Electromagnetic Compatibility (EMC) Design and Test Case Analysis

Electromagnetic Compatibility (EMC) Design and Test Case Analysis

Junqi Zheng is the Vice Chairman of IEC/CISPR, secretary-general of the Chinese National Radio Interference and Standardization Technical Committee, and deputy director of the EMC Center of Shanghai Electric Apparatus Research Institute. This famous EMC expert in China has long been engaged in the research of theory and engineering, and he has rich practice experience in the EMC engineering field. His research achievements include EMC design methods, test methods and limits, and EMC diagnostic methods in various types of products, including medical, industrial, military, and automotive. He is the founder of the EMC design risk evaluation approach, which is the design methodology that can be adopted by an R&D department.

The book features:

EMC Testing and Design Analysis Cases, 2006
EMC Design Analysis and Risk Evaluation of Electronics Products, 2008

This edition first published 2019 by John Wiley & Sons Singapore Pte. Ltd under exclusive licence granted by Publishing House of Electronics Industry for all media and languages (excluding simplified and traditional Chinese) throughout the world (excluding Mainland China), and with non-exclusive license for electronic versions in Mainland China.

Registered Offices
John Wiley & Sons, Inc., 111 River Street, Hoboken, NJ 07030, USA
John Wiley & Sons Singapore Pte. Ltd, 1 Fusionopolis Walk, #07-01 Solaris South Tower, Singapore 138628

Editorial Office
1 Fusionopolis Walk, #07-01 Solaris South Tower, Singapore 138628

For details of our global editorial offices, customer services, and more information about Wiley products visit us at www.wiley.com.

Wiley also publishes its books in a variety of electronic formats and by print-on-demand. Some content that appears in standard print versions of this book may not be available in other formats.

Library of Congress Cataloging-in-Publication Data

Names: Zheng, Junqi, 1975– author.
Title: Electromagnetic Compatibility (EMC) Design and Test Case Analysis
Other titles: Electromagnetic compatibility design and test case analysis
Description: Hoboken, NJ : Wiley, 2019. | Includes bibliographical references and index. |
Identifiers: LCCN 2018046968 (print) | LCCN 2018056375 (ebook) | ISBN 9781118956854 (Adobe PDF) |
ISBN 9781118956830 (ePub) | ISBN 9781118956823 (hardcover)
Subjects: LCSH: Electromagnetic compatibility. | Electromagnetic interference.
Classification: LCC TK7867.2 (ebook) | LCC TK7867.2 .Z44 2019 (print) | DDC 621.38–dc23
LC record available at https://lccn.loc.gov/2018046968

Cover design: Wiley
Cover image: © Brostock/Getty images

Set in 10/12pt Warnock by SPi Global, Pondicherry, India
Printed in Singapore by C.O.S. Printers Pte Ltd

10 9 8 7 6 5 4 3 2 1

Contents

Preface

The majority of domestic electromagnetic capacity books have a common defect, which is the lack of connections between design and testing. The discussion of the approach and techniques of EMC design should be based on EMC testing, not only because the first challenge of EMC design is the EMC test but also because those key factors like interference source, receiving antenna, and equivalent radiated antenna, which are critical to EMC analysis, will only exist during the EMC test. Taking the conducted emission test as an example, its essence is the voltage across a resistor in the line impedance stabilization network (LISN), when the resistance is fixed, the level of conducted disturbance depends on the current passing through the LISN resistor. EMC design is to reduce the current flow through the resistor. Possible tests include the typical immunity test, electrical fast transient/burst (EFT/B) test, big current injection (BCI) test, and electrostatic discharge (ESD) test, which is a typical common mode immunity test. The source of disturbance is a common-mode disturbance, referred to the reference ground plane, i.e. the reference point of these disturbance sources is the reference ground plane used in the test, which means that the current generated by the disturbance will eventually return to the reference ground plate. This is the basic starting point to analyze such disturbance problems.

Imagine, for the above-mentioned conducted disturbance test, that during the product testing, that the disturbance current does not flow through the LISN resistor, and at the same time, for the immunity test, that this disturbance current never passes through the product circuit, it is certainly very favorable for this product to pass the EMC tests, and this is what product design needs to consider. Therefore, the EMC design must be started from the EMC test. *Electromagnetic Compatibility (EMC) Design and Test Case Analysis*, as a project reference book, makes a close connection with the EMC test substance, EMC design principles, and specific product design to narrate EMC design methodology. Highly integrating the practical and theoretical contents is the biggest characteristic of this book.

The book is divided into seven chapters, in which the basic EMC knowledge is described in Chapter 1, mainly served for the 2–7 chapters. When readers read those later sections, if some basic concept is vague and not clearly explained, it can be easily consulted and checked from Chapter 1. Chapters 2–7 includes cases, which are typical and representative. Case descriptions use the same format: [Symptoms], [Analyses], [Solutions], and [Inspirations]. By analyzing each case, we introduce the practical information about EMC design and diagnostic technology to the designers to reduce the mistakes made by the designer in the product design and the diagnostic of EMC problems, and achieve good product EMC performance. At the same time, illustrating the design principles through EMC cases enables readers to achieve better understanding on the origin of the design. [Inspirations] section actually sums up the problem and highlights related issues. It can be used as a checklist of product EMC design. The cases are divided into the following six categories:

1) *Products' structural framing, shielding, and grounding versus EMC.* For most devices, shielding is necessary. Especially with the increasing frequency in the circuit, relying solely on the circuit board design often fails to meet EMC standards. Proper shielding can greatly strengthen EMC performance, but an unreasonable shielding design can not only fail to play its desired effect but also oppositely cause additional EMC problems. In addition, grounding will not only help solve the safety problem but is also very important for EMC. Many EMC problems are caused by an unreasonable grounding design, as the ground potential is a reference potential of the entire circuit. If the ground is not properly designed, the ground potential may be unstable, which leads to failed circuits. It may also generate additional EMI problems. The purpose of the grounding design is to ensure that the ground potential is as stable as possible, to reduce the voltage drop on the ground, thereby eliminating the interference.
2) *Cables of products, connectors, and interface circuit versus EMC.* Cable is always the path, which gives rise to radiation or bringing in the major disturbance. Because of their length, the cable is not only the transmitting antenna but also a good receiving antenna. And the cable has the most direct relationship, with the connector and interface circuit. Good interface circuit design not only can make the internal circuit noise well suppressed, so that there is no driving source for the transmitting antenna, but can also filter out the cable disturbance signal received from outside. Proper connector design of cable and interface circuit provides a good matching path.
3) *Filtering and suppressing.* For any devices, filtering and suppressing are key techniques to resolve electromagnetic interference (EMI). This is because the conductor of the device is acting as a highly efficient receiving and radiating antenna, and therefore, most of the radiation generated by the device is achieved through a variety of wires, while the external disturbance is often received by the conductor first, then brought into the device. The goal of filtering and suppressing is to eliminate these interfering signals on the wire, to prevent circuit interference signals being transferred onto the wire and then radiated through the wire, and also to prevent the conductors receiving the disturbance and taking them into the circuit.
4) *Bypass, decoupling, and energy storage.* When the device is operating, the signal level of the clock and data signals pins changes periodically. In this case, decoupling will provide enough dynamic voltage and current for the components when the clock and data are changing in normal operation. Decoupling is accomplished by providing a low-impedance power supply between the signal and power planes. As the frequency increases, before reaching the resonant point, the impedance of the decoupling capacitors will decrease, so that the high-frequency noise is effectively discharged from the signal line. Then the remained low-frequency RF energy will not be affected. Best results can be achieved through storage capacitors, bypass capacitors, and decoupling capacitors. These capacitance values can be calculated and obtained by specific formula. In addition, the capacitor insulation material must be correctly selected, rather than randomly selected based on the past usage and experience.
5) *PCB design versus EMC.* Whether the device emits electromagnetic interference or is affected by outside disturbance, or generates mutual interference between the elects, PCB is the core of the problem (the component layout or the circuit routing of the PCB), and will have an impact on the nature of the product overall EMC performance. For example, a simulated interface connector position will affect the direction of common mode current flows in, and the path of the routing will affect the size of the circuit loop, these are the key

factors of EMC. Therefore, a properly designed PCB is important to ensure good EMC performance for the product. The purpose of PCB design is to reduce the electromagnetic radiation generated by the circuit on the PCB and susceptibility to outside interference, and to reduce the interaction between the PCB circuits.

6) *Components, software, and frequency jittering technique.* Circuits are composed of components, but the EMC performance of the components is often overlooked. In fact, the packaging, the rising edge, the pinout of the component, and the ESD immunity of the device itself have a huge impact on the performance of a product's EMC performance. Although the software does not belong to EMC academic areas, in some cases software fault-tolerant technique can be used to avoid the impact on the products from the outside interference. Frequency Jittering is a popular technique to reduce the conducted and radiated emission from circuits in recent years, but the technology is not foolproof. This chapter will give details to the substance of the case and precautions for frequency jitter technique.

In fact, EMC design rules are just like traffic regulations. Noncompliance will certainly not result in a traffic accident, but the risk is bound to increase. EMC design is in accordance, noncompliance of some rules may also be able to pass the test, but the risk is bound to be increased. So there is an urgent need to introduce the product design risk awareness to the industry. The purpose of EMC design is to minimize the risk of EMC test, as only for those products complying with all the EMC, and traffic rules have the lowest EMC risk. Most of the listed problems in this book are originated from EMC problems encountered in practical work, each case is originated from the experience of these cases, which come from the accumulation of a large number of typical EMC cases the author encountered. For those classic cases, there are more detailed theoretical analysis. Each of the results of those cases is formed with one or more of EMC design rules, and it is worth learning and referring. As the engagement of the author is limited, this book may not contain all kinds of EMC issues in electronics and electrical products.

If readers discover any mistakes due to the author's incomprehensive knowledge, leading to unreasonable or inaccurate descriptions or even critical mistakes, please feel free to contact me. In addition, it must be mentioned that cases in this book are taken from specific products, for the convenience to readers, including the component symbols, codes, graphics, etc., and are not normalized in accordance with GB Standard.

I would like to express my gratitude to those who mentioned comments and suggestions for the book, beginning with Professor Wu Qinqin, PhD; I would also like to thank Professor Alain Charoy, Renzo Piccolo, Beniamino Gorini who is chairman of IEC/CISPRA, as well as those of my colleagues who mentioned comments about this book. Great thanks to the Electronics Industry Publishing House deputy editor Zhang Rong, Niu Pinyue and the Wiley Publishing House editors.

Junqi Zheng
Vice Chairman of IEC/CISPR

Exordium

Electromagnetic Compatibility (EMC) Design and Test Case Analysis has received more attention from readers since first publication in 2006. A number of defects in the version first published in the PHEI edition have been modified in this book, and on the basis of the original case analysis, the following important principles of design key points and specific treatment measures of EMC have been clarified by adding more cases.

1) Regarding the essence of EMC testing, the essence of various EMC test items defined in the standard is analyzed.
2) The design method of the filtering circuit on the power supply port has been clarified, including the selection of the filtering circuit and the parameters of the filtering components.
3) The EMC design method for digital-analog mixed circuit has been clarified, including not only the crosstalk between analog circuit and digital circuit but also how to consider the EMC problem from a system perspective – especially how to deal with the analog ground and the digital ground, which puzzles the majority of designers.
4) The advantages and disadvantages of segregating the ground plane in the printed circuit board (PCB) have been clarified.
5) The method and principle of the interconnection between various working ground in PCB and the metallic casing for the product with metallic casing have been clarified, concerning whether the interconnection between them is needed; how to connect; and where the connection shall be positioned and other problems.
6) The sensitive traces, sensitive components, clock traces, or clock components and so on have been more fully described, including why they can't be positioned at the edge of the PCB. Concrete solution and remedy measures have been detailed.
7) The relationship between the stack-up design and the EMC problems in the design of multilayer PCB has been clarified.
8) The magnitude of the differential mode radiation induced by loop has been clarified.

In China, EMC design got off to a late start but is developing rapidly. More and more companies and engineers have come to understand EMC, also gradually mastering a number of EMC design rules and applying them to guide the product design after several years of development. However, in China, with the rapid development of electronic design technique, there are many misconceptions about the nature of EMC in the process of product design. Eliminating these misunderstandings can help the reader to solve the inevitable EMC problem. These misconceptions are mainly reflected in the following aspects.

1 Grounding

The word "grounding" has entered the vision of the vast number of electronic product designers. Before we get to EMC, the most familiar ground is the earth. For safety reasons, certain metal conductors in electronic and electrical products must eventually be connected to the earth (called protective earth), generally connected to the earth through the building structure or a dedicated ground bar. For EMC, EMI radiation of the product can be reduced to the maximum extent by grounding, and the external disturbance going into the product can also be reduced to a maximum extent through grounding. However, is it necessary to connect the product to the earth? How can we correctly understand the grounding in EMC? These issues are addressed in Case 13, "The Metallic Casing Oppositely Causes the EMI Test Failed," Case 14, "Whether Directly Connecting the PCB Reference Ground to the Metallic Casing Will Lead to ESD," and Case 69, "Detailed Analysis Case for the PCB Design of Analog-Digital Mixed Circuit."

To a certain extent, the answer to the above question is given, to control product EMC, it is not necessary to connect the product to the earth of the natural world, grounding is to guide the flow of common mode current for EMC. Actually, for EMI, the reference point of the disturbance source of disturbance is the ground of the PCB. In order to avoid the disturbance source, through various channels, going into the antenna (such as cable in product), the correct grounding point should be a certain point in the ground of the PCB. So, this grounding from the flow of EMI disturbance current, should occur before the antenna (such as the cable); for most of the product's high frequency immunity, the reference point of disturbance source is the ground reference plate in the test, the correct grounding position should be the ground reference plate, to the purpose of this grounding is not to allow the external common mode current to be injected into the product. So, this grounding, from the perspective of the disturbance flow, occurs before the product's circuit. Regarding the grounding design of a product, the first thing needs to be considered is not selecting or designing "single point grounding" or "multipoint grounding," but the location of the grounding and the grounding ways. If a product has a metallic casing, the above two kinds of grounding ways can be well realized with the help of the metallic casing or other parasitic parameters, this is why the equipment with metallic casing is more easily to pass the EMC tests, these two kinds of grounding are more difficult for the equipment with nonmetallic casing to pass the EMC tests.

[Loop and differential mode radiation] How is the radiation produced in the product? This is an indisputable fact that a PCB signal loop will produce the differential mode radiation.

The formula $E(\mu V/m) = 1.3 \times I(A)S(cm^2)F(MHz)/D(m)$ gives the order of magnitude of the differential mode radiation.

> $E(\mu V/m)$: Radiation field strength, in unit of $\mu V\, m^{-1}$
> $I(A)$: Current at a certain frequency in the loop, in unit of A
> $S(cm^2)$: Signal loop area, in unit of cm^2
> $F(MHz)$: Signal frequency of radiation in the loop, in unit of MHz
> $D(m)$: The distance from the test point to the loop, in unit of m

According to this formula, if the voltage amplitude of a clock signal is 3.3V, its frequency is 20 MHz (the working current is 3.3V/100Ω = 33mA. The virtual value of the third-harmonic current at 60 MHz is 0.005A), the loop area is 1 × 1 cm (it is generally considered that this is a very poor PCB design). At the harmonic frequency 60 MHz, the radiation field strength at 3 m test distance from this clock circuit is $7\, \mu V\, m^{-1}$. This value is a far below the standard radiated emission limit. That is to say, when the loop area is not large (with the popularization of multilayer PCB board technology, the loop area can be designed to be smaller), this radiation will not exceed the radiated emission limit specified in the current EMC standards. But it is worth noting that the resulting common-mode radiation problem by loop area is increased, and Case 25, "The Excessive Radiated Emission Caused by the Loop," gives an analysis. So, do not place undue emphasis on the differential mode radiation and neglect the more important common-mode radiation.

2 Shielding

Assuming the differential-mode radiation already mentioned is more than the standard limit (or the equivalent antenna, which causes the radiation to exceed the standard is inside the shield), as long as a metallic casing with an opening that is not large is used for shielding, the radiation problem can be solved. At this time, the metallic casing does not need to be connected with the PCB. However, with the elimination of the above misunderstanding, and with the equivalent antenna that causes the radiation to exceed the standard also outside the shield (such as the cable), at this time, the necessity of shielding with the metal casing has gradually declined. Case 13, "The Metallic Casing Oppositely Causes the EMI Test Failed," is a typical case of this misunderstanding. Using metallic casing to achieve better EMC performance, it is because the metallic casing provides a better grounding path or bypassing path. If you want this path to become more direct, a reasonable interconnection between the PCB and the metallic casing will be needed. Designers must get rid of this misunderstanding when you want to add a shield to the product, and you have to be responsible for the consequences of this. We must take into account the physical location of the equivalent antenna of the radiation for product shielding design. If you cannot put them inside the shield, we must consider making a reasonable interconnection between the PCB and the metallic casing to achieve the transformation between shielding and bypassing.

3 Filtering

Capacitance and inductance are the basic components of the filtering circuit. The inductor behaves as inductive reactance and the inductive reactance increases with the increased frequency. The capacitor behaves as capacitive reactance and the capacitive reactance decreases with the increasing frequency. When the original circuit is in series with an inductor or parallel with a capacitor, the voltage divider network formed by the inductance and capacitance can reduce the interference voltage on the load. It seems that we could say, "It would be good to add more inductors in series or more capacitors in parallel." In fact, as inductor and capacitor are energy storage components, there is a phase difference between the voltage and current on them, an extreme performance of the filtering network consisting of an inductor and a capacitor creates resonance. An LC filtering circuit occurs when the resonance occurs. The interference signal is not attenuated and oppositely amplified; it's horrible. This misunderstanding must be eliminated to design a good filtering circuit, and the resonant point of the filtering circuit must be far from the EMC test frequency. In the case of filters, more is not better.

By Junqi Zheng

Introduction

In this book, we use electromagnetic capacity case analyses as the main avenue for discussing EMC issues and introduce EMC technique in product design by describing and analyzing the cases. Our purpose is to explain the related EMC practical design techniques and diagnostic techniques during product design processes and to reduce designers' mistakes when designing the products and diagnosing the EMC problems. The EMC cases in this book refer to various aspects such as structure, shielding and grounding, filtering and suppressing, cabling, routing, connectors and interface circuits, bypassing, decoupling and energy storage, PCB layout, components, software, frequency jittering technique, and more.

This book is based on practical applications, to explain the complicated principles with representative cases and avoid unnecessarily long theory. This book can be used as a necessary EMC reference book in the electronic product design departments, as well as a basic EMC training and reference material for electrical and electronics engineers, EMC engineers, and EMC counselors.

1

The EMC Basic Knowledge and the Essence of the EMC Test

1.1 What Is EMC?

Electromagnetic compatibility (EMC) is the capability of electronic and electrical equipment and systems to operate as designed in the forecasted electromagnetic (EM) environment, which is an important technique performance of the electronic and electrical equipment and the systems. EMC has the following two implications:

1) Electromagnetic interference (EMI), that is, the disturbance produced by the equipment and systems that operate in a certain environment, shall not exceed the limit required in their corresponding standard regulations. And the corresponding test items depend on the product classifications and the standards, for residential devices, ISM (industrial, scientific, and medical) devices and railway devices. The basic EMI test items include the following:

   - Conducted Emission (CE) Test on Power Line
   - Conducted Emission (CE) Test on Signal and Control Line
   - Radiated Emission (RE) Test
   - Harmonic Current Test
   - Voltage Fluctuation and Flicker Test

   For military device, basic EMI test items include the following:

   - CE101: Conducted Emissions, Power Leads, 30 Hz to 10 kHz
   - CE102: Conducted Emissions, Power Leads, 10 kHz to 10 MHz
   - CE106: Conducted Emissions, Antenna Terminals, 10 kHz to 40 GHz
   - CE107: Conducted Emissions, Power Leads, Spike, Time Domain
   - RE101: Radiated Emission, Magnetic Field, 30 Hz to 100 kHz
   - RE102: Radiated Emission, Electric Field, 10 kHz to 18 GHz
   - RE103: Radiated Emission, Antenna Spurious and Harmonic Outputs, 10 kHz to 40 GHz

   For vehicles, as well as vehicle electronic and electrical products, the basic EMI test items include the following:

   - Vehicle Radiated Emission Test
   - Conducted Emission Test for Vehicle Electronic and Electrical Parts/Modules
   - Radiated Emission Test for Vehicle Electronic and Electrical Parts/Modules
   - Transient Emission Test Vehicle Electronic and Electrical Parts/Modules

2) Electromagnetic susceptibility (EMS), that is, in normal operation, the devices and system in a certain environment can withstand the EM disturbance specified in their corresponding

Electromagnetic Compatibility (EMC) Design and Test Case Analysis, First Edition. Junqi Zheng.

standard regulations. Similarly, according to the product classification and standards for residential devices, ISM (industrial, scientific, and medical) devices, and railway devices, the basic EMC test items include the following:

- Electronic Static Discharge (ESD) Immunity Test
- Electrical Fast Transient Burst (EFT/B) Immunity Test
- SURGE Immunity Test
- Radiated Susceptibility Immunity Test (RS)
- Conducted Susceptibility Immunity Test (CS)
- Voltage Dip and Voltage Interruption Test

For military device, the basic EMI test items include the following:

- CS101: Conducted Susceptibility, Power Leads, 30 Hz to 50 kHz
- CS103: Conducted Susceptibility, Antenna Port, Inter–modulation, 15 kHz to 10 GHz
- CS104: Conducted Susceptibility, Antenna Port, Rejection of Undesired Signals, 30 Hz to 20 GHz
- CS105: Conducted Susceptibility, Antenna Port, Cross-modulation, 30 Hz to 20 GHz
- CS106: Conducted Susceptibility, Power Leads, Spike, Time Domain
- CS114: Conducted Susceptibility, Bulk Current Injection, 10 kHz to 40 MHz
- CS115: Conducted Susceptibility, Bulk Current Injection, Impulse Excitation
- CS116: Conducted Susceptibility, Damped Sinusoidal Transients, Cables and Power Leads, 10 kHz to 100 MHz
- RS101: Radiated Susceptibility, Magnetic Field, 30 Hz to 100 kHz
- RS103: Radiated Susceptibility, Electric Field, 10 kHz to 40 GHz
- RS105: Radiated Susceptibility, Transient Electromagnetic Field

For vehicles, and vehicle electronic and electrical products, the basic EMS test items include:

- Conducted Coupling/Transient Immunity Test at Power Port Conforming to ISO7637-1/2
- Coupling/Transient Susceptibility Test on Sensor Cable and Control Cable Conforming to ISO7637-3
- RF Conducted Susceptibility Test Conforming to ISO11452-7
- Radiated Susceptibility Test conforming to ISO11452-2
- Radiated Susceptibility Test in Transverse Electric and Magnetic (TEM) Cell Conforming to ISO1145-3
- Big Current Injection (BCI) Susceptibility Test Conforming to ISO11452-4
- ISO11452-5: Strip Line Susceptibility Test
- Parallel Plate Susceptibility Test Conforming to ISO11452-6
- Electrostatic Discharge Immunity Test Conforming to ISO10605

3) Electromagnetic environment: the operation environment for the systems or devices.

1.2 Conduction, Radiation, and Transient

The moment an air conditioner is switched on, a nearby indoor fluorescent lamp dims transiently. This is because a large amount of current flows to the air conditioner and the supply voltage drops quickly, affecting the fluorescent lamp supplied by the same power network. Also, when someone turns on a vacuum cleaner, the radio will make a "para-para" noise because the weak (low amplitude but high frequency) voltage/current change generated by the motor of the cleaner is transferred into the radio through the power line. So, when the voltage/current generated by a device is transferred through the power line or signal line and then affects other devices, this voltage/current change is called the *conducted disturbance*. In order to solve this problem, the general method is to install a filter for the noise source and the power line of the victim to prevent the transmission of the conducted disturbance. Alternatively, when the noise appears on the signal line, the coupling path can be cut off by changing the signal cable to the optical fiber.

When using a cellphone, a nearby computer's LED display might flash. The reason is that, in normal operation, the EM signal generated by the cellphone is transferred into the LED display through space. When a motorcycle drives past a house, the TV screen image might be affected because the pulse current generated by the motorcycle's ignition device transmits EM waves, which will propagate into space, then to the nearby television antenna, and then to the circuit, thus disturbing the voltage/current. In situations like this, the harmful disturbance, which is transferred through the space and causes the undesired voltage/current for the other device's circuit, is called *radiated disturbance*. The occurrence of the radiation phenomenon must be accompanied with the antenna and the source. Since the route of this transmission is space, shielding is the effective way to solve this problem.

So to summarize, the root cause of the disturbance is the undesired voltage/current change. If the change is transferred directly through the wire into the other devices, it is called *conducted disturbance*. If the change is transferred through the space in the form of EM waves into the other devices and causes the undesired voltage/current on the circuit or the line, this harmful disturbance is called *radiated disturbance*. However, in reality, the categorization is not so simple.

For example, for the disturbance source of the computing devices, such as computers, although the voltage/current of the digital signals are transferred through the circuit inside the devices, the interference is conducted out of the device through the power line or signal line and directly transferred to the other devices. At the same time, the EM wave generated from these wires can be harmful for the nearby devices in the form of the radiated interference. In addition, the circuit inside the computing device can also generate the EM wave and affect the other devices.

The occurrence of the radiated disturbance phenomenon is always inseparable from the antenna. According to the operation principle of the antenna, if the length of a wire is equal to the wavelength, it easily generates the EM wave. For example, a several-meter-length power line can radiate the EM wave at VHF frequency band (about 30–300 MHz). At frequencies lower than this frequency band, because of the longer wavelength of those frequencies, when the same current flows through the power line, strong EM wave cannot be generated and radiated. So, at frequency bands below 30 MHz, conducted interference is the primary problem. However, the disturbing magnetic field can be generated around the power line affected by the conducted disturbance and can then disturb amplitude moderation (AM) radio and others. In addition, as mentioned above, since the leakage disturbance from the power line at the very high frequency (VHF) band can be transformed into the EM wave and the scattered into space, the radiated disturbance becomes a more importance problem than the conducted disturbance. At higher frequencies, the circuit inside the devices of which the size is smaller than that of the power line can generate the radiated disturbance, which is harmful for other devices.

Thus, on the one hand, when the size of a wire and device is smaller than the wavelength, the main problem is the conducted disturbance; On the other hand, when the size is bigger than the wavelength, the main problem is the radiated disturbance.

There are some transient high-energy impulse disturbances in the operation environment that are harmful for electrical devices. Generally, this kind of disturbance is called *transient disturbance*. Transient disturbance can be transferred into the devices not only through the wire in the conducted mode but also through the space in the broadband radiated mode, such as the disturbance to the radio, which is generated from the motor ignition circuit and the brushes of the DC motor. The main root cause of the transient disturbances generally includes lightning, electrostatic discharge, switching on or off the load (especially the inductive load) of the power line, and the nuclear electromagnetic pulse, for example. It can be seen that the transient disturbance is the EM disturbance with high amplitude but short duration. Common transient disturbance (the immunity of the devices needs to be verified by test) is divided into three groups, various electrical fast transient/burst (EFT/B), various surge, and various electrostatic discharge (ESD).

1.3 Theoretical Basis

1.3.1 Time Domain and Frequency Domain

Any signal can be converted from time domain to frequency domain through Fourier transform, as shown in the following formula:

$$H(f) = \int_{0}^{T} x(t) e^{-j2\pi \cdot f \cdot t} dt \tag{1.1}$$

where $x(t)$ is the time-domain waveform function of the electrical signal; $H(f)$ is the frequency function of this signal; $2\pi f = \omega$, ω is angular frequency and f is the frequency.

The frequency spectrum of a trapezoidal pulse function, as shown in Figure 1.1, comprises the main lobe and countless minor lobes. Although every minor lobe has its maximum value, the general trend is to decline linearly with the increased frequency. The peak value curve in the frequency spectrum of the trapezoidal pulse, which has the rising time t_r and the pulse width t, includes two turning points, $1/\pi t$ and $1/\pi t_r$. In the frequency spectrum, the peak value is constant at low frequency, but it drops with the slope of −20 dB/10 dec after the first turning point and then drops with −40 dB/10 dec slope after the second turning point.

Thus, when designing the circuit, under the condition that the normal logic function is ensured, we should increase the rising time and the falling time as much as possible, which contributes

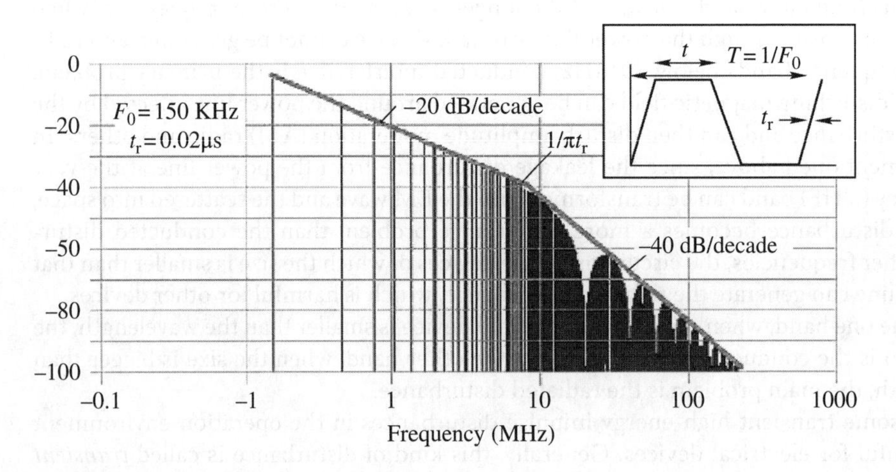

Figure 1.1 Spectrum of trapezoidal pulse function.

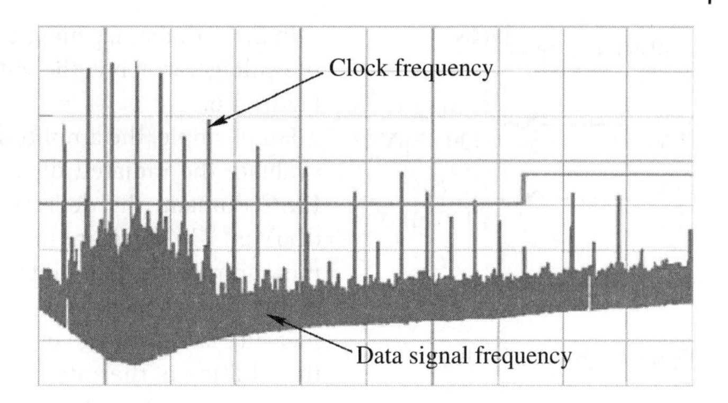

Figure 1.2 Spectrum of clock noise and data noise.

to reducing high-frequency noise. But, because of the existence of the first turning point, the periodic signal with a steep rising edge and low frequency does not include higher-order harmonic noise with a high level (for the calculation of each harmonic amplitude, you can refer to my book *Electronic Product Design – EMC Risk Evaluation*, published by PHEI in 2008).

Since at every sampling of the periodic signal, the frequency spectrum is the same, the frequency spectrum is dispersed but its amplitude is high at each frequency, which generally is the narrow-band noise. For a nonperiodic signal, since the frequency spectrum at every sampling is different, the frequency spectrum is very wide and its amplitude is small, which generally is the wide-band noise. In an ordinary system, the clock signal is a periodic signal, but the signal on the data line and the address line generally is nonperiodic, so the root cause for the radiated emission beyond the standard limit is always the clock signal. The frequency spectrum of the clock noise and the data noise is shown in Figure 1.2.

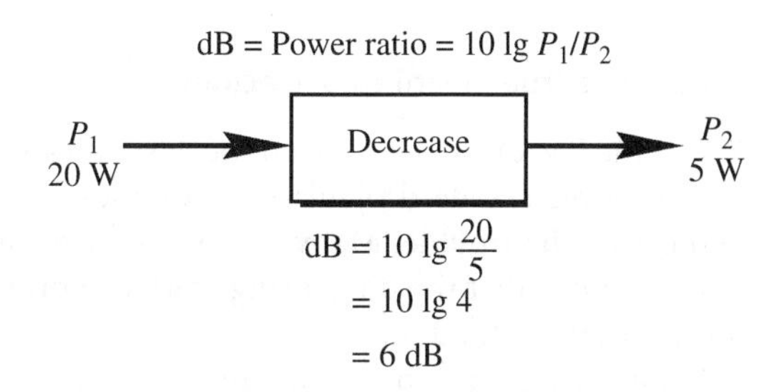

Figure 1.3 The concept of dB.

dBm = Power ratio = Value relate to 1mV

$$\text{dBm} = 10\ \text{lg}\ \frac{P(\text{watts})}{0.001}$$

P_1 → $P_{1'}$
1 W → 0 dBm

P_1 → $P_{1'}$
10 W → 40 dBm

Figure 1.4 The dB of power value.

1.3.2 The Concept of the Unit for Electromagnetic Disturbance, dB

Electromagnetic disturbance is generally measured in decibels. The original definition of *decibel* is the ratio of two powers. As shown in Figure 1.3, the decibel is a logarithmic unit used to express the ratio of two powers.

The unit dBW is often used to denote a ratio with 1 W reference power, and similarly dBm for the 1 mW reference power, as shown in Figure 1.4.

The decibel scale of the voltage can be calculated from the decibel scale of the power, as shown in Figure 1.5. (The precondition is, $R_1 = R_2$, and they are generally 50 Ω.)

$$(\text{W}) = \frac{U^2}{R} \qquad P_1 \quad P_2 :$$

$$P_1 = \frac{(U_1)^2}{R_1} \text{ and } P_2 = \frac{(U_2)^2}{R_2}$$

$$\text{dB} = 10\ \text{lg}\ \frac{(U_1)^2}{(U_2)^2}$$

$$\text{dB} = 20\ \text{lg}\ \frac{(U_1)}{(U_2)}$$

Figure 1.5 The concept of decibels-voltage.

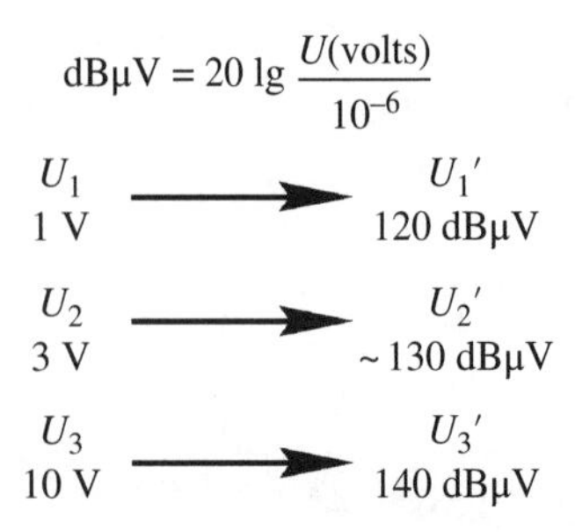

Figure 1.6 The dB of voltage value.

In EMC domain, generally, dBμV is used directly to express the voltage, which is the voltage relative to 1 μV, as shown is Figure 1.6.

For example, the amplitude of the electrical field is used to evaluate the radiated disturbance, and its unit is $\mathrm{V\,m^{-1}}$. In EMC domain, the decibel unit is generally used, which is $\mathrm{dB\mu V\,m^{-1}}$. When combining the antenna and the disturbance measurement instrument together to measure the amplitude of the disturbance field strength, what the instrument can measure is the voltage on the antenna port. The voltage with the addition of the antenna coefficient is the field strength of the measured disturbance.

$$E\left[\mathrm{dB\mu V\,m^{-1}}\right] = U\left[\mathrm{dB\mu V}\right] + \mathrm{antenna\_factor}\left[\mathrm{dB}\right]$$

Note that the cable attenuation is not taken into account.

1.3.3 The True Meaning of Decibel

When the EM disturbance from the devices cannot meet the limit requirement specified in the relative EMC standard regulations, we must analyze the root cause of the excessive emission and clarify the trouble. Many engineers have tried to accomplish this, but troubles still exist. One reason is that the diagnosing work can drop into an endless loop. The following example illustrates this situation.

Assume that the conducted emission is excessive when testing a system, and it means that the system cannot meet the CLASS B limit required in EMC standard CISPR22, as shown in Figure 1.7. After preliminary analysis, there are four possible reasons:

1) Conducted emission produced by the transformer
2) Conducted emission produced by the switching transistors in the switching mode power supply
3) Conducted emission produced by the PCB design defects
4) Conducted emission produced by the auxiliary equipment

When locating the problem, the factors related to the transformer are removed first to reduce the conducted emission. The result is that there is no obvious reduction, and the constitution and the amplitude of the conducted emission at the power port after those factors are removed is shown in Figure 1.8. So it is concluded that the transformer is not the main reason for the excessive conducted emission but we request the change for the transformer. Then we dispose of the switching transistor in the switching mode power supply. Remove its adverse factors to reduce the conducted emission at the power port, and then it is found that the test result is not obviously improved. The constitution and the amplitude of the conducted emission at the power port after removing these factors is shown in Figure 1.9. It is then concluded that the switching transistor is not the main reason, either.

Then check the printed circuit board (PCB). If we improve the PCB to remove the original defects, we discover that the signal amplitude on the frequency spectrum is barely decreased. The constitution and the amplitude of the conducted emission at the power port after improving the PCB is shown in Figure 1.10. So, it is concluded that PCB is not the main reason, either. From the change of the relative amplitude, it seems that the PCB can be ignored.

Hereto, the conducted emission problem of this product still exists because the test engineers ignored the signal amplitude on the frequency spectrum, expressed in dB scale. Let's have

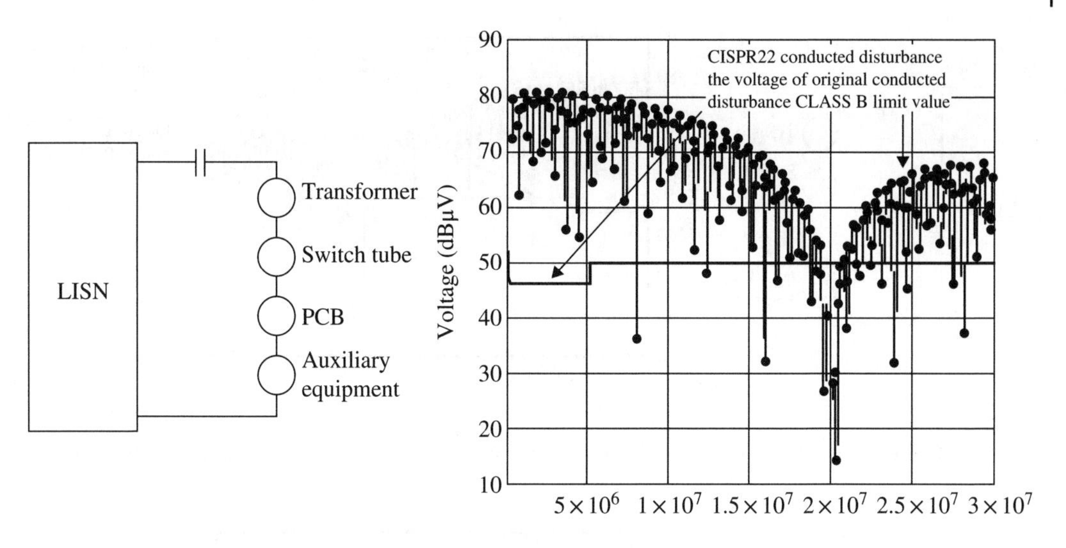

Figure 1.7 The composition and level of the conduction noise of a product's power port.

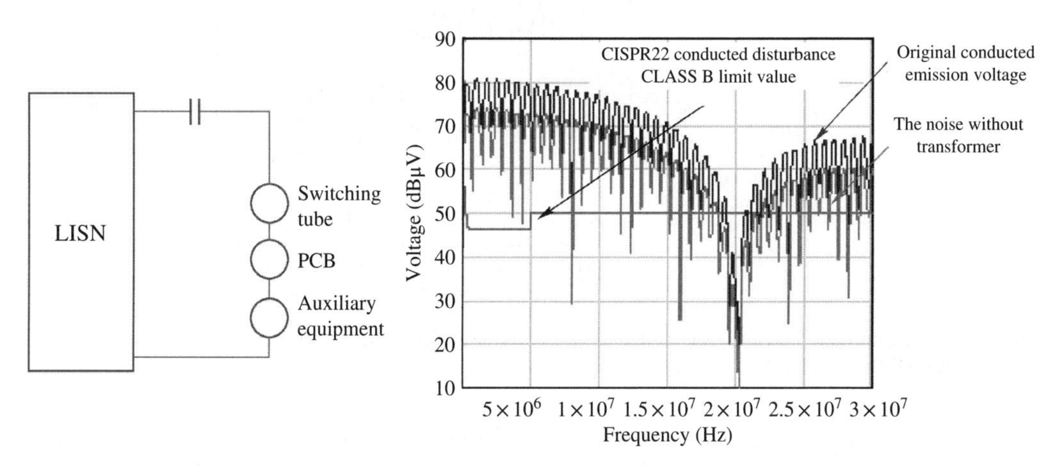

Figure 1.8 The composition and level of the conduction noise of a product's power port without a transformer.

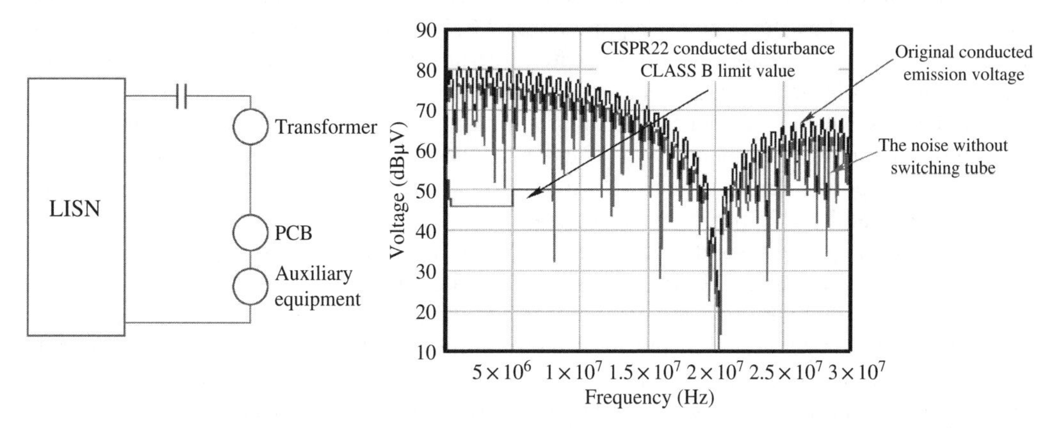

Figure 1.9 The composition and level of the conduction noise of a product's power port without the switching tube.

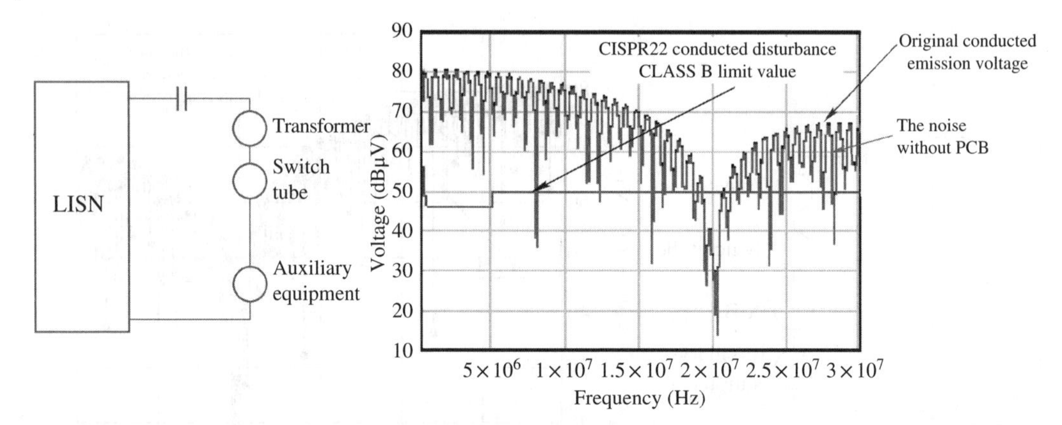

Figure 1.10 The composition and level of the conduction noise of a product's power port without the PCB.

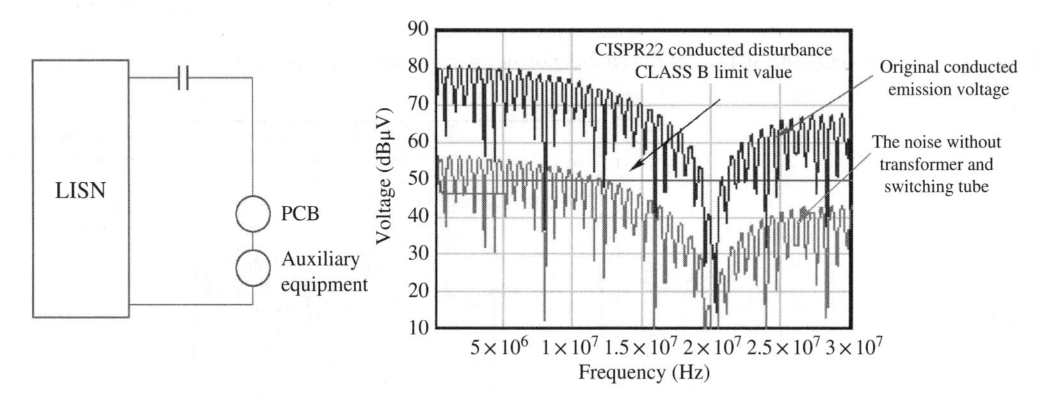

Figure 1.11 The composition and level of the conduction noise of a product's power port without transformer and switching tube.

a look at the cause of this phenomenon. Assume the amplitude of the conducted emission generated by the transformer is V_n, the amplitude of the conducted emission generated by the switching transistor in the switching mode power supply is $0.7V_n$, the amplitude of the conducted emission generated by PCB design defects is $0.1V_n$, and the amplitude of the conducted emission generated by the auxiliary equipment is $0.01V_n$. On this condition, if we remove the factors related to the transformer and the switching transistors at the same time, the test results show an obvious improvement, as shown in Figure 1.11. On this basis, we remove the PCB relevant factors, which were considered to have nothing to do with this problem formerly, and then the result show a big change, too. The constitution and the amplitude of the conducted emission at the power port after removing the factors related with the transformer, the switching transistors, and the PCB is shown in Figure 1.12.

Actually, although the absolute value of the PCB contribution is only $0.1V_n$, which is a small value compared with the amplitude of the conducted emission generated by transformer V_n and that of the switching transistors $0.7V_n$, it is a high value relevant to the amplitude of the conducted emission generated by the auxiliary equipment $0.01V_n$. Thus, when the factors relevant to the transformer and the switching transistors exist, the effectiveness of the PCB factors is insignificant. However, when the factors relevant to the transformer and the switching transistors are removed, the effectiveness of the PCB factor becomes decisive.

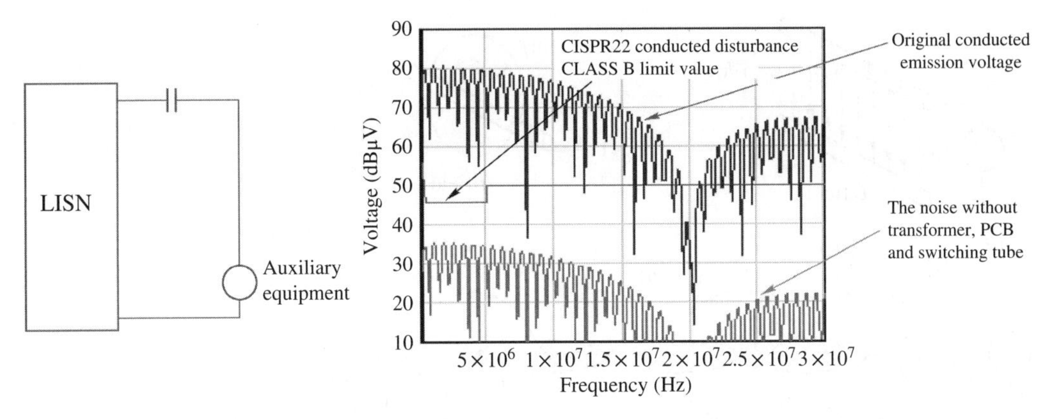

Figure 1.12 The composition and level of the conduction noise of a product's power port without transformer, PCB and switching tube.

In conclusion, the correct EMI diagnostic method is that, even if there is no obvious improvement, keep the former suppression measure to a possible disturbance source, and add more suppression measures to the other possible sources. If the disturbance amplitude declines a lot and passes the test when a certain measure is taken, it does not mean that this source is the main reason but only that this source is big in magnitude relative to the previous sources, and can be the last one to be solved.

Above all, assume that when a countermeasure to a certain disturbance source is taken, all sources in this product are removed in 100%, and then the reduction of the EM disturbance should be infinitely great after the last source is removed. Actually, this is impossible. Any countermeasure cannot remove 100% of the disturbance. It can remove 99, 99.9, or even above 99.99%, but not 100%, ever. So after the last disturbance source is removed, the improvement is really good but still limited.

Finally, when the device meets the relevant regulations completely, to reduce the product cost and remove the unnecessary components or designs, we can take out the measures one by one. The first one to be considered should be the high-cost components or material, as well as the measures that are not easy to be implemented. If the radiated emission is not excessive after removing the factors, then this measure can be removed in order to minimize the product cost.

1.3.4 Electric Field, Magnetic Field, and Antennas

1.3.4.1 Electric Field and Magnetic Field

Electric field (E-field) exists between two conductors with different electric potential. Its unit is $\mathrm{V\,m^{-1}}$. The electric field strength is proportional to the voltage between conductors and inversely proportional to the distance between conductors. A magnetic field (*H* field) exits around current-carrying conductors and the unit is $\mathrm{A\,m^{-1}}$. Magnetic field is proportional to the current and inversely proportional to the distance away from the conductors. When alternating current (AC) is generated by the alternating voltage on conductors, the EM wave will be produced. *E* field and *H* field will propagate mutually and orthogonally at the same time, as shown in Figure 1.13.

The propagation velocity of the EM field is decided by the propagation medium. In free space, the speed is equal to the velocity of light $3\times10^{8}\,\mathrm{m\,s^{-1}}$. Near the radiation source, the geometric distribution and strength of the EM field is decided by the characteristics of the interference source, and only at the far distance, the EM field is mutual orthogonal. When the

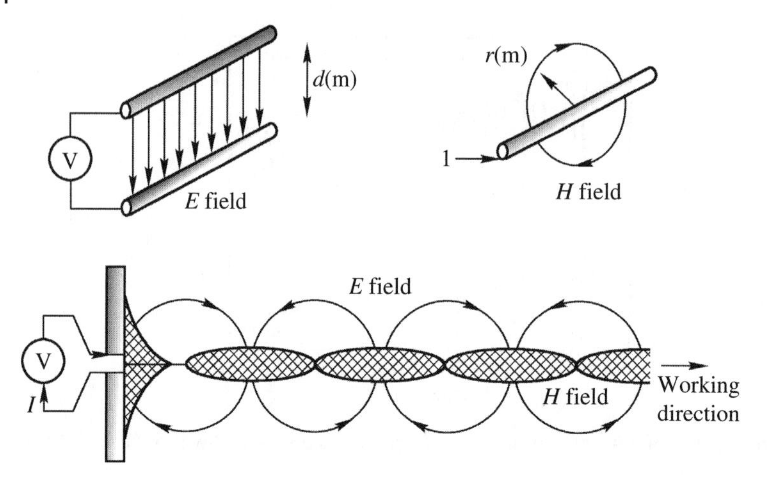

Figure 1.13 The *E* field and the *H* field propagate at the same time and are orthogonal to each other.

frequency of the interference source is high and the wavelength is smaller than the structure size of the victim, or the distance from the interference source to the victim meets $r \gg \lambda/2\pi$, the disturbance can be regarded as a radiated field – namely, the far field. It can radiate the EM energy in the form of the plane wave and get into the circuit of the victims. Through the insulation supports (including air), the disturbance can be coupled into circuits, devices, and systems; when it passes through the common impedance of the circuit, the circuit may be disturbed. If the frequency of the disturbance source is low and the wavelength λ is bigger than the structure size of the victim, or the distance from the interference source to the victim meets $r \ll \lambda/2\pi$, the disturbance can be regarded as near field interference, which can be induced into the circuit. The near-field coupling can be expressed by the capacitance and the inductance in the circuit. The capacitance represents the electric field coupling and the inductance represents the magnetic field coupling. So, the radiated disturbance can be directly conducted into the circuits, the devices, and the systems. Figure 1.14 shows the relationship between the far field, the near field, the magnetic field, the electric field, and the wave impendence, versus the distance away from the disturbance source.

At 30 MHz, the turning point of the plane wave is 1.5 m, at 300 MHz, the turning point of the plane wave is 150 mm, and at 900 MHz, it is 50 mm.

1.3.4.2 Using an Antenna to Detect Signals

An antenna has two transforming functions: one is to transform the EM wave to the usable voltage and current in the circuit, the other is to transform the voltage and the current to the EM wave propagating to the space. Signal is transmitted to the space in the form of EM wave, and the EM wave comprises the magnetic field and the electric field, which are measured, respectively, by $\mathrm{A\,m^{-1}}$ and $\mathrm{V\,m^{-1}}$. The structure of an antenna depends on the type of the measured field. As shown in Figure 1.15a, the antenna used to detect the electric field is constituted by a rod and a metallic plate, and in Figure 1.15b, the antenna used to detect the magnetic field is a wire loop.

Sometimes a part of the electronic and electrical products (such as the cable, the PCB trace, etc.) possesses these characteristics unconsciously and becomes the antenna. One of the important tasks in EMC design is to find and remove these unconscious antennas.

The electric field ($\mathrm{V\,m^{-1}}$) around the antenna will induce a voltage to ground ($m \cdot V/m = V$) along the direction of the antenna length. The receiver connected to the antenna can detect the

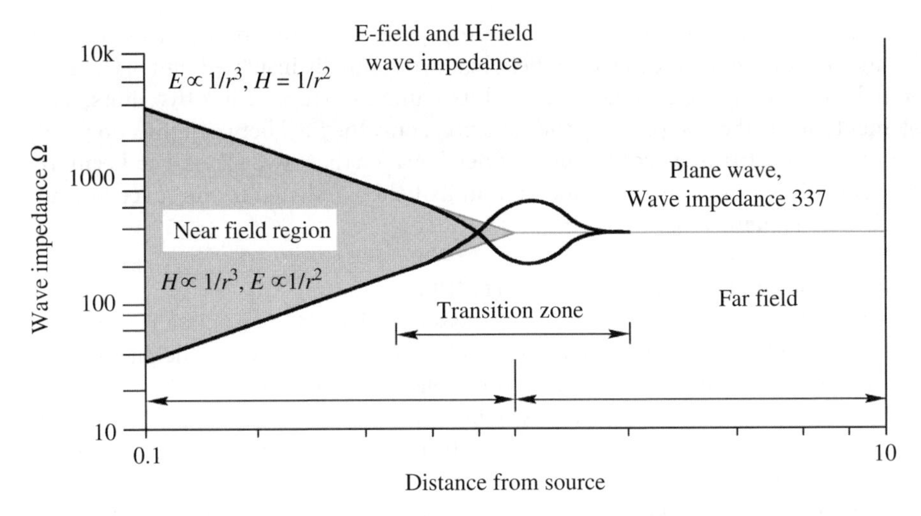

Figure 1.14 The relationship between near field, far field, magnetic field, electric field, and wave impedance in the radiation field.

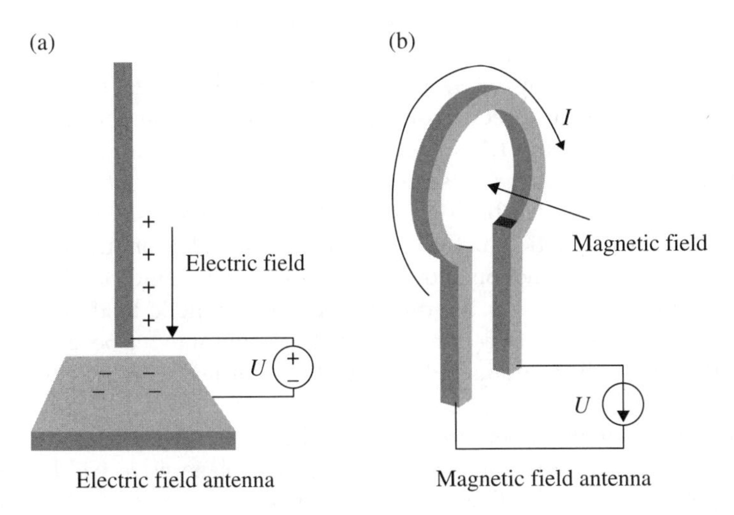

Figure 1.15 The shape of the antenna for picking up electric and magnetic fields.

voltage between the antenna and the ground. This antenna model can also be equivalent as a lead wire of the voltmeter, which is used to measure the potential in the space, and the other lead wire of the voltmeter is the ground of the circuit.

1.3.4.3 The Significance of the Antenna's Shape

Some antennas are made up of coils. These antennas are used to detect the magnetic field rather than the electric field, and they are *H*-field antennas. The current will be induced while the magnetic field penetrates the coil, just as the current flowing through the coil can produce a magnetic field that will pass through the coil. The two ends of the magnetic field antenna are fixed on a receiving circuit, so that the magnetic field can be detected by measuring the current of the loop antenna. The magnetic field is generally perpendicular to the propagation direction of the field, so the torus should be parallel with the propagation direction of the wave to detect the field. The antenna of the radiated electric field has two mutually insulated units. The simplest electric field antenna is a dipole antenna and its name hints that it has two units. Two conductor elements act as the two plates of the capacitor; only the field between the two plates of the capacitor is radiated to the space but not confined between these two plates. Furthermore, the magnetic-field coil is similar as the inductor, and its field is radiated to the space but not confined to a closed magnetic loop.

1.3.4.4 Formation of Antennas and the Radiation of EM Field

As mentioned earlier, the electric field antenna can be associated with the capacitor. Figure 1.16a shows a simple parallel plate capacitor. The electric field is generated between the plates while the electric charge is accumulated on the plates. If the plate is spread and placed on the same plane, the electric field between the plates will be extended to the space. As shown in Figure 1.16b, the same situation occurs on the electric field dipole antenna. The electric charge on each part of the antenna will generate a field, which will be transmitted to the space, and there is an inherent capacitor between the two poles of the dipole antenna, as shown in Figure 1.16c. The dipole arms need to be charged by current, which flows on each portion of the antenna in the same direction, and this electric current is called as an antenna mode current. This condition is quite special, because it leads to the generation of the radiation. When the signal applied on the antenna poles oscillates, the *E* field is kept continuously alternating and transmitted to the space.

The electric charge and the *E* field generated by the electric current are perpendicular to each other. Applying the voltage on the antenna, the direction of the electric field *E* is from the positive pole to the negative pole. As shown in Figure 1.17a, the alternating current on the antenna generates a magnetic field *H*, of which the direction surrounds the metallic conductor and meets the right-hand law, as shown in Figure 1.17b. God created this law; when the electrons move along the metallic conductor, the magnetic flux that surrounds the metallic conductor is created. If the right-hand thumb directs the direction of the current, the direction indicated by the fingers surrounding the metallic conductor is the direction of the magnetic field. The inductance of the antenna is caused by the surrounded magnetic field. So, the antenna behaves as a capacitance formed with the charge distribution and an inductance formed with the current distribution.

As shown in Figure 1.17c, the *E* field and the *H* field are perpendicular to each other. They are linked with each other and extended from the antenna to the space. A plane wave will be formed while the signal oscillates on the antenna. The transverse electromagnetic wave (TEM)

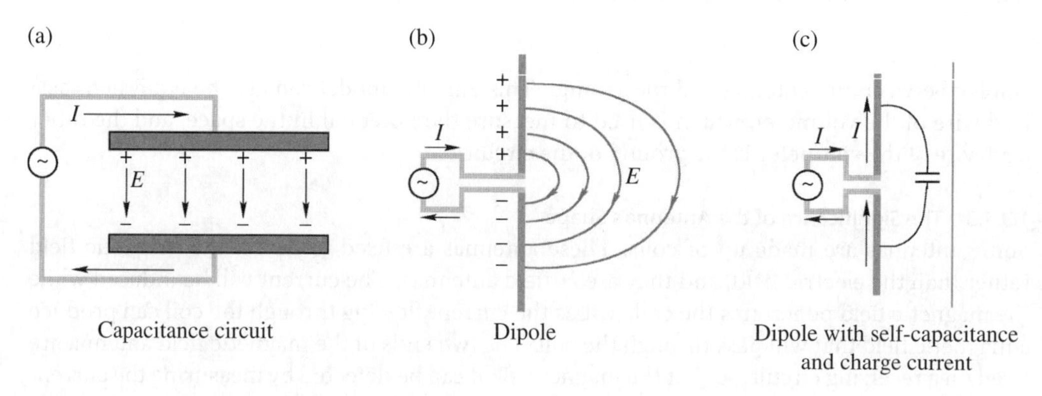

Figure 1.16 The principle of electric field antenna.

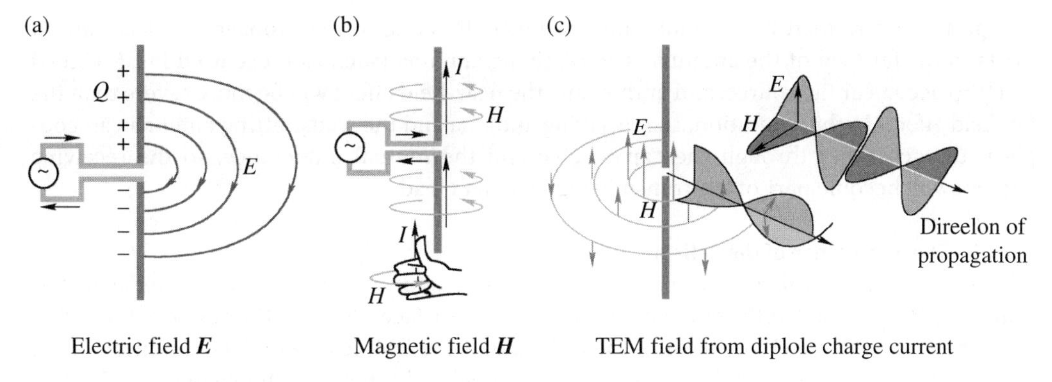

Figure 1.17 Schematic diagram of electric field antenna radiation.

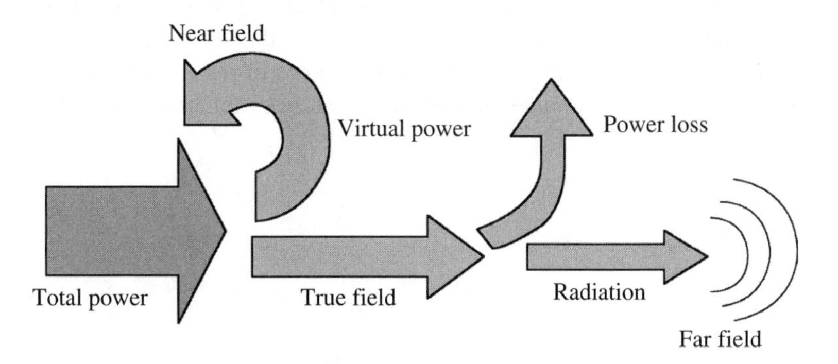

Figure 1.18 The power flow of radiation.

is generated while the E and H are perpendicular to each other. The antenna can also transform a TEM wave through the reciprocal theory back to the current and the voltage, as the antenna has the complementary nature of transmitting and receiving. The radiation situation of the antenna is shown in Figure 1.18. The antenna's reactance sections store the energy in the electric field and magnetic field around the antenna. The reactive power in the antenna is exchanged backward and forward between the power supply of the antenna and the reactive components of the antenna.

As there is a 90° phase difference between the voltage and the current in the L–C circuit, a 90° phase difference exists between the E field (generated by the voltage) and the H field (generated by the current) of the antenna if the antenna's resistance can be ignored. In a circuit, active power can only be consumed when the load impedance has an active component and the current and the voltage are in phase. This situation also applies to the antennas. The antennas have some little resistance, so the power consumption caused by these active power components exists in thc antenna. In order to produce the radiation, E field and H field must be in phase, as shown in Figure 1.17c. For the antennas behaving as a capacitance and inductance, how is the radiation produced? The in-phase component is the consequence of the propagation delay. The EM wave from the antenna does not simultaneously arrive at all points in the space, but propagates at the speed of light. At the distance far away from the antenna, this propagation delay puts the E field and the H field in phase.

Thus, the E field and the H field have different components, which contain a field of energy storage (the imaginary part) and the radiation part (the real part). The imaginary part is determined by the antenna's capacitance and inductance, which mainly exists in the near field. The real part is determined by the radiation resistance. It is due to the propagation delay, and it exists in the far field of the antenna. The receiving antenna (such as those used in EMC test) can be placed near the source, and at this time the near-field effect will be more severe than the far-field effect. In this condition, the receiving antenna and the transmitting antenna can couple with each other through the capacitance and the mutual inductance, so the receiving antenna will become part of the transmitting antenna's load.

1.3.4.5 The Importance of the Reflection

When you look in a mirror, you can think of the reflection effect as as being similar to EM radiation. Why is wave reflected back from the metal surface? What is the result of the radiation when it is reflected? The basis of the reflection is the EM field boundary condition of the metal surface. The field boundary conditions for *E* field and *H* field are shown in Figure 1.19. In the interior of the metal, due to the electric field, the charges will move freely. When the time-varying magnetic field exists, the current will be generated. The charge near the metal will cause the displacement of the charge on the metal surface. Any tangential component of the *E* field will cause movement of the charge until the tangential component of the *E* field is zero. The result is that there are the equivalent images or the virtual charges located below the

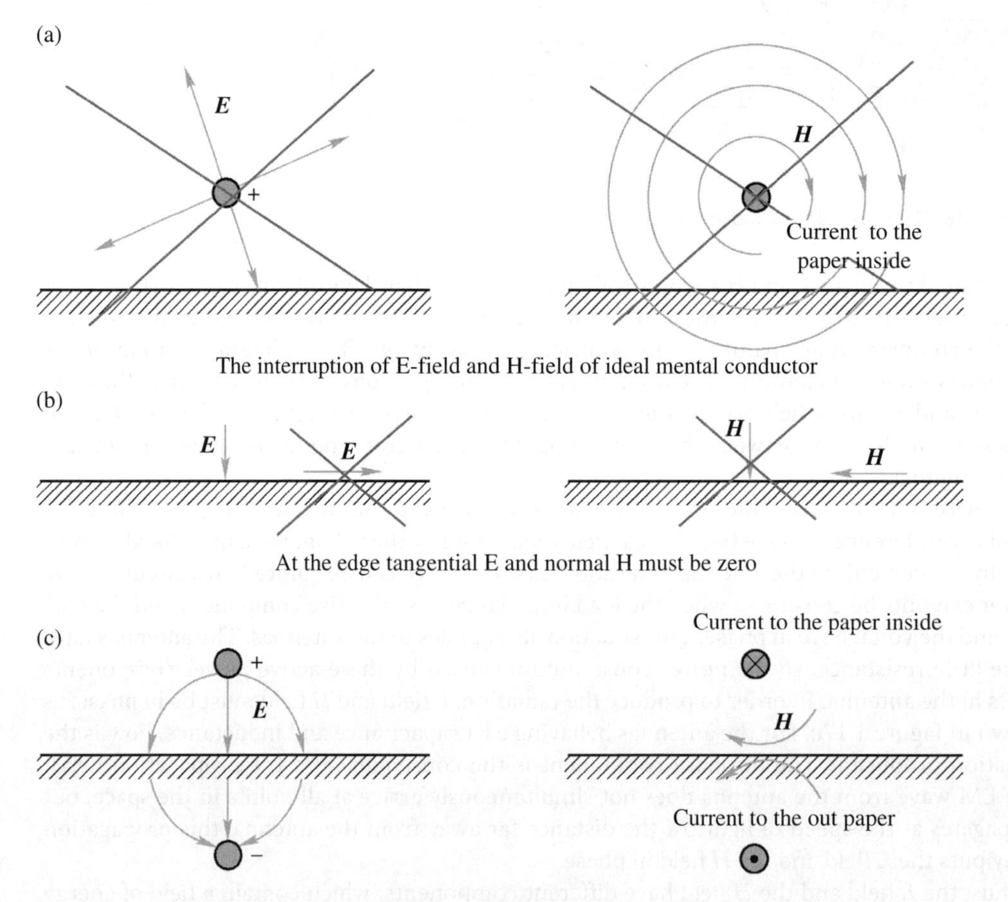

Figure 1.19 The diagram of electric field reflection.

surface of metal, as shown in Figure 1.19c. The image charges do not actually exist, but are the equivalent charge characterization of the actual results. The time-varying magnetic field induces an electric current in an ideal conductor. The current would resist the variation in the magnetic field, therefore, there will be no tangential component penetrating the metal surface. Thus, as shown in Figure 1.19c, the current image causes the normal component of H field to disappear on the metal surface.

The image effect is very important, because the antenna is often near the conductive surface, such as earth, car, or the metallic plates, the reference ground plane of the circuit board, the metallic housing of the products, the ground reference plate in the EMC test, and so on. The field radiated to the space is the sum of the radiated field from the antenna and that from the image field. If we look at the E-field of the dipole, it is very easy to see this effect. Figure 1.20a shows the dipole and its image, which is parallel with the conductor. When the dipole is perpendicular to the ground plane, the image of the dipole with the reverse charges exists below it, as shown in Figure 1.20b. In these two examples, the field at some points in the space is the sum of the field from the dipole and that from its image field. When the field is radiated to the metal from the dipole, as shown in Figure 1.20c, the reflection can be interpreted as the transmitting wave from the image.

According to this analysis, the monopole antenna shown in Figure 1.21 can be equivalent to the dipole antenna, and the monopole antenna with half-length of the dipole antenna has the equivalent length of the dipole antenna due to the effect of image on the ground plane. It has the equivalent antenna length as the dipole antenna, i.e. the length of the dipole antenna is two times the length of the monopole antenna.

1.3.4.6 The Relationship Between Antenna Impedance and Frequency

Antenna impedance is a function of frequency. The current and charge distribution on the antenna varies with the frequency. The current on the dipole is generally a sinusoidal function of the position along the antenna, and its frequency is fixed. Since the wavelength of the signal depends on the frequency, the length of the antenna is a fraction of the wavelength at a frequency. When the length of the dipole is half the wavelength of the frequencies, the current on the dipole is shown in Figure 1.22a. When the length of the dipole is a wavelength of the frequencies, the current on the dipole is shown in Figure 1.22b. If the length of the dipole is half the wavelength (one-fourth the wavelength for the monopole antenna), the current on the excitation source is the largest; therefore the input impedance of the antenna is the minimum at this frequency, and is equal to the resistance of the antenna (the actual resistance + the radiation resistance).

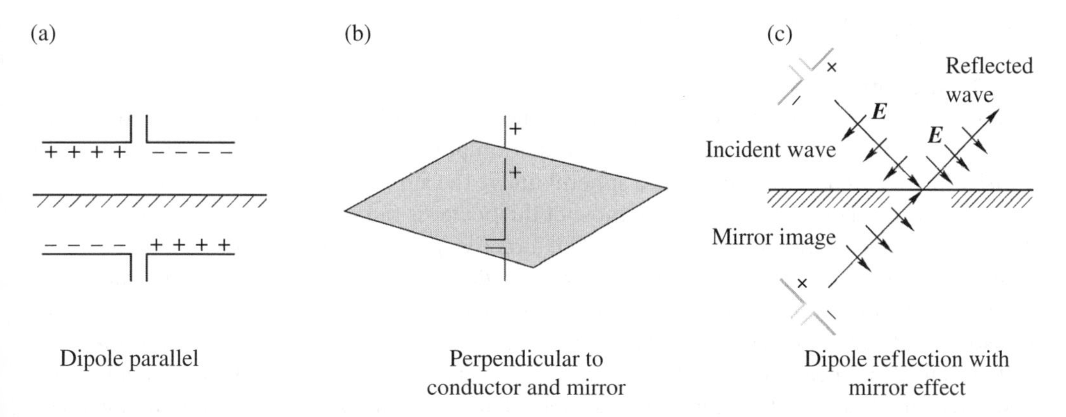

Figure 1.20 Image theory of dipole antenna.

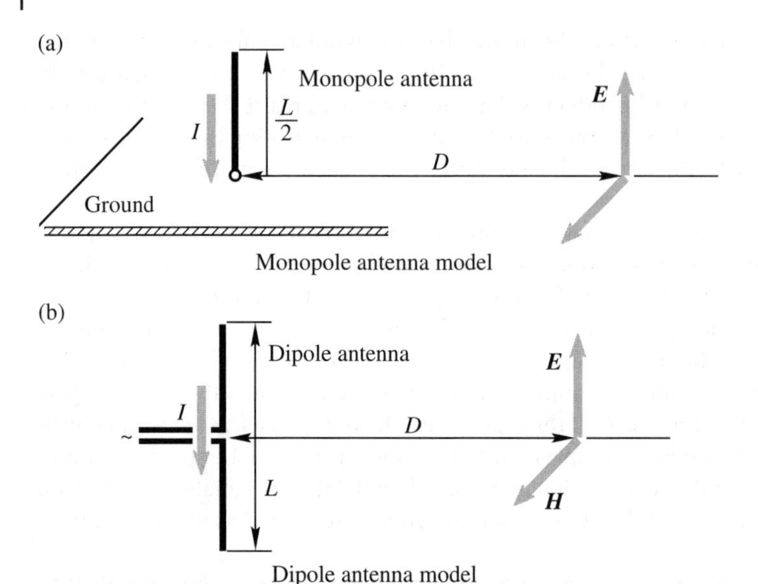

Figure 1.21 Radiation model of monopole antenna and dipole antenna.

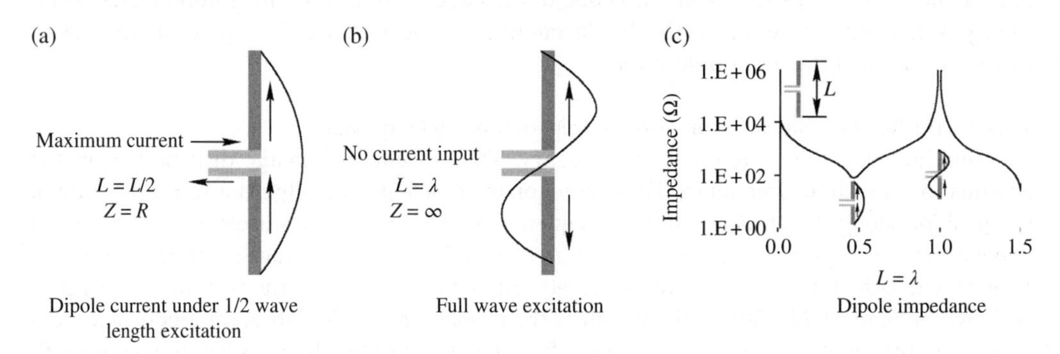

Figure 1.22 Antenna impedance frequency characteristic.

For the case where the dipole length is one wavelength, the current of excitation source is zero, therefore, the input impedance is infinite. The relationship between the impedance and the frequency is shown in Figure 1.22c.

1.3.4.7 The Antenna Radiation Direction

The radiated power plot from one antenna is not uniform in all directions. The gain of the antenna is characterized by the ratio of the radiation power in a given direction and the power density (distributed in the surface of a sphere) under the situation that the radiation is uniform in all directions. For a dipole antenna, most of the power is radiated in the vertical direction of the antenna axis, as shown in Figure 1.17. The directivity of the antenna is the gain in the direction with the maximum power, which is perpendicular to the dipole axis. The gain is measured with dBi = 10*log (gain). A three-dimensional or two-dimensional antenna directivity pattern is also known as the power directivity pattern, the power plot, or the power distribution.

It visually describes how the antenna receives or transmits within a specific frequency range, and it usually describes the scenario at far field. The radiation pattern of the antenna is influenced mainly by the geometric size of the antenna, also impacted by the surrounding terrain or

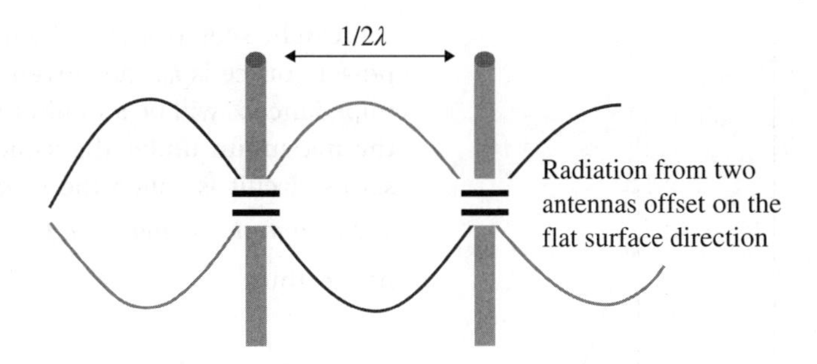

Figure 1.23 The side view of a half-wave dipole array.

other antennas. Sometimes the antenna directivity can be changed by using multiple antennas in the antenna array. As shown in Figure 1.23, using two antennas with the same feeder can eliminate the field on the antenna plane if the interval between them is half the wavelength.

1.3.5 Resonance of the RLC Circuit

1.3.5.1 RLC Series Resonant Circuit

The resistor inductance capacitor (RLC) series circuit is shown in Figure 1.24.

The relationship between the AC voltage U and the AC current I (both are RMS value) can be expressed as follows, according to AC Ohm's law:

$$I = \frac{U}{Z} = \frac{U}{\sqrt{R^2 + \left(\omega L - \frac{1}{\omega C}\right)^2}} \tag{1.2}$$

The phase difference between the voltage and the current is

$$\varphi = \text{arctg}\frac{\omega L - \frac{1}{\omega C}}{R} \tag{1.3}$$

In formula (1.2), $Z = \sqrt{R^2 + \left(\omega L - \frac{1}{\omega C}\right)^2}$ is the impedance of the AC circuit, $X_L = \omega L$ is the inductive reactance, $X_C = \frac{1}{\omega C}$ is the capacitive reactance, the angular frequency $\omega = 2\pi f$, and f is AC frequency. It can be seen from formulas (1.2) and (1.3) that I and ϕ are the function of the angular frequency ω (or f), and the two formulas, respectively, show the relationship between the amplitude and the frequency and the relationship between the phase and the frequency of the circuit. Starting from formula (1.2), if the voltage U remains constant, and R, L, C is fixed, let's study the relationship between the current I and the changing frequency f. The relationship curve between I and f in RLC series is drawn in Figure 1.25.

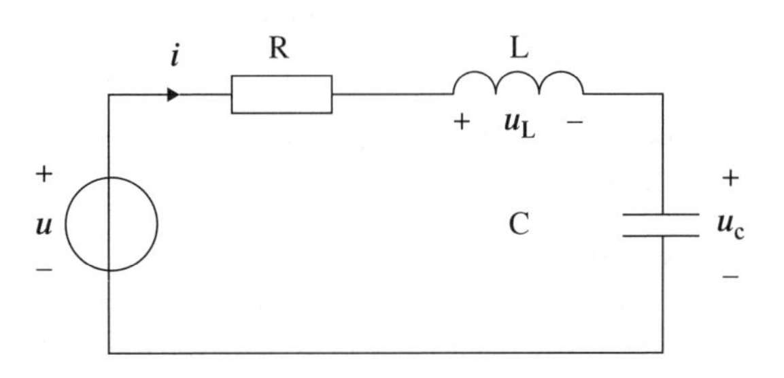

Figure 1.24 RLC series resonant circuit.

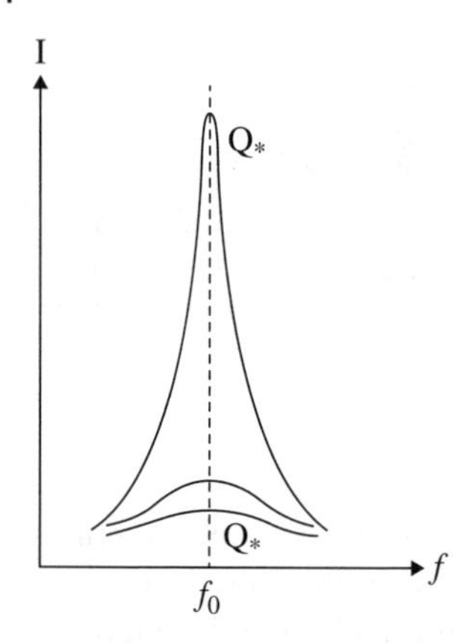

Figure 1.25 The changing condition of current I with the frequency of changing in RLC series circuit.

It can be seen from this figure that when the frequency of the power source is f_0, the current I has a maximum value, i.e. the impedance Z will be a minimum value, because the value of I is the maximum under the same voltage. This state of the RLC series circuit is called the *series resonance* and f_0 is called the *self-resonant frequency*. $\omega L - \frac{1}{\omega C} = 0$, when it works in resonance; thus:

$$f_0 = \frac{\omega_0}{2\pi} = \frac{1}{2\pi\sqrt{LC}} \tag{1.4}$$

Formula (1.4) shows the circuit composed of the inductance L and the capacitance C has a certain natural frequency f_0 (or the intrinsic angular frequency ω_0). When the frequency of the external power source is the same as the natural frequency of the circuit, the resonance phenomena will occur. From Formula (1.3), $\varphi = 0$, i.e. the phase angle between the current and the voltage when the resonance occurs at f_0, we can determine that the circuit is resistive and the voltage on the inductance L and the capacitance C, respectively, are as follows: $U_L = IX_L = \frac{U}{R}\omega_0 L$ and $U_C = IX_C = \frac{U}{R}\frac{1}{\omega_0 C}$, since $\omega_0 L = \frac{1}{\omega_0 C}$, $U_L = U_C$. The quality factor Q is used to measure the performance of the resonant circuit, which is defined as the ratio of the voltage on the inductor (or the capacitor) U_L (or U_C) and the voltage on the resistor when the resonance occurs:

$$Q = \frac{U_L}{U} = \frac{U_C}{U} = \frac{\omega_0 L}{R} = \frac{1}{R\omega_0 C} \tag{1.5}$$

The voltage on the inductor or the capacitor U_L or U_C is Q times the voltage on the resistor when the resonance occurs. When R is much larger than X_L (or X_C), $Q \gg 1$, so that the voltage on the inductor or capacitor can be much larger than the voltage on the resistor, which is also the supply voltage U when the resonance operates at f_0, and series resonance is also known as the voltage resonance.

The circuit shown in Figure 1.24 is often used in the filtering circuit. R is the equivalent series resistance of the LC filtering circuit, which is very small, L is the filtering inductance, C is the filtering capacitor, U is the interference source, and the load is in parallel with C So at or near the resonance frequency of the LC filtering circuit, the filtering circuit plays the role of enlarging the interference source, which should be considered in the filtering circuit design. In addition, when considering the actual frequency characteristics of a capacitor, the equivalent circuit between the two terminals of the capacitor is the series circuit with LCR, while the designers should be more concerned about the voltage between the LCR terminals at this time, not the voltage between the terminals of C Thus, the series resonance can achieve better EMC effects.

Example 1.1 The circuit is shown in Figure 1.24, $U(t) = 10\sqrt{2}\cos\omega t$ V, $R = 1\,\Omega$, $L = 0.1$ mH, $C = 0.1\,\mu$F.

Questions

1) At what frequency does the circuit operate when the resonance of the circuit occurs?
2) What is the magnitude of U_L and U_C when the resonance of the circuit occurs?

Solution

The frequency of the voltage source should be:

$$f = f_0 = \frac{1}{2\pi\sqrt{LC}} = \frac{1}{2\pi\sqrt{10^{-4}\times 10^{-8}}} = 159(kHz)$$

The quality factor of the circuit is:

$$Q = \frac{2\pi f_0 L}{R} = 100$$

Then,

$$U_L = U_C = QU_S = 100\times 10(\mathrm{V}) = 1000(\mathrm{V})$$

1.3.5.2 RLC Parallel Resonant Circuit

RLC parallel resonant circuit is shown in Figure 1.26.

A parallel circuit consists of the resistance R, the inductance L, and the capacitance C. Its total impedance Z, the phase difference between the current I and the voltage U are, respectively:

$$Z = \frac{R^2 + (\omega L)^2}{\sqrt{R^2 + \left[\omega CR^2 + \omega L\left(\omega^2 LC - 1\right)\right]^2}} \quad (1.6)$$

$$\varphi = \mathrm{arctg}\left[\frac{\omega L - \omega CR^2 - \omega^3 L^2 C}{R}\right] \quad (1.7)$$

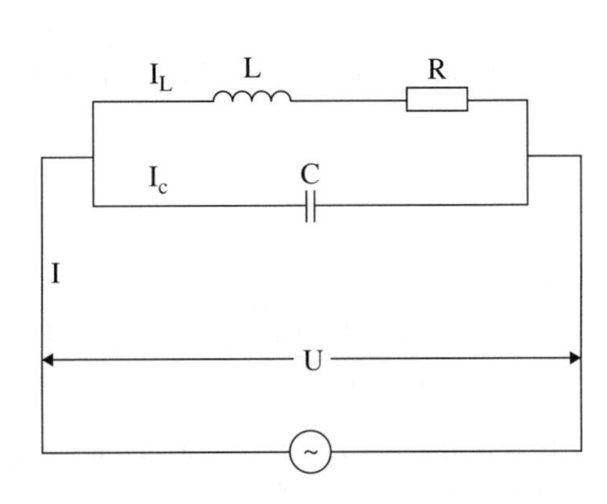

Figure 1.26 RLC parallel resonance circuit.

Here, $\omega = 2\pi f$

According to formula (1.6), the relationship curve of the total current I and the total impedance Z versus the frequency is shown in Figure 1.27. The frequency at which the maximum impedance in the figure corresponds to the condition is $[\omega L - \omega CR^2 - \omega^3 L^2 C] = 0$. In this case, Z is the maximum, I is the minimum, $\varphi = 0$, and the circuit presents resistive. This state is called the parallel resonance, and the resonant frequency is:

$$f = \frac{\omega}{2\pi} = \frac{1}{2\pi\sqrt{LC}}\sqrt{1 - \frac{CR^2}{L}} = f_0\sqrt{1 - \frac{CR^2}{L}} \quad (1.8)$$

f_0 is the self-resonant frequency of the series circuit, and when $(CR^2)/L \ll 1$, it can be ignored, so $f = f_0$. This is the same as the series resonant circuit – the larger the quality factor Q of the circuit, the better the circuit selectivity will be. When the resonance occurs, the currents in the two circuit branches I_L and I_C are approximately equal, and they are equal to Q times the current in the resistor branch I, so the parallel resonance is also known as the current resonance.

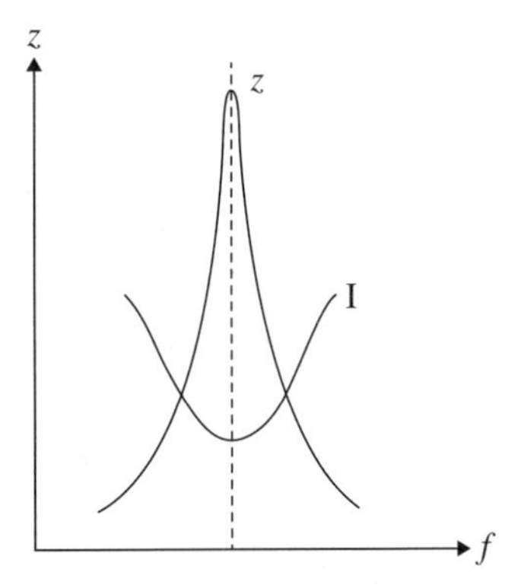

Figure 1.27 Relationship between current and frequency in RLC parallel resonance circuit.

At high frequency, the parasitic capacitance between the inductor terminals and the equivalent series resistance of the inductor must be considered, and the equivalent model of the inductor is the RLC parallel resonant network shown in Figure 1.26. At this time, the inductor can achieve good EMC effect.

Example 1.2 The model of the parallel circuit consisting of the inductor and the capacitor is shown in Figure 1.26. $R = 1\,\Omega$, $L = 0.1\,\text{mH}$, $C = 0.01\,\mu\text{F}$. Try to solve the resonant angular frequency and the impedance of the resonant circuit.

Solution

Write the conductance equation at the driving point according to the phasor model:

$$Y(j\omega) = j\omega C + \frac{1}{R + j\omega L}$$

$$= \frac{R}{R^2 + (\omega L)^2} + j\left[\omega C - \frac{\omega L}{R^2 + (\omega L)^2}\right]$$

$$Y(j\omega) = \frac{R}{R^2 + (\omega L)^2} + j\left[\omega C + \frac{\omega L}{R^2 + (\omega L)^2}\right]$$

Set the imaginary part of the formula equal as zero:

$$\omega C - \frac{\omega L}{R^2 + (\omega L)^2} = 0$$

Then:

$$\omega_0 = \frac{1}{\sqrt{LC}}\sqrt{1 - \frac{CR^2}{L}} = \frac{1}{\sqrt{LC}}\sqrt{1 - \frac{1}{Q^2}}$$

In this formula, $Q = \frac{1}{R}\sqrt{\frac{L}{C}}$, which is the quality factor of the series RLC circuit.
When $Q \gg 1$,

$$\omega_0 = \frac{1}{\sqrt{LC}}$$

Substitute the numerical data and get:

$$\omega_0 = \frac{1}{\sqrt{10^{-4} \times 10^{-8}}}\sqrt{1 - \frac{10^{-8}}{10^{-4}}}\left(\text{rad}\,\text{s}^{-1}\right) = 10^6\left(\text{rad}\,\text{s}^{-1}\right)$$

The impedance at resonance:

$$Z(j\omega_0) = \frac{1}{Y(j\omega_0)} = R + \frac{(\omega_0 L)^2}{R} = R\left(1 + Q^2\right)$$

When $\omega_0 L \gg R$,

$$Z(j\omega_0) = \frac{(\omega_0 L)^2}{R} = \left(10^6 \times 10^{-4}\right)^2 \Omega = 10(k\Omega)$$

1.4 Common Mode and Differential Mode in the EMC Domain

The voltage and current consist of two modes when they are transmitted through conductors, namely, the common mode and the differential mode. A product may have power supply lines, signal lines, and other communication lines to exchange with other devices or peripheral equipment, and there will be at least two conductors serving as the round-trip routes to transmit the power or signal. In addition to these two conductors, there is usually a third conductor, which is the ground. The disturbance voltage and current can be divided into two types. One is that these two conductors serve as the round-trip routes, respectively; the other one is that these two conductors serve as the forward path, and the ground conductor serves as the return path. The former is called *differential mode* and the latter is called *common mode*. As shown in Figure 1.28, the power supply, the signal source, and the load are connected by two conductors. The currents through these two conductors are the same in amplitude and opposite in direction. In fact, the source of the disturbance is certainly not connected between the two conductors. Since there are various forms of the noise sources, there is also a voltage difference between these two conductors and the ground. As a result, the disturbance currents flowing through these two conductors are different in amplitude.

As shown in Figure 1.29, with the driving of the disturbance voltage applied between these two conductors, the currents on these two conductors are the same in magnitude but opposite in direction (differential-mode currents). But when the disturbance voltage is applied between these two conductors and the ground, the currents on these two conductors are the same both in amplitude and direction. These currents (common mode) go together to the ground and flow back through the ground. The differential-mode disturbance current and the common-mode disturbance current on one conductor are in the same direction, so they shall be added together; while the differential-mode noise and the common-mode noise on the other conductor are in

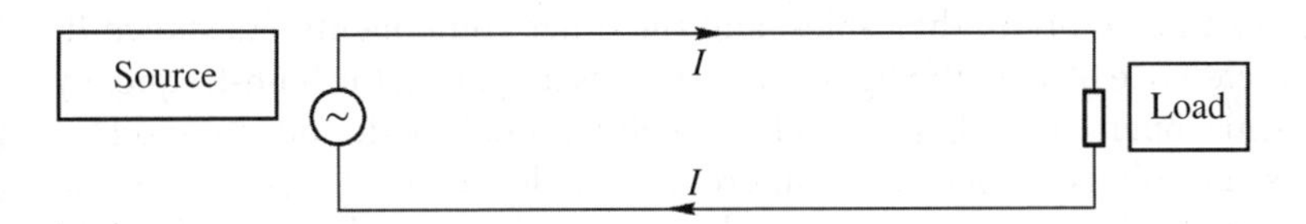

Figure 1.28 Differential-mode signal.

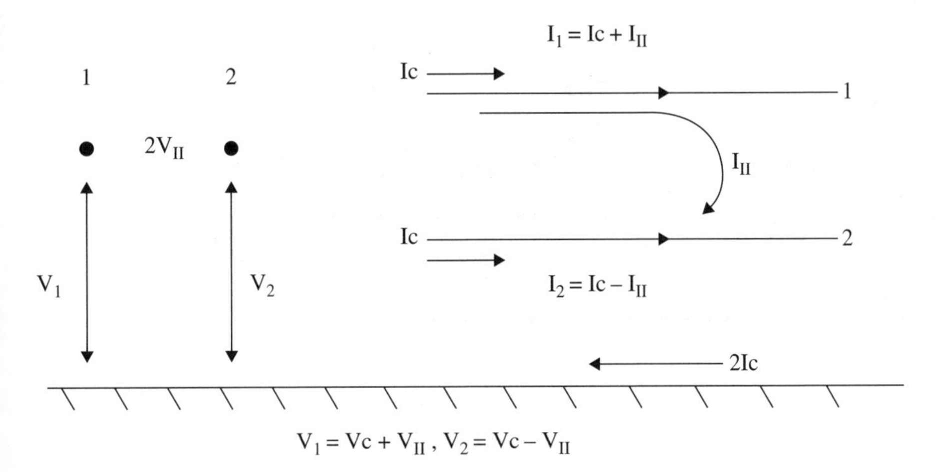

Figure 1.29 The relationship of voltage/current to the ground and the differential and common-mode voltage/current.

the opposite direction, so they shall be subtracted. Therefore, the currents flowing through these two conductors are different in amplitude.

Take the voltage to the ground into account. As shown in Figure 1.29, for the differential-mode voltage, the voltage on one conductor is the positive half of the voltage between these two conductors, while the voltage on the other conductor is the negative half of the voltage between these two conductors, so they are balanced. However, the common-mode voltages on these two conductors are the same, so when these two modes exist at the same time, the voltage to the ground on these two conductors will be different.

Thus, when the voltage or the current to the ground on these two conductors are not the same, the components of these two modes can be calculated by the following method:

$$U_N = (U_1 - U_2)/2 U_c = (U_1 + U_2)/2$$

$$I_N = (I_1 - I_2)/2 I_c = (I_1 + I_2)/2$$

In the actual circuit, the common-mode disturbance and the differential-mode disturbance are always mutually transformed; the impedance (for this impedance, the impact of the distribution parameters shall be considered) exists between the terminals of these two conductors and the ground. Once the impedance of these two conductors is unbalanced, the mutual transformation of the modes will appear on the terminals. When one mode transferred on the conductor is reflected at the terminal, part of this mode will be transformed into another mode. Furthermore, the interval between these two conductors is generally small, and the distance between the conductors and the ground conductor is large. Therefore, if the interference radiated from the conductors is considered, compared with the radiation generated by the differential-mode current, the strength of the common-mode current radiation is larger.

Then consider what common-mode or differential-mode signals exist in the products. The electronic products in the EMC test environment are usually composed of the PCB, the shell (nonmetallic shell can be ignored), the cables, and the other components. As shown in Figure 1.30, the ground reference plate in the figure is the necessary part in the high-frequency EMC test. U_{AC} is the normal operating voltage, which is the differential-mode voltage, and U_{BD} is also the normal operating voltage, which is the differential-mode voltage. U_{CD} exists on the reference ground of the PCB, which is a common-mode voltage. U_{DE} exists between the PCB reference ground and the metallic shell, which is also a common-mode voltage. U_{EF} exists

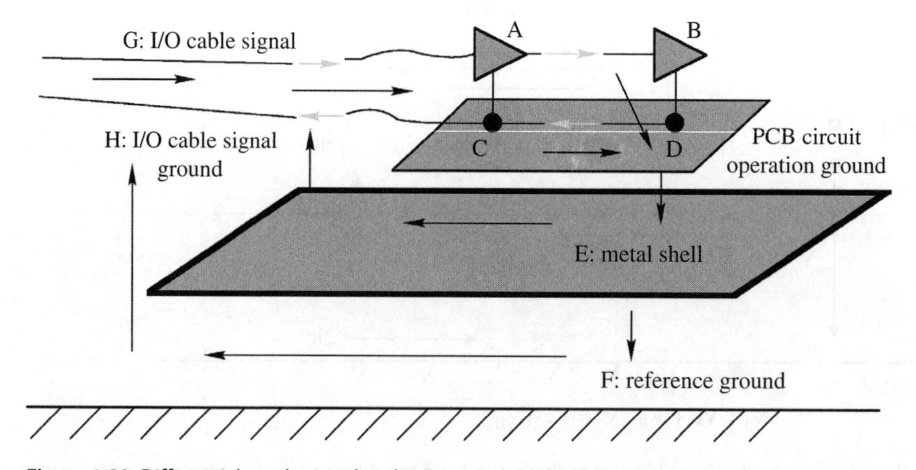

Figure 1.30 Differential-mode signal and common-mode signal in product.

between the metallic shell and ground reference plate. It is also a common-mode voltage. U_{FH} exists between ground reference plate and the cable. It is also a common-mode voltage. U_{EH} exists between the metallic shell and the cable. It is also a common-mode voltage. The current I_{AB} flows inside the PCB. It is a normal operation signal, which is a differential-mode current. I_{DC} is the return current of the signal with I_{AB}, with the same amplitude as I_{AB}. It is also a differential-mode current. I_{CD} is generated by the voltage drop U_{CD} when I_{DC} (the return current of the signal with I_{AB}) flows through the reference ground in the PCB. It will flow to the metallic shell or the ground reference plate, and will then return from the cable or the other conductors. It is a typical common-mode current. I_{DE}, I_{EH}, I_{EF}, and I_{FH} are all common-mode currents.

Thus, for a product, the common-mode current always flows to the ground reference plate or the metallic shell. It is an undesirable current signal. The differential-mode current always flows inside the PCB or between PCBs. It is a useful and desirable current signal. The common-mode voltage is the voltage causing the common-mode currents, and the differential-mode voltage is the voltage causing the differential-mode currents.

1.5 Essence of the EMC Test

1.5.1 Essence of the Radiated Emission Test

The essence of the radiated emission test is to test the radiation generated by two equivalent antennas in the product. The first one is the signal loop equivalent as the antenna. The signal loop is the equivalent antenna producing the radiation, and the source of this radiation is the current flowing through the loop (it is generally the normal operation signal, which is a differential-mode signal, such as a clock signal and its harmonics), as shown in Figure 1.31. If the current with the amplitude I and the frequency F flows through the loop of which the cross-section area is S, then, in the free space, the radiated field strength at the distance D away from the loop is:

$$E = 1.3SIF^2/D_0 \tag{1.9}$$

In this formula, E is the electric field strength, in units of $\mu V\,m^{-1}$; S is the loop area, in units of cm^2; I is the current amplitude, in units of A; F is the signal frequency, in units of MHz; and D is the distance, in units of m.

The loop exists in any signal transmission circuit of the electronic products, if the signal is alternating, and then the signal loop will generate the radiation. When the current amplitude and the frequency of the signal in the product are determined, the radiation strength generated by the signal loop is related with the loop area. Therefore, controlling the signal loop area is important for controlling the EMC problem.

Another equivalent antenna model that generates the unintentional radiation in the product is the monopole antenna (shown in Figure 1.21a), or the symmetric dipole antenna (shown in Figure 1.21b), the conductor of which can be equivalent, as the monopole antenna or asymmetric

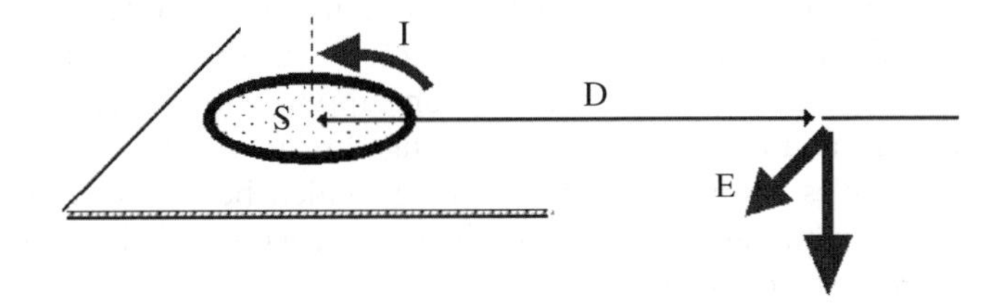

Figure 1.31 The loop is equal to radiation antenna.

dipole antenna is usually the cable in the product or the other conductors with long length. The source of such radiation is the common-mode current flowing through the cable or the other conductors with long length (the equivalent antennas). It is usually not the useful working signal in the cable or the long conductors, but a parasitic "useless" signal.

The amplitude study of this common-mode current that generates the common-mode radiation is the key to analyzing the radiated emission problem. As shown in Figure 1.21, if the signal is flowing in the antenna with the current amplitude I, frequency F, then the field strength of the radiation at the distance D away from the antenna is

When $F \geq 30\,\text{MHz}$, $D \geq 1\,\text{m}$ and $L < \lambda/2$,

$$E \approx 0.63ILF/D \tag{1.10}$$

When $L \geq \lambda/2$,

$$E \approx 60 \times I/D \tag{1.11}$$

In formulas (1.10) and (1.11), E is the electric field strength, in units of $\mu\text{V m}^{-1}$; I is the current amplitude, in units of μA; F is the signal frequency, in units of MHz; D is the distance, in units of m; and L is the cable length, in units of m.

In the electronic products, in addition to the information expressed in the circuit schematic that describes the function of the product, there are many unknown details, such as the parasitic capacitance and inductance between the signal lines, the parasitic capacitance between the signal lines and the ground reference plane, the lead inductance of the signal line, and so on. These parameters are frequency-dependent, and their values are very small, so they are often ignored by the designers in the DC or low-frequency applications. However, for high-frequency radiated emission, these parameters are of increasing importance. These factors also generate the undesirable parasitic common-mode current on the equivalent antenna (the cable or the long conductor) in the products. Its current magnitude is small (usually below the level of mA or μA), but it is the main cause of the radiation (how the common-mode currents are produced is described in the later chapters).

From formula (1.11) it can also be seen that, when the length of the equivalent antenna is greater than half the wavelength of the signal, the radiation strength produced by the antenna only depends on the amplitude of the common-mode current. So studying the amplitude of the common-mode current on the cable or the long conductor in the products is significant for controlling the radiated emission of the products.

Many practices show that the radiated emission problems in most products are caused by such equivalent monopole antennas or dipole antennas. Especially with the use of multilayer PCB technology, the signal loop area is controlled to be smaller and smaller, and the radiation generated by the normal operation signal loops is much more limited. The case 25 "The excessive radiated emission caused by the loop" is an example that helps eliminate the misunderstanding of the loop radiation. In contrast, recently engineers have increasingly focused on the radiation caused by the equivalent monopole or dipole antennas as the products become more complex.

For the radiated emission test of the military-used products and the vehicle-electronics-related products, the standards will require the ground reference plate routed on the test bench, and the EUT and the cables connected with the EUT (the equivalent radiation transmitting antenna) are placed on an insulation support with 5 cm height from the ground reference plate this ground reference plate has a huge impact on the results of the radiation. In theory, when the cables connected with the EUT, which becomes the equivalent radiation antenna, are placed with the height h above the reference ground plane (shown in Figure 1.32), the different scenarios are shown as below.

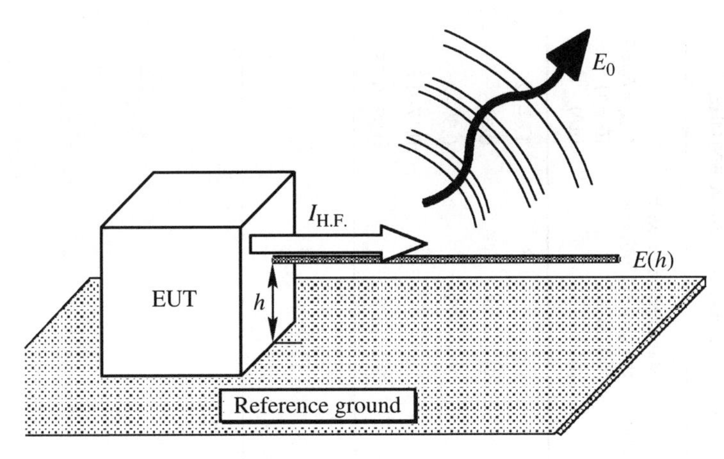

Figure 1.32 The cable radiation emission on the reference ground is decayed by the ground floor.

When $h \le \lambda/10$,

$$E(h) \approx E_0 \times 10 \times h/\lambda \tag{1.12}$$

When $h > \lambda/10$,

$$0 \le E(h) \le 2E_0 \tag{1.13}$$

In formulas (1.12) and (1.13), h is the height from the cable, which is equivalent as the radiated emission antenna to the ground reference plate, in units of m; E_0 is radiation strength from the cable in free space, in units of V m^{-1}; $E(h)$ is the radiation strength emitted to the space when the cable is placed with the height h above the ground reference plate, in units of V m^{-1}; and λ is the wavelength, in units of m.

According to the radiation principle of the monopole and dipole radiation model, since the cause of the monopole and dipole antenna radiation is the common-mode current on the antenna (namely the common-mode current on the cable or long conductor in the products), then under the radiated emission test conditions prescribed by the military standard and vehicle-electronics-related standards, the radiation generated by the common-mode current is lower than that under the conditions required by other standard. The relationship between the radiated emission limits prescribed by CISPR25 and the common-mode current amplitude of the equivalent monopole and dipole antennas to emit the radiation that exceeds the standard limit is shown in Table 1.1.

The relationship between the radiated emission limits prescribed by EN55022 or CISPR22 and the common-mode current amplitude of the equivalent monopole and dipole antenna to emit the radiation that exceeds the standard limit is shown in Table 1.2.

1.5.2 Essence of the Conducted Emission Test

The linear impedance stabilization network (LISN) is the key equipment in the power port conducted emission test. It can be seen from Figure 1.33, the receiver is connected between the 1 kΩ resistor in the LISN and the ground. After the receiver and the LISN are interconnected, the 50 Ω impedance of the receiver's signal input port is in parallel with the 1kΩ resistor in the LISN, and the equivalent impedance is close to 50 Ω, then it can be seen that, the essence of the conducted disturbance test on the power port is to measure the voltage across this 50 Ω impedance

Table 1.1 The relationship between the radiation emission limits specified in the CISPR25 standard and the common-mode current size required by the product of the equivalent monopole antenna and dipole antenna.

	30–54 MHz ($5.6\,m < S < 10\,m$)				70 MHz–1 GHz ($0.5\,m < L < 4.3\,m$) (not suitable for above 600 MHz)			
Level	Standard value (@1 m) $dB\mu V\,m^{-1}$	Linear standard value $E(h)$ (@1 m) $\mu V\,m^{-1}$	Standard value with the ground influence E_0 (@1 m) $\mu V\,m^{-1}$	$I_{COM} = E_0/60$ ($D = 1$) (μA)	Standard value (@1 m) $dB\mu V\,m^{-1}$	Linear standard balue (@1 m) $\mu V\,m^{-1}$	Standard value with the ground influence (@1 m) $\mu V\,m^{-1}$	$I_{COM} = E_0/60$ ($D = 1$) (μA)
1	46	200	2240–4000	37–67	36	83.3	83.3–716	1.4–12
2	40	100	1140–2000	19–34	30	41.6	41.6–358	0.7–6
3	34	50	570–1000	9.5–17	24	20.8	20.8–179	0.35–3
4	28	25	285–500	4.75–8.5	18	10.4	10.4–89.5	0.17–1.5
5	22	12.5	142.5–250	2.4–4.8	12	5.2	5.2–44.7	0.09–0.75

$E_0 = 60 I_{COM}/D; D = 1\,m;$

$F_{max} = 600\,MHz$, when $\lambda_{max} = 0.5\,m, h \le \lambda/10, h = 0.05\,m$

h is the height from cable to ground.

Table 1.2 The relationship between the radiation emission limits specified in the EN55022 or CISPR22 standards and the common-mode current required by the product of the equivalent monopole antenna and dipole antenna.

Level	37.5–230 MHz ($1.3\,m < \lambda < 10\,m$)				230 MHz–1 GHz ($0.3\,m < \lambda < 1.3\,m$)			
	Standard value (@1 m) $dB\mu V\,m^{-1}$	Linear standard value $E(h)$ (@1 m) $\mu V\,m^{-1}$	Standard value with the ground influence E_0 (@1 m) $\mu V\,m^{-1}$	$I_{COM} = E_0/60$ ($D = 1$) (μA)	Standard value (@1 m) $dB\mu V\,m^{-1}$	Linear standard value (@1 m) $\mu V\,m^{-1}$	Standard value with the ground influence (@1 m) $\mu V\,m^{-1}$	$I_{COM} = E_0/60$ ($D = 1$) (μA)
B	30	32	16	2.7	37	72	36	6
A	40	100	50	8.3	47	225	112.5	18.75

$E_0 = 60I_{COM}/D$; $D = 10\,m$;
When above 37.5 MHz, $h > \lambda/10$ is always correct, $0 \le E(h) \le 2E_0$ and
h is the height from the cable to reference plate, which is stipulated in EN55022 and CISPR22 standard as 0.8 m in bench device.

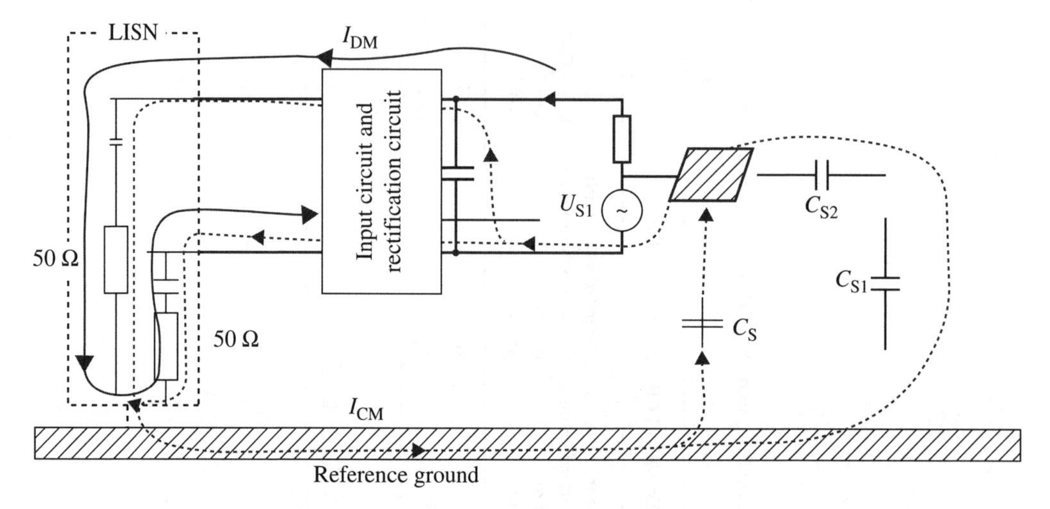

Figure 1.33 The current that causes the power port conducted disturbance.

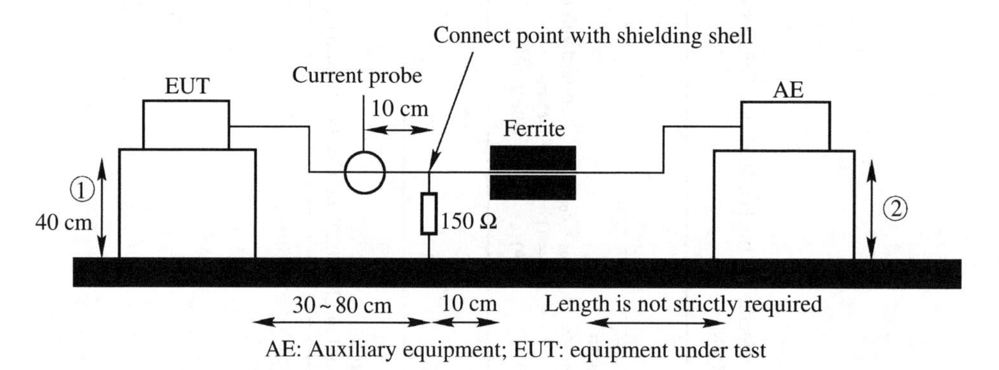

Figure 1.34 Configuration diagram of signal-to-port-conducted emission test.

(this impedance is constituted by the 1 kΩ resistor in the LISN in parallel with the input impedance of the receiver). When the 50 Ω impedance is constant, the essence of the conducted disturbances on the power port can be understood as the amplitude of the current flowing through the 50 Ω impedance. Two currents may flow through the 50 Ω impedance in the real products. One is I_{DM} shown in Figure 1.33, and the other is I_{CM} shown in Figure 1.33. Whether it is I_{DM} or I_{CM}, the tested value will be exhibited on the receiver, and the receiver itself cannot differentiate them. It needs the designer to control and analyze them. Controlling the disturbance current not flowing through the 50 Ω impedance is the key to solve the conducted disturbance problem on the power port. According to a lot of practices, it is proved that most of the conducted disturbance problems on the power port are caused by I_{CM}, which is the common-mode current, so analyzing its path and amplitude is very important.

The current probe is a key equipment to measure the conducted disturbances on the signal port. Figure 1.34 shows the conducted disturbance test configuration for the signal port. It can be clearly seen from Figure 1.34 that the current probe essentially measures the common-mode current on the cable of the EUT. Of course, just as with the radiation model of the monopole antenna or the dipole antenna, the common-mode current leading to the conducted disturbances on the signal port is usually not the normal operating current on the signal port, but some undesired common-mode current. So, the test essence of the conducted disturbance on the signal port is the same as that of the radiated emission generated by the cable or the long conductor, which is equivalent as the monopole or the dipole antenna in the radiated emission test, but the test frequency band will not be the same.

1.5.3 Essence of the ESD Immunity Test

It can be seen from the configuration description of the ESD test that during the ESD test, the grounding cable of the ESD gun is required to be connected with a ground reference plate (the ground reference plate is connected with the protective ground), and the EUT is placed above the ground reference plate (on a table or support with 0.1 m height). The electrode of the ESD gun approaches various parts in the EUT that are accessible or the vertical and the horizontal coupling plates, it determines the ESD test is an immunity test mainly in common-mode, since the ESD current always flows to the ground reference plate finally.

The ESD disturbance principle can be considered in two ways. Firstly, when the ESD discharge phenomenon occurs on the tested part of the EUT, along with the ESD discharge current, analyzing the ESD discharge current path and the current amplitude is extremely important. It is worth noting that the rising time of the ESD discharge current waveform is shorter than 1 ns, which means that the ESD discharge is a high-frequency phenomenon. The discharge current path and the amplitude of the ESD current are not only determined by the internal connections (this part is mainly represented in the schematic), but also affected by the distribution parameters. The distribution path of the ESD discharge current in the ESD test is shown in Figure 1.35. The C_{P1}, C_{P2}, and C_{P3} in Figure 1.35 are the parasitic capacitances between the discharge point and the internal circuit, between the cable and the ground reference plate, between the EUT shell and the ground reference plate, respectively. The magnitude of the capacitance will affect the amplitude of the ESD current on each path. Thinking about that, if there is an ESD current path containing the internal operation circuit of the product, then the impact will be enormous in terms of the ESD discharge when the ESD test is performed for the

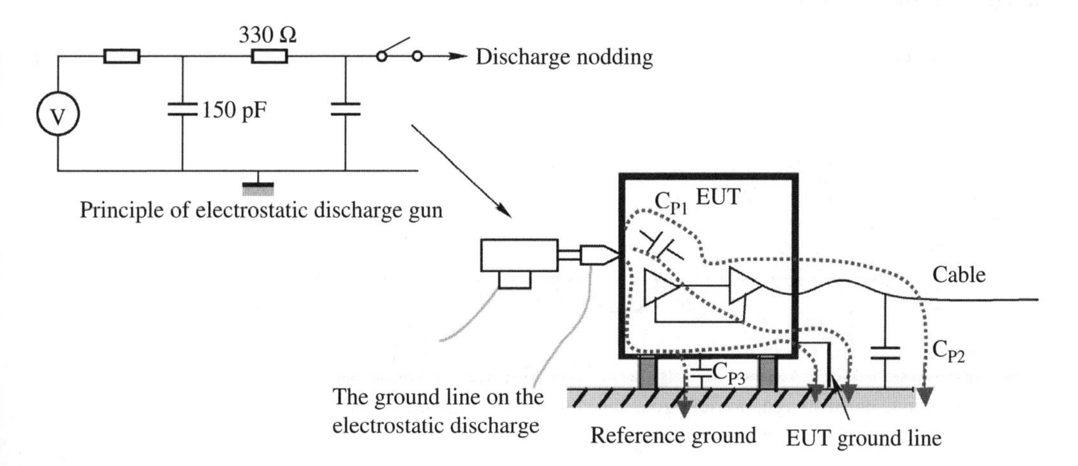

Figure 1.35 ESD discharge current distribution path when doing ESD test on a certain product.

product; On the contrary, the product will more easily pass the ESD test. So, if the product is designed to avoid the ESD common-mode current flowing through its internal circuit, then the ESD immunity design of this product is successful, the ESD immunity test essentially comprises a common-mode transient current (the ESD current) flowing through the products (the principle of the common-mode transient current disturbs the normal operation of the circuit is described in Section 1.5.5).

Second, the ESD current generated in the ESD test is accompanied with the transient magnetic field. When this time-varying magnetic field penetrates any circuit loop, an induced electromotive force will be generated in the loop, which will affect the normal operation of the circuit loop. For example, the area of a circuit loop $S = 2\,\text{cm}^2$, the distance from the loop to the discharge current in the ESD test $D = 50\,\text{cm}$, the maximum transient current $I = 30$ A during ESD test, then the magnetic field at 50 cm distance away from the ESD transient current can be calculated by formula (1.14):

$$H = I/(2\pi D) = 30/(2\pi \times 0.5) \approx 10\ \text{A m}^{-1} \tag{1.14}$$

The transient induced voltage U in the loop with the loop area S can be calculated by formula (1.15):

$$U = 0.0002 \times 4\pi \times 10^{-7} \times 10 / 1 \times 10^{-9} \approx 2.5\,\text{V} \tag{1.15}$$

In this formula, $\Delta t = 1$ ns is the rising time of ESD current, μ_0 is the air permeability. It can be seen from the result of the calculation that 2.5 V is a dangerous disturbance voltage compared with the normal operating voltage in the circuit.

1.5.4 Essence of the Radiated Immunity Test

The radiated immunity test is essentially a test procedure reciprocal to the radiated emission test. In the PCB, the signals come from the driving source, and are transferred to the load, and then return to the driving source, which forms the signal current closed loop, namely, each signal transmission path contains a closed loop. When the external EM field passes through this loop, the induced voltage is generated in the loop, as shown in Figure 1.36.

The induced voltage while the magnetic field penetrates a single-turn circuit can be calculated by formula (1.15). Since,

$$\Delta B = \mu_0 \Delta H \tag{1.16}$$

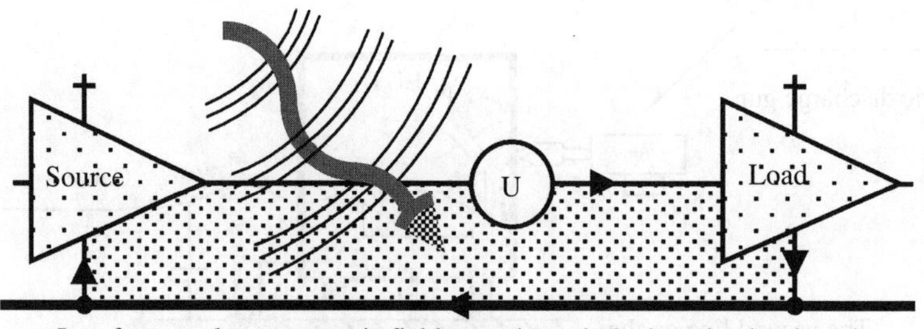
Source
U
Load
Interference electromagnetic field goes through the loop in circuit and generates induced voltage

Figure 1.36 Flux passes through the loop and generates induced voltage.

Then formula (1.15) can be transformed to formula (1.17),

$$U = S \cdot \Delta B / \Delta t \tag{1.17}$$

In formula (1.16) and (1.17), U is the induced voltage, in units of V; H is the magnetic field strength, in units of $\mathrm{A\,m^{-1}}$; B is the magnetic induction strength, in units of T; μ_0 is permeability of free space, $\mu_0 = 4\pi \cdot 10^{-7}\ \mathrm{H\,m^{-1}}$; and S is the loop area, in units of $\mathrm{m^2}$.

When the plane wave passes through the loop, the induced voltage will also appear, and can be calculated as follows:

$$U = S \times E \times F / 48 \tag{1.18}$$

In this formula, U is the induced voltage, in units of V; S is loop area, in units of $\mathrm{m^2}$; E is the strength of the electric field, in units of $\mathrm{V\,m^{-1}}$; and F is the frequency of the electric field, in units of MHz.

For example, there is an electric circuit with the loop area $20\,\mathrm{cm^2}$ in the PCB, when the radiated immunity test is carried out with $30\,\mathrm{V\,m^{-1}}$ electric field, the induced voltage U_1 in the circuit at the 150 MHz can be calculated by formula (1.18) as follows:

$$U_1 = SEF/48 = 0.0020 \times 30 \times 150 / 48 \approx 200(\mathrm{mV})$$

This is one of the reasons why the circuit is disturbed when the radiated immunity test is carried out. From the previous result, it is found that the disturbance voltage is not so high. It can also be found that the disturbance produced by this principle is not common in practice. Another phenomenon is more common, i.e. the opposite process to the essence of the radiated emission test, which corresponds to the monopole antenna or the symmetrical dipole antenna model essence. When the EUT is placed in the radiated immunity test environment, the cables or the other long conductors in the EUT will become the antenna receiving the EM field. The induced voltage will appear on the ports connected with these cables or long conductors. At the same time, there will be induced currents on the cables or long conductors, which are generally the common-mode voltage and the common-mode current. For example, an EUT with an L long cable is placed in the free space, the electric field strength of the free space is E_0, and when $L \leq \lambda/4$, the common-mode current I is induced on the cable,

$$I \approx \frac{E_0 L^2 F}{120} \tag{1.19}$$

When $L \leq \lambda/2$,

$$I \approx \frac{1250 E_0}{F} \tag{1.20}$$

In formulas (1.19) and (1.20), I is the induced current (mA); E_0 is the field strength in the free space ($\mathrm{V\,m^{-1}}$); F is the frequency (MHz); L is the cable length that is equivalent as a dipole antenna (or twice the cable length equivalent as the monopole antenna); and λ is the wavelength (m).

As with the radiated emission, when the cables, which become the equivalent receiving antenna in the EUT, are placed with the height h above the ground reference plate (as shown in Figure 1.37), we get the following relational expressions:

When $\mathrm{h} \leq \lambda/10$

$$E(\mathrm{h}) \approx E_0 \times 10 \times \mathrm{h}/\lambda \tag{1.21}$$

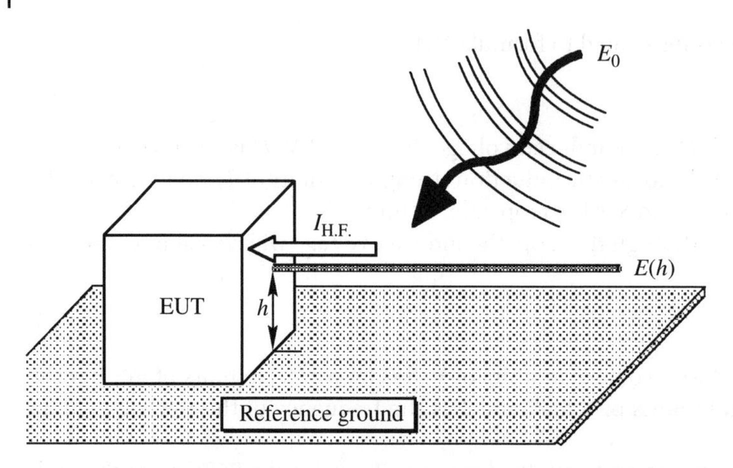

Figure 1.37 Cable induces common-mode current in electromagnetic field.

When $h > \lambda/10$

$$0 \le E(h) \le 2E_0 \tag{1.22}$$

In formulas (1.21) and (1.22), h is the height of the cables above the ground reference plate, which is equivalent as the receiving antenna in the EUT and placed above the ground reference plate, in units of m; E_0 is the electric field strength in the free space, in units of V m^{-1}; $E(h)$ is the strength of the equivalent electric field, which is attenuated by the ground, in units of V m^{-1}; and λ is the wavelength (m).

This means that the signal traces and the signal cables of the products will suffer less radiated disturbance when they are close to the cabinet wall or the ground reference plate.

The common-mode current induced by the cable will flow into the product along the cable port and the cable, as well as the internal circuit. The principle of this common-mode current disturbing the normal operation of the circuit is the same as that of the other transient common-mode current disturbing the circuit. Please refer to the description in Section 1.5.5.

1.5.5 Essence of the Common-Mode Conducted Immunity Test

There are many kinds of common-mode conducted immunity tests. Such as IEC61000-4-6, the conducted immunity test prescribed by ISO11452-7, the electrical fast transient burst (EFT/B) test prescribed by IEC61000-4-4, the surge test specified by IEC61000-4-5, the BCI test prescribed by ISO11452-4, and the CS114, CS115, CS116 test prescribed by GJB152A. EFT/B and BCI tests are the most typical common-mode immunity tests.

In the common-mode immunity test, the common-mode voltage is superimposed to the various power ports and signal ports of the products. Then, the noise voltage flows into the internal circuit (the mechanical structure of the product greatly impacts the path and the amplitude of the common-mode current to be measured in the EFT/B test, which can be referred to my reference book (*Electronic Product Design – EMC Risk Assessment*, published in 2008). The noise current can also directly flow into the internal circuit of the product in common mode. The common-mode current will be transformed into the differential voltage and disturb the normal operation of the internal circuit (the power supply voltage of the product is differential mode). For a single-ended transmission signal, just as shown in Figure 1.38, when the common-mode noise is injected simultaneously on the signal line and ground (GND) flows

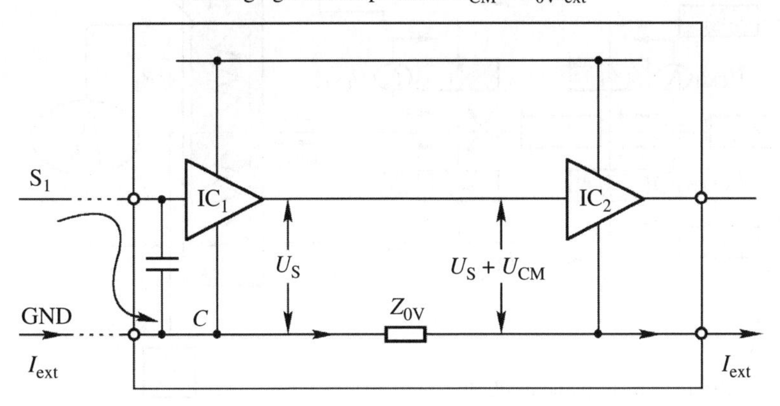

Figure 1.38 Voltage drop caused by common-mode interference current flowing through ground impedance.

into the circuit, at the port of the IC_1, due to the common-mode impedances of S_1 and GND are not the same (the impedance of S_1 is higher than that of GND), the common-mode disturbance will be transformed into the differential-mode signal and the differential-mode signal exists between S_1 and GND. In this way, the interference will first disturb the input port of IC_1. Because of the filtering capacitor C, the input port of IC_1 will be protected. The differential-mode noise between the input port of IC_1 and the reference ground will be bypassed by the capacitor C (if there is no capacitor C, the noise will directly affect the input signal of IC_1). Then, most of the noise current will flow from one end to the other end of the low-impedance PCB ground plane. The noise voltage will appear when the current flows through the ground plane. (We ignore the factor of crosstalk. The crosstalk will make the disturbance current path more complex. Crosstalk control in the EMC design is also very important.) Among them, Z_{0V} in Figure 1.38 is the ground impedance between the two integrated circuits on the PCB. U_s is the voltage of the signal transmitted from the integrated circuit IC_1 to the integrated circuit IC_2.

When the common-mode noise current flows through the ground impedance Z_{0V}, there will exist a voltage drop $U_{CM} \approx Z_{0V}I_{ext}$ across the two ends of the impedance Z_{0V}. For the integrated circuit IC_2, this voltage drop will be superimposed on the voltage U_S, which is the output of IC_1. The disturbance voltage is related not only to the current amplitude of the common-mode transient disturbance but also to the magnitude of the ground impedance Z_{0V}. When the noise current is in a certain amplitude, the amplitude of U_{CM} will be determined by Z_{0V}. For example, with a complete (no hole, no slot) ground plane, at 100 MHz, its impedance is about 3.7 mΩ. Even when the transient current flowing through the 3.7 mΩ impedance is 100 A, the voltage drop is only 0.37 V. For the 3.3 V transistor-to-transistor logic (TTL) circuit, this voltage drop is acceptable. Because the 3.3 V TTL logic circuit will switch its logic when the voltage is higher than 0.8 V, the circuit has quite enough anti-interference ability. For another example, consider a ground plane through which the EFT/B noise current flows has a 1 cm long slot. The slot will result in 1 nH inductance on the ground plane. When a 100 A EFT/B noise current flows through this ground plane, the voltage drop on the ground plane is:

$$V = |L \times dI/dt| = 1\,\text{nH} \times 100\,\text{A}/5\,\text{ns} = 20\,\text{V}$$

For the 3.3 V TTL circuit, the 20 V voltage drop is very dangerous. It is visible that the importance of anti-interference capability is the ground impedance. From experience, we know that

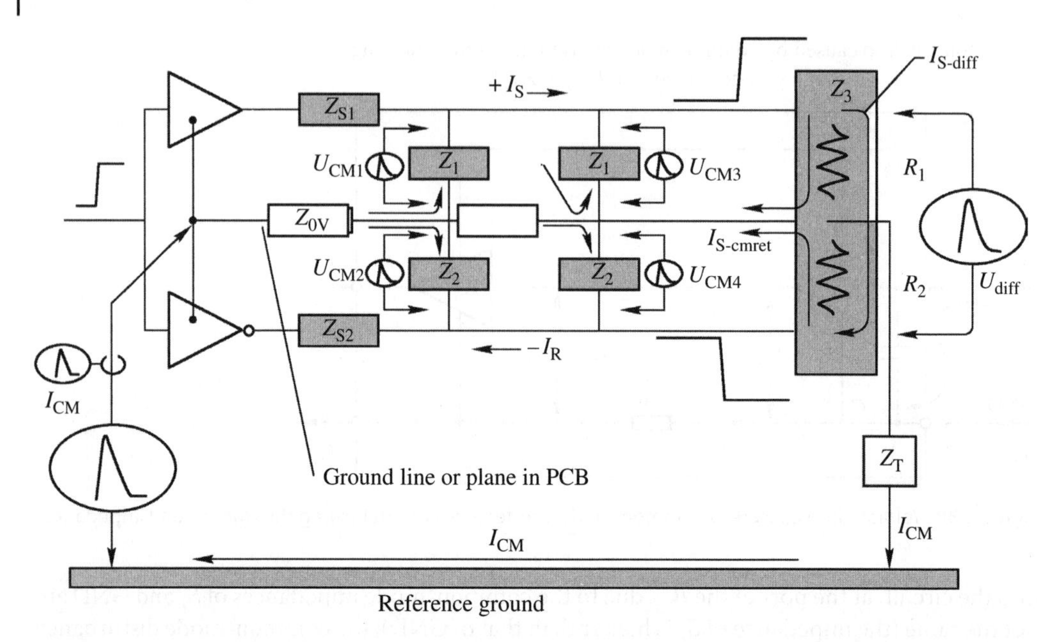

Figure 1.39 The principle of disturbance from common-mode interference current on a differential circuit.

for the 3.3 V TTL logic circuit, a voltage drop caused by the common-mode disturbance current on the ground plane is clearly safe at less than 0.4 V, and voltage higher than 2.0 V is dangerous. For the 2.5 V TTL logic circuit, the voltage threshold will be even lower (0.2 V and 1.7 V). The 3.3 V TTL circuit has more anti-interference ability than the 2.5 V TTL circuit.

For the differential transmission signal, when the common-mode current I_{CM} flows through the ground plane, the current will cause a voltage drop at the two ends of the ground impedance Z_{0V}. When the common-mode current I_{CM} is fixed, the larger the ground impedance is, the bigger the voltage drop will be. As in the principle by which the single-ended signal is disturbed, the voltage drop is applied between the signal line and the reference ground, just like the $U_{CM1}U_{CM2}U_{CM3}U_{CM4}$. Because the impedances Z_1, Z_2 are not always the same, which is the same for the impedance Z_{S1}, Z_{S2}, $U_{CM1}U_{CM2}U_{CM3}U_{CM4}$ will not be the same. The different part is transformed to the differential-mode noise voltage U_{diff} and disturbs the differential circuit. It is clear that, for the differential circuit, the impedance of the ground plane is equally important. And the parasitic parameters of the differential pairs must be balanced and routed as symmetrically as possible in PCB, which is also very important (Figure 1.39)

1.5.6 Essence of the Differential-Mode Conducted Immunity Test

In EMC test, the low-frequency conducted immunity tests are usually dominated in differential modes, such as the CS101 and CS106 tests specified by the military standard GJB152A, the line-to-line surge test specified by the standard IEC61000-4-5, and the pulse immunity test, such as P1, P2a, P2b, P4, P5a, P5b specified by ISO7637-2.

The test principle of the differential-mode conducted immunity is very simple. In the test, the differential-mode noise voltage is superimposed on the normal operation circuit directly, and the operation state of the circuit is monitored. This is because the single differential-mode conducted immunity test is usually in low-frequency range, and is usually related to the transient noise, which makes the analysis relatively easy (small parasitic parameters cannot cause great impact on the transmission of the low-frequency signal).

1.5.7 Differential-Mode and Common-Mode Hybrid Conducted Immunity Test

The differential-mode and common-mode mixed conducted immunity test mainly refers to the test that, in the conducted immunity test, is performed not only in the differential mode but also in the common mode.

The typical differential-mode and common-mode mixed conducted immunity test is P3a and P3b, prescribed by ISO7637-2. Whether the product is mounted directly on the frame or not, the noise will flow into the ground reference plate through the parasitic capacitance between the grounding cable, the EUT, or the cable and the ground reference plate. Therefore, in this kind of test, these two kinds of disturbance are directly injected on the port of the EUT. The pulse immunity tests P1, P2a, P, P4, P5a, and P5b are specified in the standard ISO7637-2, which are required for the products directly mounted on the vehicle frame are also the differential common and common-mode mixed conducted immunity test. Although the disturbance source is in the low-frequency range, because of the existence of the grounding cable between the product and the ground reference plate, the current will inevitably flows into the ground reference plate (in the test, the negative terminal of the disturbance source is directly connected to the ground reference plate).

For the surge test specified in IEC61000–4-5, because of the difference between the line-to-line test and the line-to-ground test, as a whole, this test is also the differential-mode and common-mode mixed conducted immunity test.

The surge test is a low-frequency EMC test, which is determined from the microseconds' rising time of the surge test voltage/current waveform. Most of its energy is distributed below several tens of kHz. However, this is one immunity test with high energy for the EUT. When the port of the device is injected with the surge disturbance, the system might malfunction and the components might be damaged. Surge testing is clearly classified as both the common-mode (line-to-ground) test and the differential-mode test. Due to the low-frequency spectrum characteristics of the surge, test problem analysis is relatively easy. We do not need to consider too many parasitic parameters, such as the parasitic capacitance. If you use software simulation, you can also get results close to the actual results.

2

Architecture, Shielding, and Grounding Versus EMC of the Product

2.1 Introduction

2.1.1 Architecture Versus EMC of the Product

Architecture is an important part of the product, although it is not the sole cause of electromagnetic compatibility (EMC) problems. However, addressing the architecture is a good place to begin to solve the problem. It's necessary to have an appropriate architecture for the electromagnetic shielding, a good grounding system, and coupling prevention.

As to EMC, the structure design is a kind of system-level design, and the design of structural morphology. In the product design, the designs of shielding, grounding, and filtering can't exist independently. The position of the signal input/output connectors, the layout of all kinds of circuits, the placement of cables, and the selection of grounding positions all greatly affect the EMC. In general, the EMC design should prevent the common-mode disturbance current from flowing through the sensitive circuit or the high impedance ground path. Also, in the architecture design, we should avoid additional capacitive coupling and inductive coupling, and pay attention to providing a good, low-impedance bypass path for the transient disturbance.

Figure 2.1 shows a product with the EMC design problem. This product has some severe EMC problems. First, the grounding point is too far away from the power input port, which will lead to a longer grounding path and worsen the grounding effect. Second, the grounding point of the product is too far away from the power input cable and the signal cable 1, greatly increasing the risk of EMC, because when the disturbance current is applied on the power input cable and the signal cable 1, the common-mode current will flow through the whole circuit of the PCB and the high impedance interconnecting ribbon.

If for some reason we can't change the architecture, we should try to solve the problems caused by the high impedance, the filtering, and the loop in the path through which the common-mode current flows. We need to design a good ground plane in a printed circuit board (PCB) and a reasonable filtering circuit. It's also necessary that we should design a low-impedance metallic plane near the ribbon. Third, as the signal cable 1 and signal cable 2 are at the two opposite extremities of the PCB, in the typical electrical fast transient/burst (EFT/B) test on the signal cable 1, the common-mode current flows through PCB board 2, and flows into the ground reference plate through the distributed capacitance ($50\,\text{pF m}^{-1}$) between signal cable 2 and the ground reference plate or through the grounding point, as shown in Figure 2.2. In this situation, the PCB board 2 must have a good ground plane and no holes and slots.

A good structure design needs a good EMC architecture design. For the example shown in Figure 2.1, it would be better if the signal cables, the power cables, and the grounding point are

Electromagnetic Compatibility (EMC) Design and Test Case Analysis, First Edition. Junqi Zheng.

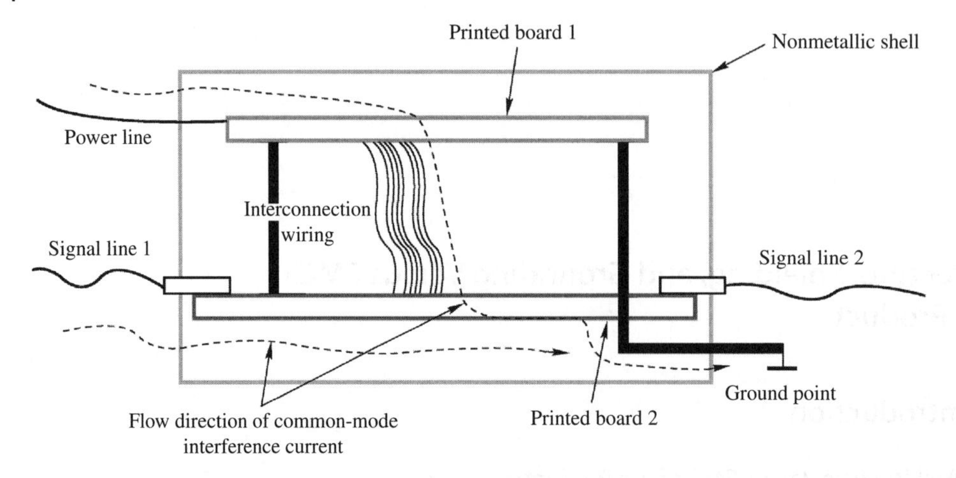

Figure 2.1 A product design example with EMC problem.

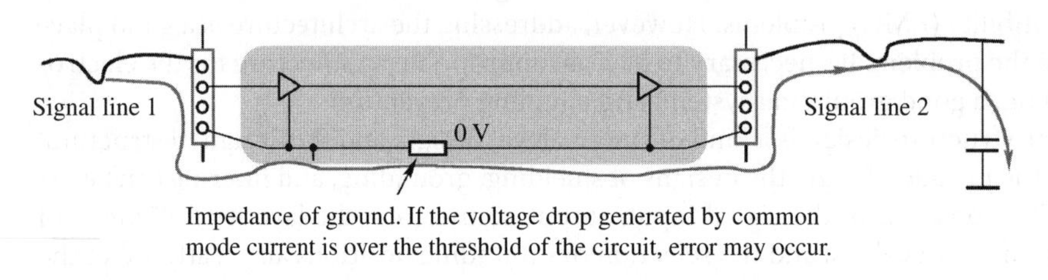

Figure 2.2 Schematic diagram of common-mode current flow direction.

in one PCB board. It's necessary that we choose the shielded cables as the signal cables, and the additional metallic ground plane is designed in the architecture.

2.1.2 Shielding Versus EMC of the Product

Shielding isolates two spaces by a metallic shield in order to control the induction and radiation of the electric field, the magnetic field, and the electromagnetic wave from one space to the other space. The shields are used to enclose the disturbance source, such as the components, circuit, constructional frame, cable, or the whole system in order to prevent the scattering of the internal electromagnetic field. They enclose the receiving circuit, devices, and systems to prevent them from being disturbed by the external electromagnetic field. The shields can absorb (eddy-current loss), reflect (the reflection of the electromagnetic wave on the shield surface), and counteract (an inverse electromagnetic field produced in the shield due to the electromagnetic induction) the electromagnetic energy from the external wires, cables, components, circuits, and system and from the internal circuit, so the shield has the function of attenuating the disturbance. In most product applications, the reflection principle is used.

Figure 2.3 shows the operating principle of the shielding explained from the electromagnetic field and the transmission lines perspectives. The design of shielding is always associated with the contact. The contact is to construct a low-impedance electrical connection between two metal surfaces. If the structure of both surfaces can't contact very well, a gasket is needed. Gasket includes conductive elastomers, wire mesh, metal spring finger, spiral ribbons, multiple conductive elastomers, and fabric over foam, for example. When we choose an electromagnetic gasket, we should consider the requirements of the shielding effectiveness, the

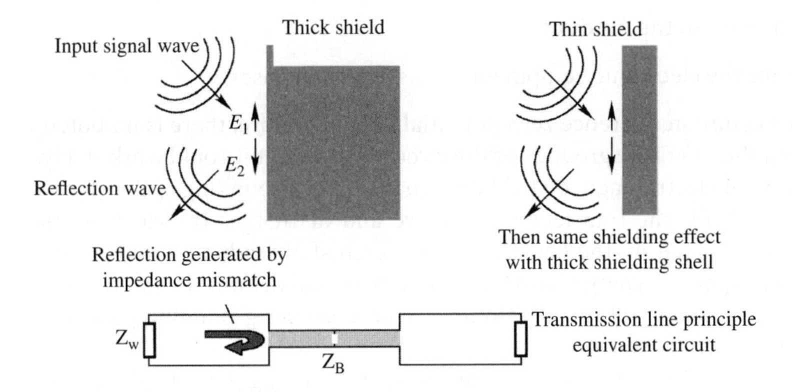

Figure 2.3 Electric field shielding principle diagram.

Table 2.1 Characteristics comparison of different liner materials.

Liner type	Advantages	Disadvantages	Occasion applicable
Conductive rubber	Provide environment encapsulation and electromagnetic shielding, high shielding efficiency at high frequency	Need high pressure, high price.	Need environment encapsulation and high shielding efficiency
Metal wire mesh	Low cost; difficult to damage	Low shielding efficiency at high frequency, unsuitable for occasions at over 1 GHz, no environment encapsulation	Interference frequency under 1 GHz
Finger strip	High shielding effectiveness allows sliding contact, big deformation range	High price, no environment encapsulation	Needs high shielding performance and sliding contact
Spiral tube	High shielding effectiveness, low prices; compound type can also provide environment encapsulation and electromagnetic shielding	Easily damaged under excessive compression	Need high shielding performance; need good compression limit; need environment encapsulation and high shielding efficiency
Multiple conductive rubber	Good elasticity, low price, can provide environment encapsulation	The surface conducting layer is thin, and easy to fall off with repeated friction	Need environment encapsulation and common shielding efficiency; cannot provide large pressure
Conductive fabric liner	Soft, low pressure, and low cost	Easily damaged in hot and humid environment	Cannot provide large pressure

environment sealing, the installation structure, and the cost. Table 2.1 shows the feature of different gaskets.

The key to the shielding design is the electrical continuity. A fully sealed single metal shell has optimal electrical continuity. In a practical application, considering the existing of ventilation holes, cable entry, and attachable conductor, the key point of the shielding design is to reasonably design the contact of the ventilation holes, the cable entry, and the attachable conductor. By coordinating the relationship between the size of holes and seams, the wavelength of signal, the propagation direction, and the contact impedance, the shielding can be well designed.

2.1.3 Grounding Versus EMC of the Product

Grounding is important for the electronic equipment. It has three purposes:

1) Grounding provides a common reference zero potential, and, therefore, there is no potential difference between the reference grounds of the circuits. The circuit could work stably.
2) It can prevent the external electromagnetic field disturbance. The grounding of the chassis provides a discharge path for the transient disturbance and discharges the electrostatic charges, which are accumulated on the chassis due to electrostatic induction; otherwise, these charges may cause spark discharge inside the equipment and disturb the internal circuit. In addition, for the shielding of the electronic circuit, a proper grounding point can help improve the shielding effectiveness.
3) The system can work safely with a proper grounding point. Grounding prevents damage to electronic equipment when there is a power surge, such as during a lightning strike. Grounding also protects operators from electric shock when the power supply input is connected with the chassis directly due to faulty insulation or for some other reasons. In addition, grounding is necessary for medical equipment connected to the body of patients, as it can be very dangerous when the chassis is energized to 110 or 220 V.

So, grounding is an important way to suppress the noise and prevent the disturbance. Grounding can be understood as an equipotential point or surface, which is the reference potential of the circuits and systems, and it does not always connect to the earth. To avoid the potential damage caused by lightning and a danger to operators, the chassis of the electronic equipment and the metallic parts in the machine room must be grounded and the grounding resistance should be very small and not more than a specified value. Although the ground can be the earth, an isolated ground, or a floating ground, the grounding structure must exist. This ground is often confused with the signal ground providing the signal current return path. In practice, the portion of the grounding problem is related with PCB. These problems can be summarized to provide a reference connection between the analog circuits and the digital circuits, and the high frequency connection between the PCB and the metallic casing.

Although the grounding is most important in EMC design, the grounding problem is not easily understood and is often difficult to be modeled and analyzed. Grounding is influenced by many uncontrollable factors, which creates confusion. In fact, every circuit must have its reference ground, which can't be avoided. The grounding design should be first considered in the circuit design. Grounding is an important way to minimize undesired noise and disturbance and to partition the circuit. Appropriately applying the PCB grounding method and the cable shield can avoid many noise problems. One advantage of a well-designed grounding system is that it prevents unwanted disturbance and emission with a very low cost. However, its meaning is different for different technique engineers, and it's a relatively broad concept in this book. For the logic circuit, the ground refers to the reference voltage for the logic circuits and components, and it is not necessary to be connected with the earth. The ground is the reference for the logic voltage, a typical value of the potential difference on the ground must be less than mV level. As shown in Figure 2.2, if the transistor-transistor logic (TTL) operating level is 3.3 V, the ground voltage drop caused by the common-mode current flowing through the ground is more than 0.4 V, and the EMC test may be not passed. In addition, if the ground plane of the high-speed digital circuits is not complete, as shown in Figure 2.4, the cable connected to the ground will cause the EMI problem because it is driven by the noise. For the systems and structures, grounding means connecting them to the metallic casing or the framework of the circuit. The grounding is to connect the equipment under test (EUT) to the ground reference plate in the EMC test.

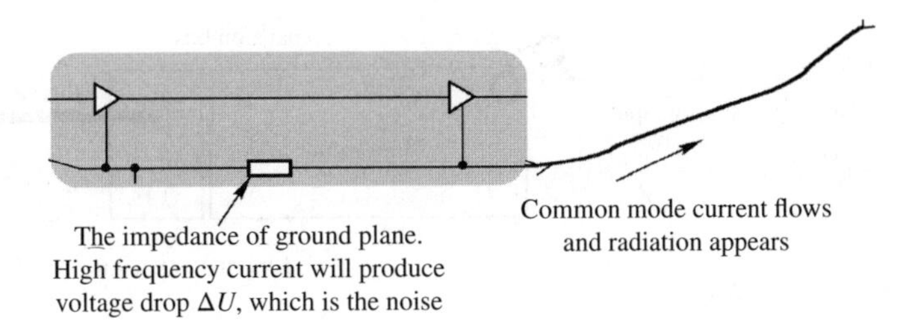
The impedance of ground plane.
High frequency current will produce
voltage drop ΔU, which is the noise
Common mode current flows
and radiation appears

Figure 2.4 Incomplete ground plane causes EMI problem.

When discussing the grounding current, we must know two basic concepts:

1) Once the current flows through the finite impedance, a certain voltage drop will be produced. As set forth in Ohm's law, in circuit, there is never 0 V, and the voltage or the current may be in micro volts or micro amps level, but there must be a small finite value.
2) The current always returns to its source. There may be many different paths in circuit, and there may be different current amplitude on each path, which is related to the impedance of the path. We do not want some current flow on a path, because there may be no suppression measures for this path.

When designing a product, taking the grounding into account is the most economical way. In a well-designed grounding system, we could prevent the radiation from the product and protect the product against the disturbance, not only for the PCB but also for the system. If there is no careful consideration on the grounding system in the design phase, or the grounding system is not redesigned when redesigning an existing product, it means that the system may fail the EMC test.

2.2 Analyses of Related Cases

2.2.1 Case 1: The Conducted Disturbance and the Grounding

[Symptoms]

The configuration diagram of a device during conducted emission test is shown in Figure 2.5.

The conducted emission test results with the configuration shown in Figure 2.5 are shown in Figure 2.6. From the spectral plot of test results, the conducted emission of the device on the power supply port cannot pass the requirements of Class B limit.

- P_1: The 0 V point of the base module is used for grounding, which is connected to the metallic frame under test.
- P_2: The grounding point of ethernet module, inside ethernet module, it is connected to 0 V through a capacitor, and it is not connected to the metallic frame under test.
- P_3: Three separated points of the metallic frame, as these three points are all in the same metallic plate and the impedance between each other is approximately zero. Hence, the three points can be approximately regarded as the same point in the schematic.
- Expansion bus: The interconnection bus between the base module and ethernet module, through which the 0 V of base module and 0 V of ethernet module are connected.

It is found that, during testing, the form of grounding is changed from that shown in Figure 2.6 to that shown in Figure 2.7, i.e. to connect the point P_2 with the point P_1 and disconnect the previous interconnection between P_1 and P_3, then test again, the test is passed. The test result is shown in Figure 2.8.

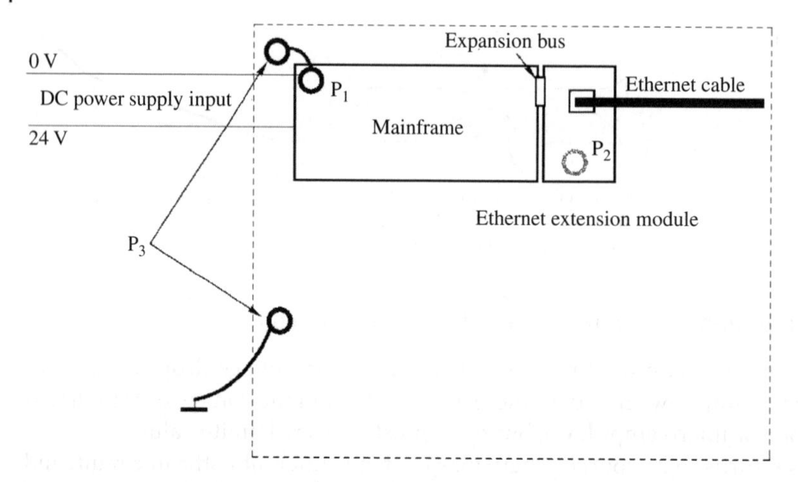

Figure 2.5 The configuration diagram in the conducted emission test. Note: the dotted line shown in the figure is the metallic frame, which constitutes the EUT together with the base module and the ethernet module. The EUT is grounded through the frame. The base module and the ethernet module are powered by 24V power supply.

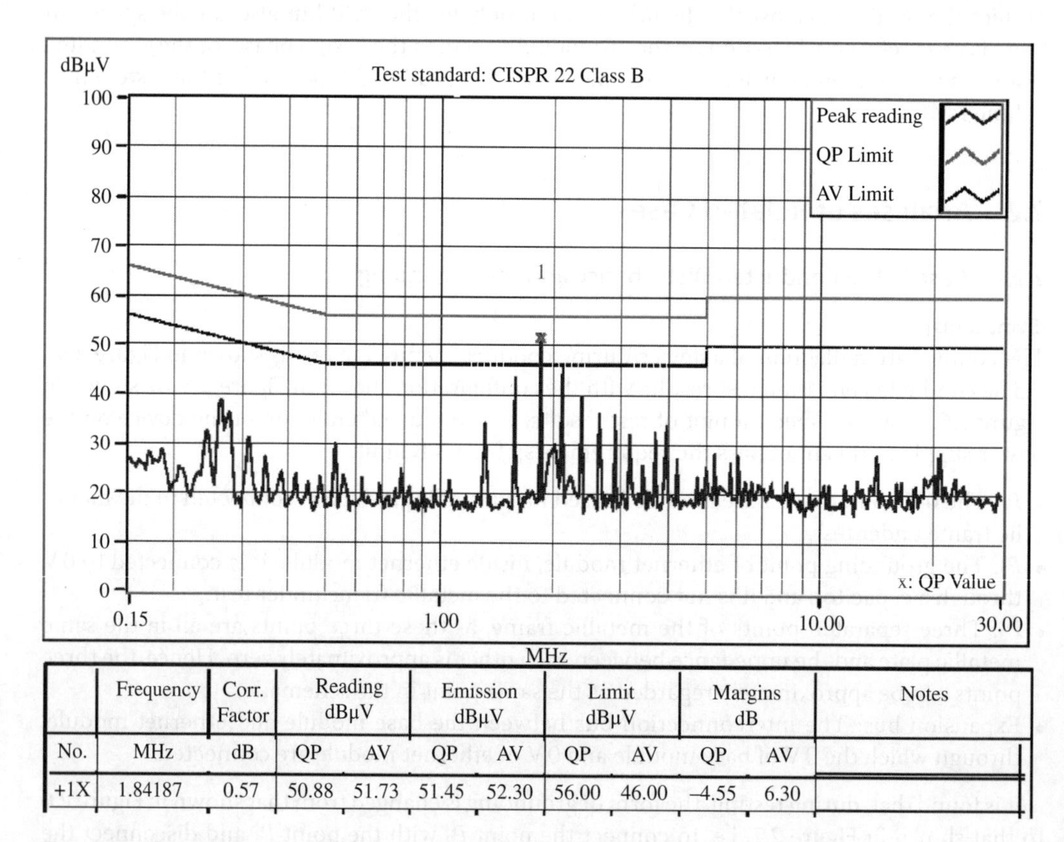

No.	Frequency MHz	Corr. Factor dB	Reading dBμV QP	Reading dBμV AV	Emission dBμV QP	Emission dBμV AV	Limit dBμV QP	Limit dBμV AV	Margins dB QP	Margins dB AV	Notes
+1X	1.84187	0.57	50.88	51.73	51.45	52.30	56.00	46.00	−4.55	6.30	

Figure 2.6 The conducted emission test results with the configuration shown in Figure 2.5

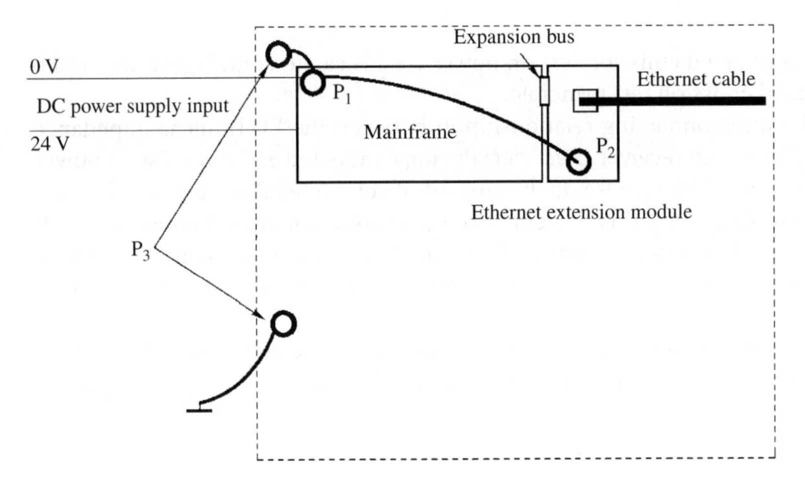

Figure 2.7 Configuration diagram that passes the test.

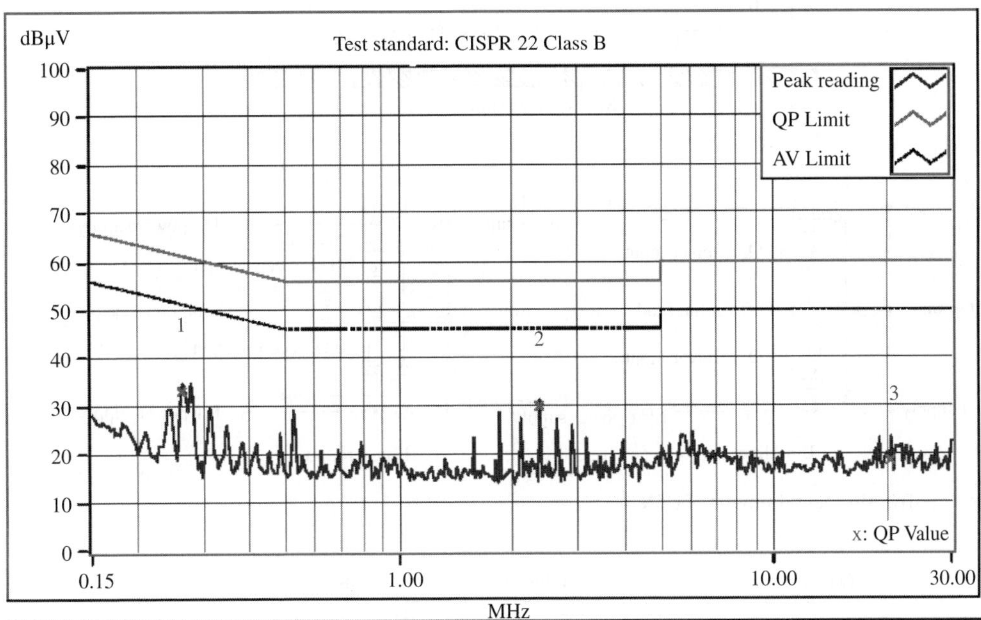

No.	Frequency MHz	Corr. Factor dB	Reading dBμV QP	Reading dBμV AV	Emission dBμV QP	Emission dBμV AV	Limit dBμV QP	Limit dBμV AV	Margins dB QP	Margins dB AV	Notes
1	0.26231	0.80	31.94	32.75	32.74	33.55	61.36	51.36	−28.62	−17.81	
+2	2.36905	0.56	28.79	29.45	29.35	30.01	56.00	46.00	−26.65	−15.99	
3	20.36162	1.16	17.42	15.31	18.58	16.47	60.00	50.00	−41.42	−33.53	

Figure 2.8 The results of test that passed.

[Analyses]

First, let's see how the conducted emission test on power cord is carried out, Figure 2.9a and b can explain the conducted emission test principle.

Figure 2.9a shows the interconnecting relationship in between the EUT, linear impedance stabilization network (LISN) and receiver (Receiver) during conducted emission test on power cord. The arrow line in Figure 2.9b represents the current of conducted disturbance. The voltage drop on 50 Ω resistor caused by this current is the measured conducted emission result expressed by voltage. The left diagram in Figure 2.9b is the differential-mode conducted emission measurement principle; the right one is the common-mode conducted emission measurement principle.

In this example, as for the grounding way with which the conducted emission test of EUT is not passed, its equivalent schematic is shown in Figure 2.10, where C_1, C_2, C_3 are the respective

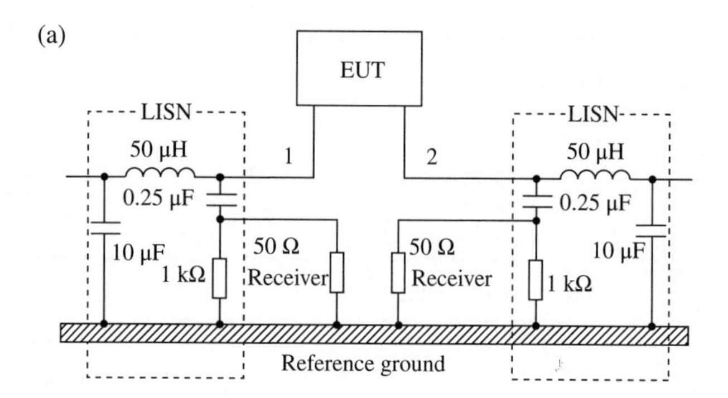

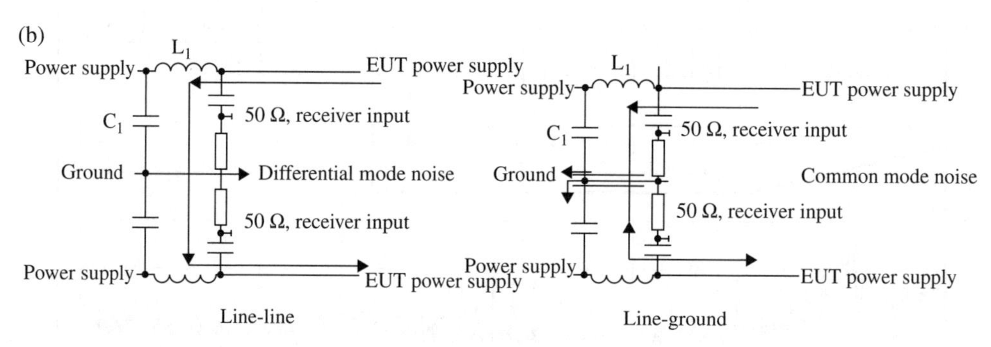

Figure 2.9 The internal schematic of LISN.

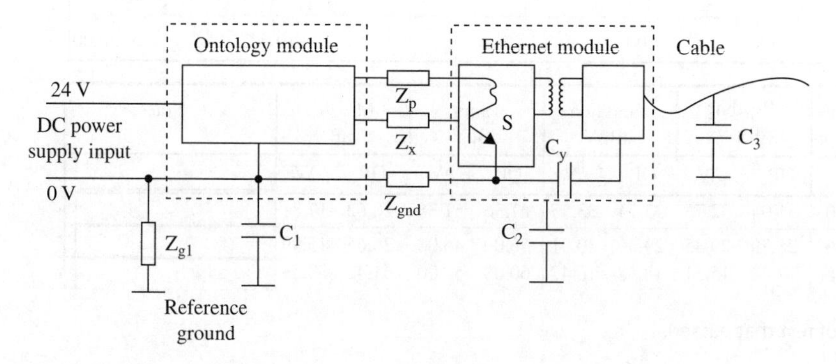

Figure 2.10 Topological diagram of the failed test.

distributed capacitance between base module, ethernet module, ethernet cable, and the reference ground; C_y is the bypass capacitor in PCB between 0 V and ethernet module grounding terminal. Z_p is the impedance of the 24 V link of the connector interconnecting base module and ethernet module; Z_x is the impedance of the signal bus link of the connector interconnecting base module and ethernet module; Z_{gnd} is the impedance of 0 V link of the connector interconnecting base module and ethernet module, and the impedance of 0 V plane in each module; Z_{g1} and Z_{g2}, respectively, represent the grounding impedance between these two ground terminals of EUT and the reference ground plate; S represents the switching-mode power supply inside the ethernet module, which is the main disturbing source of the conducted emission test. A high pulse current will flow through the power switch of the switching-mode power supply when it is switched on, the common-mode noise across 0 V will appear due to this current. For example, the current waveform of forward, push-pull, and bridge converter while driving resistive load is approximately a rectangular wave, which includes rich high-order harmonic components. In addition, when the power switch is switched off, the switch current will be changed abruptly. Due to this abrupt current change, the leakage inductance of the high-frequency transformer could generate interference, too.

In common mode, the conducted emission schematics caused by 0 V common-mode noise are shown in Figure 2.11.

In Figure 2.11, the circle symbol is the common-mode noise across 0 V. The arrow line indicates the flow of conducted disturbing current, which directly determines whether the test can be passed. The part inside the dotted line represents a line impedance stabilization network.

Taking a look on the connections of EUT under which the conducted emission test is passed, its equivalent EMC analysis diagram is shown in Figure 2.12.

The line directly connected between C_1 and C_2 is the interconnecting line between grounding terminal of ethernet module and 0 V of base module. In common mode, the schematic in Figure 2.12 may be simplified as the schematic shown in Figure 2.13.

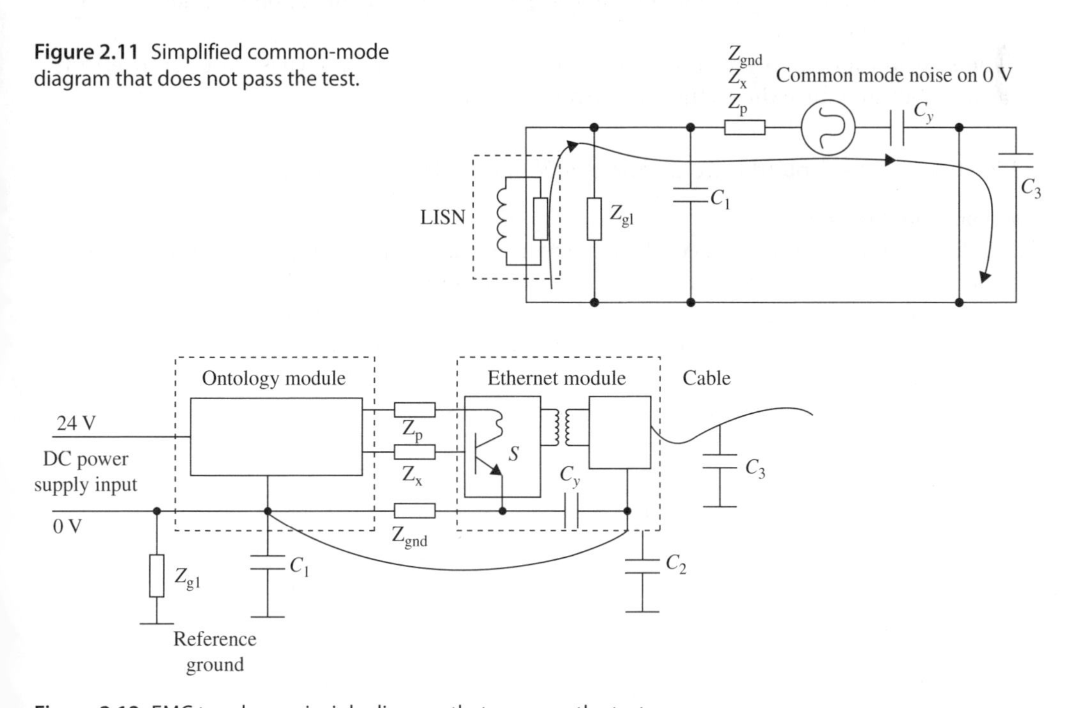

Figure 2.11 Simplified common-mode diagram that does not pass the test.

Figure 2.12 EMC topology principle diagram that can pass the test.

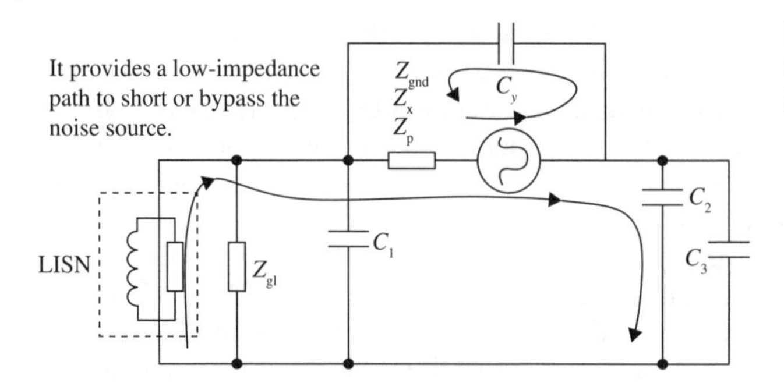

Figure 2.13 Common-mode current simplified schematic diagram that can pass the test.

Comparing the differences between Figures 2.13 and 2.10, it can be seen that, when C_y is connected to the 0 V of base module, a low-impedance path is provided so that part of the common-mode current is bypassed, thus the current flowing into the LISN is reduced, and finally the test is passed.

[Solutions]

From this analysis we can conclude that the following major solutions may be referred by other similar products. There is a need to provide a structure that helps to realize the internal equipotential connection between 0 V of the base module and the grounding of the ethernet module. This structure should ensure that its impedance is low enough, which is the precondition for EUT to use a single grounding point.

[Inspirations]

1) Grounding is very important for EMC. A grounded product will greatly reduce the risk of failure for EMC test.
2) For EUT with multiple grounding points, the equipotential connection between each other is very important for EMC.
3) The solution for the problem of conducted emission is not to bypass interference to the ground but to reduce the disturbing current flowing into LISN.

2.2.2 Case 2: The Ground Loop During the Conducted Emission Test

[Phenomenon Description]

One information technology device has an external signal cable and power supply cable. For the conducted emission test on power supply port, the grounding cable of EUT is connected to its adjacent reference ground plate. The test configuration is shown in Figure 2.14, and the test result is shown in Figure 2.15. From Figure 2.15, we can see that the conducted emission of the power supply port of this device is not within the standard limit shown in Figure 2.15. We need to analyze the reason of such high-conducted emissions.

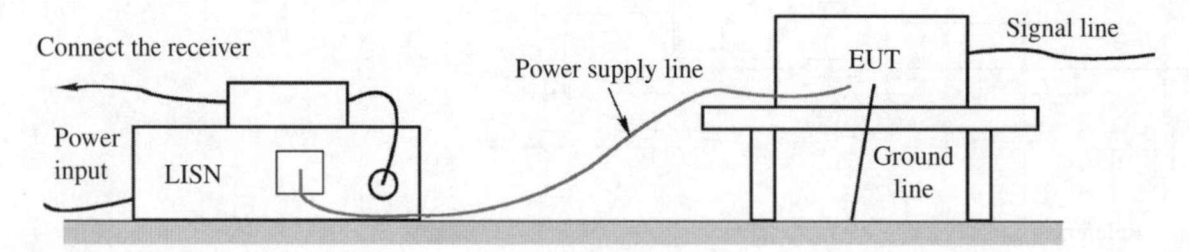

Figure 2.14 The configuration diagram of the conduction emission when EUT ground line connects to the reference ground plane nearby.

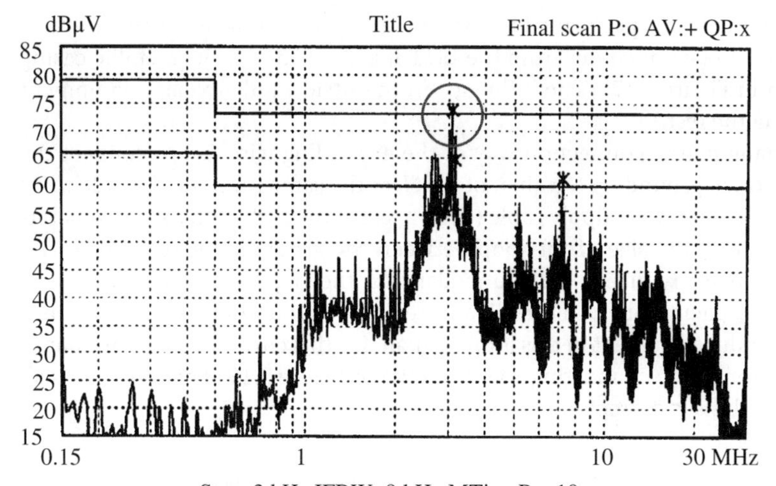

Figure 2.15 Spectrum of the initial conduction emission.

[Analyses]

Regarding the principle of conducted emission test on power supply port, we can refer to Case 1. The principle of the test configuration shown in Figure 2.14 can be expressed by Figure 2.16.

Figure 2.16 shows the relationship between the EUT, the LISN, and the receiver when the conducted emission test is performed on power supply port. The arrow line in Figure 2.16 represents the common-mode conducted disturbing current. The voltage drop across 25 Ω resistance or two parallel 50 Ω resistors due to this current is the measured common-mode conducted emission (the differential-mode conducted emission is irrelevant to this case, which is not shown in Figure 2.16). Just as shown in Figure 2.16, with this kind of test configuration, a large loop (shown in the dotted line in Figure 2.16) is formed by the power cord, LISN, EUT, the grounding cable of EUT, and the reference ground plate. The significance of the loop in EMC has been mentioned in other cases of this book (e.g. Cases 16 and 65).

According to the electromagnetic theory, a loop can be an antenna that is the necessary condition of emission, and it can also be a receiving antenna that receives interference. When the magnetic flux passing through the loop changes, a current will be induced in the loop. The magnitude of the current is proportional to the area of the loop. For a loop with a certain size, the receiving antenna with the loop will resonate at a specific frequency. If the induced current exists in the loop shown in Figure 2.16, the current flowing through the 25 Ω resistance in LISN will increase and LISN will detect more conducted disturbance.

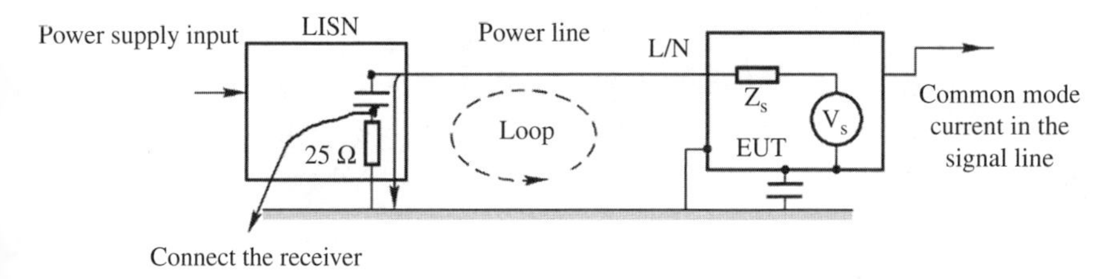

Figure 2.16 The configuration diagram of the conduction emission when EUT ground line connects to the reference ground plane directly.

As the conducted emission test is performed in a shielded room, from where is the interference received by the loop? In fact, it comes from the EUT itself, radiated by its metallic casing and signal cables shown in Figure 2.17. (The influence from outside and auxiliary equipment has been excluded during the test.)

During the test, by changing the connection way of the grounding wire, i.e. to connect the grounding wire of EUT to the grounding terminal of LISN, at the same time side-by-side routing the grounding wire and the power supply cable (shown as Figure 2.18) with a small separating distance (less than 5 mm), the size of the loop between the grounding wire of EUT and reference ground plate is greatly reduced. As the impedance of the loop formed by power cord, LISN, EUT, the parasitic capacitors between EUT and reference ground plate, and reference ground plate is high, the induced current in this loop is small (i.e. this current is not dominant). The test result after modifying the connecting way is shown in Figure 2.19. The test is passed, further confirming the analysis is correct.

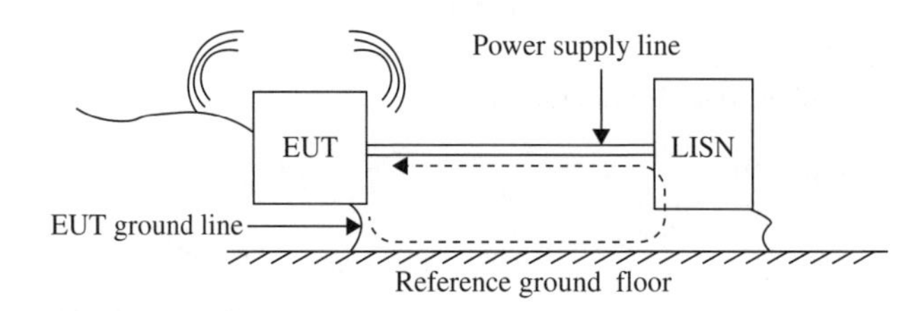

Figure 2.17 A schematic diagram of the radiation of the cable and the shell when doing conduction emission test.

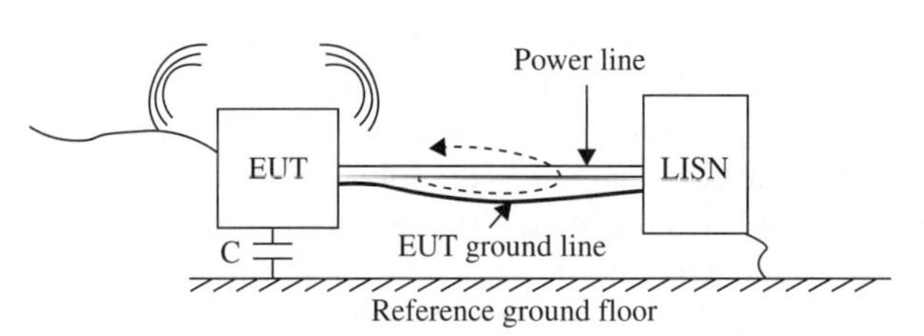

Figure 2.18 A schematic diagram showing connections of the ground terminal of the EUT ground wire to the LISN line.

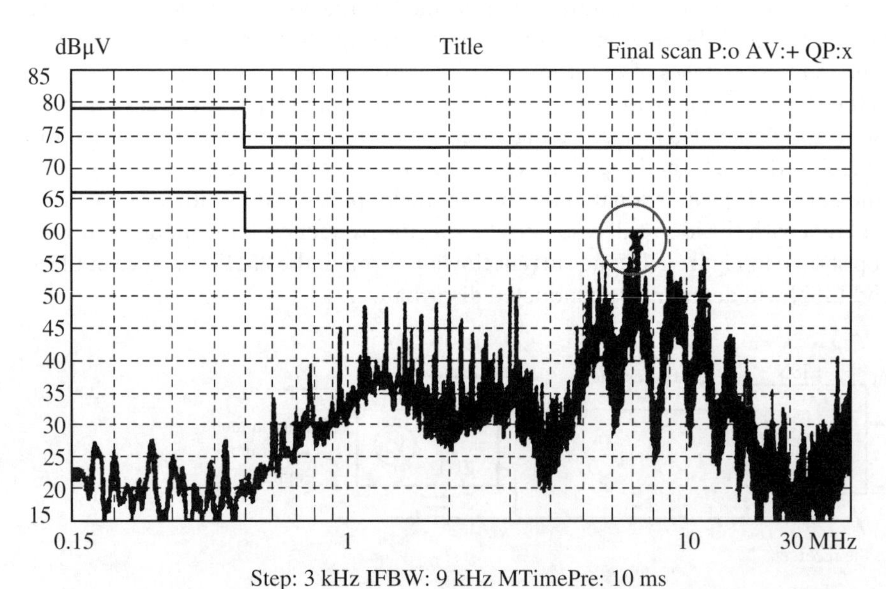

Figure 2.19 The test results of connected EUT to the end of the ground wire of to LISN.

[Solutions]

This case is not a design problem but a problem caused by test configuration. Therefore, the best solution is to connect the EUT grounding wire to the grounding terminal of LISN, and at the same time route the grounding wire and the power cord side by side with a small separation (less than 5 mm) and decrease the area of the loop.

[Inspirations]

1) The grounding problem in this case is a common problem during conducted emission testing. The problem is also worth being paid attention. In conducted emission test, the grounding cable must be routed with power cord side by side instead of "grounding in the neighborhood," in order to avoid forming a large loop and receiving unexpected interference. For the device with high emissions over low-frequency range (150 KHz–30 MHz), such as the device with signal cables, we must pay particular attention to the loop composed by the power cord, LISN, EUT, the grounding wire of EUT and the reference ground plate.
2) The problem in this case can be solved through other means without changing the area of the grounding loop (e.g. filtering on power supply input). As the test result is always a comprehensive result, it is possible to make the result meet the test requirement by removing one part of the influencing factors, but the unreasonable test configuration is mostly needed to be eliminated.

2.2.3 Case 3: Where the Radiated Emission Outside the Shield Comes From

[Symptoms]

A device has modules and a backplane, and each module has a connector. When the module is inserted into the backplane, the connector of the module is connected to the corresponding connector of the backplane, which is also the only connection position between the module and backplane. Each module is designed with shielding. The backplane is also designed with shielding. The backplane is used to fix each module and interconnect the signals between each module. In the radiated emission test, it is found that the radiated emission exceeds the required limit of the product standard. The frequency at which emission exceeds the standard limit is 350 MHz, and the test spectral plot is shown in Figure 2.20. The limit is marked with a thick line. Here, a magnetic near-field probe is used for disturbing source locating test. (Using the magnetic near-field probe to detect the emission leakage of metallic casing of the device is one of the most effective and economical ways. The near-field probe group developed by Shanghai Ling Shi Electric Ltd. is a cost-effective and efficient choice.) According to such a test, we can identify that the radiated emission is due to the emission leakage from the connector between one module and the backplane. The disturbance source at this frequency is the 50 MHz crystal inside the PCB of this module.

[Analyses]

As this device needs to work outdoor, it must be waterproof, including the interconnection position between the module and the backplane. The connection between the module and the backplane must be sealed using a waterproof rubber gasket. However, the waterproof gasket is not conductive, so there must be "seam" that makes the radiated emission possible.

By removing the module, powering the module alone, and measuring the radiated emission of the module with near-field probe, we found that 350 MHz radiation exists around the connector interconnecting the module and the backplane. The PCB already takes EMC into account. There is a shielded metallic cover on the right of the PCB trace without solder mask,

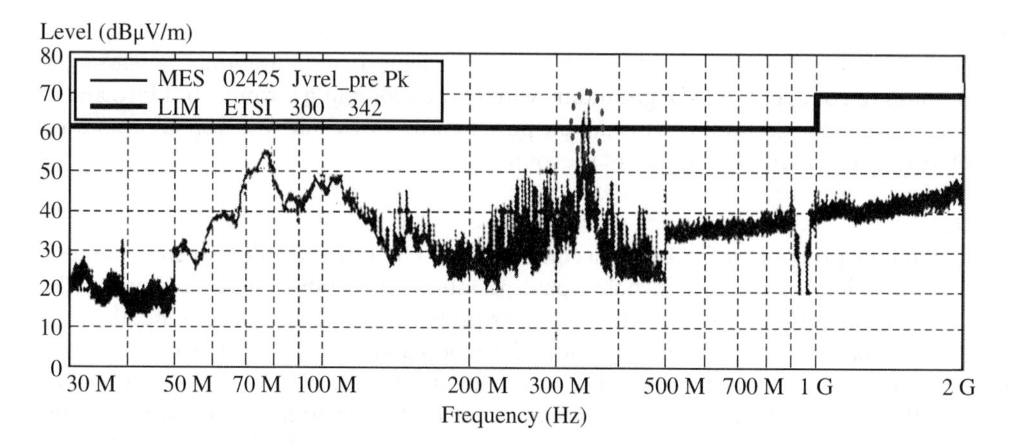

Figure 2.20 Spectrum of radiation emission.

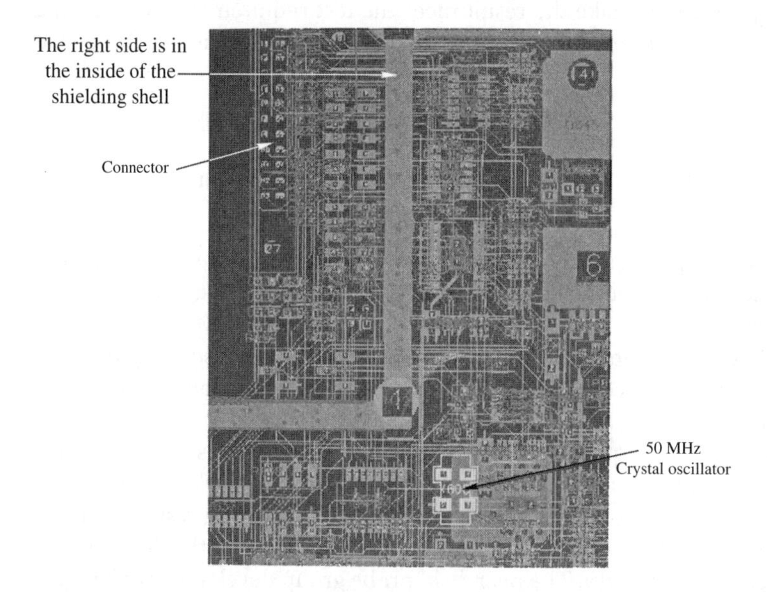

Figure 2.21 The PCB layout with the connector.

which is shown in Figure 2.21. The path through which the shell of crystal radiates directly to space has been isolated by the shielded cover. The only possible radiated emissions may come from the connector J1 and the PCB traces connected to it, which probably couple the noise from the crystal or the clock traces. They could be regarded as the radiating antenna and driven by the noise. Then, after checking the module itself, part of its PCB layout with the connector is shown in Figure 2.21.

The following problems exist in the PCB layout:

1) The local ground plane has not been routed in the external layer of PCB under the crystal. The local ground plane under the crystal and clock circuits can provide a return path for the

common-mode radio frequency (RF) current generated by the crystal and its related circuits, thereby, the RF emission can be limited to the minimum. In order to sustain the common-mode RF current that flows to local ground, it must be connected to local ground plane and the system ground plane with multiple points with the ground plane on the top layer relative to the internal ground plane, implementing low impedance to the ground plane.

2) The signal traces linked to the connector are routed under the 50 MHz crystal, which not only destroys the effect of local ground plane but also couples the noise caused by 50 MHz crystal to the signal traces under the crystal through capacitive coupling way. Common-mode voltage noise exists in these signal traces, and these signal traces extend outside the shield of PCB and take the noise out of the shield. This is a typical common-mode radiation model. Its schematic is shown in Figure 2.22.
3) The position of the crystal is too close to the interface.

It can be seen that the two necessary conditions of radiation, the driving source and antenna, have been formed. The driving voltage is formed due to the coupling between the signal traces connected with the interface and the crystal. The signal traces connected to the interface or the ground layer of PCB are driven by the noise and become the radiating antennas.

[Solution]

Three approaches can be used to solve this radiation problem:

1) Shield all the parts related to the radiated emission (including the antenna and the driving source). This reduces the shield defects. Change the waterproof gasket between the module and the backplane to the conductive gasket to provide continuity of the shielding for the entire device.
2) Reduce the driving voltage, i.e. dealing with this in PCB. Reduce the coupling between crystal and the signal traces connected to the interface connector. Adjust the position of 50 MHz crystal and its surrounding traces, move the crystal to the middle area of the board as possible avoid routing traces under the crystal body, and keep traces more than 300 mil away from the crystal. Fill the area under crystal body on PCB external layer with ground plane and ensure the integrity of the ground plane under the crystal.
3) Remove the radiating antenna is difficult to be achieved. It is necessary to connect signal traces to the interface connector.

Note: The reason for keeping the integrity of the ground plane under the crystal is that the application of multilayer PCB board technology in this device makes the signal return current in opposite direction and the same amplitude as the signal itself. These will reduce the magnetic amplitude of the signal current. This characteristic can guarantee good signal integrity and EMC performance. However, when the local ground plane is not perfect, the current in return path cannot be canceled by the current of signal itself (in fact this current imbalance is inevitable), which will cause a small part of common-mode current with same amplitude and in same direction, especially for crystals with high-level noise. The common-mode voltage transmits along the adjacent reference plane, and the coupled small voltage drives the peripheral structure,

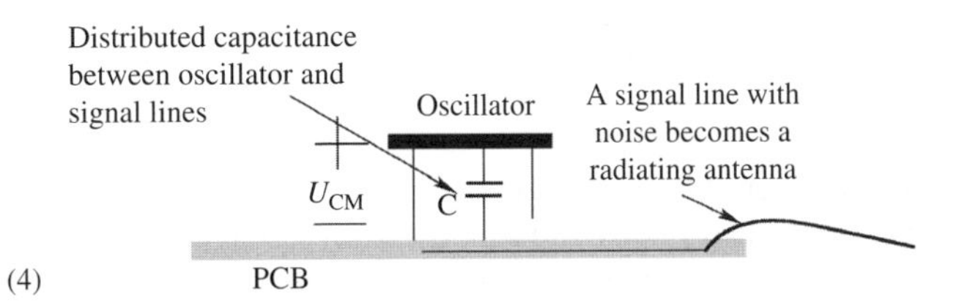

Figure 2.22 The principle diagram of common-mode radiation driven by the coupling voltage.

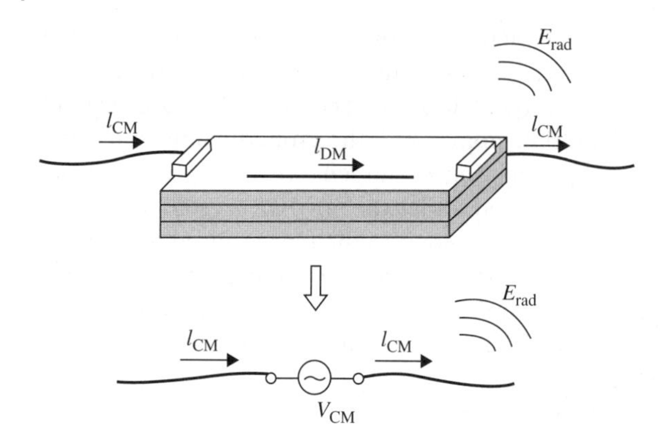

Figure 2.23 Equivalent model of common-mode radiation caused by incomplete ground plane.

which make it a radiating antenna. Figure 2.23 shows the common-mode radiation model of PCB routing on an external layer (the reference plane).

From a high-frequency perspective, the reference ground plate becomes the path of return current, where there may be a high-frequency AC voltage. A common-mode voltage drop U_{CM} will appear across the reference ground plate routed under signal traces, which is shown in Figure 2.23. This voltage drives the big peripheral structure and generates common-mode current I_{CM}.

After treatments (1) (2), the radiated emission measurement around interface connector is performed again with near-field probe. We find that the radiation at 350 MHz frequency is significantly reduced by more than 10 dB.

[Inspirations]

1) Waterproofing and shielding should be considered at the same time; do not neglect the shielding integrity for the sake of waterproofing.
2) The signal traces cannot be routed under the crystal, especially for the signal traces that are directly connected to the external interface.
3) The plane below the crystal and clock circuits can provide the return path for the common-mode RF current generated by crystal and its related circuits; therefore, the RF emission is significantly reduced.

2.2.4 Case 4: The "Floating" Metal and the Radiation

[Symptoms]

A certain device provides 24 10/100 Mb s^{-1} ethernet ports and two expansion slots. In actual networking, different clipping boards can be configured per the needs. In the radiated emission test, the expansion slot is configured as a gigabit optical interface (LC connector) for remote communication. All the 24FE electrical interface cables are self-loop connected as well as LC optical interface. It is found from the test results that there are several high frequencies (625, 687.5, 812.5, 875 MHz) at which the emissions exceed the standard requirements of CLASS B limit, the test spectral plot is shown in Figure 2.24.

[Analyses]

Radiated emission problem of shielded device generally is related with three aspects, the leakage from power cord, the leakage from signal cables, and the leakage from structural shielding. However, such high-frequency radiation is generally not from the power cord (power cord

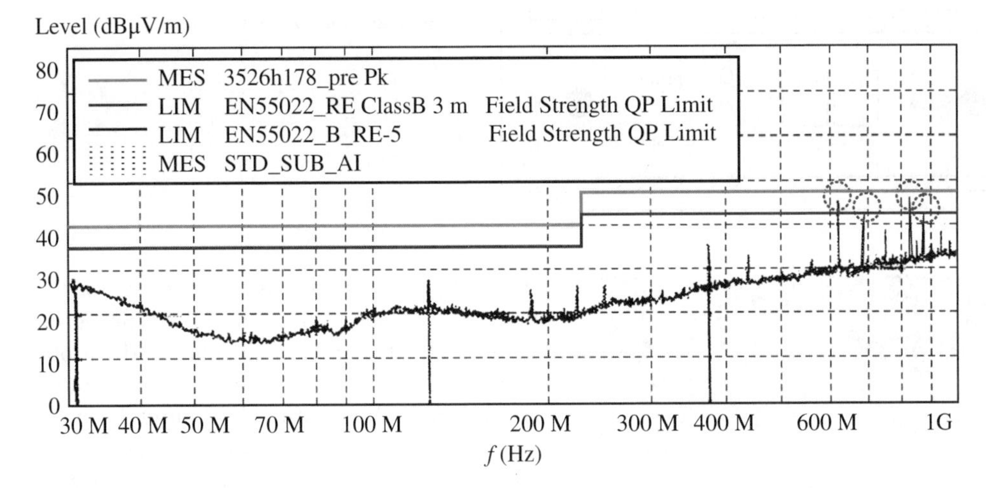

Figure 2.24 Initial test results.

radiation is generally below 230 MHz), but only from the leakage of signal cables and structural shielding, therefore, the following locating measures are performed based on this analysis.

At first, the test results shows only a small reduction in emissions after the 24 100 Mbps ethernet cables are removed. Then when the emissions from a device with near-field probe are scanned, it is found that they are very high on both sides of the clipping boards installed on expansion slots. Locating the frequencies, it is found that emission levels are very high on both sides of hte clipping boards installed on expansion slots. The frequencies are the same as the frequencies in which the emissions exceed the standard limit. So it is suspected the contact between the frame of clipping boards and the cabinet is not good. Opening the contact, there are respective spring piece in the upper side and down side of the clipping board which makes a good contact with the cabinet. But there is no spring piece on both vertical sides, even though the length of the seam is not more than 1.5 cm due to the existence of screws for fixation. More details about the structure are shown in Figures 2.25 and 2.26

Does such short seam cause a large emission leakage overt the frequency range below 1 GHz? Try to seal the seam with conductive copper foil, and test with a near-field probe, as expected, the emission on both sides of the panel of clipping board disappeared. It is thought that the problem should be solved, and then move equipment into the laboratory for testing and find that those frequencies at which the emissions exceed the standard limit with high emissions still exist, and there is almost no decline in amplitude (in which case there are only the optical

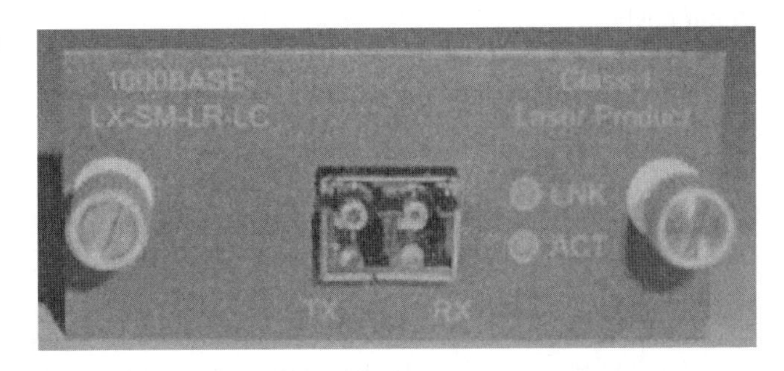

Figure 2.25 Buckle positive physical map.

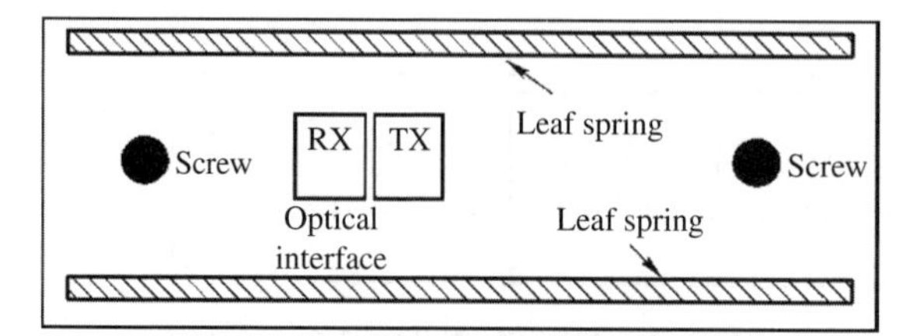

Figure 2.26 Back of the board.

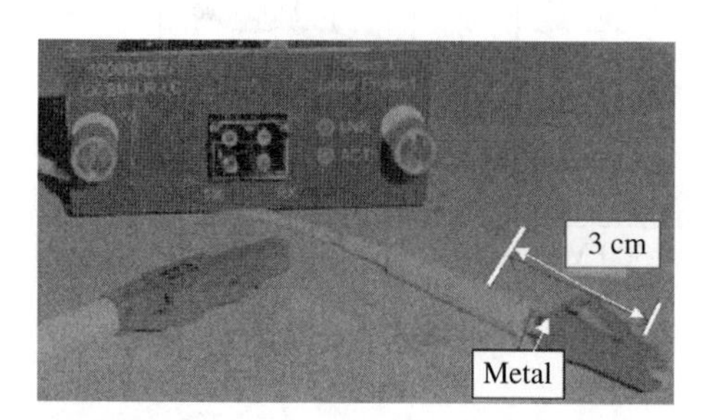

Figure 2.27 LC optical fiber and optical module panel interface.

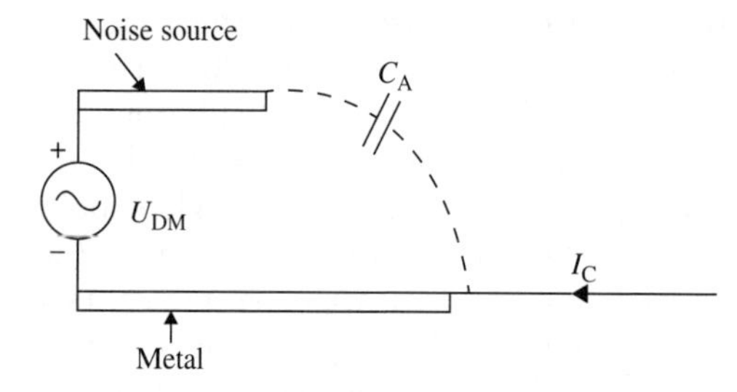

Figure 2.28 Principle of common-mode radiation caused by floating metal.

fibers and power cords connected to the device). We have to use a near-field probe for scanning the test once again, choose the probe with smaller loop area in order to make more accurate problem locating. It is found the emission from the outlet of optical fiber is very high by accident. After carefully looking at the outlet of optical fiber and its handle bar, it is only a square hole with 1 cm × 1 cm, and there is the shielded shell all around it. LC connector is used for optical fiber, is shown in Figure 2.27.

The outlet of optical fiber is always thought small, and optical fiber just transmits optical signal without radiated emission. After carefully checking, it is found that there is a metal reinforced rib inside the optical fiber connector, about 3 cm long, which is floating (without any connection to any metallic frame). The TX transmitter in optical module interface is metallic, although there is no direct contact between two metals when optical fiber is plugged in, due to the close distance, there will be a high-frequency noise coupled to the reinforced rib of optical fiber by capacitive coupling, which makes a driven-driving relationship between the metal reinforced rib and the internal noise sources. As a result, the reinforced rib becomes a monopole antenna. The monopole antenna radiates emission driven by a common-mode voltage. This is another radiated emission example driven by common-mode voltage, the principle is shown in Figure 2.28.

The U_{DM} in Figure 2.28 is the driving source, and the metallic reinforced rib is the *radiating antenna*.

[Solutions]

1) Remove the metal reinforced rib.
2) Make a good contact between the metal reinforced rib and the metallic parts of the interface panel, to bypass U_{DM}.

As removing the metal reinforced rib can cause application problems of the optical fiber, solution (2) is actually used. Testing it again after modifications, the results are shown in Figure 2.29.

[Inspirations]

1) High near-field emission doesn't mean that far-field emission is also large, which is mainly related to the radiation efficiency of antenna and the radiating path.

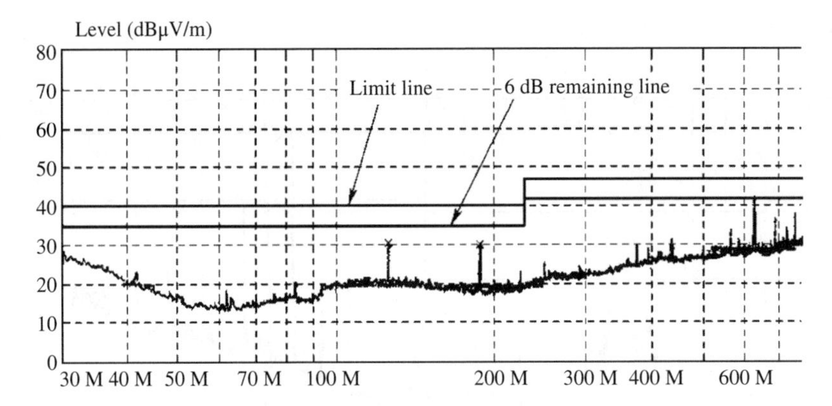

Figure 2.29 Modified test spectrum.

2) Prevent floating metallic parts, which will cause high common-mode voltage between noise source and floating metallic part. Floating metallic parts driven by the common-mode voltage will radiate strong emissions. In particular, a metallic part with large area has larger distributed capacitance and easily captures electric field coupling. Any metallic parts may cause common-mode radiation if there is voltage difference between them, so they must be well grounded locally. Heatsink, shielding metallic shell, metallic frame, and the unused areas of the metallic parts in PCB should be grounded.
3) The integrated circuit (IC) on the PCB sometimes has some unused pins, which are equivalent as a small antenna that can receive or transmit interference, so they should be connected to an adjacent ground plane or power plane.
4) Change the position of common-mode source relative to the antenna and reduce the parasitic capacitance C_A.

2.2.5 Case 5: Radiated Emission Caused by the Bolt Extended Outside the Shield

[Symptoms]

One telecommunication device is under radiated emission test. Its emission at 891 MHz frequency is more than the required margin limit, and the margin is less than 5 dB. The test result is shown in Figure 2.30. (The emission at 363 MHz also exceeds the standard limit, which is caused by the cable and is not summarized in this case.)

[Analyses]

Scanning the panel of the enclosure, the interface connector and the seam of enclosure with near-field probe, 891 MHz frequency is found. Although the emission at 891 MHz is high, is only confined to a small area around the heat sink of the enclosure. Open the enclosure and observe it. The internal structure of this area is shown in Figures 2.31 and 2.32.

There is a metallic bolt on the enclosure reinforced. This metallic bolt is connected to upside cover plate and bottom plate of enclosure with plastic bolts. Then it becomes a floating bolt. This bolt extends out of the shield with 5 cm length around. This floating bolt is driven by the 33 MHz clock in PCB board and behaves as a radiating antenna (891 = 33 × 27). The sectional view of the enclosure is shown in Figure 2.33.

The outermost layer of the enclosure is made of plastic material. The intimal layer is a metal mesh shield (as shown in Figure 2.31). There are heat-dissipating holes with 5 mm diameter in the shield, and the plastic bolt integrated with the enclosure passes through the shield. The metallic bolt is fixed on the plastic bolt and extends out of the shield in order to ensure that the enclosure can

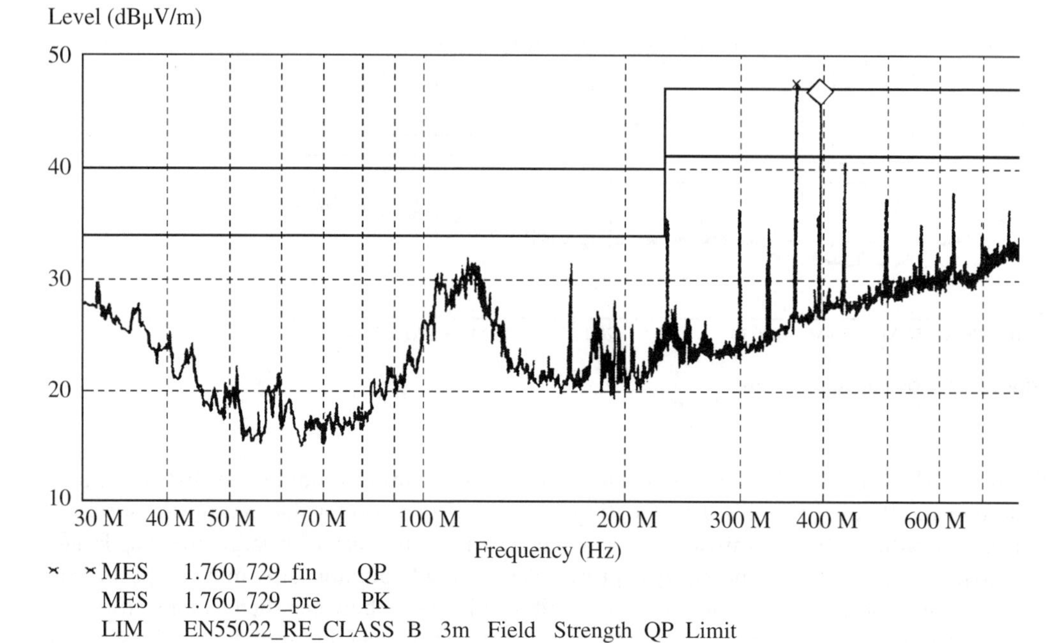

MEASUREMENT RESULT: "1760_729_*fin QP*"

7/29/02 5:01 PM

Frequency	Level	Transd	Limit	Margin	Height	Azimuth	Polarisation
MHz	dBV/m	dB	dBV/m	dB	cm	deg	
891.000000	44.20	4.9	47.0	2.8	147.0	0.00	

Figure 2.30 The failed test spectrum.

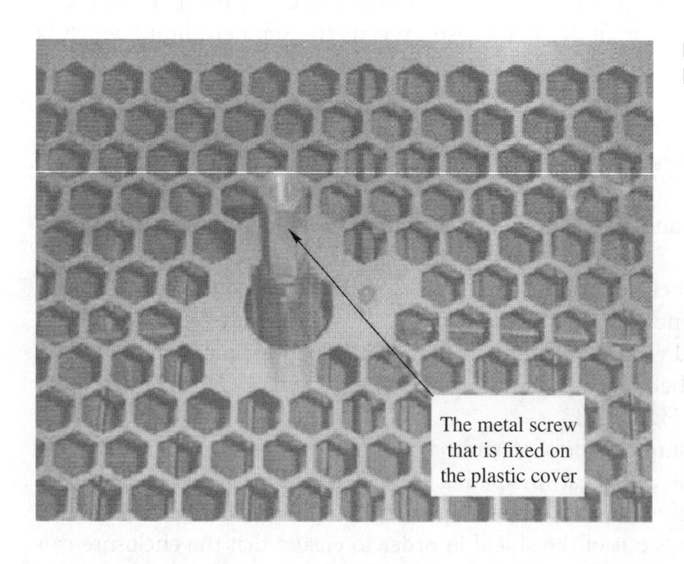

Figure 2.31 Upper cover plate of box body.

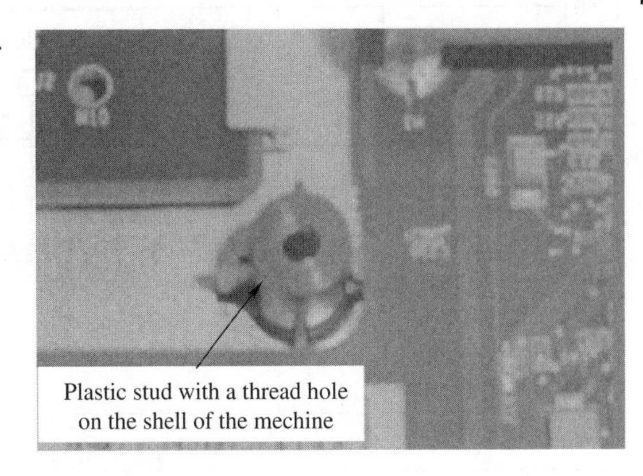

Figure 2.32 Chassis plate.

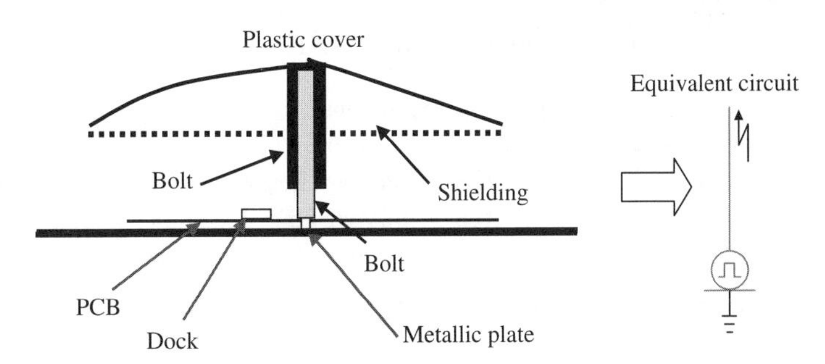

Figure 2.33 Section view of the shell.

withstand a certain mechanical stress, and there is no any electrical connection with the mechanical structural parts. Figure 2.33 also describes the radiation mechanism. In fact, the metallic bolt becomes a monopole antenna radiating as an electric field source. The voltage coupled between clock circuits and metallic bolt is the source, which becomes the driving source of common-mode emission. In high frequency, the common-mode current will flow through this bolt driven by the common-mode voltage. The change in the charge is caused by the current flowing along the length of the monopole antenna, which is transformed to radiated electromagnetic energy.

[Solutions]

In the countermeasure seeking test, we connect the bolt to ground (i.e. connect with the shield of the enclosure) by copper foil, then perform the test; the test is passed. The test result is shown in Figure 2.34.

Per the test result, it is confirmed that the emission at 891 MHz is indeed related to the bolt. However, this method in actual application is not feasible, because the bolt needs to be connected to the shield. If we choose the welding process, its processability is not good.

Therefore, we consider the following two solutions:

1) Change the metallic bolt to a plastic bolt. In this case, the mechanical strength of the structure will need to be further verified.
2) Shrink the metallic bolt toward the inner side. The metallic bolt extends the shield, which is equal to break the continuity of the shield. Shrinking back the bolt inside the shield (Figure 2.35) can ensure the continuity of the shield and does not impact the mechanical strength of the structure.

After EMC verification with these two solutions, both of them can meet the requirement. As the first one takes insufficient mechanical strength, it cannot be used. The second solution is finally chosen. The test result is shown in Figure 2.36.

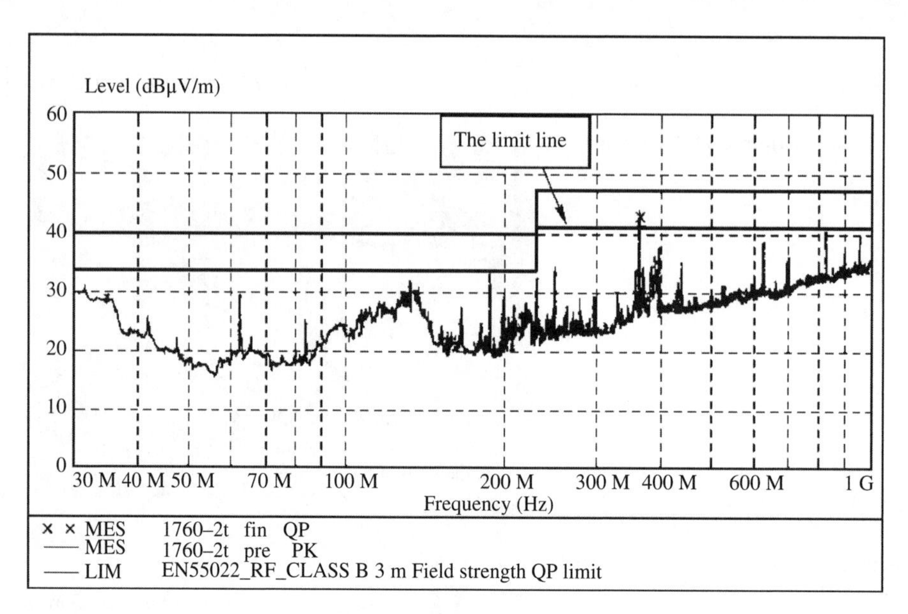

MEASUREMENT RESULT: "1760–2T_fin QP"

8/13/02 7:51 PM

Frequency	Level	Transd	Limit	Margin	Height	Azimuth	Polarisation	
MHz	dBV/m		dB	dBV/m		dB	cm	deg
363.000000	42.10		–2.0	47.0		4.9	101.0	266.00

Figure 2.34 Test spectrum after modification.

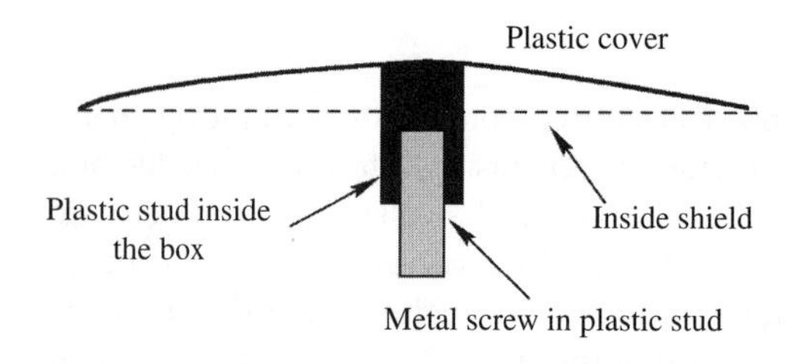

Figure 2.35 The diagram of screw retracting into the shield.

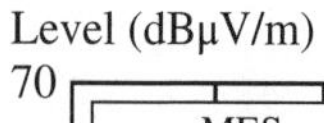

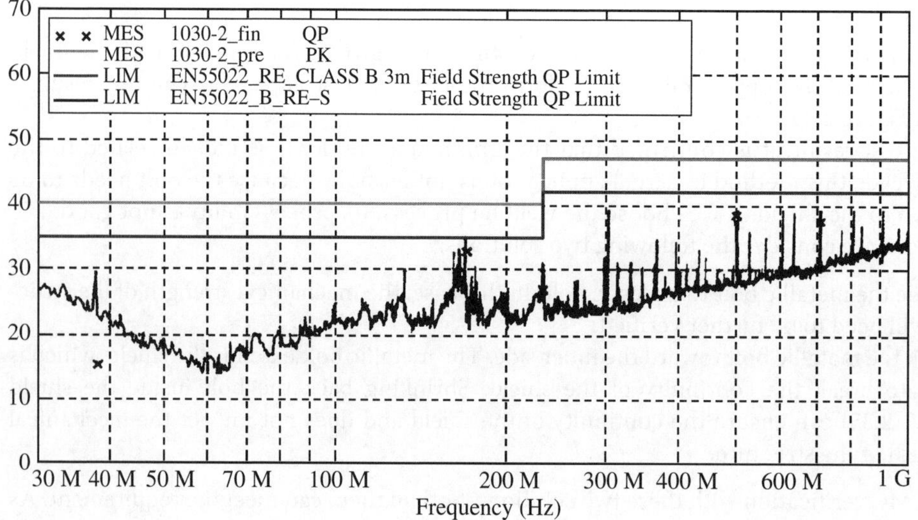

Figure 2.36 Final test result.

[Inspirations]

1) Avoid the existence of floating metallic parts. Floating metallic parts must be grounded or connected to 0 V.
2) Floating metallic parts could become a radiating antenna. Even though the floating metallic part cannot become an antenna, it could become a very good coupling path.

2.2.6 Case 6: The Compression Amount of the Shield and Its Shielding Effectiveness

[Symptoms]

The radiated emission of one device used in residential environment is required to meet EN55022 Class B limit. It is found in the test that there are multiple frequencies near 120 MHz exceeding the standard limit, which is shown in Figure 2.37.

[Analyses]

To identify whether this emission is from cable, pulling out all the external cables, the radiated emission results does not change significantly, still exceeding the standard limit. As this device is shielded with shielding structure, the emission must be radiated from the seam of the mechanical structure. It is the best way to use a near-field probe to locate the emission source radiated by the seam. A near-field probe detects that the maximum emission is located around the interface connector and that the module is installed on the shielded backplane (Figure 2.38).

At the beginning of the design, the interface connector between the module and backplane is shielded with conductive rubber, and is contacted a full 360° with the metallic parts of the

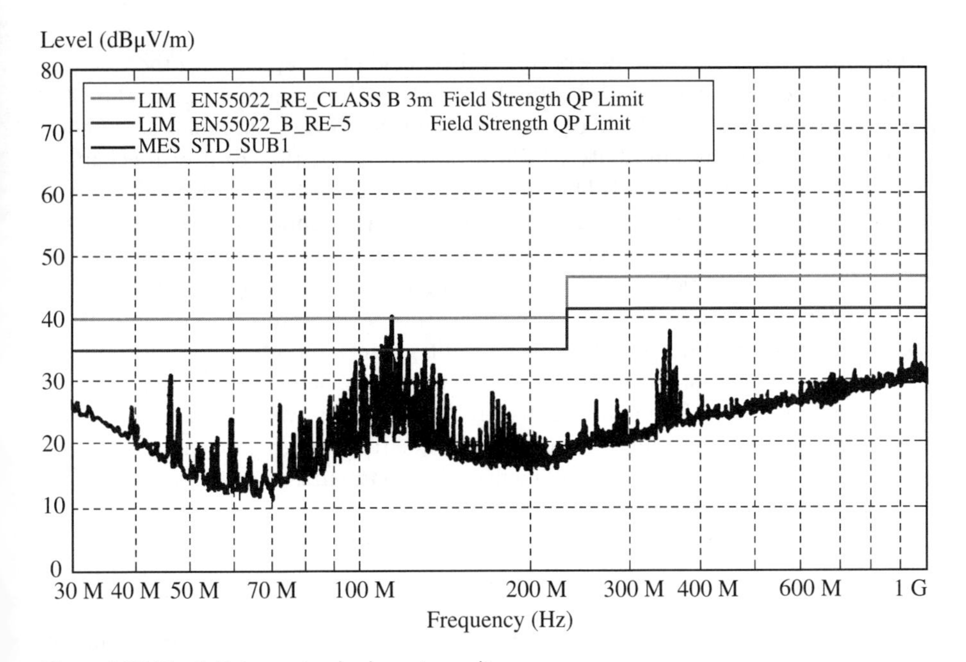

Figure 2.37 The initial overstandard spectrum diagram.

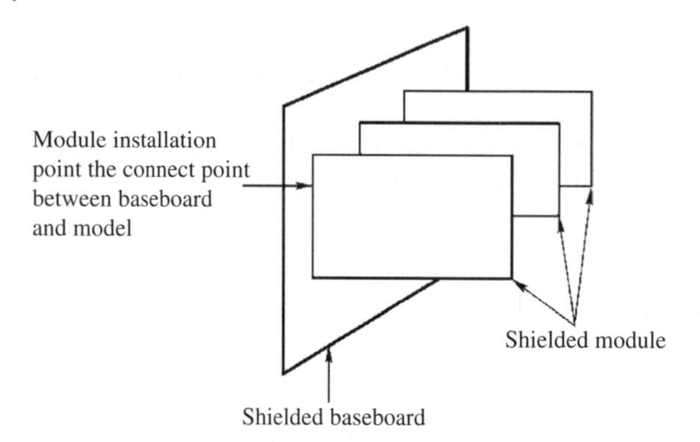

Figure 2.38 Product structure diagram.

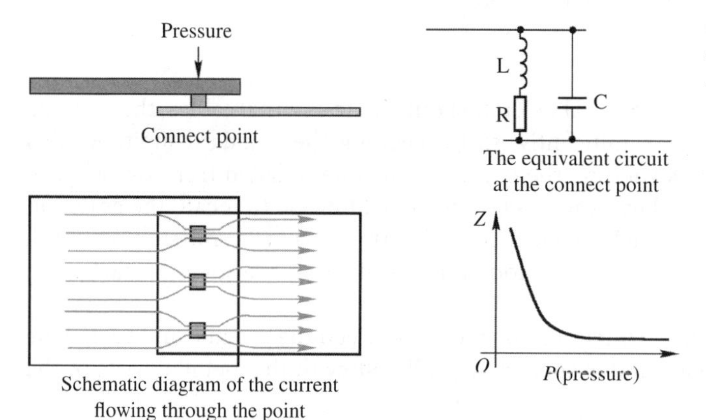

Figure 2.39 Relationship between the pressure and the impedance at the lap point.

backplane, which provides good continuity between the shield of backplane and the shield of the modules. Such high emission is detected with a near-field probe because the contact in 360° between conductive rubber and the shield of backplane and the shield of modules is not good enough, which results in the discontinuity of shield impedance or the seam of mechanical structure. Using a thicker conductive rubber during the problem-locating test, the test is passed. It indicates that there was not enough pressure between the conductive rubber and the shield of modules or that between conductive rubber and the shield of backplane because the conductive rubber was not thick enough; therefore, there was not enough compression on the conductive rubber (see Figure 2.39). The naked eye cannot perceive seams or impedance discontinuity (it may measure a low resistance with multi-meter), but after amplification, we can see there is a seam, or high impedance of the contact interface between conductive rubber and the shield of modules in high frequency (such as 100 MHz). Thus, when the common mode or induction current flows through the contact interface, due to this high impedance, a voltage drop across this contact interface appears, and it will drive the antenna acted by the seam, such that the emission is radiated.

Figure 2.40 shows the details of the installation relationships between backplane and modules.

[Solutions]

According to analysis and the test results with the locating way, it can be seen that replacing the original conductive rubber with a thicker one can increase the amount of compression on it,

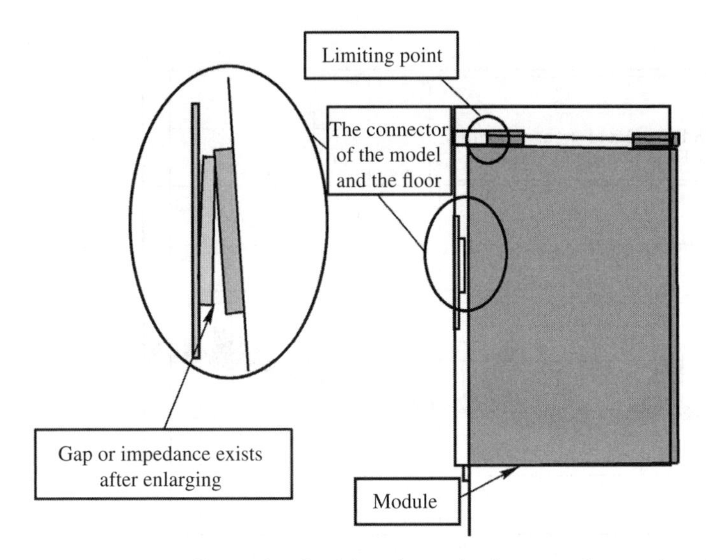

Figure 2.40 Installation details of the relationship between floor and module.

which makes the test result meet the requirements. Another option, rather than changing the conductive rubber, is to elevate the PCB of the backplane, as shown in Figure 2.41. This method can increase the amount of compression on conductive rubber as well, and realize 360° contact between conductive rubber and the shield of modules in its true sense, therefore ensuring conductive continuity. After modification, the test result as shown in Figure 2.42 is good.

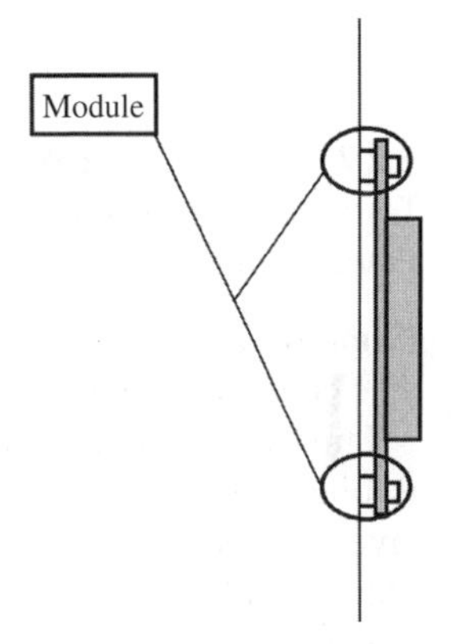

Figure 2.41 Diagram of block-up PCB.

[Inspirations]

1) While using the shielded material like conductive rubber and conductive gasket, we should not only ensure good conductivity for the contact surface (remove all the paintings on the contact surface) but also guarantee a certain amount of compression. However, we must pay attention to the compression limit problem of conductive rubber and conductive gaskets. Any gasket will be damaged when it is excessively compressed, and after it is damaged, its elasticity becomes poor and its sealing effect will be lost.
2) While using shielded material like conductive rubber and conductive gaskets, we should pay attention to cleaning the contact surface in order to prevent the gasket corrosion. Otherwise, the conductivity of contact surface is decreased, and the shielding effectiveness is reduced, too. The necessary conditions for electrochemical corrosion to occur between the gasket and the shield of backplane are moisture and corrosive gasses. Thus, a method of preventing corrosion is to isolate the electromagnetic sealing gasket through environmental sealing.
3) The seam is also an antenna.
4) Generally, from the EMC perspective, the system in which the contact resistance is less than $2\,\Omega$ and the resistance between any locations of the expected equipotential system is also less than $25\,\Omega$ can be regarded as a equipotential bonding system.

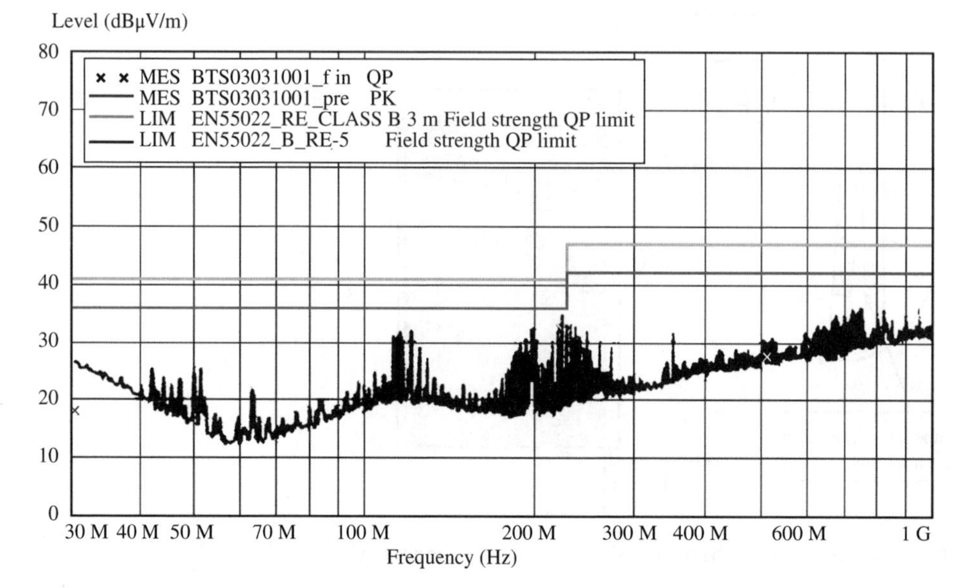

Figure 2.42 Test spectrum after changing.

2.2.7 Case 7: The EMI Suppression Effectiveness of the Shielding Layer Between the Transformer's Primary Winding and Secondary Winding in the Switching-Mode Power Supply

[Symptoms]

The configuration of a switching-mode power supply is shown in Figure 2.43.

The transformer shown in Figure 2.43 is designed with the shielding. The shielding layer is placed between the primary winding and the secondary winding. And the shielding layer is connected to the 0 V of the primary circuit through wire, which is shown in Figure 2.44.

The radiated emission and conducted emission test results of this power supply are, respectively, shown in Figures 2.45 and 2.46.

From these test results, we can see that the switching-mode power supply meets the Class B requirement, which is specified in Standard EN55022.

Changing the transformer of the power supply to a nonshielded transformer, i.e. to remove the shielding foil between the primary winding and secondary winding, we perform the conducted emission and radiated emission test. The test results are, respectively, shown in Figures 2.47 and 2.48.

From the test results, we can see that the product cannot meet the Class B requirement, which is specified in the Standard EN55022.

[Analysis]

For the switching-mode power supply, the EMI generated by switching circuits is one of the main disturbance sources of the switching-mode power supply. The switching circuit is the critical part of the switching-mode power supply, which is mainly composed of a switching transistor and a high-frequency transformer. The dU/dt generated by the switching circuit is a

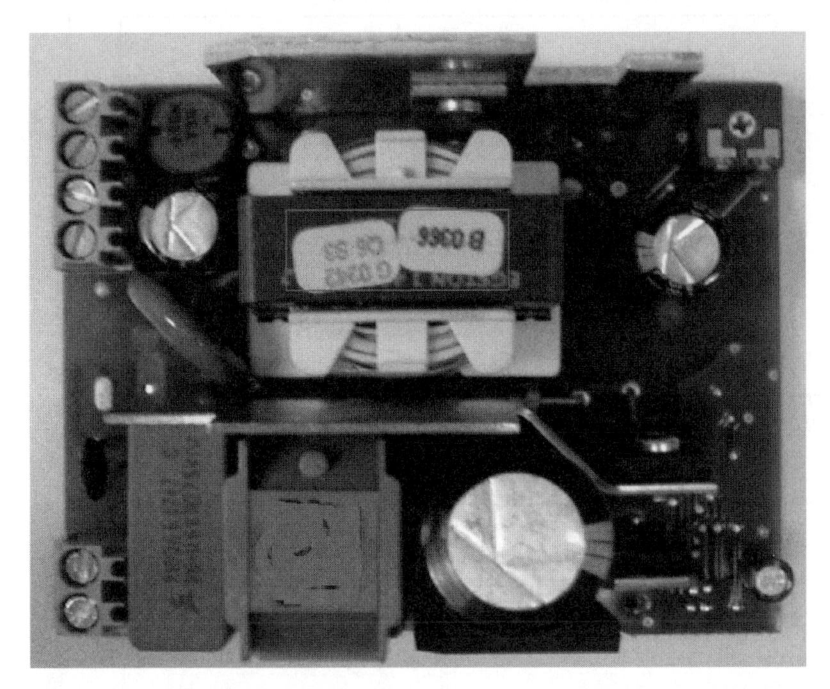

Figure 2.43 Picture of switch power supply.

Figure 2.44 Picture of transformer.

pulse with large amplitude, which has a wide frequency band and rich harmonics. The disturbance propagation schematic is shown in Figure 2.49.

There are two main reasons that the pulse disturbance is produced:

1) The load of the switching transistor is the primary winding of the high frequency transformer, which is inductive. While the switching transistor is switched on, a huge inrush current is produced in the primary winding, and a high peak voltage occurs across both ends of the primary winding. While the switching tube is switched off, due to the leakage magnetic flux of the primary winding. A part of energy cannot be transmitted from the primary winding to the second winding.

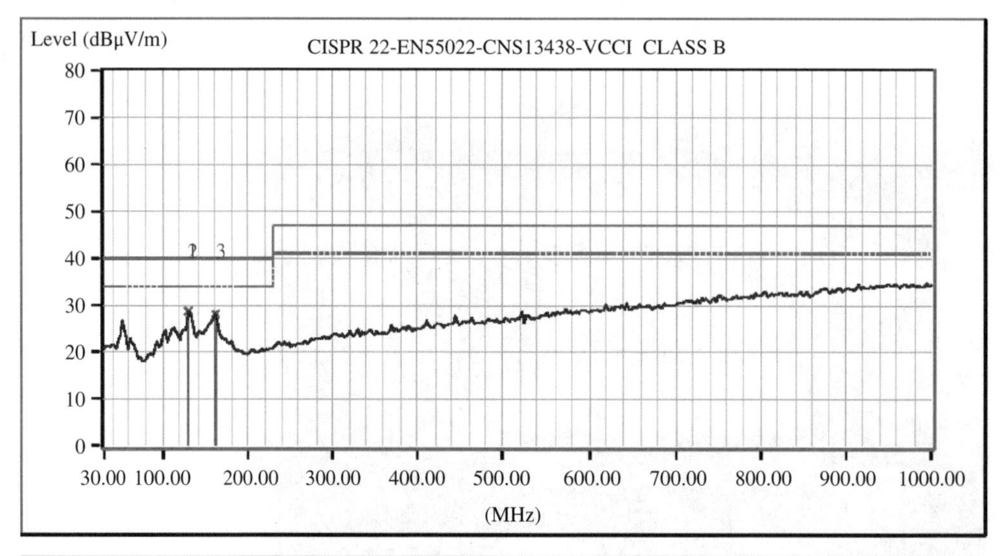

No.	Frequency MHz	Factor dB	Reading dBµV/m	Emission dBµV/m	Limit dBµV/m	Margin dB	Tower/Table cm	deg
1	129.43	15.33	13.38	28.70	40.00	–11.30	100	19
2	129.43	15.33	13.38	28.70	40.00	–11.30	100	19
3	160.95	16.98	11.05	28.02	40.00	–11.98	100	19

Figure 2.45 Radiate emission test result of switch power supply with shielded transformer.

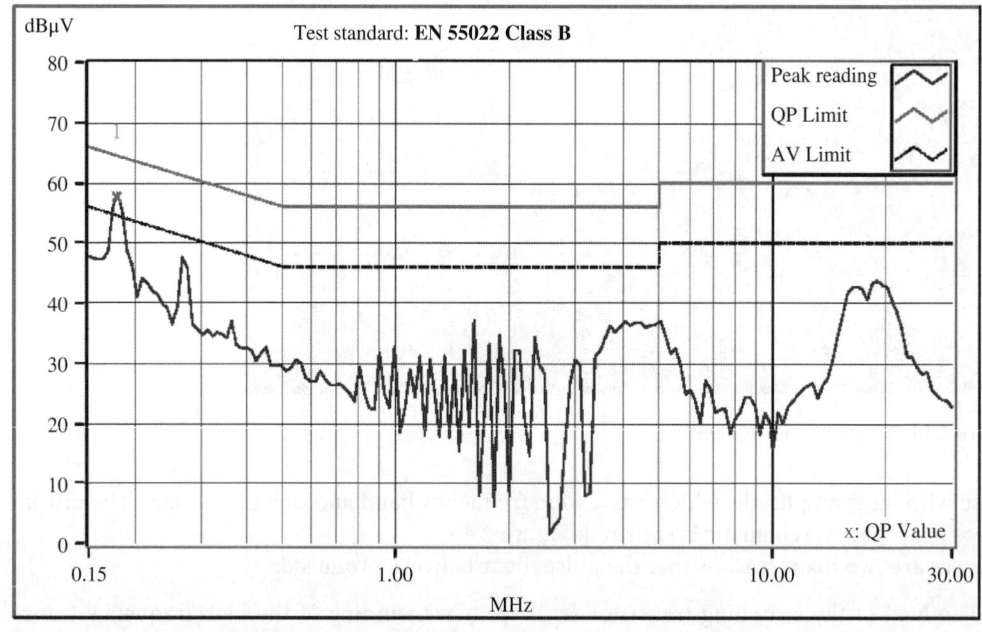

No.	Frequency MHz	Corr. Factor dB	Reading dBµV QP	Reading dBµV AV	Emission dBµV QP	Emission dBµV AV	Limit dBµV QP	Limit dBµV AV	Margins dB QP	Margins dB AV	Notes
+1	0.18015	0.50	57.22	47.89	57.72	48.39	64.48	54.48	–6.76	–6.09	

Figure 2.46 Conduct emission test result of switch power supply with shielded transformer.

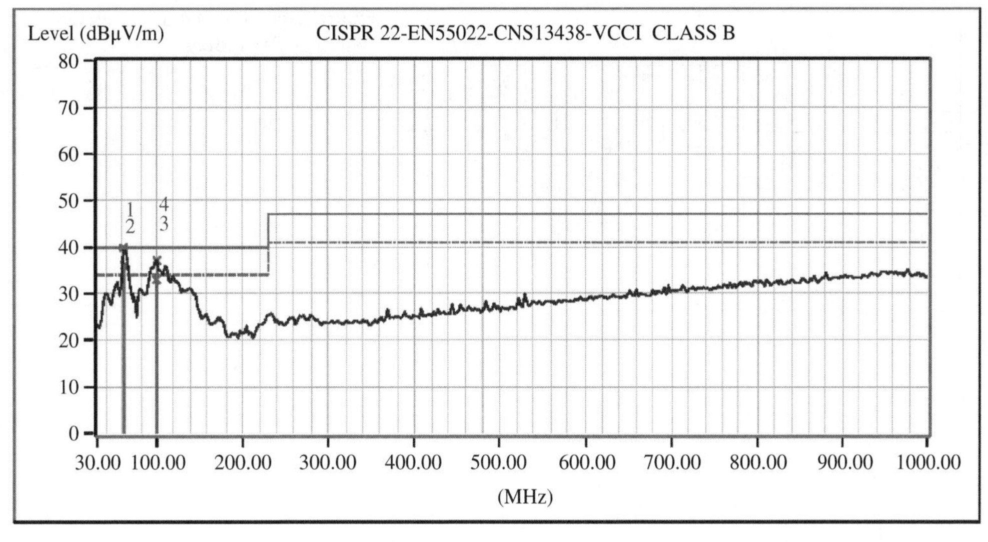

	No.	Frequency MHz	Factor dB	Reading dBµV/m	Emission dBµV/m	Limit dBµV/m	Margin dB	Tower/Table cm	deg
*	1	61.52	14.51	25.43	39.93	40.00	–0.07	---	---
	2	62.42	14.34	21.73	36.07	40.00	–3.93	97	18
	3	99.90	12.55	20.49	33.04	40.00	–6.96	97	106
	3	160.95	12.59	24.53	37.13	40.00	–2.87	---	---

Figure 2.47 Radiate emission test result of switch power supply with unshielded transformer.

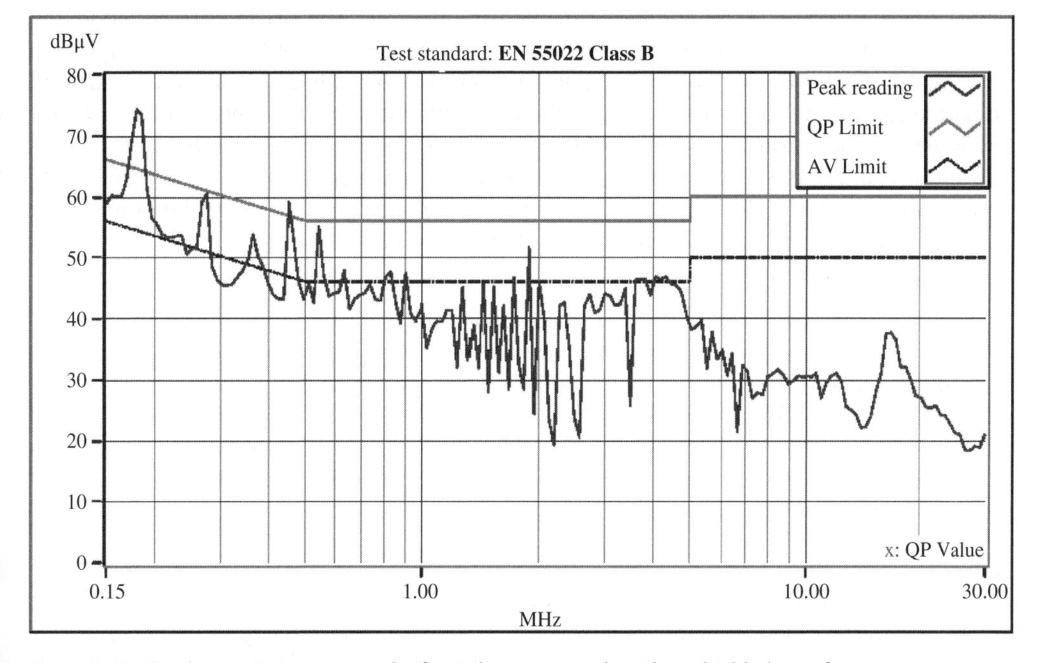

Figure 2.48 Conduct emission test result of switch power supply with unshielded transformer.

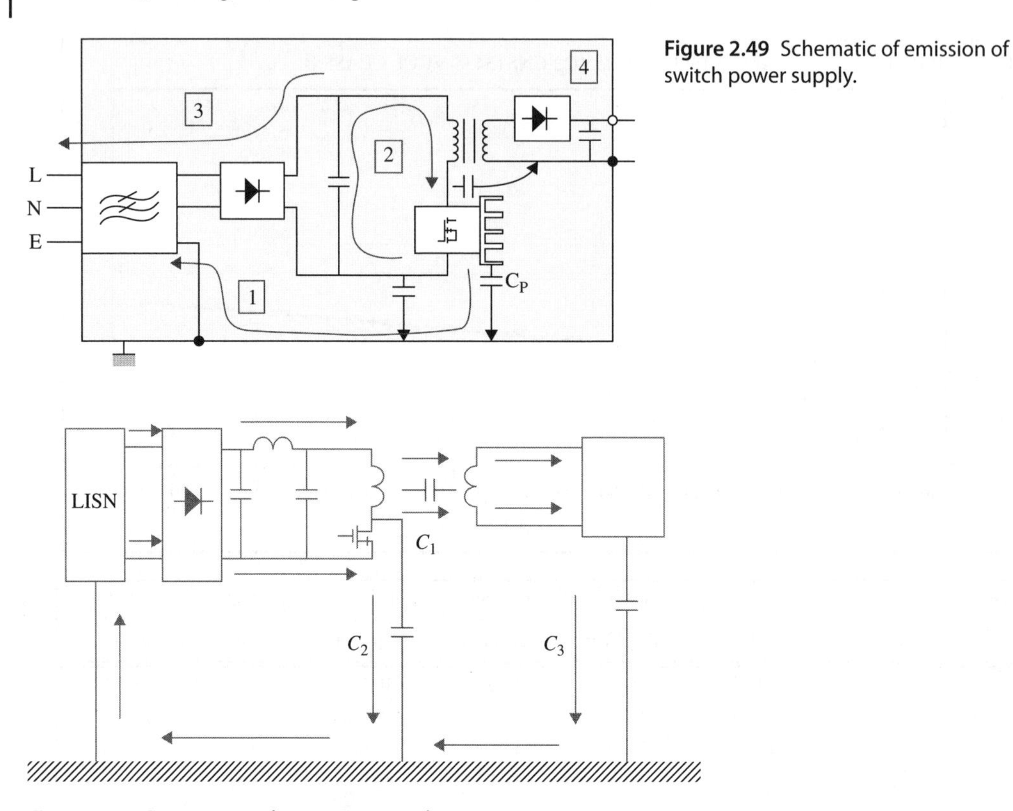

Figure 2.49 Schematic of emission of switch power supply.

Figure 2.50 Common-mode current in conduct emission test.

2) The high-frequency switching current loop that consists of the primary winding of the high-frequency transformer, the switching transistor, and the filtering capacitor may produce great space radiation and cause the radiated disturbance. If the filtering capacitance of the capacitor is not enough or its high-frequency characteristics are not good, the high-frequency current flowing through the high-frequency impedance of the capacitor will be transferred to the AC power supply in differential mode and cause the conducted disturbance. At the same time, there is the distributed capacitance between the primary winding and the secondary winding of the transformer, which causes the disturbance produced by primary circuit be transferred to the secondary circuit. As shown in Figure 2.50, on the one hand, the disturbance propagation loop is increased, and on the other hand, there will be more current flowing into the LISN. The EMI characteristic will be deteriorated, as shown in Figure 2.50.

The equivalent circuit of the circuit in Figure 2.50 is shown in Figure 2.51. Adding the shielding layer in the transformer and connecting it with the 0 V of primary circuit is equivalent to cutting off the path through which the disturbance propagates backward, which is shown in Figure 2.52. From the equivalent circuit shown in Figure 2.53 we can see that the conducted disturbance source is confined in a small loop. Therefore, the conducted disturbance and the radiated disturbance are restricted. (Note: The point A in Figure 2.52 is the point A in the equivalent circuit shown in Figure 2.53.)

[Solution]

One of the main reasons for the EMI problems in the switching-mode power supply products is that the common-mode noise in the primary side of the transformer propagates to the secondary

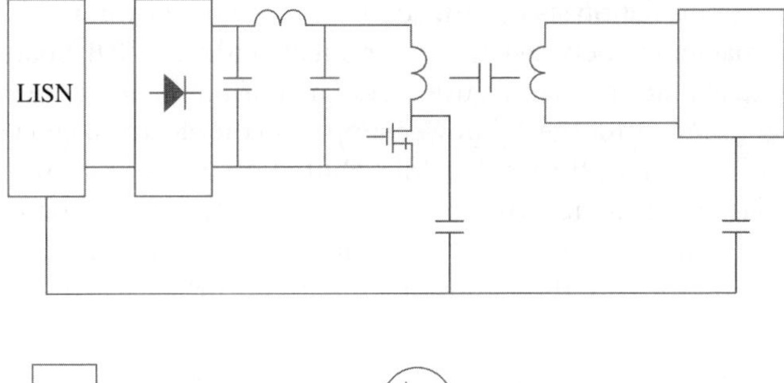

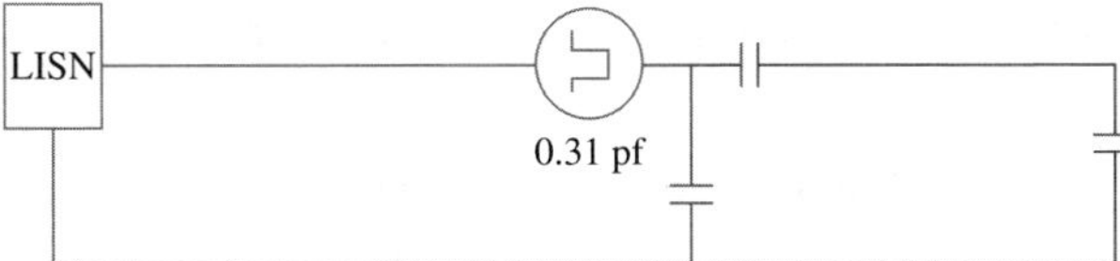

Figure 2.51 Equivalent circuit diagram of Figure 2.50.

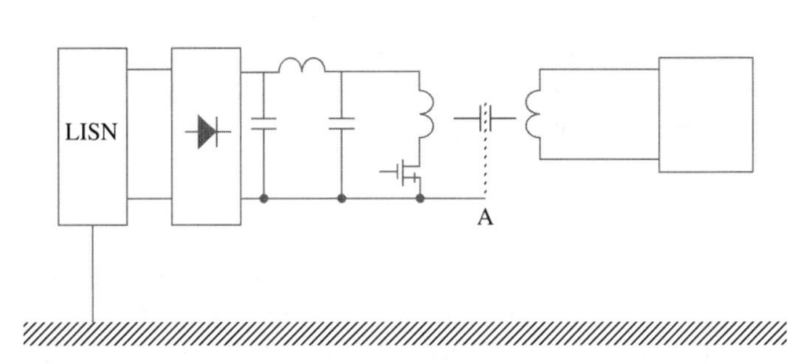

Figure 2.52 Position of the shield layer in the schematic diagram.

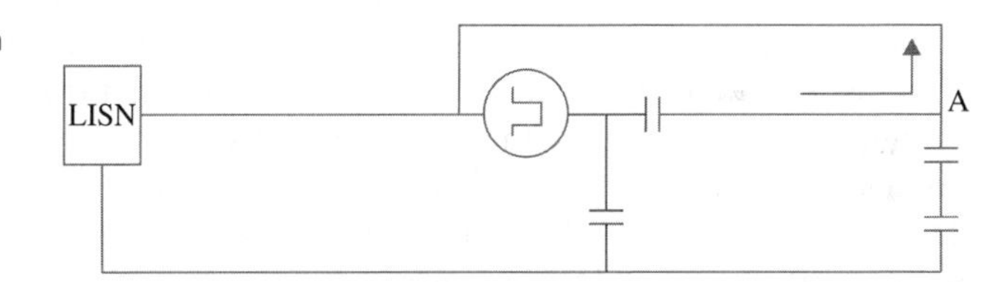

Figure 2.53 Equivalent circuit diagram of Figure 2.52.

side of the transformer. To cut off the transmission path, when we make the transformer, we need to create a shielding layer between the primary winding and the secondary winding, and connect the shielding layer to the ground or the DC high voltage point in the primary side.

In order to ensure the noise isolation function of the shielding layer, *zero impedance* must be ensured between the shielding layer to the ground or the DC high-voltage point in the primary circuit. This is the key of a good shielding. The practice proves that a single metal conductor whose ratio of length to width is less than 5:1 has extremely low impedance.

[Inspirations]

1) Using the shielding technology in the transformer, we can effectively control the common-mode noise coming from the switching-mode power supply propagating to the post-stage circuit. This kind of shielding is not a general electromagnetic shielding but a kind of electrostatic shielding. The shielding layer must be connected to the ground (the shielding layer is connected to the ground or the DC bus). The ungrounded shielding conductor will cause a *negative electrostatic shielding* effect.
2) Similar to this kind of shielding technology, if the power losses of the power switching transistor and the output diode are high, a heatsink must often be installed or the switching components directly mounted on the bottom plate for heat dissipation.

An insulation sheet with good heat conductivity must also be used, which creates distributed capacitance between the component and the PCB board or the heatsink. The distributed capacitance is C_P in Figure 2.49. The bottom plate of the switching-mode power supply is the ground line for the AC power supply. So the electromagnetic interference can be coupled to the AC line input through the distributed capacitance between the components and the bottom plate and produces the common-mode disturbance. The solution to this problem is inserting a shielding plate between two insulation layers. Connect the shielding sheet to the ground and cut off the path through which the RF disturbance is transmitted to the power supply input.

2.2.8 Case 8: Bad Contact of the Metallic Casing and System Reset

[Symptoms]

In the ESD immunity test for a product, when performing the ESD test (−4 kV contact discharge) on the DB connector of an interface PCB board (pure analog circuits), reset phenomenon appears on the PCB board connected (through the motherboard) with this PCB board. After checking the DB connector of this board, it is found that the shell of the DB connector is not well connected with the metallic panel. After interconnecting them with the conductive glue, and then performing the test once again (−6 kV contact discharge), all the circuits work well.

[Analysis]

ESD is a high-energy, wide-spectrum electromagnetic disturbance. It disturbs the device mainly in two ways. One is the direct energy, i.e. a large transient contact current may cause the malfunction or damage of the internal circuit. Another is the space coupling, since the leading edge of the ESD current is very short, approximately 0.7 ns, of which the frequency spectrum can reach hundreds of megahertz. Using slightly longer wires can create effective coupling.

In the test, after careful examination, it is found that the contact between the metallic panel of the PCB and the shell of the DB connector was not very good. There was no fixed electrical connection between them – obviously a great gap existed. From the circuit perspective, this gap is the impedance. Therefore, due to the existence of the impedance, the ESD current starting from the shell (e.g. represented as the dashed line B in Figure 2.54) will lead to a high voltage drop ΔV across the impedance, as shown in Figure 2.54.

There exists distributed capacitances between the shell of the DB connector and the metallic casing, and the reference ground plane and the signal traces of the internal circuit, among them. The distributed capacitance between the shell of the DB connector and the metallic casing, and the PCB ground plane is the largest, as the capacitance shown C_p in Figure 2.54. This

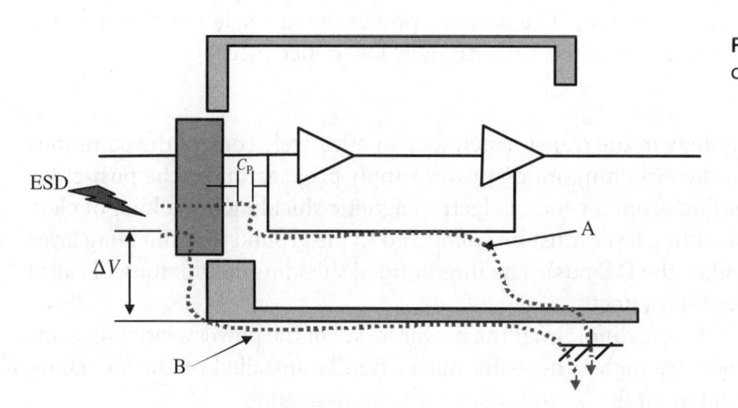

Figure 2.54 Schematics of ESD current.

distributed capacitance can't be ignored in the case of high-frequency ESD disturbance. In the presence of ΔV, it will inevitably lead to some ESD current flows to the reference ground plane through the distributed capacitance C_p, and finally flows to the earth, as the dashed line A shown in Figure 2.54 (Note: The reference ground in this product is connected to the casing somewhere, in fact, even if there is no connection between them, the ESD current will also flow to the earth through the distributed capacitance, so it is not feasible to solve this problem by disconnecting the circuit from the casing.).

In fact, the reference ground plane is also not very complete (the impedance of a complete, ground plane without via can be considered as $3\,\text{m}\Omega$), it has certain impedance, due to the existence of the slot caused by the via holes, as shown in Figure 2.55. When the current flows through the reference ground plane, due to the existence of the impedance, there will be a voltage drop ΔU_1, which will disturb the circuit (for more detailed analysis, refer to Case 49).

It is also worth noting that the presence of ΔU also provides the possibility of the radiation, which directly affects the internal signal lines through space.

Through this analysis, we understand that the disturbance signal is difficult to quickly release. It is coupled to the circuit through stray capacitor between the shell of the DB connector and the PCB, as the PCB is composed of some transformers and some analog devices. The product reset and crashed; other similar phenomena don't exist in the test. But the abnormal phenomenon as described above appears in some relevant digital circuit boards (which are connected to the same backplane), when the disturbance caused by ESD discharge is coupled to these boards. After using conductive adhesive to interconnect the shell of the DB connector shell and the metallic panel, on the one hand, due to the characteristics that the ESD disturbance will be discharged to its nearby objects, in the ESD discharge, the ESD disturbance is discharged quickly to the earth through the casing, which has no chance to go inside the PCB. On the other hand, after using the conductive adhesive to interconnect the shell of the DB connector and the metallic panel, the casing has better shielding effectiveness, which can shield the electromagnetic field generated in the ESD discharge outside the casing, so the PCB is protected and maintains its normal operating state.

[Solutions]

In order to ensure a good electrical connection between the DB connector and the metallic panel, the DB connector is fixed above the metallic panel by a screw, which can provide the tight connection and good electrical continuity between the DB connector and the metallic panel. This way not only improves the shielding effectiveness of the whole product but also

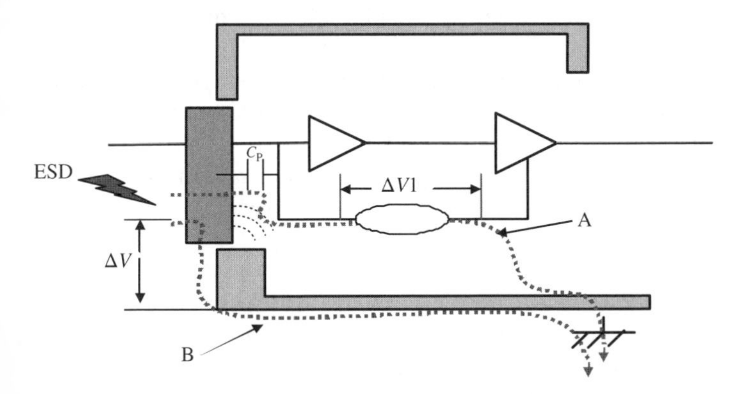

Figure 2.55 Disturbance voltage induced by ESD current flow ground plane of PCB.

quickly discharges the ESD disturbance current through the metallic panel; therefore, the problem is solved.

[Inspiration]

An effective way to prevent ESD disturbance directly coupled into the PCB is to guide the ESD disturbance directly to the ground through a grounded metal. Therefore, if there isn't a conductive integral (by means of mechanical connection) between the discharge point and the ground, it is necessary to pay special attention to the electrical continuity of the connection.

2.2.9 Case 9: ESD Discharge and the Screw

[Symptoms]

During the ESD test on the shell of the router, we found that the router will cause the system to crash when the test level is higher than ±4 kV. We locate the problem after carefully checking the equipment. The screw on the wide-area network (WAN) port of the router is not installed, and there is some insulating paint on the inner layer of the casing, which causes bad electrical continuity between different parts of the casing; thus, the router is susceptible to the ESD disturbance.

[Analysis]

During the ESD discharge, the electronic equipment is usually impacted by the following four ways:

1) The initial electric field can be coupled to the net with large surface area through capacitance, and it produces high voltage at the distance 100 mm away from the ESD discharge point.
2) The charge and current injected by the electric arc may cause the following damage and failure. (i) The charge will flow through the thin insulating layer inside the component and damage the gate of the MOSFET and CMOS component. (ii) The trigger in the CMOS device will be locked. (iii) The reverse biased diode will short circuit. (iv) The positively biased diode will short circuit. (v) The charge will melt the welding wire or the aluminum wire inside the active device.
3) The current will cause a voltage pulse on the conductor ($U = L^* \, di/dt$).
4) These conductors may be the signal, the power supply or the ground. The voltage pulse will enter all the components, which are connected to these nets.
5) The electric arc will produce a strong magnetic field with a frequency range of 1–500 MHz, and the electric arc will be coupled to each wiring loop. The electric arc will produce magnetic field that is tens of ampere per meter at the point that is 100 mm away from the ESD arc. The electromagnetic field produced by the electric arc will be coupled to the signal line. And the signal line acts as the receiving antenna.
6) We can refer to the case "bad contact and reset" to explain the problem in this case. Just a screw exists in this case, but a DB connector exists in the referenced case.

[Solutions]

Remove the insulating paint, and keep the electrostatic discharge path with good electrical continuity. Practice proves that, for a conductor, if the ratio of its length to width is less than 5:1, and there is no via hole and slot on it, it has good electrical continuity.

[Inspirations]

1) The continuity design of a grounding conductor is very important for the improvement of the ESD immunity of the system.
2) While locating the ESD problem, if the problem exists at the metal contact position, we should check whether the contact is good.
3) Painting usually disrupts the electrical continuity, which is a common problem in the structure design and process management. From this aspect, the EMC performance is not only related with the design, but also related to the technology, production, process, etc.

2.2.10 Case 10: Heatsink Also Affects the ESD Immunity

[Symptoms]

One device uses metallic casing. While ESD is injected on it, it is found that one area with a screw is extremely sensitive to ESD. When 3 kV contact discharge is injected on the screw, the reset of circuits appears for one PCB of this device. After observation and analysis, it is found there is one chip close to the area with screw, on which there is a 2 cm height heatsink without any grounding connection. After temporarily removing the heatsink, the ESD immunity of the area with screw can be improved to ±6 kV.

[Analyses]

During electrostatic discharge, tens of amperes discharge current appear in a very short time, and the rising time of the discharge current pulse is less than 1 ns. According to the formula of the highest-order harmonic frequency of pulse waveform,

$$f = \frac{1}{\pi T_r} \left(T_r \text{ is rising time} \right)$$

It can be seen that the electrostatic discharge is a process of high-frequency energy discharged and transmitted. All sensitive electronic circuits and components along the transmitting path can be disturbed, and they cause abnormal operation of the device.

In this case, due to the wide spectrum of the electrostatic discharge signal, some parasitic capacitance between structures cannot be neglected. Figure 2.56 shows the transmitting path of ESD disturbance.

In Figure 2.56, C_0 represents the parasitic capacitance between the injection point and the heatsink, C_2 is the parasitic capacitance between the heatsink and the chip. ESD disturbance is

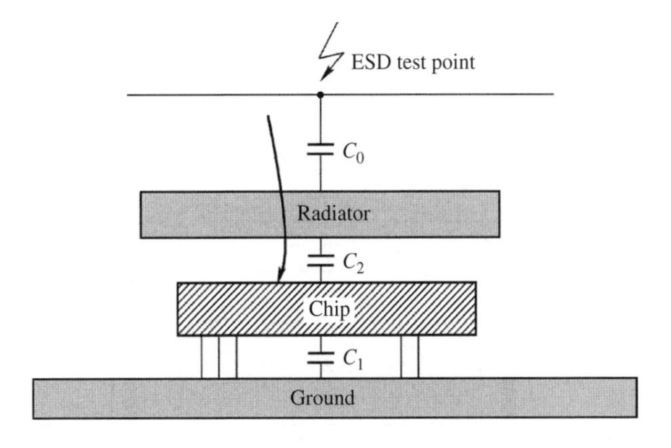

Figure 2.56 The diagram of ESD transferring path.

started from the injection point, through C_0 and C_2, then goes into the internal circuits of the chip, consequently disturbing the system. The existence of the heatsink significantly increases the capacitive coupling between the injection point and the chip, since the surface area of the heatsink is larger than that of the chip, and its existence reduces the distance between the injection point and the chip. Therefore, after removing the heatsink, the ESD immunity of the device is improved.

[Solutions]

According to above analysis, by grounding the heatsink, the discharge path of ESD can be altered, and thus the chip is protected. The arrowed curve in Figure 2.57 shows the new ESD discharge path after grounding the heatsink.

[Inspirations]

The metallic parts on the PCB have to be directly or indirectly grounded instead of floating. In addition, the sensitive circuits or chip should be placed far away from ESD discharging points during PCB layout.

2.2.11 Case 11: How Grounding Benefits EMC Performance

[Symptoms]

The architecture of one device is shown in Figure 2.58.

When an EFT/B test is performed on this device, with ±2 kV on power port and ±1 kV on signal port, meanwhile, P_1, P_2, and P_3 are all grounded, and the tests fail. If only P_1 is grounded, the EFT/B test on power port can pass, while they fail on signal ports with both cable 1 and cable 2. If P_1 and P_2 are grounded and P_3 is floating, the EFT/B tests on power port and signal cable 1 pass, while they fail on signal cable 2. If P_1, P_3 are grounded and P_2 is floating, the EFT/B tests on power port and signal cable 2 pass, while they fail on signal cable 1.

According to the above test results, none of those grounding ways works well on all ports during EFT/B tests.

[Analyses]

First, the nature and characteristics of EFT/B disturbance should be well known. The EFT/B is generated when the inductive loads are switched off. Instead of a single pulse, EFT/B is composed by a group of pulses. Figure 1.12 shows the waveform of EFT/B, the rising time of a single

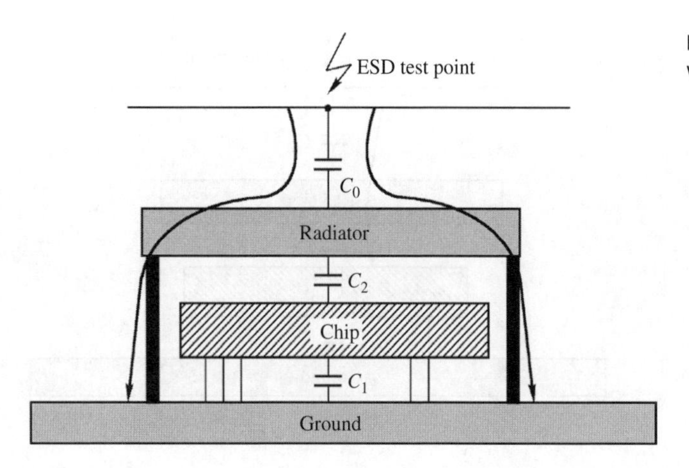

Figure 2.57 The ESD discharge path with grounded heatsink.

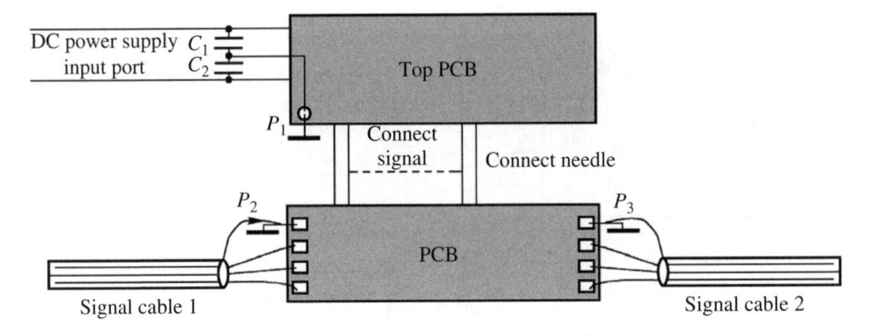

Figure 2.58 Product general structure diagram.

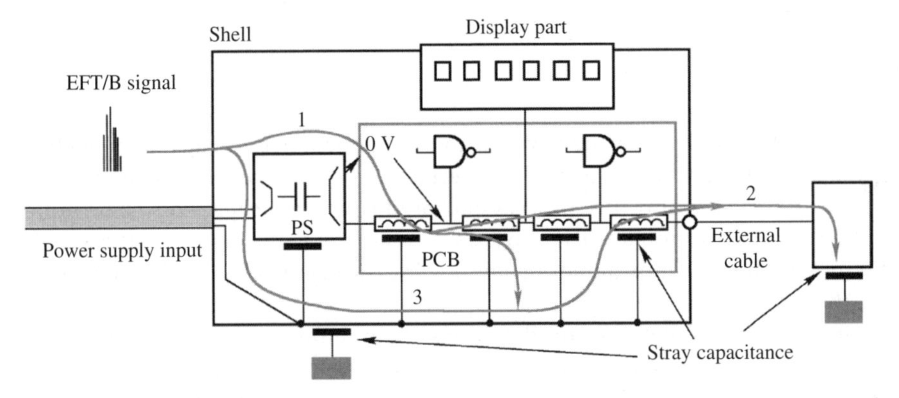

Figure 2.59 EFT/B interference affects the device circuit.

pulse t_r can be as short as 5 ns and its half-width duration T is 50 ns, which assures that the pulse group includes rich harmonic components. The maximum frequency of harmonic components can be $1/\pi t_r$, which is around 60 MHz. The parasitic capacitances exist between power cable, EUT, signal cable, and reference ground plate, which provide a high-frequency path for EFT/B current. Therefore, during the EFT/B test, the disturbance current can be injected everywhere through the parasitic capacitance in common mode, as shown in Figure 2.59, and significantly impacts on the electronic circuits.

The impact of pulse sequence is accumulated at the injection point, which makes the disturbance amplitude exceed the noise margin of electronic circuits. Based on this mechanism, the shorter the repetitive period of the pulse is, the more significant its impact on the circuit. As the interval between two adjacent pulses of the pulse sequence is short, there is not enough time to discharge the input capacitance of circuits before the next charge arrives, which easily raises the voltage level on the circuit input. If this voltage level is high enough to affect the normal operation of the circuit, the system will be disturbed.

The principle of the EFT/B test is shown in Figure 2.60. In Figure 2.60, EFT is the disturbance source. During the test, the disturbances are applied on the DC power port, signal cable 1, and signal cable2. C_1 and C_2 are the Y type capacitors at DC power supply input; C_3 and C_4 are the capacitances between signal cables and ground; P_1, P_2, and P_3 are three points where could be grounded. The PCB on top side and the PCB on bottom side are connected through pins of hard-link connector; Z_1–Z_n are the impedance of the signal pins of the connector; Z_{g1} is the impedance of its ground pins; Z_{g2} is the ground impedance of the interconnections between P_2 and P_3 on PCB.

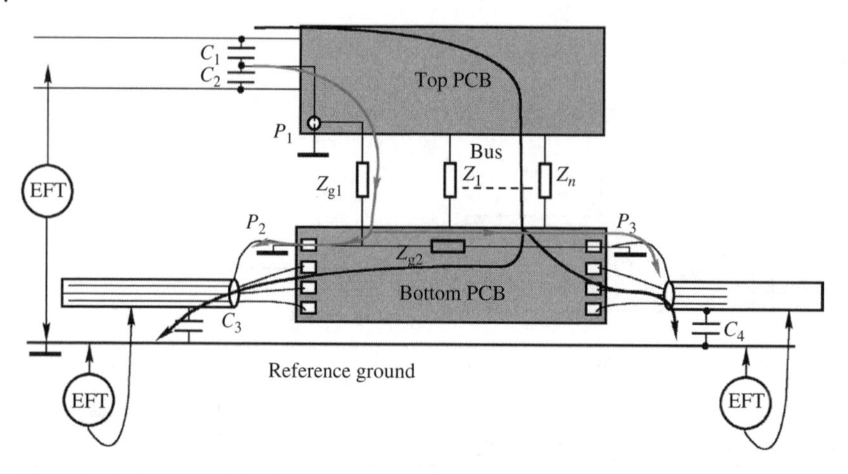

Figure 2.60 Test principle diagram.

The mechanism that EFT/B disturbs the device is to charge the junction capacitors of electronic circuits. When its energy is accumulated to a certain high level, the circuits could operate abnormally (even for the system). The process of charging junction capacitors is the same as the process that the common-mode current passes through the EUT; therefore, the amplitude and the duration of the common-mode current flowing through EUT directly determines the result of EFT/B test.

The arrow line in Figure 2.60 shows the flow directions of the common-mode currents. It can be seen that the common-mode current easily passes through the internal circuits if the grounding position is at the far end of EUT from the injection point. When the common-mode current passes through the internal circuits, the impedance of the current path is critical to judge the impact on the circuits from disturbance. If the impedance is large, a high-voltage drop may occur, which means more significant impacts on EUT (generally, the through hole component on PCB has520 μH distributed inductance for each pin of it; a double-row 24-pin integrated circuit socket contributes 4–18 μH distributed inductance for each pin of it). The three grounding points of Figure 2.60 are connected by a relatively narrow PCB copper track, so there is high impedance between them. In this regard, one side, the EUT, must be grounded at a single point to reduce the common-mode current to pass through the internal circuit. On the other hand, according to impedance analysis and test results, the three grounding points are different, so the impedances between the three grounding points are high. These impedances should be reduced by a certain way, in order to reduce the voltage drop on the ground while the common-mode current passes through the internal circuits, which will increase the success probability of the EFT/B test.

Regarding the impedance of the reference ground, the following supplementary should be noted:

When talking about the malfunction caused by voltage difference between anywhere of ground due to non-neglectable ground impedance, many people think it is hard to imagine. While measuring ground impedance with an Ohm meter, the ground impedance is in the order of milliohm. How can such significant voltage drop on the ground be generated while the common-mode current passes through so such small ground impedance?

To figure out this problem, we have to differentiate the concept of the resistance and the reactance. Resistance represents the impedance of a conductor when direct current (DC) current flows through, while reactance is its impedance when alternating current (AC) current flows though. The reactance of the conductor is mainly caused by its inductance. All conductors have inductance; in high frequency, the reactance of conductor can be much higher than its resistance. Table 2.2 shows this trend. In real circuits, the disturbances are usually pulses, which contain rich high-frequency components. As a result, a relatively high voltage drop is generated. The frequency of the interference from digital circuits is much higher; therefore, the ground impedance is important for digital circuits.

Note: D is the diameter of conductor, and L is its length. If the impedance at 10 Hz is approximately considered as resistance, it can be seen that when the frequency is up to 10 MHz, for 1 m length cable, its impedance is 1000 to 100 000 times its DC resistance. Therefore, a large voltage drop appears when RF current passes through the ground conductor.

Table 2.2 also shows that increasing the conductor's diameter helps to reduce it DC resistance, but is less helpful for reducing its AC reactance. In EMC tests, people are mostly concerned with the AC reactance. To reduce AC reactance, planes are generally used in PCB, like the perfect power plane or ground plane designed in PCD layout. Moreover, via holes and slots should be avoided on the planes as much as possible. Metallic structures can also be used as supplements to reduce the impedance of non-perfect planes. Normally, the impedance between anywhere on a perfect plane without via holes is considered as 3 mΩ. With this type of ground plane, TTL circuit can sustain at least 600 A pulse current (i.e. 1.8 V voltage drop will appear when 600 A current passes through it). Whereas, the maximum current of 4 kV EFT/B is only 80 A, which is limited by the 50 Ω, internal impedance of EFT/B generator. However, in real cases, vias cannot be avoided on ground plane. If there are slots caused by juxtaposed vias on a ground plane as Figure 2.61 shows, and 1 cm long slot contributes around 10nH inductance, the voltage drop with 80 A EFT/B peak current can be calculated as:

$$U = LdI/dt = 160\,\text{V}$$

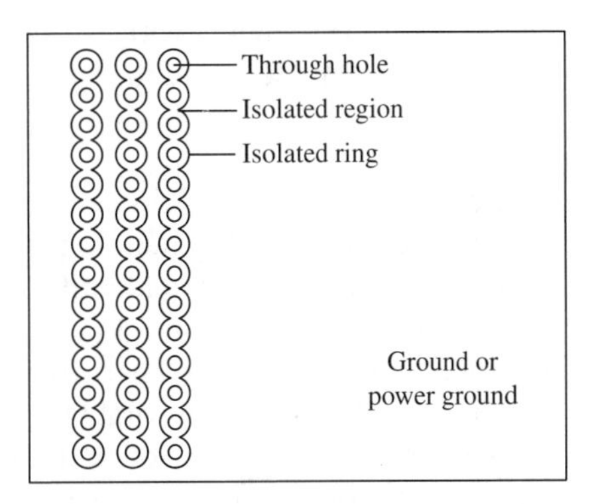

Figure 2.61 An example of a ground plane with slot.

Table 2.2 Wire impedance (Ω).

Frequency (Hz)	D = 0.65		D = 027		D = 0.065		D = 0.04	
	L = 10 cm	L = 1 m	L = 10 cm	L = 1 m	L = 10 cm	L = 1 m	L = 10 cm	L = 1 m
10	51.4 μ	517 μ	327 m	3.28 m	5.29 m	52.9 m	13.3 m	133 m
1 k	429 μ	7.14 m	632 μ	8.91 m	5.34 m	53.9 m	14 m	144 m
100 k	42.6 m	712 m	54 m	828 m	71.6 m	1.0	90.3 m	1.07
1 M	426 m	7.12	540 m	8.28	714 m	10	783 m	10.6
5 M	2.13	35.5	2.7	41.3	3.57	50	3.86	53
10 M	4.26	71.2	5.4	82.8	7.14	100	7.7	106
50 M	21.3	356	27	414	35.7	500	38.5	530
100 M	42.6		54		71.4		77	
150 M	63.9		81		107		115	

Here, L is the inductance of the slot, 10 nH cm^{-1} dI is the EFT/B current with the maximum value of 80 A, dt is the rising time of EFT/B current, which is around 5 ns. Obviously, 160 V is a dangerous voltage for TTL circuit. Therefore, grounding, filtering, or metallic plate must be applied to solve the problem caused by EFT/B disturbance. It can be seen that a perfect ground plane is important to improve the immunity performance. Specifically, for devices without grounding, a perfect ground plane becomes even critical.

[Solutions]

According to this analysis, the following major solution can be applied:

1) Ground at a single point instead of multiple points. In other words, for on-site application and EMC test, P_2 and P_3 are only connected to the cable shield, and the ground connections at P_2 and P_3 are removed; only the grounding connection at P_1 is kept.
2) Interconnect P_1, P_2, and P_3 with a metallic plate, and make sure the ratio of length-to-width between any two of them is smaller than 3:1, which guarantees that their mutual impedance is relatively low.

After the improvements with (1) and (2), perform the test again; all tests pass. The power port can pass ±2 kV test, and the signal ports can pass ±1 kV test.

[Inspirations]

1) In high-frequency EMC domain, while grounding at multi-points, the equipotential connection between each grounding point is quite critical. One reliable way to justify the equipotential connection is to identify if the ratio of length-to-width of the conductor between each point is smaller than 5:1 (or 3:1 for even better performance).
2) Grounding at the far end apart from EFT/B injection point is bad for the immunity performance of EUT, which inevitably lets the common-mode disturbance current pass though the internal ground plane of circuits.
3) A perfect ground plane is critical not only for EMS but also for EMI.
4) The main concerns for proper grounding include:
   - Place the high-frequency components carefully to reduce the area of current loop, or minimize it.
   - Separate the high-frequency wide-band circuits from low-frequency ones by segmenting PCB or system into different zones.
   - Avoid disturbance current passing through the common ground path to affect other circuits in PCB or system design.
   - Select grounding points properly in order to minimize loop current, ground impedance, and transfer impedance.
   - Consider the current passing through grounding points as the injection current or the noise out of circuits.
   - Connect the sensitive (low noise margin) circuits to a stable reference ground with minimum ground impedance.

2.2.12 Case 12: The Heatsink Shape Affects Conducted Emissions from the Power Ports

[Symptoms]

The test results of conducted emission test on power port of one charger is shown in Figure 2.62.

It can be seen that the power port of this charger cannot meet the requirement of EN55022, Class B limit.

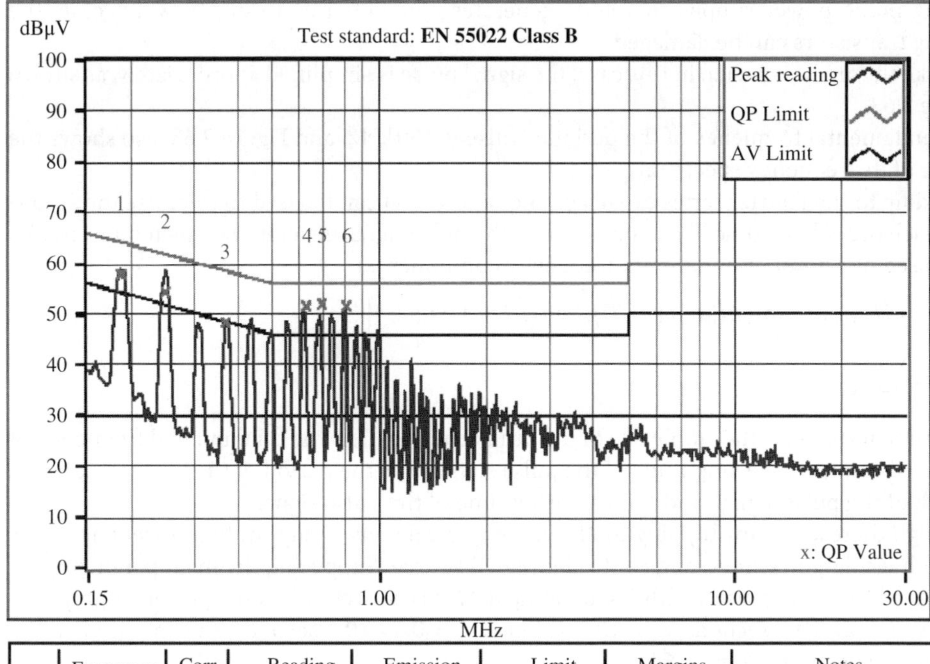

No.	Frequency MHz	Corr. Factor dB	Reading dBμV QP	Reading dBμV AV	Emission dBμV QP	Emission dBμV AV	Limit dBμV QP	Limit dBμV AV	Margins dB QP	Margins dB AV	Notes
1	0.18516	0.50	57.23	51.45	57.73	51.95	64.25	54.25	−6.52	−2.30	
2X	0.24766	0.50	53.78	51.79	54.28	52.29	61.84	51.84	−7.56	0.45	
3	0.36875	0.50	47.96	43.32	48.46	43.82	58.53	48.53	−10.07	−4.71	
+4X	0.62734	0.50	51.26	46.14	51.76	46.64	56.00	46.00	−4.24	0.64	
5	0.69766	0.50	51.43	42.82	51.93	43.32	56.00	46.00	−4.07	−2.68	
6	0.81016	0.50	51.22	43.89	51.72	44.39	56.00	46.00	−4.28	−1.61	

Figure 2.62 Conduct emission test results.

[Analyses]

The previous chapter has described the EMC nature of switching-mode power supply. In other words, for switching-mode power supply, the electromagnetic disturbance generated by switching circuits (which are mainly composed of switching transistors and high frequency transformer) is the main source of interference. The switching circuits generate a pulse signal, which is caused by the inductive load, the primary winding of high-frequency voltage transformer. When the switching transistor is turned on, the primary winding is magnetized, which induces a relatively high surge peak voltage across transformer's primary winding. When the switching transistor is turned off, due to the leakage inductance of the primary winding of transformer, part of the energy cannot be transferred from the primary winding to the secondary winding. The energy stored in leakage inductance drives the resonance between leakage inductance, parasitic capacitance and resistance, which will generate a damped oscillation with voltage spikes. The switching-off high-voltage spike is caused by the damped oscillation wave superposed on the normal operating voltage at the time of switching-off. This voltage spike will magnetize transformer primary winding as it is magnetized at switching-on, and generates transient magnetizing pulse current that could be transferred to the inputs and outputs of switching-mode power supply; therefore, generates conducted emission. In worst case, the switching transistors can be damaged.

To make the analysis easier, in this case, the signal pulse is simplified appropriately, as shown in Figure 2.63.

The fundamental frequency of the periodic pulse is 150 kHz, and Figure 2.63 also shows the envelope of its frequency spectrum.

According to the Fourier series expansion, Equation (2.1) can be used to calculate the amplitude of each order harmonic. It can be seen that the pulse signal generated by switching circuits is composed of a number of different frequency components.

$$A_n = 2U_0 t_w / T\{\sin(nF_0 T)/(nf_0 T)\}\{\sin(nF_0 t_r)/(nf_0 t_r)\} \tag{2.1}$$

$$n = 1,2,3,\ldots$$

In this equation, A_n is the amplitude of the nth harmonic; F_0 is the fundamental frequency of the pulse signal; U_0 is the amplitude of the pulse signal; T is the period of the pulse signal; t_w is the width of the pulse signal; and t_r is the rising time of the pulse signal.

Although there are many topologies of switching-mode power supply, their essential part is a high-voltage, large-current, and controlled signal source. Supposing the main parameters of switching pulse in a pulse-width modulating (PWM) switching-mode power supply are: U_0=500 V, $T = 2 \times 10^{-5}$ seconds, $t_w = 10^{-5}$ seconds, $t_r = 0.4 \times 10^{-6}$ seconds, the amplitude of its harmonics is shown in Figure 2.64.

The harmonics of switching pulses shown in Figure 2.64 are the EMI source for other electronic equipment. The amplitude of harmonics can be measured through conducted emission (from 0.15 to 30 MHz) and radiated emission (from 30 to 1000 MHz) tests. In Figure 2.64, the amplitude at fundamental frequency is around 160, and 30 dBμV at 500 MHz. Therefore, it is a certain difficulty to control the EMI of switching-mode power supply below the limit specified in the EMC standards.

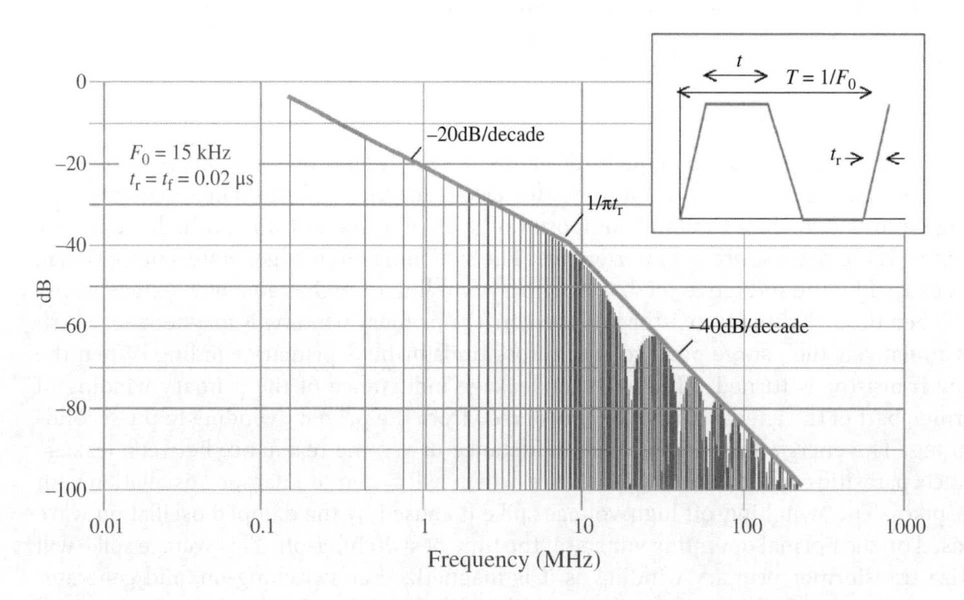

Figure 2.63 Frequency spectrum of periodic signal generated by switching power supply.

As we know now, the switching circuits are the essential part of switching-mode power supply, and it generates high dU/dt pulses, and their amplitude is high, frequency-band is wide and rich harmonics are included. It is difficult to change this fact, but the interference transmission path is controllable. Therefore, for switching-mode power supply, controlling the transmission path of interference is an important aspect of its EMC design.

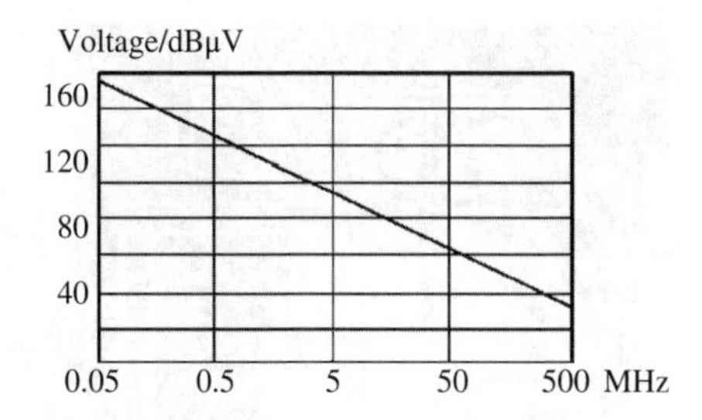

Figure 2.64 Harmonic of switching power supply.

The power switch is the critical component to generate the switching pulses as already mentioned, which usually consumes high power by itself. For heat dissipation, a heatsink is installed and connected to the drain (or collector) of the switch (or coupled through the distributed capacitances even if not directly connected. In this case, the heatsink becomes a part of key interference source of switching-mode power supply. Figure 2.65 shows the impact of heatsink on EMI. Due to the large surface area of the heatsink, high capacitance will appear between the surface of the heat sink and the printed wirings on PCB, components, power cords, and reference ground plate, etc., and becomes the root cause of conducted emission. Over the frequency range concerned in EMC domain, the impact of small parasitic capacitances should never be neglected.

Taking the case shown in Figure 2.65a as an example, if C_{S1} = 0.1 pF (a very small capacitance value), U_{S1} = 300 V, at 150 kHz, the conducted interference voltage detected on LISN is 1400 μV, which is far beyond the Class B limit defined in EN55022 (630 uV at 150 kHz). And then with the example shown in the Figure 2.65b, if C_{S2} = 0.1 pF, U_{S2} = 300 V, at 150 kHz, the measured conducted interference voltage is 700 μV, which is also over than the Class B limit defined in EN55022.

*NOTE: the coupling capacitance between two parallel planes can be calculated as followed:

$$C_S = C_i + C_p,$$

C_i (intrinsic capacitance in pF) =35 · D (the diagonal in meter);

C_p (capacitance between parallel plates in pF) =9 · $S(m^2)/H$ (the distance between two plates in meter)

For example: The areas of two parallel metallic plates both are 10 cm × 20 cm. Therefore:

$$D = 0.22m, S=0.02m^2;$$

The distance between them, H = 10 cm;

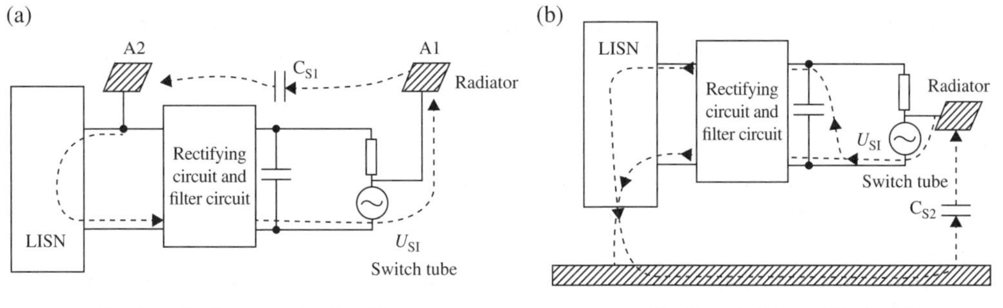

Figure 2.65 Couple from radiator.

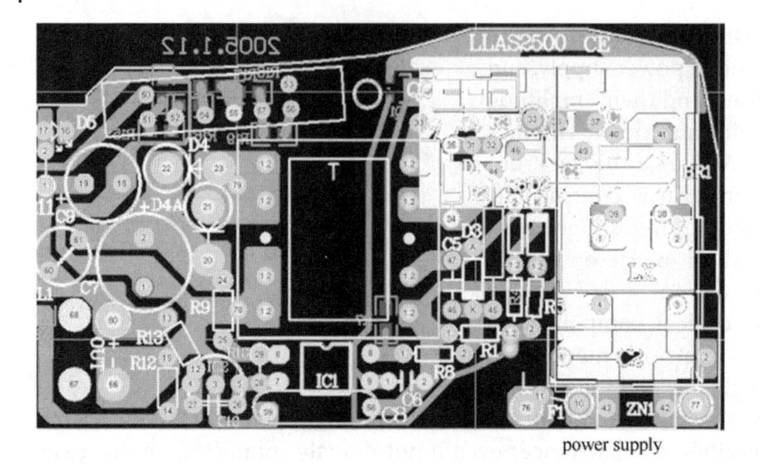

Figure 2.66 Switching power supply device layout diagram.

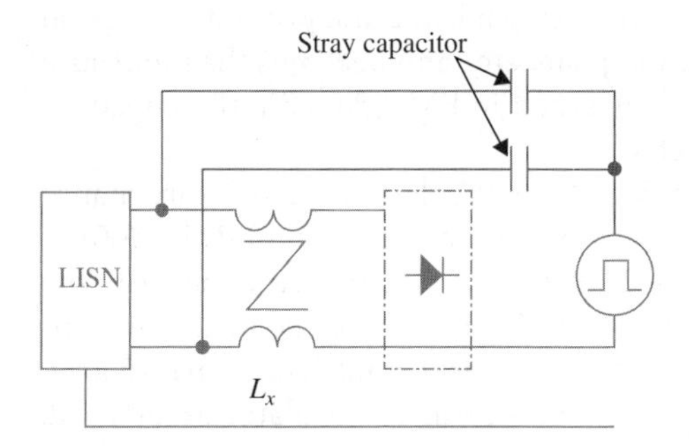

Figure 2.67 Schematics of coupling path of noise.

Then, $C_i = 35 \times 0.22 = 7.7\,\text{pF}$; $C_p = 9 \times 0.02/0.1 = 1.8\,\text{pF}$

The coupling capacitance between two parallel plates:$C_S = 9.5\,\text{pF}$.

During the design of switching-mode power supply, to avoid the heatsink becoming floating metallic plates and causing unnecessary coupling, or becoming monopole antenna and to bypass the noise in a small loop with lower resistance, in the design of switching-mode power supply, the heatsink needs to be grounded or connected to 0 V to in general.

In the charger described in this case, although the heatsink has been connected to 0 V, its size is relatively large, and it is extended to the area of the power supply input. As shown in Figure 2.66, the area with light color on the right is covered by the metallic fins heatsink.

It can be seen that in the power supply input circuits, the common-mode choke L_X, filtering capacitor, etc., are all close to the heat sink; thus, the parasitic capacitance or coupling capacitance between them is relatively high. Figure 2.67 shows the noise coupling principle between heatsink and power supply input circuits due to the parasitic capacitance.

From Figure 2.67, it can be seen that in the capacitive coupling between the interference source and the power supply input circuits, the noise source skips over the designed filtered components, which bypasses the filtering effectiveness as they are intended. Therefore, the test results are not good.

[Solutions]

According to this analysis, the shape of the heatsink is modified to cut off the capacitive coupling path between the interference source and power supply input circuits. The top view of the device after the installation with the modified heat sink is shown in Figure 2.68.

This installed heatsink does not have relatively high capacitive coupling with the filtering components including the common node choke L_X, filtering capacitor C_2, etc. After modification, the conducted emission test is performed again, and the measured spectral plot is shown in Figure 2.69.

From the plot and data shown in Figure 2.69, we can see the test is passed.

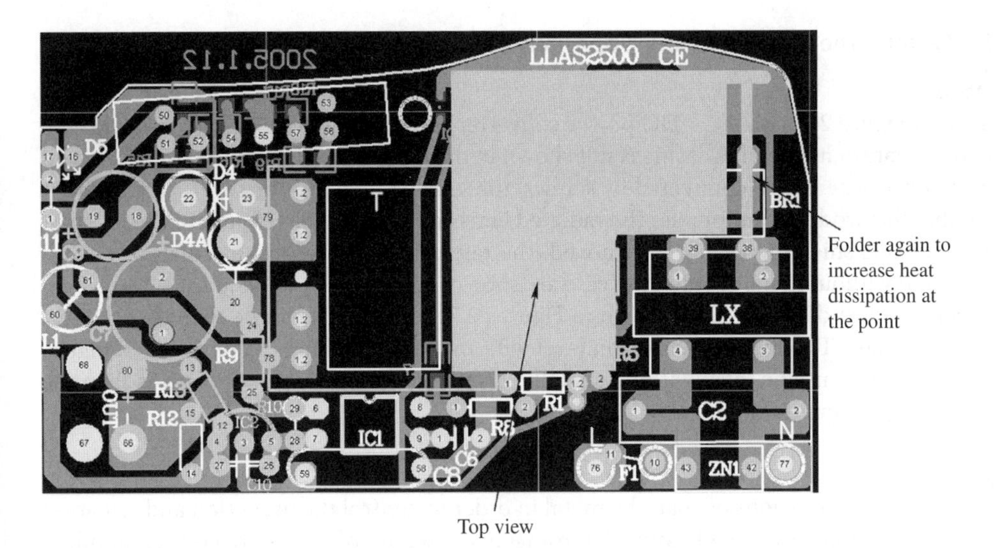

Figure 2.68 Top view after modified installation.

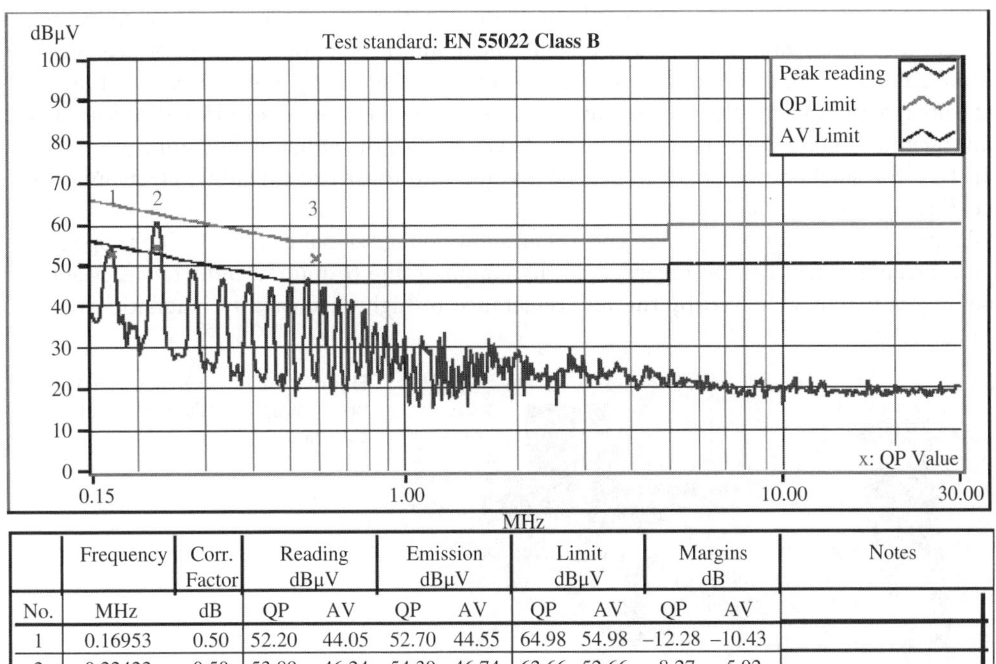

No.	Frequency MHz	Corr. Factor dB	Reading dBμV QP	Reading dBμV AV	Emission dBμV QP	Emission dBμV AV	Limit dBμV QP	Limit dBμV AV	Margins dB QP	Margins dB AV	Notes
1	0.16953	0.50	52.20	44.05	52.70	44.55	64.98	54.98	−12.28	−10.43	
2	0.22422	0.50	53.89	46.24	54.39	46.74	62.66	52.66	−8.27	−5.92	
+3	0.58516	0.50	50.95	44.32	51.45	44.82	56.00	46.00	−4.55	−1.18	

Figure 2.69 Conduct emission test result after modification.

[Inspirations]

Although the heatsink is not an electronic device and cannot generate signal interference on its own, it can often become the transmitter or receiver of noise. Especially in switching-mode power supply design, the design of a heatsink can impact the EMC test result significantly. Power supply design engineers should carefully consider how to design their proper shape and installation.

2.2.13 Case 13: The Metallic Casing Oppositely Causes the EMI Test Failed

[Symptoms]

As shown in Figure 2.70, an AC / DC power converter using its metallic casing as shield (the upper cover plate of its metallic casing is not shown in the figure, and the upper cover plate and the lower cover plate is well connected by screws, the spacing between the screws are 5 cm). It is found that the product cannot pass the radiated emission test. It is also found in the test that after the metallic shielding casing is removed, the test can be passed. The radiated spectral plots with and without metallic shielding case are shown in Figure 2.71 and Figure 2.72, respectively. From the test data and plots shown in Figures 2.71 and 2.72, it can be seen that the test results vary greatly. The radiated emission level with metallic shielding casing is much higher than the radiated emission level without metallic shielding cover, which seems to violate the theory of electromagnetic shielding.

[Analyses]

Shielding isolates two regions of space by metal in order to control the induction and radiation of electric field, magnetic field, and electromagnetic wave from one region to another. Specifically, shielding encloses the components, circuits, subassemblies, cables, or the noise source of the entire system in order to prevent electromagnetic field to propagate outward. As the shield can absorb (eddy current loss), reflect (reflection while electromagnetic wave intrudes the shield), and multi-reflect (inside shield, multi-refection on the interface between air and shied and the interface between shield and air) the energy of either external incident electromagnetic wave or the internal induced electromagnetic wave of printed wires, cables, components, circuits or system, etc., the shield can reduce the radiated emission.

In fact, the shielding mechanism can be divided into magnetic shielding, electromagnetic shielding, and electric field shielding.

Magnetic field is induced by high current and low voltage signal in circuit. The propagation of the magnetic field can be regarded as the coupling due to mutual inductance. Magnetic shielding is mainly achieved by the low reluctance of high permeability magnetic material,

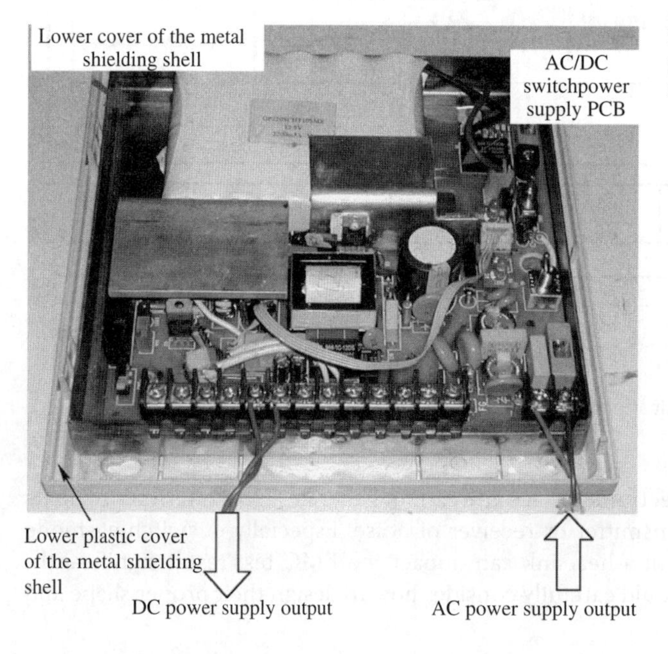

Figure 2.70 Physical map of a AC/DC power supply product with metal shielding shell.

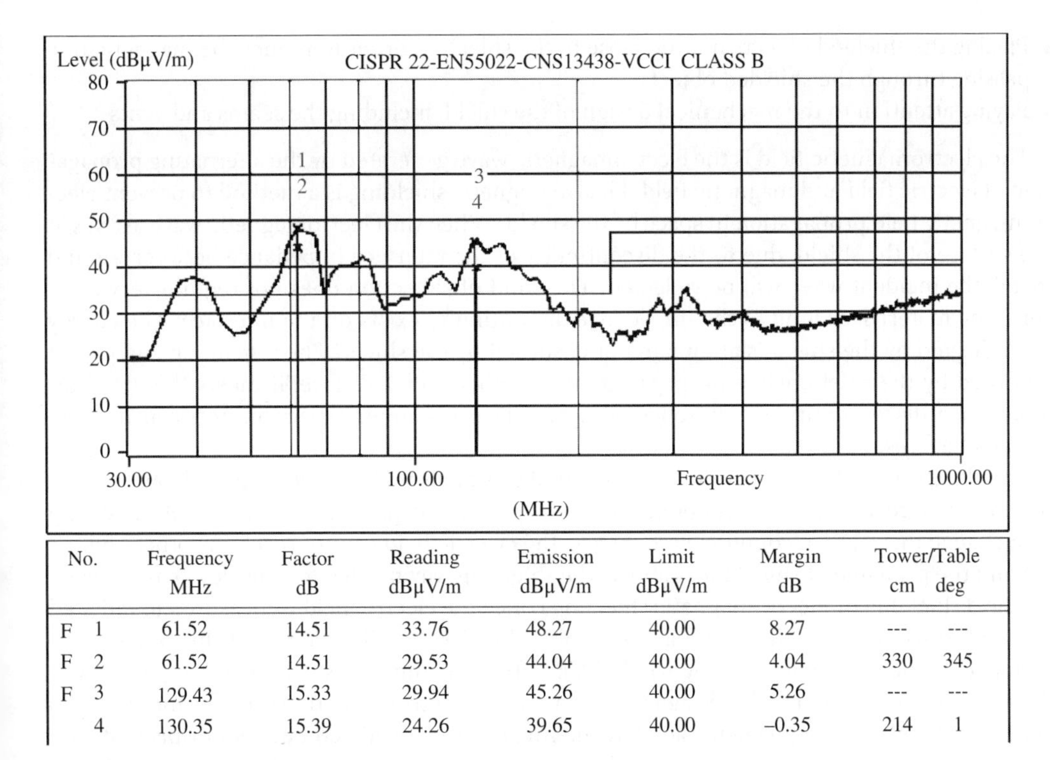

No.	Frequency MHz	Factor dB	Reading dBμV/m	Emission dBμV/m	Limit dBμV/m	Margin dB	Tower/Table cm	deg
F 1	61.52	14.51	33.76	48.27	40.00	8.27	---	---
F 2	61.52	14.51	29.53	44.04	40.00	4.04	330	345
F 3	129.43	15.33	29.94	45.26	40.00	5.26	---	---
4	130.35	15.39	24.26	39.65	40.00	–0.35	214	1

Figure 2.71 The spectrum of the radiation emission test results of the metal shielding shell.

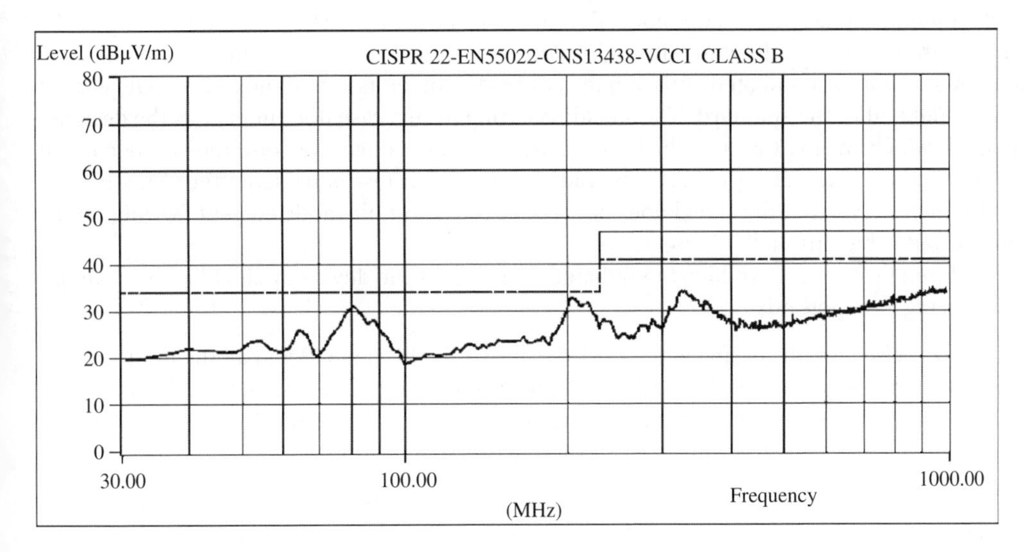

Figure 2.72 The spectrum of the radiation emission test results without metal shielding shell.

which can shunt the magnetic flux, and significantly reduces the internal magnetic field strength of the protected area by shield. For shield design, these approaches may decrease the shield's effectiveness:

- Generally selecting a high permeability magnetic material, such as permalloy
- Increasing the thickness of the shield

- Placing the shielded objects not too close to the shield, in order to reduce the magnetic flux passing through the shielded objects
- Paying attention to the mechanical design of the shield, including the seams and vents

The electromagnetic field is the electromagnetic wave generated by the alternating propagation of electric field and magnetic field. Electromagnetic shielding is a method to prevent electromagnetic field propagation in space by the shield. When the electromagnetic wave arrives at the surface of the shield, due to the discontinuity of the interface impedance between air and metal, the incident wave will be reflected. This kind of reflection does not require very thick shielding material, as long as the impedance discontinuity exists on the interface. The energy not reflected by the shield is attenuated or absorbed by the shield. The remaining energy not absorbed by the shield; while propagating to the other metal surface, it meets the metal-air impedance discontinuity and is reflected again. There is multi-reflection between the two metal interfaces.

The electrical field around circuits is generated by signals with high voltage and low current, which can be regarded as the coupling caused by parasitic capacitance. Electric field shielding is to change the original coupling path; hence, the electric field cannot reach the other end.

From the principle of shielding as above, the shielding way needed for this device is the electric field shielding or electromagnetic shielding (since the electric field is measured in radiated emission test). It seems that there is no problem for the shielding design of this power converter. The shield has enclosed the whole PCB within the metallic shielding casing. However, there is one point ignored by designer, which is, for electromagnetic emission measured in anechoic chamber, its equivalent radiated antenna is not a certain component or printed wire on PCB. Instead, it is the input/output cables of this power converter (if longer than 1 m). The reason is that only the cable length is comparable to the wavelength of emissions. The cable is a real "antenna" radiating the emissions. Practice and theory both show us that, as long as the common-mode current flowing through cables is more than several microamperes over the frequency range of radiated emission test, the radiation emission of the cable would exceed the limit defined in the standard. The shielding casing in this case does not act as the designer expected, which means that the shielding casing did not reduce the common-mode current flowing on the input/output cables. The radiated emission test is passed after removing the shield, which shows that the shield does not reduce the common-mode current flowing on the input/output cable. Instead, it increases it.

Therefore, if we want to reduce the radiated emission of the device by shielding, we can use the following two methods:

- Method 1: Use metallic casing and shielded cable to shield the PCB and all the input/output cables, and the shield of shielded cable and the shield casing of PCB shall be well connected.
- Method 2: Reduce the common-mode current flowing on the input/output cables by connecting metallic casing to ground of PCB at a reasonable position.

Obviously, method 1 is not feasible for the power converter because the input/output cables generally are not shielded cable. Therefore, we can only use Method 2, i.e. to take the profit of the shielded casing of PCB, through reasonable connection to reduce the common-mode current flowing on the input/output cables, and ultimately reduce the radiated emission from cables. Actually, it is a kind of electrical field shielding, i.e. to shield electric field generated from PCB inside the metallic casing. If the electric field is coupled to the reference ground or the input/output cables, the shielding effectiveness is lost.

According to the previous analysis, the equivalent radiating antenna of small switching-mode power supply is its input/output cables, and the root cause of radiated emission from cables is that there is common-mode current flowing on those cables. This common-mode current has two main causes. The first kind of common-mode current is caused by the capacitive coupling between the high *du*/*dt* circuits on the transformer primary side and the reference ground plate, as I_1 shown in Figure 2.73; The second kind is caused by the capacitive coupling between primary high *du*/*dt* circuits and secondary circuits, and the capacitive coupling between the secondary circuits and the reference ground plane, which is shown as I_2 in Figure 2.73. If there is no additional path, then after flowing into the reference ground plane, these two kinds of common-mode currents will flow back to the input cables of the switching-mode power supply.

If common-mode filtering capacitors exist, these two kinds of common-mode current will be bypassed (or shunted) by C_{Y1}, C_{Y2} before flowing back to the input cables of switching-mode power supply (as shown in Figure 2.74), thus the common-mode current I_{cable} flowing into the input cables of switching-mode power supply is reduced (if common-mode choke exists at the input of power supply, the common-mode current I_{cable} will be further reduced), and ultimately reduce radiated emission.

Therefore, when a metallic shielding casing is used in the switching-mode power supply, since the capacitive coupling from high *du*/*dt* source of switching-mode power supply to the metallic casing is always before the capacitive coupling to the reference ground plate, as long as

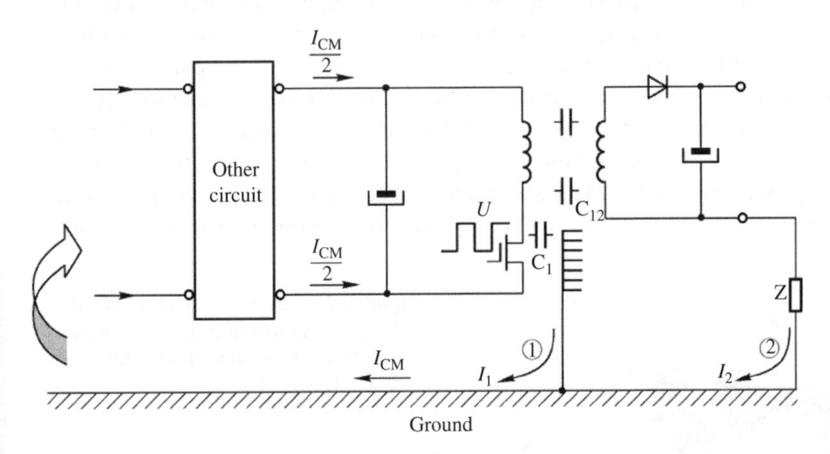

Figure 2.73 Schematic diagram of common-mode current on the switch power input cable.

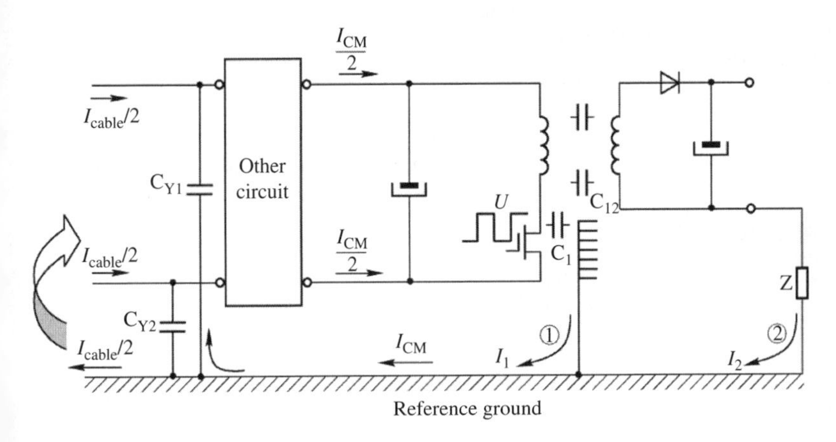

Figure 2.74 Common-mode filter capacitor bypass common-mode current.

the circuits of switching-mode power supply are connected properly to the metallic casing, the common-mode current shown in Figure 2.74 can be prevented from flowing into the reference ground plate. And if the common-mode current can be prevented from flowing into the input of the switching-mode power supply, then the radiated emission can be reduced. Figure 2.75 shows the schematic of using a metallic casing to prevent common-mode current from flowing into the reference ground plane.

If we take a look at Figure 2.76, when there is no connection between the metallic shielding casing and the PCB of switching-mode power supply, the common-mode current flowing into the input cable is not reduced. Furthermore, after carefully analyzing Figure 2.76, with the metallic shielding casing, the internal noise coupling of the switching-mode power supply circuits is changed. Taking the example of the coupling between switching point (with high dU/dt) on PCB of switching-mode power supply and its power input cables, in the absence of the metallic shielding casing, the coupling mechanism is shown in Figure 2.77. In this case, as long as the PCB layout is reasonable, in Figure 2.77, the parasitic capacitance between the switching point on PCB of switching-mode power supply and power supply inputs, C_{S1} and C_{S2},are relatively small (a few tenths of pF in general). In the existence of the metallic casing, due to the presence of a metallic plate, the parasitic capacitance, between the switching point (with high dU/dt) on PCB of switching-mode power supply and power input lines, equals the series capacitance of C'_{S1} and C'_{S2} as shown in Figure 2.78. Here, C'_{S1} is the parasitic capacitance between switching point (with high dU/dt) and the metallic casing; $C'_{S2}$is the parasitic capacitance between power supply input lines and the metallic casing. Because the distance (several centimeters) from the switching point on PCB (with high dU/dt) of switching-mode power supply and its power input lines to the metallic casing is much lower than the distance (about 1 m) to the reference ground plane, C'_{S1}, C'_{S2} are much bigger than C_{S1}, C_{S2}; i.e. the coupling between switching point (with high dU/dt) on PCB of switching-mode power supply and its

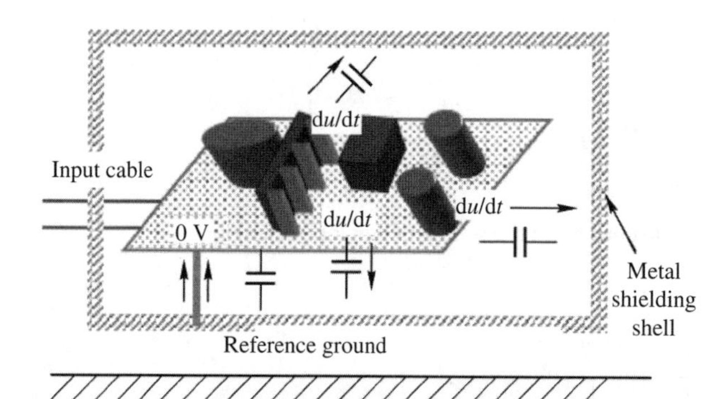

Figure 2.75 Schematic diagram of the principle of preventing common-mode current from flowing into the reference ground floor.

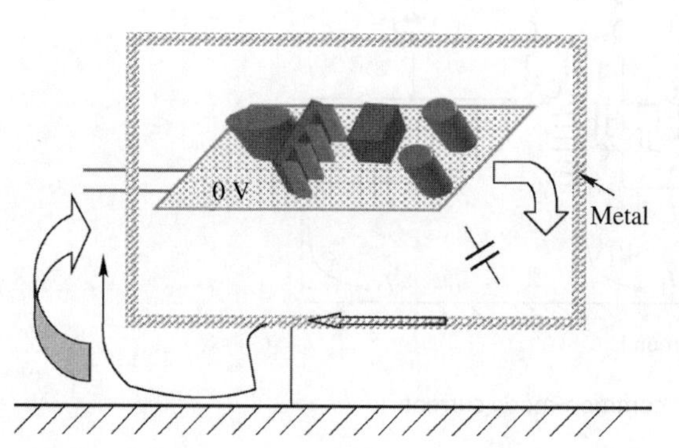

Figure 2.76 The metal housing does not reduce the common-mode current in the cable.

power input lines are greatly increased; therefore, the common-mode current flowing into the power input lines is greatly increased, and then radiated emission is also greatly increased. This is why a metallic shielding casing unexpectedly leads the radiated emission test to fail in this case. When the metallic casing and 0 V of PCB are connected, although the parasitic capacitance C'_{S1}, C'_{S2}shown in Figure 2.78 still exists, the potential of the interconnection point between C'_{S1} and C'_{S2} is zero, which leads the common-mode current coming from C'_{S1}not flowing to C'_{S2}. Thereby, the common-mode current flowing to its power input cable is reduced, and then the radiated emission is reduced.

[Solutions]

According to the above analysis, the main reason the metallic casing causes the radiated emission test to fail is that there is no connection (directly or through capacitors) between the metallic casing and PCB reference ground. To fully utilize the metallic casing, the following connections should be implemented:

1) Connect the capacitor C of the filter to the metallic casing near the placement position of capacitor C And placing C after the diode bridge is more effective. The primary 0 V of DC/DC switching-mode power supply can be directly connected to the metallic casing.
2) Connect the secondary 0V of switching-mode power supply to the metallic casing directly or by capacitor C, as shown in Figure 2.79.

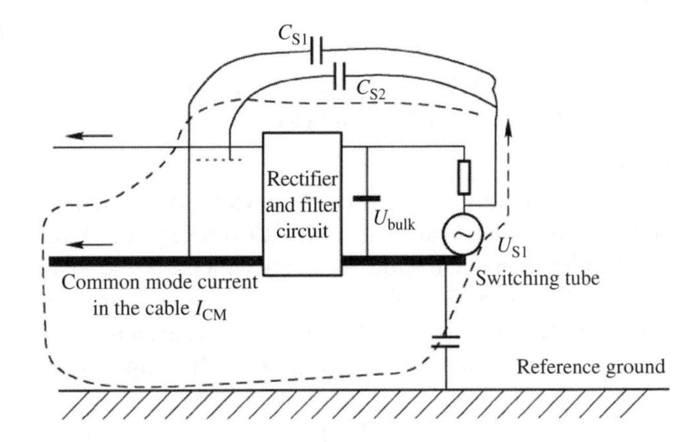

Figure 2.77 The coupling relationship without metal shielding shell.

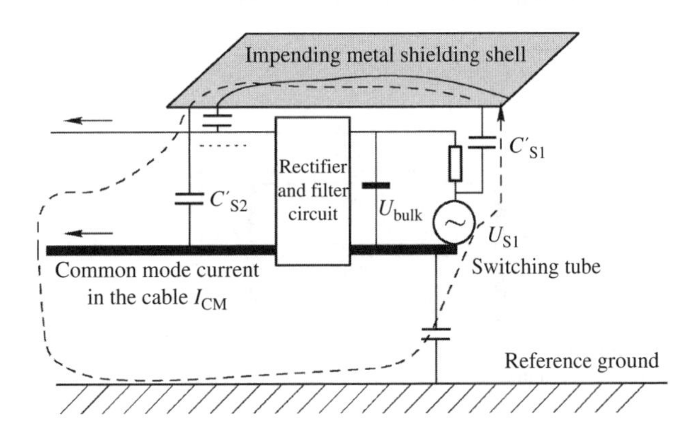

Figure 2.78 The coupling relationship with metal shielding shell.

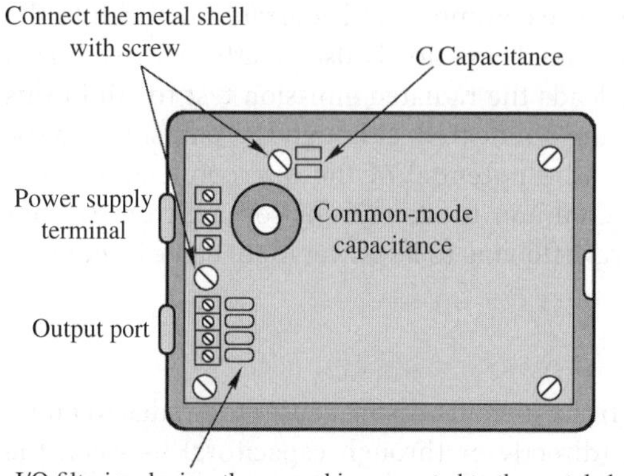

Figure 2.79 The connection between the PCB and the metal shell.

[Inspirations]

It can be seen that the metal cover is not always secured. The connection between metallic casing and the reference ground of PCB, and the location selection of this connection is extremely important. Adding the metallic casing arbitrarily may even worsen the EMC performance of device. The reduction of radiated emission from cables can be achieved by decreasing the common-mode current flowing on cables, instead of simply grounding. Analyzing common-mode current is the critical method for EMC analysis. The goal of shielding is to prevent common-mode current from flowing into LISN. In the design the shielding, cables must be taken into account.

2.2.14 Case 14: Whether Directly Connecting the PCB Reference Ground to the Metallic Casing Will Lead to ESD

[Symptoms]

The basic structure of an electronic device used on vehicle is shown in Figure 2.80. The dimension of this device is about20 mm × 20 mm × 10 mm. From Figure 2.80 it can be seen that the product has metallic enclosure, inside which there are two PCBs. These two PCBs are fixed on the metallic casing by bolts. There is no connection between the bolt and the PCB reference ground or the circuits of PCB. The ribbon is used to interconnect these two PCBs. There is an I/O connector on the PCB2 to connect with a bundle of cables inside, which are power supply, input/output signals, and other control signals.

When we performed ESD testing for this device according to its relevant standard ISO10605, we found that as long as the test voltage is higher than ±2 kV, system errors will appear.

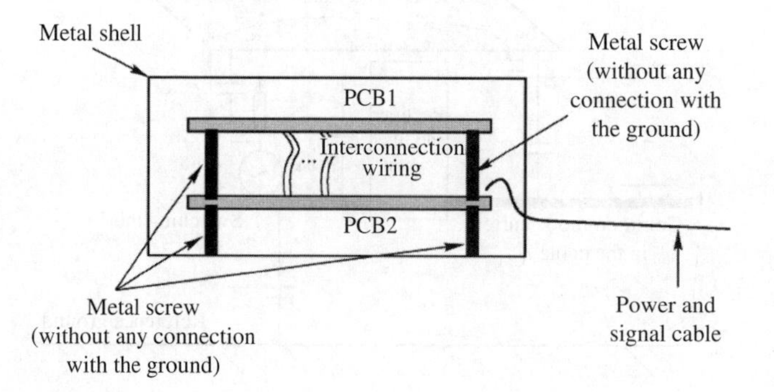

Figure 2.80 Product frame diagram.

[Analyses]

There exists an obvious EMC flaw on the structure design of this device. Through analysis of ESD disturbance path for this device, we can see where the flaw is. Figure 2.81 shows the ESD common-mode disturbance analysis when the electrostatic discharge is injected on the metallic casing.

There are two ESD common-mode interference current paths. The first one is indicated with thick arrow (current I_{CM1} is in this path). The second one is indicated with thin arrow (current I_{CM2} is in this path).

In Figure 2.81, the second path (the path of current I_{CM2}) is passed by the ESD common-mode disturbance current which flows through PCB1, the ribbon, PCB2 and the attached cables. This is a "undesired" ESD disturbance current path. In this path, if the ESD disturbance current is higher, the device will disturbed more easily. The following interpretation can help understanding this phenomenon.

Figure 2.82 shows the simplified equivalent circuits of that shown in Figure 2.81 (when we analyze common-mode current disturbance path, the parasitic inductance, resistance and other parameters of it can be temporally ignored).

C_{p1} is the parasitic capacitance, around 10 pF, between the metallic casing of device and PCB1 reference ground. Parasitic capacitance could exist between metallic casing of the device and

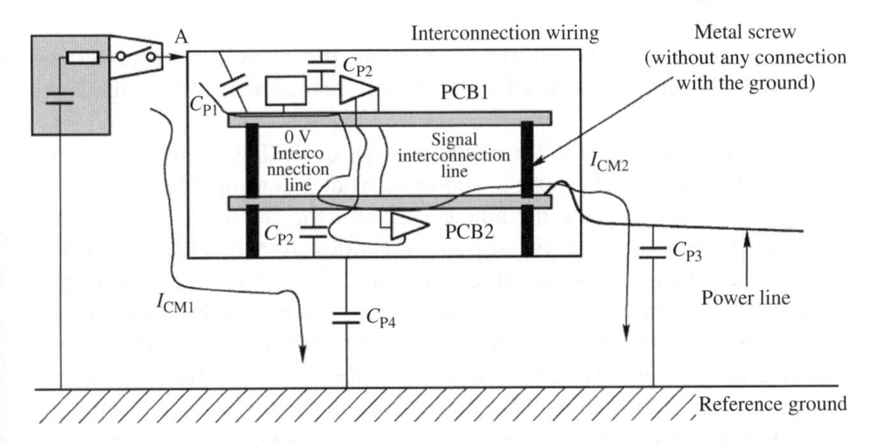

Figure 2.81 Analysis of ESD common-mode interference path when the static discharge point is on the metal shell.

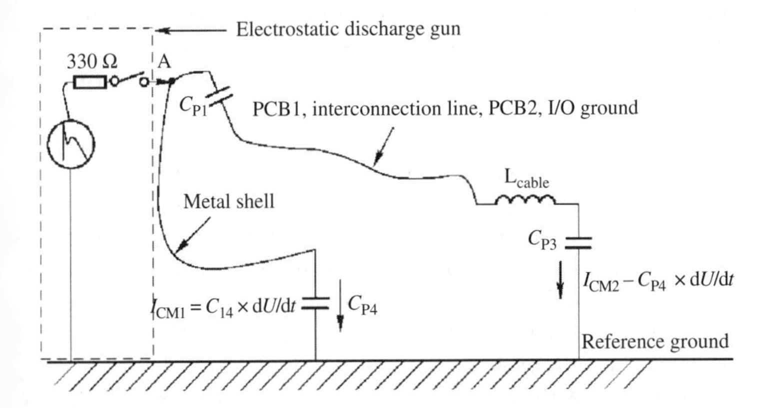

Figure 2.82 Simplified equivalent circuit diagram of Figure 2.81.

the components, printed wires, ground plane and power plane in PCB1. And the parasitic capacitance between metallic casing of the device and the ground plane and the power plane is the largest. C_{p3} is the parasitic capacitance between reference ground plate and the bundle of cables. During the test, these cables are placed above reference ground plate with 25 mm height. This capacitance can be estimated as 60 pF m^{-1} (cable configuration is defined in ISO10605). In this case, the length of cables is 2 m. C_{p3} is can be estimated about 120 pF. C_{p4} is the parasitic capacitance between metallic casing of the device and reference ground plate. The device is placed above the reference ground in the test. The device is separated from the reference ground plate with a insulation whose relative dielectric constant is less than 1.4. The height of the insulation is 25 mm, this capacitance is about 30 pF. (Note: the estimation of parallel-plate capacitance can refer to Case 12.)

C_{p2} is the parasitic capacitance between PCB2 and metallic casing. In the schematic, C_{p2} is in series with C_{p4} and then the serial capacitance is in parallel with C_{p3}, and $C_{p2} << C_{p3}$, so C_{p2} can be neglected. You can see the voltage between ESD injection point A and the reference ground plate is not zero (when the metallic casing is well connected to the reference ground plate, the voltage between point A and the reference ground plate is close to zero). Assuming, in 4 kV contact discharge test, the transient voltage at point A is 1 kV, which will cause the common-mode current I_{CM2}:

$$I_{CM2} = C_{p1} \times dU/dt = 10\,pF \times 1\,kV/1\ ns = 10\,\text{A}$$

Note: Over the frequency range of ESD disturbance current, the equivalent characteristic impedance caused by cable L_{cable} and C_{p4} *is* about 150 Ω, which is smaller than the impedance of C_{p1}.

In fact, for the ESD common-mode disturbance current I_{CM2} of ESD, we can also say that when the electrostatic discharge occurs, because the equipotential is not possible to be implemented between the injection point on the metallic casing and the reference ground plate (grounding devices will be better), the potential at point A is not zero, which finally leads to the common-mode ESD current I_{CM2} flowing through PCB1, the interconnecting ribbon between PCB1 and PCB2, PCB2 and cables attached to PCB2. I_{CM2} primarily flows through the 0 V ground of the PCB1, the 0 V ground wires of the interconnecting ribbon between PCB1 and PCB1, the 0 V ground of PCB2 and the 0 V ground cables (because the path impedance of 0 V ground is the minimum). The ESD common-mode disturbance current I_{CM2} caused by an ESD common-mode voltage relative to the ground does not directly impact the operation of the electronic circuits. It is a common-mode current, but inside the device, the transmitted operating signals of circuits are voltages. The voltage signals exist between the chip or the ports of the circuits and the internal reference ground; i.e. they are differential-mode voltage signals. To disturb the normal operating signals of circuits of the device that are transmitted in differential mode, this kind of common-mode voltage or current must be transformed.

Figure 2.83 shows the common impedance coupling principle when ESD common-mode transient current flows through the interconnection connector, which explains the transformation principle. U_0 is the normal operating voltage signal transmitted between PCB1 and PCB2 in Figure 2.83. When there is no ESD common-mode disturbance current flowing through the interconnecting ribbon, U_o can be normally transmitted from PCB1 to PCB2. However, when the ESD common-mode current flows through the internal 0 V ground, because the parasitic inductance L (about 10 nH cm^{-1}) exists on the ribbon interconnecting PCB1with PCB2, this ESD common-mode current will cause a voltage drop exists across this ribbon. This voltage drop is $\Delta U_{Z0V} = |L_{0V} \times dI_{CM2}/dt|$. In this case, the length of ribbon between PCB1 and PCB2 is about 10 cm, the parasitic inductance can be estimated about 100 nH, so $\Delta U_{Z0V} = |LdI_{CM2}/dt|$

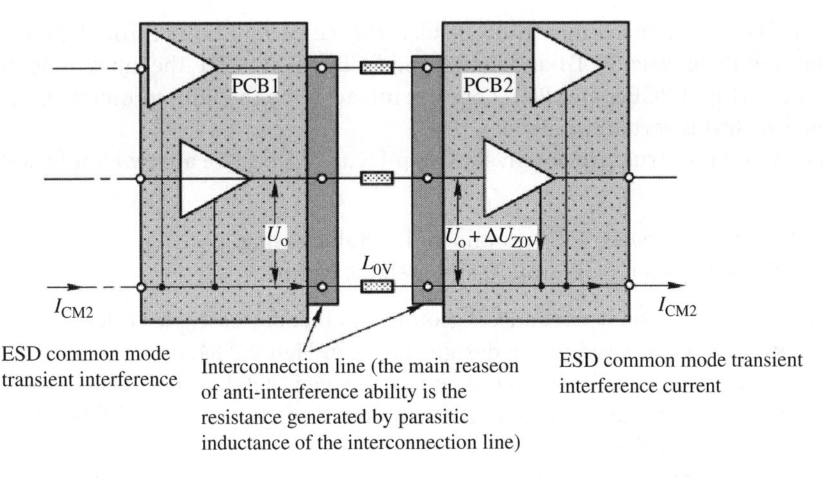

Figure 2.83 Common impedance coupling principle of ESD common-mode transient current flowing through interconnect connector.

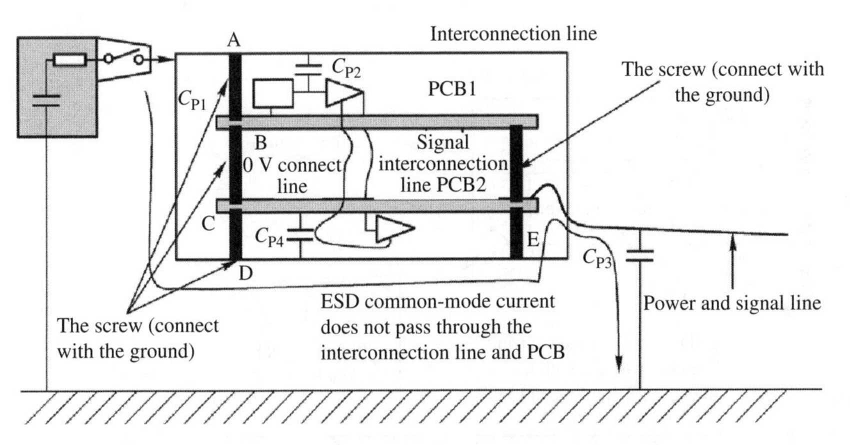

Figure 2.84 ESD interference current path analysis for PCB and metal case.

= 100 nH × 4 A/1 ns = 400 V. This voltage drop is far beyond the noise margin of the circuit itself. This causes the described phenomenon in this case.

However, the amplitude of the common-mode disturbance current in Figure 2.81 can be changed through the design of the structure. Figure 2.84 presents a solution from a mechanical structure perspective. This solution is the most reliable and effective solution. The changes are mainly focused on the interconnection between the internal 0 V ground of PCB and the metallic casing, the interconnection positions and the choice of interconnection ways. As shown in Figure 2.84, if the impedance continuity from point A to point D on the metallic casing is good enough (if the length-to-width ratio of the metallic plane between point A and point D is less than 3:1 and there is no hole or slot on the metallic casing, then the impedance is less than 11 mΩ at 100 MHz frequency and the impedance is less than 20 mΩ at 300 MHz), then the equipotential (voltage difference in between point A, point B, point C and point D when the ESD transient disturbance current flows through them is less than 400 mV) can be kept in between point A (the mounted position of the bolt interconnecting the metallic casing and PCB1), point B (the mounted position of the bolt on PCB1), point C (the mounted position of the bolt on PCB2), and point D (the mounted position of the bolt interconnecting the metallic casing and PCB2). This results in no common-mode disturbance current flows through PCB1, the interconnecting ribbon between PCB1 and PCB2, and PCB2, then PCB1, the interconnecting ribbon between PCB1 and PCB2, and PCB2 are not impacted by the ESD common-mode disturbance current, the test is certainly passed.

For the improvement on the structure design shown in Figure 2.84, there are two important points:

1) The interconnection point between PCB ground and metallic casing.
2) The contact impedance between PCB ground and metallic casing.

As for the selection of the position between PCB ground and metallic case, please look at the design shown in Figure 2.85. Compared to the design shown in Figure 2.84, only the position near the cable bundle is removed for the interconnection between PCB2 ground and metallic casing, with which the ESD common-mode current will flow through the entire PCB2, And PCB2 will be highly disturbed by ESD.

As for the impedance between the PCB ground and metallic casing, please look at analysis diagram of the ESD disturbance current path when PCB2 and the metallic casing are linked together with a high-impedance connection as shown in Figure 2.86. In the design shown in

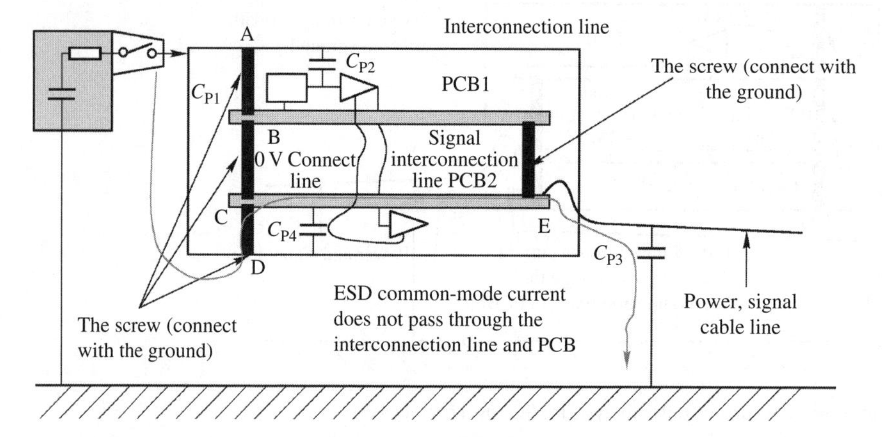

Figure 2.85 ESD current with no connection between metal flame and GND plane of PCB at point *E*.

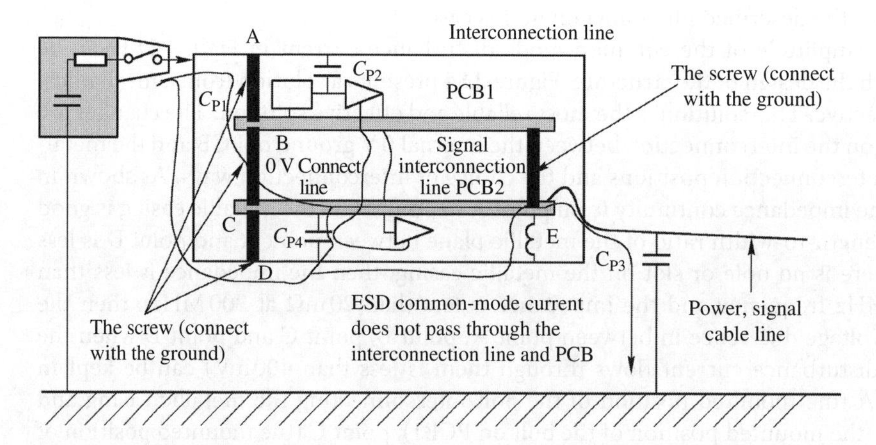

Figure 2.86 ESD current with high impedance connection between PCB2 and J metal case.

Figure 2.86, due to the bad contact between point E of PCB2 and metallic casing, there is remained common-mode voltage on metallic casing. Through C_{P4} or the bolt between point C and point D, this ESD common-mode voltage will go inside PCB2 and greatly disturb PCB2. It can be imagined that, if the bolt between PCB1 ground and PCB2 ground is not mounted well enough, ESD common-mode disturbance current will flow into PCB1 and the interconnection ribbon between PCB1 and PCB2, the disturbing will be even worse.

Figure 2.87 shows one of the simplest scenarios of the interconnection between PCB ground and metallic casing. In this case, only if the connections between point A and point B, between point E and the metallic casing are good enough, meaning that their contact impedance is low, an obvious EMC performance improvement can be achieved. Because, at this moment, there is almost no ESD common-mode current flowing through PCB1 ground, the ground of the interconnecting ribbon between PCB1 and PCB2, and PCB2 ground, the critical circuits of the device are not disturbed.

The prior choice of interconnection point between PCB ground and metallic casing should be near the I/O cables. The elsewhere additional interconnection points between PCB ground and the metallic casing must make the metallic casing and the interconnection ribbon between PCBs in parallel. If this kind of parallel connection can reduce the loop area (between two parallel branches), the effectiveness will be even better. The preferred interconnections are the metallic parts whose length-width ratio is less than 5:1. The mounting way prefers screwing, welding, riveting, contacting with reeds, and other intentional connection. If for some reason, such as avoiding low-frequency disturbance due to ground loop, safety reasons, or some other reasons, the PCB ground cannot be directly connected with metallic casing, we can also use capacitors to interconnect them. The typical capacitance is around 1–10 nF.

[Solutions]

We will change the product structure according to Figure 2.84.

[Inspirations]

As long as the interconnection between metallic casing and PCB ground is correct, the external disturbance cannot come into the PCB. On the contrary, the low-impedance characteristic of the metallic casing could bypass the disturbance, which potentially passes through PCB into the metallic casing.

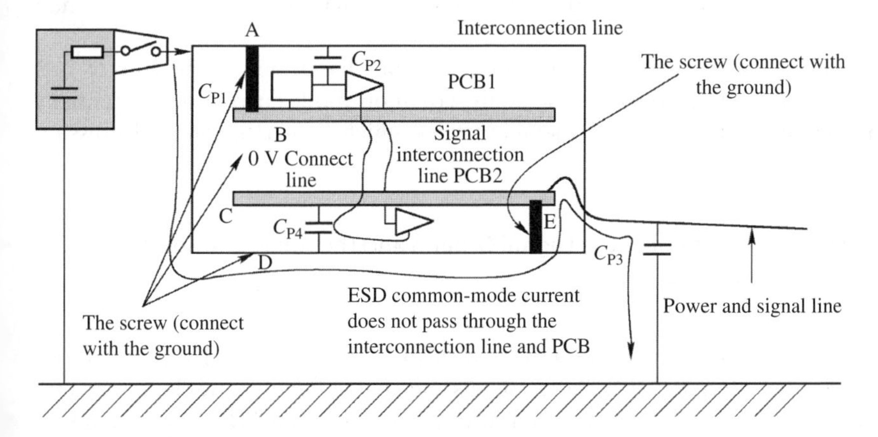

Figure 2.87 The simplest PCB working with a metal shell.

When the PCB ground interconnects with the metallic casing at multi-points, between the interconnection points, the impedance of metallic casing must be lower than the impedance of PCB ground and the ground impedance of ribbon.

The case described in this section is relevant to immunity testing; in fact, it is not only a matter of immunity but also a matter of EMI. Let's look at the following estimation example.

Assuming that PCB1 and PCB2 shown in Figure 2.18 are both multilayer boards with perfect ground (0 V) plane, and the length of the interconnect ribbon is 10 cm. The parasitic inductance in each wire of ribbon is $L \approx 100\,\text{nH}$ (estimated with 10 cm in length, and $10\,\text{nH}\,\text{cm}^{-1}$). The number of wires which is defined as 0 V ground is 10. A voltage pulse with amplitude of 3.3 V and frequency of 10 MHz passes though the ribbon. The characteristic impedance of the signal wires is 100 Ω. The length of the cables attached on the PCB2 is 3 m.

The harmonic amplitude of this voltage pulse at 30 MHz (the third order harmonic) is U:

$$U = 0.7\ \text{V}$$

The harmonic current at 30 MHz is I:

$$I = U/Z = 0.7/100 = 7(\text{mA})$$

The parasitic inductance L caused by the 10 parallel ground wires is:

$$L \approx 100\,\text{nH}/10 = 10\,\text{nH}$$

The common-mode voltage drop while the 30 MHz harmonic flowing through the ground wires U_{CM} is:

$$U_{CM} = L2\pi FI = 10\,\text{nH} \times 2 \times \pi \times 30\,\text{MHz} \times 7\,\text{mA} \approx 12.4\ \text{mV}$$

The common-mode characteristic impedance between the cables attached on PCB2 and reference ground plate Z_{cable} is about:

$$Z_{\text{cable}} \approx 150\,\Omega$$

The actual common-mode current I_{CM} in the cables attached on PCB2 at 30 MHz is:

$$I_{CM} = U_{CM}/Z_{\text{cable}} = 12.4\,mV/150\,\Omega = 0.083\,mA$$

It can be estimated that, at 30 MHz, the radiated emission of the device is far beyond the limits of EN55025, CISPR25 or EN55022 (this current shall be typically less than 0.003 mA).

Using the connecting ways shown in Figure 2.84, on the one hand the voltage drop U_{CM} is shunted by the metallic casing. On the other hand, the common-mode current that potentially flows into the cables is also bypassed by the metallic casing, which makes the radiated emission test passed successfully.

2.2.15 Case 15: How to Interconnect the Digital Ground and the Analog Ground in the Digital-Analog Mixed Devices

[Symptoms]

One device is wireless communication equipment with the hybrid of digital signals and RF signals, in which the conversion between digital signals and analog signals is implemented by a digital/analog converter (DAC) and analog/digital converter (ADC). The ADC has two kinds of power supply pins, namely the power pins to respectively supply digital parts and analog part of ADC, and the ground corresponding to each power supply is also divided into digital ground (DGND) and analog ground (AGND), whose structure is shown in Figure 2.88.

Abnormal operation appears for this product during 6 kV contact electrostatic discharge test, which is passed as a result. After checking, *t* is caused by the fact that ADC does not work properly (the inputs and outputs of ADC during the test are monitored, and the abnormal operation is found).

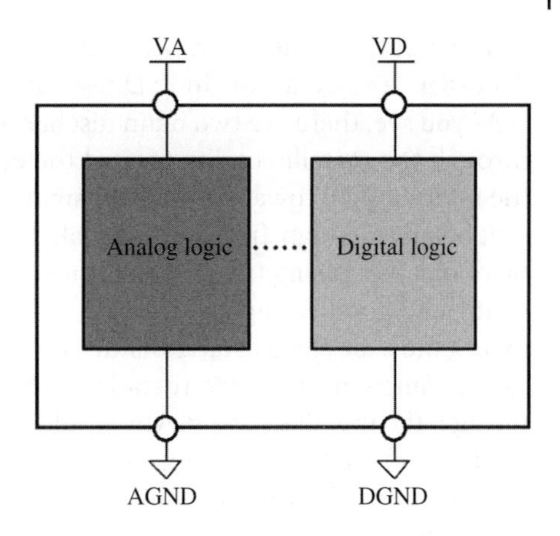

Figure 2.88 The logical structure diagram of ADC.

[Analyses]

First, let's describe the overall structure of the device, as shown in Figure 2.89. This device uses the metallic shielding structure, and the power supply input goes into the internal circuits after filtering. The RF interface and communication interface both are equipped with the connectors with metallic shell, which implements the contact in 360° with the cabinet. The RF interface uses 50 Ω coaxial connectors, theoretically whose outer sheath is the operating ground of analog circuit, in other words, analog ground. The PCB inside metallic casing includes digital signals and analog signals. ADC and DAC are placed across the border line between digital circuits and analog circuits. PCB is fixed on the bottom of the metallic casing through screws, and the operation ground (digital ground) in digital circuit of PCB is directly connected with the bottom plate of the metallic casing through the screw. Similarly, the operation ground (analog ground) in the analog circuit of PCB is directly connected with the bottom plate of metallic casing through the screws in the same way.

The ESD injection point where abnormal operation happens is on the metallic casing near the RF connector. The nature and disturbing principle of electrostatic discharge has been described in detail in other cases, which will not be mentioned here. But figuring out the flow of electrostatic discharge current will greatly help in analyzing this issue. As the grounding

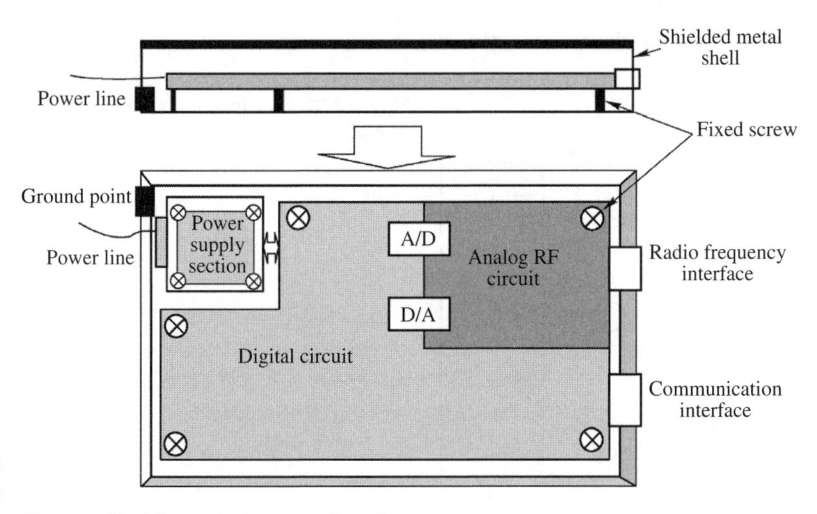

Figure 2.89 Schematic diagram of product.

position of the device is in the vicinity of the power supply input, the electrostatic discharge disturbance current flow in ESD test can be shown in Figure 2.90.

As you see, there are two main discharging paths for electrostatic disturbance current: one is through the metallic casing toward the earth (most of the ESD current); the other is through the internal PCB toward the earth (small part of the ESD current). Electrostatic discharge current belongs to high-frequency signal, then the skin effect and the low-impedance characteristic of metallic casing (the contact impedance of metallic casing is checked during the test, and confirmed good enough. If it is not, it will lead to extra disturbance, as described in Case 49, makes most of electrostatic disturbance current flowing into the earth through the metallic casing. Since most of electrostatic discharge disturbance currents have flowed into the earth through the metallic casing, why would the abnormal operation of ADC appear? for the design weakness of ADC circuits must exist. When checking the circuits, ADC has analog ground and digital ground. In order to avoid disturbing analog circuits by the interference generated from the digital circuits in the circuit design, a ferrite bead is inserted between digital ground and analog ground to isolate each other. The schematic of ADC and its layout is, respectively, shown in Figures 2.91 and 2.92.

Because of the existence of ferrite bead between analog ground and digital ground, there will be a voltage drop ΔU on the ferrite bead when the high frequency electrostatic discharge disturbance current passes through it (Figure 2.93).

What kind of impact will this voltage drop ΔU cause on ADC? Maybe someone thinks that digital circuits and analog circuits in ADC are separated from each other, and their voltage levels are not mutually referred. In fact, the existence of the parasitic capacitance and ferrite bead between the digital circuits and analog circuits has correlated the two parts of the circuit; therefore, the existence of ΔU must impact the normal operation of the ADC. During the test,

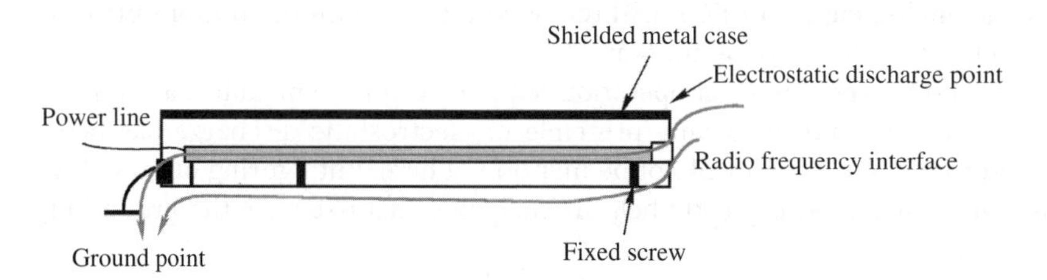

Figure 2.90 Flow of interference current in electrostatic discharge.

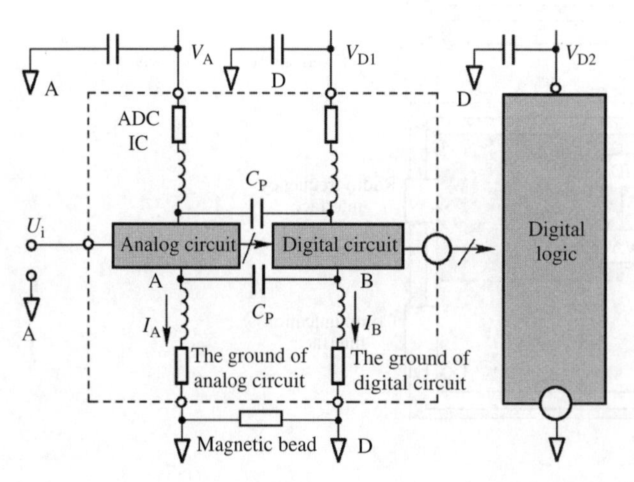

Figure 2.91 Principle block diagram of ADC.

connecting the digital ground and analog ground of ADC with a conductor and at a single point, this device can pass the ±6 kV electrostatic discharge test, which is due to the fact that the voltage drop ΔU is greatly reduced after the single point interconnection.

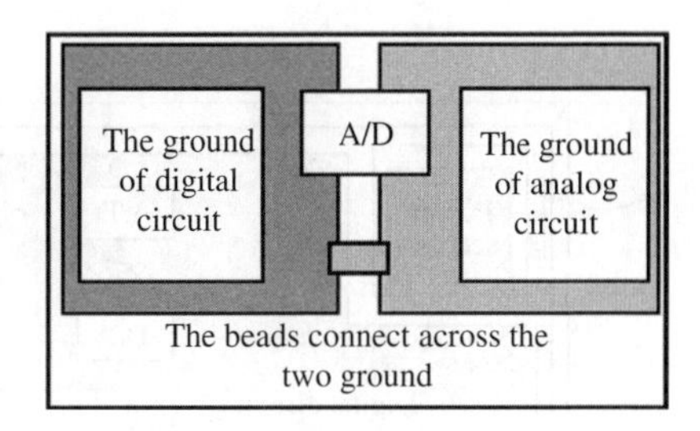

Figure 2.92 PCB layout diagram of ADC.

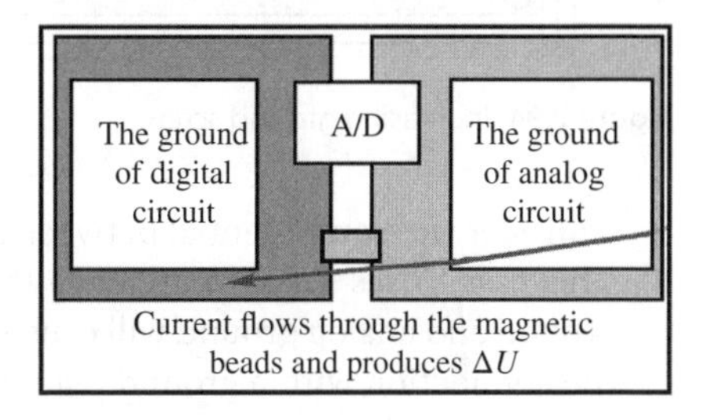

Figure 2.93 The pressure drop generated by electrostatic discharge interference current flowing magnetic bead.

Interconnecting ADC's digital ground and its analog ground by a ferrite bead is a mistake that engineers often make in the circuit design, and these mistakes are derived from the naming of ADC's digital ground pins and analog ground pins. The naming of analog ground pins and digital ground pins indicates the role of the internal circuits themselves, but it does not mean that the external circuit should be designed in accordance with the internal functions. There are two parts, analog circuits and digital circuits, inside ADC, as shown in Figure 2.91, its analog ground and digital ground are usually separated in order to avoid the digital signal coupled to analog circuit. However, the lead joint inductance and resistance from the soldering pads of the die to the pins of its package are not specifically added by the IC designers. The high slew rate digital currents cause a voltage at point B, which inevitably coupled to the point A of analog circuitry through the parasitic capacitance. It can be seen that the inserted ferrite bead between digital ground and analog ground can't reduce the noise transmitted from digital circuits to analog circuits, and any additional external impedance in serial with the digital ground pin will cause a large noise at point B, and then, the large noise will be coupled to the analog circuits by stray capacitance. The problem of the coupling due to parasitic capacitance inside IC shall be considered by IC designers during its manufacturing process. In order to prevent further coupling, the analog ground pins and digital ground pins shall be connected to the same low-impedance ground plane with the shortest wire outside IC. In this case, the bottom plate of the metallic casing actually acts as a low-impedance ground plane, but the connection point between the PCB and the metallic plate is too far from the ADC. In high frequency, the low-impedance grounding for the two kinds of ground pins of ADC is not implemented. If two additional grounding points are added in the vicinity of ADC, one is in analog circuits, the other is in the digital circuits, and the grounding effectiveness will be greatly improved, as shown in Figure 2.94.

[Solutions]

According to this analysis and test results, the direct connection at a single-point should be applied between the analog ground and digital ground. If more grounding points can be added as shown in Figure 2.94, the EMC performance of this device can be further improved.

[Inspirations]

1) Ferrite beads are usually recommended to be used on power lines or signal lines to enhance the decoupling effect, but we must be careful when using it between grounds (maybe in some cases you can use like this), especially when those currents like electrostatic discharge disturbance currents or EFT/B disturbance currents pass through it.

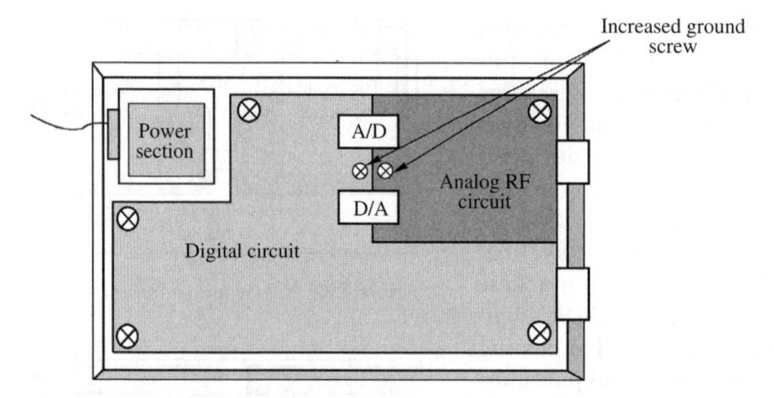

Figure 2.94 Increasing ground screw.

2) Consider the equipotential between separated grounds as well.
3) This case raises a question: Since the voltage drop on the impedance between the digital ground and analog ground will cause EMC problems similar as that of this case, why not use the connection with a ground plane? ADC, DAC are both analog-digital mixed devices, so which ground plane is more appropriate to connect to? The answer is, if we do not separate the analog ground and digital ground on the PCB and connect all of the analog ground pins to the digital ground plane, then in structure design of this case, the analog input signal will be superimposed by the digital noise. The reason is that there are some digital ground noises flowing into the analog ground. It will be a little bit better if the analog ground and digital ground pins are connected together to the analog ground plane, because superposing several hundred millivolts noise on the digital signals is significantly better than superposing the same noise to the analog signals. For 16-bit ADC with 10 V input voltage, the voltage of the least significant bit is only 150 μV! The digital currents on digital ground pin are virtually impossible to be worse than this, otherwise they will make the analog part inside ADC fail! If high-quality high-frequency ceramic capacitors (0.1 μF) are connected between the ADC power pins and analog ground plane to bypass the high-frequency noise, these noise currents will be isolated to a very small range around the integrated circuit and minimize its impact on the rest of the system. Although the analog signal can also affect the digital circuits and reduce the digital noise margin, if less than a few hundred millivolts, it is often acceptable for the TTL and CMOS logic circuits. But pay attention on the plane continuity for mirror currents of digital signal (usually, in this case, digital signal mirror current plane is not perfect).

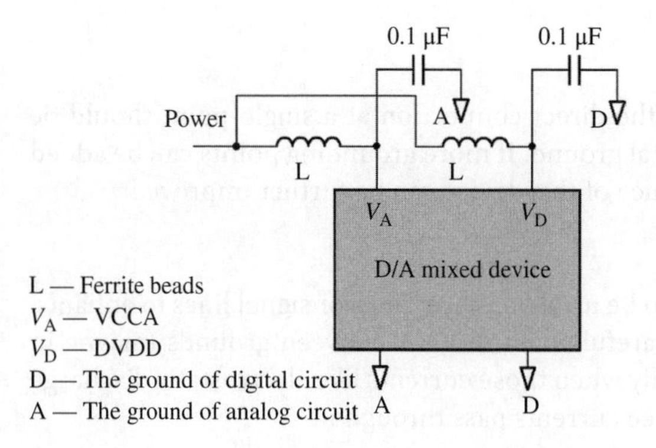

Figure 2.95 Isolation of analog and digital power supply by ferrite.

4) What requirements should analog and digital power supply meet as for the analog-digital mixed devices like ADC, DAC? Which power supply on earth shall we choose, separated analog power supply and digital power supply or the same power supply? The answer is, it is possible to use the same power supply, but the digital power supply decoupling must be good enough, such as each power supply pin of chip shall be decoupled with one 0.1 μF ceramic capacitor, and a ferrite bead is used to further separating analog power supply and digital power supply, which are more important than ground connection. A right connection is shown in Figure 2.95. A more secure way, of course, is to use a separated power supply. Further analysis on analog-digital mixed circuits can be Case 69.

3

EMC Issues with Cables, Connectors, and Interface Circuits

3.1 Introduction

3.1.1 Cable Is the Weakest Link in the System

When the radiated emission test is performed for a device alone, it can pass the test; however, when it is connected with the cable, the system might not pass the test. This is due to the cable radiation. Practice shows that the shielding effectiveness of the metallic casing designed in accordance with the shielding design specifications generally reach 60–80 dB quite easily, but because of the improper disposal of the cable, severe electromagnetic compatibility (EMC) problems may appear in the system. About 90% of the EMC problems are caused by the cables. This is because the cable is a highly effective antenna to receive and radiate the electromagnetic wave, and it is also a good path for disturbance.

The radiation generated by the cable is particularly severe. The cable can radiate the electromagnetic emission because the common-mode voltage existing at the cable port drives the cable as a monopole antenna, as shown in Figure 3.1.

The electric field generated by the radiation is

$$E = 12.6 * 10^{-7}(fIL)(1/r) \tag{3.1}$$

In this equation, I is the amplitude of the common-mode current caused by the common-mode voltage driving the cables; L is the length of the cable; f is the frequency of the common-mode current; r is the distance from the observation point to the radiation source. It can be seen that, to reduce the cable radiation, we can reduce the amplitude of the high-frequency common-mode current and shorten the cable length. The cable length is often not possible to be arbitrarily reduced, and the best way to control the common-mode radiation by the cable is to reduce the amplitude of the high-frequency common-mode current. The high-radiation efficiency of the high-frequency common-mode currents is the main factor causing the excessive radiation by the cable.

The shielded cable may be used to solve the cable radiation problem, but the shielded cable must be appropriately grounded. The pigtail, the improper grounding position, and the other issues will all cause EMC problems.

In addition, the cable positioning also has a significant impact on the product EMC performance. The coupling between the cables and the cable loop caused by cabling both are important for EMC cable design.

Electromagnetic Compatibility (EMC) Design and Test Case Analysis, First Edition. Junqi Zheng.

3.1.2 The Interface Circuit Provides Solutions to the Cable Radiation Problem

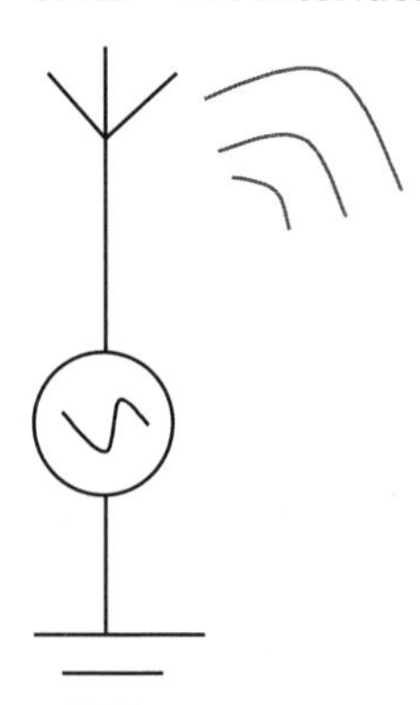

Figure 3.1 Cable common-mode radiation-model.

An effective way to reduce the common-mode high-frequency current on the cable is to reasonably design the interface circuit of the cable port or use a low-pass filter or suppression circuit at the cable port to filter out the high-frequency common-mode currents on the cable, as shown in Figure 3.2. Visibly, the importance of the filtering circuit for electromagnetic interference (EMI) is the same as that for the immunity.

The interface circuit is directly connected with the cable in the circuit, and whether the effective EMC design is implemented in the interface circuit is directly related to whether the whole system can pass the EMC tests. The EMC design of the interface circuit includes the design of the filtering circuit and the design of the protection circuit. The purpose of the filtering design in the interface circuit is to reduce the radiation from the interface and the cable, which is produced inside the system, and suppress the external radiated and conducted disturbance on the whole system. The interface protection circuit is designed so that the circuit can withstand a certain degree of the overvoltage and overcurrent impact.

The design of the interface filtering circuit and the protection circuit should comply with the following basic design rules:

1) The impact of the filtering circuit and the protection circuit on the interface signal quality should meet the requirements.
2) The filtering circuit and the protection circuit should be designed based on the actual needs, do not just copy the other design.
3) When we need both the filtering circuit and the protection circuit at the same time, the protection circuit shall be firstly ensured, then the filtering circuit.
4) The interface chips, including the corresponding filtering, protection, isolation devices, etc., should be placed in the direction of the signal flow, along the straight line and as close to the interface connector as possible.
5) The filtering, protection, and isolation devices for the interface signal should be placed close to the interface connector and the appropriate signal routings must be as short as possible (the shortest distance in accordance with the manufacturing process).
6) The interface transformer should be placed at the nearest place near the connector, usually within 3 cm from the corresponding interface connector.

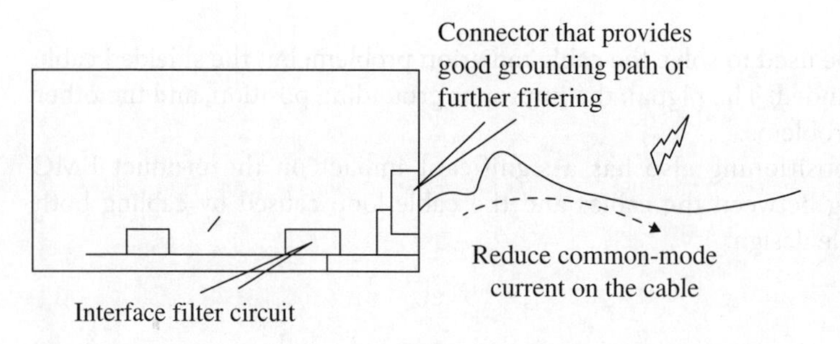

Figure 3.2 Common-mode low-pass filter on circuit board.

7) Between the analog interface and the digital interface, between the low-speed logic signal interface and high-speed logic signal interface (distinguish them per the susceptibility and the interference emission level), a certain interval should be ensured. When there is the possibility that mutual interference exists between the connectors, the isolation, shielding, and other measures must be taken.
8) When different types of signals are present in the same interface connector, we should isolate these signals with the ground pin, especially for some sensitive signals.
9) The width of interface signal trace should always be the same. For high-speed signal trace, if the trace must be bent somewhere, we should use a smoothly curved arc trace.
10) Prohibit other signal traces routed between the differential signal traces and the return path. Route differential pairs in parallel, closely, and on the same layer. The length of the differential pair shall be the same as possible.
11) If the interface signal traces are too long (the distance from the driver and the receiver to the interface connector is more than 2.5 cm), treat them as transmission lines. Their routings must meet the characteristic impedance requirement.
12) All signal traces can't be routed across the different reference plane, unless they have been isolated or filtered.
13) The interface signal trace and the interface chip must be treated in terms of the impedance matching, filtering, isolation, and protection to meet the requirements from the manufacture or the product standard.
14) All signals should be filtered, otherwise, so long as there is one signal trace without filter, the high-frequency common-mode current on it will couple into the other signal inside the same connector or cable, and produce radiation.

3.1.3 Connectors Are the Path Between the Interface Circuit and the Cable

The main role of the connector is to provide a good interconnection between the cable and the interface circuit, and to ensure a good grounding. A bad connector may ruin the function of the forestage filtering circuit. Electrostatic discharge (ESD) performance, impedance matching, the pin definition, the contact characteristics of the grounding, and so on need to be considered for the connector. If the connector is packaged with plastic material, it is necessary to ensure that there are sufficient air gaps between the surface of the seam and the metallic conductors inside the connector.

Sometimes, when an interface filtering circuit is mounted on the circuit board, the route of the signal line after it is filtered by the filtering circuit is too long inside the cabinet, which may result in cable radiation by inducing the interference signal and producing a new common-mode current.

The induced signals are from two sources, the electromagnetic waves inside the cabinet can be induced on the cable, and the interference signal before the filtering circuit may be directly coupled to the cable port through the parasitic capacitance. The solution to this problem is to minimize the wire length exposed inside the casing after the signal is filtered by the filtering circuit. The connector with a filtering function is ideally suited to solve this problem. There is a low-pass filter on each pin of the connector, and the common-mode current flowing through the pins can be filtered. These connectors with the filtering function are identical in shape and size with the conventional connectors, and can directly replace the conventional connectors. Since the connector is mounted on the port through which the cable goes inside the cabinet, the disturbance cannot be coupled on the signal wires after they are filtered, as shown in Figure 3.3.

If you select a connector with a filter, it is necessary to ensure that the filter connector has good grounding characteristics. Especially for the filtering connectors containing the bypass

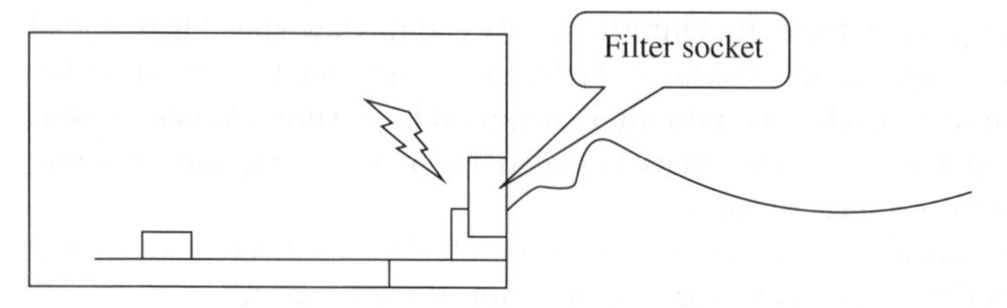

Figure 3.3 The filter connector can prevent the interference of the wire from inducing interference again after filtering.

capacitor (most of these connectors contain the bypass capacitor), most of the disturbance on the signal line is bypassed to the ground, so there will be a great disturbance current flowing through the connection point between the filter and the ground. If the contact impedance between the filter and the ground is large, there will be a large voltage drop across this impedance, leading to severe EMC problems.

The following points are the basic rules for choosing the connector:

1) Choose a shielded connector with metallic shell, especially for the high-frequency signal connector.
2) Good electrical continuity between the metallic shell of the connector and the casing should be guaranteed. For the connector that can be connected with a 360° contact, it must be connected with a 360° contact, and the connection impedance usually shall be smaller than 1 mΩ.
3) If the connector can't be connected with a 360° contact, it is recommended to use the connector with some reeds upward the enclosure on its metallic shell, and the size and the performance (the elasticity) of the reed must be sufficient to achieve a good electrical connection to the casing.
4) The filtering connectors can greatly help improve the EMC performance of the product, but its cost is relatively high. When, with the internal filter and the shielded cable, the problem can be solved, we will not use the filtering connectors. The filtering connectors are typically used in some special cases, such as the strict military applications and the harsh industrial applications with small quantities, and some of the special applications (such as the structural size is limited, etc.).
5) The shield layer of the shielded cable shall be connected to the metallic shell of the connector with a 360° contact. For the interface that can't reach this, there are usually other corresponding measures to ensure the EMC performance of the interface.
6) If the connector is mounted on the circuit board and connected to the cabinet through the grounding trace in the circuit board, we have to pay attention to provide a clean ground connection, and this ground shall be separated from the reference ground of the signal in the circuit board. There is only one connection point, and the good connection to the cabinet must be guaranteed.

3.1.4 The Interconnection Between the PCBs Is the Weakest Link of the Product EMC

EMI problems often become more complex because of the interconnecting of the high speed and high rising edge signal, which can also result in the crosstalk and the voltage difference between the reference grounds. For an interconnection connector without a shield or good ground plane, the crosstalk between the signals on it is larger than that in a multilayer printed circuit board (PCB). The ground impedance between the different subsystems is high due to the parasitic inductance of the interconnection connector pin, therefore the voltage difference between the 0 V reference grounds is larger than that of the PCB (since the voltage drop exists between the 0 V reference points [grounds] in different structures, as a common reference voltage, it shall be limited to a certain extent. This voltage drop on the same PCB board, compared with that on the different PCBs, which are interconnected by a cable, is much easier to be controlled, because there is higher induction to the external electromagnetic signals for the PCBs interconnected by a cable).

Therefore, before designing a product with the interconnection, as a designer, you should ask yourself, "Can this product feature be achieved without the interconnection? Can these subsystems that need the interconnection be concentrated to one PCB?" Using a subsystem (PCB) system is better than connecting several small PCB boards together with the cables.

In the products system with interconnections, the crosstalk and the ground (0 V) impedance of the interconnecting connector is the key to the EMC design:

1) If the number of the ground pins is less, the RF circuit loop will be large, which will cause a greater differential-mode radiation (although sometimes the differential-mode radiation is not the main factor leading to the excessive radiation of the products).
2) If it can't be guaranteed that at least one ground pin is accompanied to the signal line, the cross talk between signals caused by the capacitive coupling and inductive coupling can also be intensified.
3) If the number of ground pins is less, the overall equivalent parasitic inductance of its ground pins is larger, the RF return current will cause a higher common-mode voltage drop on the ground pins, which means that there will be a high-frequency RF voltage between the two interconnected PCBs (unless there are some other additional measures), and this high-frequency RF voltage will cause a common-mode current in between devices, and then lead to the common-mode radiation in the current drive-mode, increasing the overall radiated emissions and conducted emissions of the product system.
4) Even if the ground pins are enough, the EMC weakness still exists. See the interconnection schematic of the product with the structure of backplane and inserted card in Figure 3.4. In this mechanical structure, the high-speed bus is usually located in the backplane and connected with the inserted board. The common-mode voltage U_{CM} will be induced between the inserted board and the backplane. This common-mode voltage will be the main reason for causing the common-mode radiation, as shown in Figure 3.5.

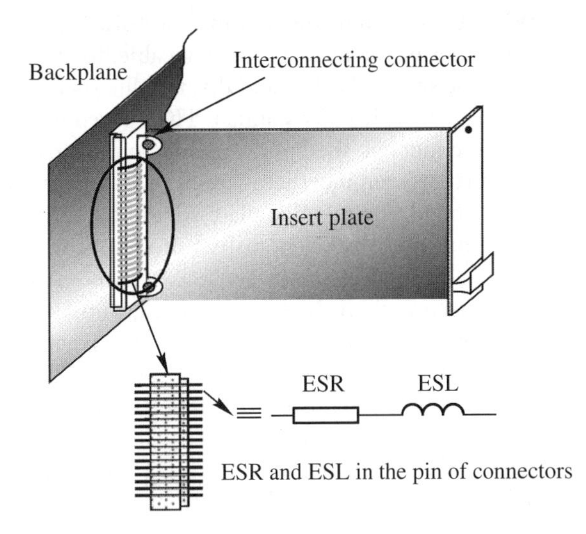

Figure 3.4 Plug-in board structure product interconnect schematic diagram.

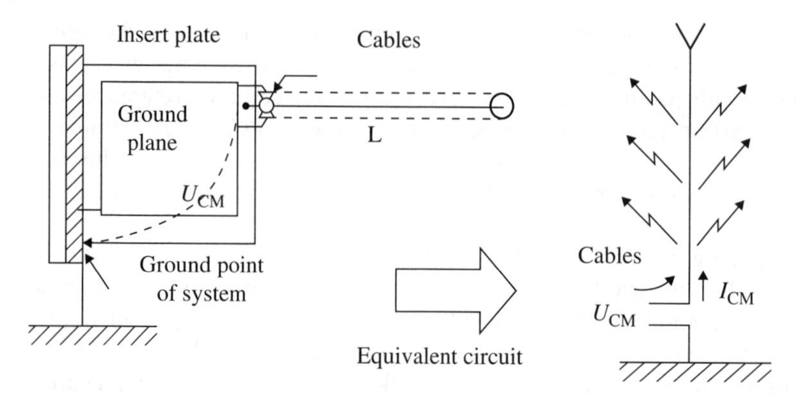

Figure 3.5 The principle diagram of the interconnection of common-mode radiation.

The interconnecting connector or the interconnecting cables inside the product are also the main reason affecting the immunity performance of the product because of the high impedance caused by the parasitic inductance of the interconnecting connector or the interconnecting cable in high frequency. When performing the test, such as big current injection (BCI), electrical fast transient/burst (EFT/B) and ESD immunity test, the common-mode transient disturbance current produced in the test flows through the ground (0 V) lines of the interconnecting connector or the interconnecting cable, and due to the impedance of the ground lines in the interconnecting connector or the interconnecting cable, there will be a common-mode voltage drop on the interconnection. If the voltage drop ΔU_{Z0V} of the ground lines in the interconnecting connector or the interconnecting cable is more than the noise margin of the circuits at both ends of the interconnection, an error will occur (the schematic is shown in Figure 2.83).

Therefore, when designing the product, avoiding the common-mode interference current flowing through the interconnecting connector or the interconnecting cable is the first step to solve the EMC immunity problem due to the interconnection inside the product. When the product mechanical architecture design can't prevent the common-mode interference current flowing through the interconnecting connector or the interconnecting cable, the following points shall be considered for the interconnection design of the product:

1) When the common-mode transient disturbance current flows through the interconnecting connector or the interconnecting cable, we recommend using the connector with metallic shell, the shielded cable, and the metallic shell, and connecting the cable shield to both ends of the cable with a 360° contact. The interconnecting 0 V ground shall be directly connected with the metallic shell of the connector at the signal input/output terminals on the PCB. If we can't interconnect them directly, we can use the bypass capacitor. For the equipment that is grounded, the metallic plate should be connected to the earth. The aim is to make the common-mode transient disturbance current flow through the shell of the interconnecting connector and the shield of the cable, and avoid the common-mode disturbance current flowing through the high impedance ground lines of the interconnecting connector and the interconnecting cable and causing the transient voltage drop.
2) If we only use the interconnecting connector with nonmetallic housing and the unshielded cables (such as the unshielded ribbon), it is recommended that you use an additional metallic plate to connect both ends of the interconnection (you can also use the existing metallic casing of the product, as shown in Figure 5.20), and directly connect the interconnected 0 V ground to the metallic plate at the PCB signal input/output terminals. If we can't

interconnect them directly, we can use the bypass capacitor. For grounded equipment, the metallic plate should connect to the earth.

3) If the methods described in (1), (2) are not feasible, all the interconnected devices must be filtered.

3.2 Analyses of Related Cases

3.2.1 Case 16: The Excessive Radiation Caused by the Cabling

[Symptoms]

It is found, in the radiated emission test for a product, that it can't meet the requirements, the specific phenomenon is that the emission level over the frequency range from 100 to 230 MHz is severely more than the standard limit, and the maximum emission at one frequency exceeds the Class B limit more than 20 dB. The test spectral plot is shown in Figure 3.6.

[Analyses]

After checking, the DC power supply cable of the equipment is arranged as shown in Figure 3.7. The DC power supply cable goes into the equipment from the power supply input connector and the power supply filter, and then it is divided into two ways and connected to the internal

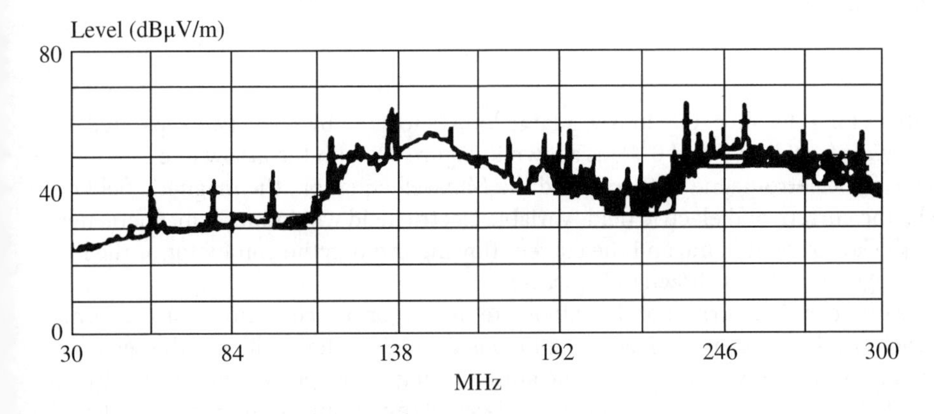

Figure 3.6 Spectrum of radiated emission test.

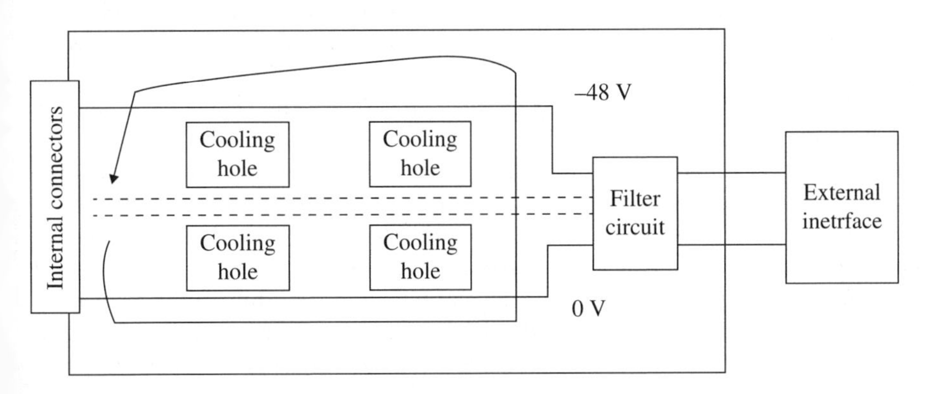

Figure 3.7 DC power line layout.

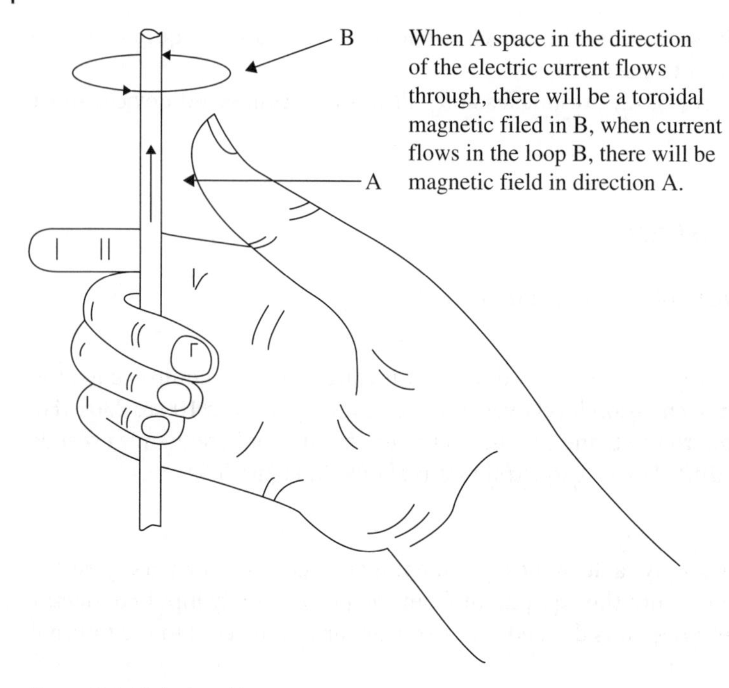

B
A
When A space in the direction of the electric current flows through, there will be a toroidal magnetic filed in B, when current flows in the loop B, there will be magnetic field in direction A.

Figure 3.8 Right-hand law.

connector after a long distance. Obviously, a large loop is formed by these two separated DC power supply wires (−48 line V and 0 V), as shown in Figure 3.8 with the arrow line.

According to the electromagnetic theory and the Maxwell equation, the magnetic field can be induced by the current and electric field. Variable electric field will induce a magnetic field, according to the law of right hand, and the current flowing through the conductor or the loop will induce a magnetic field, as shown in Figure 3.8.

It can be clearly seen that current and the loop are the important conditions for the formation of radiation. In this case, a relatively large loop is formed by the cabling of the separated positive and negative power wires, in which the noise flowing on the power wires is the driving source of radiated emission. If the closed loop area is small (much smaller than the wavelength of the signal frequency or the frequency we care), then the field strength is proportional to the area of the closed loop. The bigger the closed loop area, the lower frequency observed at the antenna terminal side. For a certain physical size, the antenna will resonate at a certain frequency.

For the differential-mode radiation generated by the loop antenna, the far-field electric field strength varies linearly with the loop area. The bigger the area of the closed loop, the more severe radiation can be generated by the differential-mode current. In addition, for the same area of the closed loop, if the loop shape varies, instead of a square structure, the radiation effect will vary simultaneously, and will even lead to considerable differences. As the frequency increases, the radiated emission generated by the fixed-shape closed loop will be enhanced, and the radiation energy from the differential-mode current will gradually shift to the front right above the loop. More importantly, with the gradually changing closed loop from a square into more and more narrow rectangular, the radiated emission generated by the differential-mode

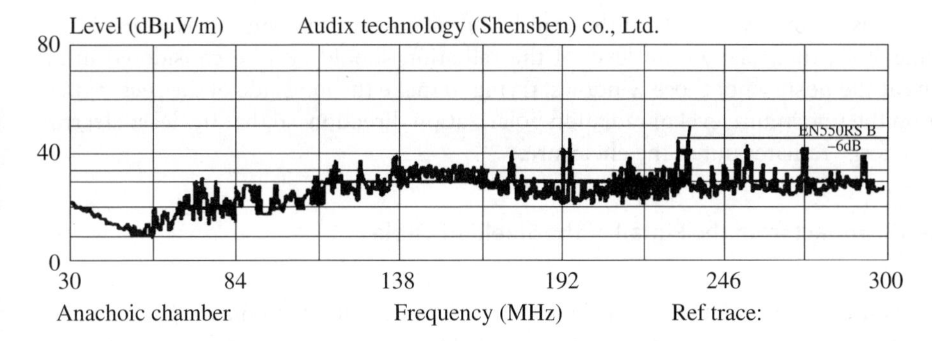

Figure 3.9 The radiation emission result after change the power supply wiring.

current will be significantly reduced. That is to say, even the area of the closed loop is the same, changing its shape properly to be narrower. The strength of the radiated emission caused by the differential-mode current can also be reduced. The radiated emission produced by the differential-mode current flowing through the closed loop has different distribution in each polarization direction.

The polarized component of the radiation mainly focus on both sides of the front right above the loop, and the heat dissipating holes in this case is just located above the loop, which is also the direction with the strongest radiation.

In view of the above simulation analysis, it should be easy to understand the radiation mechanism in this case.

[Solutions]

According to the above analysis result, this problem can be solved just by routing the ground wire and the power wire in the middle of the heat dissipating holes (such as the dotted line in Figure 3.7), with this way the loop area of power supply line is greatly reduced compared with the original one, after testing, the radiated emission is reduced by nearly 20 dBμV m^{-1} over the frequency range from 111 to 165 MHz, as shown in Figure 3.9.

[Inspirations]

1) Controlling loop area is a necessary mean to decrease the radiation, not only to minimize the loop area in the PCB routing but also to pay attention on the size of the current loop in cabling.
2) It is also effective to reduce the radiated emission caused by differential-mode current by changing the shape of the closed loop to be narrow as possible.
3) As the radiation level from differential-mode current is different in each polarization direction, the electrical length of the loop formed by adjacent cable, the traces of PCB or components should be as minimized as possible in the polarization direction, on which the radiation is higher, and this guarantees less electromagnetic energy coupled to them.
4) For the cabling design inside the enclosure, we should ensure that the minimum electrical length of cables on the polarization direction in which higher emission is radiated, so that the cable is coupled with the least electromagnetic energy.
5) Be sure to determine the location and the structure of the ventilation window or observation window that have the minimum effect on the radiation from inside to outside of enclosure. The ventilation or observation window should be placed at the position where

the radiation is as low as possible. If the ventilation or observation window is formed by a rectangular hole, the polarization level of the radiation should also be considered in all directions at the position of those windows, trying to make the long side of the rectangular hole not on the maximum level of radiation polarization direction, so that the least electromagnetic energy radiates out of the enclosure.

3.2.2 Case 17: Impact from the Pigtail of the Shielded Cable

[Symptoms]

An industrial controller product uses shielded cable to connect with its signal output port, and it is found that, in the radiated emission test, the radiated emission is under Class B limit but with insufficient margin. The tested spectral plot is shown in Figure 3.10.

Considering the test uncertainty, there must be some improvement so that the test results have more than 6 dB margin below the limit.

After the preliminary locating test, the radiated emission is very low without signal output cable, and meets the Class B requirements with 6 dB margin.

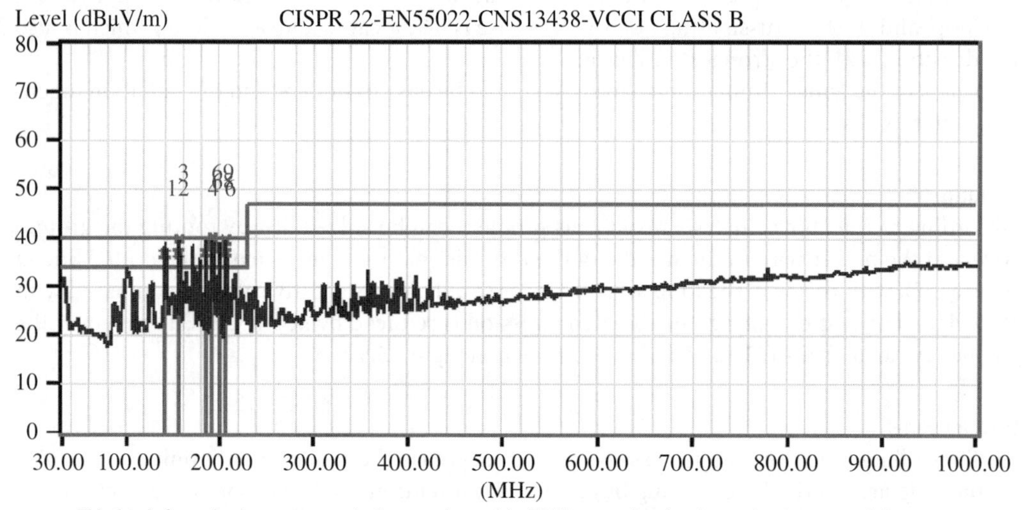
Level (dBμV/m)
CISPR 22-EN55022-CNS13438-VCCI CLASS B
80
70
60
50
40
30
20
10
0
30.00
100.00
200.00
300.00
400.00
500.00
600.00
700.00
800.00
900.00
1000.00
(MHz)
This data is for evaluation purposes only. It cannot be used for EMC approvals unless it contains the approved signature.
If you have any questions regarding the test data, you can write your comments to service@mail.adt.com.tw

	No.	Frequency MHz	Factor dB	Reading dBμV/m	Emission dBμV/m	Limit dBμV/m	Margin dB	Tower cm	Table deg
	1	140.06	16.22	20.42	36.64	40.00	−3.36	237	40
	2	154.81	17.02	19.74	36.76	40.00	−3.24	176	19
	3	156.10	17.03	22.74	39.77	40.00	−0.23	–	–
	4	184.29	14.12	22.80	36.92	40.00	−3.08	164	0
	5	191.66	13.36	24.66	38.02	40.00	−1.98	144	13
*F	6	192.47	13.32	26.77	40.09	40.00	0.09	–	–
	7	199.03	13.00	24.89	37.89	40.00	−2.11	100	349
	8	206.41	13.06	23.82	36.88	40.00	−3.12	99	0
	9	207.03	13.07	26.44	39.51	40.00	−0.49	–	–

Figure 3.10 Test result that is over the limit line.

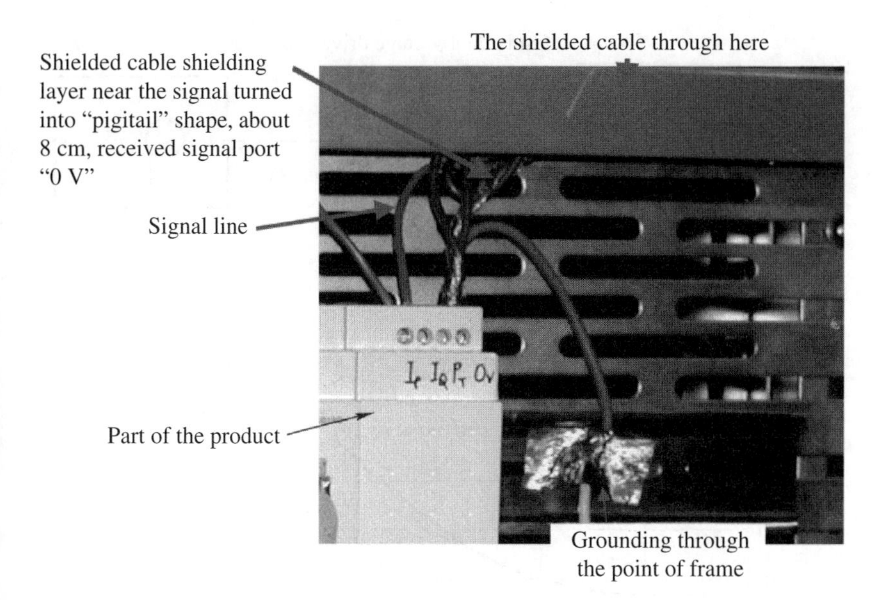

Figure 3.11 Signal terminal cable connection mode.

[Analyses]

From the test results, we know that frequencies with high radiated emission are focused on the range from 150 to 230 MHz, and due to the small size of the product, only the length of cable can be comparative with the wavelength of the frequencies with high radiated emission. Therefore, such higher radiated emission of this product is certainly related to the cable. The connection way of the shielded cable is shown in Figure 3.11.

It is shown in Figure 3.11 that, even though the product uses the shielded cable, the shield of the shielded cable is twisted into a pigtail near the product signal connector with about 10 cm length. Pigtail is a common EMC problem when shielded cable is used. The principle of the EMC problem regarding the impacts on product from pigtail can be explained in Figure 3.12, which shows that pigtail acts as a common-mode voltage ΔU, and the ΔU drives the shield of the signal cable with pigtail on its shield, the radiation happens finally.

The concept of transfer impedance can help further explain the principle of the impact from pigtail. The transfer impedance is, when RF current is injected on the shield of the shielded cable, the ratio of the voltage on its inner conductor to the injected current on its shield. For a given frequency, lower Z_T means that a lower voltage exists on its inner conductor when RF current is injected on its shield, that is, good shielding effectiveness from the external disturbance. It also shows the induced current on the shield will be lower when there is a voltage on the inner conductor, that is, it has a higher effectiveness of shielding the interference from the inner conductor. If the Z_T of a shielded cable over the entire frequency range is only a few milliohms, the shielding effectiveness is good. At the same time, lower transfer impedance of shielded cable means better effectiveness of shielding external interference and internal radiated emission from itself. However, the presence of pigtail can be equivalent as that a tens of nH inductor is in serial with the shield, which can cause a common-mode voltage while there is current flowing through the shield of the cable. With the increasing of frequency, the equivalent transfer impedance of Pigtail will increase rapidly, which will make the shielding effectiveness of shielded cable completely failed. It shows, in Figure 3.12, that the principle of the radiated emission caused by the

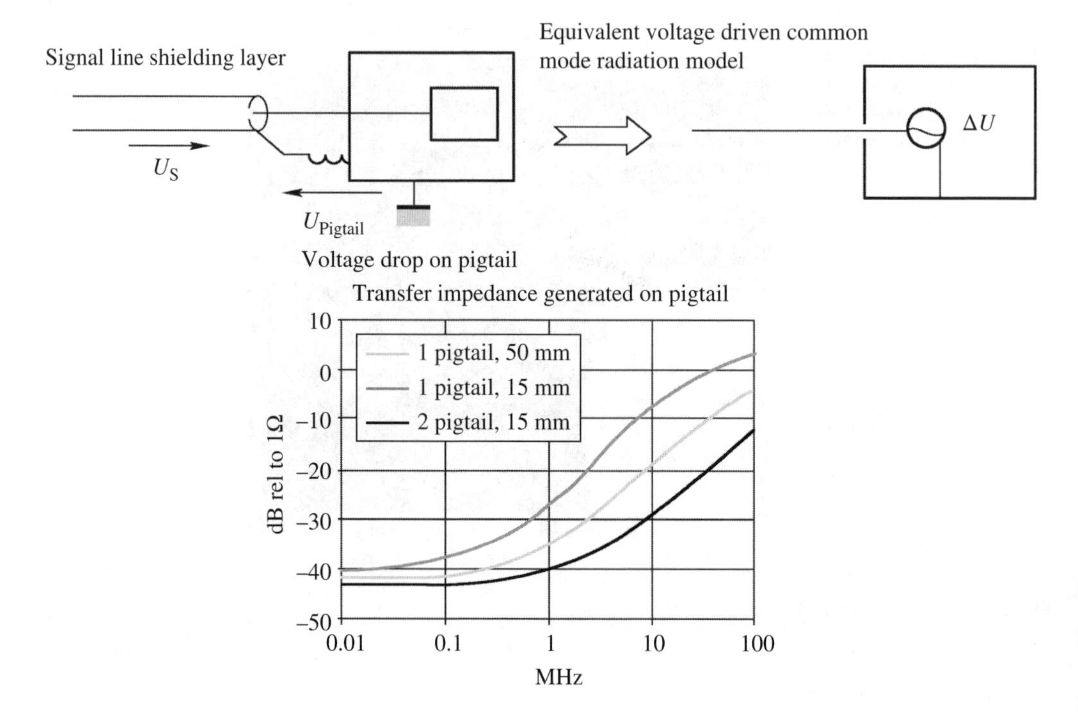

Figure 3.12 Radiation forming principle.

shielded cable with pigtail. The existence of pigtail causes large impedance, and causes a large voltage drop U_{pigtail}, which is the driving source of the radiation, and the shield becomes the antenna.

[Solutions]

After shortening the pigtail to 1 cm, the test results are shown in Figure 3.13.

[Inspirations]

1) The shield of shielded cables must be bonded in 360°.
2) From the concept of risk evaluation, according to the experience, below 30 MHz, if cable shield has zero-length pigtail, there is no risk; there is 30% risk if the length of the pigtail is 1 cm; 50% for 3 cm pigtail; 70% for 5 cm pigtail.
3) All cables are affected by their parasitic resistance, capacitance, and inductance. Look at the following simple examples, which can help readers have a perceptual awareness:
   - At 160 MHz, the impedance of a wire with 1 mm diameter is more than 50 times the DC resistance. This is the result of the skin effect, which forces 67% of the current of the conductor, at this frequency, flow within the area with 5 μm depth from the outer layer.
   - The wire with 25 mm length and 1 mm diameter has about 1 pF parasitic capacitance. It seems not worth mentioning, but it presents around 1 kΩ impedance at 176 MHz. If this wire is driven in free space by an ideal square wave signal whose peak to peak voltage is 5 V and frequency is 16 MHz, then at the 11th-order harmonic, that is at 16 MHz, only driving this wire, a 0.45 mA common-mode current will be generated, which is a dangerous current causing radiated emission test failure.
   - For the pins of the connector having a length of about 10 mm, and the diameter of 1 mm, each pin has about 10nH self-inductance. This too seems scarcely worth mentioning, but

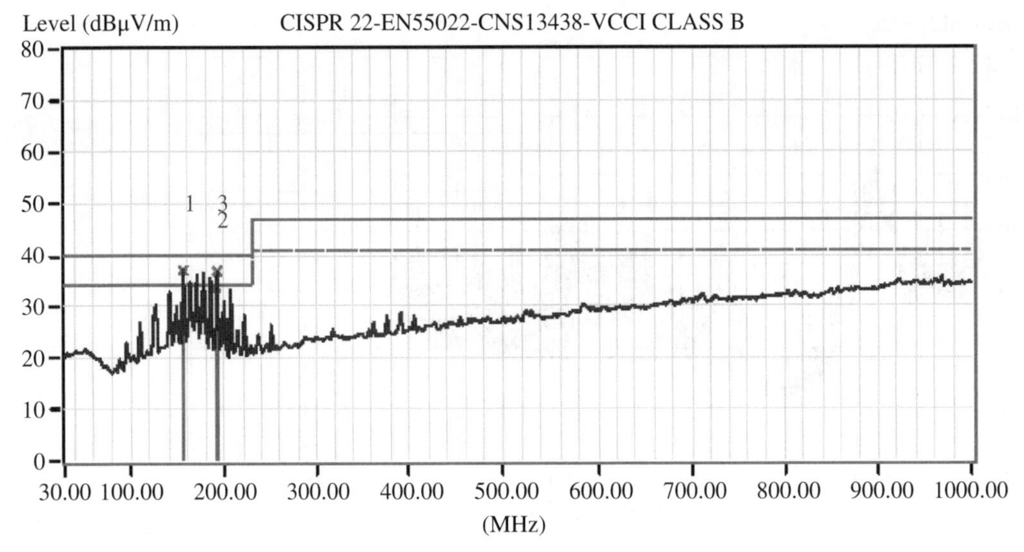
Level (dBμV/m)
CISPR 22-EN55022-CNS13438-VCCI CLASS B
80
70
60
50
40
30
20
10
0
1
3
2
30.00 100.00 200.00 300.00 400.00 500.00 600.00 700.00 800.00 900.00 1000.00
(MHz)
This data is for evaluation purposes only. It cannot be used for EMC approvals unless it contains the approved signature.
If you have any questions regarding the test data, you can write your comments to service@mail.adt.com.tw

	No.	Frequency MHz	Factor dB	Reading dBμV/m	Emission dBμV/m	Limit dBμV/m	Margin dB	Tower/Table cm	deg
*	1	156.10	17.03	19.73	36.76	40.00	–3.24	–	–
	2	191.67	13.36	20.00	33.36	40.00	–6.64	129	190
	3	192.47	13.32	23.26	36.58	40.00	–3.42	–	–

Figure 3.13 Test result after modification.

when the 16 MHz square wave signal passes through it to the motherboard for bus transmission, if the driving current is 40 mA, the drop voltage on this connector is around 40 mV, which is large enough to cause serious signal integrity problems and EMC problems.

- The 1 m wire has around 1 μH self-inductance. If used as the grounding of the equipment, it will affect the effectiveness of surge protector and filter.
- Since the self-inductance of 100 mm grounding wire of filter can reach about 100 nH, it will cause filter failure when the frequency is higher than 5 MHz.
- The empirical data: For the wire with the diameter smaller than 2 mm, its parasitic capacitance and inductance, respectively, are: 1 pF in.$^{-1}$ and 1 nH mm^{-1} (not to unify unit, but it is easier to remember them). The simple impedance calculation formula is

$$Z_C = \frac{1}{2\pi fc} Z_L = 2\pi fl$$

3.2.3 Case 18: The Radiated Emission from the Grounding Cable

[Symptoms]

A product is found not meeting the requirements in the radiated emission test, in which the specific phenomenon is that the emission seriously exceeds the standard limit over the frequency range from 30 to 300 MHz. The maximum emission exceeds Class B limit by more than 20 dB. The spectral plot is shown in Figure 3.14.

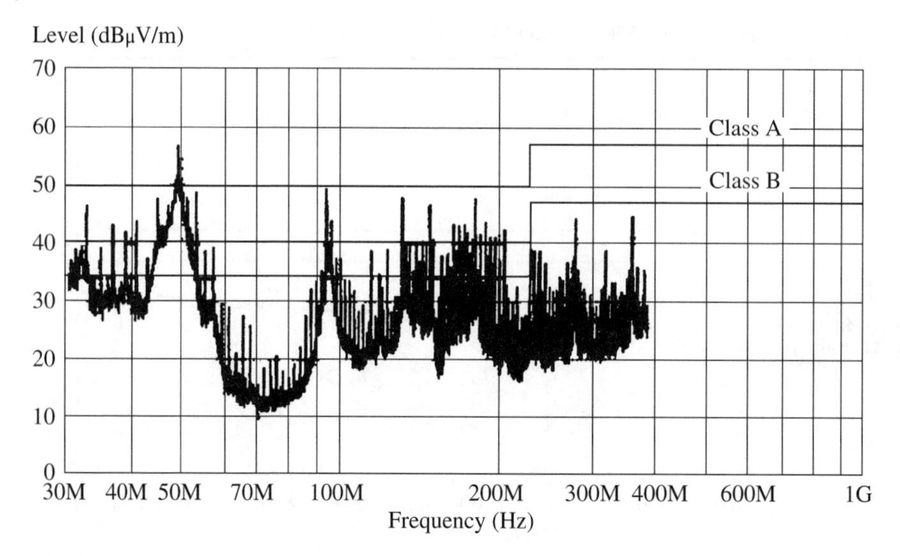

Figure 3.14 Spectrum of radiated emission test.

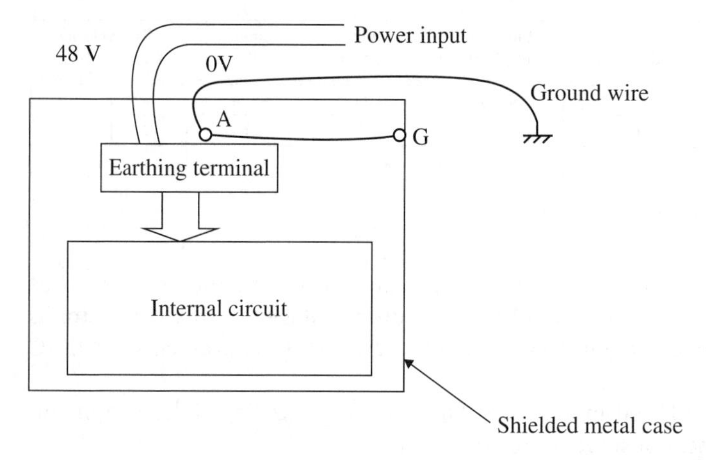

Figure 3.15 Internal cable layout.

[Analyses]

After checking, it is found that the cabling of the DC power supply power wires of the equipment is shown in Figure 3.15, the DC power wires (including the 0 V wire and 48 V wire) and the grounding wire go inside the device through the metallic casing, and then they are connected to the connectors. If the grounding wire is out of the shielded casing from point A, and from which it is connected to the point G of the casing, then a grounding wire for the whole system is connected to system protective earth from point A (the bold black curve in this figure). Regarding the common-mode radiation, it has been described in other cases, there are two necessary conditions forming common-mode radiation:

1) Common-mode driving source, such as the existence of RF voltage difference between two positions of the metallic casing.
2) Common-mode antenna, i.e. one pole of antenna is an external terminal device, the other one is the internal PCB ground of the device.

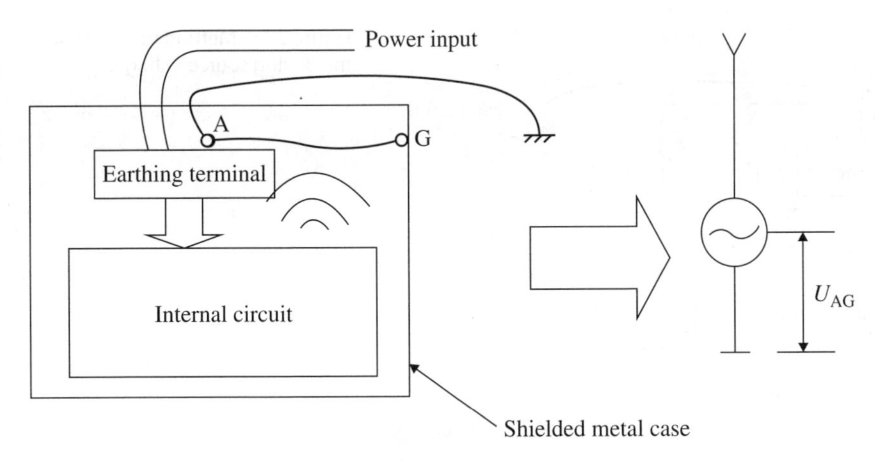

Figure 3.16 Common-mode radiation formation diagram.

In fact, it already has the necessary conditions forming common-mode radiation in this case, as shown in Figure 3.16.

Since the interconnecting wire between point A and point G (the bold black line as shown in Figure 3.16) is inside the equipment, and the working circuits inside the device are full of noise, the capacitive coupling or inductive coupling exists between the cable interconnecting point A and point G and the noise source in the internal circuit, so that the cable interconnecting point A and point G is actually not pure sense of the grounding wire, after picking up the noise, it becomes a polluted wire. Across the both ends of this wire, there is a common-mode voltage drop U_{AG}. The diagram on the right side of Figure 3.16 is an equivalent circuit that a common-mode radiation is formed, the cable with common-mode voltage U_{AG} is the common-mode driving source, and the grounding wire (the bold black wire in Figure 3.16) for whole system is the driven radiating antenna.

[Solutions]

From the analysis, to solve the problem of common-mode radiation, we can start from the two necessary conditions forming the common-mode radiation. Since the wire interconnecting point A and point G in this case is the driving source, and the relatively long one is the antenna, if one of these two is canceled or the correlation between them is broken, we can solve this problem. Two kinds of solutions are shown in Figures 3.17 and 3.18. The former method is to

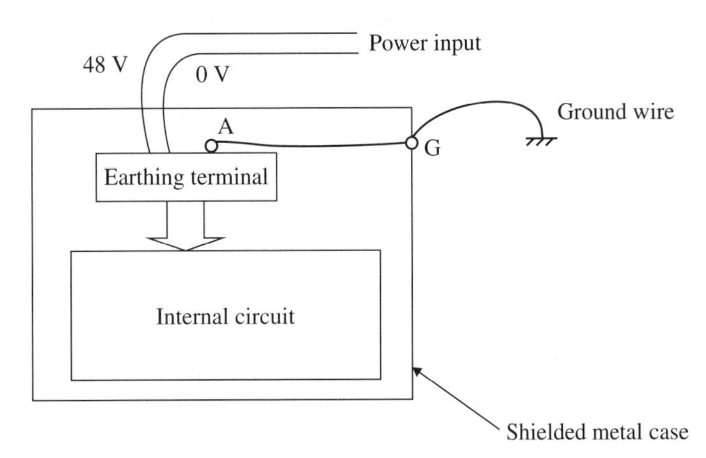

Figure 3.17 A scheme for the solution of the connection relationship between the driving source and the antenna.

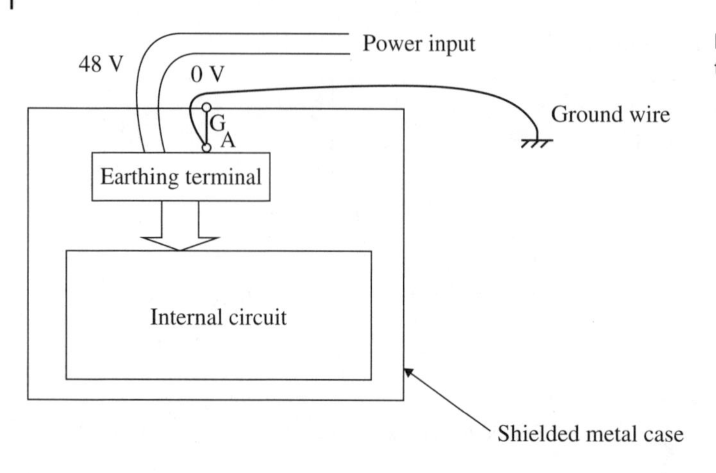

Figure 3.18 Method for reducing the driving source voltage.

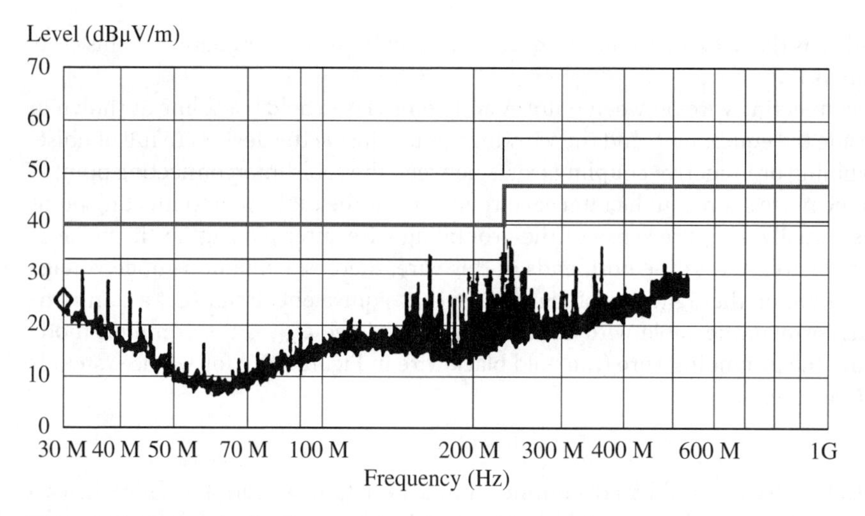

Figure 3.19 Test result of radiate emission after modification for grounding.

break the correlation between the common-mode noise source and the antenna, since the potential at point G is zero. The grounding wire cannot be driven by the noise source of U_{AG}, so the radiation can't be formed. The other method is to reduce the voltage of the driving source, changing the wire interconnecting point A and point G to a very short wire, which causes U_{AG} to be 0, thus the radiation can be reduced.

Figure 3.19 shows the new test results with the solution of breaking the correlation between the driving source and the antenna, in other words, connecting the grounding wire to the cabinet. As seen from the plot, the radiated emissions over low-frequency range are significantly reduced and they are under the Class B limit with around 6 dB.

[Inspirations]

1) Three elements for EMC design is shielding, filtering, and grounding, among of them, grounding is the most critical. Improper grounding will result in lower filtering and shielding performance.

2) For the equipment with the design of shielding the entire device, grounding should be connected with the shield, and make sure the equipotential between the shield and the grounding point.
3) To solve the problem of cable radiation, we can start from the two necessary conditions of the formation of radiation.

3.2.4 Case 19: Is the Shielded Cable Clearly Better than the Unshielded Cable?

[Symptoms]

A product uses ethernet communication interface. The ethernet cable is shielded cable. In the test, we found that the radiated emission is over the standard limit (Class B), and the emission is related to the ethernet cable. The relevant test spectral plot is shown in Figure 3.20.

From Figure 3.20, we can see, the emission at 150 MHz exceeds the limit of Class B. When we change the ethernet cable to unshielded ethernet cable, we unexpectedly found that the emission meets the limit of Class B, and there is a certain margin. The radiated emission spectral plot when we use unshielded ethernet cable is shown in Figure 3.21.

Generally speaking, the shield of shielded cable has the function of isolating the internal signal from being transmitted outside, so it has advantage in the EMI test. So in the design of many products, taking the performance of EMC into account, we choose the shielded cable by increasing the cost a little bit. But why, in this case, does a "thankless task" phenomenon appear?

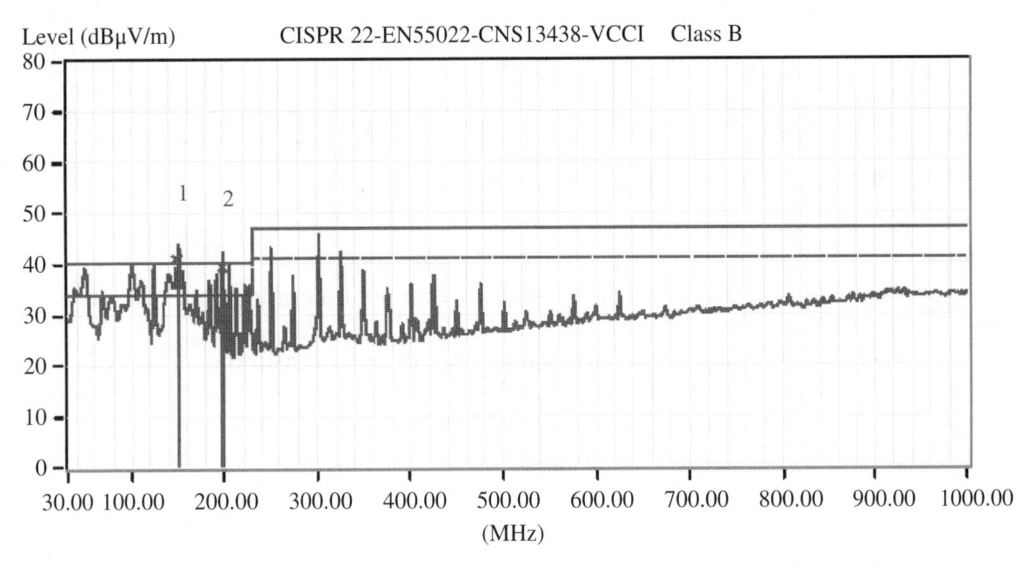

	No.	Frequency MHz	Factor dB	Reading dBμV/m	Emission dBμV/m	Limit dBμV/m	Margin dB	Tower/Table cm	deg
*F	1	150.01	16.97	23.79	40.76	40.00	0.76	99	235
	2	199.06	13.00	26.22	39.21	40.00	–0.79	99	271

Figure 3.20 Radiation emission spectrum using shielded cable.

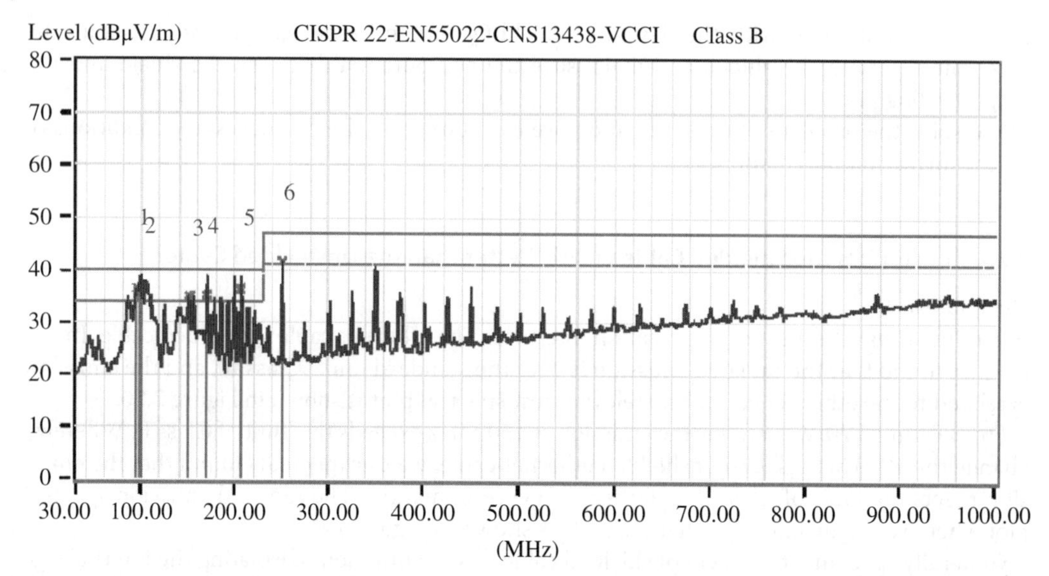

	No.	Frequency MHz	Factor dB	Reading dBμV/m	Emission dBμV/m	Limit dBμV/m	Margin dB	Tower/Table cm	deg
*	1	95.47	12.20	24.06	36.26	40.00	−3.74	–	–
	2	100.01	12.56	22.30	34.86	40.00	−5.14	218	232
	3	151.25	16.98	17.77	34.76	40.00	−5.24	–	–
	4	169.57	16.15	18.88	35.03	40.00	−4.97	173	213
	5	206.43	13.06	23.16	36.22	40.00	−3.78	179	230
	6	250.68	14.83	26.77	41.60	47.00	−5.40	–	–

Figure 3.21 Radiation emission spectrum using nonshielded cable.

[Analyses]

Firstly, we should look at the layout of the ethernet communication interface. The layout is shown in Figure 3.22.

The ethernet communication interface adopts the ethernet transformer. The PCB trace between the shell of RJ45 connector and the earth terminal is about 6 cm long, which is shown in Figure 3.22. As we can see in Figure 3.22, a certain problem exists in the grounding trace. The reason is that the impedance of the 6 cm PCB trace is very high when the frequency is very high. Because of the limitation of product structure, we must do like that.

The principle of the radiation in this product is shown in Figure 3.23.

The amplitude of the common-mode current I_{CM} determines the amplitude of the radiated emission. A part of the common-mode current comes from the unbalance transmission and coupling of ethernet signal line, and the other part comes from RC common-mode suppressing circuit, which is connected with the center tap of the transformer. The thick arrow line in Figure 3.23 shows the flowing direction of the common-mode current. The amplitude of the common-mode current is totally controlled by the common-mode voltage drop U_n (U_n is caused by the grounding impedance of the shielded cable). So, to a certain extent, U_n determines the success or failure of the radiated emission test. The common-mode current flowing on the cable shield is, $I_{CM} = U_n/150$, assuming the impedance between shielded cable and ground is 150 Ω.

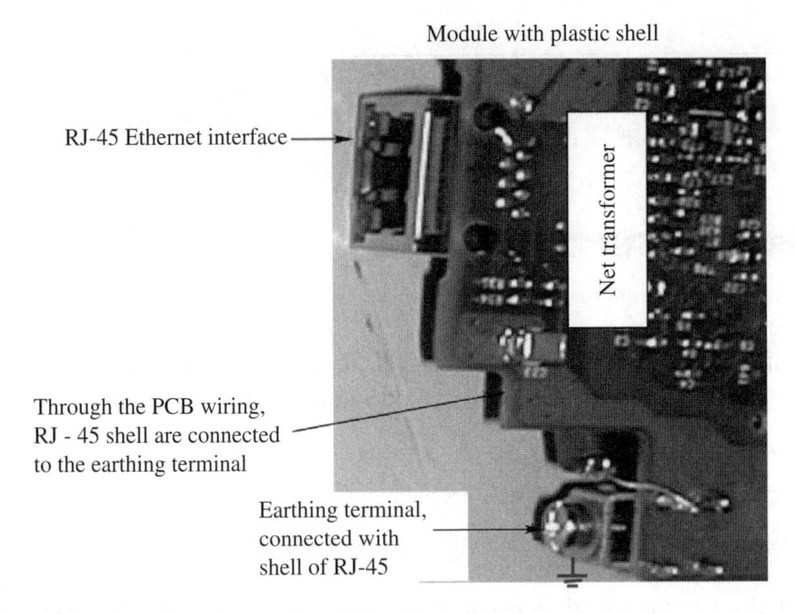

Figure 3.22 Layout of the ethernet communication interface part.

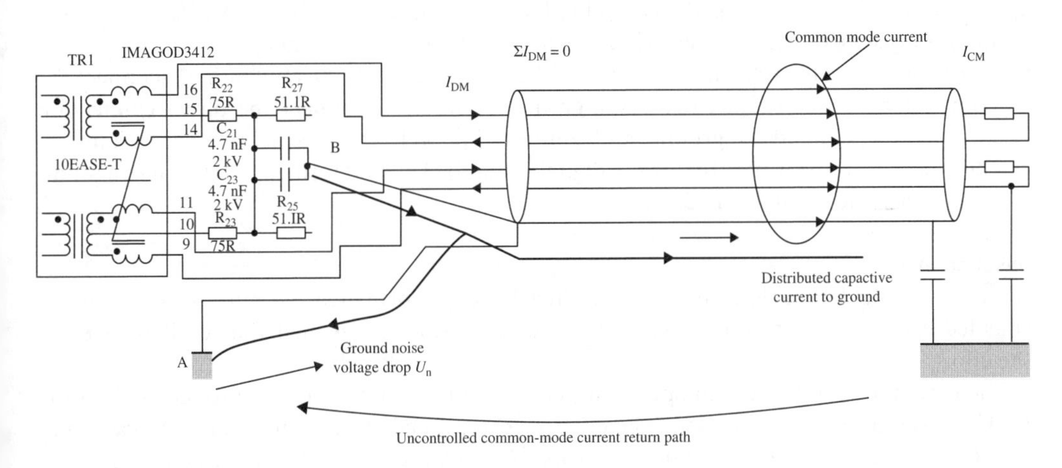

Figure 3.23 The forming principle of radiation.

In this case, due to the limitation of mechanical structure, the impedance of the grounding path connected to the metallic shell of RJ45 connector is relatively high. The shield of shielded cable or the metallic shell of RJ45 connector cannot be well grounded, which results in high grounding impedance. When the ethernet transformer and the relevant common-mode suppressing circuits (including C21, R22, etc.) are activated, a common-mode current is generated and it flows through the grounding wire connected to the metallic shell of RJ45 connector (the wire between A and B shown in Figure 3.23), then there will exist a high voltage drop U_n. The shielded ethernet cable is driven by U_n, there is a large current existing on the shied of the shielded ethernet cable. The shield which the common-mode current flows through becomes the carrier of radiation, which is the antenna. This is a typical-model of the radiating antenna driven by the common-mode voltage. Figure 3.24 shows the schematic of the radiation driven by common-mode voltage.

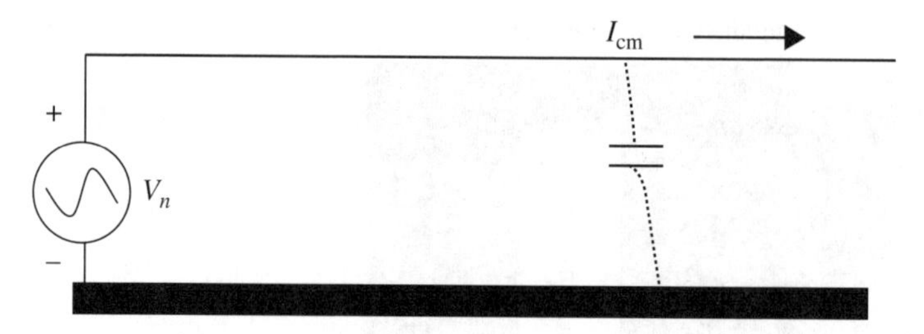

Figure 3.24 Schematics of common-mode emission.

When the shielded cable is changed to an unshielded cable, the common-mode voltage U_n still exists but the radiation carrier antenna disappears, so the radiated emission is reduced.

[Solutions]

Four approaches can improve the radiated emission problem:

1) Remove the radiating antenna. If the grounding effectiveness cannot be improved, change the shielded cable to unshielded cable.
2) Cut off the path of the common-mode current: Disconnect the connections between C_{21}, C_{23} and the ground.
3) Reduce the grounding impedance, and reduce the common-mode voltage, U_n: Replace the grounding wire of the cable shield routed in PCB with a metallic sheet.
4) The product finally uses the way of cut off the common-mode current path: Disconnect the connection between C_{21}, C_{23},and the ground. That is, disconnect the direct connection between C_{21}, C_{23} and the ground, and connect C_{21}, C_{23} to the internal reference ground of the ethernet transformer (that is the digital ground). The test results with the above modification are shown in Figure 3.25.

[Inspirations]

If we use a shielded cable to improve product EMC performance, we must ensure that the shielded cable is well grounded. Otherwise, it is possible to get half the result with twice the effort.

The ethernet circuit is a high-speed interface circuit. A lot of ethernet products fail in the EMC test because of the radiation from ethernet cable. In addition to grounding the cable and selecting a shielded ethernet connector, properly designing ethernet interface circuits is also important. The following is a summary of the EMC design of 10 M/100 M ethernet interface circuits:

Schematic design

Figure 3.26 shows the common ethernet interface circuits. This part is used for the impedance matching and the suppression of EMI.

The transformer in this circuit integrates common-mode choke in the transmitting port, not in the receiving port. If the ethernet transformer does not integrate common-mode choke, we need insert an external common-mode choke. Of course we can also choose the ethernet transformer, which integrates common-mode chokes not only in transmitting port but also at receiving port, such as H1012. If we choose the ethernet transformer with integrated common-mode chokes, the PCB layout method described below is still applicable.

R_9,R_{10} are the differential-mode impedance matching resistors in the receiver port. They provide common-mode impedance matching effect via a capacitor connected between the

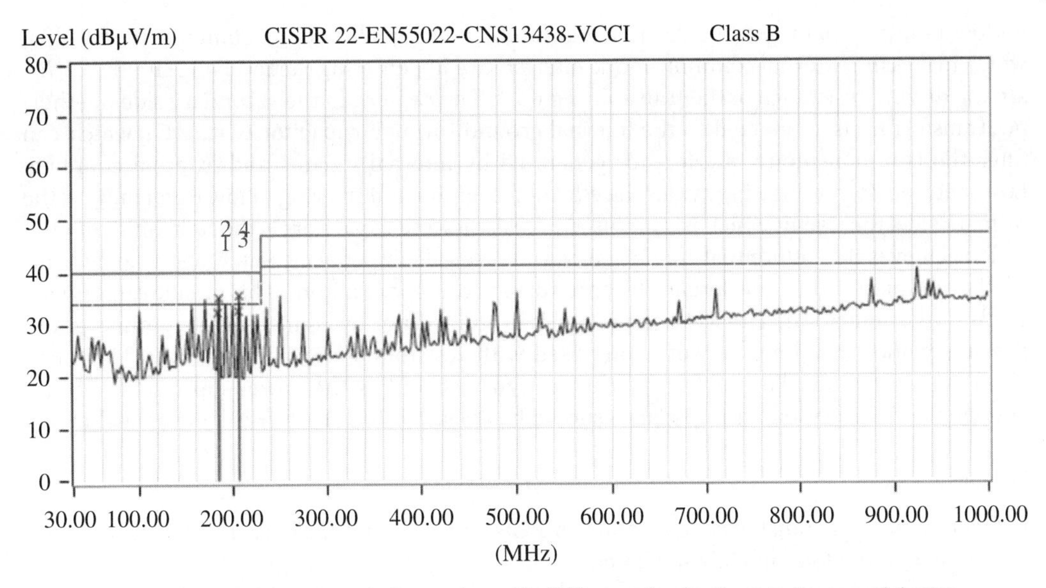

This data is for evaluation purposes only. It cannot be used for EMC approvals unless it contains the approved signature.
If you have any questions regarding the test data, you can write your comments to service@mail.adt.com.tw

	No.	Frequency MHz	Factor dB	Reading dBμV/m	Emission dBμV/m	Limit dBμV/m	Margin dB	Tower/Table cm	deg
	1	184.18	14.14	18.16	32.30	40.00	−7.70	100	22
	2	185.20	14.02	21.10	35.12	40.00	−4.88	–	–
	3	206.28	13.06	19.56	32.62	40.00	−7.38	99	90
*	4	207.03	13.07	22.52	32.59	40.00	−4.41	–	–

Figure 3.25 Test results after modification.

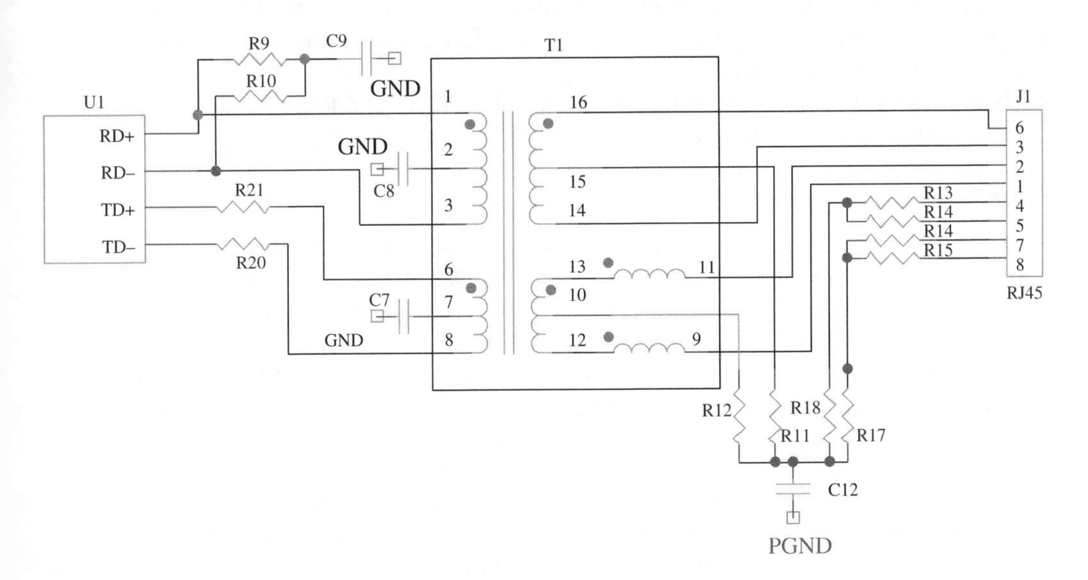

Figure 3.26 Schematics of ethernet interface circuits.

middle point of them and the reference ground, and they also act as common-mode filter, which makes the external common-mode interference not entering the receiving circuit. R_{20} R_{21} are the driving resistance of the transmitting port. The center tap the secondary side of ethernet transformer is connected to the internal ground through capacitor C_7 and C_8, which can filter the common-mode interference generated by internal circuits and external coupling. Ethernet transformer itself provides the effect of isolation and filtering in low frequency. In the primary side, the circuit which is composed of resistors and capacitors is the Bob Smith circuit, which achieves the differential-mode and common-mode impedance matching. It can also filter the common-mode interference by grounding through capacitors. This circuit can provide 10 dB EMI attenuation. The RJ45 connector is grounded not by connecting its ground pin to this impedance matching network composed with resistors and capacitors, in order not to generate interference. For the ferrite bead used for decoupling the power supply of the interface chip and the crystal, its impedance should be at least 100 Ω (100 MHz) or even higher.

PCB layout

- The direction of transformer's location in PCB board should make its primary circuit and secondary circuit completely separated.
- The distance L_1 between transformer and RJ45 connector, and the distance L_2 between the interface chip and transformer should be limited within 1 in.
- The receiving port of the interface chip should face directly to the transformer, to keep the inherent A/D separation of the interface chip. At the same time, the shortest path can easily achieve a balanced PCB routing it can reduce the interference coupled to PCB board, and prevent the transformation from the common-mode current to the differential-mode current, which can affect the signal integrity at the receiving port.
- The differential-mode and common-mode impedance matching resistors and capacitors at the receiving port shall be placed near the interface chip; these two resistors shall be placed symmetrically and the capacitor shall be connected to the central position of their common node. The series resistors at the transmitting port shall be placed near the interface chip.
- The common-mode filtering capacitors of the secondary side of the transformer shall be placed near the transformer. Bob Smith circuit shall be placed near the RJ45 connector.
- In Figure 3.27, the circuit in area A is near the interface chip and the circuit in area B is close to the ethernet transformer.
- The distance between the signal traces TX+ and TX− (RX+ and RX−) should be maintained within 2 cm.

PCB routing

- The segmentation line of PGND is right below the transformer. The width of segmentation area should be above 100 mil, just as L_5 shown in Figure 3.27. The transmitting/receiving traces should be well separated, just as L_4 in Figure 3.28. Separation effect can be achieved by fill GND plane between them, just as shown in Figure 3.29.
- Except PGND layer, we should hollow out all the plane layers vertically below the circuit at Ethernet transformer primary side, just as the area enclosed by the white square, inside which J1 is located (Figure 3.30 is a PCB layout with well EMC designed ethernet interface circuit). So we suggest that in this area, the design of the pad and via in PGND layer should comply with the following restrictions: The diameter of antirelief pad and the thermal relief pad should be bigger than the diameter of the regular pad by more than 70 mil.
- The critical signal traces which should be handled firstly are TX+ and TX− (RX+ and RX−), just as the nets which are highlighted and marked with the TPI and TPO shown in Figure 3.30.

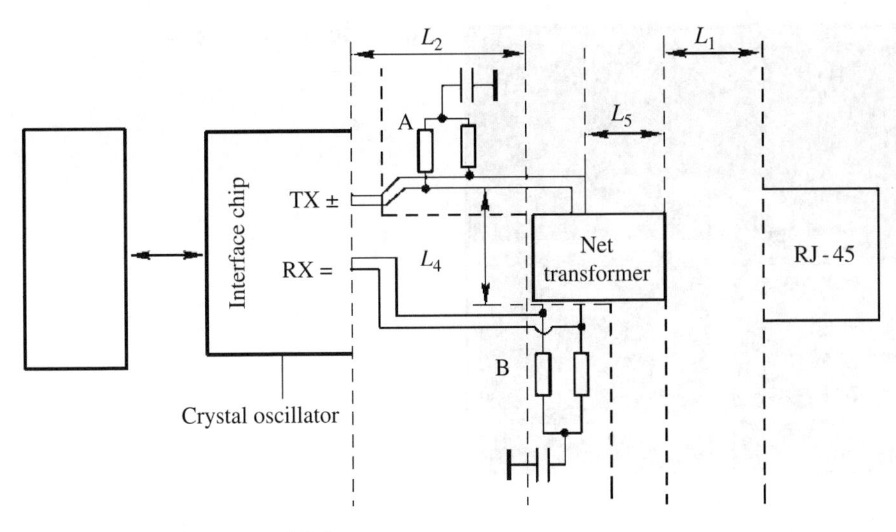

Figure 3.27 Interface circuit PCB layout.

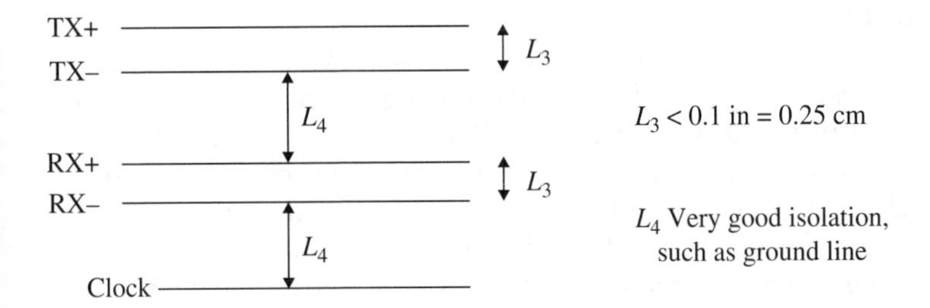

Figure 3.28 Trace routed in PCB.

TX+ and TX– (RX+ and RX–) should be routed in differential pairs. In order to improve the receiving performance and prevent the radiated emission from transmitting side, balanced and symmetric routing is the most important. The interval of the differential pairs shall be not more than 100 mil (just as L_3 in Figure 3.28). Signal traces shall be routed tightly adjacent to the ground plane. It is recommended that the signal traces are interconnected only on top layer without any via. The second layer adjacent the top layer is the ground plane. High-speed signals cannot be placed near them, especially the digital signal. In order to improve the immunity performance, the width of the wiring is recommended to be 20 mil. If the space is enough, we can use ground guard trace along those critical traces for protection. The ground guard trace must be connected to the internal ground plane with via holes at certain intervals.

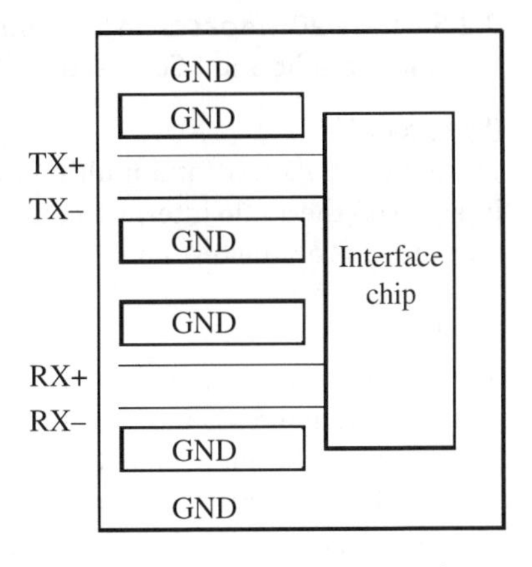

Figure 3.29 Input/output trace with GND isolation.

- The interface chip supply that the digital power supply and the analog power supply should be separated, as shown in Figure 3.30. Each analog power supply pin must be connected with a high-frequency capacitor. The analog power supply is separated from digital power supply

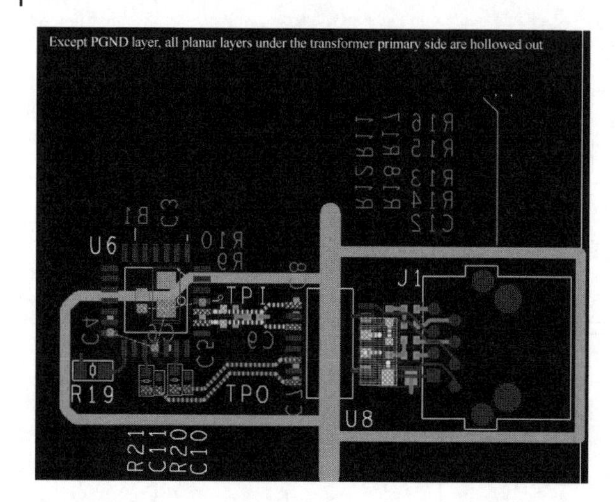

Figure 3.30 Layout of interface.

in the power supply layer, just as the left area enclosed by the rectangle in Figure 3.30. The border width is 50 mil; The digital power supply cannot be extended to be near the signals TX+ and TX− (RX+ and RX−).

- Other high-speed signals cannot pass through the DC bias resistor (R_{19} in Figure 3.30).
- The transmitting and receiving signal traces between ethernet transformer and RJ45 connector are handled the same as that of the secondary traces TX+ and TX− (RX+ and RX−).
- We should widen the traces of Bob Smith circuit. The net of the node between the resistors and capacitors (the highlighted white net in Figure 3.30) is routed with a copper shape in inner layer, like the highlighted white copper shape shown in Figure 3.30.
- It is better that there are no via holes on the TX+ and TX− traces. The RX+ and RX− traces must be routed on the same layer with the components.

3.2.5 Case 20: Impacts on ESD Immunity of the Plastic Shell Connectors and the Metallic Shell Connector

[Symptoms]

During the ESD test for a multimedia product with metallic casing, when 2 kV ESD voltage is injected on the audio interface connector, mosaic and image solidification phenomenon appear on the monitor, the test fails.

[Analyses]

Through observation, we find that the shell of audio interface connector is plastic, and the joint of audio signal lines is more on the outside. So the ESD disturbance can be directly coupled to PCB through the audio signal lines. The operation of the equipment can be disturbed.

From the prior introduction of ESD disturbance, we know that ESD is a kind of high-voltage energy discharging. In the ESD test, the ESD disturbance has the characteristic of discharging the energy to the nearby low-potential point. When we use the audio interface connector with a plastic shell, the audio signal lines are the nearest conductors to the electrostatic discharge injection point. ESD disturbance is discharged to the nearest audio signal lines. Such a high-voltage ESD disturbance goes into the device through the signal lines, and the abnormal operation or the damage of the device is difficult to be avoided, just as Figure 3.31 shows.

If the electrostatic discharging point is not on the signal lines, the estimated situation may be better. In order to ensure the device smoothly passes the electrostatic discharge test, the

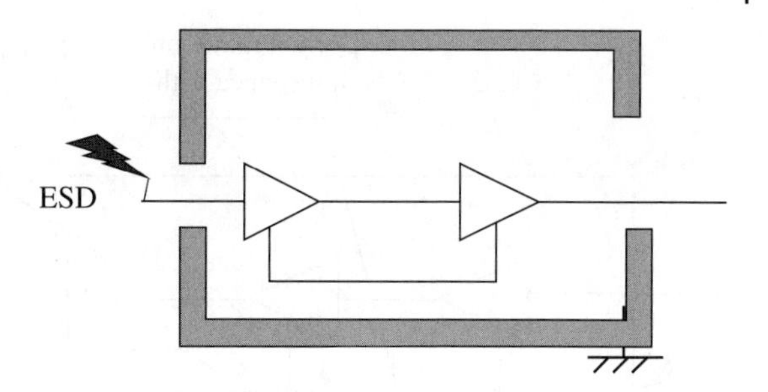

Figure 3.31 Principle of static electricity failure.

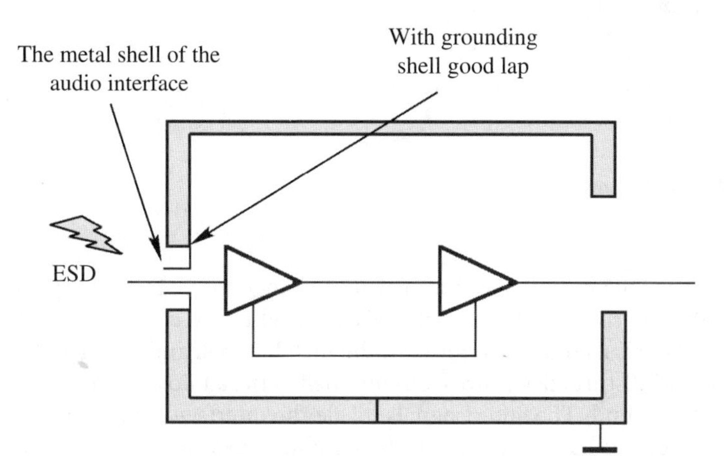

Figure 3.32 Use a metal shell connector.

best way is to make sure the ESD energy is discharged through a good grounding path. Then, inside the device, all the circuits, components, and signals will not be directly influenced from ESD energy. For this equipment, to achieve this goal we must change the energy discharging path at the audio interface connector (see Figure 3.32). If we use audio interface connector with a metallic shell and connect the shell to the ground, the electrostatic discharging path can be well changed.

As a design with good immunity against ESD, it is necessary to add some supplements. Just as Figure 3.33 shows, when the electrostatic discharging point is on the metallic casing of the device, because bad bonding and seams exist on the metallic casing, when the electrostatic discharge current flows through the bad bonding and seams, a voltage drop ΔU appears. This voltage drop will directly impact the grounding circuit. Even the internal circuits without grounding (i.e. without the bold lines shown in Figure 3.33), will also be impacted because of capacitive coupling (i.e. the part shown with red line is replaced by parasitic capacitance). At the same time, if the seam size is close to the wavelength of the frequencies of ESD disturbance, the seam will act as a slot antenna and transmit the electromagnetic energy.

Therefore, the ESD discharge path must be kept with low impedance. Otherwise, the discharge arc may pass through the much lower impedance path composed of the electronic circuits. Because of the skin effect, in high frequency, the impedance is even high. We can increase the surface area to alleviate this problem. What is low impedance in EMC? The practice proves that a complete (no seam, no opening) metallic plane with the ratio of length to width less than 3:1 can well satisfy the ESD discharging.

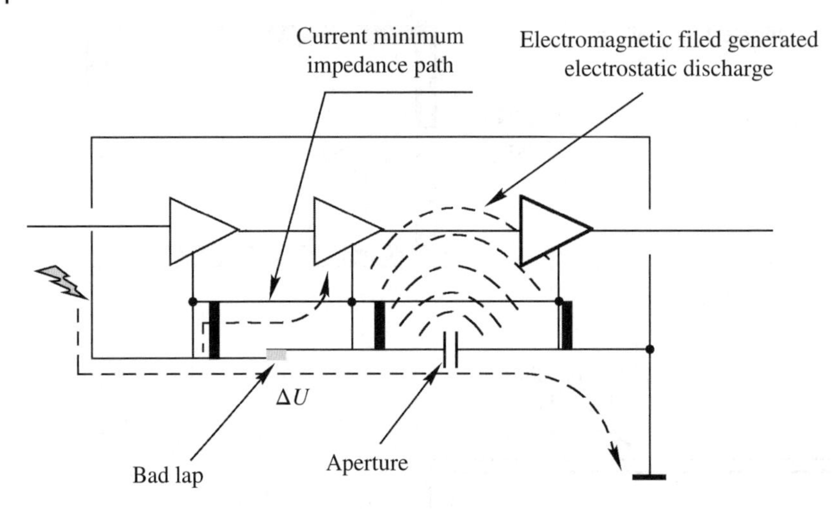

Figure 3.33 The radiation caused by static current.

[Solutions]
We can use the characteristic that ESD is discharged nearby it to change the ESD discharging path. Then we change the audio interface connector with plastic shell to the connector with metallic shell and keep a good electrical continuity between the metallic shell and the metallic casing of the device. Then the ESD disturbance flows from the shell of the audio connector via the casing of the device shell, to ground. The audio signals inside the interface connector are protected. Through the test, this device has the ability of withstanding ±8 kV air discharge and ±6 kV contact discharge. The problem in this case can be solved:

[Inspirations]

1) For the signal interface connector in the device, the design of the structure and the selection of the connector should take account avoiding ESD disturbance directly coupled to the signal lines.
2) For the selection of the connection, if the plastic shell cannot reach the required distance to avoid discharge happening, we must select the connector with a metallic shell. In the structure design, the shell of the connector must be well grounded.

3.2.6 Case 21: The Selection of the Plastic Shell Connector and the ESD Immunity

[Symptoms]
An industrial product (Figure 3.34) needs to pass an 8 kV ESD air discharge test. The shell of connector of the product is plastic. Air discharge is needed at the position around the connector. In the test, we find that ESD air breakdown appears on the plastic shell of the connector. The test fails

[Analyses]
During the air discharge test, the round tip of the discharge electrode should approach and touch the equipment as fast as possible (without causing mechanical damage). Every time, after discharging, the discharge electrode of ESD generator should be removed from the EUT. Then trigger the ESD generator, and perform the discharges again. This procedure should be repeated until the discharge test is completed.

Figure 3.34 Product front view.

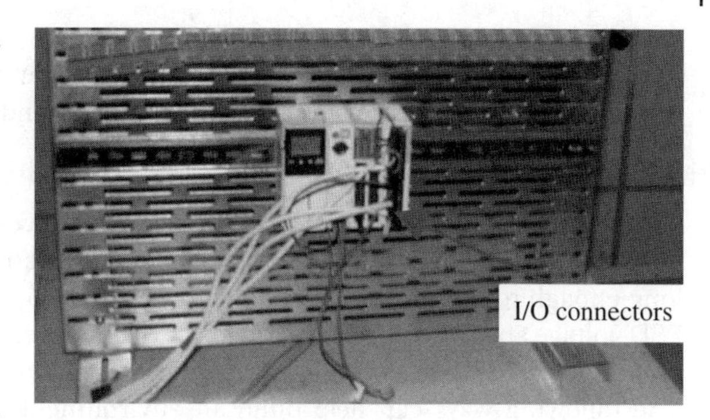
I/O connectors

Figure 3.35 Connector views.

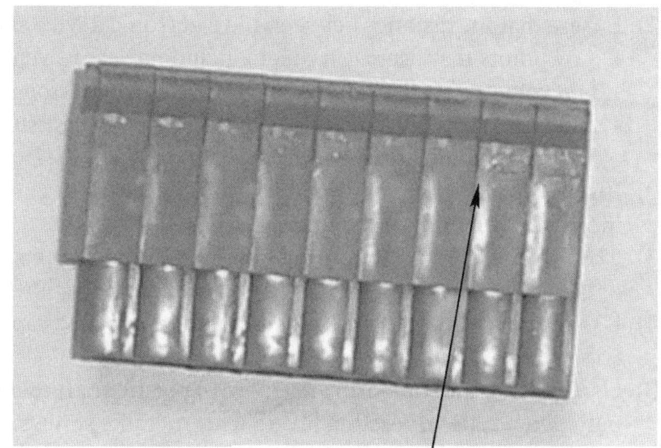
Gap in the connector, and the interior
is a metal conductor

For the air discharge test, itis essential that when a charged object approaches a conductor whose potential is not equal to the charged object or a grounded conductor, the charge on the object will be discharged through another conductor or a grounded conductor. This is the ESD air discharge phenomena. When the discharge phenomenon occurs, because the ESD waveform has the characteristic of high amplitude and short rising time, it will generate strong and wide-spectrum electromagnetic field and cause electromagnetic disturbance on the equipment, circuit or components. The rising time depends on the inductance in the discharge path. The waveform of the discharge current shown in Figure 1.8 is the discharge waveform while ESD is discharged on human body. According to Fourier transformation, the rising time of the pulse is 1 ns, the bandwidth is 300 MHz.

For the product in this test, when the round tip of the discharge electrode approaches and contacts the test point (the plastic surface of the connector), and if a low-potential conductor or grounded conductor exists away from the injection point with less than the air breakdown distance (such as, when the voltage is 8 kV, the air breakdown distance is 6 mm), the discharge phenomenon occurs. After investigating the connector used in the test, we found that the distance between the plastic shell of the connector and the inner conductor is less than 3 mm, just as the Figure 3.35 shows.

When the discharge phenomenon occurs, the disturbance will also influence on the internal circuit. For some products, this discharge phenomenon may not make the test failed, but we have to say that this phenomenon is a great threat.

[Solutions]

According to the analysis, we should use new connector and make sure that the distance between the surface of the new selected connector and the inner conductor is above 6 mm.

[Inspirations]

For the device with plastic casing or the selection of connector, we should also pay attention to that the clearance between the surface of the plastic structure and the inner conductor shall be long enough to prevent air breakdown by ESD voltage. Any clearance can be broken down by ESD voltage and the discharging arc may exist between ESD injection point and the conductors or circuits inside the electronic equipment. We can use distance to protect the internal circuit. The following ways can help build an environment against ESD discharging, in which the breakdown voltage is larger than the ESD test voltage:

1) Ensure that the distance between the electronic device and the following parts is long enough.
   - Any points that users can touch, including seams, vents, and mounting holes. Under a certain voltage, the discharge arc can travel farther through the device surface than through air.
   - All the ungrounded metal that users can access, such as fasteners, switches, operating levers, and indicators.
2) Install the electronic equipment into the groove or the notch of the casing of the device to increase the length of the seam.
3) Cover the seams and mounting holes with polyester inside the device. This extends the edge of the seam and holes, and increases the length of the discharging path.
4) Use the metallic cap or shielded plastic dustproof cap to cover the unused or rarely used connectors.
5) Use switches and operating levers with plastic shaft to increase the length of the discharging path. Do not use operating levers with metallic screws.
6) Put LED and other indicators into the inner hole of the equipment, and cover them with straps or caps. Then we can extend the edge of the hole or use a light guide to increase the length of the discharging path.
7) Extend the edge of the membrane covering on the touching keyboard to be far from the metallic border with enough distance. For example, for 8 kV air to avoid ESD, it needs a distance of more than 6 mm.
8) The heatsink must be placed close to the seams of the casing. The edge and the corner of the metallic parts of vents or mounting holes must be circular, in order to avoid tip discharge.
9) In the plastic casing, the metallic fixing parts are close to the electronic circuits or are not grounded cannot be exposed outside of the casing.
   - For the touching keyboard with membrane, we must ensure that the wiring is compact and we must extend the membrane to increase the length of the discharging path.
   - Coat the electronic circuits of the membrane-covered touching keyboard with adhesives or sealing compound.
   - At the seams of the casing, we should use high-pressure resistant silicone resin or washer to implement sealing, protecting against ESD, waterproofing, and dust-proofing.

3.2.7 Case 22: When the Shield Layer of the Shielded Cable Is Not Grounded

[Symptoms]

When performing radiated emission test for a product, we find that the product cannot meet the requirements. The specific phenomenon is that when the frequency is above 700 MHz, a large number of high-order harmonic emissions of 50 MHz appear. The spectral plot is shown in Figure 3.36.

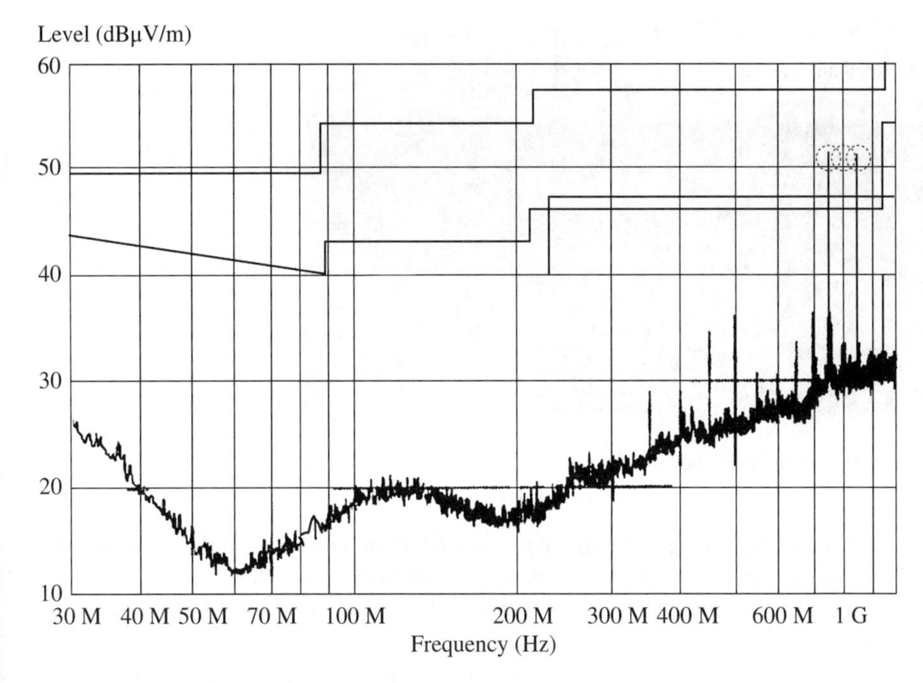

Figure 3.36 Radiate emission test result.

[Analyses]

Since the frequencies at which the emission is over the limit concentrates in the high-frequency range, the first suspect is that the structure shield is not good and emission leakage exists. From the front panel of the casing, we can see several seams are wide. We open the casing and seal the seam with conductive copper foil. Then we test it again. The result is the same as the prior result. The high-order harmonic emission is still over the limit. We search for remaining hidden seams that could cause the emission leakage, but after careful examination, we did not find any.

Unplug all the cables, leaving only the power supply cable and the grounding wire. We test it again, and the test result is quite good. All the high frequencies at which there were high emissions previously now disappear. This shows that the shield of the structure has no problem. The emission at 50 MHz is radiated by the cable.

The device has four interfaces. Two 485 interfaces and the 485 interface uses DB connector. One alarm output interface and the alarm output interface also uses DB connector. One ethernet port and this port uses an RJ45 connector with a metallic shell, which is bonded with the casing of the device through metallic reed. According to the product configuration, all the interface cables are shielded cables.

Plug the cable and test it one by one. Once the ethernet cable is plugged, the radiated emission will exceed the standard limit. Other cables have almost no effect on the radiated emission (because the decibel value is a relative value, and the plugging sequence has been tried in many ways). To verify if the ethernet cable is shielded cable, open its insulation layer, it shows that it is shielded cable which is copper meshed and aluminum-foil wrapped. The meshed density of the copper meshed layer is also quite high, which shall be higher than 80%. Practice has proven that the shielded cable with more than 80% meshed density has good shielding effectiveness. Returning to the grounding of the shield of shielded cable, if the shield of shielded cable is not well grounded, the cable shield can easily become the antenna driven by common-mode current or common-mode voltage. With speculation, we open the casing and find that there

Figure 3.37 Ethernet front-end ports layout.

already exist edgefold on the casing right above the ethernet connector. It can be seen that the reed is well bonded with the structure of the casing (as shown in Figure 3.37). We test the DC resistance between the casing and ethernet connector with multimeter. The DC resistance is in the milliohm level. So, we can say that there is no problem for grounding.

In order to find out the shielding effectiveness of this shielded ethernet cable, we plug an unshielded cable into the device and test the device. Then we found that above 700 MHz the radiated emission is reduced slightly. Then we plug the shielded cable and test it; the result is kept the same as before. According to experience, we think the problem lies in the grounding of the shield of the shielded cable. One possibility is that the shield is not bonded to the RJ45 connector in 360°. Plugging the cable and testing the resistance between the casing and the cable shield with multi-meter, we found that this path is unexpectedly open. So the problem is basically clear. This shielded cable actually is in name only. When we replaced it with other ethernet cable well connected to the casing, the test result is good. The high frequencies with high radiated emission disappear. It is proved that if the shield of the shielded ethernet cable is not connected with the casing, the expected shielding effectiveness will be lost and the shield will behave as antenna and generate even high radiated emission.

For the shielded cable, the signal wires are tightly coupled with its shield, and the distributed capacitance is large. The harmonics for which frequencies are higher than 700 MHz of the signal are coupled to the shield through distribution capacitance (driving voltage source). The shield transmits the signals into the power supply input through the distributed capacitance between the earth and the cable shield (common-mode current is generated). This is the common-mode radiation driven by voltage. The shield is equivalent to a monopole antenna at this time. From the test results, the ungrounded shield also behaves as an antenna whose bandwidth of frequency response is narrow and gain is high.

[Solutions]

Bonding the shield of the ethernet shielded cable with the RJ45 connector in 360°.

[Inspirations]

1) This case reminds us: We must carefully check and control the component bought from outside before the test.
2) For the shielded cable, its grounding is the most important. Do not think that the shielded cable is always better than the unshielded. A shielded cable with good shield but poor grounding might be worse than an unshielded cable.

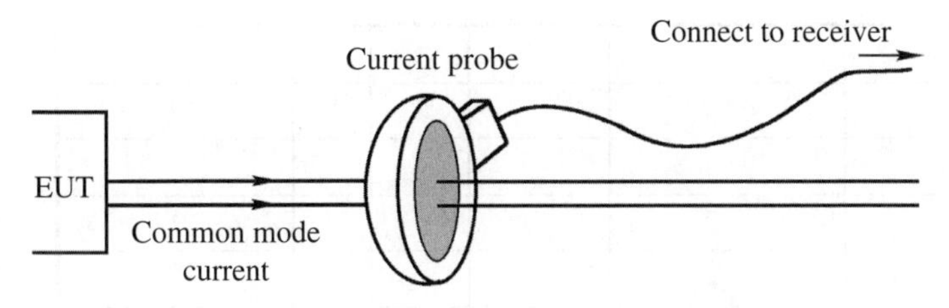

Figure 3.38 Using current probe to predict cable radiation.

3) The radiated emission from cable is mainly caused by common-mode current. Before the official radiated emission test, if we can test the common-mode current flows through the cable, we can predict whether the equipment can pass the radiated emission test. This will help enterprises save lots of expenses that are consumed on the corrections if the test in chamber cannot be passed. Table 1.1 gives the method evaluating whether the radiated emission from cable can meet the Class A limit specified in the CISPR 25 by testing the common-mode current flowing on the cable. Table 1.2 gives the method evaluating whether the cable radiated emission from cable can meet the EN55022 limit by testing the common-mode current flowing on the cable.

The common-mode current shown in Tables 1.1 and 1.2 (Chapter 1) can be tested by current probe and spectrum analyzer.

The test configuration is shown in Figure 3.38. The current probe LEP613 is produced by Shanghai Ling Electronics Co. (the price is about RMB 10000). In the test, the resolution bandwidth of the spectrum analyzer is recommended to be set as 120kHz. The status of spectrum analyzer is set as "MAX HOLD." During the test, we should move the current probe slowly along the cable to find out the frequencies with maximum current.

3.2.8 Case 23: The Radiated Emission Problem Brings Out Two EMC Design Problems of a Digital Camera

[Symptoms]

One kind of digital camera has a USB port. Its casing is plastic material and the internal control circuit PCB is double layer board. When performing radiated emission test, the digital camera is connected to the computer USB connector, and communicates data with computer to simulate the actual operating situation. The test results at a 3m (the distance from antenna to EUT) semi-anechoic chamber are shown in Figures 3.39 and 3.40. Figure 3.39 shows the tested spectral plot when the polarization of the receiving antenna for the radiated emission test is in horizontal. Figure 3.40 is the tested spectral plot when the polarization of the receiving antenna is in vertical.

It can be seen from the Figures 3.39 and 3.40, for this kind of camera, when the polarization of the radiated emission receiving antenna is in horizontal, the emission at one frequency (148.34MHz) exceeds the Class B limit specified in EN55022 standard, and the emission at another frequency (194.9MHz) is below the limit with only 0.38dB margin. The emissions at a few frequencies are below the limit with small margin when the polarization of the radiated emission receiving antenna is in vertical.

[Analyses]

USB port can provide duplex, real-time data transmission, with the advantages of plug and play, hot plug, and low price. Now it has become the first choice for computer and electronic product such as digital camera to connect to their peripherals. Popular USB 2.0 has

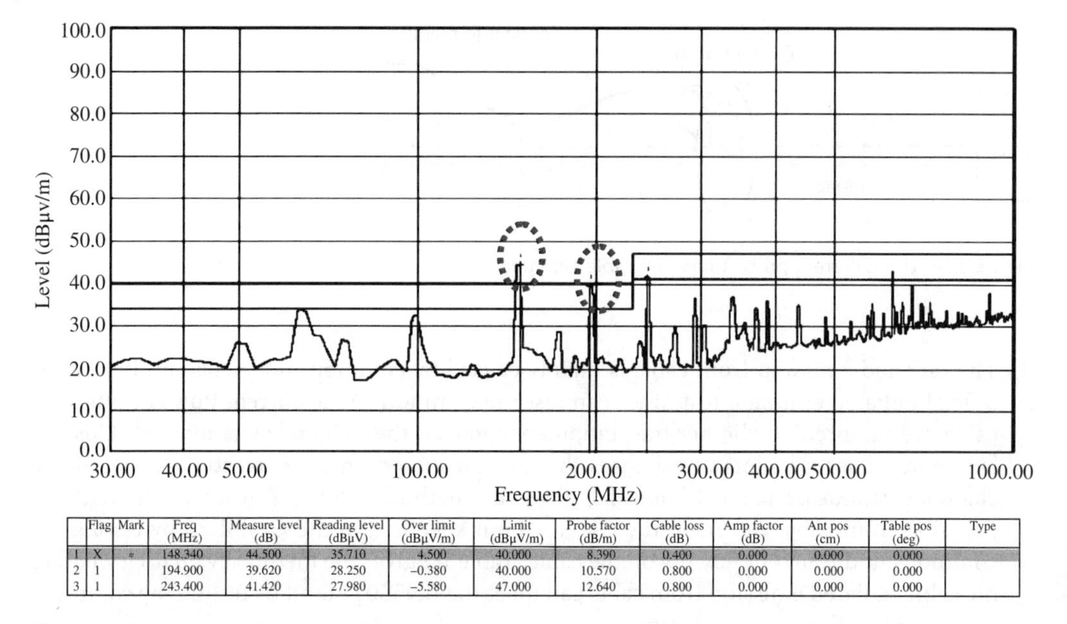

	Flag	Mark	Freq (MHz)	Measure level (dB)	Reading level (dBµV)	Over limit (dBµV/m)	Limit (dBµV/m)	Probe factor (dB/m)	Cable loss (dB)	Amp factor (dB)	Ant pos (cm)	Table pos (deg)	Type
1	X	*	148.340	44.500	35.710	4.500	40.000	8.390	0.400	0.000	0.000	0.000	
2	1		194.900	39.620	28.250	–0.380	40.000	10.570	0.800	0.000	0.000	0.000	
3	1		243.400	41.420	27.980	–5.580	47.000	12.640	0.800	0.000	0.000	0.000	

Figure 3.39 Test result of radiate emission in horizontal.

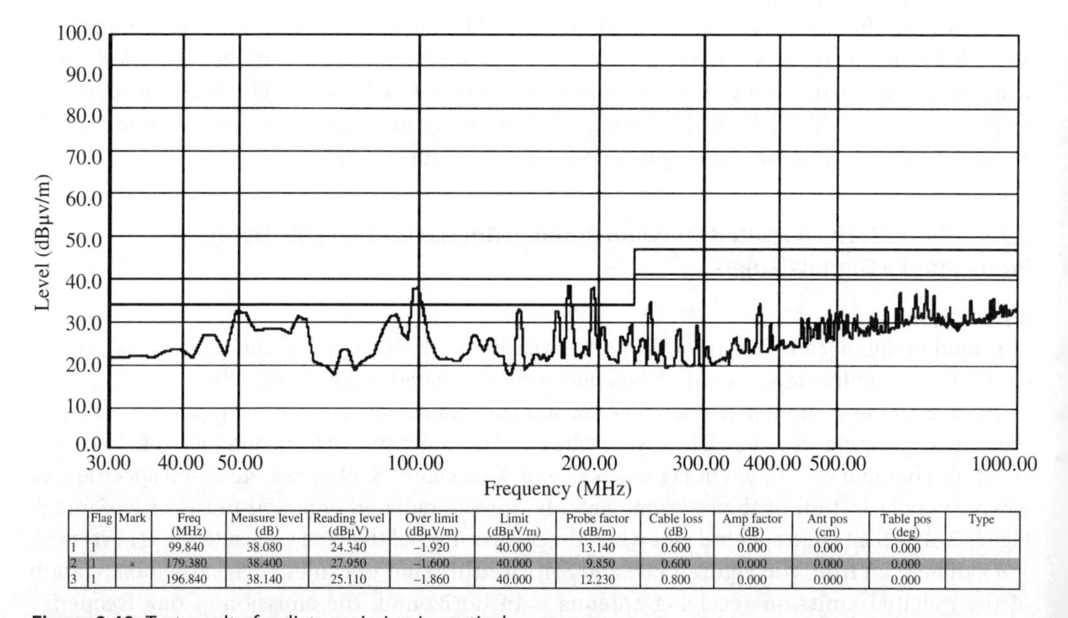

	Flag	Mark	Freq (MHz)	Measure level (dB)	Reading level (dBµV)	Over limit (dBµV/m)	Limit (dBµV/m)	Probe factor (dB/m)	Cable loss (dB)	Amp factor (dB)	Ant pos (cm)	Table pos (deg)	Type
1	1		99.840	38.080	24.340	–1.920	40.000	13.140	0.600	0.000	0.000	0.000	
2	1	*	179.380	38.400	27.950	–1.600	40.000	9.850	0.600	0.000	0.000	0.000	
3	1		196.840	38.140	25.110	–1.860	40.000	12.230	0.800	0.000	0.000	0.000	

Figure 3.40 Test result of radiate emission in vertical.

up to 480 Mb s^{-1} transmission speed, and is fully compatible with full speed USB 1.1 whose transmission speed is 12 Mb s^{-1} and low-speed USB 1.0 whose transmission speed is 1.5 Mb s^{-1}. This makes digital image editor, scanner, video conference cameras, and other consumer products can do data transmission with computer with high speed and high

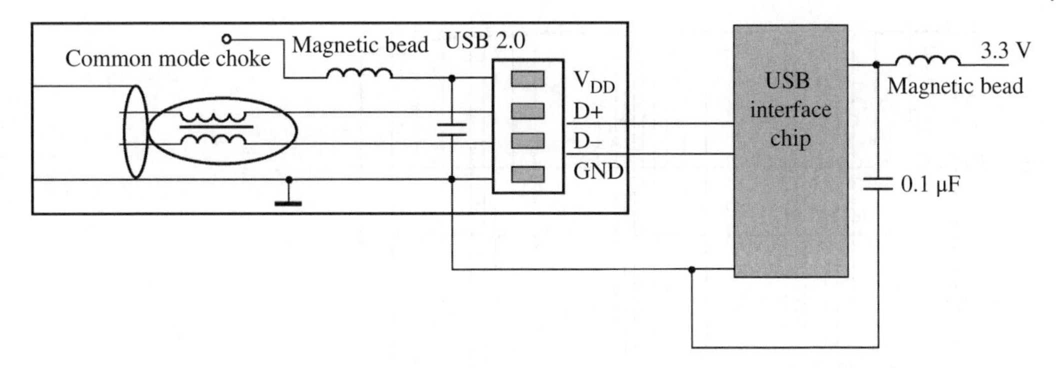

Figure 3.41 EMI noise suppression measured at USB interface.

performance. In addition, it is worth mentioning that the strengthened version of USB 2.0 – USB OTG can realize data transmission between USB equipment when there is no USB host. For example, a digital camera can be directly connected to the printer and prints photos, personal digital assistants (PDAs) can proceed data transmission and exchange file with other brands of PDA.

The transmission speed of USB is very high, its periodic signal and its harmonics can produce radiated emission through USB cable. In addition, when the control chip and the interface chip transmit signals, the voltage noise will exist between chip's ground and its power supply with the transmission of the signals (such as what is described in Case 47). Therefore, the following four methods are generally used to suppress the EMI noise of USB interface, as shown in Figure 3.41:

1) Use a shielded cable as the USB cable.
2) Put ferrite core on USB cable.
3) Add a common-mode choke in series with the differential pair. The common-mode choke includes two windings which are circled on the magnetic core in the same direction, common-mode choke can provide high impedance due to the superposition of the magnetic fluxes induced by these two windings when common-mode current passes through it, and low impedance due to the cancellation of the magnetic fluxes induced by these two windings when difference-mode current passes through it. As a type of common-mode choke, SDCW2012-2-900 has only about 4.6 Ω differential-mode impedance at 100 MHz. As shown in Figure 3.42. From the attenuation characteristics as shown in Figure 3.42, we can also see that the common-mode choke in the USB interface circuit won't impact the differential signals, while selectively and mainly attenuates the common-mode current.
4) Good decoupling for the power supply of USB interface circuit and control circuit's is also important to reduce EMI noise of USB interface circuit.

Checking the digital camera in this case, it is found first that, the shield of the shielded cable at the side of the digital camera is connected to the cable's metallic connector with pigtail connection, namely the shield near its metallic connector is twisted to about 3 cm long wire and then welded on the cable's metallic connector. This is an obvious connection defect, such as what is described in Case 17. However, the existence of the pigtail is equivalent to adding an inductance of tens of henries in series with the shield, which can cause a common-mode

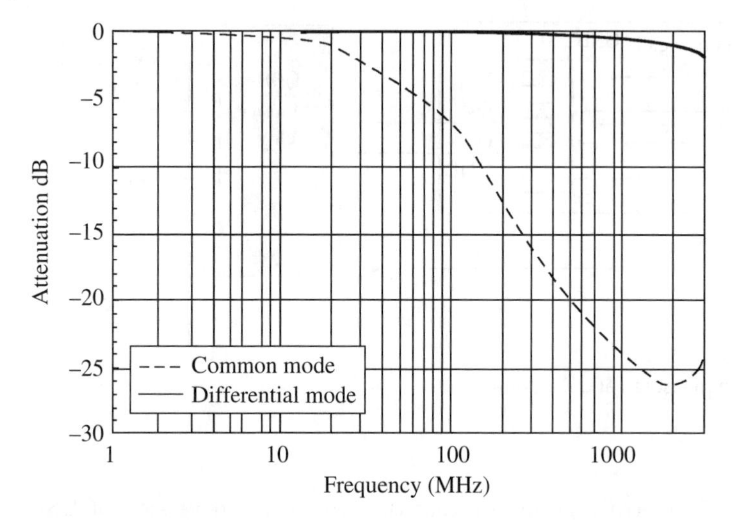

Figure 3.42 Common-mode inductance frequency attenuation characteristic curve.

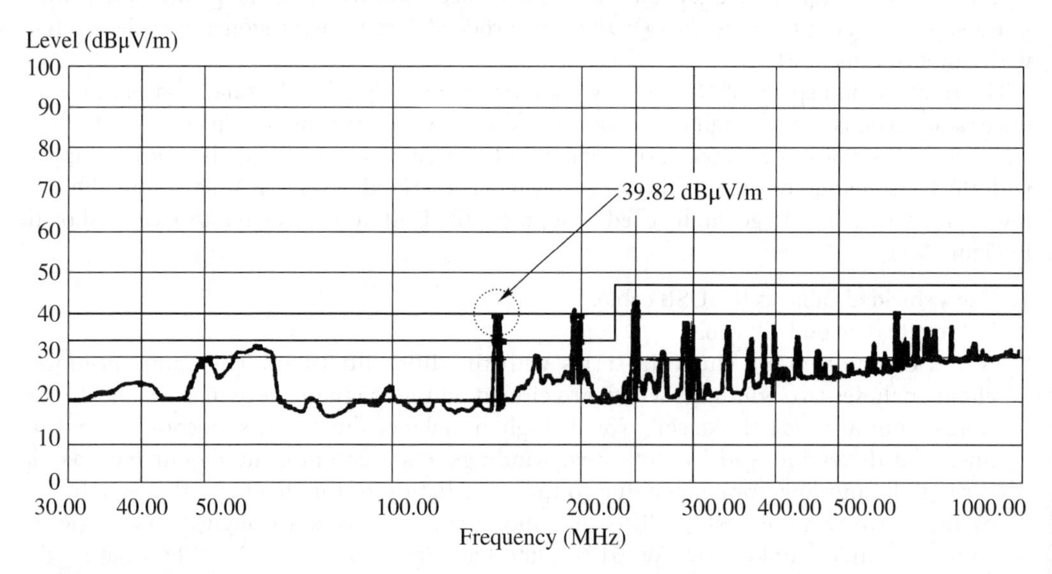

Figure 3.43 Test spectrum diagram of the receiving antenna of horizontal polarization in emission test after changing.

voltage on the cable shield due to the common-mode current flowing through the shield. With increased frequency, the equivalent transfer impedance of pigtail connection will also increases quickly, which will not only make the shielding effectiveness of the shielded cable completely lost, but also may generate additional disturbance.

Change the connection way between the shield and the metallic connector of the shielded cable, namely that the cable shield is bonded to its metallic shell connector in 360°. After modifications, the tested spectral plots of radiated emissions when the polarization of receiving antenna is in horizontal and in vertical are respectively shown in Figures 3.43 and 3.44, respectively. At 148.34 MHz, the radiated emission fell by nearly 4.5 dB, but the margin from the limit was small.

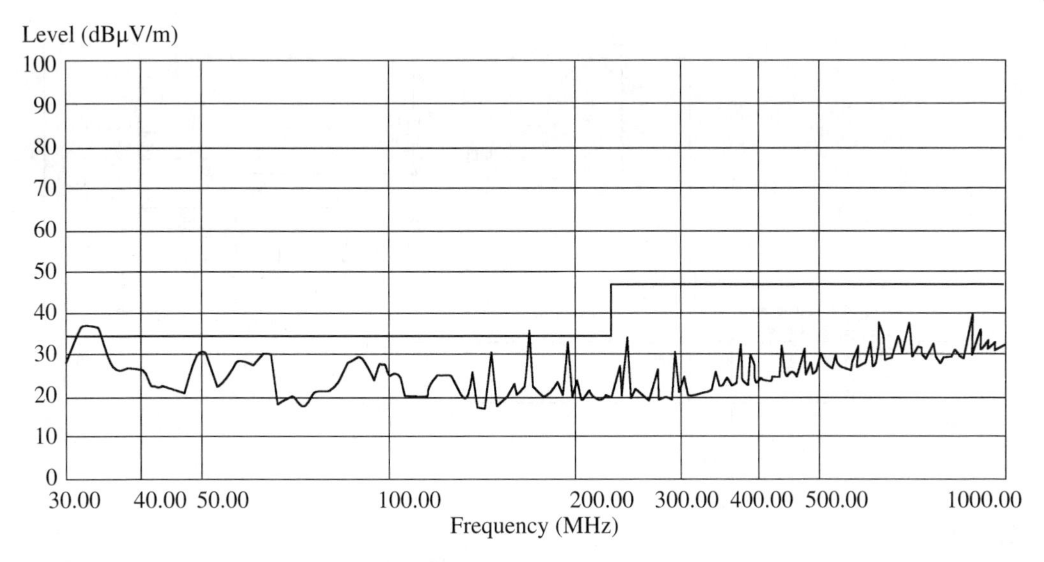

Figure 3.44 Test spectrum diagram of the receiving antenna of vertical polarization in emission test after changing.

Further checking the schematic of the printed circuit boards in the digital camera, we found the ferrite bead and capacitors are used for decoupling the power supply of the control chip, and here, capacitor C28 is 0.1 μF, as shown in Figure 3.45.

Actually 0.1 μF SMD capacitor is not good for decoupling in the frequency more than 100 MHz, the reason mainly lies in two aspects: one is, the capacitor has parasitic inductance on its own; Another is that there is parasitic inductance of the decoupling loop. For an ideal power supply, its impedance is zero. At any point of the power plane, the potential is constant (equal to the system's supply voltage). However, the actual situation is not in this case, there is a lot of noise, which possibly affects the normal operation of the system. The decoupling capacitor is used to reduce the impedance of power supply and guarantee that the power supply near the components fluctuates within a small scope. About the decoupling capacitor and why the decoupling is needed, the relative descriptions already exist in Case 47 and in Section 5.1. The self-resonant frequency of 0.1 μF ceramic capacitor is 10 MHz commonly, that is to say, 0.1 μF ceramic capacitor can keep the impedance of the power supply at the lower level only in the frequency range near 10 MHz, which is far from the frequency at which the radiated emission of the digital camera is over the limit in this case. Figure 3.46 shows the reason why there is high emission at 148.34 MHz when the interface chip uses 0.1 μF decoupling capacitors, the arrow line in Figure 3.46 shows the noises, such as the noise at 148.34 MHz, which is not well decoupled by the 0.1 μF capacitor are transferred to the power supply, and due to the higher impedance of the power supply at this frequency, there is a high-voltage drop between power supply and ground. In this way, it is equivalent to forming a 148.34 MHz voltage source between power supply and ground, and because the digital camera is a floating system, the cable shield connected to the ground becomes a radiating antenna.

After checking the impedance-frequency characteristics of capacitor, we learn that, for the SMD capacitor with around 1000 pF capacitance, if its lead inductance is guaranteed to be the minimum (if lead inductance is high, it will cause the capacitor out of it function), and because the 1000 pF capacitor's self-resonant frequency is about 150 MHz, it can suppress high-frequency noise well, as shown in Figure 3.47, it is equivalent to that the voltage noise source

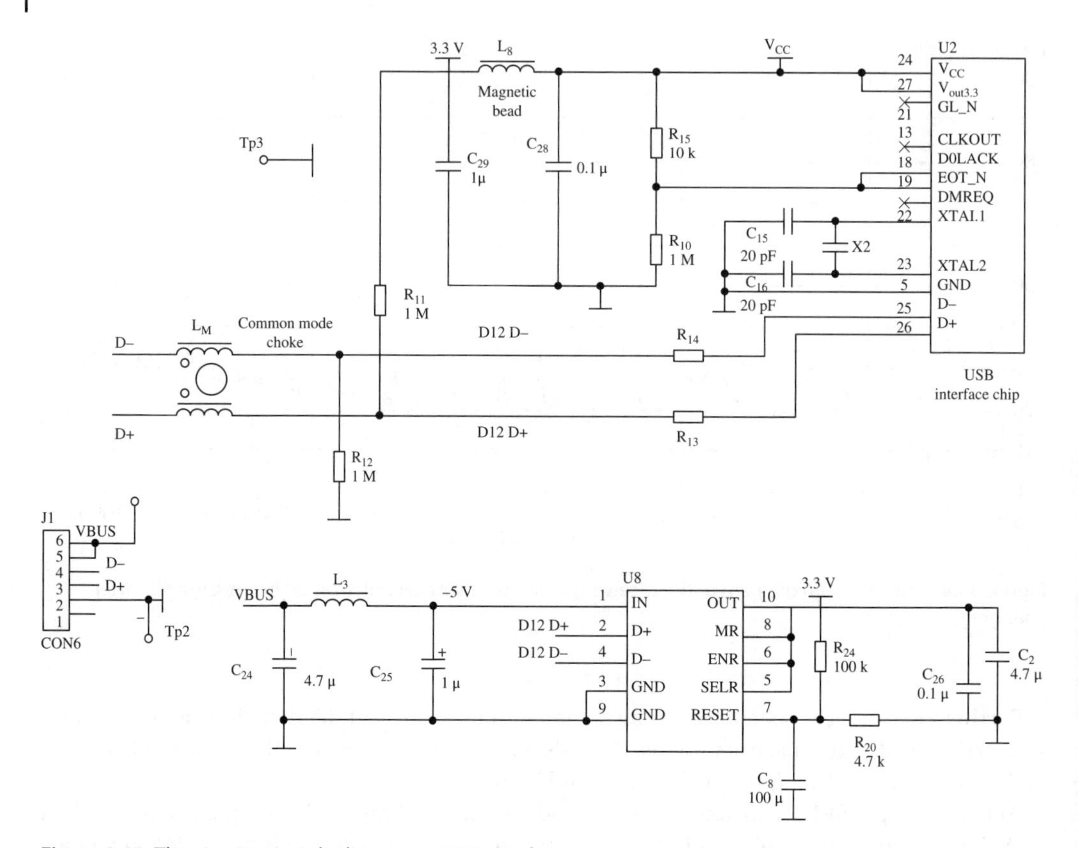

Figure 3.45 The circuit principle diagram at USB interface.

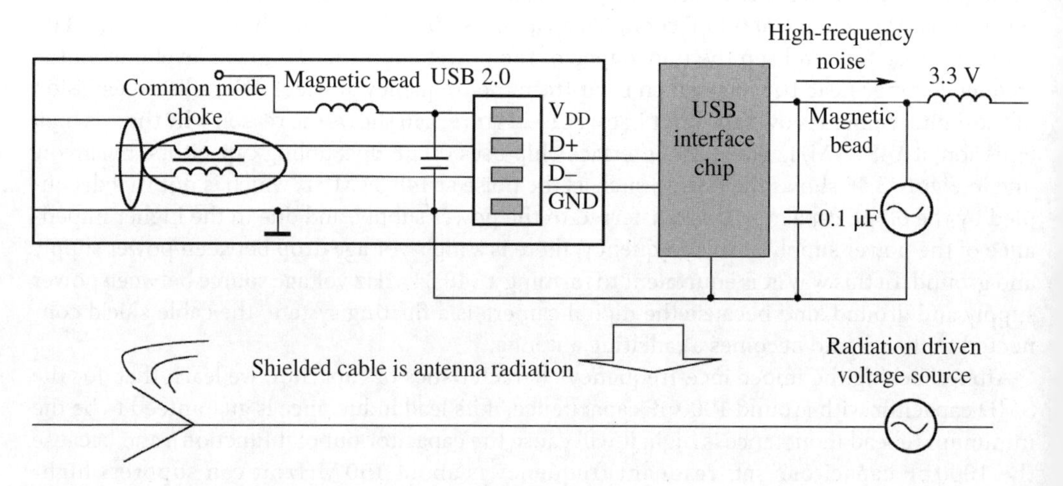

Figure 3.46 Decoupling of the formation of radiation principle.

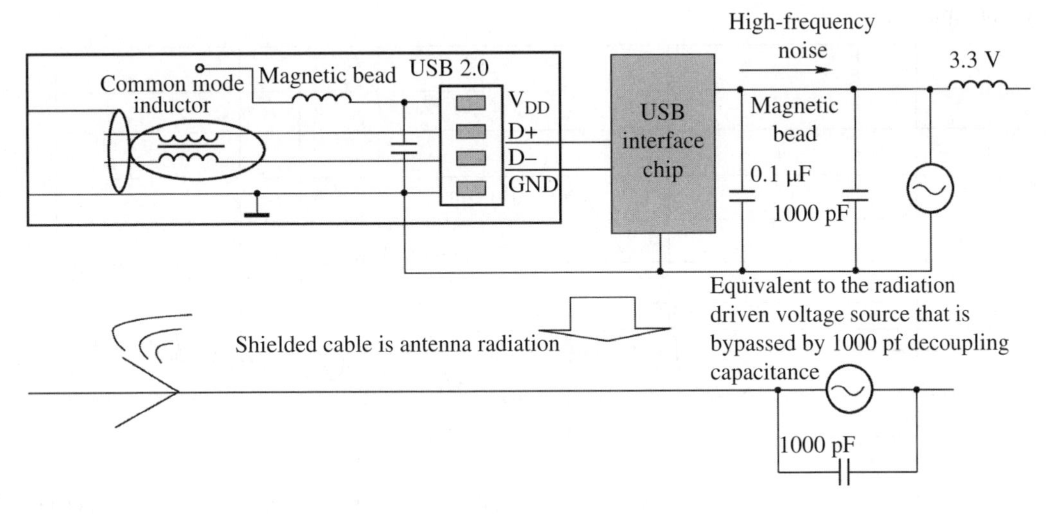

Figure 3.47 1000 pF capacitor in parallel mechanism action principle.

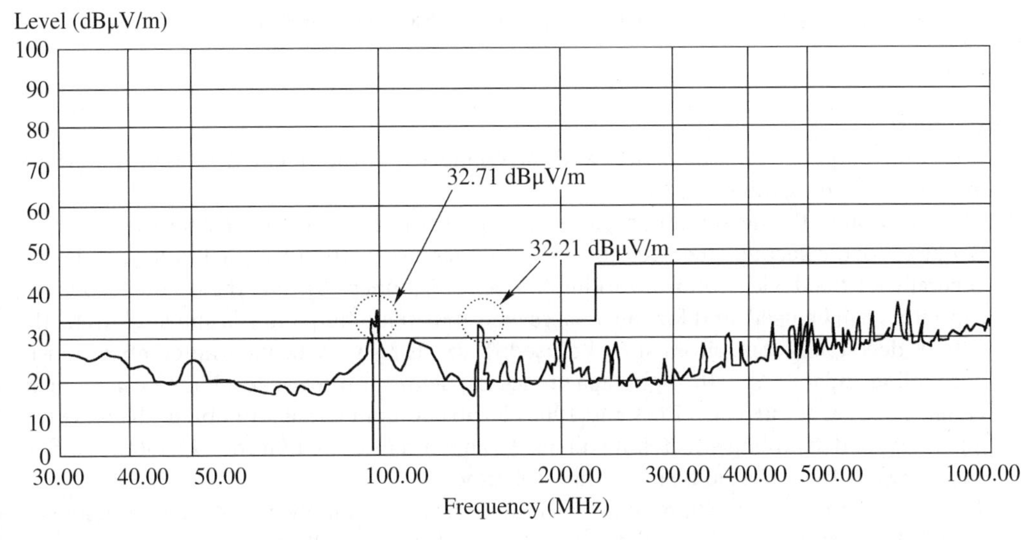

Figure 3.48 The test spectrum of the antenna after paralleling 1000 pF when the antenna is horizontally polarized.

driving the radiation is bypassed, finally the amplitude of the voltage noise is reduced, so the radiated emission is reduced naturally.

According to the principle analysis, try to use 1000 pF in parallel with the power pins of the USB interface chip, namely in parallel with 0.1 μF decoupling capacitors, and test it again, as shown in Figures 3.48 and 3.49, the tests are passed, and it validates correctness of the analysis.

[Solutions]

1) Change the connecting way between the shield and the metallic connector of the shielded cable, remove the original pigtail, and achieve bonding in 360°.
2) Add 1000 pF decoupling capacitors for power pins of the interface chip, and place it close to the power supply pins on the PCB layout.

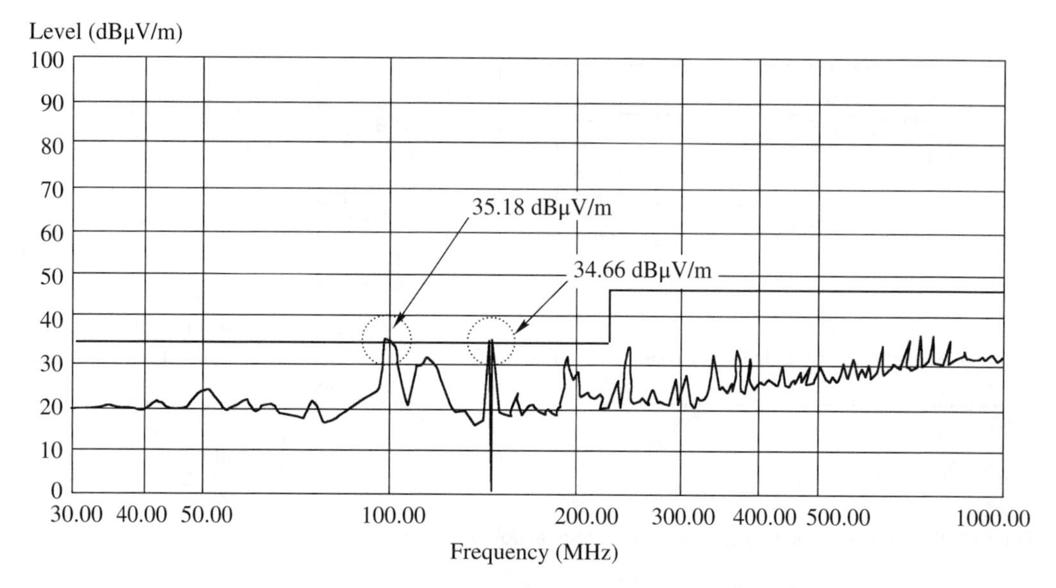

Figure 3.49 The test spectrum of the antenna after paralleling 1000 pF when the antenna is vertically polarized.

[Inspirations]

1) The connection between the shield and the connector of the shielded cable is important; make sure bonding them in 360°.
2) The selection of the decoupling capacitors on power pins needs to consider the operating frequency of the decoupled device and its harmonics, not use 0.1 μF capacitor for any device, generally for the devices whose working frequency is under 20 MHz, the 0.1 μF decoupling capacitors can be used, and for the devices whose working frequency is above 20 MHz, the 0.01 μF decoupling capacitors shall be used. We can also try to use the combination of decoupling capacitors with high-capacitance capacitor and low-capacitance capacitor in parallel, such as 0.1 μF capacitors and 1000 pF capacitors in parallel to obtain the wide frequency band decoupling effect, but still need to pay attention to that the capacitance difference between them should be more than 100 times.
3) The decoupling on power supply can help reduce the power supply impedance, and decrease the noise of power supply and the noise of ground. As a result, it also helps suppress the radiated emission, especially for the decoupling on the power supply of the interface circuit, because the cable near the interface circuit is the radiating antenna.
4) For the floating equipment, the decoupling of the power supply, and the integrity of power plane and ground plane are more important for EMC.

3.2.9 Case 24: Why PCB Interconnecting Ribbon Is So Important for EMC

[Symptoms]

An industrial product is a controller, composed of main module and extension module, the main module and the extension module are equipped with I/O cables. The I/O cable is shielded cable; Figure 3.50 shows the EMC test configuration of the product.

The radiated emission test results of the product are shown in Figures 3.51 and 3.52. Figure 3.51 is the radiated emission spectral plot when the antenna polarization is in horizontal. Figure 3.52 is radiated emission spectral plot when the antenna polarization is in vertical.

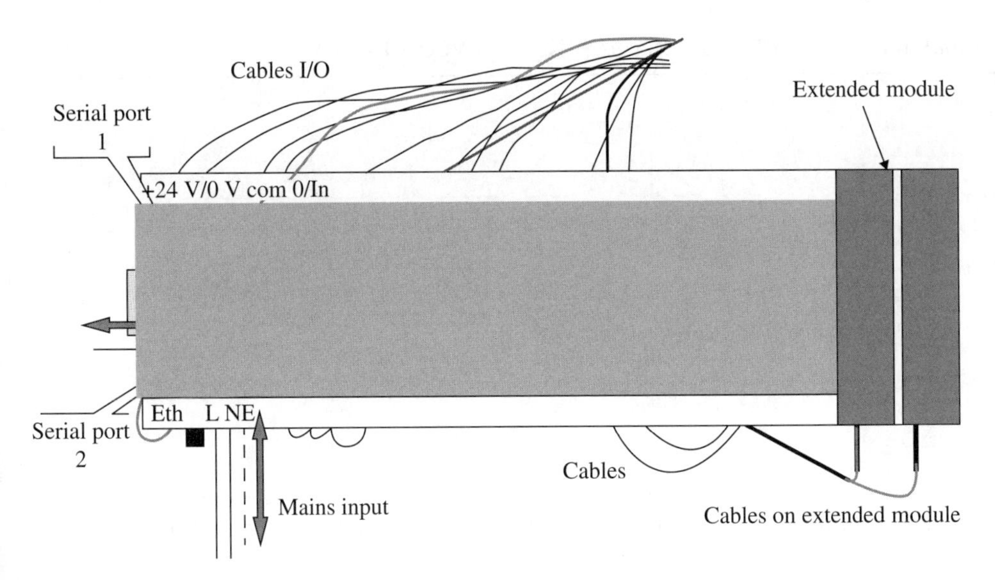

Figure 3.50 Product EMC test configuration diagram.

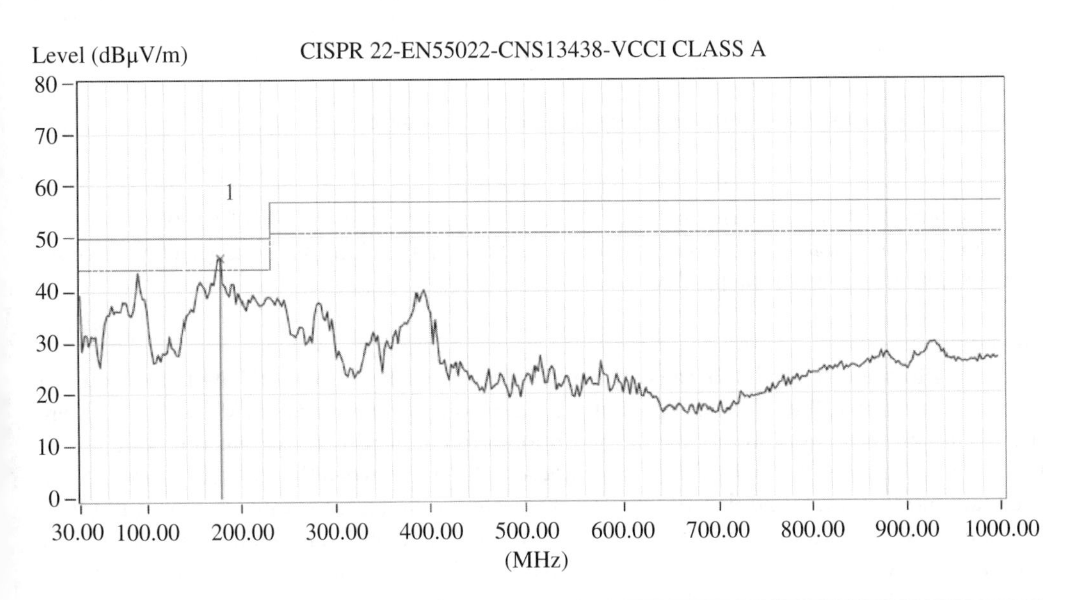

	No.	Frequency MHz	Factor dB	Reading dBµV/m	Emission dBµV/m	Limit dBµV/m	Margin dB	Tower/Table cm	deg
*	1	177.56	–23.57	69.61	46.04	50.00	–3.96	–	–

Figure 3.51 The radiated emission spectral plot when the antenna polarization is in horizontal.

From the spectral plots shown in Figures 3.51 and Figure 3.52, it can be seen that the product cannot pass the Class A radiated emission limit specified in the standard EN55022, the frequencies with excessive emissions are 177.56 and 30 MHz, respectively. During the test, it is also found that if the extension module is removed, the main module can pass the Class A limit specified in the standard EN55022.

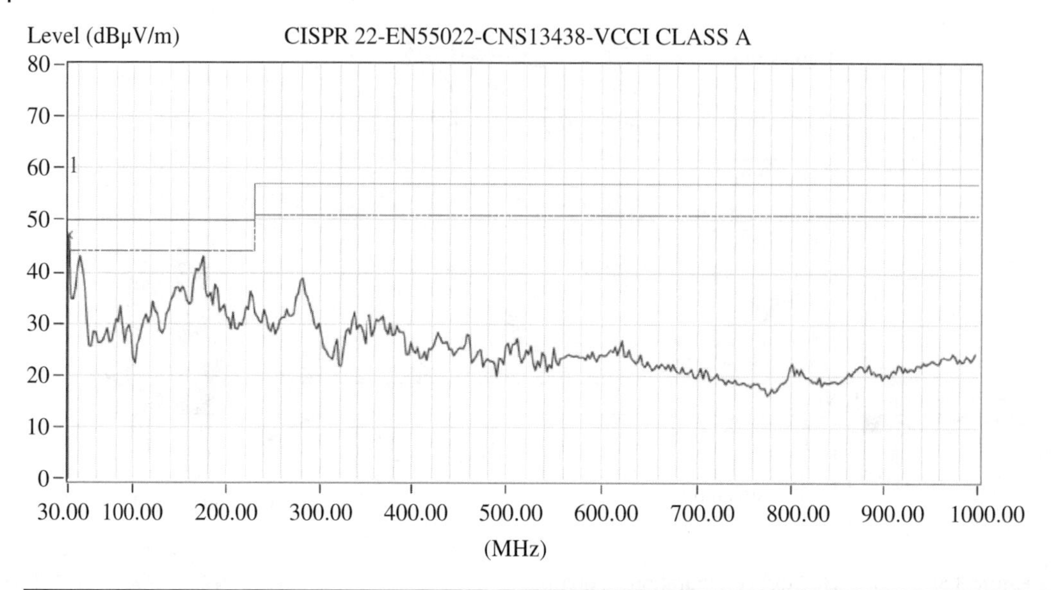

	No.	Frequency MHz	Factor dB	Reading dBμV/m	Emission dBμV/m	Limit dBμV/m	Margin dB	Tower/Table cm	deg
*	1	30.00	–20.15	67.10	46.95	50.00	–3.05	–	–

Figure 3.52 Radiated emission spectral plot when the antenna polarization is in vertical.

At the same time, when performing EFT/B test defined in the standard IEC61000-4-4 on the I/O cable of this product's extension module, we found that as long as the test voltage is more than 1 kV, malfunction of the product will appear.

[Analyses]

According to the test results, because the main module itself can pass radiated emission test and the passed EFT/B level on I/O port of the extension module is only ±1 kV, so the problem usually lies on the extension module or the ribbon interconnecting the extension module with the main module. By analyzing the interconnection design between the extension module and the main module, we found that the PCB in the main module is interconnected with the PCB in the extension module via a common flexible flat cable. The signal sequence in the flexible flat cable is: ground-ground-signal-signal-signal-signal-ground-signal-ground-power, respectively. The highest frequency of the signals in the interconnecting ribbon is about 3 MHz, and the rising time of the signal is less than 1 ns. Figure 3.53 shows photos of the actual interconnection part between the PCB in the main module and the PCB in the extension module.

Thus, we can find a very important design problem of EMC; namely, the interconnection between the main module and the extension module is the flexible flat cable as shown in Figure 3.53. Figure 3.54 shows the schematic how the radiated emission is caused, and Figure 3.55 shows the schematic of the EFT/B immunity test problem. From Figure 3.54 we can see that a radiated emission problem is due to the high-impedance Z_{gnd} of the interconnecting wires between the ground of PCB1 and the ground of PCB2, and the length of the interconnecting wire is about 10 cm. A 10 cm long, ordinary cable has about 150 nH parasitic inductance (assuming the wire diameter is 0.65 mm, at 177 MHz, its impedance is about 100 Ω, and at 30 MHz, it is about 15 Ω); in high frequency, the parasitic inductance dominates the impedance

PCB 1 in mainframe
PCB2 to extended module
Interconnected flat cable about 10 cm long
Plastic casing

Figure 3.53 PCB and the extension of the PCB in the extension of the part of the physical photos.

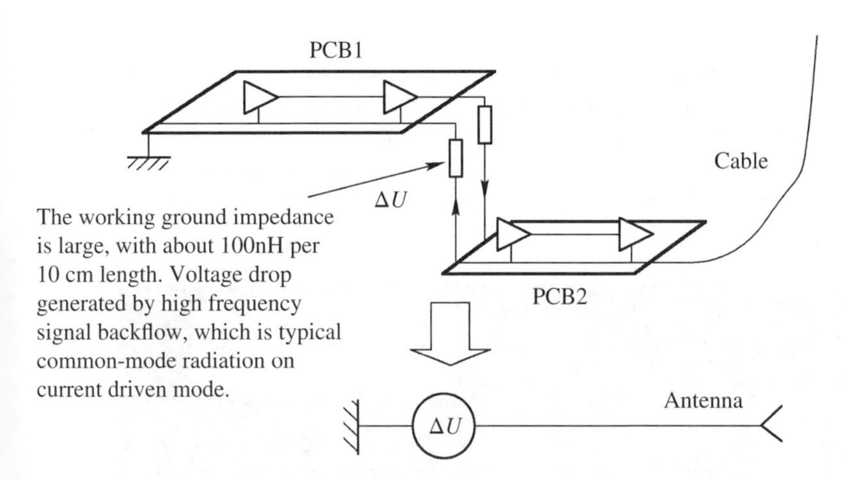
PCB1
ΔU
Cable
The working ground impedance is large, with about 100nH per 10 cm length. Voltage drop generated by high frequency signal backflow, which is typical common-mode radiation on current driven mode.
PCB2
Antenna
ΔU

Figure 3.54 The principle diagram of the radiation.

of the interconnecting wire. So when the return current of signals in the interconnecting ribbon goes through the ground wires, it will cause a voltage drop, which is a typical current-driving common-mode radiated emission.

Figure 3.55 shows the schematic of the EFT/B immunity test problem, from which we can clearly see, as the grounding wire of the product is on the left side of the main module, when EFT/B common-mode disturbance is injected on the I/O cable of the extension module, the common-mode current must go through the main module, the extension module, and their interconnections, finally flow into the ground. As the impedance Z_{gnd} of the interconnecting wire between the ground of PCB1 and the ground of PCB2 is large, and the length is about 10 cm, when the common-mode disturbance current injected into the I/O cable of the extension module passes through the interconnecting ground wires, a voltage drop $\Delta U = |LdI/dt|$ is produced across the interconnecting ground wires. When ΔU exceeds the noise margin of the

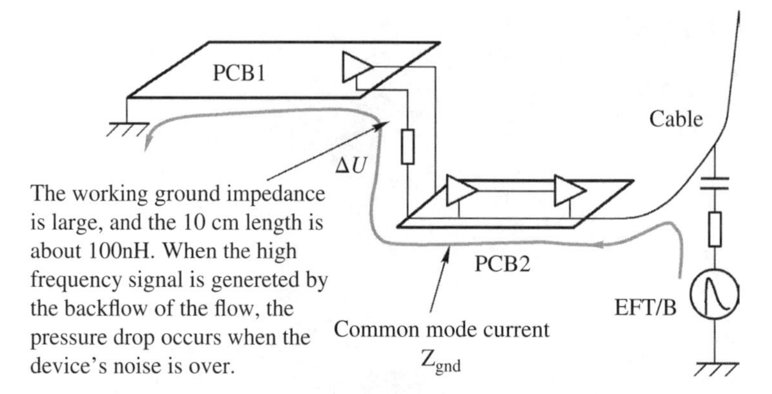
PCB1
Cable
ΔU
The working ground impedance is large, and the 10 cm length is about 100nH. When the high frequency signal is genereted by the backflow of the flow, the pressure drop occurs when the device's noise is over.
PCB2
EFT/B
Common mode current
Zgnd

Figure 3.55 EFT/B immunity principle diagram of the testing problems.

device, it will disturb the device. (From the formula, $\Delta U = |LdI/dt|$, we know that, even 1 A EFT/B transient common-mode current flows through the ground wires, the voltage drop ΔU is, $\Delta U = |LdI/dt| = 150\ \text{nH} \times 1\ \text{A}/5\ \text{ns} = 30\ \text{V}$.)

[Solutions]

From this analysis, reducing the ground impedance Z_{gnd} of the signals between PCB1 and PCB2 can solve this problem. In actual product application, changing the interconnecting ribbon to PCB interconnection, namely change the original flexible flat cable between PCB1 and PCB2 to PCB and use the four-layer board for the PCB interconnection to ensure that it has a complete ground plane. The PCB layout of each layer is shown from Figures 3.56 to 3.60.

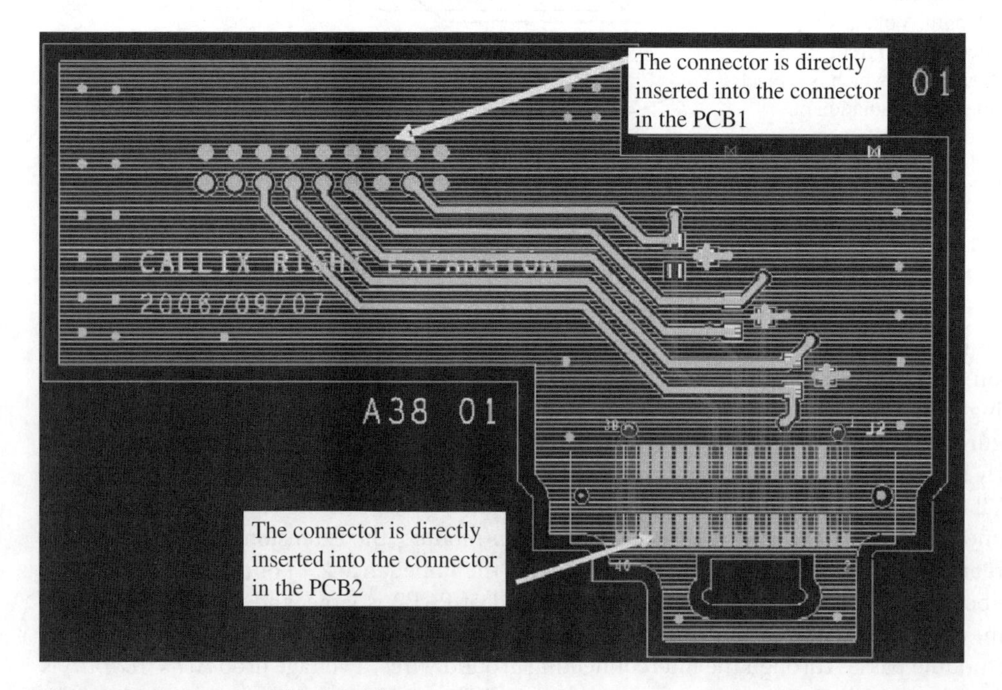
The connector is directly inserted into the connector in the PCB1
01
A38 01
J2
The connector is directly inserted into the connector in the PCB2

Figure 3.56 PCB layout of the global figure.

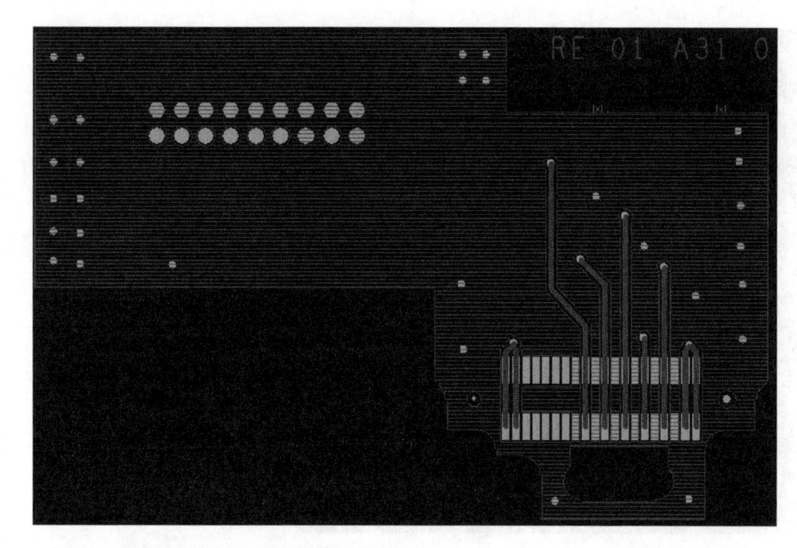

Figure 3.57 Layout of the top layer.

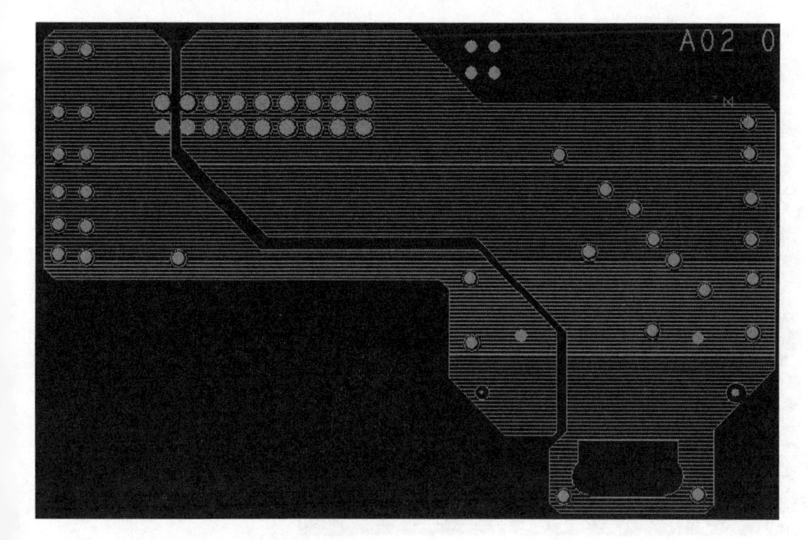

Figure 3.58 Layout of the second layer.

After using the above PCB as the interconnection, under the same test configuration for the product, the radiated emission test results are shown in Figures 3.61 and 3.62. The immunity level of the EFT/B on signal cable is also improved from original ±1 to ±2 kV.

[Inspirations]

In product design, the interconnection between different PCBs often exists. On the interconnection, not only the crosstalk is caused since the signals on it are close to each other, but more importantly, and often overlooked, the ground impedance of the interconnecting connector is much higher than that of the multilayer PCB using ground plane design (in engineering, we can use $10\,\text{nH}\,\text{cm}^{-1}$ to estimate the parasitic inductance of each pin in the connector), it is like in PCB, when the external common-mode disturbance current flows by, the operating signals on the interconnection will be disturbed.

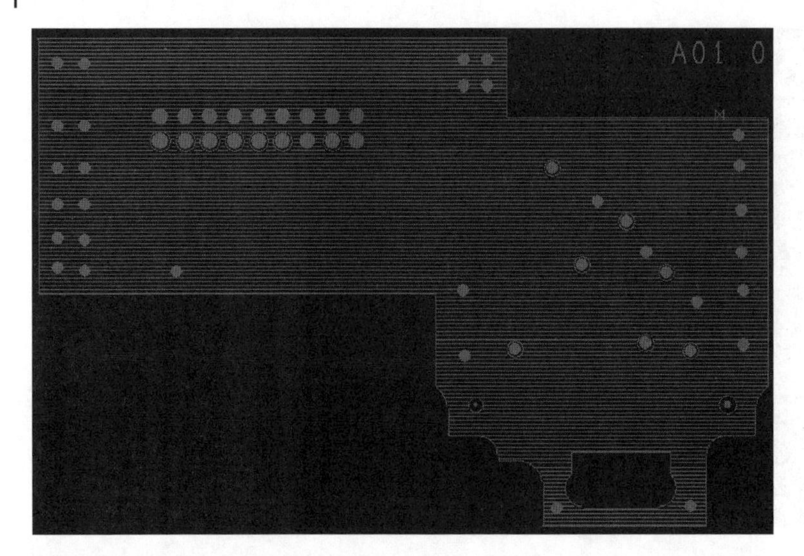

Figure 3.59 Layout of the third layer.

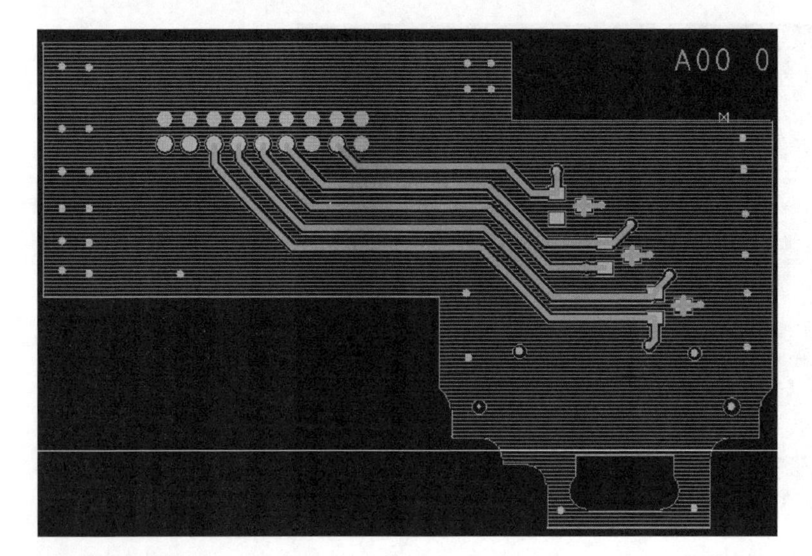

Figure 3.60 Layout of bottom layer.

When the return current of the signals on the interconnection flows back this high-impedance ground interconnecting wire, the common-mode EMI problems will appear.

Focusing on the interconnection between PCBs in products, which is the bottleneck of EMC problems in the product, it is the best way to solve this problem to avoid the common-mode current flowing through the interconnection between PCBs when designing product structure.

3.2.10 Case 25: Excessive Radiated Emission Caused by the Loop

[Symptoms]

The radiated emission test results of one residential product are shown in Figure 3.63.

The spectral plot in Figure 3.63 shows that, the frequencies with excessive radiated emission are 125 and 170 MHz.

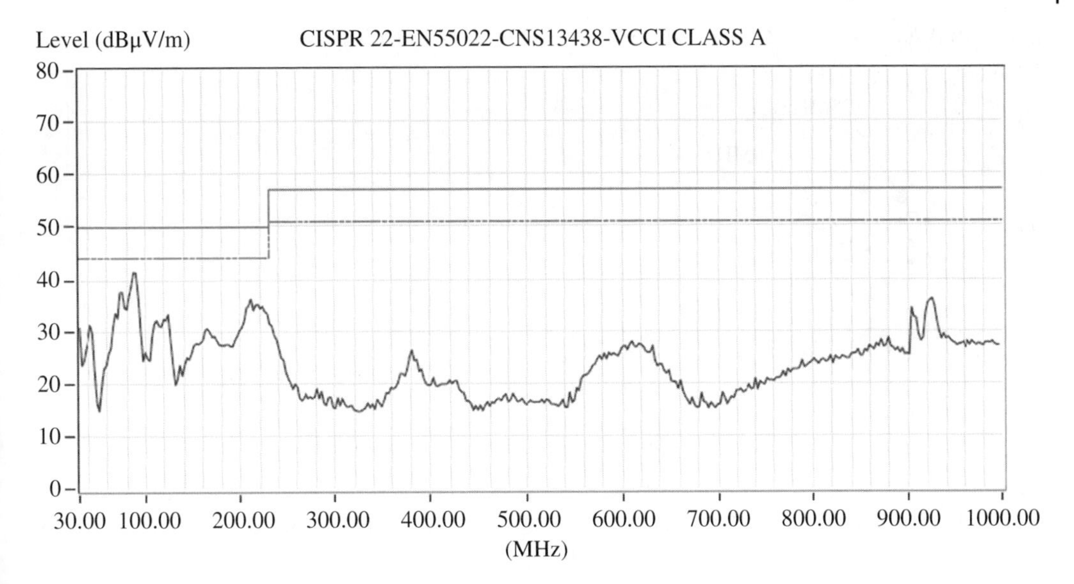

Figure 3.61 Modified horizontal polarization test spectrum.

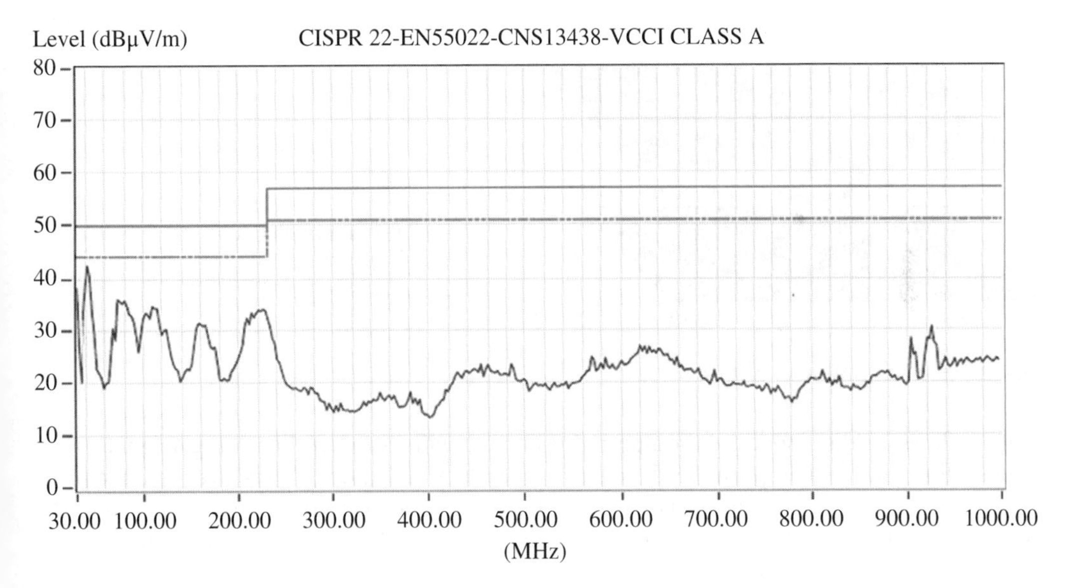

Figure 3.62 Modified vertical polarization test spectrum.

[Analyses]

The product structure is shown in Figure 3.64, the product is mainly composed of two pieces of PCBs, a connector interconnecting two PCBs and a communication cable, the PCB1 and PCB2 in the product are interconnected by the interconnecting connector. The highest frequency of the signal transferred in the interconnecting connector is 25 MHz, which voltage level is 2.5 V, and its working current is about 25 mA. The pin pitch of the connector is 2 mm, and the length of the pin of the connector is 2 cm; the loop area has a signal pin and its return current pin in the connector is 0.4 cm^2. For high-speed signal, it doesn't seem to be a small loop, according to the basic principle of electromagnetic field, a current will create a magnetic field

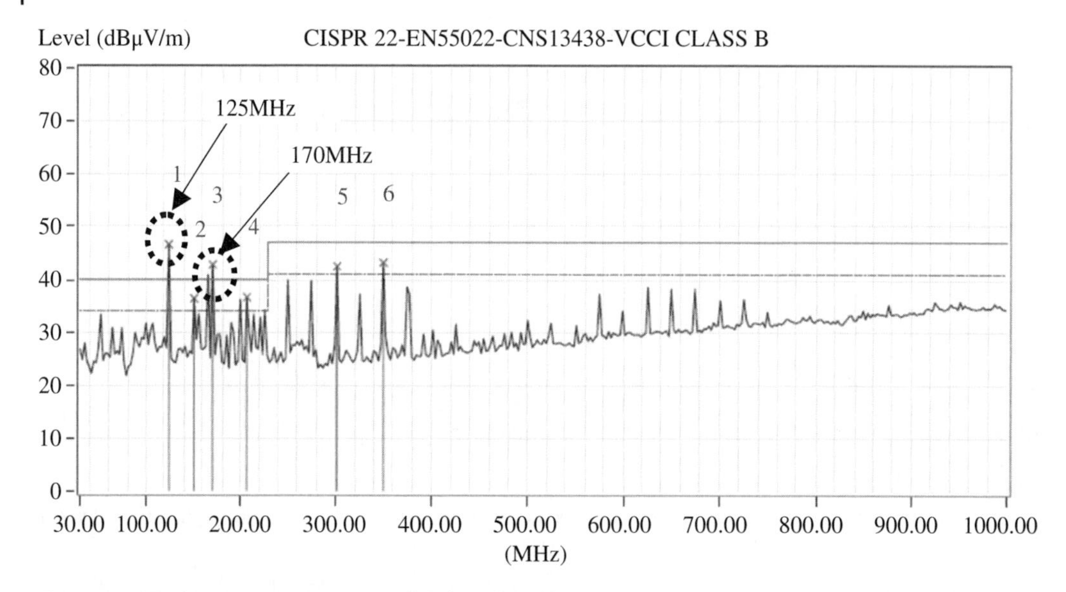

Figure 3.63 Radiation emission test results of a civil product.

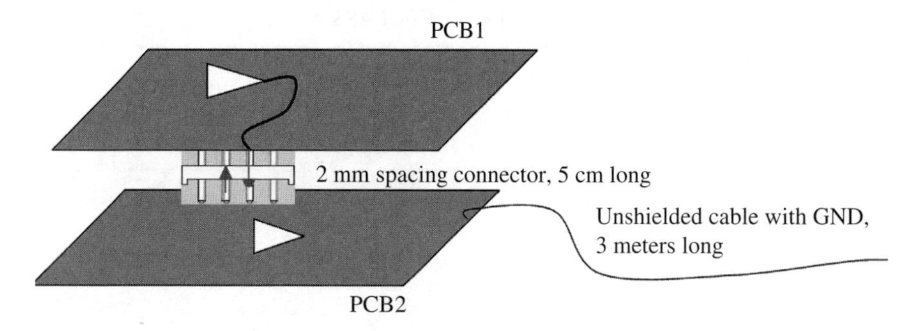

Figure 3.64 Product structure diagram.

when it flows inside a loop, and the alternating current will create an alternating magnetic field when it flows inside a loop, and leads to electromagnetic radiation. This is a very simple principle. Thus, it can be seen that this loop will produce a certain radiation. This is a defect of the product design, and the field strength radiated by this kind of loop in free space can be calculated using formula (1.9) or the formula in Figure 3.65.

In the formula, $E_{\mu V m^{-1}}$ is the electric field strength at the point away from the loop by D_m in free space, with the unit μV m^{-1}; D_m is the distance away from the loop, with the unit m; F_{MHz} is the frequency of the current inside the loop, with unit MHz; S_{cm^2} is the loop area, with the unit cm^2; I_A is the current amplitude inside the loop, with the unit A.

According to some basic information of the product, we can get the RMS value of the harmonic current I_{5A} at 125 MHz of the 25 MHz operating signal current (square wave):

$$I_{5A} = 0.45 \times 25mA/5 = 2.25(mA)$$

The RMS value of the harmonic current I_{7A} at 170 MHz is:

$$I_{7A} = 0.45 \times 25mA/7 = 1.60(mA)$$

Figure 3.65 Current flows through the loop caused by radiation.

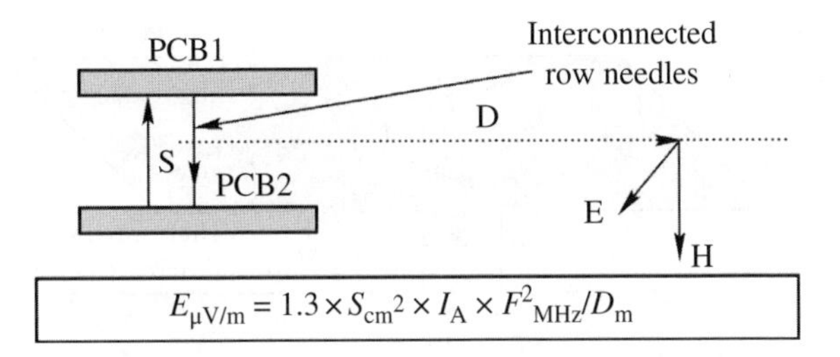

Note: the reference book of the calculation method of harmonic current is *EMC Risk Assessment of Electronic Products Design*, mentioned in Chapter 1.

To calculate the differential-mode radiated emission strength at the position with 3 m away from the loop at 125 and 170 MHz, respectively:

The radiated emission strength at 125 MHz is:

$$E_{5\mu V\,m^{-1}} = 1.3 \times S_{cm^2} \times I_A \times F_{MHz} \times F_{MHz}/D_m = 1.3 \times 0.4 \times 0.00225 \times 125 \times 125/3 = 18.2\left(\mu V\ m^{-1}\right)$$

If this 18.2 μV m^{-1} is converted into dB scale, its value is 25.2 dBμV m^{-1}.

The radiated emission strength at 170 MHz is:

$$E_{7\mu V\,m^{-1}} = 1.3 \times S_{cm^2} \times I_A \times F_{MHz} \times F_{MHz}/D_m = 1.3 \times 0.4 \times 0.0016 \times 170 \times 170/3 = 24\left(\mu V\,m^{-1}\right)$$

If this 24 μV m^{-1} is converted into dB scale, its value is 27.6 dBμV m^{-1}.

From the above calculation result, we learn that even considering the superposition effect from the electromagnetic wave reflected by laboratory ground during radiated emission test, the radiated emission caused by the loop still doesn't beyond the limit defined in the product standard (40 dBμV m^{-1} limit for 3 m distance away from the EUT). So although the 2 cm-long, 2 mm-pin pitch connector directly transferring the 5 MHz square wave signal is a EMC design defect, the radiated emission caused by it is not over the limit.

Then what caused the excessive radiated emission – is it related to this loop? In fact, in addition to the differential-mode radiation caused by the above loop, another kind of radiated emission needs more attention, the common-mode radiation. Figure 3.66 is the schematic that the common-mode radiation is generated.

ΔU in Figure 3.66 is the voltage difference between the ground of PCB1 and the ground of PCB2, which is caused by the return current of the 25 MHz rectangular wave high-speed signal of the interconnecting connector. Its value can be calculated by $\Delta U = 2\pi FLI$, F is the frequency of the current. In this case, it is the fifth-order harmonic frequency and the seventh-order harmonic frequency of the 25 MHz rectangular wave, respectively, $F_5 = 125$ MHz and $F_7 = 170$ MHz; L is the parasitic inductance of each pin of the 2 cm-long connector, about 20 nH (estimated per 10 nH cm^{-1}); I is the harmonic current of the 25 MHz rectangular wave, the fifth-order harmonic current, and the seventh-order harmonic current, respectively, I_5 = 2.25 mA and I_7 = 1.6 mA. So, at 125 and 170 MHz, the voltage difference between the ground of PCB1 and the ground of PCB2, ΔU_5 and ΔU_7 is, respectively:

$$\Delta U_5 = 2\pi F_5 L I_5 = 2\pi \times 125\,MHz \times 20\,nH \times 2.25\,mA \approx 35\,mV$$
$$\Delta U_7 = 2\pi F_7 L I_7 = 2\pi \times 170\,MHz \times 20\,nH \times 1.60\,mA \approx 34\,mV$$

The common-mode current path is started from ΔU_5, ΔU_7, via PCB1ground, PCB2 ground, the ground of the interconnection between PCB1 and PCB2, the cables, then the characteristics

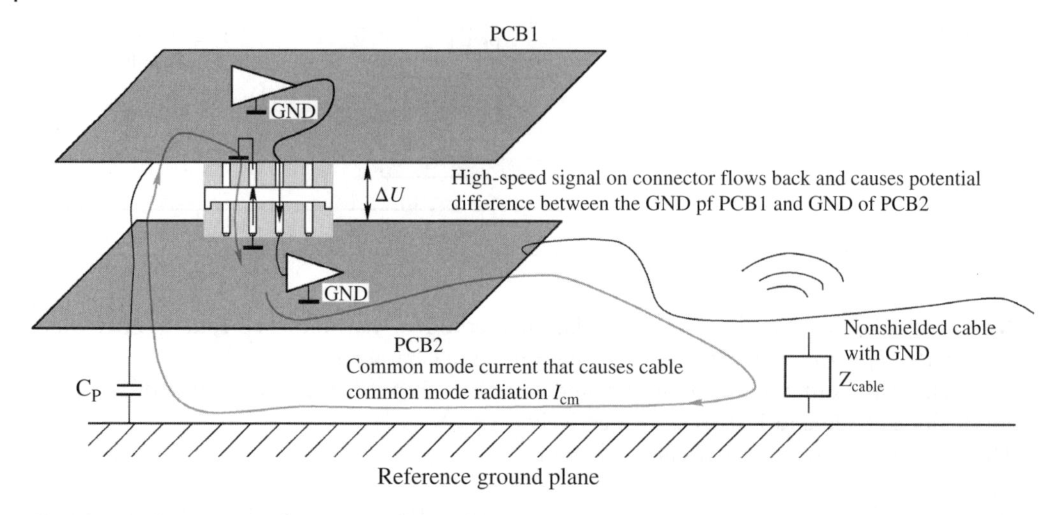

Figure 3.66 The principle of common-mode radiation.

impedance from cable to the reference ground plate (which is formed by the parasitic inductance of the cable and the parasitic capacitance between the cable and the reference ground plate) to the parasitic capacitance between the ground plane of PCB1 and the reference ground plate, C_p, and the common-mode current in this path flows through the cable. When the cable is long, it acts as an antenna (when the cable length is comparable to the wavelength of the signal frequency). The capacitance of C_P depends on the size of the GND plane in PCB1 and the distance between PCB1 and the reference ground plate. The size of the PCB1 GND plane in this case is about 10 cm × 8 cm, the distance from PCB1 to reference ground plate is about 0.8 m, which is defined in the radiated emission test standard for residential product, so we can estimate C_p as about 3 pF.

Z_{cable} is from 150 to 250 Ω, which depends on the cable's parasitic inductance L_{cable} and the parasitic capacitance C_{cable} between the cable and the reference ground plate, and $Z_{\text{cable}} = (L_{\text{cable}}/C_{\text{cable}})^{0.5}$ ($L_{\text{cable}} \approx 1\,\mu\text{H m}^{-1}$, C_{cable} is between 20–50 pF m^{-1}, associated with the cabling). So, at 125 and 170 MHz, the common-mode current flowing through the cable, $I_{5\,\text{cm}}$ and $I_{7\,\text{cm}}$, is, respectively, about:

$$I_{5cm} = 2\pi \times F_5 \times C_p \times \Delta U_5 = 2\pi \times 125\,MHz \times 3\,pF \times 35\,mV \approx 82\,\mu A$$

$$I_{7cm} = 2\pi \times F_7 \times C_p \times \Delta U_7 = 2\pi \times 170\,MHz \times 3\,pF \times 34\,mV \approx 112\,\mu A$$

Note: the reference book of the detailed estimation method of $Z_{\text{cable}}$and C_p is *EMC Risk Assessment of Electronic Products Design,* Chapter 2.

According to electromagnetic field theory, when the cable is longer than the half of the wavelength of the common-mode current signal frequency, the field strength of the common-mode radiated emission from the cable to the free space, E_{CM}, can be calculated referring to the formula (1.11) $E_{\text{CM}} = 60 \cdot I_{\text{CM}}/D_{\text{m}}$, E_{CM} is the radiated emission field strength caused by cable at the position with D_{m} away from the cable, with the unit μV m^{-1}; I_{CM} is the common-mode current in the cable, with the unit μA; D_{m} is the distance away from the cable, with the unit m.

At the 125 and 170 MHz, the common-mode radiated emission field strength at the position with 3 m away from the cable, $E_{5\,\text{cm}}$ and $E_{7\,\text{cm}}$ is, respectively:

$$E_{5\text{cm}} = 60 \times I_{5\text{cm}}/D_{\text{m}} = 60 \times 82\,\mu\text{A}/3\,\text{m} \approx 1640\,\mu\text{V m}^{-1}$$

$$E_{7\text{cm}} = 60 \times I_{7\text{cm}} / D_{\text{m}} = 60 \times 112\,\mu\text{A}/3\,\text{m} \approx 2240\,\mu\text{V m}^{-1}$$

If the 1640 μV m^{-1} is converted into dB scale, its value is 64 dBμV m^{-1} and the converted value of 2240 μV m^{-1} into dB scale is 67 dBμV m^{-1}, which is far beyond the limit defined in the relevant product standards.

[Solutions]

According to the principle of the common-mode radiation, to solve the radiated emission problem of this product, it mainly needs to reduce the common-mode radiated emission, and the common-mode radiated emission of this product is closely related with the impedance of the interconnecting connector between PCB1 and PCB2 and the parasitic capacitance C_{P} between GND plane of PCB1 and the reference ground plate. Therefore, the solution can also be started from the following two points:

1) Reduce the GND impedance of the interconnection between PCB1 and PCB2, such as increasing the GND pin number (it is worth noting that the impedance is not a linearly increased with the increasing of the GND pin number), or paralleling a metallic plate with low length-to-width ratio with the GND pin.
2) Reduce the parasitic capacitance C_{p} between the GND plane of PCB1 and the reference ground plate, such as adding metallic casing for the product, etc.

[Inspirations]

- Do not emphasize too much on the differential-mode radiation, rather than the more important common-mode radiation.
- For digital signal, its frequency is below 200 MHz and the loop area is less than 1 cm^2, it basically will not cause the differential-mode radiation problem. If the product radiated emission excesses the limit, we can focus on the common-mode radiated emission perspective.
- It is similar for radiated immunity test; usually, more problems lie in that the common-mode current induced by the cable exposed in the radiated field flows into the internal circuits of the products and impacts their normal operation, unless the induced disturbance is lower (i.e. less than 1 mV).
- The loop with 1 cm^2 area cannot be ignored in the ESD immunity test.

3.2.11 Case 26: Pay Attention to the Interconnection and Wiring Inside the Product

[Symptoms]

The radiated emission test result of a microcomputer is shown in Figure 3.67. It is not difficult to see that it cannot pass the Class B limit defined in the standard EN55022.

[Analyses]

Microcomputers generally consist of motherboard, power supply, and casing. In many cases, the manufacturers of the microcomputer only design the motherboard, often outsourcing the power supply and casing. If considering the EMC problem for all three components, we should put forward the requirements for each, i.e. the power supply needs to have CCC certification, and the casing needs to have certain shielding effectiveness. If a microcomputer system is composed of these three parts (motherboard, power supply, and chassis) with good EMC performance, will the EMC performance of the entire microcomputer system be good? This is not guaranteed. Another important factor to the EMC performance of the entire microcomputer system is the interconnection and the wiring between each part.

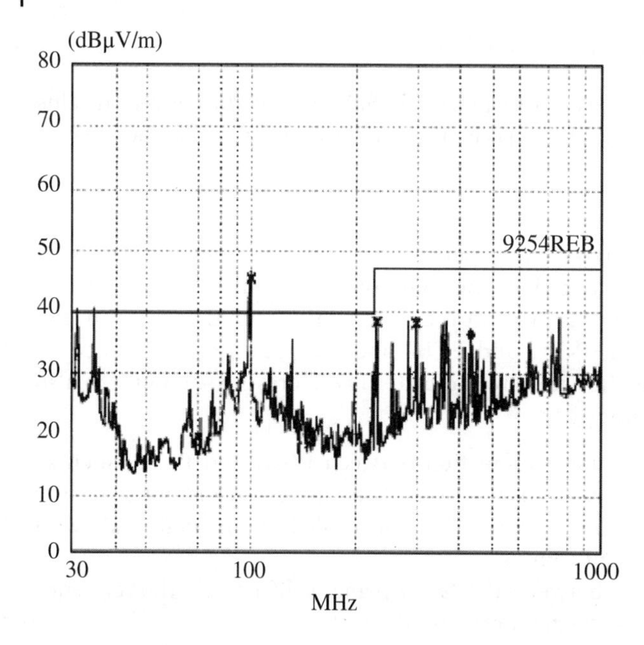

Figure 3.67 The spectrum of radiation emission test results of a microcomputer.

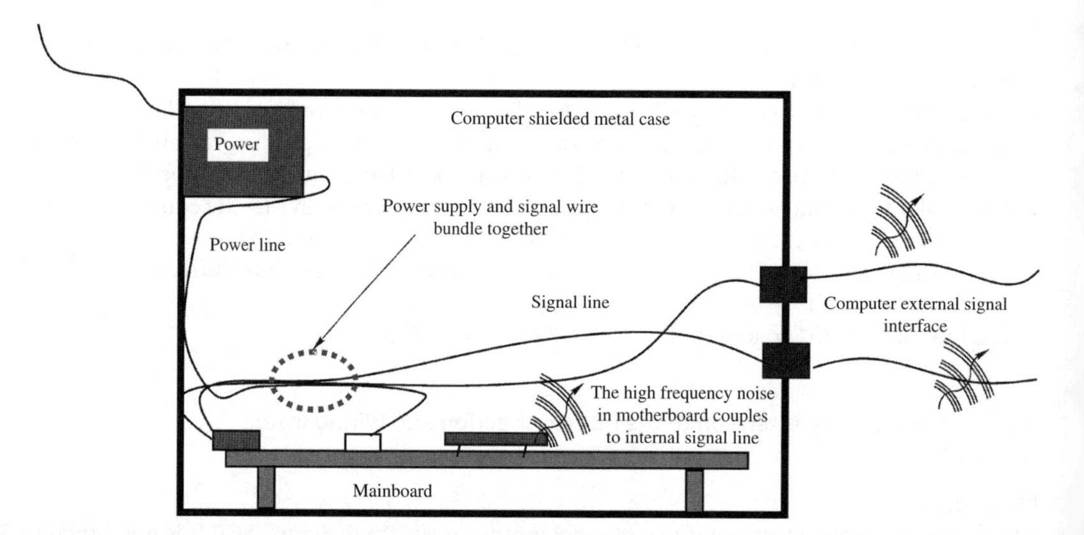

Figure 3.68 A microcomputer internal wiring diagram.

The problem in this case is actually due to defects of the interconnection and the wiring between each part, as described in Figure 3.68, which shows the internal wiring of the microcomputer. In Figure 3.68, a signal wire of the motherboard is connected to its interface connectors installed on the casing (the USB connector, the earphone connector, and the microphone connector), which has two EMC defects:

1) The power wires and signal wires are bundled together, which leads to the noise of the power wires transferred to the signal wires (USB signal wires, earphone signal wires, and microphone signal wires).

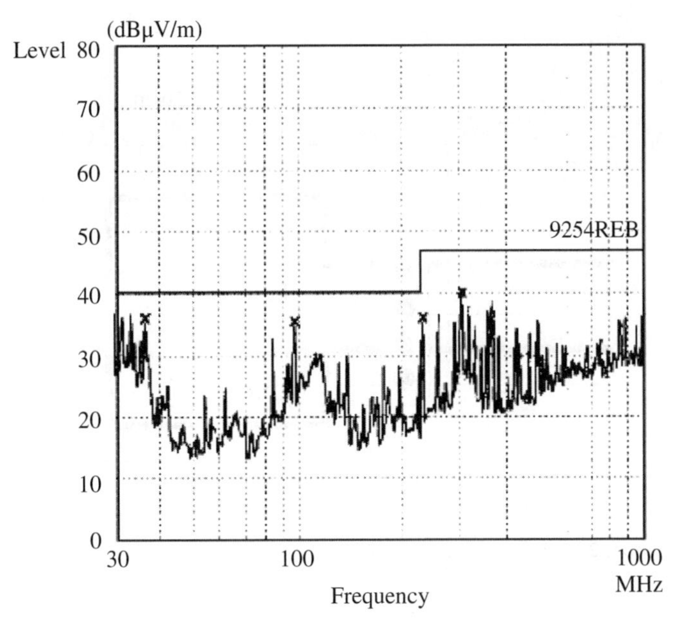

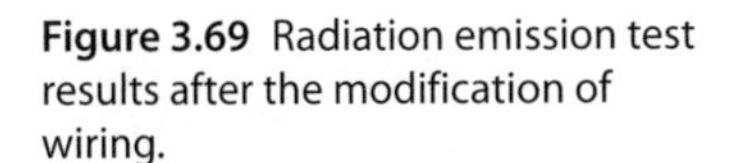
Figure 3.69 Radiation emission test results after the modification of wiring.

2) When signal wires (USB signal wires, earphone signal wire, and microphone signal wire) are connected from the motherboard to the interface connectors on the casing, they pass over the whole motherboard, with a short distance above the surface of the motherboard (about 2 cm), which led to high-frequency noise on the motherboard being coupled to the signal wires (USB signal wires, earphone signal wires, and microphone signal wires).

The USB signal wires, the earphone signal wires, and the microphone signal wires are polluted by the noise on the power wires and the high-frequency noise on the motherboard, and this noise is brought out of the casing shields during EMC test and eventually leads to excessive radiated emission.

[Solutions]

According to the above analysis, the signal wires (USB signal wires, earphone signal wire, and microphone signal wires) are distributed along the casing independently (wiring along the casing can reduce the coupling between wires and the radiated emission from wires), rather than bounded together with the power wires; they do not cross over the motherboard. The radiated emission test result after the modification of the wiring is shown in Figure 3.69.

[Inspirations]

For product design in the system level, we should not only control the EMC performance of each part that constitutes the system but also deal well with the interconnection and the wiring between each part; both are indispensable.

3.2.12 Case 27: Consequences of the Mixed Wiring Between Signal Cable and Power Cable

[Symptoms]

During an EFT/B test in the case of 1 kV test voltage on power port, a DC amplifier product loses its function, due to the saturation of the amplifier.

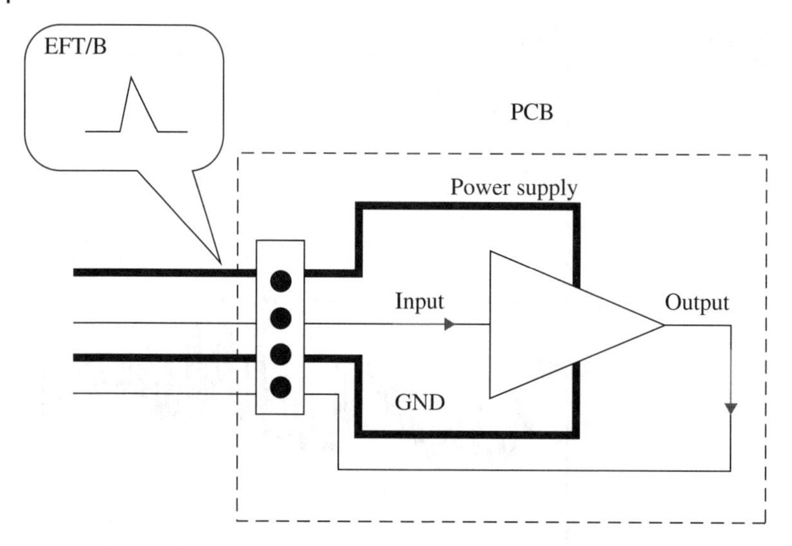

Figure 3.70 Schematics of amplifier.

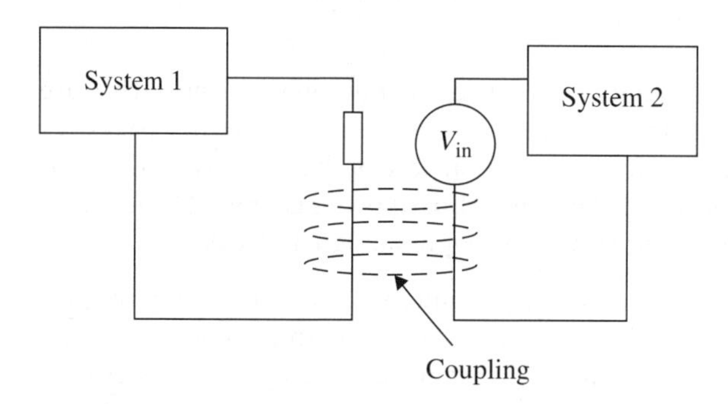

Figure 3.71 Inductive coupling between two loop.

[Analyses]

This DC amplifier is installed on a PCB. To facilitate the installation, the whole PCB is connected with other modules by one cable, as shown in Figure 3.70. So the input/output wire of the amplifier, the power wire and the ground wire, is bundled together and placed into one cable.

According to electromagnetic induction principle, the AC current I_L flowing in the conductor will generate magnetic field. This magnetic field can be coupled to its nearby conductor and induce voltage on it (as shown in Figure 3.71).

The induced voltage in the victim conductor can be calculated by formula (3.2).

$$U = -MdI_L/dt \tag{3.2}$$

In this expression, M is mutual inductance, and the unit is a henry.

M depends on the loop area, the direction, the distance, and the existence of magnetic shielding between disturbance source and the victim circuit. Generally, the mutual inductance between two nearby short conductors is about 0.1–3 H. The magnetic coupling is equivalent to that of the disturbance voltage source in series with the victim circuit. It is worth noting that, generally, whether two circuits have a direct connection has no effect on the coupling, and the induced voltage is the same whether these two circuits are isolated to ground or connected to ground.

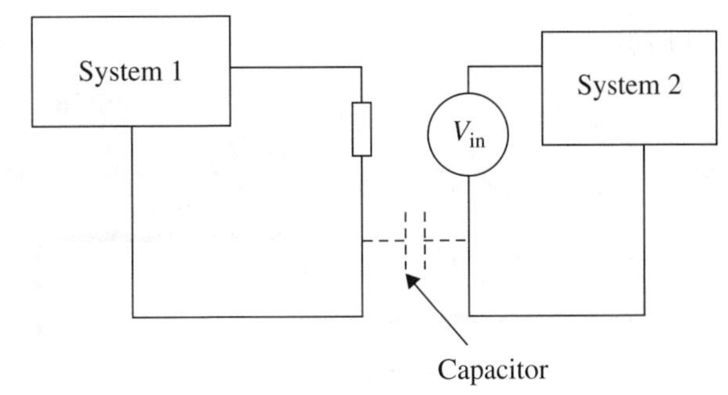

Figure 3.72 Capacitive coupling.

In the meantime, the AC voltage U_L on the conductor creates electric field, which is coupled by the nearby conductor and induces voltage on it (as shown in Figure 3.72).

The induced voltage on the victim conductor can be calculated from formula (3.3).

$$U = C \times Z_i \times dU_L/dt \tag{3.3}$$

In this formula, C is the parasitic capacitance; Z_i is the impedance between the victim circuit and ground.

It is assumed here that the impedance of the parasitic capacitance is much higher than the circuit impedance. The noise seems to be injected from a current source, with the current, $C \times dU_L/dt$. The value of C is related to the distance between conductors, the effective area, and the presence of shielding materials. A typical example is that, for two parallel insulated conductors with the 0.1 in. interval, the parasitic capacitance between them is about 50 pF m^{-1}; And the parasitic capacitance between the primary side and secondary side of an unshielded medium-power power transformer is about 100–1000 pF.

The parasitic capacitance and the mutual inductance are both influenced by the physical distance between the disturbance source and the victim conductor. The influence of the distance between two parallel conductors in free space on the parasitic capacitance and the mutual inductance between two conductors above a ground plane (which provides the return current path for each power) are shown in Figure 3.73.

In this amplifier product, the parasitic capacitance and the mutual inductance between conductors is higher, because the input/output wires of the amplifier, the power wires and the ground wires are bundled into one cable and the cable is longer.

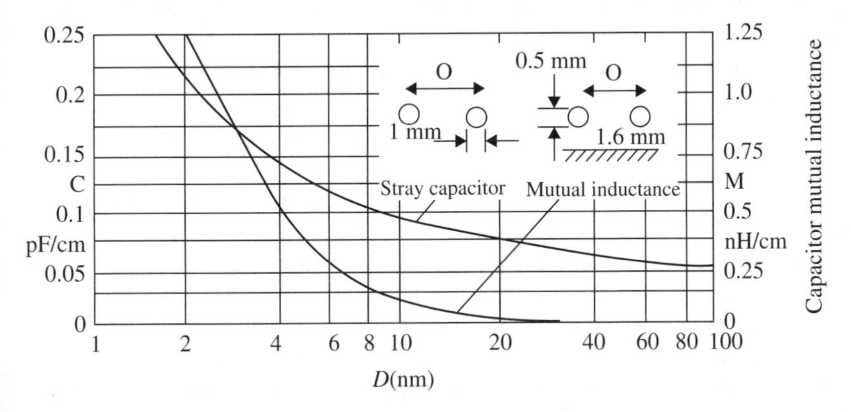

Figure 3.73 The influence of the distance between two parallel conductors in free space on the parasitic capacitance and the mutual inductance between two conductors above a ground plane.

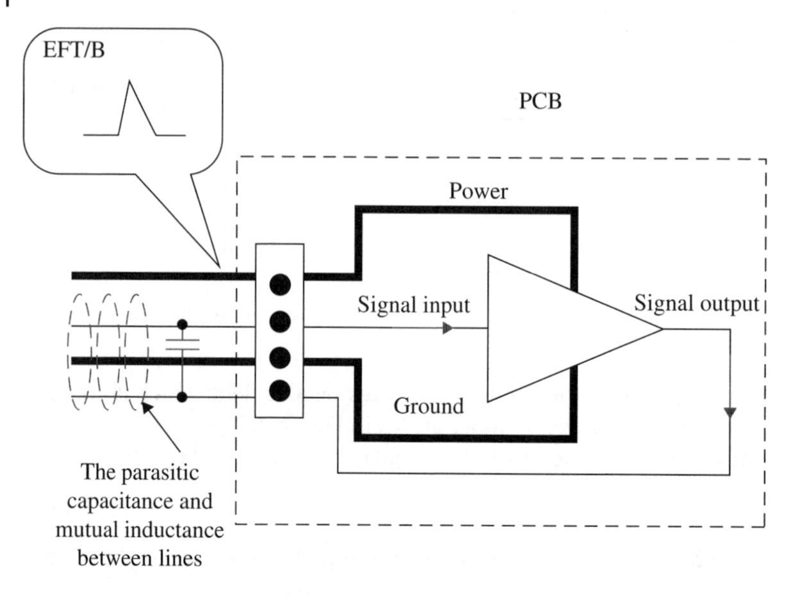

Figure 3.74 The principle of the saturation of the amplifier, caused by parasitic capacitance.

During EFT/B test, since a lot of high-frequency components are included in EFT/B signal, the disturbing energy will be coupled to the input of the amplifier via the parasitic capacitance and the mutual inductance between conductors. Although this is a DC amplifier, no limit has been set on the bandwidth of the amplifier by designer. As a result, the amplifier amplifies the high-frequency signal, which is coupled to the input. Big parasitic capacitance and mutual inductance also exist between the input wire and the output wire of the amplifier, so the input signal after amplification is coupled to the input again, and then a positive feedback appears, which causes the saturation of the amplifier. Figure 3.74 shows the principle of the saturation of the amplifier, caused by parasitic capacitance.

[Solutions]

There are two ways to solve this problem. The first one is to separate conductors in order to decrease the parasitic capacitance and the mutual inductance between conductors. Especially the input wires and output wires of the amplifier are well separated from the power wires, and the input wires are well separated from the output wires, which will avoid the test pulse coupled to the amplifier input.

The other way is to reduce the amplifier bandwidth so the amplifier does not respond to the high-frequency signal that will be coupled to the amplifier input. Since it is a DC amplifier, it should have an amplification effect only on DC and low-frequency signal, but not on a high-frequency signal.

But considering the actual situation, it will be inconvenient to separate the conductors. So we can only use one cable to connect the amplifier to the system. If we replace it with an amplifier with narrower bandwidth, this component change may result in a longer product development cycle, even though the problem can be solved. So in order to work this out, the filtering components should be installed at the periphery of the amplifier and reduce the amplifier bandwidth to make the amplifier have no response to high-frequency signal which will be coupled to the input. The coupling efficiency of low-frequency signal is lower, so it won't cause amplifier saturated. The circuit after improvement is shown in Figure 3.75. A low-pass filter is installed on the amplifier input. It is equivalent to reduce the bandwidth of the amplifier. After this improvement, the amplifier successfully pass ±1 kV EFT/B test.

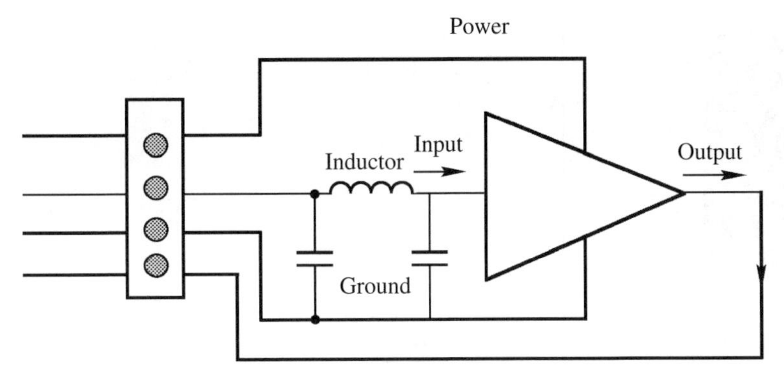

Figure 3.75 Amplifier circuit after transformation.

[Inspirations]

1) When designing circuit with amplifier, reduce amplifier bandwidth as much as possible on the premise of function assurance. Do not use a bandwidth more than the necessary one.
2) In the design of product wiring, crosstalk, and coupling problems among different signal wires should be considered.

3.2.13 Case 28: What Should Be Noticed When Installing the Power Filters

[Symptoms]

A product didn't pass the conducted emission test at AC power port. It can be seen from the conducted emission spectral plot at the power port (as shown in Figure 3.76). The conducted emission at 13 and 21 MHz is higher and exceeds the limit defined in standard.

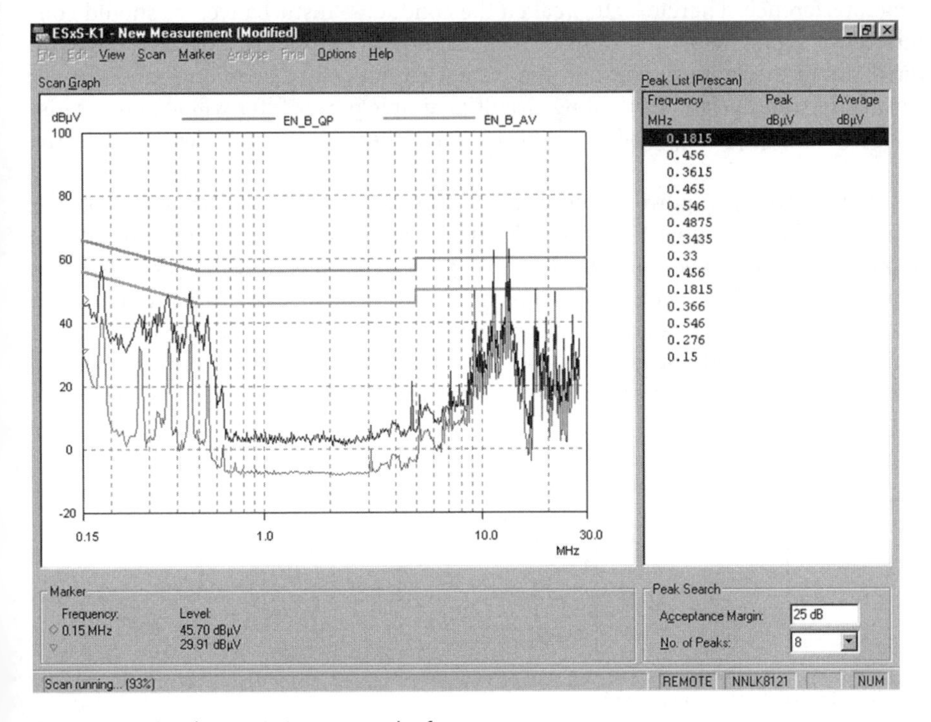

Figure 3.76 Conduct emission test result of power port.

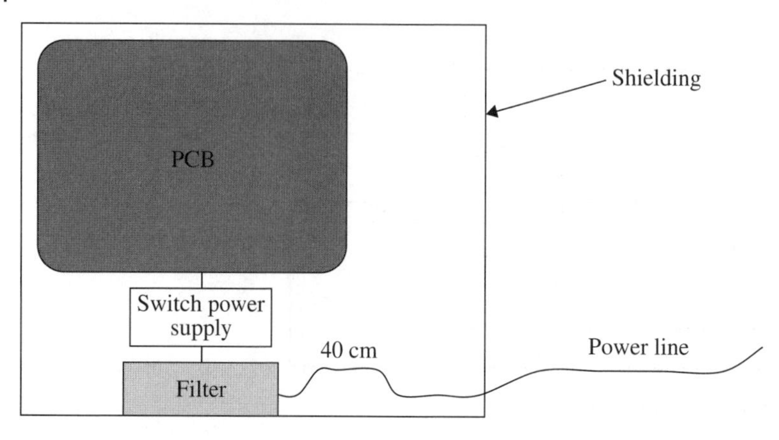

Figure 3.77 Architecture of product.

[Analyses]

This product is designed by using structure shield and the power port is filtered by power filter. The structure block is shown in Figure 3.77.

Filtering is an effective measure to suppress the interference and disturbance, and is especially efficient for suppressing the conducted disturbance and radiated disturbance of the EMI signal from switching-mode power supply. All the conducted disturbance signals on the power line can be expressed by differential-mode signal and common-mode signal. Differential-mode interference transfers between two wires, which is symmetrical interference; common-mode interference transfers between wires and ground (product casing), which is dissymmetrical interference. In general, differential-mode interference has low amplitude, has low frequency, and causes small interference; common-mode interference has high amplitude, high-frequency, and causes big interference. Therefore, to weaken the conducted disturbance, we should control the interference signal below the limit of EMC standards. The most effective way is to install power filters to the input circuit and the output circuit of the switching-mode power supply. Choosing the appropriate decoupling circuit or simple power filter will lead to a satisfied result. The operating principle of the power filter's internal circuit is described in the Section 4.1.1.

Because of the reciprocity of power filter, in practical usage, the power filter can filter the interference not only from the power port but also from the inside of the product. In order to effectively suppress the disturbance and interference signal, we must choose the network structure and the parameters of the power filter according to the impedance of the disturbance and interference signal source and the load impedance, which will be connected to the power filter. When the impedances of both sides of the power filter are mismatched, namely, $Z_S \neq Z_{in}$, $Z_L \neq Z_{out}$ shown in Figure 3.78, the disturbance and interference signal will be reflected at the

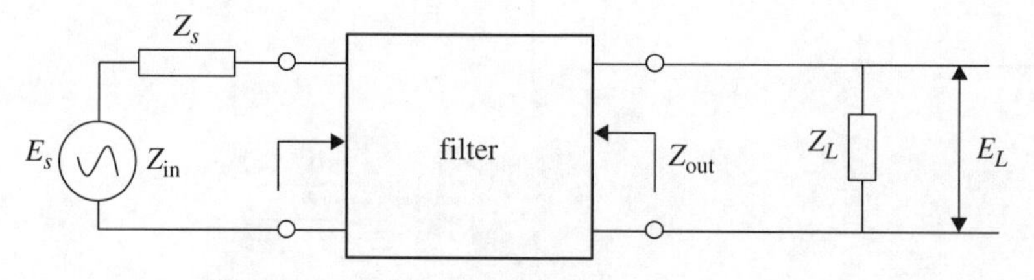

Figure 3.78 The principle of the power supply filter.

input and the output of the filter, the attenuation of the disturbance and interference signal is increased. The relationship between the attenuation of the signal (A) and the reflection (Γ) is,

$$A = -10\lg\left(1 - |\Gamma|2\right)$$

The purpose to design a power filter in the product is to choose the component parameter as appropriate as possible, under the principle that the network structure complies with the maximally mismatching, in order to get the biggest attenuation of the disturbance and interference signal.

Since the principle of power filter is that all impedances are mismatched, and the power filter will reflect the disturbance and interference signal, which are transferred to the power filter circuit. If we want the power filter to be effective as expected, we must ensure that the disturbance and interference signal are transferred to the power filter circuit. That is to say, while the power filter is being used, especially at high frequency, the disturbance and interference signal must be avoided to be transferred through the space and over the power filter. In fact, the product in this case has the possibility that the disturbance and interference signal at high frequency could transfer over the power filter.

As shown in Figure 3.79, after the power cable passes though the shield of the product, there is still some distance (about 40 cm) from its entrance to the power filter. Through the coupling in space, the high-frequency signal from PCB or switching-mode power supply will be coupled into this segment of the power cable, which makes the power filter act unexpectedly. There are clock generation circuits working at 13 and 21 MHz in the actual product, and their routings are widespread in PCB; therefore, the radiation is stronger. If we want the power filter to have an excellent filtering performance at high frequency, the problem of high-frequency interference signal transferring through the space and coupling into the power input lines should be solved. Installing the power filter near the power line entrance, as shown in Figure 3.80, can avoid the interference coupled through the space into the power input cable.

In the state of the installation way, as shown in Figure 3.80, since the shielding structure is used in this product, the input power cable (the power cable on the right side of the power filter in Figure 3.80) of the power filter is out of the shield, and the shielding structure are also used for the power filter. It has a good overlap with product shielding structure. So the radiated interference signals from the inside of the product will not be coupled with the internal circuit of power filter or its power supply input cable. Even if the radiated interference signals coupled to the links between the power filter and the switching-mode power supply, it would be filtered by the power filter.

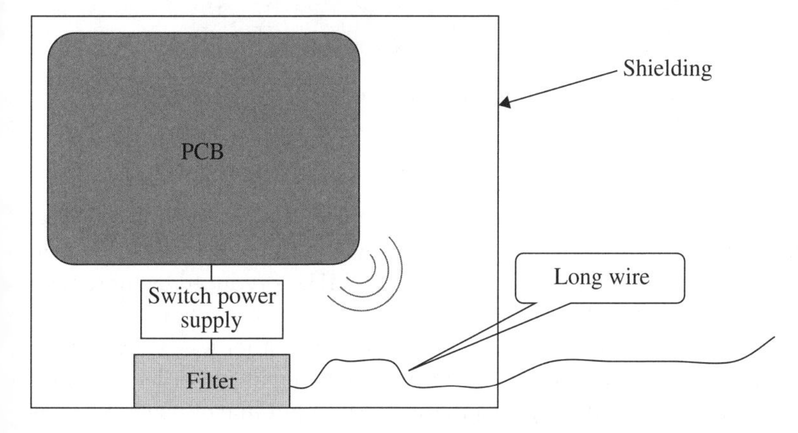

Figure 3.79 The disturbance signal is transmitted over the power supply filter.

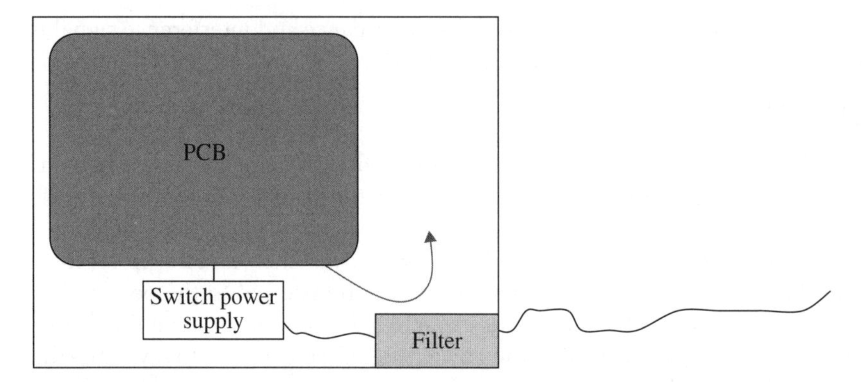

Figure 3.80 Power filter installed at the entrance of the shield of power cable.

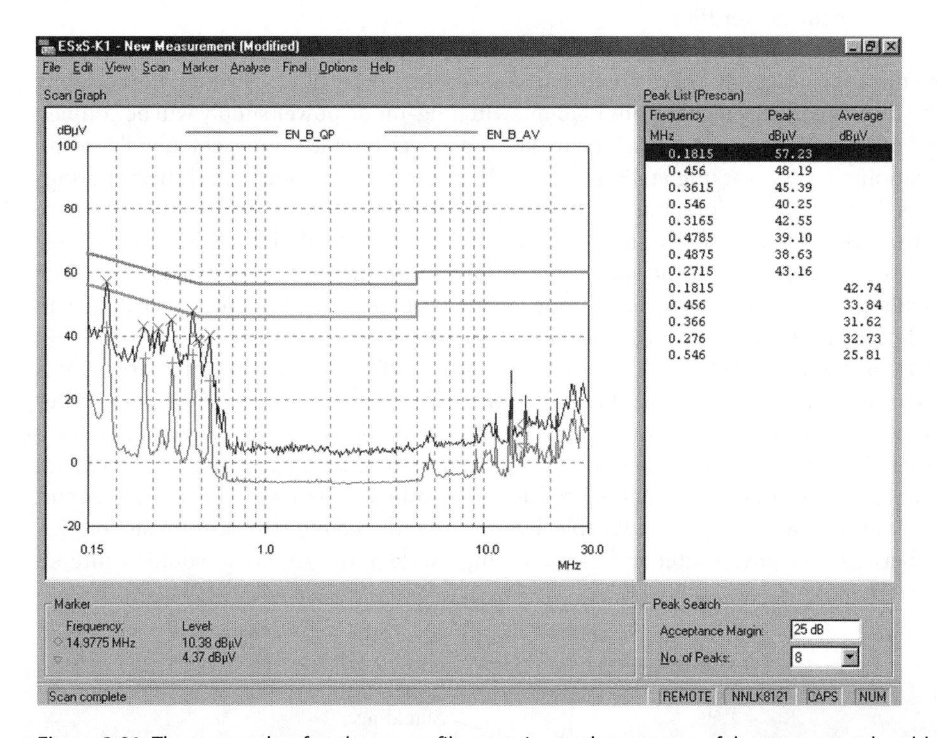

Figure 3.81 The test results after the power filter moving to the entrance of the power supply cable.

[Solutions]

According to the above analysis, we move the power filter to the entrance of power cable, as shown in Figure 3.80, and then we test it again. The result is shown in Figure 3.81. It can be seen that the prior excessive conducted emission between 10 and 20 MHz is reduced by more than 20 dB, which confirms the correctness of the analysis.

As a reference, to solve the problem in this case, we can also take another way, shielding the 40 cm power cable as shown in Figure 3.77, i.e. to shield the inner subpart, in order to cut off the coupling between the inner interference signal and the power input line of power filter, as shown in Figure 3.82.

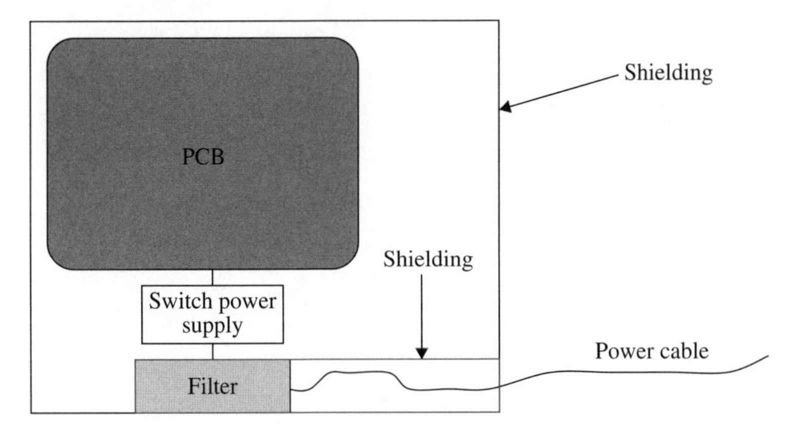

Figure 3.82 Coupling between disturbance signal and power supply line of the subshielded body.

[Inspirations]

1) No doubt, it is important to choose a suitable power filter per the interference and disturbance characteristic of the product, but it is more important to properly arrange the wires at both sides of the power filter.
2) For the installation of the power filter, it is necessary to note the separation between the input and output signals of the power filter. The separation includes not only the separation between the wires at two sides of the power filter but also the separation between the circuits at two sides of the power filter.
3) Signal filter is the same as the power filter. When installing the signal filter, the separation between the input and output signals of the signal filter must be noted.

4

Filtering and Suppression for EMC Performance Improvement

4.1 Introduction

Optimal EMC performance hinges on proper filtration and surge protection. This chapter discusses these elements in detail.

4.1.1 Filtering Components

The filtering system is comprised of five components: resistors, capacitors, inductors and common-mode inductors, ferrite beads and ferrite ring cores, and filters.

4.1.1.1 Resistors

Resistors are the most commonly used device on printed circuit boards (PCB). The use of the resistor is limited for electromagnetic interference (EMI). The application limitation in frequency domain depends on the resistor's material (compound carbon, carbon film, mica and wire-wound, etc.). The wound wire results in the parasitic inductance, so the wire-wound resistor is not suitable for high-frequency application. Thin film resistors also contains some inductance, but due to the lower lead inductance, they are sometimes used for high-frequency application.

The overall characteristics of the resistor are related to the package size and the parasitic capacitance. The parasitic capacitance exists between the two ends of the resistor. For extremely high-frequency design, especially GHz frequency design, the parasitic capacitance will cause a destructive effect. For the majority of applications, the resistor lead is more important than the parasitic capacitance between the pins.

For the resistor, the possible overvoltage stress is mainly concerned. Injecting electrostatic discharge (ESD) voltage on the resistor belongs to this kind of situation. If there is a surface-mounted resistor, the arc discharge will be observed; If there is through-hole resistor, the ESD voltage will encounter a high-impedance path, and the inductive and capacitive characteristics hidden behind the resistive one will block the ESD into the circuit.

4.1.1.2 Capacitors

Capacitor is usually used for the decoupling of the power supply bus, filtering, bypassing, and voltage regulation. Below its self-resonant frequency, the capacitor keeps capacitive. Above the self-resonant frequency, the capacitor presents as inductive. Its impedance can be described as $X_C = 1/2\pi fC$. Among them, X_C is the capacitive reactance, the unit is Ohm (Ω); f is the frequency, the unit is Hertz (Hz); C is the capacitance, the unit is farad (F). The capacitive

Electromagnetic Compatibility (EMC) Design and Test Case Analysis, First Edition. Junqi Zheng.

reactance of a 10 μf electrolytic capacitor at 10 kHz is 1.6 Ω. At 100 MHz, it drops to 160 μΩ. So the short circuit exists at 100 MHz, which is favorable for EMI. However, the electrolytic capacitor has high-ESL and electrostatic resistance (ESR), and these parameters confine its application below 1 MHz frequency. In addition, the lead inductance for the use of the capacitor is also an important aspect that needs to be considered. In short, the parasitic inductance of the capacitor leads makes the capacitor work as an inductance when the frequency is above its self-resonant frequency instead of the capacitance.

4.1.1.3 Inductors and Common-Mode Inductors

Inductors are also commonly used to control the EMI. With increased frequency, the impedance of the inductor increases linearly, which can be described by the formula $X_L = 2\pi fL$. For example, the impedance of an ideal 10 mH inductor is 628 Ω at 10 kHz, and it will increase to 6.2 MΩ at 100 MHz, which looks like an open circuit. So it is very difficult for a 100 MHz signal to pass it. Similar to the capacitor, the parasitic capacitance of the inductor winding limits the application frequency not to be infinite.

The common-mode inductor is also known as a common-mode choke. Why can the common-mode inductor prevent the common-mode EMI? To understand this, the structure of the common-mode inductor needs to be analyzed first.

Figure 4.1 shows the schematic and the magnetic field distribution of the common-mode inductor. L_a and L_b are the common-mode inductances. These two coils are wound on the same magnetic core. The number of turns and the phase of them are the same (the winding direction is reverse). In this way, when the normal current flows through the common-mode inductor, the currents in the in-phase wound coils produces the reverse magnetic fields, which cancel each other. At this time, the normal signal current is mainly affected by the coil resistance (and a small leakage inductance). When the common-mode currents flow through the coils, since the common-mode currents are in the same direction, the magnetic fields with the same direction are produced in the coils and the coil impedance is increased, so that the coil acts as high-impedance and causes strong damping effect to attenuate common-mode current and achieve the purpose of filtering.

In fact, one end of the common-mode inductor is connected to the disturbance source, and the other end is connected with the disturbance device. It is usually used together with the capacitor to constitute a low-pass filter so that the common-mode EMI signal can be controlled at a very low level. The circuit can not only suppress the external EMI signal but also attenuate the EMI signal generated from the internal operation circuit, thus effectively reducing the strength of EMI.

For the ideal inductor-model, when the coil is wound, all the magnetic flux is concentrated in the center of the coil. But in the general case, the circular coil will not be wound as a complete circle or the coil is not tightly wound, which will cause the leakage of the magnetic flux. The

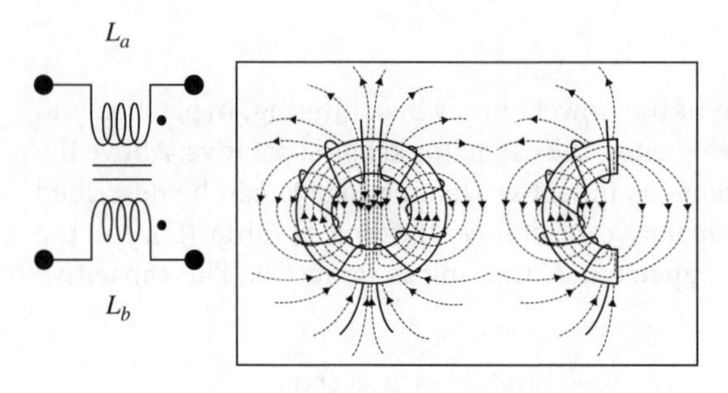

Figure 4.1 Schematic diagram of differential-mode in common-mode ring magnetic core.

common-mode inductor has two windings, and there is a large gap between them, which will cause the leakage of the magnetic flux and the differential-mode inductance is produced. Therefore, the common-mode inductor generally is able to attenuate the differential-mode interference.

In the design of the filter, the leakage inductance can be used. If only one common-mode inductor is installed in the ordinary filter, the leakage inductance of the common-mode inductor can be used to contribute certain amount of the differential-mode inductance to attenuate the differential-mode current. Sometimes, the leakage inductance can also be artificially increased, in order to increase the differential-mode inductance and achieve the better filtering effect.

4.1.1.4 Ferrite Beads and Ferrite Ring Cores

When the inductor cannot be used in high frequency, what should be used? The use of ferrite bead or ferrite ring core is an option. Ferrite material is an alloy of ferromagnetic or iron-nickel. This material has a high-frequency magnetic permeability and high-frequency impedance, and the capacitance between windings is the smallest. Ferrite bead is usually used for high-frequency application. At low-frequency, its low-inductance results in small line loss. At high-frequency, its impedance is basically reactive and related with the frequency, as shown in Figure 4.2. In fact, ferrite bead is a high-frequency attenuator for the radio frequency (RF) energy.

Ferromagnetic material can be expressed by the chemical molecular formula MFe_2O_4. In this formula, M stands for the divalent metal ion, such as manganese, nickel, zinc, copper, etc. Ferrite is manufactured by sintering a mixture of these metal compounds. Its main characteristic is that the resistivity is much larger than that of the metallic magnetic material. This inhibits the generation of the eddy current, which makes the ferrite magnetic material suitable for the high-frequency application. First, according to a predetermined compound ratio, the high-purity powder oxide (such as Fe_2O_4, Mn_3O_4, ZnO, NiO and others) is mixed uniformly, after sintering, grinding, graining, and molding under high-temperature (1000–1400 °C). The sintered ferrite products can be shaped to the finished product by mechanical processing. Different ferrite material is chosen for different application. According to the different frequency ranges of its application, the application frequency band of the ferrite can be divided into the middle- and low-frequency range (20–150 kHz), the middle high-frequency range (100–500 kHz) and ultra-high-frequency range (500 kHz–1 MHz).

In fact, the ferrite bead can be better explained by the paralleling of the inductance and resistance. At low frequency, the inductance shorts the resistance, but at high frequency, the inductance is high, so that the current can only pass through the resistance. According to the different frequency of the disturbance to be suppressed, the ferrite materials with different magnetic permeability can be chosen. The higher the permeability of the ferrite materials is, the higher the impedance is at low-frequency, and the smaller the impedance is at high-frequency. In addition, the ferrite material with high-permeability usually has higher dielectric constant, when the conductor passes through it, and a large parasitic capacitance appears, which also reduces its high-frequency impedance.

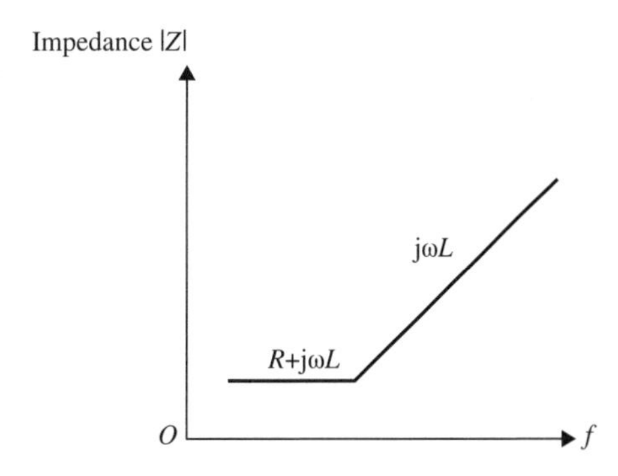

Figure 4.2 High-frequency characteristics of ferrite beads.

The determination of the ferrite size, the larger the difference between the inner diameter and the exterior diameter of the circular is, and the longer the axial is, and the larger the impedance becomes. But the inner diameter must wrap the wire tightly. Therefore, in order to obtain high attenuation, the large volume of magnetic ring should be used.

The turn number of the common-mode choke, the impedance, at low frequency, can be increased through increasing the number of the turns circled on the magnetic ring core, but the impedance at high frequency is decreased since the parasitic capacitance is increased, so it is a common mistake to blindly increase the number of turns to increase the attenuation. When the interference frequency band to be suppressed is wide, the turns with different number can be wound on the two rings. For example, there are two radiation frequency points over which the emissions exceed the standard limit in a device. One is 40 MHz, the other is 900 MHz After checking, it is determined that these two frequency points are caused by the common-mode radiation of the cable. A magnetic ring (one turn) is positioned on the cable, and then the emission at 900 MHz is significantly reduced and not over the standard limit, but the emission at 40 MHz is still excessive. When the ferrite ring core is wound with three turns, the emission at 40 MHz is decreased and not over the standard limit, however the emission at 900 MHz exceeds the standard limit again. This is because that increasing the number of the ferrite ring core on the cable can only increase the impedance at low frequency, but the impedance at high frequency will be decreased. The reason for this phenomenon is the increased parasitic capacitance. Due to the effectiveness of the ferrite ring core depends on the impedance of the original common-mode loop, if the impedance of the original loop is lower, the effectiveness of the ferrite ring core will be more obvious. So when the filter with the capacitor is installed at both ends of the original cable, the loop impedance is very low, so that the effect is more obvious.

Ferrite bead belongs to the *power dissipation component*. It consumes high-frequency energy in the form of heat. This can only be explained by the characteristics of the resistance rather than the inductance.

4.1.1.5 Filters

A filter is a two-port network. It has the characteristics of frequency-selecting, namely certain frequencies can pass it, and other frequencies are blocked. The basic circuit of the filter for the power supply line is shown in Figure 4.3.

The filter is a passive low-pass network composed of an inductor, a capacitor, and a common-mode inductor. Among them, the L_1 and L_2 can constitute the common-mode inductor; the inductance range of the common-mode choke is from 1 mH to tens of mH, depending on the disturbance frequency to be filtered. For the lower frequency filtering, larger inductance is needed; L_3 and L_4 are the independent differential-mode inductors. If one end of the filter is connected to the disturbance source and the load end is connected with the disturbed equipment,

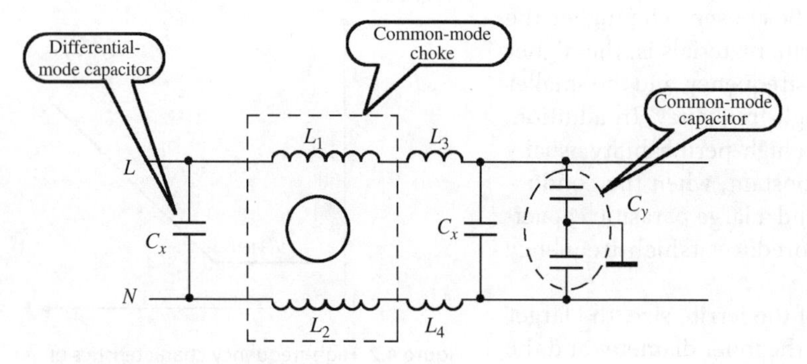

Figure 4.3 The basic circuit of power line filter.

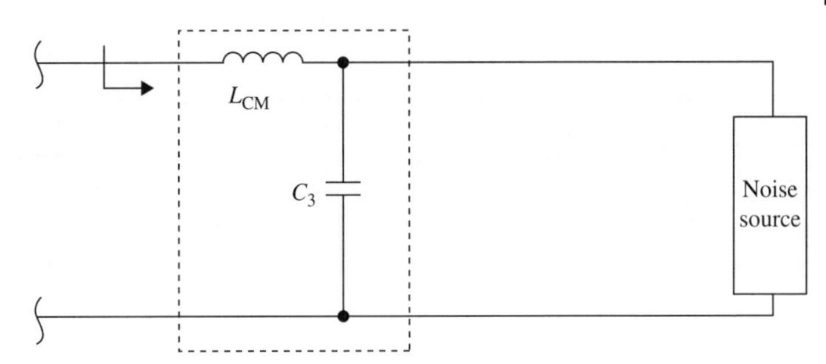

Figure 4.4 Equivalent circuit of common-mode filter network.

then L_1 and C_y, L_2 and C_y separately constitute the low-pass filter from L to earth and from N to earth, which is used to suppress and attenuate the common-mode EMI disturbance existing on the power supply line, it is controlled to a very low level. Among them, the capacitance of C_y cannot be too large, otherwise the leakage current flowing through it will exceed the leakage current limit (3.5 mA) regulated in some safety standards, generally it shall be below the 10 000 pF. In medical equipment, the requirement for leakage current is stricter. In medical equipment, the capacitance of the capacitor is much smaller, even null. The equivalent circuit of the common-mode filtering network is shown in Figure 4.4, which is composed by L_{CM} and C_y. Due to various reasons, for the common-mode inductor, the magnetic material is impossible to be absolutely uniformed, and the two coils could not be wound completely symmetrically, so that the inductance of L_1 and L_2 does not equal each other (sometimes artificially increasing the leakage inductance of the common-mode choke to increase the differential-mode inductance), so the differential-mode inductance L_{DM} is formed. L_{DM} combined with the independent differential-mode inductor L_3 and L_4, and the capacitor C_x, constitutes the low-pass filter between L and N, which is used to suppress the differential-mode EMI signal existing on the power supply line.

Figure 4.5 shows an equivalent circuit of the EMI signal filtering network. $L_{DM}$is the differential-mode inductance, which consists of the independent differential-mode inductors and the leakage inductance of the common-mode choke. C_{LL} and C_{LL2} are the equivalent parallel capacitances of the capacitor C_x in Figure 4.3. Compared with Figure 4.5a and b, C_{LL2} is added, and its capacitance selection determines whether the filtering network is mismatched with the load.

For example, when the power filter is installed in the system, it can effectively suppress the external interference signals passing into the device, and it also greatly attenuates the disturbance signal transmitted to the power network, which is generated when the device is operating.

The schematic shown in Figure 4.3 is a passive filtering network. It has the reciprocity characteristic. Choosing the appropriate filter, and taking a good installation, grounding and wiring, you can get satisfactory results. After the power supply filter is installed in the product, it can effectively suppress the external interference signals passing into the device, such as the transient disturbance signal EFT/B and so on, and also greatly attenuate the interference transmitted to the power network, which is generated when the device is operating, for example, for the conducted disturbance and the radiated disturbance form the switching-mode power supply products. The interference should be controlled below the limit level regulated in the relevant EMC standards.

In a typical filter, the main function of the common-mode choke is to filter out the low-frequency common-mode interference. At high frequency, due to the presence of the parasitic capacitance, the common-mode choke has a small suppression effect on the interference, and the common-mode filtering capacitors are mainly relied on for high-frequency filtering.

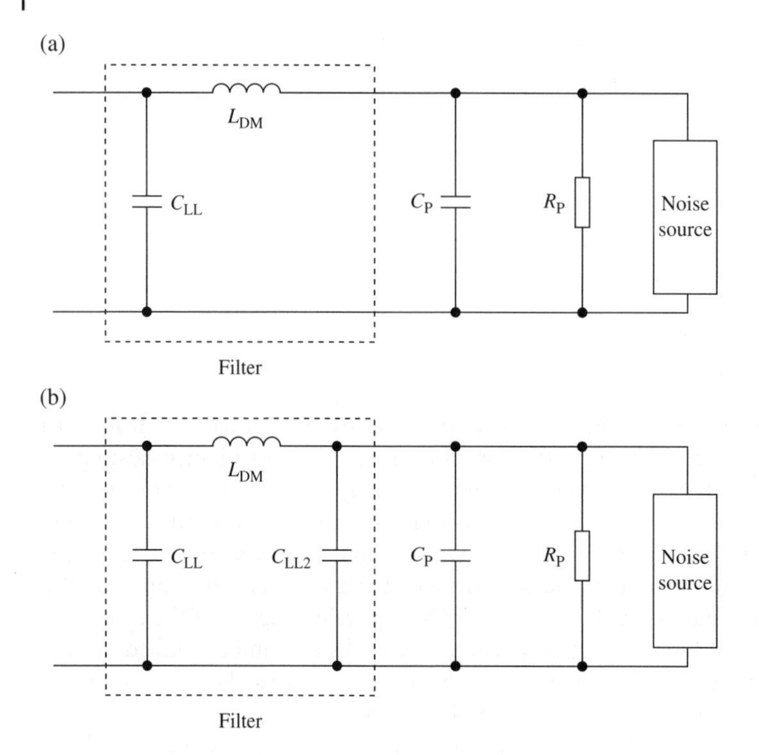

Figure 4.5 Equivalent circuit of differential-mode filter network.

For the medical equipment due to the leakage current limit, sometimes the common-mode filtering capacitor cannot be used, so we need to improve the high-frequency characteristics of the common-mode choke.

The filtering effectiveness of the basic circuit is very limited, and it only suits the case with the minimum requirements. To improve the filtering effectiveness of the filter, we can increase the number of components on the basis of the basic circuit. Here are two some common circuits:

1) Strengthen the differential-mode filtering, such as two differential-mode inductors are in series with the common-mode choke to increase the differential-mode inductance; on the right of the common-mode filtering capacitor, add two differential-mode inductors, while on the right of the differential-mode inductance, add a differential-mode filtering capacitor.
2) Strengthen the common-mode filtering, on the right of the common-mode filtering capacitor, adding a common-mode choke to constitute a T-type filter for the common-mode disturbance. To strengthening the common-mode and differential-mode filtering, add a common-mode choke on the right of the common-mode choke, plus a differential-mode capacitor. Note: Generally, increasing the capacitance of the common-mode filtering capacitor is not used to enhance the common-mode filtering effectiveness, because poor grounding can make the filtering effectiveness even worse.

The power line filter is commonly used to meet the EMC requirements. There are many kinds of the power line filters on the market, and choosing the filter can indeed be a headache. The filter insertion loss is the most important indicator for the filter. Since both common-mode interference and differential-mode interference are on the power line, the insertion loss of the filter is also divided into the common-mode insertion loss and the differential-mode insertion loss. The bigger insertion loss is better. The ideal power line filter should have a greater attenuation for the signals over all the frequencies except the power frequency, so the frequency range of the effective insertion loss should cover the entire frequency range of the potential interference. But the datasheet of almost all power line filters gives only the attenuation characteristic below 30 MHz. This is because the test frequency of the conducted emission test regulated in EMC standards is only up to 30 MHz (10 MHz regulated in military standard), and the actual attenuation effectiveness of most of the filters, at more than 30 MHz, begins to deteriorate, but in practice, the high-frequency characteristics of the filter is very important. The main reason for the poor high-frequency characteristics of the power line filter includes two aspects, the internal parasitic coupling caused by space, and the less-than-ideal filtering components. Therefore, the ways to improve the high-frequency characteristics include two aspects:

1) In the internal structure, the lines connected to the filter are positioned in one direction. If the is enough space, a certain distance between the inductor and the capacitor shall be maintained. If necessary, place an isolation plate to reduce the space coupling.
2) For the filtering components, the parasitic capacitance of the inductor shall be controlled. If necessary, you can use multiple inductors in series. The leads of the differential-mode filtering capacitor should be as short as possible, and the leads of the common-mode capacitor should be as short as possible. The understanding and precautions of this requirement is the same as that of the differential-mode capacitor.

4.1.2 Surge Protection Components

There are several aspects to protecting products from electrical surges. This section briefly reviews them, discussing both the components and the functions of components in surge protection.

4.1.2.1 Gas Discharge Tubes

Gas discharge tube is a switch-type protection component. The schematic symbol of gas discharge tube (GDT) is shown in Figure 4.6.

The operating principle of the GDT is based on gas discharge. When the voltage between the two electrodes is high enough, the gap between electrodes will be broken down, and it is transformed from the original insulation state to the conduction state, which is like a short circuit. Then the voltage between the two electrodes is very low, generally between 20 and 50 V, so it can play a protection role for the post-stage circuit. The main parameters of GDT are the response time, the DC breakdown voltage, the impulse breakdown voltage, the current handling capacity, the insulation resistance, the capacitance between electrodes, and the interruption time for the follow current.

The response time of the GDT can be from hundreds of nanoseconds to several seconds. It is the slowest responding transient protection component. When the lightning overvoltage on the cables breaks down the GDT in the surge protection device, the initial breakdown voltage

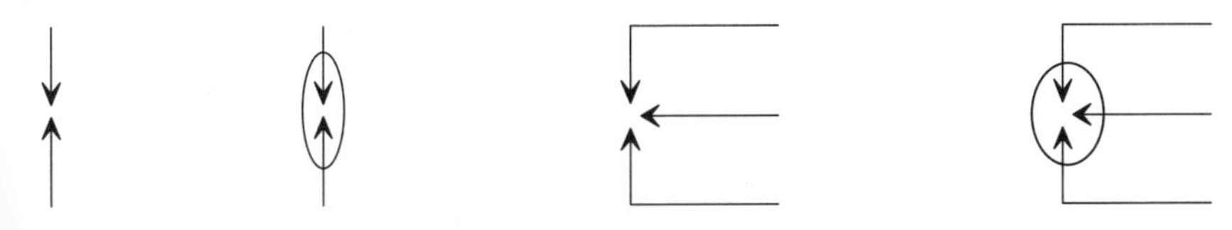

Figure 4.6 Schematic symbols of the gas discharge tube.

basically is the impulse breakdown voltage of GDT, more than 600 V in general. After the discharge tube is broken down, the voltage between the two electrodes is decreased to 20–50 V. In addition, the current handling capacity of the GDT is higher than that of the varistor and the transient voltage suppressor (TVS). When the GDT and TVS or other protection components are used together, most of the current will pass through the GDT, therefore the GDT is generally used in the most front stage of the protection circuit, followed by the post-stage protection circuit composed of the varistor or TVS.

The response time of these two devices (varistor and TVS) is very fast, and their protective effectiveness on the circuit is better. The insulation resistance of the GDT is very high, which can be in the mega-ohm range. The capacitance between the electrodes is very small, generally below 5 pF. The leakage current between the electrodes is very small, only in the nA level. Therefore, connecting the GDT between line and line will not cause any impact on the circuit.

Interrupting the flow current of the GDT is a problem that needs to be considered in the circuit design. As mentioned before, the residual voltage of the GDT when it is in the conduction state is 20–50 V. For the application in the circuit with DC power supply, if the voltage between these two lines exceeds 15 V, the GDT can't be directly used between these two lines. When used in the 50 Hz AC power supply circuit, although there is zero crossing for the AC voltage, the flow current of the GDT can be interrupted, but after the GDT is repeatedly broken down and conducted, the flow current interruption ability will be greatly deteriorated. After long-term using, the zero crossing in the alternating current circuit can't interrupt the flow current. Therefore, in the AC power supply circuit, the GDT cannot be directly connected between the phase line and the protective earth line, between the neutral line and the protective earth line. For the protection between the above mentioned lines, the GDT in series with a varistor can be connected between them. Between phases and neutral, the GDT is basically not used.

We should pay attention to the selection of the GDT, regarding the DC breakdown voltage and the impulse breakdown voltage, current handling capacity and other parameters in the design of lightning protection circuit. The discharge tube used in the general AC power supply lines shall not be activated in the normal operation voltage range of the lines plus the allowable fluctuation, and its DC breakdown voltage shall meet $\min(u_{fde}) \geq 1.8U_P$. In this formula, u_{fdc} is the DC breakdown voltage; $\min(u_{dfc})$ is the minimum DC breakdown voltage; U_P is the peak voltage of the normal operating voltage of the lines.

The GDT is mainly used for the protection between the AC power supply phase and neutral lines and the protective earth line, between the DC power supply port and the protective earth, between signal line and ground, between the inner wire and the shield layer of the RF feeder.

The failure-mode of GDT in most cases is open circuit. Due to the circuit design or other reasons, the GDT is in short circuit for a long time and then it burns out, which can also cause the short-circuit failure mode for the GDT. The life span of the GDT is relatively short. After repeated current shock, its performance will decline. Therefore, the maintenance and replacement of the lightning protection device composed of a GDT is necessary after a long time.

4.1.2.2 Varistors

Varistor is a kind of voltage-clamping protection device. Figure 4.7 shows the schematic symbol of the varistor. Using the nonlinear characteristic of the varistor, when overvoltage appears in between the electrodes of the varistor, the varistor can clamp the voltage to a relatively fixed value, in order to achieve the protection for the post-stage circuit. The main parameters of the varistor are the voltage, current-handling capacity, junction capacitance, and response time.

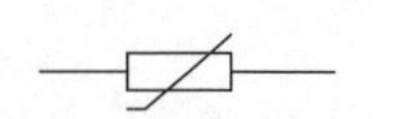

Figure 4.7 Schematic diagram of the pressure sensitive resistance.

The response time of the varistor is in the ns level, which is faster than that of the air discharge tube, slight slower than that of the TVS. It is generally used for the overvoltage protection of the electronic circuits; the response speed can meet the requirements. The junction capacitance of the varistor is generally in the order of hundreds pF to thousands pF, and it is not suitable for the protection of high-frequency signal lines in many cases. In the protection of the AC circuit, its leakage current is increased because of its large junction capacitance, so it needs to be fully considered in the design of the protection circuit. The current handling capacity of the varistor is big, but smaller than that of the GDT.

The varistor voltage (U_B) and the current handling capacity are the key parameters to be considered in the design of the circuits. In the DC circuit, U_B should be (1.8–2) U_{dc}. In this formula, U_{dc} is the DC operating voltage in the circuit. In the AC circuit, U_B should be (2.2–2.5) U_{ac}. In this formula, U_{ac} is the AC operating voltage in the circuit. The selection rule is mainly to ensure a proper safety margin when the varistor is applied in the power supply circuit. In the signal circuit, U_B should be (1.2–1.5) U_{max}. U_{max} is the peak voltage in the signal loop. The current handling capacity of the varistor should be determined according to the requirement for the lightning protection circuit design. Generally speaking, the current that the varistor can withstand twice without damage shall be larger than the designed current handling capacity of the lightening protection circuit. The varistor is mainly used for the DC power supply, AC power supply, low-frequency signal line, and the antenna feeder with power supply feeding. The failure mode of the varistor is mainly short circuit. When the current passing through it is too large, an open circuit may appear because the chip has exploded. The service life of the varistor is short, and the performance can decline after repeated current shock. As a result, the maintenance and replacement for the lightning protection device composed of the varistor is necessary after a long time.

4.1.2.3 Voltage Clamping Type Transient Suppression Diode (TVS)

TVS is a kind of voltage clamping protection device. Figure 4.8 shows the schematic symbols of the TVSs with several packages. Its function is similar to that of the varistor. Using the nonlinear characteristics of the device, the overvoltage can be clamped to a lower voltage to protect the post-stage circuit. The main parameters of the TVS are the reverse breakdown voltage, the maximum clamping voltage, the transient peak power, the junction capacitance, and the response time.

The response time of TVS can reach the level of ps, which is the fastest one of the overvoltage protection devices. Its response speed is satisfied when it is used to protect the circuit. According to the unction capacitance of the TVS, and the different manufacturing process, TVS can be roughly divided into two types, the TVS with high-junction capacitance, which is in 1 pf to tens of pF range; The TVS with low-junction capacitance, which is in the tens of pF range. Generally, the through-hole TVS has high-junction capacitance, and the surface-mounted TVS includes both types. For the protection of the high-frequency signal line, the low-junction capacitance TVS should be used.

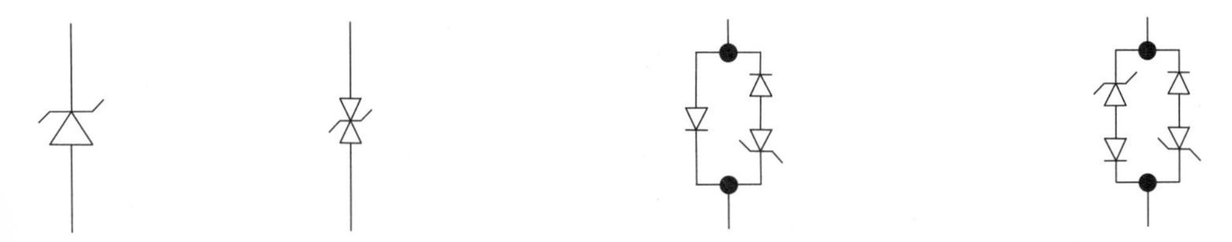

Figure 4.8 TVS schematic diagram of several packages.

The nonlinear characteristic of TVS is better than that of the varistor. When the current of the TVS is increased, the clamping voltage of the TVS is increased slower than that of the varistor. Therefore, the residual voltage of the TVS can be more ideal than that of the varistor. TVS is a good choice in many electronic circuits that need to be carefully protected. The current handling capacity in the voltage clamping protective device is the smallest, and it is generally used for the refined protection of the last stage circuit. Because of its small current handling capacity, it is not generally used in the AC power line protection. For the lightning protection of the DC power supply, the combination of TVS and varistor or other high-current handling capability components can be used. TVS is easy to be integrated, so it is suitable for the use on the PCB board.

Another advantage of TVS is that it can be used in a flexible way, which includes the unidirectional TVS and bidirectional TVS. In the unipolar signal circuit and the DC power supply circuit, the residual voltage of the TVS can be lower than that of the varistor by 50%.

The reverse breakdown voltage and the current handling capacity of the TVS are the key to be considered in the design of the circuits. In the DC circuit, the reverse breakdown voltage of the TVS should be (1.8–2.0) U_{dc}. In this formula, U_{dc} is the DC operating voltage in the circuit loop. In the signal circuit, the reverse breakdown voltage of the TVS should be (1.2–1.5) U_{max}, U_{max} is the peak voltage of the signal loop.

TVS is mainly used for the lightning protection of the DC power supply, the signal lines, and the cable lines. The failure-mode of TVS is short circuit. But when the current is too large, it may also be open circuit because the TVS may be broken down. TVS's service life is relatively long.

4.1.2.4 Voltage Switching-Mode Transient Suppression Diodes (TSS)

Voltage switching-mode transient suppression diode is the same as TVS, and is also the voltage clamping device with the semiconductor technology. But the working principle is similar to the GDT, which is different from that of the varistor and TVS. Figure 4.9 shows the schematic symbol of TSS. When the voltage across the TSS exceeds the breakdown voltage of TSS, the voltage will be clamped to a lower voltage, which is close to 0 V. After that, the TSS will keep this short-circuit state until the current of TSS drops below a critical value, and then the TSS is recovered to the open-circuit state.

TSS has the same characteristics as the TVS regarding response time and junction capacitance. It is easy to be manufactured into a surface-mounted device, which is suitable for a PCB. After the activation of the TSS, the overvoltage is drained from the breakdown voltage to the voltage close to 0 V. At this time the voltage drop of the diode is small, so for the protection of the high-signal level lines (such as the analog subscriber line, ADSL, etc.), its current handling capacity is better than that of the TVS, the protective effectiveness is also better than that of TVS. TSS is suitable for the protection of the signal with high-voltage level.

One thing that we should pay attention to while using TSS is, when TSS is broken down due to the overvoltage, only if the current flowing through the TSS drops below the critical value, TSS can be recovered in the open state. When TSS is used to protect the signal line, the normal current of the signal line should be lower than the critical recovery current of the TSS.

The breakdown voltage and current handling capacity of the TSS should be considered in the circuit design. In the signal circuit, the breakdown voltage of TSS should be (1.2–1.5) U_{max}. U_{max} is the peak voltage of the signal loop.

TSS is mostly used for the lightning protection of the signal line.

The failure mode of TSS is short circuit. But when the current is too large, the TSS may be burned out to open circuit. TSS's service life is relatively long.

Figure 4.9 TSS schematic diagram.

4.1.2.5 Thermistors (PTC)

PTC is a current limiting protection device. Its resistance can be dramatically changed with the increasing of the current passing through the PTC, which is generally used in series with the line for overcurrent protection. When the external cable suffers overcurrent, the self-impedance of the PTC is rapidly increased to limit the current. PTC can be used in the signal line and power line. The response speed of the PTC is slow, generally above millisecond, due to its nonlinear resistance characteristics. PTC will not play its role while the surge overcurrent passes through it. In this situation, only its normal state resistance can be used to estimate its current-limiting effectiveness. The role of thermal resistance is more reflected in the situation where there is long-term overcurrent due to the short circuit between power lines or ground, and it is commonly used in the protection of the user's lines. It is in the open circuit when PTC fails.

At present, there are mainly two kinds of PTC, the polymer PTC and the ceramic PTC. Among them, the overvoltage capability of the ceramic PTC is better than that of the polymer PTC. PTC is generally used in the most prior stage of the protection circuit in PCB board, and the ceramic PTC is preferred.

4.1.2.6 Fuse Tubes, Fuses, and Circuit Breakers

Fuse tubes, fuses, and circuit breakers belong to the protection device. When short circuits, overcurrent, and other faults occur inside the equipment, they can disconnect the load, preventing electrical fires and ensuring the safety features of the equipment.

Fuse tubes are generally used for protection in the PCB board, but fuses and circuit breakers can be used for the protection of the whole machine. The use of the fuse tube is briefly introduced as follows.

For the protection circuit composed by the air discharge tube, varistor, and TVS in the power supply circuit, the fuse tube must be equipped with this protection circuit, in order to avoid the safety problem of the equipment when the protection circuit is damaged. The fuse tube used in the power supply protection circuit is better to be in series with the protective device, so that the protective device can be protected against damage. After the fuse tube is fused, the main power supply will not be impacted. For the protection of the signal lines and the antenna feeder without the power supply lines, it is not necessary to use the fuse tube.

The main characteristics of the fuse tube are the rated current, the rated voltage, etc.

The rated voltage marked on the fuse tube represents that the fuse can completely and reliably interrupt the rated short circuit current of the circuit when the operating voltage of the circuit is not more than its rated voltage. The series of the rated voltage rating series is included in the NEC regulations, and is also a requirement for the manufacture's laboratory and a protective measure to prevent the fire hazard. For most of the small size fuse tubes and miniature fuse tubes, the standard rated voltage adopted by the fuse tube manufacture is classified as, 32, 125, 2.5, 600 V.

For the equipment in which the short circuit current of the power supply is limited by the circuit impedance to be lower than 10 times of the rated current of the fuse tube, a common practice is that the fuse with the rated voltage of 125 or 250 V can be used in the secondary circuit with 500 V or even higher operating voltage.

Generally speaking, the fuse tube can be used in the circuit with any operating voltage lower than its rated voltage, and its fusing characteristics are not deteriorated. The rated current can be determined according to the current handling capacity of the protective circuit. For the fuse tube used in the protective circuit, it is advised to use explosion-proof type and slow acting fuse tube. The slow acting fuse tube is also known as the time delay fuse tube, due to the time delay characteristic, the circuit can be kept intact while nonfaulty pulse current appears and the fuse

Table 4.1 Some protective tube rated value and measured value of bearable surge current.

Fuse type	SST1	SST2	SST5	SMP500	SMP1.25
Rating value	1	2	5	0.5	1.25
Nominal melting heat (A2sec)	1.2	5	38	1.4	14
Surge current peak (A)	300	580	1300	300	830

tube can provide the protection for the long time overload. The general fuse tube cannot sustain high-surge current, if the general fuse tube is used in the circuit which needs surge protection circuit, it might not meet the test requirements. Also, if the large size the fuse tube is used, the overload protection may be lost. The fusing part of the time delay fuse tube is made by a special processing, which has the function of absorbing energy, and adjusting the quantity of the absorbed energy, the fuse tube can withstand the impulse current and protect the circuit against overload. The related standards regulate the time delay characteristics. Table 4.1 lists some ratings of the fuse tube and the measured surge currents which the fuse tube can withstand, for reference only.

4.1.2.7 The Functions of Inductors and Resistors in Surge Protection

Inductors, resistors, capacitors, and wires themselves are not the protective devices, but in the protection circuit composed of many different protection devices, they can play the role in protecting against surges.

Among the protective devices, the features of the GDT are that its current handling capacity is high, but its response time is slow, and its impulse breakdown voltage is high. The current handling capacity of TVS is small, its response time is the fastest, and the characteristic of voltage clamping is the best. The characteristics of the varistor are in between these two. When a protection circuit is required for a high-current handling capacity and refined protection, protection circuits often need these protective devices cooperating together to achieve the relatively ideal protection characteristics. But these protective devices cannot be used in parallel. For example, if varistor with high-current handling capacity is directly in parallel with TVS with small current handing capacity, under the condition of overcurrent, the TVS will be first damaged, so the varistor cannot play its advantage of high-current handling capacity. Therefore, in the use of combining several protective devices, the inductor, the resistor, and other components are often inserted between two protective components. These elements are described as the following:

- *Inductors:* In series with the DC power supply protection circuit, a great voltage drop on the power supply lines is unacceptable, so the air core inductor can be used between two protective components, as shown in Figure 4.10. When the designed current handling capacity (higher than that of the TVS) of the protection circuit is reached, the overcurrent on the TVS shall be lower than its rated current handling capacity; therefore, the inductor needs to provide enough current-limiting capability for the surge current. Taking the circuit shown in Figure 4.10 as an example, the inductance of the air core inductor is, with 8/20 μs impulse current as the reference, the measured varistor residual voltage is U_1 at the designed current

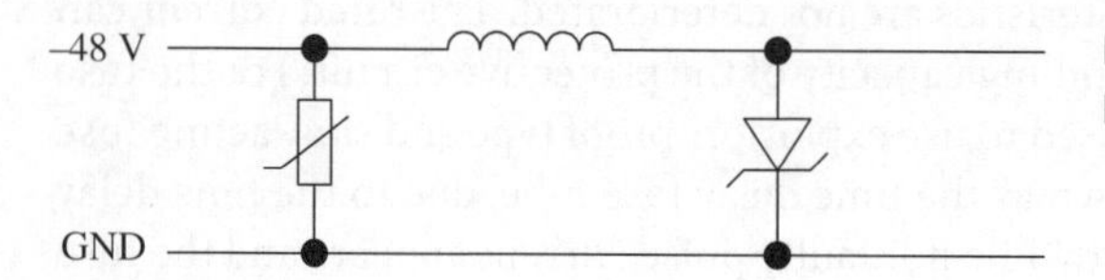

Figure 4.10 Inductance in series between the pressure sensitive resistance and TVS.

handling capacity. Checking the datasheet of the TVS, with the 8/20 μs impulse current waveform, the maximum current that the TVS can withstand is I_1 and the maximum clamping voltage of the TVS is U_2, for the 8/20 μs impulse current waveform. Its rise time is $T_1 = 8$ μs, the pulse duration time $T_2 = 20$ μs, the minimum inductance is

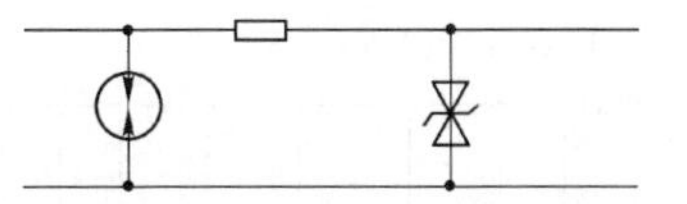

Figure 4.11 Resistance in series between the gas discharge tube and TVS.

$$L = (U_1 - U_2) * (T_1 - T_2) / (I_1/2)$$

In this formula, the voltage unit is *V*, the time unit is *s*, the current unit is *A*, the inductance unit is *H*. In the power supply circuit, for the inductor design, we should pay attention to several problems:

- The inductor should work normally at the maximum working current of the equipment and without overheating.
- The air core inductor should be used as much as possible. The inductor with magnetic core will be saturated under the condition of overcurrent, so the inductance required in the circuit can only be calculated based on the air core inductor.
- *Resistor:* In the signal line, the components in series with the line shall attenuate the normal high-frequency signal as little as possible, so the resistor can be inserted between two protective components, as shown in Figure 4.11.
  The function of the resistor is basically the same as that of the aforementioned inductor. Take Figure 4.11 as an example: The resistance calculation method is the measured impulse breakdown voltage of the GDT, U_1. Check the datasheet of TVS; for 8/20 μs impulse current, the maximum current the TVS can withstand is I_1 and the highest clamping voltage of TVS is U_2. Then the minimum resistance is

  $$R = (U_1 - U_2) / I_1$$

  While using resistor in the signal circuit, we should pay attention to several issues:
- The resistor's power should be large enough to avoid the resistor is damaged in the condition of overcurrent.
- Use the linear resistor as possible so that the resistor impacts the normal signal transmission will be small as possible.

4.2 Analyses of Related Cases

4.2.1 Case 29: The Radiated Emission Caused by a Hub Exceeds the Standard Limit

[Symptoms]

A communication device uses the cabinet structure design. In the radiated emission test, it is found that the radiated emission exceeds the limit. The initial test spectral plot is shown in Figure 4.12.

As we can see, a lot of continuous noise appears from 30 to 100 MHz. The amplitude of the noise is very high. (The spectral curve with lower amplitude is the ambient noise in Figure 4.12.)

[Analyses]

According to the experience, the low-frequency excessive emission comes from the power supply cable (because of the operating frequency range of the power supply and the long length of the power supply cable). So first, the localization of the radiation source is the power module.

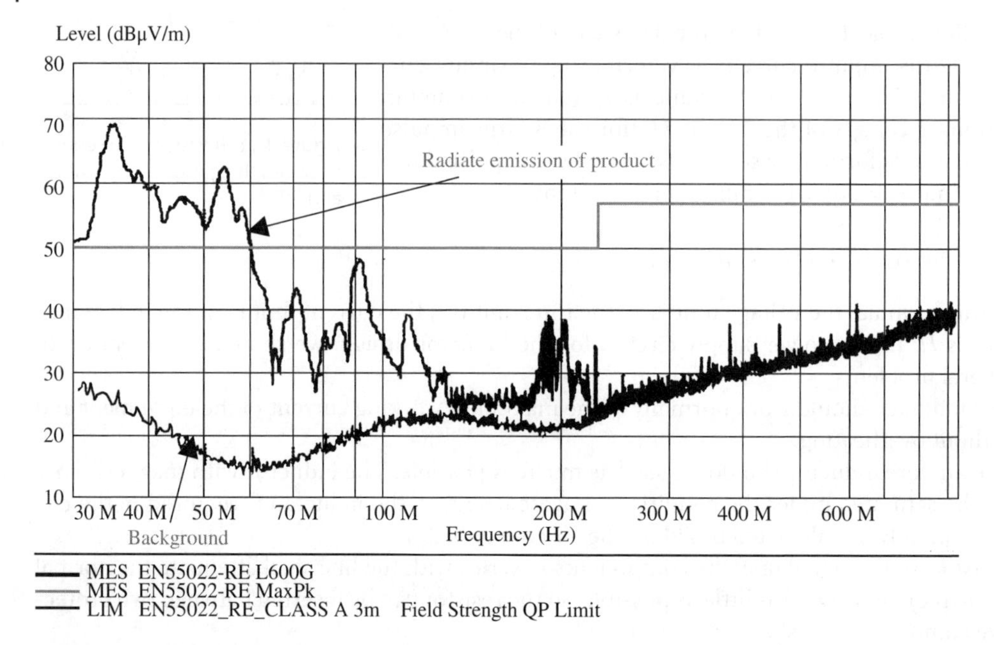

Figure 4.12 Initial test spectrum.

Table 4.2 Outage order table.

Power cut	Compared with initial test results
Cut the power of subrack 1	Almost unchanged
Cut the power of subrack 1 and 2	Almost unchanged
Cut the power of subrack 1, 2 and 3	Almost unchanged

The power supply of the EUT comes from a top power supply box. After lightning protection, filtering, combining and monitoring, the power supply is, respectively, connected to three subracks. There is a 710 μH common-mode choke at the power supply entrance in each sub rack. Trying to connect an additional filter on the power supply cable and test it again, we cannot see any improvement. It can be concluded that the problem is not in the power supply box. After examining the power supply structure of the whole cabinet, we found the cabinet needs three subracks, which need to be supplied. The PCBs vary in different subrack, and their power consumptions are also different. Even the excluding way cannot help completely identify whether the radiated emission is strong or weak as the DB is a relative value unit, but it can be used as a reference for problem locating. Following the procedure in Table 4.2, we turn off the power supply of the subrack, in turn, to locate the radiation source.

During the test, there is almost no change for the test results at low frequency. A hub is installed in the bottom of the cabinet and the height of the cabinet is 2.2 m, shown in Figure 4.13. The function of the hub is to interconnect the signals from the three subracks. If we turn off the power supply of the hub and perform the test, the radiation at low frequency disappears immediately. The spectral curve drops to the curve of the ambient noise. Powering on all the subracks (the hub is still not powered on) and testing it again, the result is also very good. Thus, we can conclude that the radiation is caused by the hub.

The hub operates with 7.5–12 V DC power supply. Because we only have −48 V DC power supply, there is a dedicated power supply board designed for the hub. The −48 V DC power supply starts from the power supply box, passes through the 2.2 m height cabinet, and then arrives at the bottom of the cabinet, where it is connected to the power supply board of the hub. The output voltage of the power supply board is 10 V DC. The length of the power supply wire between the power supply entrance of the hub and the input of the hub's power supply board is 30 cm.

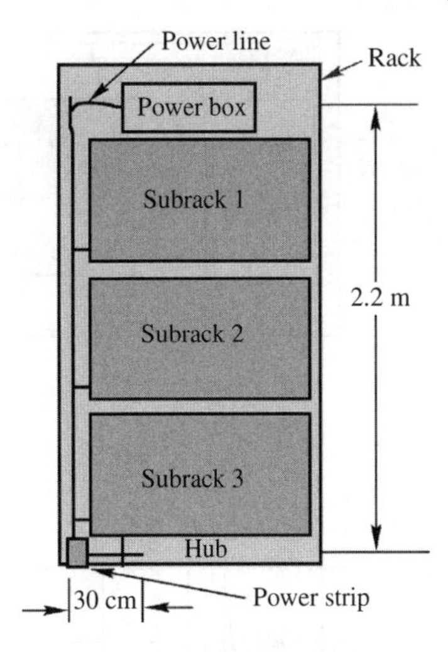

Figure 4.13 Device structure block diagram.

In order to further locate the radiation source and find out whether the radiation is mainly generated by the hub or the power supply board. We can use a linear power supply (because of its linear operating characteristics, the linear power supply have no radiation) to replace the power supply board (it is a switching-mode power supply). Then we test it, the result is shown in Figure 4.14.

In Figure 4.14 we can see that the radiated emission of the hub is compliable with the requirement. It is visible that if we solve the radiation problem in the power supply board of the hub, the radiation problem of the whole equipment can be solved. The circuit of the power supply board is very simple, as shown in Figure 4.15.

The DC/DC module is a radiation source. There is a 100 μH common-mode choke, two 1500 pF Y capacitors, and a 0.22 μF differential-mode capacitor in the −48 V power supply input port. There is no filter in the output stage. The length of the power supply cable between the output of the hub power supply board and the hub is 30 cm. The test results can meet the requirements if we disconnect the power supply wire, as shown in Figure 4.16.

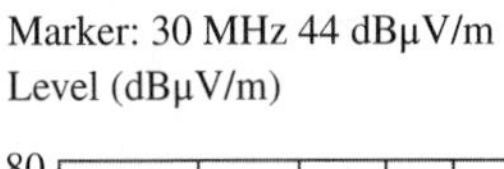

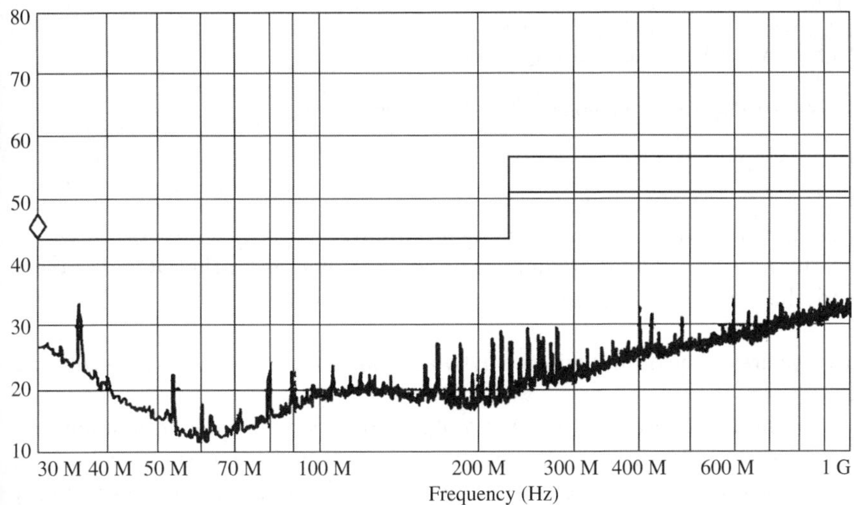

Figure 4.14 Test spectrum diagram of linear power supply instead of switching power supply.

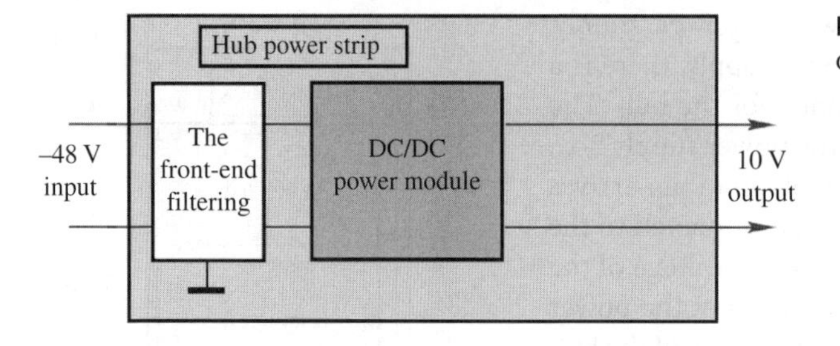

Figure 4.15 Power plate principle block diagram.

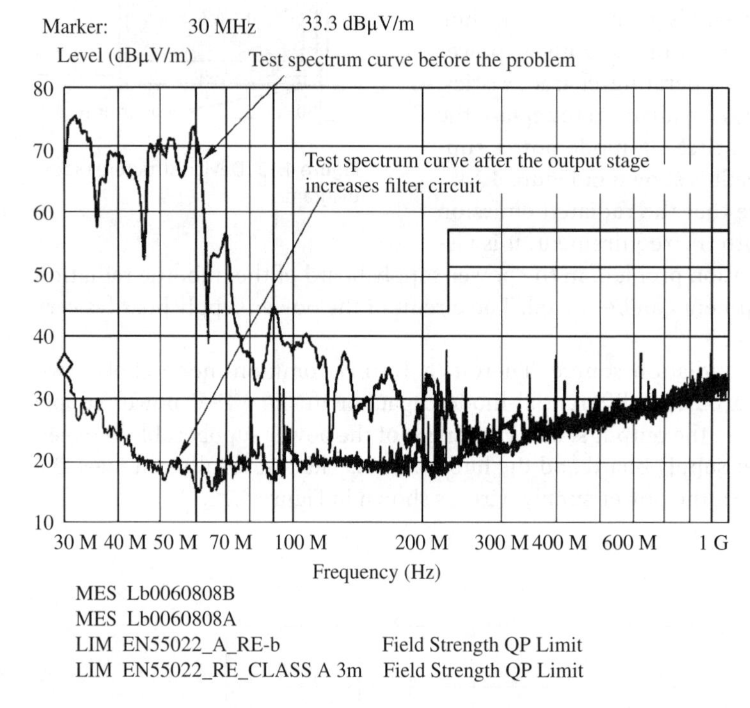

Figure 4.16 Test spectrum diagram after unplugging the output power line.

In the test, we found that when we connect the 30 cm power supply wire, even if we do not connect it to the hub, meaning that there is no load connected to the DC/DC power supply board and the output is floating, the emission is still quite large, as shown in Figure 4.17.

Add the ferrite ring core on the output wire (with two turns) and take the test. The results are shown in Figure 4.18.

It can be seen from the above test that the 30 cm output power supply wire without any disposal is the radiation antenna. The switching-mode power supply in the power supply board is the noise source. Obviously, if we filter and control the noise of the output of the power supply board, we can solve the problem of the radiated emission. Per the suppression effectiveness, the ferrite ring core is a good choice. At the same time, the flexibility of the ferrite ring core brings some convenience to locate the EMC problem. However, the ferrite ring core is only a

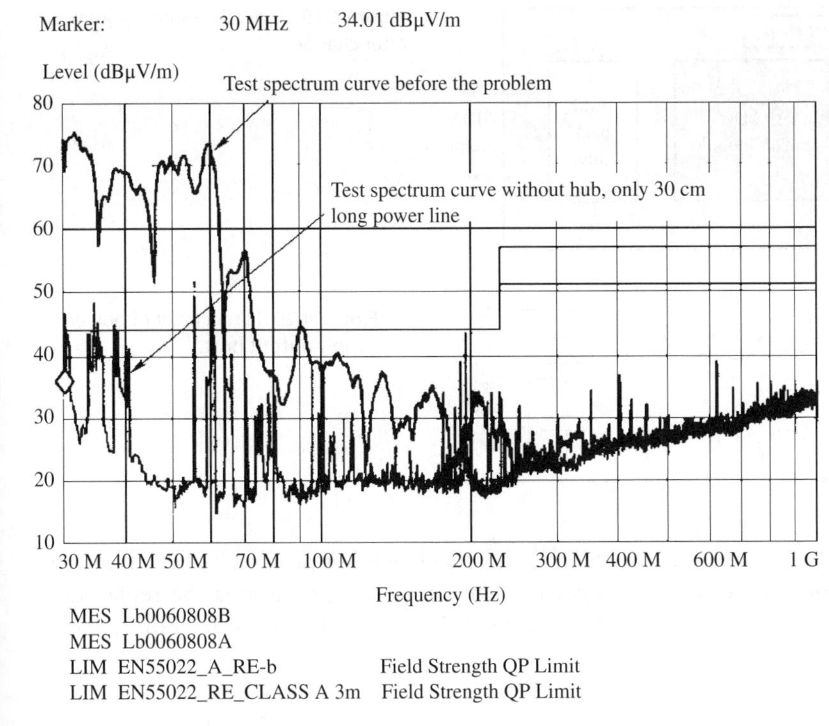

Figure 4.17 After the output power line of the test spectrum.

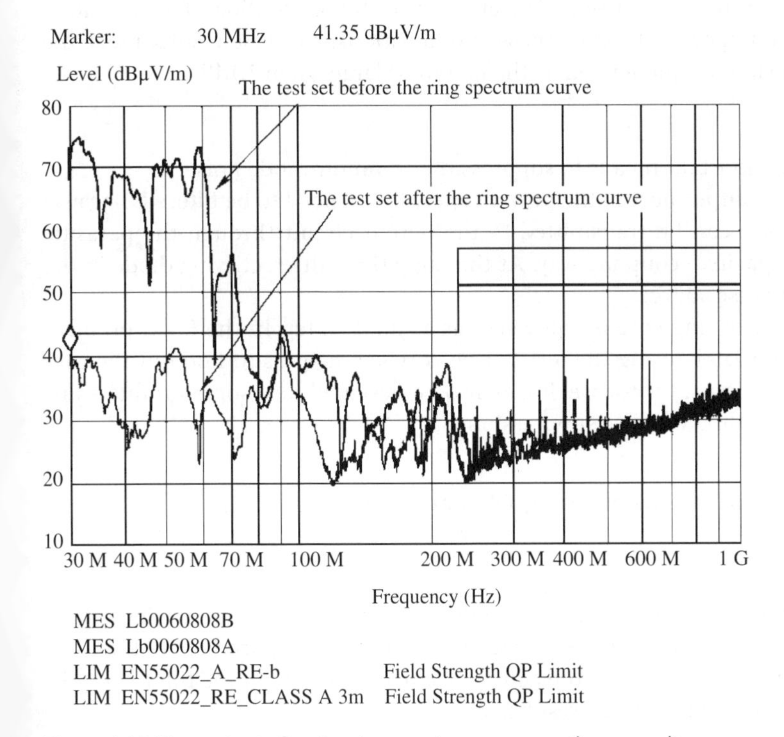

Figure 4.18 The test set after the ring spectrum curve on the power lines.

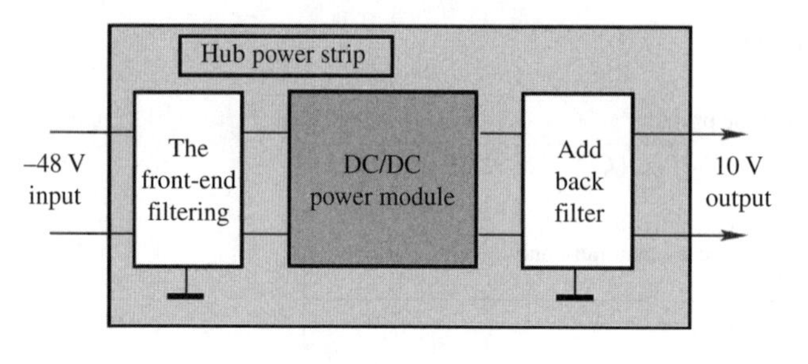

Figure 4.19 Principle frame diagram after change.

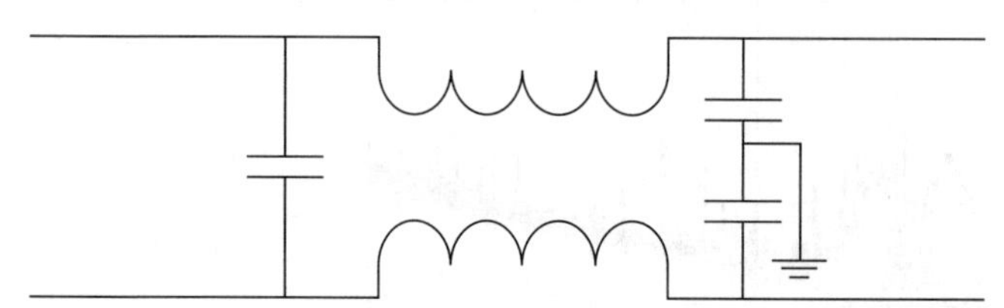

Figure 4.20 Filter circuit of power supply output port.

temporary measure for locating the problem, as there will be other problems of installing and fixing it inside the product cabinet in this case. The ferrite ring core must be replaced by other specific circuits, such as the LC filter. The modified schematic block diagram is shown in Figure 4.19.

[Solutions]

According to this analysis, we add the filtering circuit for the output port of the power supply board, shown in Figure 4.20. We add a 100 nF X capacitor, a 1.3 mH common-mode choke, and two 2200 pF Y capacitors on the output port. We also ensure the grounding of Y capacitors is good (this step is very important). After these modifications, the test results meet the requirement, and over the low-frequency range, the margin is larger than 10 dB.

[Inspirations]

1) Good grounding is an important means to suppress the common-mode noise.
2) Both the input port and output port of the power supply board need to be filtered. Because the common-mode noise can be transmitted to the output circuit through the parasitic capacitance and the capacitive coupling way. At the same time, the rectifying diode in the output circuit is also a noise source.
3) In certain cases, the ferrite ring core is equivalent as a simple filter. In the EMC problem locating process, appropriately using the ferrite ring core can help us to find out the root cause of some EMC problems. Of course, it can also be designed into some products just like some circuits as a method of noise suppression.

4.2.2 Case 30: Installation of the Power Supply Filter and the Conducted Emission

[Symptoms]

The conducted emission test results of a certain household appliance A are shown in Figure 4.21.

We can see that the conducted emission test result cannot meet the requirement of the Class B limit. During the test, we also found that replacing the current filter with the filter with different insertion loss has little impact on the results.

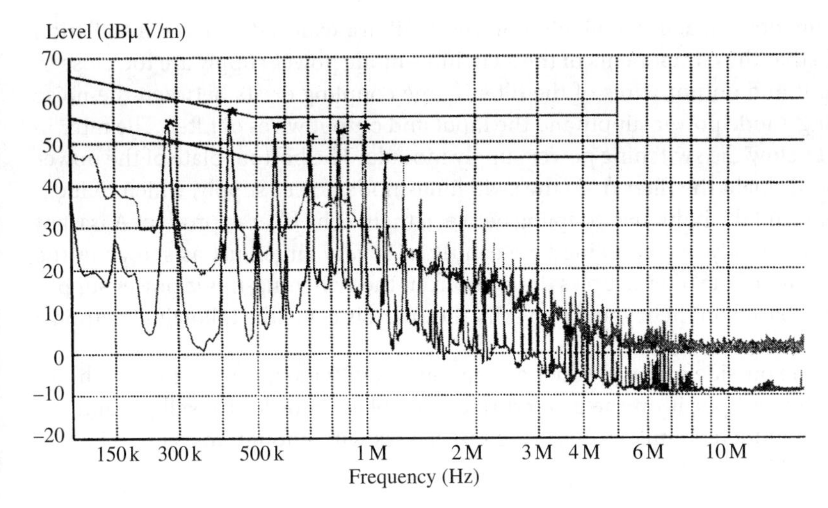

Figure 4.21 Test results of conducting disturbance.

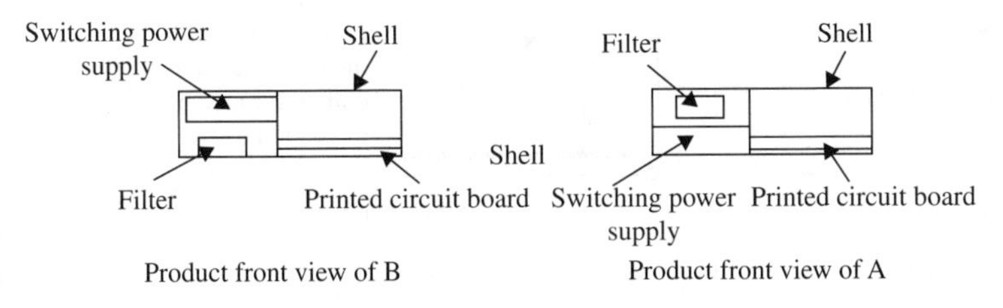

Figure 4.22 Product sketches in the face of A and B.

[Analyses]

We doubt that the bad grounding of the filter leads to the bad filtering performance. Then we put the casing on the ground and use short wire to ground the filter. The test result is the same. Then we doubt that the filtering performance of the filter is not enough to filter the conducted disturbance level generated from the power supply itself and to pass the test, and this leads to that the conducted emission exceeds the standard limit. But, when this power supply is used inside another product B, and the same filter is used, the conducted emission test of product B can be passed. We can say that the power supply cooperating with this filter can meet the requirement of conducted emission test. Then we compared the product A with B, and found that the installation positions of the power supply filter are different. The power supply filter in product B is installed below the power supply and the filter in product A is installed above the power supply, as the Figure 4.22 shows.

The reason why the noise in the switching-mode power supply is generated has been described clearly in other cases (e.g. Case 7). We will not describe it here.

As for the structure features of the AC/DC power supply, we found that only one side shell of the AC/DC power supply module is not closed and the other shells (acting as the heatsink too) is closed. The components (capacitor, coil, etc.) on the PCB are exposed. Per the installation features, we conclude that the components of the switching-mode power supply are too close to the filter and the input and output wires of the filter. Large coupling exists between the noise source in the switching-mode power supply and the input and output wires of filter. The filter in product B is installed below the switching power supply module. The bottom plate of the power supply module can well shield the filter from the switching-mode power supply, which reduces the capacitive coupling and the inductive coupling between them. The filter in product A is right above the switch power supply, the coupling between them is very large. Just as shown in the Figure 4.23, the input wire of the filter couples the noise from the switching-mode power supply.

The above analysis is based on theory, and we verify the analysis using the following methods:

1) Move the filter to the outside of the product casing (put the filter away from the switching-mode power supply) and put it on the ground reference plate in order to well ground the filter. Perform the conducted emission test; the test passes with enough margins. The test results are shown in Figure 4.24.

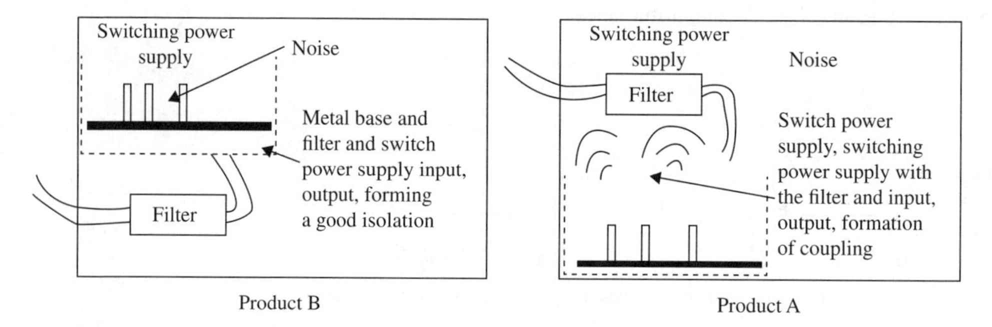

Figure 4.23 Product noise coupling diagram of A and B.

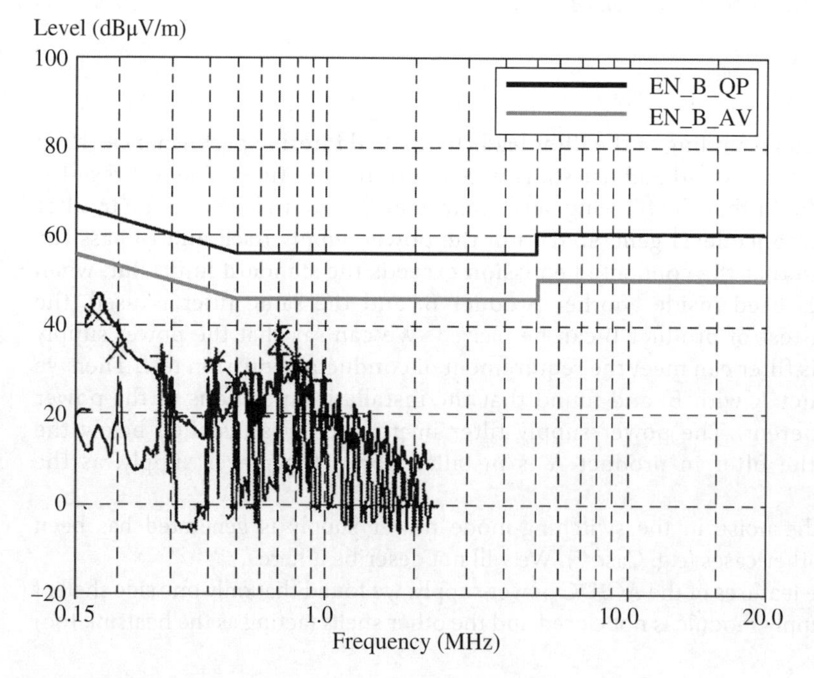

Figure 4.24 The test results after taking the filter out of the case.

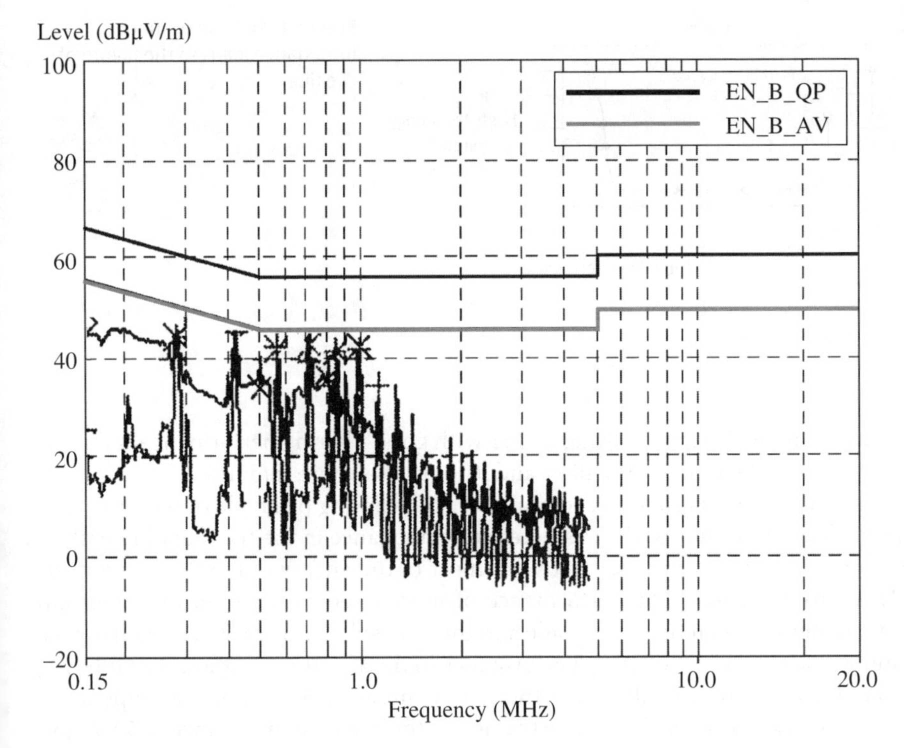

Figure 4.25 The test results were isolated by the copper foil.

2) We use the conductive copper foil to shield the coupling between the switching-mode power supply and the power supply filter and put it in the space between them. The results are shown in Figure 4.25.
3) It is clear that the small distance between the filter and the switching-mode power supply and the bad shielding between them causes the conducted emission to exceed the limit.

In addition, there is another scenario in which the filter fails, which is very common in engineering, although it does not appear in this case. The principle can also be shown in Figure 4.26. When the grounding impedance exists (the filter is connected to the ground by a long wire or a long and narrow printed wire), high-frequency noise will goes into the filter through this impedance path and cause the filter failed. Therefore, the filter should be connected to the ground by a contact point or surface with low-impedance. The wires connected to a high-frequency filter should be as short as possible. For the capacitors in filter, the ceramic capacitor with low-parasitic inductance is the best choice.

[Solutions]

The distance between the filter and the switching-mode power is too small and the shielding between them is too bad. The above characteristic leads to the conducted emission test failed. So the basic method to solve this problem is that we well shield the filter and the power supply module. The detail methods are to:

1) Design the product structure as product B. Place the filter below the power supply, and shield the filter by the bottom plate of the power supply module.
2) Add a metallic cover on the top surface of the power supply module. Consider drilling holes in the cover to help dissipate heat.

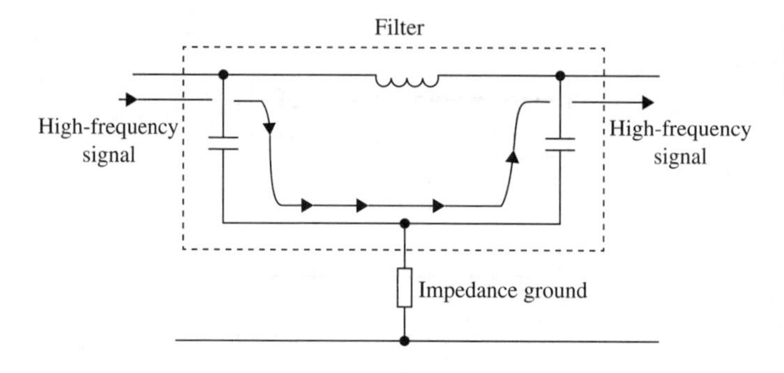

Figure 4.26 High ground impedance leads to the failure of the filter.

[Inspirations]

1) For a good power supply design, choosing a filter with good filtering performance is important, but well isolating the input of the filter and its output is more important.
2) For the filter, its impedance is also very important. The working principle of the filter is to construct large discontinuities for the characteristic impedance in the transmission path of the RF electromagnetic wave, in order to reflect most of the energy of the RF electromagnetic wave back to its source. The performance of most filters is tested in the condition when the impedance between the source side and the load side both are 50 Ω, which means the filter cannot achieve its best filtering performance in the actual application. The filtering parameters are tested in the condition that the source impedance and the load impedance both are 50 Ω, because, for most RF test instruments, their source impedance is 50 Ω, and the filter need 50 Ω load and the cable with 50 Ω characteristic impedance to operate properly. The parameters tested in this condition are perfect, but are also the most misleading. Because the filter is composed of inductors and capacitors, the filter is a resonant circuit. The performance and the resonance of the circuit mainly depend on the impedance in the source side and the load side. In fact, the performance of an expensive 50/50 Ω filter may be worse than a low-price 50/50 Ω filter.

4.2.3 Case 31: Filtering the Output Port May Impact the Conducted Disturbance of the Input Port

[Symptoms]

The structure of an AC/DC switch power supply is roughly shown in Figure 4.27. The right side of the power supply is 220 V AC input port and the left side of the power supply is its DC voltage output port. The cables connected to these two sides are more than 1 m long.

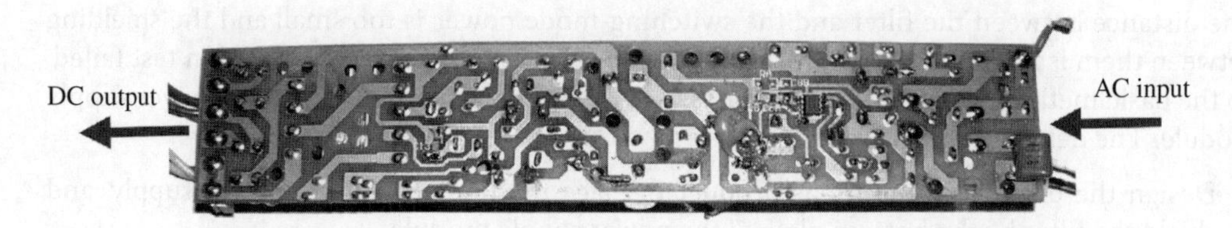

Figure 4.27 The power structure.

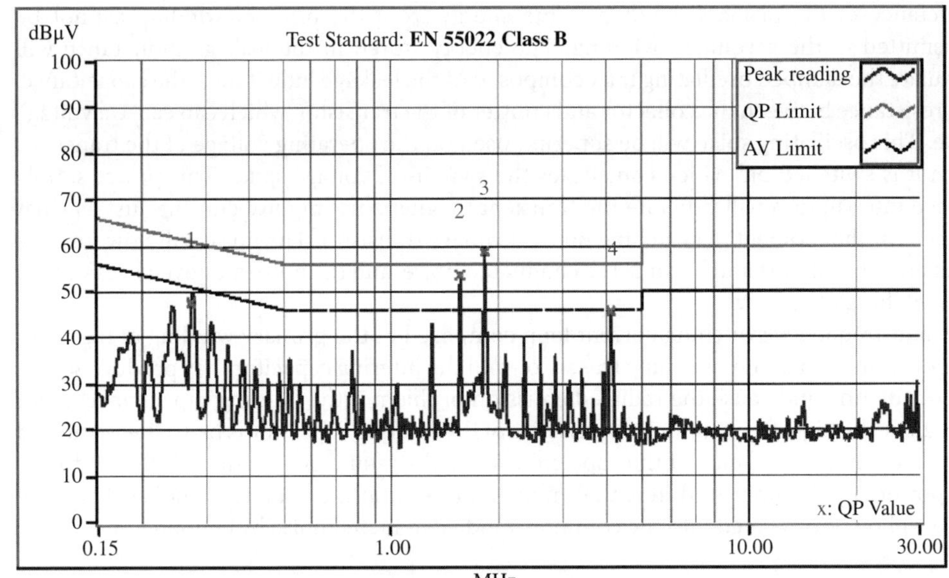

No.	Frequency MHz	Corr. Factor dB	Reading dBμV QP	Reading dBμV AV	Emission dBμV QP	Emission dBμV AV	Limit dBμV QP	Limit dBμV AV	Margins dB QP	Margins dB AV	Notes
1	0.27518	0.78	47.14	47.70	47.92	48.48	60.96	50.96	−13.04	−2.48	
2X	1.55755	0.55	52.95	53.81	53.50	54.36	56.00	46.00	−2.50	8.36	
+3X	1.81994	0.58	57.84	58.59	58.42	59.17	56.00	46.00	2.42	13.17	
4X	4.07131	0.66	44.85	45.45	45.51	46.11	56.00	46.00	−10.49	0.11	

Figure 4.28 Test results of conducting disturbance.

The spectral plot of the conducted emission test results at the AC input port are shown in Figure 4.28.

It can be seen from Figure 4.28, the result cannot meet the requirement.

[Analyses]

According to the principle of the conducted emission test on power port (the principle of the conducted emission test on power port is clearly described in Case 1, the current flowing into the LISN directly affects the test results. In order to get the appropriate result, we must control the current flowing into the LISN.

Electromagnetic disturbance source of switching-mode power supply is the switching devices, the diodes and the nonlinear passive components. The electromagnetic disturbance produced by the switching circuit is one of the main noise sources in the switching-mode power supply. The switching circuit is the core of the switching-mode power supply and it is mainly composed of the switching transistors and high-frequency transformers. The switching circuit generates high *du*/*dt* pulse with high amplitude. The pulse has wide frequency band and rich harmonics. There are two main reasons for the pulse noise:

1) The load of the switching transistor is the primary winding of the high-frequency transformer, and it is inductive load. When the transistor is switched on, there are high inrush current flowing through the primary winding, and a high-voltage spike appearing across the primary winding. At the moment the transistor is switched off, because of the leakage

inductance of the primary winding, some energy from the primary winding cannot be transmitted to the secondary winding. The energy stored in the leakage inductance will stimulate the damped oscillating tank composed of the leakage inductance, the capacitance, and resistance between the collector and emitter of the transistor, which can cause a voltage spike. This oscillating spike will be superimposed on the operating voltage of the transistor when it is switched off, which constitutes the switch-off voltage spike. This power supply voltage interruption will generate the transient magnetizing impulse current, just as what happens at the moment of switching on the primary transistor. The noise will flow into the input and output port and become the conducted noise, and the noise may even break down the switching transistor.

2) The high-frequency switching current loop composed of the primary winding of high-frequency transformer, the switching transistor, and the filtering capacitor may produce severe space radiation and cause the radiated emission problem if the filtering capacitance is not enough or the high-frequency characteristics of the filtering capacitor is bad. Due to the bad high-frequency impedance of the capacitor, the high-frequency current will flow into the AC power supply input in differential-mode and become the conducted noise. Then the noise will be transformed into the common-mode conducted noise by coupling.

In addition, the electromagnetic disturbance generated by the rectifying diode and the freewheeling diode cannot be ignored. Although in the main circuit, the $|di/dt|$ of the reverse recovery current of the rectifying diode is far below that of the freewheeling diode, as an electromagnetic disturbance source to be investigated, the reverse recovery current of the rectifying diode is already large enough and its frequency bandwidth is wide, too. As well, the transient voltage generated by the rectifying diodes is less than the transient voltage generated by switching-on and switching-off the power transistors.

The circuit schematic in this case is shown in Figure 4.29.

The freewheeling diode has RC snubber circuit. Considering the efficiency constraint, the RC snubber circuit is not added for the switching transistor and the shielding is not designed for the transformer. In the actual test, we found that changing or adding the filtering component on the power input port has no effect on the test results. But as long as an inductor (150 μH) is inserted in series with the output port, the test result can meet the requirement.

According to the principle of the conducted emission test and the common-mode coupling transmission in the switching-mode power supply, the output part (including the output power cable) of the switching-mode power supply also affects the conducted emission on the input port. Figure 4.30 shows the principle of the above phenomenon.

According to Figure 4.30, the parasitic capacitance between the primary winding and secondary winding of the transformer (the parasitic capacitance between the primary winding and the secondary winding of an unscreened medium power transformer is about 100–1000 pF) and the parasitic capacitance C_2 (50 pF m^{-1}) between the output power cable and the ground reference plate provide a transmission path for the conducted disturbance. The arrow in the figure indicates the transmission path and the direction of the conducted disturbance. It can be seen that the disturbance transmitted in the parasitic capacitance between the output power cable and ground reference plate will also flow into the LISN and impact the test results. In the test, the common-mode inductance in series with the output port suppresses the common-mode disturbance current, and then the tested conducted emission is reduced.

After further analysis, we find that, in this case, if we add an appropriate capacitor C between the output port and the 0 V ground, part of the noise can be bypassed. Part of the noise will flow into the 0 V ground, then the noise current flowing through C_2 can be reduced, that is, the noise current flowing into the LISN can be reduced. The principle is shown in Figure 4.31.

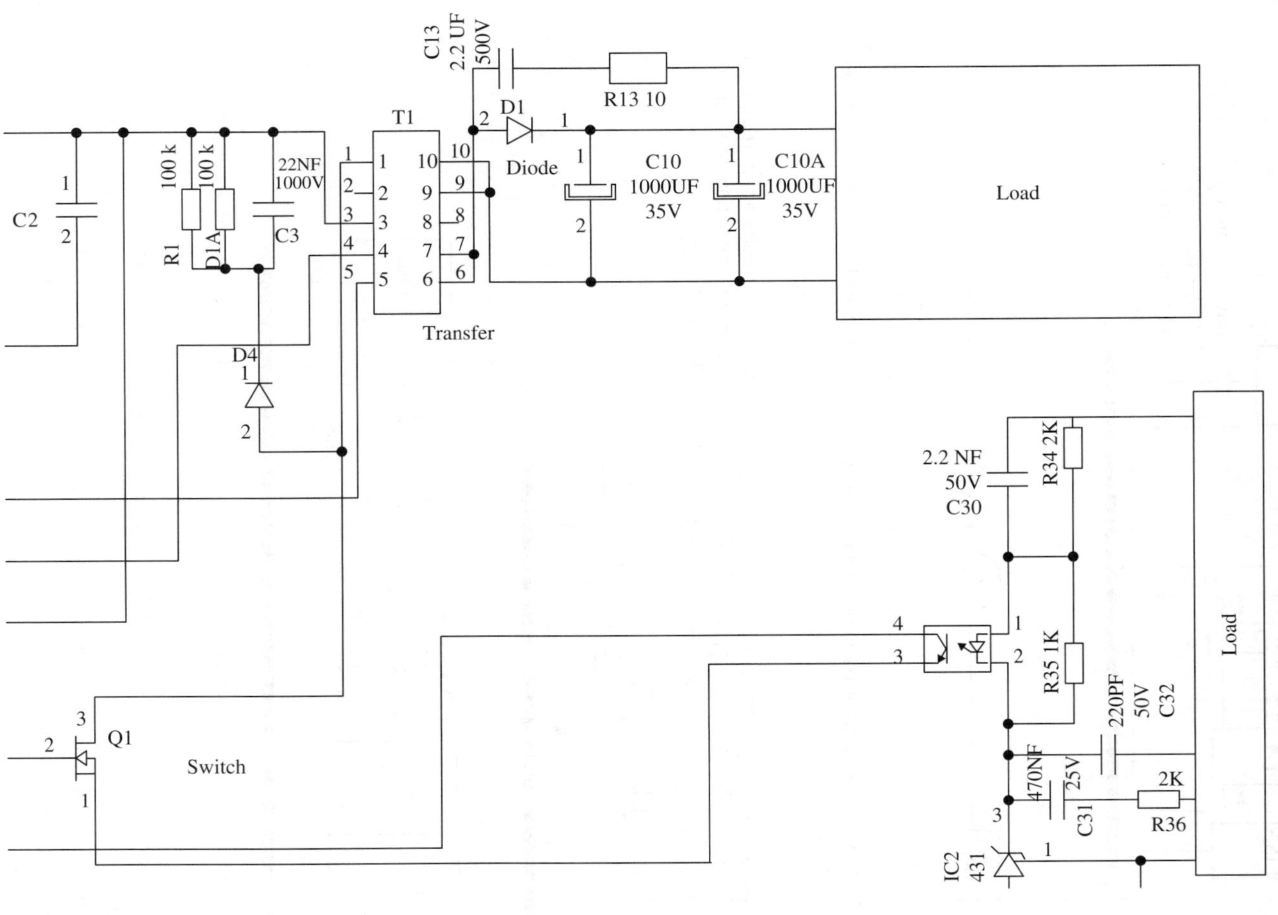

Figure 4.29 Power supply circuit schematic diagram.

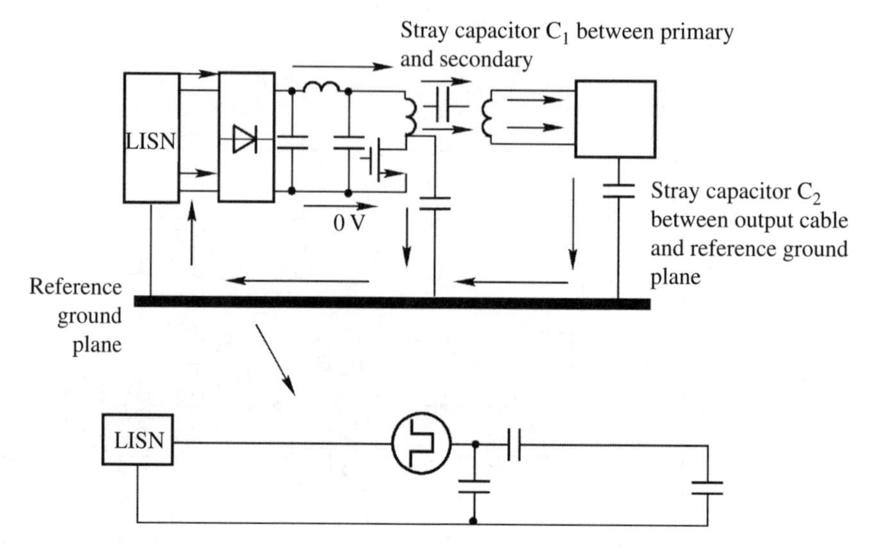

Figure 4.30 The influence principle of switching power supply output to the input port of the conduction disturbance.

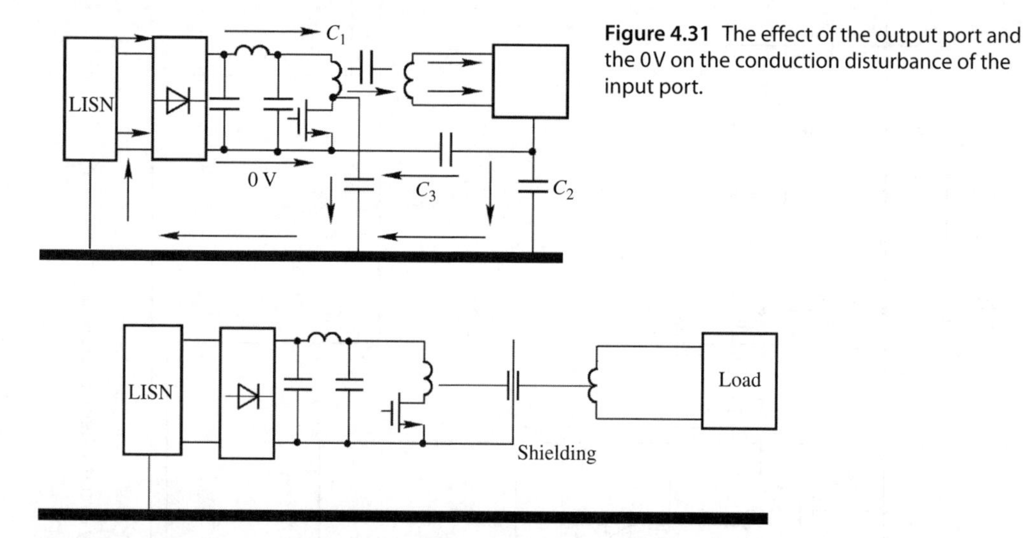

Figure 4.31 The effect of the output port and the 0 V on the conduction disturbance of the input port.

Figure 4.32 Principle of the transformer shielding layer on the input port of the conduction disturbance.

Using the transformer with a shielding layer can also help suppress the common-mode noise caused by the main switch (the transformer almost has no effect on suppressing the noise produced by the freewheeling diode). The principle is shown in Figure 4.32.

Through the above analysis, we can conclude that, for the switching-mode power supply, the output filter has great impact on the conducted emission test result on the input port.

[Solutions]

1) Connect a common-mode choke (150 mH) in series with the output port of the switching-mode power supply.
2) Connect a capacitor (2.2 nF) between the output port and the 0 V ground.

3) In the countermeasure finding tests, it is not easy to make the transformer with shielding layer at that time, but from the analysis on the problem, we can also conclude this may actively influence the test result. Adding the common-mode choke is only an alternative way to solve this problem.

[Inspirations]

1) In the design of the switching-mode power supply, if the transformer cannot be shielded due to some reason and the suppression of the noise (produced by switching components) cannot be very good. We recommend strengthening the output filtering.
2) There is always a misunderstanding on the analysis and localization of the disturbance problems, i.e. since the nonpassed conducted emission test appears on the input port, which is not related to the output port, we should just focus on the input port to settle out it. But it seems the problem is not as expected. EMC problem should be considered from the overall perspective, especially in the high-frequency scenario. The coupling path varies greatly. Only systematically analyzing the problem can help find the root cause.

4.2.4 Case 32: Properly Using the Common-Mode Inductor to Solve the Problem in the Radiated and Conducted Immunity Test

[Symptoms]

A product has the RF power-amplifying function. During the test, its input signal power is 0 dBm, and after the signal is amplified, the output signal power is 43 dBm, which is absorbed by the fixed attenuator. Through RS485 serial port, the equipment under test is configured and monitored and reports its working state to the computer. The test configuration is shown in Figure 4.33. The test level of conducted immunity test on the power port and the radiated Immunity test on enclosure respectively are 3 and $3\,\mathrm{V\,m^{-1}}$. In the test, the disturbance signal frequency at which abnormal operation of the EUT occurs is random. During the entire test, there are seven or eight times abnormal displays on the computer. The disturbance frequency at which the EUT operates abnormally cannot be obviously identified.

[Analyses]

Because the abnormal monitoring information appears when we inject the conducted disturbance on the power cable, we try to separate the power cable from the monitoring cable at first. The interface to the outside of this module includes the RF signal input and output interface connectors. The cables for configuration and monitoring and the power cable are connected to the same connector. The radio frequency cable is coaxial cable. The RS 485 monitoring cable is ordinary twisted pair cable and the power cable is ordinary power line. Inside the module, the power cable and the monitoring cable are routed in parallel with about 30 cm long. Then

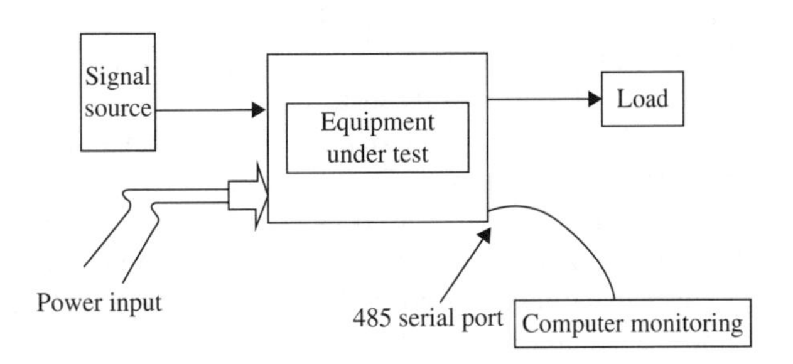

Figure 4.33 Immunity test configuration diagram.

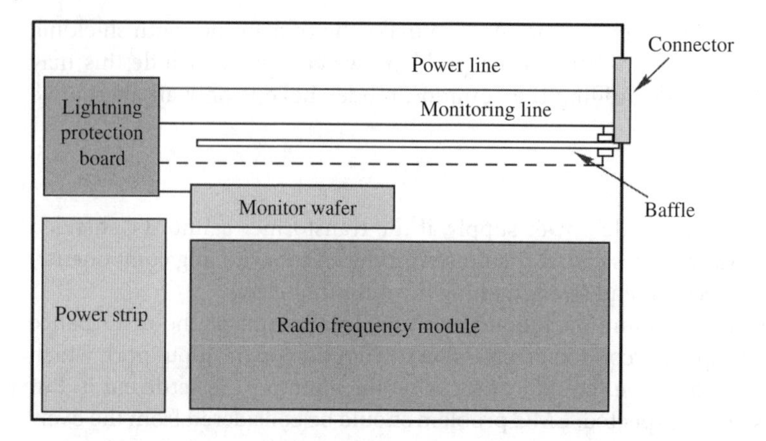

Figure 4.34 Schematic diagram of the internal structure of the device after changing the wiring mode.

the two cables are separately connected to the power supply board and the monitoring board. In order to exclude the factor of coupling between these two cables, first, we wrap a copper foil on the monitoring cable inside and outside of the module and ground it in the connector. Then perform the test again, no significant improvement occurs and the abnormal display still appears. Then, we change the monitoring cable of the module to a shielded twisted pair cable and ground it in the connector. Some improvements on the test results exist with the new modification, but the abnormal reporting still exists.

We doubt that the parallel cabling distance between the power cable and the monitoring cable inside the module is too long. So we change the cabling method inside the module. The new cabling method is shown in Figure 4.34.

The power cable and the monitoring cable are both connected to the lightning protection board after parallel cabling. Then, these cables are separately connected to the power board and the monitoring board starting from lightning protection board. We suspect that the parallel wire is too long and the parallel cabling causes the abnormal display. We change the cabling. The dashed line shown in Figure 4.34 is the new position of the monitoring cable. The monitoring cable is under the separating plate and the separated distance between these two cables is increased by the separating plate. Change the cabling and retest it. There is no obvious improvement. Abnormal information can also be reported. It is proved that increasing the separation between the power cable and the monitoring cable does not bring obvious improvement. Try to find the solution from the electronic circuit. The RS485 control circuit includes two parts, TX and RX. The reported monitoring information is from TX part. The circuit schematic is shown in Figure 4.35.

In the RS485 circuit, for redundancy, the TX part includes two branches and the relays are used to switch over them. There are bidirectional protection components and series impedance matching resistors for the differential pairs. In the actual test, we add capacitors C_1 and C_2 (0.1 μF) at the positions shown in Figure 4.35. The improvement is obvious in the new test. In each conducted immunity test, there are only two or three times abnormal monitoring information. After that, we increase the capacitance in the same position by adding two 0.1 μF capacitors in parallel with C_1 and C_2 respectively. In the new test, communication interruption exists, because the increased capacitance greatly influences the signal transmission quality. The communication speed of RS485 in this module is 10 kbps.

This capacitance has good effect, but this capacitance will affect the signal quality. Can we choose a suitable inductor?

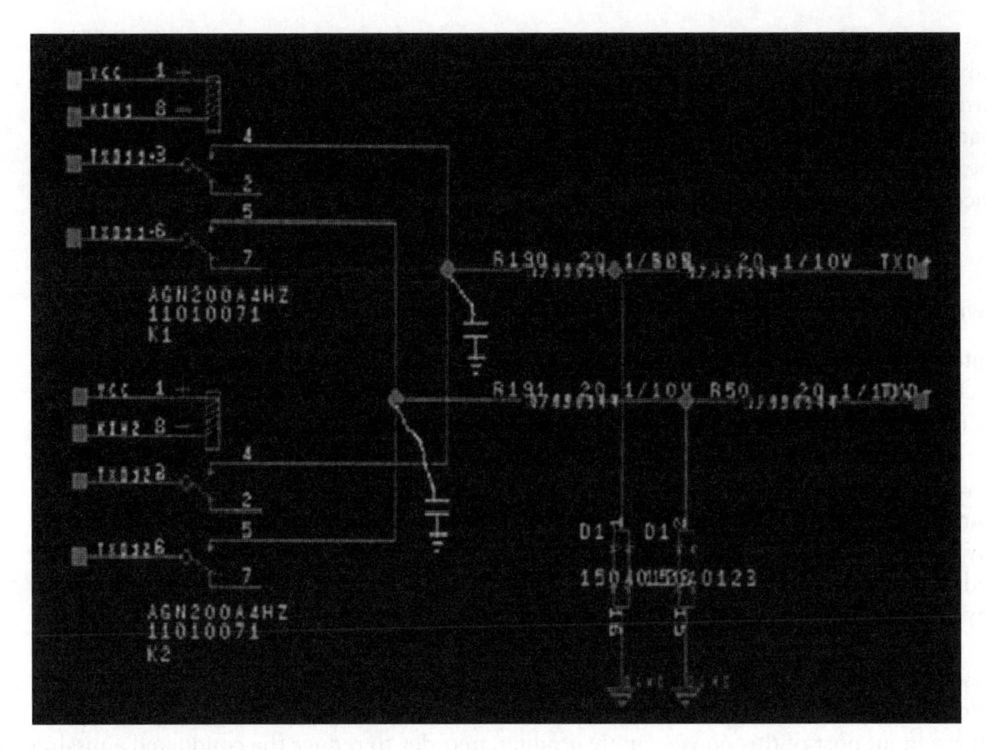

Figure 4.35 485 monitor circuit diagram.

Common-mode choke includes two windings wound on a magnetic core, whose turns and phases are the same (wound in reverse direction). So, when the operating current in the circuit flows through the two winding of the common-mode choke, the reverse magnetic fields are induced and canceled each other. At this time, the normal signal current is mainly affected by the winding resistance (and the small leakage inductance). When common-mode current flows through the windings, because the common-mode current is in the same direction, the in-phase magnetic fields are induced and then the inductance of each winding is increased. The winding impedance will be high, which contributes good attenuating effect, and then the common-mode current will be decreased. The purpose of the common-mode filtering can be achieved. The ideal common-mode choke has no impacts on differential-mode signals. Due to the existence of its leakage inductance, the common-mode choke has less impact on the differential signal than the capacitance.

In the actual test, we select a common-mode choke, its inductance is 0.5 H and its rated current is 3 A. We connect the common-mode choke in series with RS485 signals at the positions with the capacitors. There is no any abnormal phenomenon in the new test. For the radiated immunity test, there is no any abnormal phenomenon, too.

[Solutions]

Connect a common-mode choke in series with the 485 signals.

[Inspirations]

1) For the differential signal pairs, when their transmission distance are long or they are in parallel with the power wires or other polluted wires, we suggest connecting the common-mode choke in series with them in order to enhance the common-mode immunity.

2) Capacitors and inductors are filtering components, but their characteristics are different. How to choose or match them up depends on the impedances of the two sides of the LC circuit, which is composed of inductors and capacitors, namely, its source impedance and load impedance. In general, the inductor is always in series with low-impedance part and the capacitor is always in parallel with high-impedance part, in order to achieve the impedance mismatching.

4.2.5 Case 33: The Design of Differential-Mode Filter for Switching-Mode Power Supply

[Symptoms]

The schematic of a DC/DC nonisolated power supply product is shown in Figure 4.36. The IC in the figure is the main power chip.

The conducted emission spectral plot of the power input port of this power supply product is shown in Figure 4.37.

From the spectral plot, we can see that the switching frequency of the switching-mode power supply is about 50 kHz. The average value of conducted emission on the power supply port exceeds the standard limit by nearly 12 dB at 150 kHz. From Figure 4.36 we can see that there is no any filtering component on the input port of this power supply.

[Analyses]

In order to let the product pass the conducted emission test, we need to design a filtering circuit at the input port of this power supply product, in order to reduce the conducted emission and make the conducted emission test result on the power supply input port meet the standard requirement. The filtering can be classified into differential-mode filtering and common-mode filtering.

Differential-mode filtering mainly influences the conducted emission between the power supply input lines (for this product, it is the conducted emission exists between the positive power supply port and the negative one). Common-mode filtering is mainly aimed at suppressing the conducted emission between power lines and the ground reference plate (the ground reference plate used in the conducted emission test). For this product, its volume is very small (1 cm × 4 cm). That is to say, the parasitic capacitance between the product itself and the ground reference plate (used in the conducted emission test) is very small. For a switching-mode power supply whose working voltage is less than 24 V and switching frequency is about 50 kHz, as

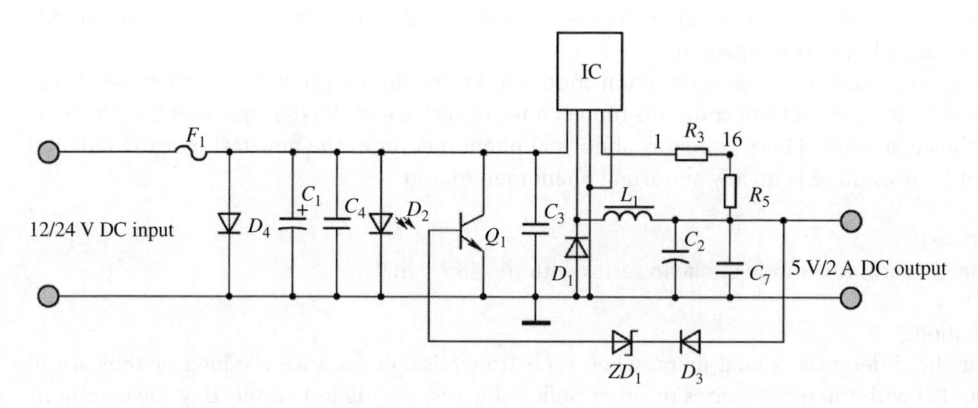

Figure 4.36 Principle diagram of a DC DC/ non isolated power supply product.

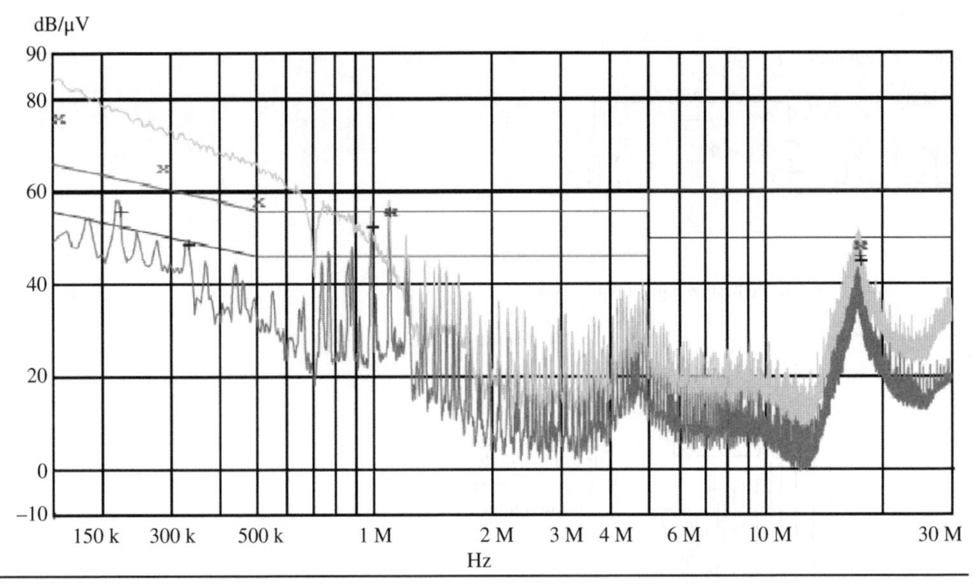

Figure 4.37 The power input port of the power supply of the conduction disturbance frequency spectrum.

long as the parasitic capacitance between the product itself and the ground reference plate is less than 1 pF (the impedance of common-mode conducted emission path is greater than 1 MΩ when the frequency is 150 kHz), the common-mode conducted emission problem basically will not occur at low-frequency. Therefore, the filtering of the product mainly lies in the differential-mode filtering.

The differential-mode conducted emission of the switching-mode power supply is mainly caused by the circuit loop with high-di/dt. In this case, when the IC is working in the switching-mode, a cyclic current signal with high-di/dt exists on the input line. If the energy storage capacitor in the product (in this case, the capacitor is C_1 just as shown in Figure 4.36) is an ideal capacitor, because the voltage of the capacitor cannot be changed abruptly, as long as the capacitance of the capacitor is large enough. The voltage across the capacitor can be controlled in a constant value. In other words, for the input port of the power supply, there will be no conducted emission. But there are always ESR (equivalent series resistance) and ESL (equivalent series inductance) in the electrolytic capacitor when it is used as the energy storage capacitor. When the cyclic current with high-di/dt flows through the energy storage capacitor, because of the ESL and ESR (if the frequency is lower than 150 kHz, the ESR is dominant for the impact) there will exist a voltage drop across the capacitor. Without a filtering circuit, this voltage drop will become the differential-mode conducted emission at the power port. And the data will be sent to the receiver through the LISN, shown in Figure 4.38, which illustrates the principle how this kind of differential-mode conducted emission is caused, and the relationship of the emission amplitude versus the frequency. The equivalent circuit of the conducted emission test without the different-mode filtering circuit is shown in Figure 4.39.

(a)

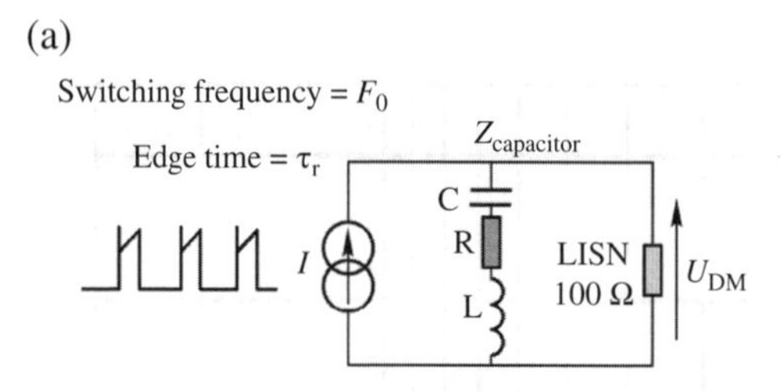

Differential mode equivalent circuit

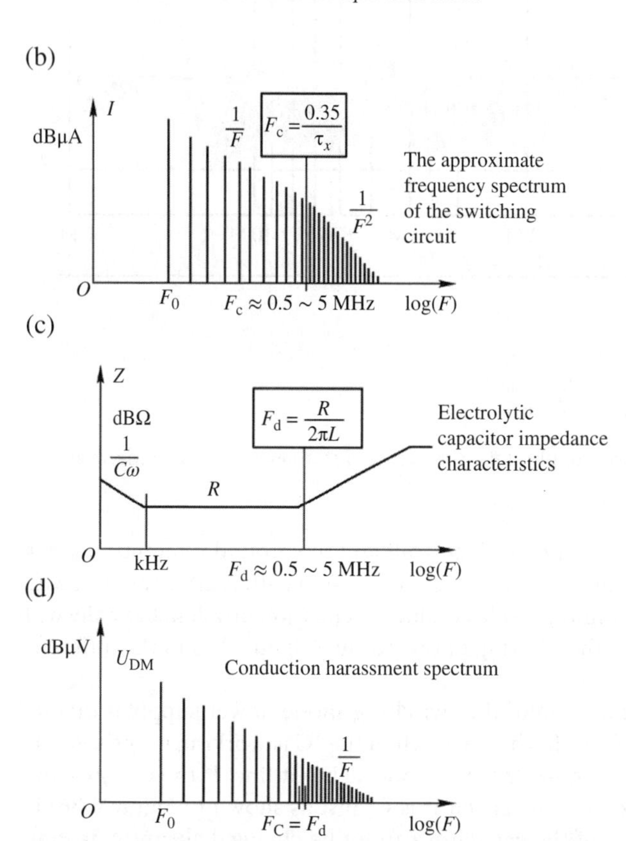

Figure 4.38 The principle for the disturbance generated by differential-mode transmission and the relationship between frequency and amplitude.

For the filtering circuit which needs to be designed, the noise shown in Figure 4.38d is the noise source to be filtered.

1) The filtering is based on the principle of impedance mismatching, namely the working principle of filtering is to create impedance discontinuity on the transmission path of RF electromagnetic wave. Then, most of the energy of the electromagnetic wave is reflected to the source of the energy, if other energy consuming components exist in the filter (such as ferrite beads and resistors), the electromagnetic energy will be transformed into thermal energy and dissipated. When the output impedance Z_0 of the filter is not equal to the load impedance R_l, reflection will appear on this port. The reflection coefficient is defined as,

$\rho = (Z_0 - R_L)/(Z_0 + R_L)$. The reflection coefficient will be high if the difference between Z_0 and R_l is big. For the disturbance signal to be controlled, when both sides of the filter are in the impedance mismatching condition, the disturbance signal will be strongly reflected in its input side and output side. For example, if a filter with a single capacitor is in parallel with the high-input impedance port, good filtering effect can be achieved, but if it is in parallel with the low-input impedance port, good filtering effect cannot be achieved. If a filter with a single inductor is in series with the low-input impedance port, a good filtering effect can be achieved, but if it is in series with the high-input impedance port, a good filtering effect cannot be achieved. Therefore, before we design the filtering circuit, we must confirm the impedance on both sides of the filter, namely the source impedance Z_s (of the conducted emission source and the load impedance Z_l (the output impedance of the conducted emission measurement equipment LISN).

2) For the filtering circuit operating in the conducted emission test, the source impedance of the filter is the source impedance Z_s of the conducted disturbance source. The value Z_s depends on the ESL and ESR of the electrolytic capacitors in the switching-mode power supply (if the frequency is low, ESR plays the main role and it is in milliohm level). The load impedance of the filter is the impedance provided by LISN and receiver, and the differential-mode impedance is 100 Ω (two 50 Ω resistors in series). By contrast, the source impedance of the filtering circuit is low, and its load impedance is high. So we can design a filtering circuit with reflection, just as the LC circuit inside the dashed box shown in Figure 4.40. L_1 and L_2 are the differential-mode inductors. C_X is the differential-mode filtering capacitor. The differential-mode inductors L_1 and L_2 can also be replaced by one inductor. Here two inductors are designed, which are respectively connected to the positive line and negative line of the power supply. This method can achieve a symmetrical design, in order that the common-mode conducted noise cannot be transformed into the differential-mode conducted noise.
3) After the relative positions of the filtering components are determined, the parameters of the differential-mode inductor and the differential-mode capacitor shall be determined, too. And these parameters can be decided based on the required insertion loss for the actual circuit. Figure 4.41 shows the insertion loss curve of the differential-mode filtering LC circuit in the actual power supply circuit. In this figure, the vertical axis represents the insertion loss, which is, in dB scale, the ratio of the differential-mode disturbance U_2 without the differential-mode filtering circuit shown in Figure 4.39 to the differential-mode disturbance U_1 with the differential-mode filtering circuit shown in Figure 4.40. The horizontal axis represents the ratio of the frequency to the resonant frequency F_0 of the filtering circuit, F/F_0.
4) From the curve shown in Figure 4.41, we can clearly see that, when the frequency F is equal to the filtering circuit's resonant frequency, F_0, the actual insertion loss is positive. That is, when the frequency F is greater than the resonant frequency F_0, the insertion loss of the differential-mode filtering circuit is increased. The increasing speed is 40 dB/decade. In this case, the desired filtering circuit must have its attenuation with more than 12 dB. The test frequency is 150 kHz. According to the curve, the resonant point F_0 of the differential-mode filtering circuit must meet the following requirements.
5) $150/F_0 > 2.5$, namely, $F_0 < 60$ kHz. $F_0 = \dfrac{1}{2\pi\sqrt{L_{DM}C_X}}$. L_{DM} is the addictive total inductance of L_1 and L_2 in the differential-mode filtering circuit in

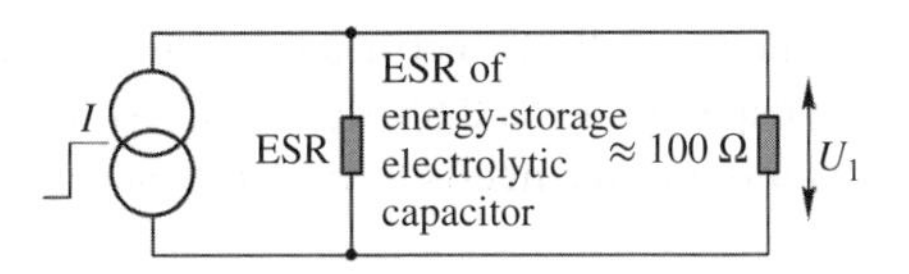

Figure 4.39 The equivalent circuit diagram for switching power supply conducted emission test without different-mode filter circuit.

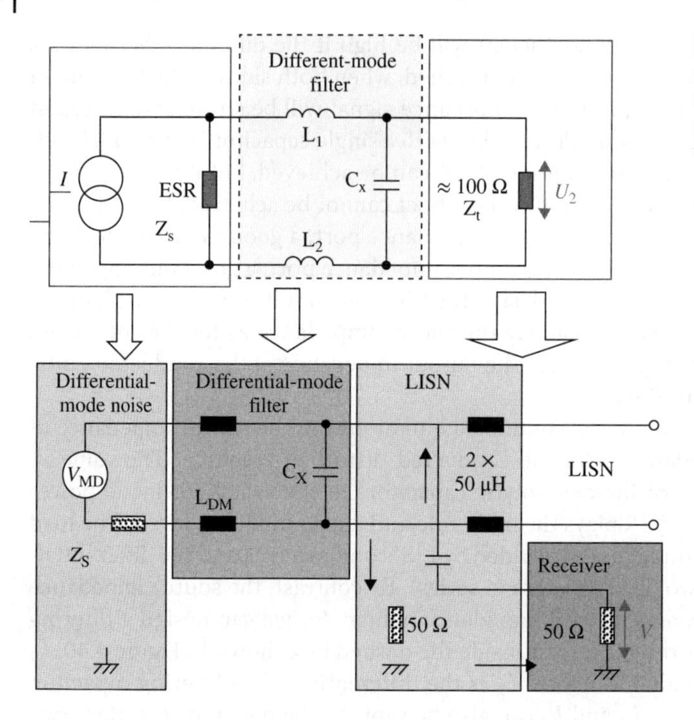

Figure 4.40 The equivalent circuit diagram of the conduction disturbance test for switching power supply noise with different-mode filter circuit.

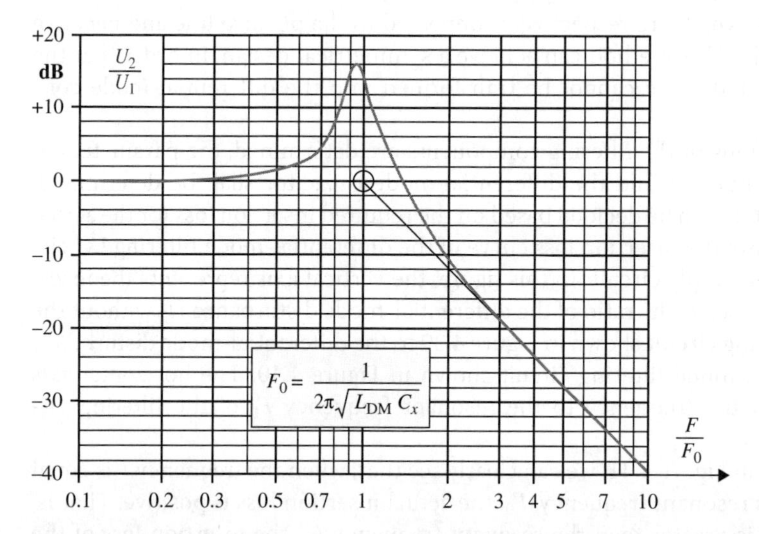

Figure 4.41 Insertion loss curve for LC different-mode filter with actual power supply.

Figure 4.40. C_X is the capacitor C_X of the differential-mode filter circuit in Figure 4.40. If the capacitance is 0.68 μF, L_{DM} must be greater than 20 μH. If the capacitance is decreased, the inductance needs to be increased.

[Solutions]

According to the above analysis, we add a 0.68 μF capacitor in parallel with the input port of the power supply and then insert 0.2 mH differential-mode inductor in series with power supply line. With this modification, the test is passed. The margin is 3 dB when the input voltage is

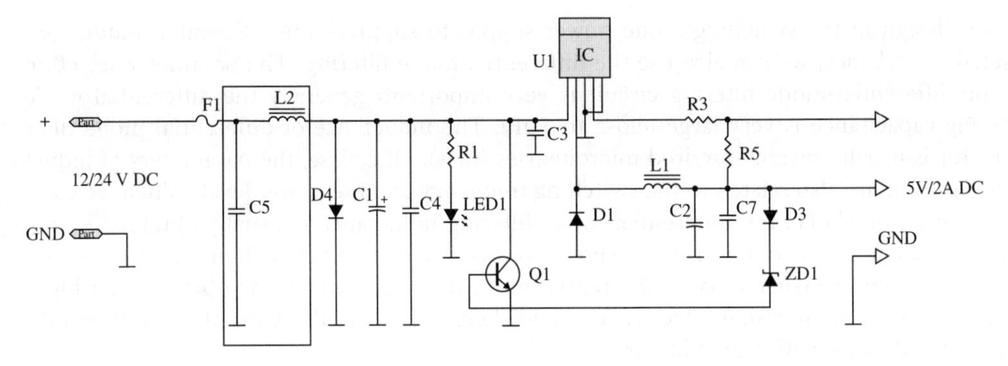

Figure 4.42 Principle diagram of DC/DC unisolated power supply products with filter circuit.

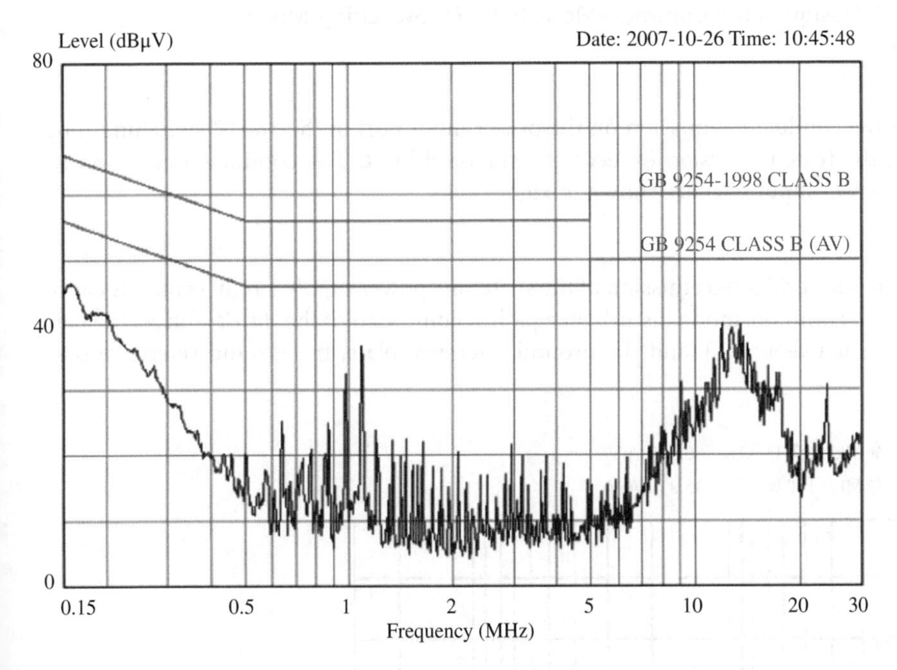

Figure 4.43 The conduction disturbance test spectrum with 12 V power supply.

24 V, and the margin is 11 dB when the input voltage is 12 V. The actual product circuit schematic is shown in Figure 4.42.

Figure 4.43 is the tested conducted emission spectral plot when the input voltage is 12 V.

[Inspirations]

From the principle of differential-mode conducted emission, we can see that the differential-mode conducted emission is produced by the current change *di*/*dt* in the switching-mode power supply circuit, and it is related to the power level of the switching-mode power supply. When the other conditions are fixed, if the current change *di*/*dt* is greater, the differential-mode conducted emission will be greater, too. That means that the filtering circuit in the power supply input port needs greater differential-mode filtering inductance or differential-mode filtering capacitance (X capacitor). The differential-mode conducted emission of switching-mode power supply occurs mainly in the low-frequency range. In addition to changing the circuit design of the switching-mode power supply, to suppress the differential-mode conducted disturbance, we can also use the differential-mode filtering. The parameter selection of the differential-mode filtering circuit is very important; generally the differential-mode filtering capacitance is very large (close to 1 μF). The inductance of differential-mode filter inductor is usually several hundred microhenries (μHs). Of course, the parameters of induct and capacitor are also related to the switching frequency and the power level. When we use a common-mode choke, the differential-mode filtering inductance is usually obtained by the leakage inductance of common-mode choke. In order to achieve better filtering effectiveness, we can use multiple (such as two) differential-mode inductors, which can obtain better high-frequency noise suppression effectiveness, and there will be a difference of more than 6 dB comparing the case with one inductor.

4.2.6 Case 34: Design of the Common-Mode Filter for Switching-Mode Power Supply

[Symptoms]

When we test the conducted emission on the power input port of the 180 W switching-mode power supply, the tested emission exceeds the standard limit. The conducted emission test results on the power input port are shown in Figure 4.44.

[Analyses]

The common-mode conducted emission of the switching power supply circuit is mainly caused by the *du*/*dt*. The common-mode disturbance path is composed of the *du*/*dt* source, the parasitic capacitance between EUT and the ground reference plate, the ground reference plate,

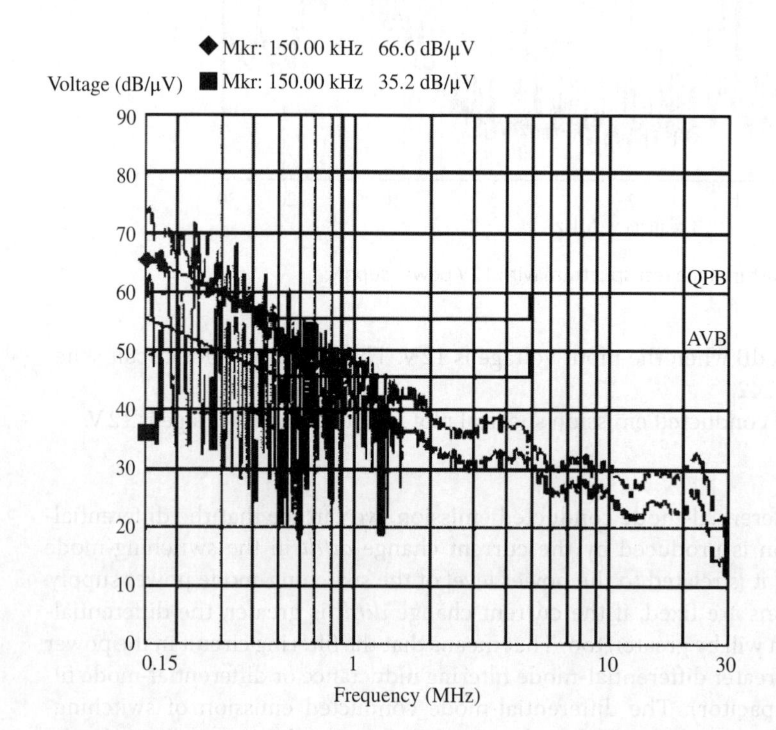

Figure 4.44 The conduction disturbance test results of the power supply input port.

LISN, and the input and output cables of the product. The common-mode current flows through the LISN. For the conducted emission problem of this product, the frequency range with excessive emission is from 150 kHz to 2 MHz. In general, this kind of emission may be differential-mode conducted emission or common-mode conducted emission. In the absence of other test equipment to figure out whether the emission is differential-mode or common-mode, the only method is to suppress both the differential-mode conducted emission and the common-mode conducted emission. For the differential-mode conducted emission suppression, we can see the description in Case 33. There are two ways to obtain the differential-mode inductance.

The first method, we use the differential-mode split inductor. The second method, we can use the leakage inductance of the common-mode choke. In general, we use 0.5–1% of the common-mode inductance.

This case mainly describes the design of common-mode filtering circuit and the parameter selection. Case 7 described the principle how the common-mode disturbance is generated in switching-mode power supply. The disturbance is generated through the parasitic path from the *du*/*dt* source of the switching transistor to the ground reference plate. Besides, the disturbance can also be generated through the parasitic path from the transformer's primary circuit and secondary circuit, to the ground reference plate. The main common-mode disturbance currents in the switching-mode power supply are I_1 and I_2.

$$I_{CM} = I_1 + I_2$$
$$I_1 = 2\pi F C_1 U_2$$
$$I_2 = YU$$

Here, $Y \approx 2\pi F C_{12}/(jZ2FC_{12}+1)$, U is the voltage at certain frequency (peak value or average value) (Figure 4.45).

In this case, the common-mode disturbance problem caused by I_1 is mainly analyzed

Similar to the analysis on the differential-mode conducted emission, the principle of the common-mode conducted emission and the relationship of the emission amplitude versus frequency can be expressed in Figure 4.46. Figure 4.47 shows the equivalent schematic of the conducted emission test for the switching-mode power supply without common-mode filtering circuit.

For the filtering circuit to be designed, Figure 4.46d shows the relationship of the amplitude of the common-mode disturbance source, which needs to be filtered versus frequency.

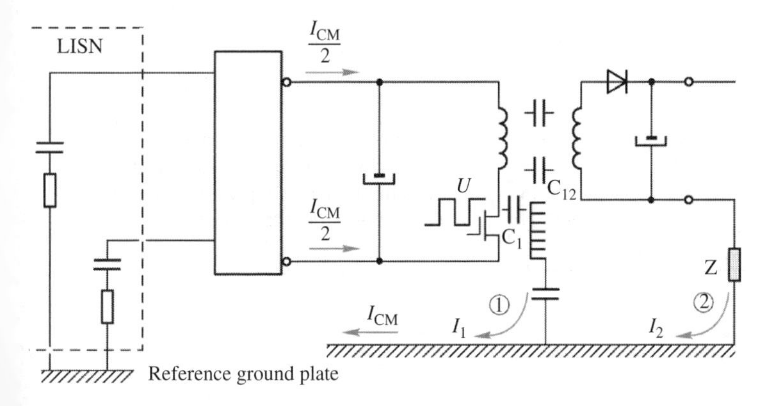

Figure 4.45 The Principle diagram of common-mode disturbance in switching power supply.

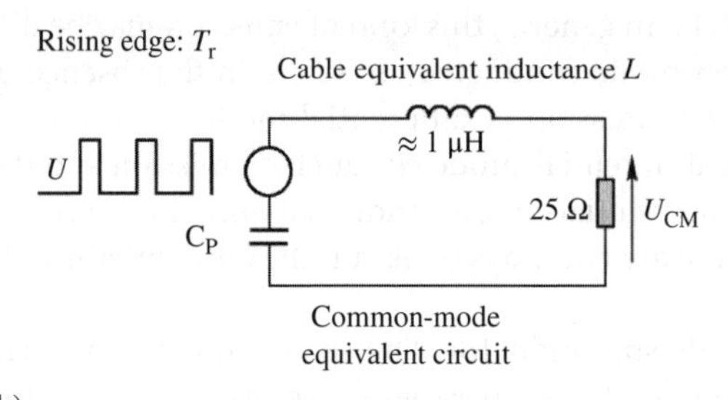

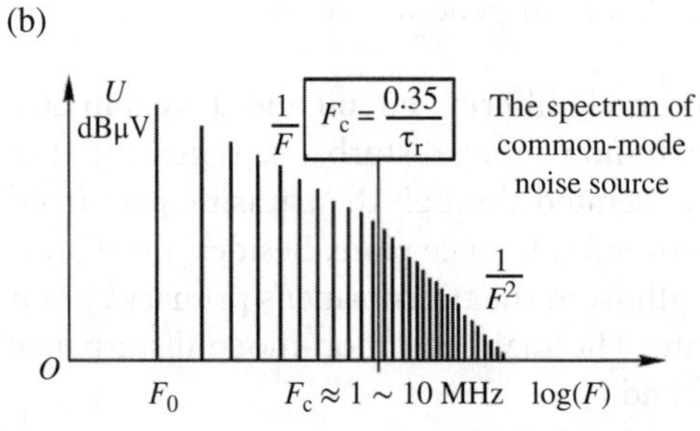

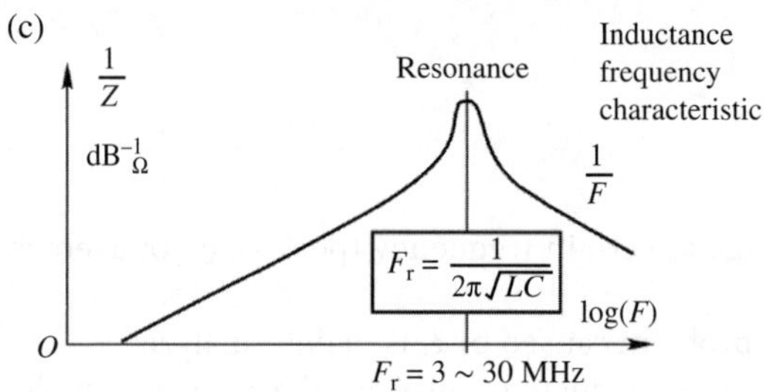

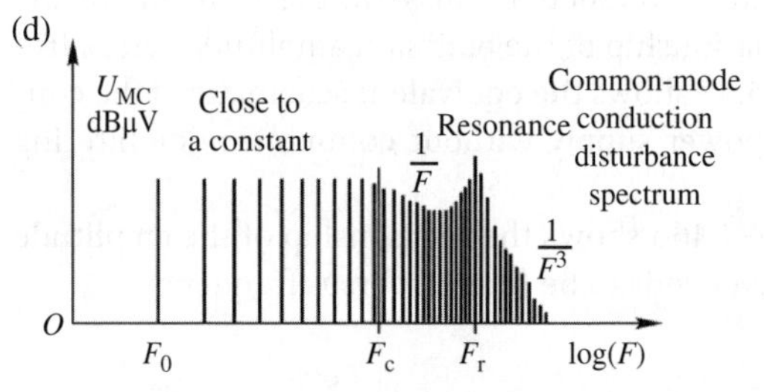

Figure 4.46 The principle of common-mode conduction disturbance and the relationship between frequency and amplitude.

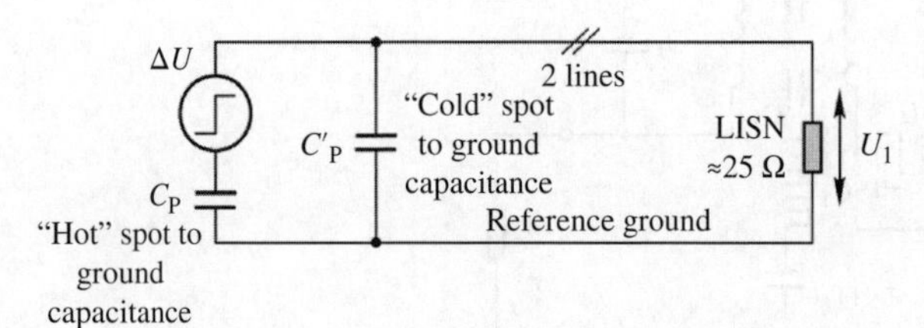

Figure 4.47 The equivalent circuit diagram of the conduction disturbance test for switching power supply noise without common-mode filter circuit.

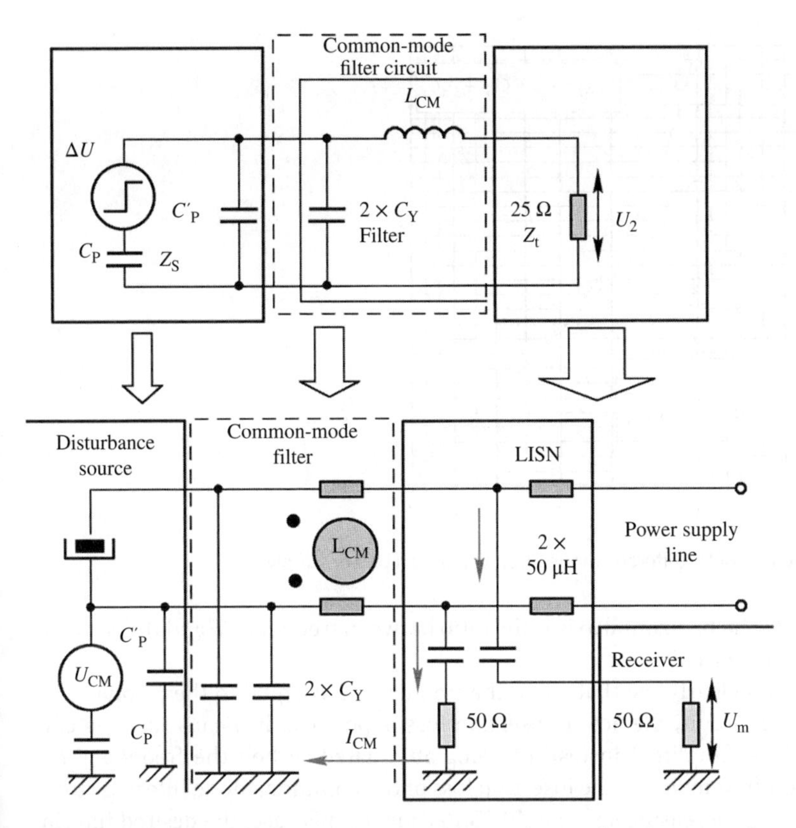

Figure 4.48 The equivalent circuit diagram of the conduction disturbance test for switching power supply noise with common-mode filter circuit.

For the common-mode filtering circuit operating in the conducted emission test, the source impedance of the filter is the source impedance of the conducted disturbance source impedance Z_s. The magnitude of Z_s depends on the impedance between the hotspot with high-du/dt in the switching-mode power supply and the ground reference plate (see Figure 4.47), which is mainly caused by the parasitic capacitance and its typical value is 30 pF–1 nF. In the low-frequency band (such as 150 kHz–2 MHz), the impedance is about 1–10 kΩ. The load impedance of the filter is the impedance provided by LISN and receiver, which is 25 Ω (two 50 Ω resistances in parallel). By contrast, the impedance of the source side of the filtering circuit is high and the impedance of the load side of the filtering circuit is low. Then, we can design a filtering circuit that can reflect the disturbance according to the principle of the filter design. For the LC circuit in Figure 4.48, L is the common-mode inductor L_{CM} and C_Y is the common-mode filtering capacitor.

Besides the relative positions of the filtering components in the filtering circuit are confirmed, we also need to determine the parameters of the common-mode filtering inductance and common-mode filtering capacitance. These parameters can be obtained from the insertion loss of the LC filtering circuit, which will be used in the actual circuit. Figure 4.49 shows the insertion loss curves of the LC common-mode filtering circuit, which will be used in the actual circuit. The vertical axis in the figure is the insertion loss, which is the ratio between U_2 and U_1 in the decibel scale. U_2 is the common-mode conducted disturbance without the filtering circuit in Figure 4.47, and U_1 is the common-mode conducted disturbance with the filtering

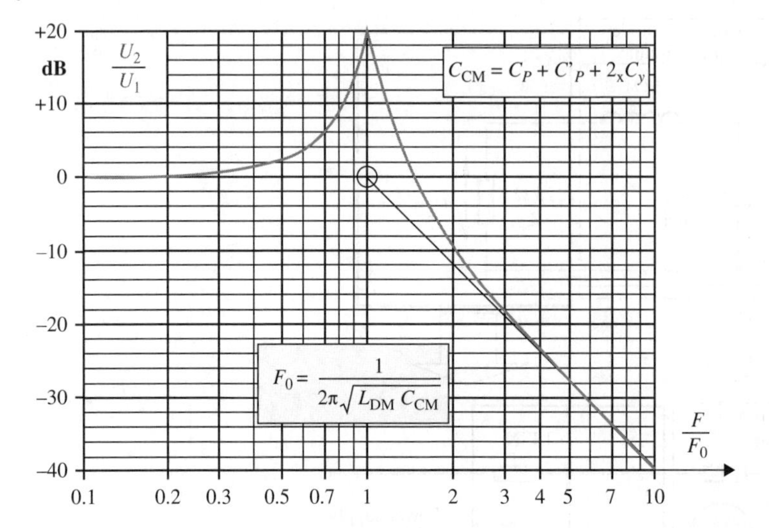

Figure 4.49 Insertion loss curve of LC different-mode filter with actual power supply.

circuit in the Figure 4.48. The horizontal axis is the ratio between frequency F and the resonant frequency F_0 of the filtering circuit.

From Figure 4.49 we can clearly see that when the frequency F is equal to the resonant frequency F_0 of the filtering circuit, the actual insertion loss is positive. It means that the conducted disturbance will be amplified instead of being attenuated. When the frequency F is greater than the resonant frequency F_0, the insertion loss of the common-mode filtering circuit increases gradually and the increasing speed is 40 dB/decade. In this case, the desired filtering circuit must have the attenuation above 40 dB near 150 kHz (we still need the attenuation above 20 dB based on the result in the Figure 4.44). The test frequency is 150 kHz. According to the curve, the resonant frequency F_0 of the common-mode filtering circuit must meet $150/F_0 > 3$, namely $F_0 < 50$ kHz, and $F_0 = 1/[2\pi(L_{CM}C_{CM})^{0.5}]$, L_{CM} is the common-mode filter inductance in the common-mode filtering circuit in Figure 4.48. C_{CM} is the sum of C_Y, C_P and C_P' in the common-mode filtering circuit in Figure 4.48, and C_Y is dominant. Due to the limitation of the leakage current, C_Y must be small. If C_{CM} is 4.7 nF, L_{CM} must be larger than 20 mH. If the capacitance is decreased, the inductance must be increased.

[Solutions]

According to the above analysis, we redesign the filtering circuit of the product. In Figure 4.50, the new filtering circuit is a low-pass filter and consists of common-mode inductor L_{CM}(24 mH), Y capacitors (3300 PF, C_{Y1},C_{Y2},), and X capacitor (C_x, 1 μF). C_{Y1}, C_{Y2} and the common-mode inductor L_{CM} filters out the common-mode disturbance. C_X and the leakage inductance of L_{CM} filter out the differential-mode disturbance.

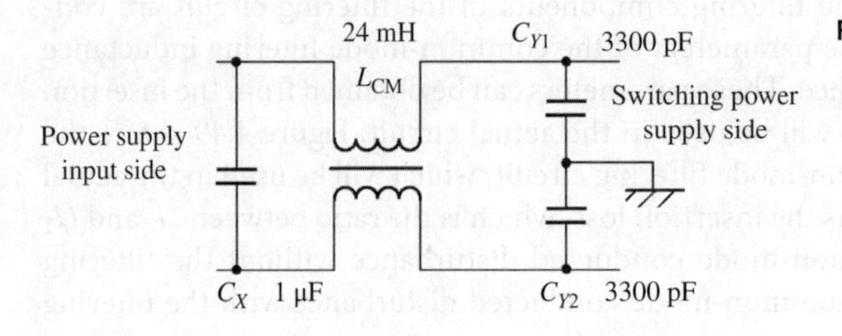

Figure 4.50 Power supply filter circuit.

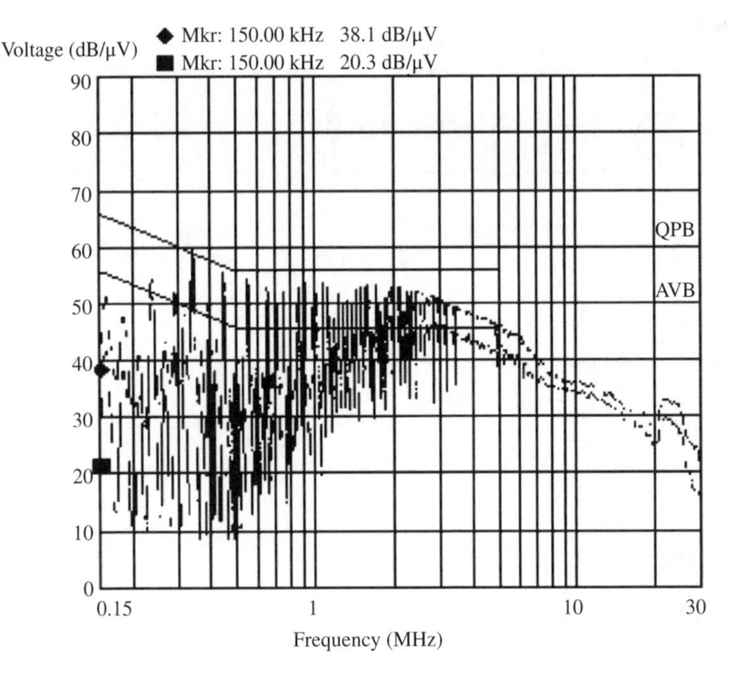

Figure 4.51 The conduction disturbance test results of the power supply input port after changing the filter circuit.

Figure 4.51 is the spectral plot of the conducted emission test with the modified power input circuit. We can see from Figure 4.51 that the emission over the middle and low-frequency range (0.15–1 MHz) is clearly decreased. But we found that the conducted emission over 1–10 MHz is increased.

The disturbance increasing over 1–10 MHz is caused by the parasitic capacitance between the turns of the large common-mode choke. Without reducing the total common-mode inductance, we use two-stage filter, which helps to reduce the parasitic parameters of the inductor and the capacitor and improves the high-frequency filtering effectiveness. Figure 4.52 shows the comparison between one-stage filter and two-stage filter. From the curve in Figure 4.52, we can see that when the frequency is higher than the frequency at point A, the two-stage filter will achieve greater attenuation.

It is obvious that as long as the frequency corresponding to point A is greater than the test starting frequency (such as 150 kHz), the two-stage filter is an alternative. The two-stage filter not only improves the high-frequency filtering effectiveness but also improves the low-frequency filtering effectiveness. The filtering circuit shown in Figure 4.53 is the improved two-stage filtering circuit. The input filter is the low-pass filtering circuit, which is composed of the common-mode inductor (L_{CM1}, L_{CM2}) of which the total inductance is 24 mH, and $C_Y$capacitors ($C_{Y1}C_{Y2}$ C_4 C_{Y4}) and C_X capacitors (C_{X1},C_{X2}). $C_{Y1}C_{Y2}$ C_{Y3} C_{Y4} and the common-mode inductors L_{CM1}, L_{CM2} filter out the common-mode disturbance. C_{X1}, C_{X2} and the leakage inductances of the common-mode inductor L_{CM1}, L_{CM2} reduce the differential-mode disturbance.

If we use the filtering circuit shown in Figure 4.53, the conducted emission test results at power port are shown in Figure 4.54. The conducted emission test passes.

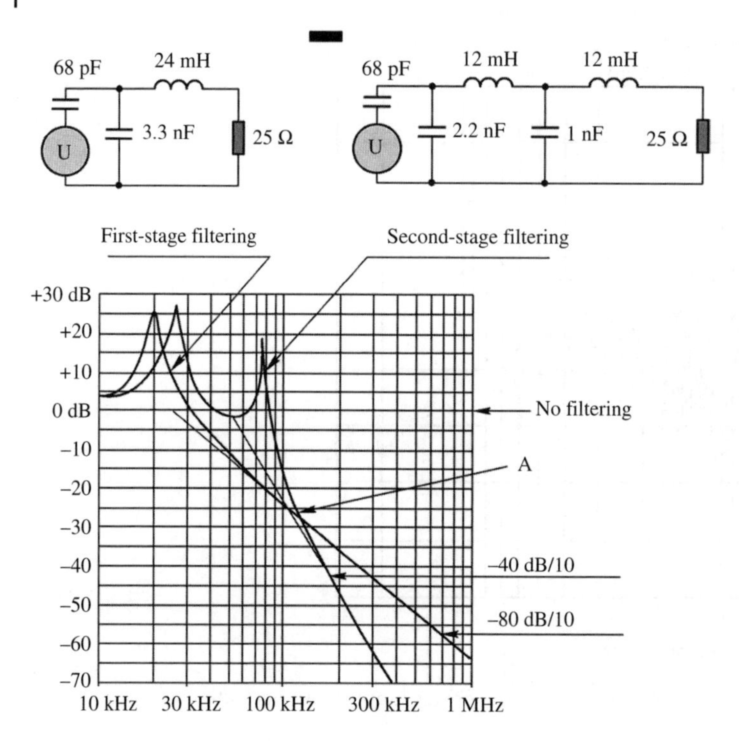

Figure 4.52 Comparison of the attenuation characteristics between the first- and second-stage filtering.

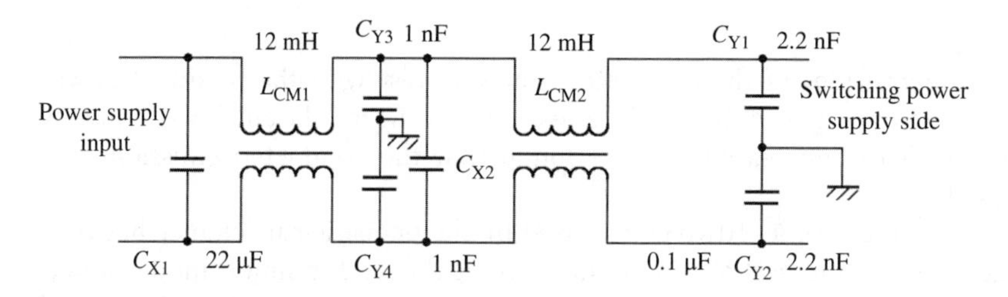

Figure 4.53 The improved two-stage filter circuit.

[Inspirations]

From the principle of the common-mode conducted emission, we can see that the common-mode conducted noise is generated by the *du*/*dt* in the switching-mode power supply circuit, which is directly relevant to the working voltage of the switching-mode power supply. The power rating of the switching-mode power supply does not have direct impact on the noise (but indirect impact). When the other conditions are the same, if the *du*/*dt* is increased, the common-mode conducted noise will be increased, which means that the filtering circuit at the power supply input port needs larger common-mode choke or common-mode filtering capacitors (Y capacitor). However, since there is a high-parasitic capacitance on the large common-mode choke, the high-frequency conducted noise will pass through the parasitic capacitance. So, a single large common-mode inductance cannot achieve good high-frequency filtering effects. If we use two common-mode chokes with the same inductance, we can get better suppression effectiveness on the high-frequency noise. Usually speaking, there will be a difference of more than 6 dB between these two ways.

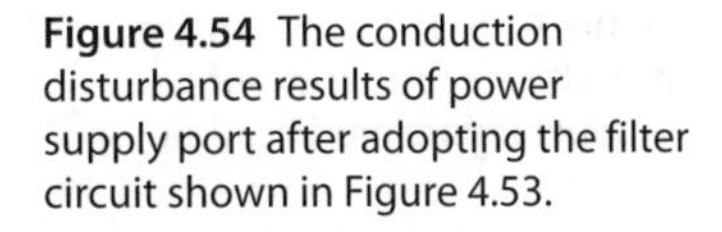

Figure 4.54 The conduction disturbance results of power supply port after adopting the filter circuit shown in Figure 4.53.

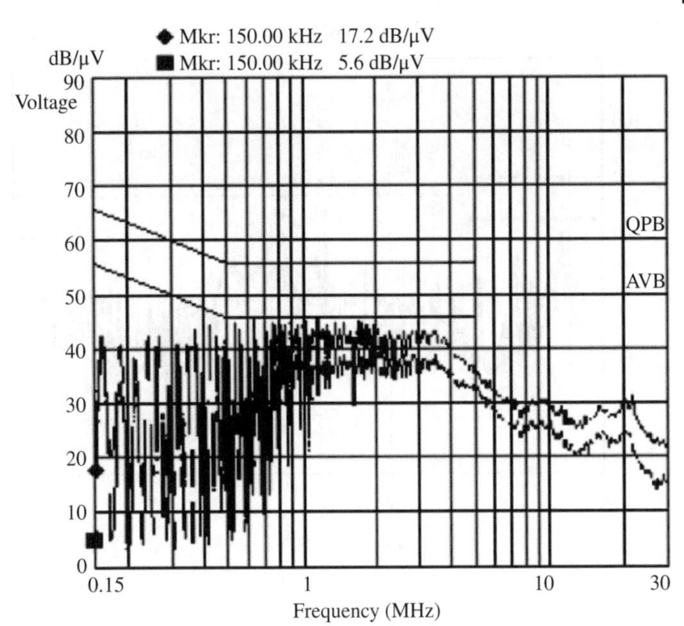

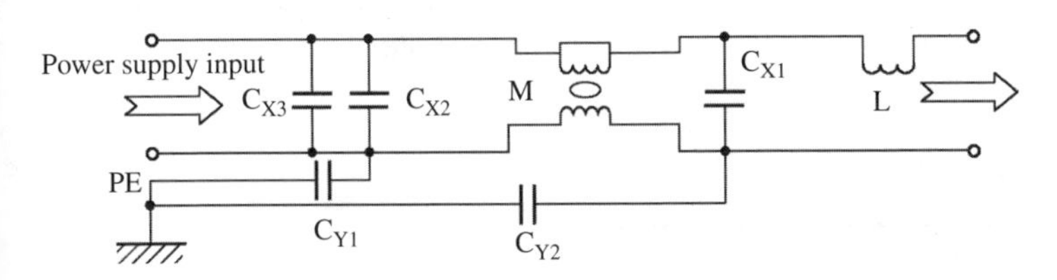

Figure 4.55 The principle of power supply filter circuit.

4.2.7 Case 35: Whether More Filtering Components Mean Better Filtering Effectiveness

[Symptoms]

For a product with plastic casing, the filtering circuit of its AC power supply input is shown in Figure 4.55. From the schematic shown in Figure 4.55, we can see that the common-mode filtering capacitors consist of two-stage Y capacitors. The Y capacitors (C_{Y2} and C_{Y1}) are respectively on the two sides of the common-mode inductance M. PE is the grounding terminal of the product. The conducted emission test frequency range is 150 kHz–30 MHz. During the conducted emission test, the product is placed on the insulation table of which the height is 0.8 m. The test results show that the conducted emission exceeds the standard limit.

In Figure 4.56, over the whole frequency range, the test results meet the limit requirements except few individual frequencies. During the test, we try to change the parameters of the inductance and the capacitance many times, but the improvements are not obvious. The excessive emissions at those individual frequencies cannot be suppressed. In addition, from the test results we know that the individual frequencies at which the emission exceeds the standard limit are in the range 2–3 MHz. Besides, these points have no obvious correlation with the CPU's crystal frequency of the device. While solving the conducted emission problem of the electronic products, we try to disconnect the link between the capacitor C_{Y1} and the earth. The test results are shown in Figure 4.57. The emission is reduced a lot in the whole frequency range. So the test is passed.

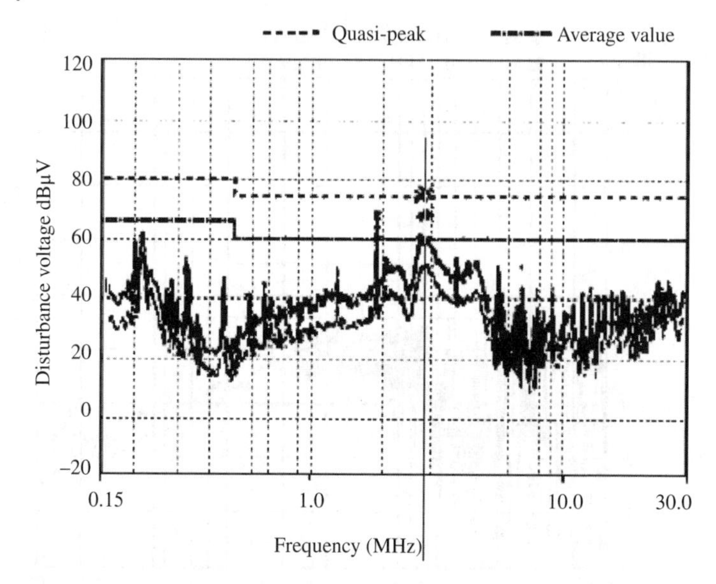

Figure 4.56 The conduction disturbance test results of a product before modification.

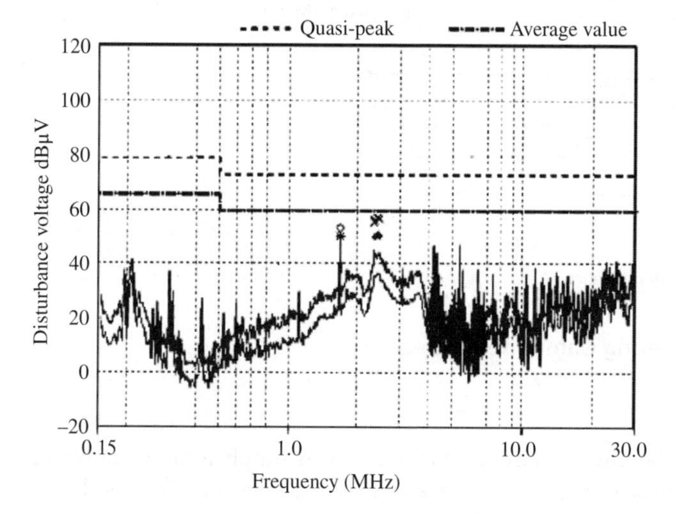

Figure 4.57 The conduction disturbance test results of a product after modification.

[Analyses]

To analyze the conducted emission problem, we can start with the essence of conducted emission test at first and then analyze the disturbance current flowing through the LISN. According to the principle of conducted disturbance of the switching-mode power supply products and the principle of the conducted emission test, the following schematic shown in Figure 4.58 describes the conducted emission problem when the capacitor C_{Y1} exists in the product with plastic casing.

According to the schematic shown in Figure 4.58, the common-mode conducted disturbance current I_{CM} produced by the switching-mode power supply are separated into two branches, one is going to the power supply input circuit, the other one is going to the measurement equipment, LISN. That is, one part flows into the PE grounding wire, I_{PE} in Figure 4.58, and the other part flows into the LISN, I_{LISN} in Figure 4.58 which directly determines the conducted emission (the common-mode emission) test result. From Figure 4.58, we can also see that the magnitude of I_{LISN} depends on the voltage difference between the PE terminal of the product

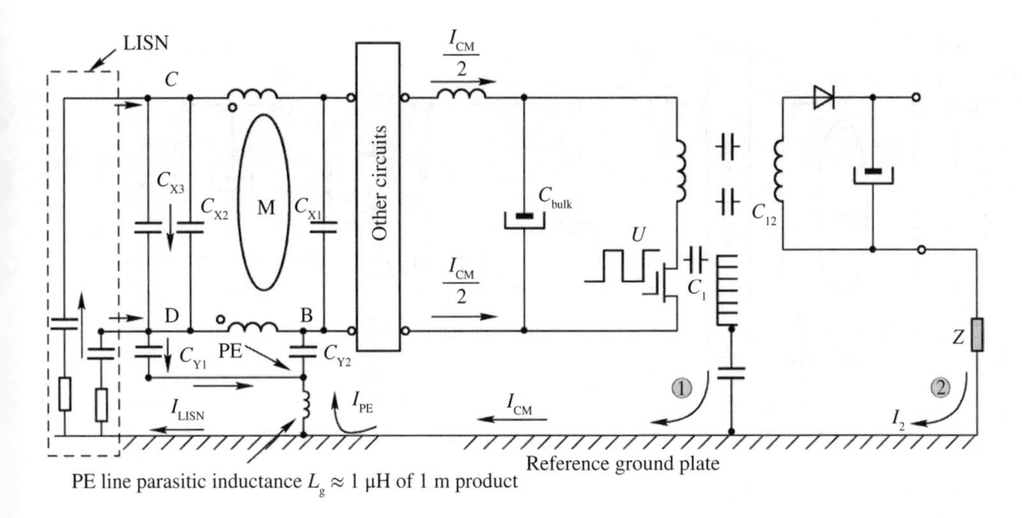

Figure 4.58 The principal diagram of conduction disturbance when Cy1 exists in the plastic products.

and the ground reference plate or the voltage difference between A/B and PE of the product, and the ground reference plate. When the voltage difference between A/B and PE of the product, and the ground reference plate is equal to zero (i.e. the impedance of PE line is zero and the filtering effectiveness of C_{Y1}, C_{Y2}filter is perfect), I_{LISN} will be equal to zero. In fact, the PE grounding wire is about 1 m in length, and its parasitic inductance is about 1 μH (the diameter of a wire has less impact on its parasitic inductance if it is relatively long, and only affects its equivalent resistance). Under this condition, when the common-mode current I_{PE} (Figure 4.58) flows through the PE line, there will exist a voltage drop ΔU on this PE line that acts as a voltage source. This voltage drop will cause a current flowing through LISN, I_{LISN} So, I_{LISN} is not equal to zero, just as Figure 4.59b shows. If the parasitic inductance of the PE wire is fixed, the magnitude of the common-mode current I_{LISN} depends on the impedance between the power supply input connected with the LISN and the PE position of the product (i.e. from C/D to PE

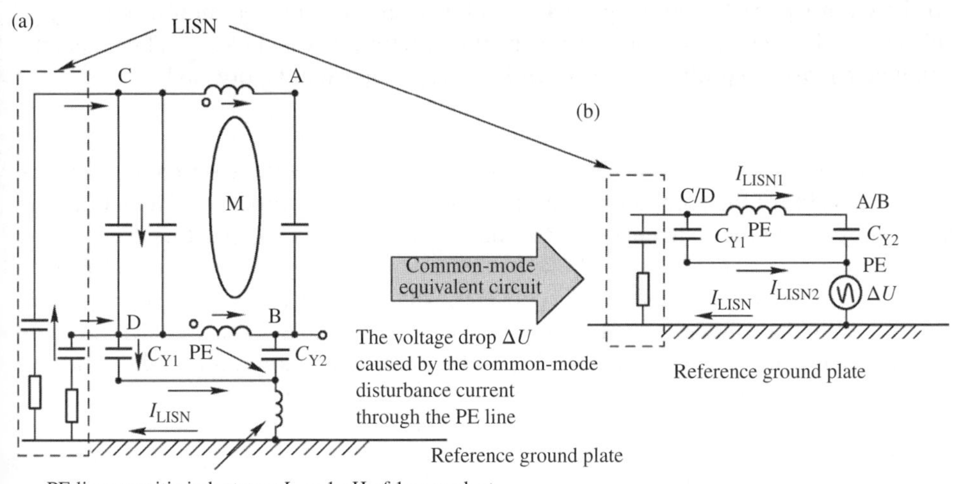

Figure 4.59 The principal diagram of common-mode current path when Cy1 exists in the plastic products.

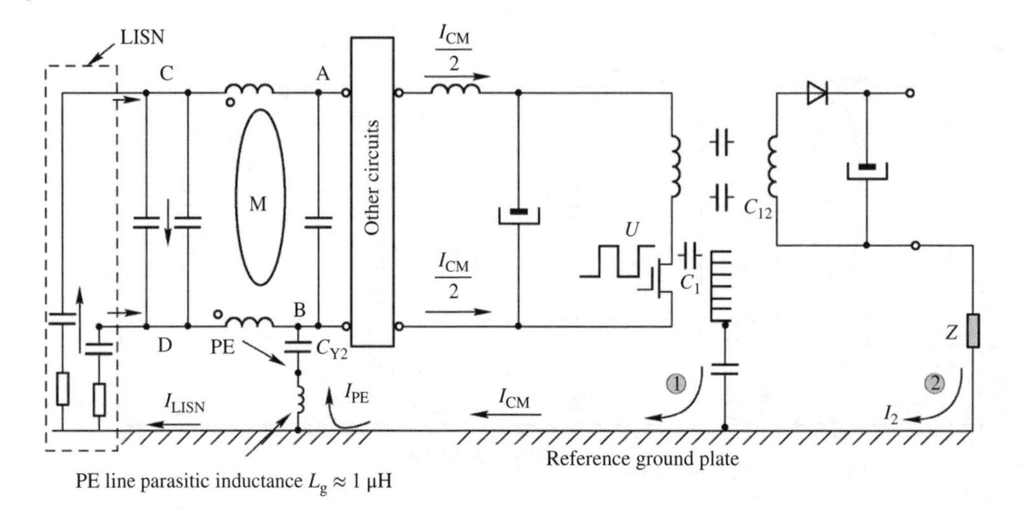

Figure 4.60 The principal diagram of conduction disturbance without Cy1.

in Figure 4.59b). For the design of the filtering circuit in this product, there are two circuit branches between the power supply input connected to the LISN and the grounding position of the product, PE. The first path is through C_{Y1} (i.e. the circuit path with I_{LISN2} in Figure 4.59b). The second path is through the common-mode inductance M and C_{Y2} (i.e. the circuit path with I_{LISN2} in Figure 4.59b). Because the impedance of the first path is much less than the impedance of the second path (i.e. if C_{Y1} is 4.7 nF, at 3 MHz, its impedance is about 10 Ω. If the common-mode inductance of M is 10 mH, at 3 MHz, its impedance is about 200 kΩ), the common-mode choke *M* is bypassed by the capacitor C_{Y1}. The common-mode current I_{LISN} ,which flows through LISN cannot be suppressed by the common-mode choke *M*, and then the tested conducted emission is very high.

If we disconnect the line between the capacitor C_{Y2} and the PE position of the product, the situation shown in Figure 4.58 and Figure 4.59 will be changed. The common-mode current I_{LISN}, which flows through LISN, is suppressed by the common-mode choke *M*. The schematic of the conducted emission problem analysis without C_{Y1} is shown in Figure 4.60.

[Solutions]

In the filtering circuit schematic shown in Figure 4.55, we can see that there is a common-mode capacitor on each side of the common-mode choke, which seems appropriate. However, from another perspective, another current path is formed due to these two capacitors. That is, the internal interference signal of the device flows back the power supply port through the capacitor C_{Y1}., and it bypasses the common-mode choke *M*, which is used to suppress the common-mode current. Then the conducted emission test fails. According to the above analysis, as long as we remove the common-mode capacitor C_{Y1} on the front of the common-mode choke, we can ensure that the conducted emission test passes with some margin.

[Inspirations]

Just grounding or increasing filtering elements are not a good way to suppress the common-mode conducted emission on the power port. The essence of conducted emission test is that the disturbance current (including the common-mode current and the differential-mode current) flows through LISN. Changing the direction of the disturbance current with the filtering circuit or grounding, avoiding the disturbance current passing through LISN, and reducing the disturbance current flowing into LISN as possible, are the right design guides for suppressing the conducted emission of the product.

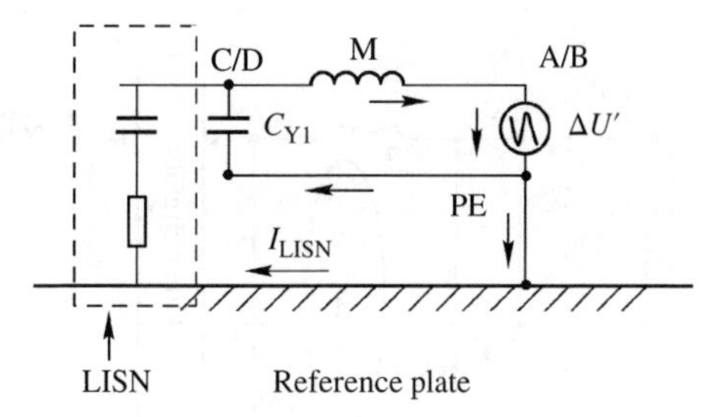

Figure 4.61 The principal diagram of conduction disturbance when common-mode current exists between the Cy2.

In this case, we solve the conducted emission problem by removing C_{Y1}, but we cannot say that this filtering circuit in Figure 4.55 is wrong. The common-mode current I_{LISN} is analyzed on the basis that the product is a soft plastic casing product and has high-grounding impedance. Besides, the filtering performance of the capacitor C_{Y2} is ideal (i.e. the voltage drop on both sides of C_{Y2} is close to zero) and the common-mode current I_{CM} flows back the inside of the product through the ground reference plate. In fact, the common-mode voltage between the position A/B in the Figure 4.58 and Figure 4.59 and the ground reference plate is related not only to the PE grounding wire itself and the common-mode current flowing on it but also to the impedance of the capacitor C_{Y2}. The actual impedance of C_{Y2} cannot be very low. That is, there exists common-mode voltage drop $\Delta U'$ at both sides of C_{Y2}. At this time, the common-mode equivalent circuit schematic shown in Figure 4.59b can be transformed into the schematic for the conducted emission problem analysis caused by the common-mode voltage drop $\Delta U'$ that exists at both sides of C_{Y2} and the voltage cause common noise problem shown in Figure 4.61.

From Figure 4.61, we can see that C_{Y1} bypasses the common-mode current, which flows into the LISN. That is to say, at this time, C_{Y1} helps the conducted emission test passed on the power input port, and this scenario is opposite to what Figure 4.59 shows.

In fact, the product with small common-mode current flowing on its PE grounding wire generally has metallic casing and good grounding connection. Figure 4.62 is the schematic of the conducted emission problem analysis when C_{Y1} exists in the metallic casing product.

For products with metallic casing, the metallic casing can bypass most of the common-mode disturbance current produced from the switching-mode power supply to the metallic casing before the current flows into the ground reference plate or LISN (the precondition is the circuit is connected correctly). In this way, the common-mode current on the PE grounding wire of this product will be very small. The common-mode voltage ΔU on the PE grounding wire will be very low, too. The conducted emission due to the existing of C_{Y1} in Figure 4.58 or 4.59 will be reduced. On the contrary, the conducted emission due to C_{Y1}, shown in Figure 4.61 will be very high. This is why, for some products, adding another Y capacitor on the power input port can help EMI test passed.

It is clear that for the product with metallic casing, the filtering circuit in Figure 4.55 can be used. At high frequency (10–100 MHz), its filtering performance is obvious. Typically, for C_{Y1}, 1000 pF is enough and the analysis schematic is shown in Figure 4.63.

At low frequency, the metallic casing of the product will not obviously help improve the suppressing effectiveness on the conducted emission at the power port. Because, at low frequency, the impedance of C_{Y1} is much larger than the 25 Ω (The equivalent common-mode impedance of LISN). The capacitance increasing of C_{Y1} will not decrease the common-mode current flowing into the LISN.

More filtering components will not always bring good effects. When we design a power supply filter, we can refer this case, Case 33, and Case 34.

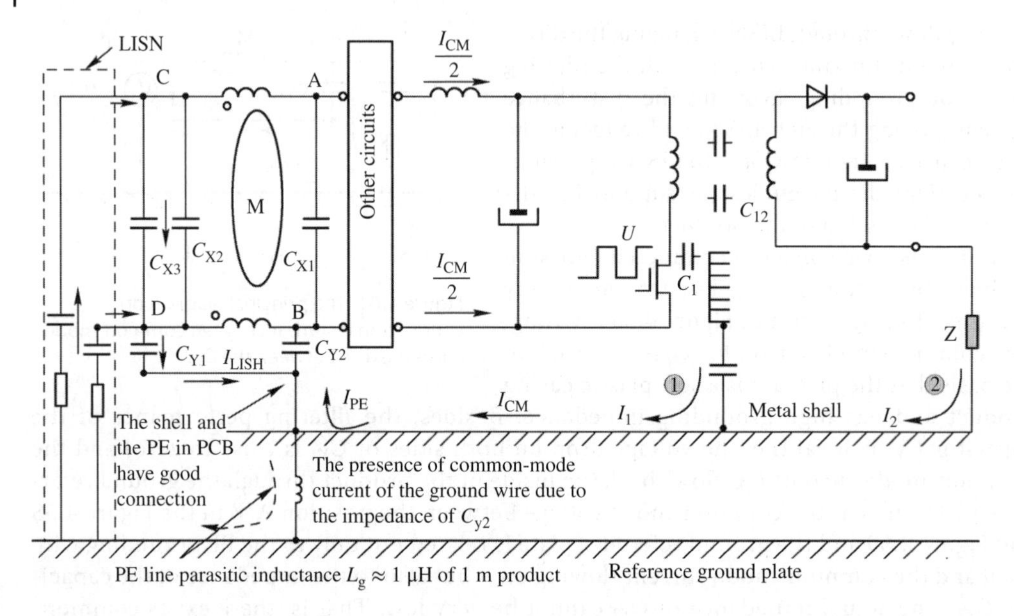

Figure 4.62 The principal diagram of conduction disturbance when Cy1 exists in the metal shell product.

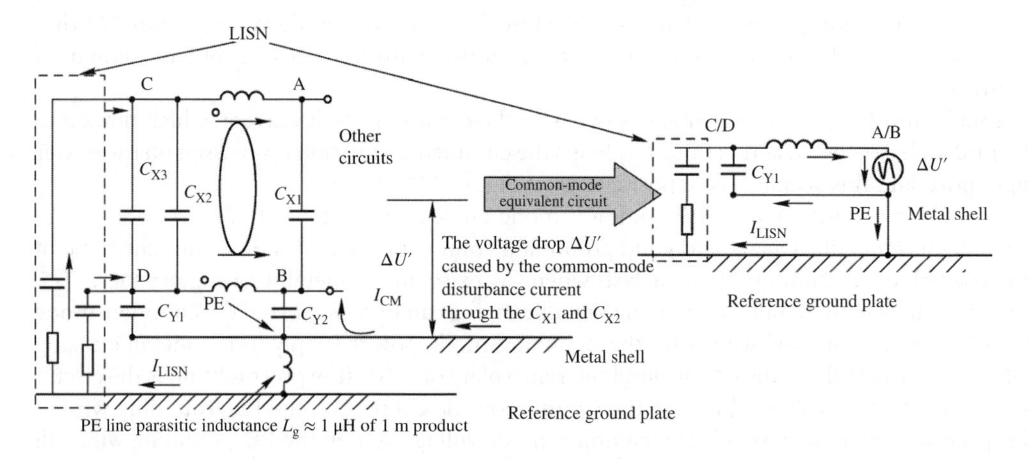

Figure 4.63 The principal diagram of conduction disturbance when Cy1 exists in the metal shell product.

4.2.8 Case 36: The Events Should Be Noticed When Positioning the Filters

[Symptom]

The radiated emission of a vehicle-mounted mini-type DC motor with electric brushes is tested. The spectral plot is shown in Figure 4.64. As illustrated, the radiated emission of the motor is over the limit.

[Analyses]

For the motor with brushes, the current in the motor coils cannot be suddenly changed when the brushes are switched. A higher counter electromotive force will appear on both ends of the coil when the coil is de-energized. The counter electromotive force can cause electric discharge phenomena in the nearby circuits. The transient discharge current has a steep rising edge,

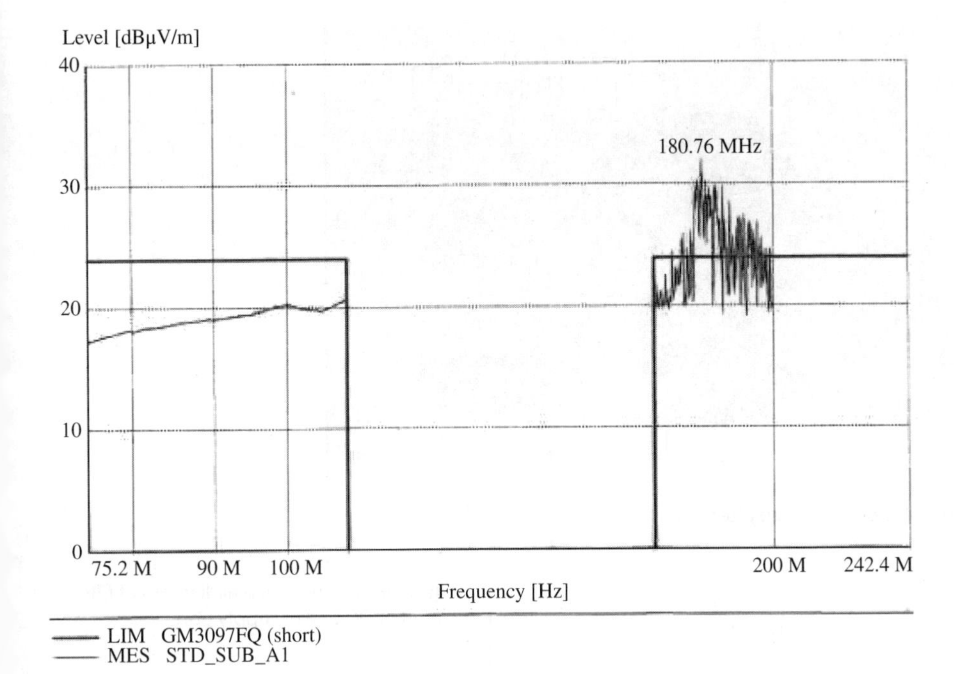

Figure 4.64 Radiation emission spectrum of motor.

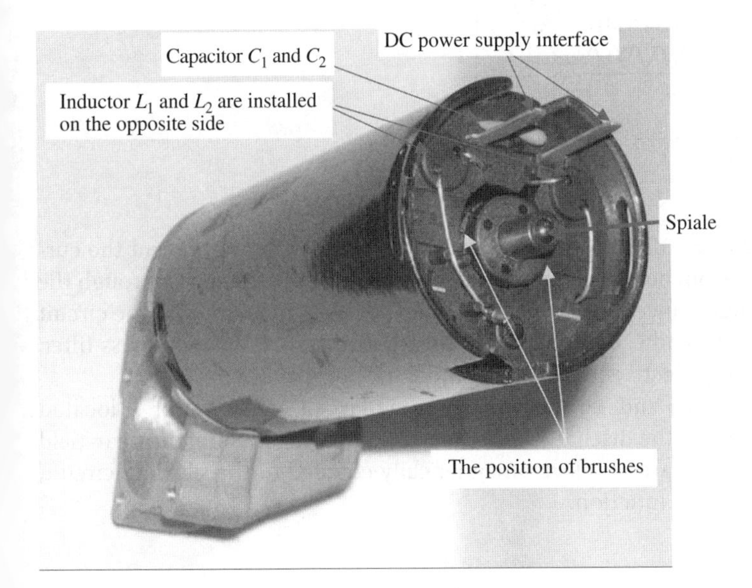

Figure 4.65 Motor physical map.

which causes high-frequency noise, and these noises are greatly random on amplitude and frequency. Radiation will appear when these high-frequency noises are coupled to the power cables or other long and unearthed conductors. Figure 4.65 shows the motor that has the radiated emission problem in this case. Figure 4.66 shows the disassembling structure of the motor, and Figure 4.67 is the schematic of the filtering circuit at its DC power port. In the filtering

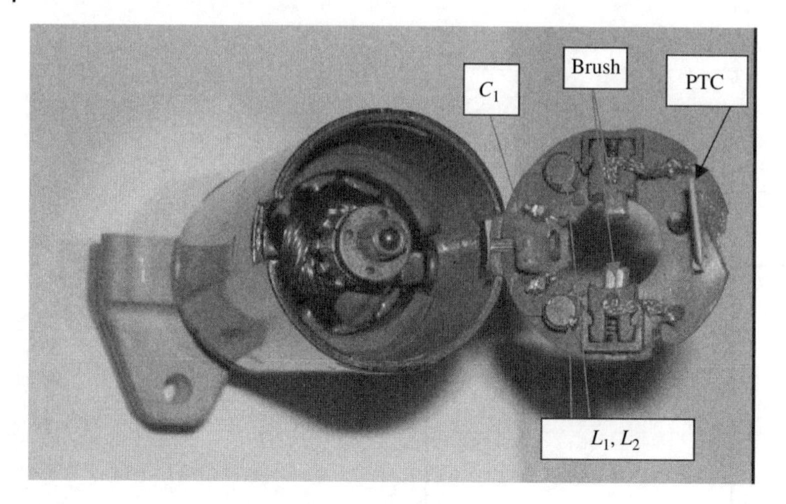

Figure 4.66 Motor physical map after opening.

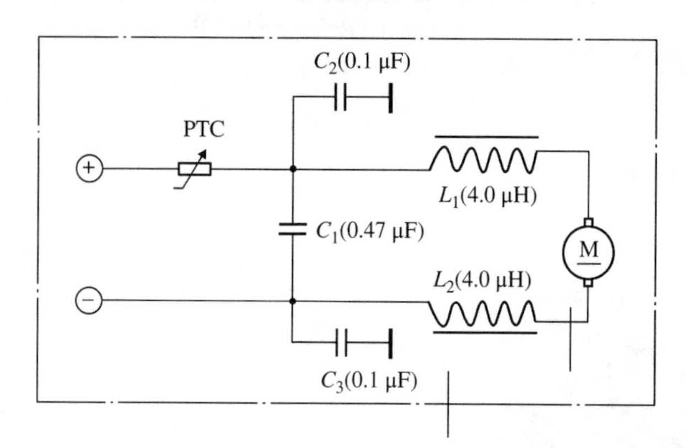

Figure 4.67 The principal diagram of filter circuit at DC power supply port.

circuit, the inductors L_1 and L_2 are directly in series with brushes, in order to prevent the current flowing into the brushes from being changed suddenly when the brushes pass through the commutator segment. The inductances are about 4 μH. The inductors in series with the circuit and the bypassing capacitors C_2 and C_3 are combined together to constitute a low-pass filter, which can improve the filtering effectiveness of single inductor or capacitor.

It can be seen from Figures 4.65 and 4.67 that the filtering circuit of the motor is located exactly among the motor brushes (the discharge disturbance source). Therefore, the near-field noise caused by brush spark discharge is electromagnetically coupled to the filtering circuit, and the filtering circuit is out of its function.

[Solutions]

- Change the position of filtering circuit and move all the components of the filtering circuit to one side of the motor brushes in order to avoid that the filtering circuit loop is in between the brushes.
- Shield the filtering circuit from the brushes to prevent the electromagnetic wave caused by brush spark discharge from being transmitted to the power cable through the space coupling.
- The leads of the capacitors are also important. Capacitor with long leads work badly, so the leads should be shortened. The link between the capacitor and the reference ground of the

noise source must be very short in order to ensure the capacitors have good filtering effectiveness. The inductance of the wire in free space is about $1\,\text{nH}\,\text{cm}^{-1}$. If the frequency of the noise caused by the brushes is 180 MHz and the wire connected with the capacitor is 4–6 cm long, even the capacitive reactance of the capacitor itself at certain frequency is not considered, only the impedance of the wire is about:

$$X_L = 2\pi fL = 6.78\Omega$$

The total impedance should include the capacitive reactance of capacitance (0.1 μF):

$$X_C = 1/2\pi fC = 0.08\Omega$$

As can be seen from this result, it is a great bypass filter if the capacitive reactance of the capacitor is the only consideration. But because of the wire inductance, the circuit does not work as a filter at all. If we shorten the wire to 1 cm long, then the impedance of the inductance is only 1.1 Ω, and the filtering effectiveness of the capacitor is increased by 20%. When using the motor shell as the grounding terminal, the lacquer on the shell must be removed to ensure that the conductor has a good contact with the ground. Even though the product shell is metallic, we must install filters directly onto the noise source instead of somewhere convenient and near the noise source or the shell. This eliminates any extra length of the lead, minimizes the impedance of the path on which the noise is circulated back the noise source, and achieves the best filtering effectiveness.

- In addition, since the voltage spike of the motor is caused by the disconnection between the brushes and the commutator segment, the amplitude of the voltage spike can be decreased by replacing the material of the brushes with a kind of softer material or boost the pressure of the brushes on the commutator segment. However, it will shorten the lifespan of the brushes and cause some other problems. It can only be taken when there is no other better way.

[Inspirations]

It is only the first step to implement the schematic design of a correct filter. To make sure that the filter can work as expected, we must pay attention to the placement of each component in the filtering circuit. While placing the components, the additional parasitic parameters shall be avoided, i.e. the additional coupling and crosstalk should be avoided.

4.2.9 Case 37: How to Solve Excessive Harmonic Currents of Switching-Mode Power Supply

[Symptoms]

There is one 180 W switching-mode power supply with plastic casing used for the desktop. Its load is a 12 V/15 A semiconductor refrigerator and the size of the power supply is 205 mm × 90 mm × 62 mm. The schematic of the switching-mode power supply is shown in Figure 4.68.

Table 4.3 is the list of the measured harmonic currents from 7th to 21th orders. The measured harmonic current of the 11th, 15th, and 17th orders are all over the standard limit.

Since most capacitors present low impedance for high-order harmonics, large current will flow through the capacitor with the high-order harmonics. When the resonant frequency of the resonant tank formed by capacitor and inductor in the system circuit is equal or close to the frequency of certain order harmonic, the harmonic current will be amplified. It will cause

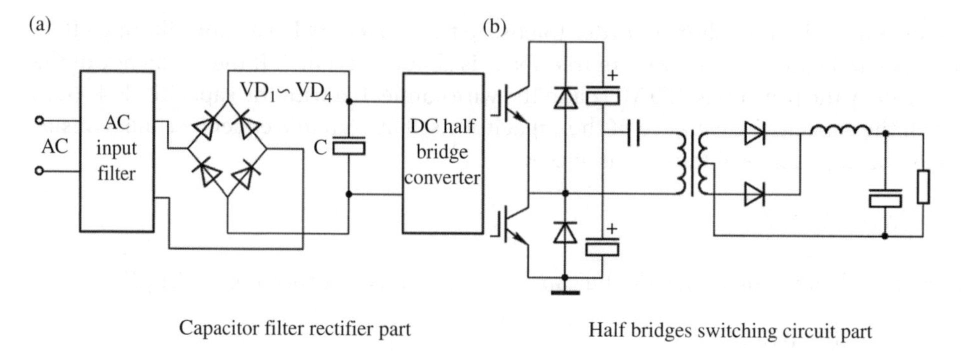

Figure 4.68 180 W switching power supply.

Table 4.3 The measured harmonic current.

Harmonic order	Measured value/A	Harmonic limits/A
7	0.694	0.770
9	0.397	0.400
11	0.334	0.330
13	0.209	0.210
15	0.165	0.150
17	0.151	0.132
19	0.101	0.118
21	0.084	0.107

overheating and overvoltage on the capacitor; then it may work abnormally, or accelerate the aging of the capacitor and shorten the lifespan of the capacitor.

The wire impedance will increase with the increasing of the frequency because of the frequency characteristic of the conductor's parasitic inductance in customer application. Furthermore, because of skin effect of conductors, harmonics will increase the additional power losses of the conductors in user's own power supply system. It is worth noting that the current of the neutral wire needs to be increased because in the three-phase power supply system, the neutral wire is overloaded and the user's electricity costs are greatly increased. Also, since it is unavoidable that there exists the distributed inductance of the transmission conductor and the distributed capacitance between transmission conductor and ground, when this kind of inductance and capacitance form a series or parallel circuit with the harmonic generation devices, series or parallel resonance will occur. Overvoltage and overcurrent caused by the resonance will harm the aforesaid related devices in return. Therefore, the harmonic problem of this product must be solved.

[Analyses]

The traditional diode-bridge rectifier and the filtering capacitor are used in this switching-mode power supply, which causes the input AC current greatly distorted and injects a large number of high-order harmonics to the power network, and then the power factor of the network becomes low. In this case, the power factor may be only about 0.6, and it causes great harmonic pollution and interference to the power network and other electronic devices. In the early 1980s, the harm from high-order harmonic current produced by this kind of device has already drawn public attention. In 1982, the International Electrotechnical Commission published the standard IEC55-2 to confine high-order harmonics (the subsequent revised standard is IEC1000-3-2), which facilitate the power electronics engineers to do more investigation on harmonics-suppression techniques. The common harmonics-suppression techniques are as follows:

1) If a PFC (power factor correction) circuit is applied into the power supply product, the electric energy utilization efficiency will be apparently increased. There are two kinds of PFCs: passive and active. In the passive PFC, inductance compensation is generally used to decrease the phase difference between the AC input fundamental current and the AC input voltage to increase the power factor. But the power factor of the passive PFC is not so high; it can only reach 0.7–0.8. The active PFC composed by inductors, capacitors, and electronic components is small and can reach a high power factor. For example, the power factor of an active PFC circuit with Boost converter can be raised above 0.99, with which the harmonic current is very low, but the cost is higher than that of the passive PFC.
2) Adding inductors to capacitance compensator can suppress harmonics. When the inductor is in series with the capacitor, a series resonant tank is formed, and at the resonant frequency of this tank, its impedance is very low (theoretically, it is zero). If the serial resonant frequency is equal to the harmonic frequency of the power network, this serial circuit simply becomes a filtering circuit. Maintaining a certain proportion between the capacitance and the inductance can filter the harmonics at different frequencies.

These two common harmonics suppression methods work well, but their circuits are complex and they cost a lot. Furthermore, the switching noise of switching transistors and high-voltage rectifying diodes in the PFC circuit will become the new interference source, which increases the difficulty for products to meet the EMI standard. The rated power of this power supply is 180 W over the AC input voltage range (AC 200–250 V). Under the condition that the voltage regulation is guaranteed, the filtering capacitance after the rectifier can be moderately reduced. So, to a certain extent, the series resistance on the power input can decrease the instantaneous peak current to meet the harmonic current limit while the filtering capacitor is charging. The power loss can also be in the acceptable range and the power-supply efficiency of the complete equipment declines only a little. So it is also a good method. The actual measured harmonic current result with this method is shown in Table 4.4.

[Solutions]
Moderately decrease the filtering capacitance after the rectifier at the power supply input according to the above descriptions. Change the 1 μF X capacitor into 0.47 μF. Cascading a 10 Ω resistor on the power supply input, to a certain extent, can decrease the instantaneous peak current to meet the harmonic current limit while the filtering capacitor is charging.

[Inspirations]
For small power products, we can cascade a resistor on the power supply input to improve the harmonic suppression. In addition, for small power products, cascading a resistor on the power supply input is also beneficial to the other EMC performance of the products.

4.2.10 Case 38: Protections from Resistors and TVSs on the Interface Circuit

[Symptoms]
How differently the relative position of TVS and resistor affects the surge protection on the interface is always a controversial problem. In order to verify that the relative position has impact on surge protection, we choose the RS485 port of one product as the

Table 4.4 Harmonic current after filtering.

Harmonic order	Measured value/A	Harmonic limits/A	Harmonic order	Measured value/A	Harmonic limits/A
1	1.120	Nan	21	0.064	0.107
2	0.004	1.080	22	0.001	0.084
3	0.990	2.300	23	0.050	0.098
4	0.004	0.430	24	0.001	0.077
5	0.812	1.140	25	0.054	0.090
6	0.003	0.300	26	0.001	0.071
7	0.594	0.770	27	0.051	0.083
8	0.002	0.230	28	0.001	0.056
9	0.379	0.400	29	0.037	0.078
10	0.002	0.184	30	0.001	0.061
11	0.212	0.330	31	0.026	0.073
12	0.001	0.153	32	0.001	0.068
13	0.142	0.210	33	0.027	0.068
14	0.002	0.131	34	0.001	0.064
15	0.141	0.150	35	0.027	0.064
16	0.002	0.115	36	0.001	0.061
17	0.132	0.132	37	0.023	0.061
20	0.001	0.092	40	0.001	0.046

experimental object. Sixteen signals are selected in this product, which, respectively, are DBUS1D+, DBUS1Dc, DBUS2D+, DBUS2D−, CLK2M+, CLK2M−, CLK8K+, CLK8K−, CFN+, CFN−, BFN+, BFN−, OBCLK+, OBCLK−, FCLK+, and FCLK−. The first eight signals are classified as the first group. The schematic of the first group is shown in Figure 4.69. The relative position of TVS and resistor can be seen from the schematic, and the TVS is positioned near the chip.

The latter eight signals are classified as the second group. The schematic of the second group is shown in Figure 4.70. The relative position of TVS and resistor can be seen from the schematic and the resistor is positioned near the chip.

The surge test configuration is shown in Figure 4.71.

The test procedures are as follows:

1) Test and record the normal operating waveform of the RS485 signals with an oscilloscope.
2) Set the surge generator to output the standard surge voltage 1.2/50 μs for 10 times continuously (5 times positive voltage and 5 times negative voltage) as a group test, with the interval at 60 seconds.
3) Test the operating waveform of the RS485 signals after the surge test on each group and compare with the normal waveform.
4) If the compared result is the same, increase the surge voltage and continue the test. Repeat (3) until the signal is distorted, and record the injected surge voltage at this time.

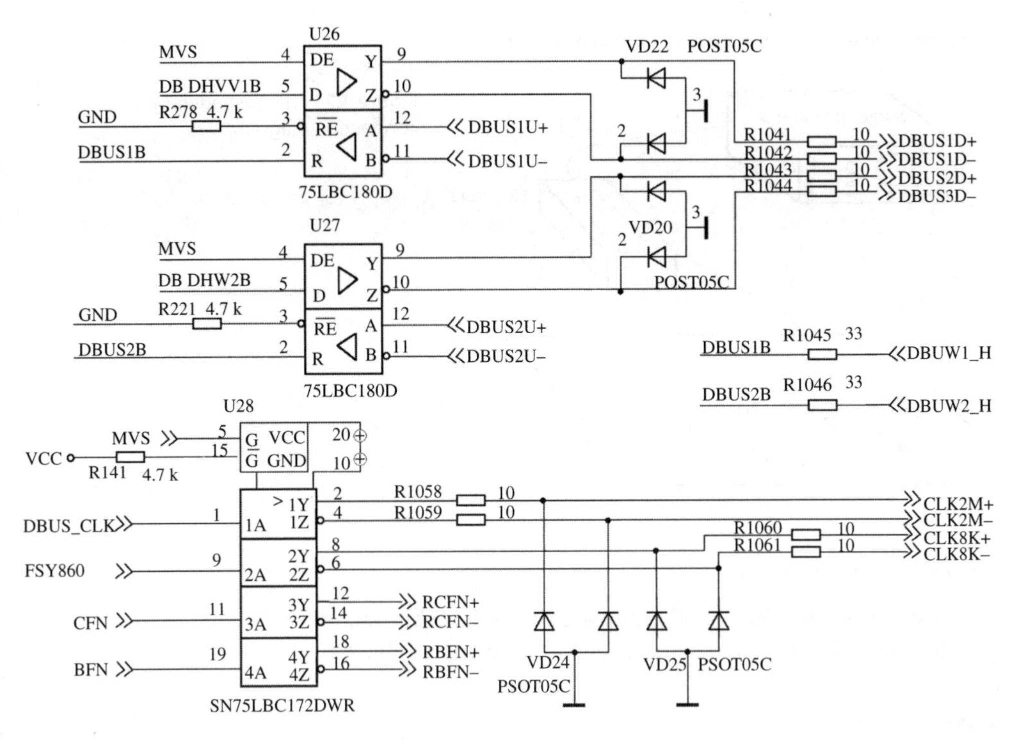

Figure 4.69 Circuit principle diagram of the first group.

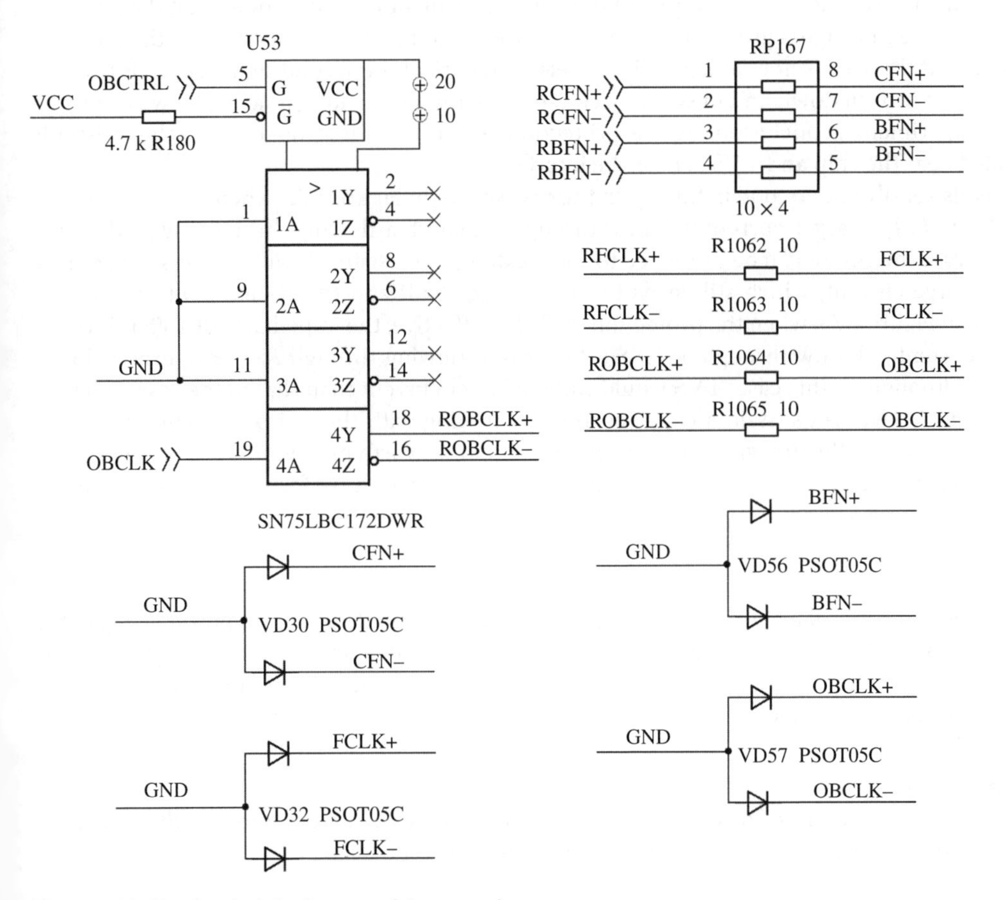

Figure 4.70 Circuit principle diagram of the second group.

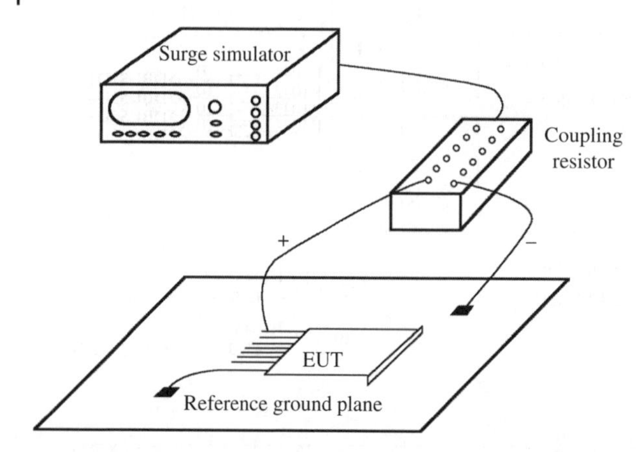

Figure 4.71 The test equipment connection diagram.

The test data are as follows:

1) The test data of the first signal group are shown in Table 4.5.
2) The test data of the second signal group are shown in Table 4.6.

[Analyses]

It can be seen from the test data that the surge protection of the first signal group is much worse that the second signal group. Test the post-experiment PCB. It is found that the interface failure of the first signal group is due to the failure of 33 Ω resistor in series with the interface circuit, but the TVS is not damaged. The interface failure of the second signal group is because the TVS is broken down and in short-circuit, but the resistor is not damaged. So we can come to an experimental conclusion that the protection performance is stronger when the resistor is located near the chip and TVS is near the interface.

It is also easily understood in theory. In the case shown in Figure 4.72, when TVS is activated, $I = I_1 + I_2$, $I > I_1$, a larger current will flow through R than that flowing through TVS. Although in this case the circuit can be protected to some extent, the resistor R_1 will be damaged because of the large current, which still leads to the interface fault and the system fault. In the case shown in Figure 4.73, when the protection of TVS is effective, the impedance of TVS is low and most current will flow through TVS. So the current flowing through R, $I_2 \ll I_1$, exactly lower that I. Although, in this case, TVS would suffer a larger current than that in the case shown in Figure 4.72, TVS can sustain a larger current comparing with the resistor R. This is why the signal interface in the first group fails more easily. But it is worth mentioning that if we increase the power rating of the resistor R enough, the surge protection performance of the circuit in Figure 4.72 is also good.

[Solutions]

When designing the surge protection for an interface circuit in the PCB, the power rating of the resistor must be considered if we want to reduce the surge current with the series resistor. If the power rating of the resistor is not enough, we suggest positioning the resistor near the chip and the TVS near the interface.

[Inspirations]

1) Resistors can be used to protect the devices against surge current and are often in series with the protected signal circuit to limit the surge current.

Table 4.5 Test data of first group.

Test signal	Test voltage	Test phenomenon	Test results
DBLSLD+	+500 V	Sparking	Broken
DBLSLD−	+200 V	None	OK
	−200 V	None	OK
	+300 V	None	OK
	−300 V	sparking	broken
OBLS2D+	+200 V	None	OK
	−200 V	None	OK
	+240 V	None	OK
	−240 V	None	OK
	+260 V	None	OK
	−260 V	none	OK
	+280 V	Sparking	Broken
DBLS2D−	+200 V	none	OK
	−200 V	None	OK
	+240 V	None	OK
	−240 V	None	OK
	+260 V	None	OK
	−260 V	Sparking	Broken
CLK2M+	+200 V	None	OK
	−200 V	None	OK
	+240 V	None	OK
	−240 V	None	OK
	+260 V	None	OK
	−260 V	None	OK
	+280 V	sparking	broken
CLK2M−	+200 V	None	OK
	−200 V	None	OK
	+240 V	None	OK
	−240 V	None	OK
	+260 V	Sparking	Broken
CLK8K+	+200 V	None	OK
	−200 V	None	OK
	+240 V	None	OK
	−240 V	None	OK
	+260 V	None	OK
	−260 V	None	OK
	+280	None	OK
	−280	None	OK
	+300	Sparking	Broken

(*Continued*)

Table 4.5 (Continued)

Test signal	Test voltage	Test phenomenon	Test results
CLK8K−	+200 V	None	OK
	−200 V	None	OK
	+240 V	None	OK
	−240 V	None	OK
	+260 V	None	OK
	−260 V	None	OK
	+280	None	OK
	−280	None	OK
	+300	Sparking	Broken

2) In the signal circuit, there are several points that need to be noticed when using a resistor:
   i) The power rating of the resistor should be enough.
   ii) Overcurrent on it should be avoided to prevent the resistor from being damaged.
3) For the above two surge protection circuits (shown in Figures 4.72 and 4.73), if the resistor's power rating isn't considered, then the voltage-dividing principle and impedance-mismatching principle can be used to explain which circuit is more suitable for the protected object. It depends on the input impedance of the protected circuit. If the input impedance is high ($R_1 \gg R$), the protection circuit in Figure 4.72 is suitable. Because, at this time, the series of R and R_1 contributes little to the current-limiting or voltage-dividing, and TVS is always in low-resistance state when it is activated, so only when the current-limiting resistor is connected in the circuit as shown in Figure 4.72 will it play a better current-limiting performance. If the input impedance is low (R_1 is similar to R or even smaller), the protection circuit in Figure 4.73 is more suitable. Because, at this time, the surge voltage after TVS will be further divided by R, the surge voltage on the protected circuit can be greatly reduced.
4) When a resistor is connected in series with the signal wire, the impact on the signal quality should be paid attention.

4.2.11 Case 39: Can the Surge Protection Components Be in Parallel Arbitrarily?

[Symptoms]

When performing surge test on the power supply input port of a certain device, each time performing the ±1 kV differential-mode surge test, (the required surge test level on this device is ±1 kV differential-mode voltage and ±2 kV common-mode voltage, and the accepted criteria is B), the rotating speed of the cooling fan in this device will be reduced and it cannot be recovered. After the test, the protection diode at the power supply input port of the operating circuit was damaged.

[Analyses]

The schematic of the surge protection circuit near the power supply input of this device is shown in Figure 4.74.

Two varistors in parallel are used for differential-mode surge protection. Two GDT in parallel are used for common-mode surge protection. During designing the operating circuits of the

Table 4.6 Test data of the second group.

Test signal	Test voltage	Test phenomenon	Test results
CFN+	+160	None	OK
	−160	None	OK
	+180	None	OK
	−180	None	OK
	+200	None	OK
	−200	None	OK
	+220	None	OK
	−220	None	OK
	+240	None	OK
	−240	None	OK
	+260	None	OK
	−260	None	OK
	+280	None	OK
	−280	None	OK
	+300	None	OK
	−300	None	OK
	+320	None	OK
	−320	None	OK
	+340	None	OK
	−340	None	OK
	+380	None	OK
	−380	None	OK
	+420	None	OK
	−420	None	OK
	+460	None	OK
	−460	None	OK
	+500	None	OK
	−500	None	OK
	+540	None	OK
	−540	None	OK
	+600	None	OK
	−600	None	OK

(*Continued*)

Table 4.6 (Continued)

Test signal	Test voltage	Test phenomenon	Test results
	+660	None	OK
	−660	None	OK
	+720	None	OK
	−720	None	OK
	+800	None	OK
	−800	None	OK
	+900	None	OK
	−900	None	OK
	+1000	None	OK
	−1000	None	OK
	+1200	None	OK
	−1200	None	OK
	+1400	None	OK
	−1400	None	OK
	+1600	None	OK
	−1600	None	OK
	+1800	None	OK
	−1800	None	OK
	+2000	None	OK
	−2000	None	OK
	+2200	None	OK
	−2200	None	OK
	+2400	None	OK
	−2400	None	OK
	+2800	None	Broken
CFN+	+2500	None	OK
	−2500	None	OK
	+2600	None	OK
	−2600	None	OK
	+2700	None	OK
	−2700	None	Broken
BFN+	+2500	None	OK
	−2500	None	OK
	+2600	None	OK
	−2600	None	OK

Table 4.6 (Continued)

Test signal	Test voltage	Test phenomenon	Test results
	+2700	None	OK
	−2700	None	OK
	+2800	None	OK
	−2800	None	Broken
BFN−	+2600	None	OK
	−2600	None	OK
	+2700	None	OK
	−2700	None	OK
	+2800	None	OK
	−2800	None	OK
	+2900	None	Broken
FCLK+	+2600	None	OK
	−2600	None	OK
	+2700	None	OK
	−2700	None	Broken
FCLK−	+2600	None	OK
	−2600	None	OK
	+2700	None	OK
	−2700	None	Broken
OBCLK+	+2600	None	OK
	−2600	None	OK
	+2700	None	Broken

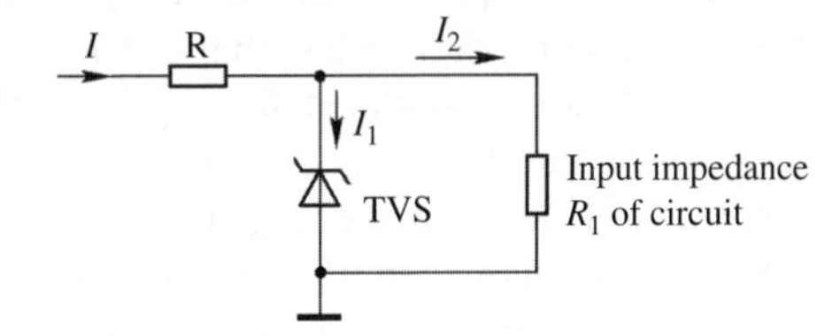

Figure 4.72 The circuit principle diagram of resistor outside the TVS.

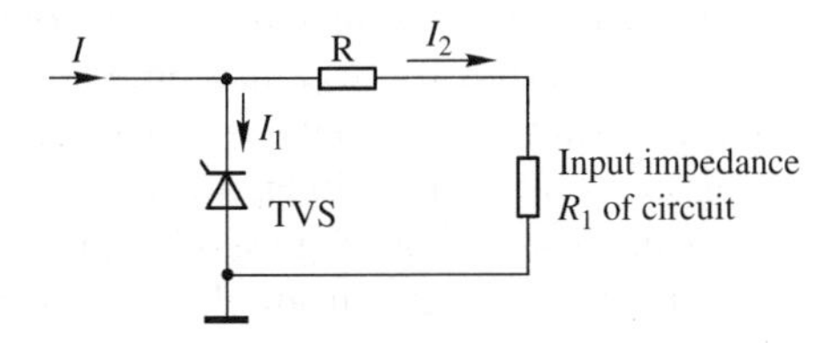

Figure 4.73 The circuit principle diagram of resistor inside the TVS.

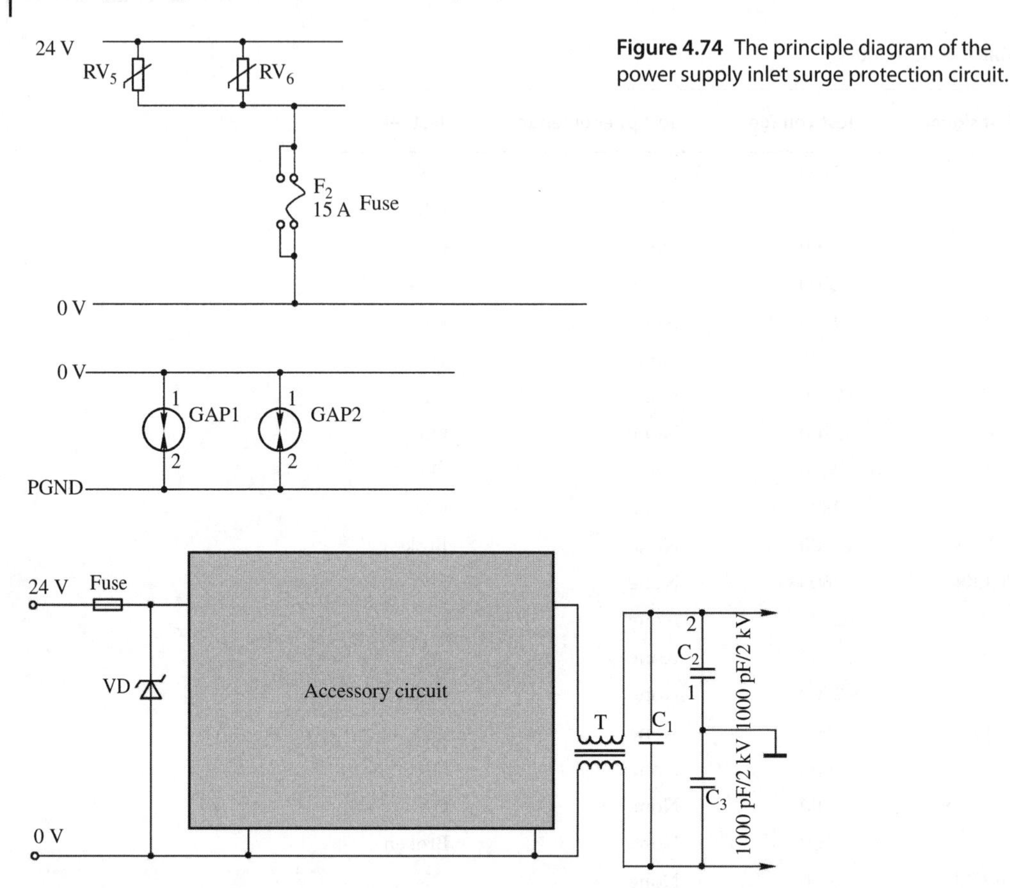

Figure 4.74 The principle diagram of the power supply inlet surge protection circuit.

Figure 4.75 The principle diagram of the power supply inlet of the fan working circuit.

fan, to get a further surge protection, a TVS is paralleled in the DC 24 V power supply input port for differential-mode protection. The schematic of the power supply input circuit in the fan is shown in Figure 4.75.

The output of the main power supply input board is directly connected to the power supply input of the fan's operation board by a cable without any component between them, and the interconnecting cable is less than 0.4 m. So, in theory, it is equivalent that the varistors RV_5, RV_6 and TVS VD are paralleled directly. Since the rated peak current of TVS is small and its responsive time is the shortest; in this case, TVS will be conducted first,. TVS will be damaged due to overcurrent when most energy flows through it. Then the cascaded circuit cannot be protected and the fan is damaged (the rotating speed is reduced). Under this condition, when performing a surge test, the residual voltage on the output of the lightning protection circuit in the main power supply input port of this device is above 150 V.

Among the surge protection components, the current handling capability of GDT is high, but its response time is long and its spark-overvoltage is high. The current handling capability of TVS is small, but its response time is fast and its clamping voltage is low. For varistor, the characteristics fall in between. The characteristics comparison of these three protection components are shown in Figure 4.76.

As can be seen from Figure 4.76, the working principle of varistor is that when the voltage on it is above a certain amplitude, its resistance will drop, and then the surge energy is discharged and the surge voltage is clamped in a certain range. Its advantage is that it can

Figure 4.76 Comparison of characteristics of three kinds of protective devices.

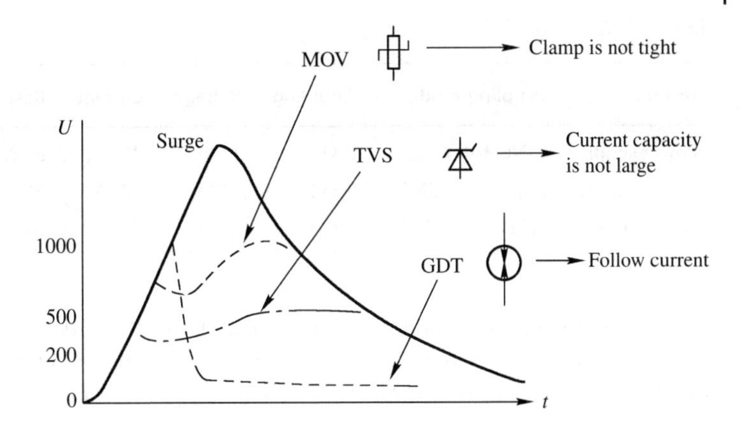

withstand high-peak current and its cost is low; But its disadvantage includes high-clamping voltage (relative to the operating voltage), long response time, big parasitic capacitance, and high-leakage current with the increasing of the suffered surge pulses. For TVS, when the voltage on it is above certain amplitude, it will be conducted immediately and discharge the surge energy, and then limit the amplitude of the surge voltage in a certain range. The advantage of TVS is fast response time and low-clamping voltage (relative to the operating voltage). However, its disadvantage is that it can only withstand low peak current, and since the parasitic capacitance of the general TVS is big, we must use specially made low-parasitic-capacitance TVS if it is applied on a high-speed data line. For GDT, its working principle is that when the voltage is above certain amplitude, it will be in short-circuit condition and discharge the surge energy. Its advantage includes high-current handling capability and small parasitic capacitance; its disadvantage is long response time, and since the maintaining voltage is very low, the follow current would exist, which means it cannot be applied in DC circuit (the discharge tube cannot be switched off). Note also that when it is used in AC circuit, the follow current may exceed the rated power of the component. A resistor can be in series with it to limit the current amplitude. The GDT can be used for around 50 times; then its spark-over-voltage begins dropping.

When the device requires the protection circuits with high-current handling capability and accurate protection, it is always necessary that these protection components cooperate to reach an ideal prevention effect. But these components cannot be simply paralleled to reach the cascading protection. If the varistor is directly in parallel with TVS (the current handling capability between them is widely different), even if the current handling capability of the varistor could satisfy the surge protection requirement of the device, TVS will still be damaged first by the high-surge current and the varistor does not play its role of high-current handling. The reason is that TVS will be conducted first, and after that, before the varistor is conducted, there is nothing to limit the large surge current, and then TVS will be damaged by this large surge current. Therefore, when designing a surge protection circuit for DC power supply input, it is always necessary that an inductor or wire is inserted between two surge-protection components when they are used to cooperate together. Inductor and wire are not surge protection components themselves, but they can be applied in the protective circuit cooperating with the other protective components. The schematic is shown in Figure 4.77.

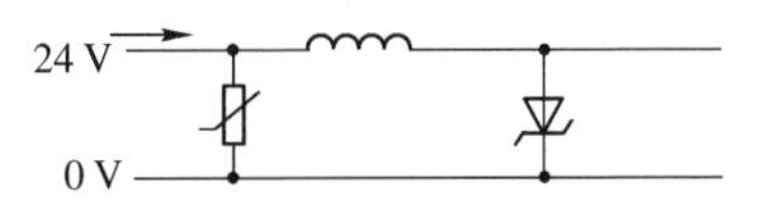

Figure 4.77 The schematic diagram of using the inductor of between piezoresistor and TVS.

The effectiveness of an inductor and wire in this protection circuit is seen in Section 4.1.2. In fact, in the practical

Table 4.7 Test results.

Test port	Coupling-mode	Coupling	Voltage	Current	Result
Power supply	L-N(24–0 V)	2 Ω	±1 kV	310 A	Normal
Power supply	LTE(24 V-PCND)	12 Ω	±1 kV	76 A	Normal
Power supply	NTE(0 V-PCND)	12 Ω	±1 kV	105 A	Normal

application, the parasitic inductance of the wire is effective. Its operating principle is the same as the inductor, and the wire can solve the problem that the DC current handling capability of the inductor is small. For the empirical value, the inductance of 1 m wire is 1–1.6 μH.

[Solutions]
Cascade an inductor at the power supply input circuit in the fan (the preceding stage of TVS). The inductance is 7 μH. The test results (shown in Table 4.7) verify the validity of this cascading inductor.

[Inspirations]
1) For GDS, varistor, and TVS, each has its own advantages and disadvantages. Each should play its own roles in the protection circuit design.
2) The surge protection components cannot be used arbitrarily. It should be carefully chosen; otherwise it would work opposite to what we expect, especially for those circuit boards developed separately.

4.2.12 Case 40: Components in Surge Protection Design Must Be Coordinated

[Symptoms]
A certain product has a 12 V power port (to supply the product) and a RS485 communication port. The 12 V power port needs to pass ±2 kV common-mode surge test because of its special application environment. It was found during the test that, the RS485 interface chip was damaged and the protection circuit connected with RS485 port was damaged, too.

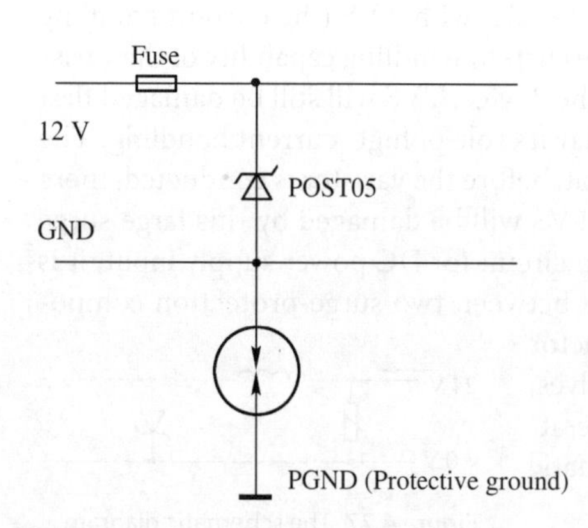

Figure 4.78 12 V power supply interface protection circuit.

[Analyses]
The circuit designs of the surge protection circuit for the 12 V power port and the RS485 communication port are, respectively, shown in Figures 4.78 and 4.79.

In the figure, protective ground (PGND) is the system grounding position of this product; GND is the working reference ground of its internal circuit. At the 12 V power port of this product, PGND and GND are not short-circuited, but GND and PGND are interconnected through GDT. The purpose of this design is to discharge the common-mode surge energy through GDT when high-surge voltage appears on the power port. After analyzing the surge protection circuit of the RS485 port, we found that the bidirectional TVSs TPN3021S are used to be the first stage differential-mode and common-mode surge protection before the common-mode choke. TVSs POST05 are used to be the second stage protections after the common-mode choke. So, in the entire product, there are two kinds of TVSs: TPN3021S and POST05 connected between GND and PGND, besides the GDT on the 12 V power port. The actual protection circuit between GND and PGND is shown in Figure 4.80.

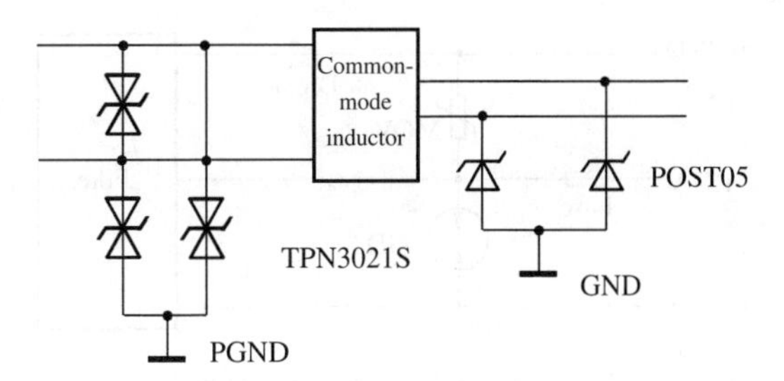

Figure 4.79 485 interface protection circuit.

So it is equivalent that there are two stages of protection between GND and PGND. One stage is composed of GDT and the other one is composed of TVSs and the common-mode choke (the characteristics of GDT and TVS are already described in other cases). Because there is no the coordinated decoupling components (such as inductor, resistor, etc.) between these two stages and the response time of TVS is faster than that of GDT, when performing the ±2 kVcommon-mode surge test on the 12 V power port (the surge voltage is applied between GND and PGND), the second-stage protection circuit existing in the RS485 interface will be conducted immediately, but the GDT is not activated yet, which means GDT doesn't play its roles. That is to say, the surge energy injected on 12 V power port was discharged through the protection circuit of the RS485, instead of the protection circuit of the power port. Besides, in the second-stage protection circuit of the RS485 port, the current handling capability of TVS is very small, so the surge current is larger than the rated peak current of TVS, which causes the TVS and the RS485 interface chip to be damaged.

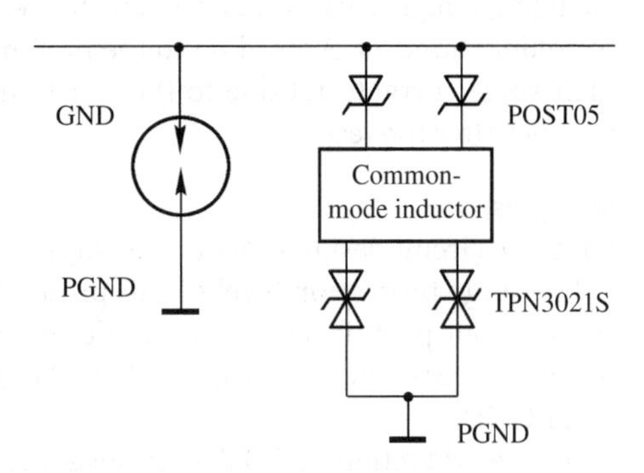

Figure 4.80 The protection circuit after the deformation.

[Solutions]

1) Remove the second-stage protection circuit of the RS485 interface circuit, i.e. disconnect the second-stage protection component POST05 in Figure 4.79. The test result proves that, after removing the second-stage protection circuit of the RS485 port, the RS485 port can pass the ±4 kV surge test and the 12 V power port can also pass the required common-mode surge voltage test.
2) We could also shorten GND and PGND to solve this problem. The test result proves that, after shortening PGND and GND, the product can also pass ±2 kV common-mode surge test.
3) Connecting the first stage protection components of the RS485 port to GND is also a solution to the problem.

[Inspirations]

This is another case about how to coordinate the surge protection components in appropriate cooperation. The protection design shall be considered for the overall product. The protection circuit for any interface cannot be designed alone with ignoring the protection for the other interfaces. That is to say, it is important to design the surge protections as a whole.

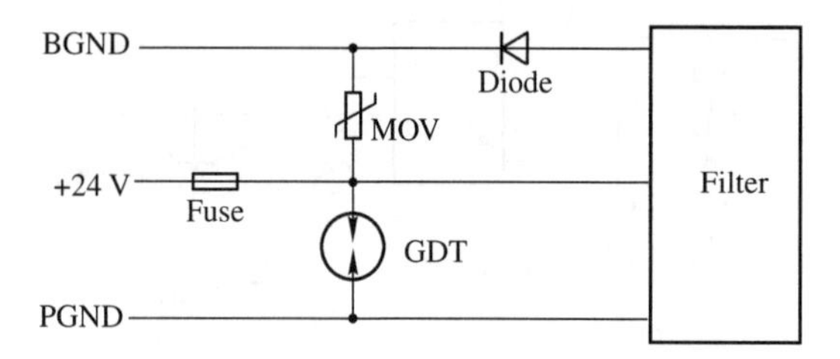

Figure 4.81 Power supply input port circuit.

4.2.13 Case 41: The Lightning Protection Circuit Design and the Component Selections Must Be Careful

[Symptoms]

A certain product is supplied by a 24 V DC power supply. After performing surge test on the DC power port, its fuse was burnout. And the phenomenon still appeared after changing a high-current fuse. Then shorten the fuse with a high-gauge wire and test it again; "smoking" appears. According to the preliminary analysis, the burnout of the fuse is not due to high-surge current, but due to the short-circuit on the 24 V power supply input circuit in the product after the test.

[Analyses]

First, the circuit design of its power supply input port is shown in Figure 4.81.

The surge protection level of this product is high, so GDT and varistor are both used in the power input port. In fact, a severe design error is obvious in the schematic: There is a GDT, whose DC spark-overvoltage is 90 V, and is directly connected between the +24 V power supply and PGND.

For the protection of GDT to its poststage, during the surge test, the air inside GDT is instantaneous broken down. Then the GDT is approximately in short-circuit state to bypass the surge current to the ground. After the transient disturbance disappears, GDT needs to be recovered to open-circuit state; otherwise, GDT can be damaged by a long-term large current that will also cause the power supply to be short-circuited at the power supply interface port of the device when surge is injected on the power input circuit, which could cause an accident or even fire. Therefore, one condition must be guaranteed for the design of a circuit with GDT is that, in the normal operation, GDT can be recovered to open-circuit state automatically after being broken down to be in short-circuit state, that is to interrupt its follow current. In this case, the remained DC voltage of GDT in the short-circuit state is 20–25 V. In other words, if the GDT is conducted, the remained voltage on it is above 20 V. The GDT will be continuously conducted until its voltage drops or the GDT is burned out. The voltage remained on GDT, which, in the short-circuit state, is called the residual voltage under the follow current.

When the surge voltage is not injected on the power input port, the DC voltage between +24 V power supply and PGND is always not over the GDT's DC spark-overvoltage (90 V) and GDT is in open-circuit state. So the interface circuit shown in Figure 4.81 can operate normally and no problem appears. After the GDT is broken down to be in the short-circuit state, there is 24 V continuous operating voltage applied on it to resist GDT from being recovered in the open-circuit state. Then the fuse is burned out, and even "smoking" appears when replacing the fuse with a heavy-gauge wire.

[Solutions]

Change the design schematic according to the above analyses. The modified power supply input port circuit is shown in Figure 4.82.

[Inspirations]

1) The lightning protection design and the components selections at the power port must be careful. Generally, the power supply can provide continuous large energy. If the lightning protection circuit is not designed properly, the electrical safety cannot be ensured and the

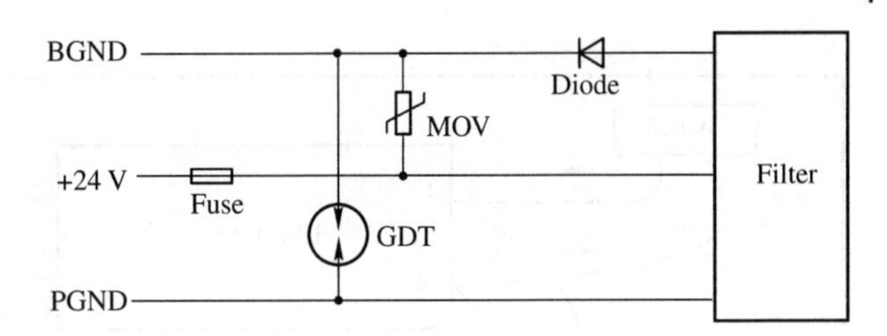

Figure 4.82 Power supply input port circuit after modified.

product can be even on fire in severe case. This is much more terrible than the consequences after lightning stroke on the device. In the power supply circuit design, safety design is always the first priority.
2) When designing the surge protection circuit, it is necessary to make clear the characteristics of the protective components; otherwise, the unintended consequence could appear.
3) Pay close attention to the follow current of GDT. That is also why GDT can only be paralleled with the low-voltage circuits.
4) In most cases, the failure-mode of GDT is in open-circuit state. However, if the GDT is burned out because of staying in long-term short-circuit state, the failure-mode can also be short-circuit.

4.2.14 Case 42: Strict Rule for Installing the Lightening Protections

[Symptoms]

For a large-sized outdoor industrial equipment, the required surge current its power supply input should sustain is 40 kA (8/20 μs surge current waveform) in differential-mode and common-mode. In order to meet this requirement, a surge protection module with nominal 40 kA surge current is paralleled on the power supply input port specially. But after performing the differential common surge test, it was discovered that many components inside the product were not well protected but damaged, such as the filter, the power supply, and the voltage regulator. However, the lightening protection device was not broken. The other equipment with this type of surge protection module can pass 40 kA surge current test successfully.

[Analyses]

The actual installation for this surge protection module inside this product is shown in Figure 4.83.

In the differential-mode surge test, the surge current flows along the direction of the arrow. It can be seen from Figure 4.83, the cables L_1 and L_2 are inside the loop through which the surge current flows. There will be 40 kA surge current flows on the cables L_1 and L_2. The actual total length of L_1 and L_2 is about 0.5 m. So, the residual voltage that the input of the filter suffers is the sum of the residual voltage of the surge protection module itself, and the voltage drop on L_1 and L_2 caused by the surge current on L_1 and L_2. The simplified schematic in the differential-mode surge test is shown in Figure 4.84. H_1 and H_2 are respectively the parasitic inductance of L_1 and L_2. In general, the cable self-inductance is about 1–1.6 μH m^{-1}. The voltage between A and B is the overall residual voltage that the post-stage circuit will suffer.

Under the 40 kA surge current, the voltage drop caused by the two cables can be estimated as:

$$\Delta U = L\times(di/dt) = 1\,\mu\mathrm{H\,m}^{-1}\times 0.5\left(40\,\mathrm{kA}/8\,\mu\mathrm{s}\right) = 2500\,\mathrm{V}$$

As can be seen from the above estimation, the 40kA surge current will cause 2500 V voltage drop when it flows through the two cables. This high-voltage drop is superimposed on the

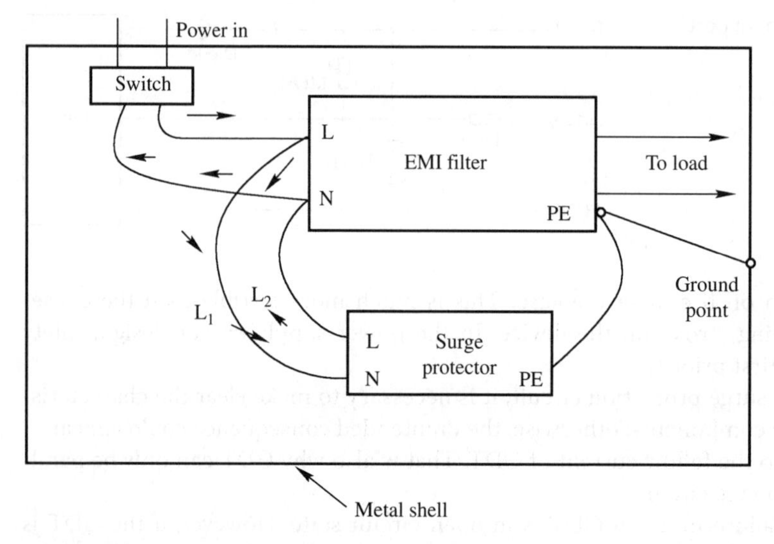

Figure 4.83 Lightning arrester wiring diagram.

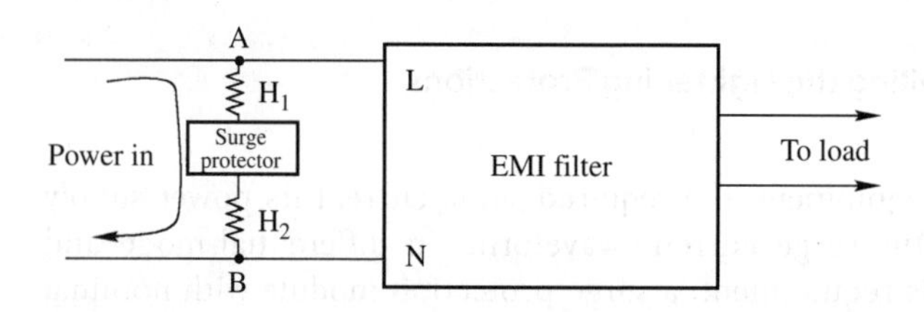

Figure 4.84 The simplified circuit principle diagram under different-mode.

1500 V residual voltage of the surge protection module itself, the overall residual voltage that the input of the filter will suffer is 4000 V. It is obvious that this excessive residual voltage will ultimately damage the filter, the power supply, the voltage stabilizer, etc.

[Solutions]

It can be seen from the above analyses that if L_1 and L_2 are shortened (see Figure 4.83), namely if the inductance H_1 and H_2 is reduced, then the lightening protection effectiveness will improve. If wiring is as shown in Figure 4.85, the residual voltage on the post-stage circuit can be reduced to the minimum, which is only the residual voltage 1500 V of the surge protection module itself.

After testing the product with the connection as shown in Figure 4.85, the test passes.

[Inspirations]

1) Under large surge current, a high-voltage drop can appear on a short cable. The length of the cable cannot be ignored.
2) For the protection components paralleled on the power supply wires or signal wires, the signals shall come into the protection components first, and then go to the post-stage circuits.
3) When designing surge protection against large surge current, the parasitic inductance of the wire itself cannot be ignored. Under the surge current, the voltage drop on the cable can

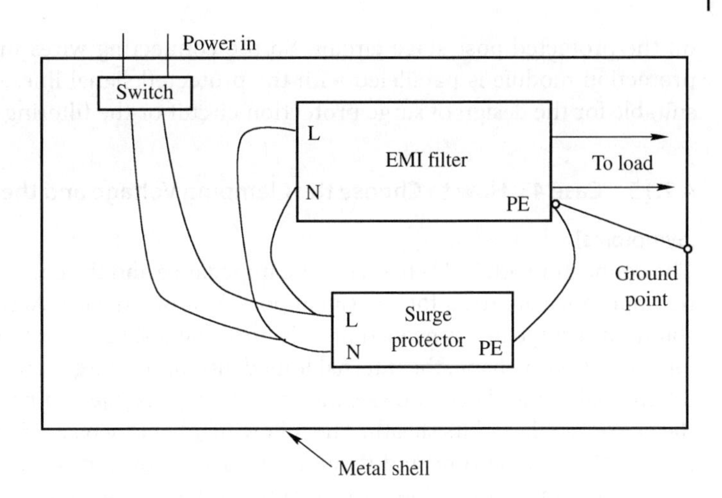

Figure 4.85 Connection-mode after modified.

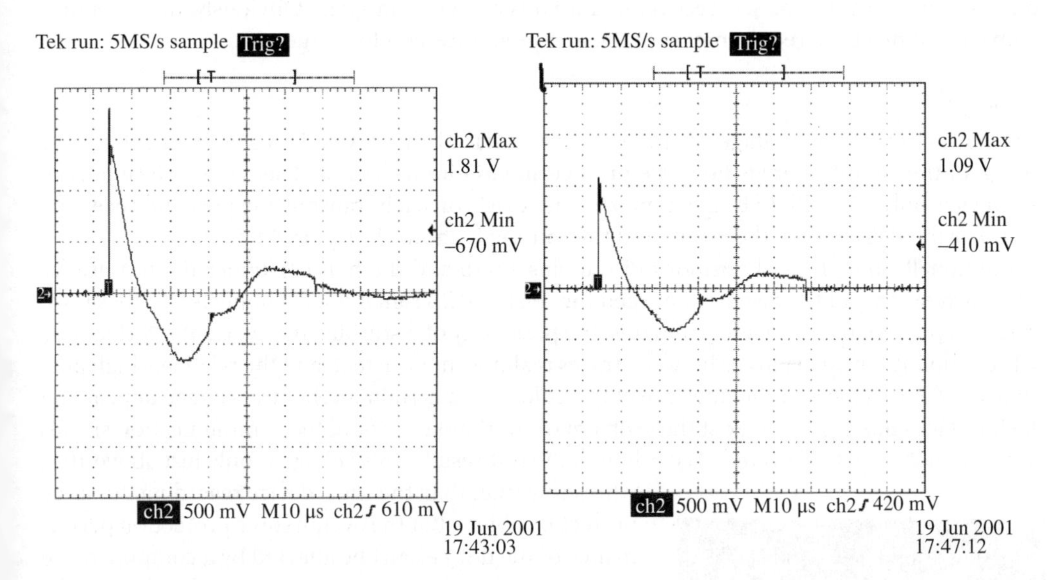

Figure 4.86 The voltage drop at the end of 8/20 μs 1 m long wire for 5 kA (left), 3 kA (right).

be estimated in theoretical calculation. A wire can be equivalent to an inductor, and when alternating current flows through the wire, the voltage drop on it is

$$\Delta U = |L \times di/dt|$$

Here L is the self-inductance of the wire. Generally the self-inductance of 1 m long wire is about 1–1.6 μH (1 μH can be used when calculating). di/dt is the slew rate of the current. It can be seen from this formula, ΔU is directly proportional to L, and L is directly proportional to the length of the wire. Figure 4.86 shows the measured voltage waveform on the 1 m long wire under the 8/20 μs surge current of 5 kA (left), 3 kA (right) (the probe attenuation is 500 : 1). So when 5 and 3 kA surge current flows through the 1 m long wire, the peak voltage on the two ends of the wire is, respectively, 900 and 550 V. Therefore, reducing the length of the wire connected to the surge protection module has a great effect on the reduction of surge voltage on the protected post-stage circuit. So the connecting wires must be short when the surge protection module is paralleled with the protected signal lines. This design principle is also suitable for the design of surge protection circuit or the filtering circuit on PCB.

4.2.15 Case 43: How to Choose the Clamping Voltage and the Peak Power of TVS

[Symptoms]

The connector with 75 Ω characteristic impedance and the coaxial cable with 75 Ω characteristic impedance are used for the signal port A of one device. According to the related product standard, a surge test is needed on this port, the test surge voltage is ±2500 V, the surge waveform is 1.2/50 μs shape, the internal impedance of the surge generator is 42 Ω, and the test is in differential mode. That is to say, the surge voltage is injected between the signal conductor of the coaxial cable and its sheath. The following phenomena are found during the test: (i) Falsely alarming for a long time and it cannot be self-recovered to the normal operation; (ii) When testing on signal port A, the same warning problem also appears on signal port B, lasting for a long time and it cannot be self-recovered until it is powered on again. Obviously, this phenomenon cannot meet the requirements of the accepted criteria B for surge test.

[Analyses]

The normal operating voltage of the interface signal of the device is 3.3 V, by checking the design of the circuit, there is no surge protection circuit on the tested signal interface port of the device, only a 1 : 1 PULSE signal transformer exists on each transmitting port and receiving port. For the PCB design, the layout and the routing is shown in Figure 4.87.

Additionally, the physical positions of the pins 14/15 and the pins 6/7 under the test are far from each other, and the distance between the related PCB signal traces connected to these two groups of pins is 0.07 in. To infer from this, the possibility of disturbing the adjacent interface due to the coupling between pins or between traces is almost nonexistent, and the relatively high possibility is the weakness of the interface chip, As long as a certain level of the voltage appears on the interface chip, there will exist the problem of continuous error of the communication signals and it cannot be self-recovered. According to the test results, this voltage is only just above 10 V, which is weaker than the usual device regarding the surge protection ability. But the overall surge protection performance of the device can't be affected by a component, so the residual surge voltage amplitude on the chip can be reduced only by adding the external surge protection. Several kinds of common surge protection components are planned to be tested, taking into account the parasitic capacitance and the signal speed, the selected component is TVS. The relationship between the clamping voltage of TVS and its current is shown in Figure 4.88.

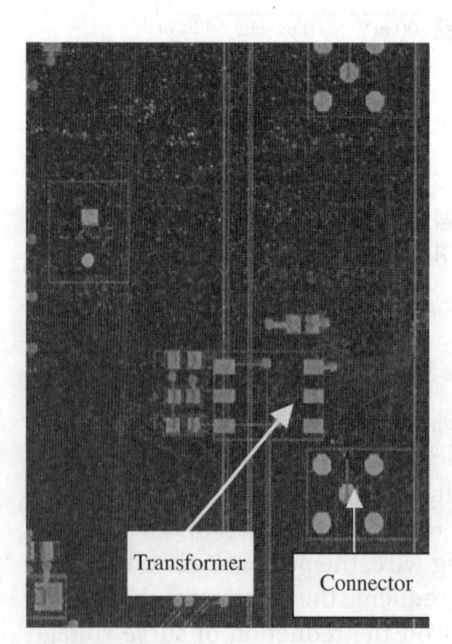

Figure 4.87 Interface PCB layout.

The clamping voltage is calculated:

$$U_c = (I_p / I_{pp}) \times (U_{cmax} - U_{BRmax}) + U_{BRmax}$$

where U_c is the intermediate value of the clamping voltage (V), measured at the current I_p, which is the peak current (A) during the experiment. I_{pp} is the maximum peak current (A) for the design; U_{cmax} is the maximum clamping voltage (V); U_{BRmax} is the maximum avalanche breakdown voltage (V).

Figure 4.88 The relationship between TVS voltage and current.

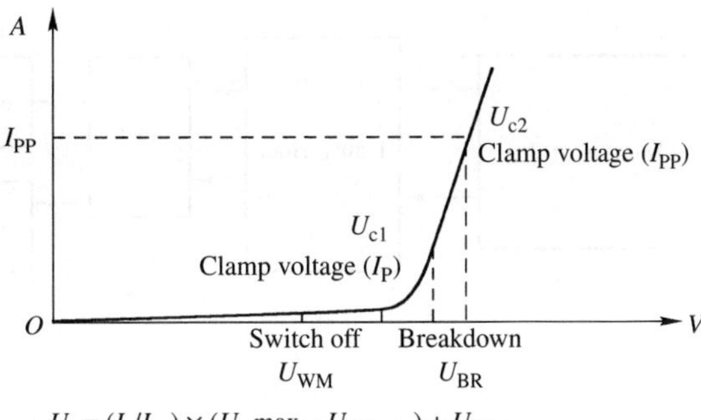

$U_c = (I_p/I_{pp}) \times (U_c \max - U_{BR\ max}) + U_{BR\ max}$

U_c = Clamp voltage (V), I_P = Peak current of surge (A), I_{PP} = Target maximal current of design (A), $U_{c\max}$ = Target maximal clamp voltage of design (V), $U_{BR\max}$ = Target maximal breakdown voltage (V)

There are several kinds of surge protection components for testing.

PSOT05L3, its rated peak power is 300 W, and the clamping voltage is 9.8 V (at 1 A)/11 V (at 5 A).

SLUV2.8-4, its rated peak power is 400 W, and its current handling ability is 24 A, and the clamping voltage is 5.5 V (at 2 A)/8.5 V (at 5 A).

LC03–3.3, its rated peak power is 1800 W, its current handling ability is 100 A (8/20 μs surge current), it can be used for differential-mode and common-mode surge protection, and its clamping voltage is 15 V (at 100 A surge current flowing from line to ground).

During the test, TVS is placed in front of the transformer, and the specific layout is shown in Figure 4.89.

Because the impedance of the test instrument is high, in order to reproduce the problems, the test voltage is increased to verify the solutions.

4.2.15.1 Test Results with PSOT05L3

At the beginning, the ±1500 V surge voltage is injected, and no problem appears during the test, the same for ±2000 V test voltage; Signal communication is blocked when the test voltage is ±2500 V, the interface PCB circuit cannot be self-recovered and, it must be powered on again. Analyzing the test results, the test surge current is already up to 15 A with ±2500 V surge voltage injection. It can be seen from the datasheet of this component, it cannot effectively clamp the surge voltage at this current, the clamping voltage is at least above tens of volts. And after this clamping voltage passes through the transformer, the same voltage amplitude appears on the pins of the chip, with which the problem described as above appears, and then the problem occurs; obviously, this component cannot solve this problem.

4.2.15.2 Test Results with SLUV2.8-4

This component is better on the champing voltage and the current handling ability than PSOT05L3, and the test is started from ±2500 V surge voltage, and no problem happens at this voltage, but the same problem happens when performing ±3000 V surge voltage test, the interface PCB circuit cannot be self-recovered, it must be powered on again. The reason is similar to PSOT05L3 after analysis, and the clamping voltage of SLUV2.8-4 is above 10 V when the fault happens. This TVS also can't solve this problem effectively.

4.2.15.3 Test Results with LC03–3.3

The test voltage is started from ±2500 V and no problem happens, and the same results for ±3000 and ± 3500 V surge voltage test, respectively. It can be seen from the datasheet of this

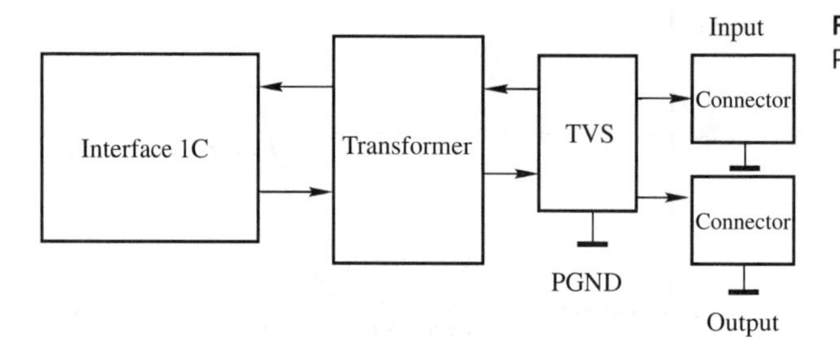

Figure 4.89 The diagram of interface PCB layout.

Table 4.8 The influence of capacitance on the signal itself.

	Low-speed interface 10–100 kb s^{-1}	High-speed interface 2 Mb s^{-1}	Low-speed CMOs	TTL
Rising time tr	0.5–1 us	50 ns	100 ns	10 ns
Bandwidth Bw	320 kHz	6 MHz	3.2 MHz	32 MHz
Total impedance	120 Ω	100 Ω	300 Ω	100 – 150 Ω
Max capacitance	2400 pF	150 pF	100 nF	30 pF

TVS that the champing voltage is about 5 V now. Repeat the test and there are still no abnormal phenomena, so this TVS can meet the requirements.

[Solutions]

Through the comparative tests with the above components, the following conclusions can be made:

(i) The clamping voltage of the selected TVS under the assigned surge current should be low, at least below 8 V. Usually in the signal circuits, the clamping voltage of TVS shall be 1.2–1.5 times the maximum normal working voltage of the signal. In fact, it is also required to use a TVS which has sufficient peak power and can withstand high-surge current; (ii) The space availability in PCB shall be taken into account for the common-mode and differential-mode protection design, and the highly integrated components should be used first. So as long as the TVS meeting the above (i) and (ii) conditions is chosen, the surge problem of this equipment can be solved. So choosing LC03-3.3 from SEMTECH can solve the problem.

[Inspirations]

1) In the selection of the clamping type surge protection components, the peak pulse power, the current handling capacity and the clamping voltage must be considered. Usually, the better the current handling capability is, the lower the champing voltage is, i.e. the better the effectiveness is. Usually in the signal circuits, the clamping voltage of TVS should be 1.2–1.5 times the maximum normal working voltage of the signal.
2) When choosing TVS for surge protection, we should not only consider TVS's clamping voltage and peak power, but also the impact on the signal itself due to the junction capacitance of TVS. Table 4.8 and Figure 4.90 can be used as a reference.

4.2.16 Case 44: Choose the Diode for Clamping or the TVS for Protection

[Symptoms]

A product has a 2 MHz operating frequency interface circuit. Due to the required application environment of the product, a surge protection is needed for this interface circuit and the surge test voltage is 1 kV. The protection circuit composed with TPN3021 and DA108S1RL is recommended

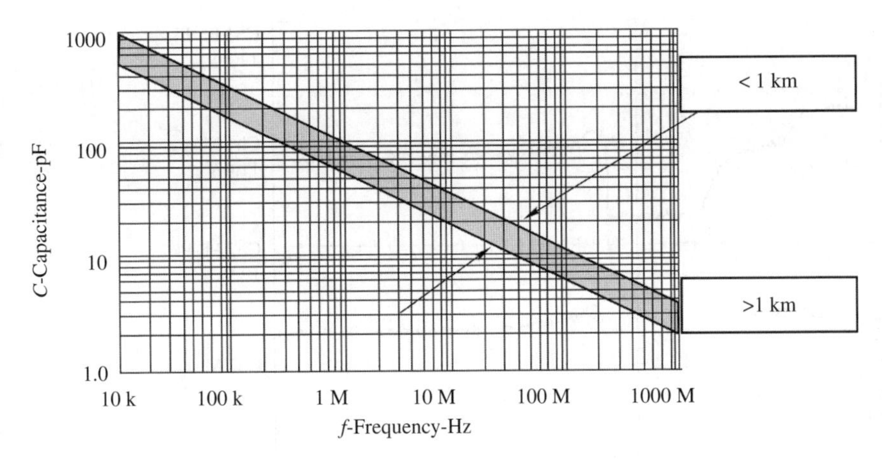

Figure 4.90 The relationship diagram between signal frequency and wire-to-wire capacitance.

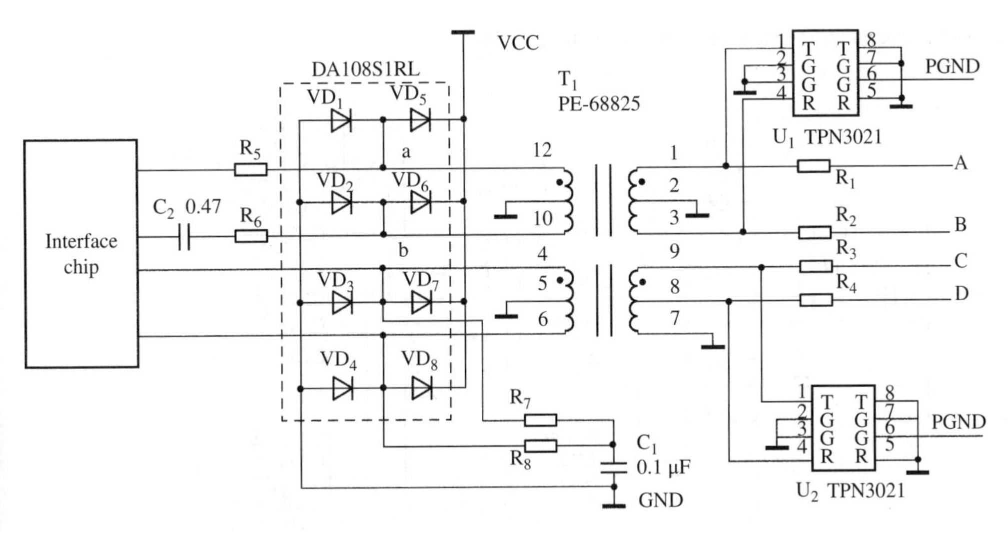

Figure 4.91 The principle diagram of interface circuit.

by SGS – THOMSON Co. for this kind of interface circuit, which is shown in Figure 4.91. During the surge test, it is found that the interface circuit cannot meet the requirement for surge test.

[Analyses]

In the schematic shown in Figure 4.91, the diode DA108S1RL is used to constitute the clamping circuit. Due to the unidirectional conduction of the diode and the fast response time accordingly, when the surge disturbance voltage is higher than the diode's forward conducted voltage, the diode will be conducted, and its post-stage circuit will be protected. R5, R6 in Figure 4.91 can limit the current. Taking point a as an example, when the potential on point a is higher than VCC (+5 V) by 0.7 V, the diode VD5 will be conducted, the potential at point a is clamped at 5.7 V; when the potential of point a is lower than GND by 0.7 V, the diode VD1 will be conducted, and the potential at point a is clamped at −0.7 V. In the ideal case, the potential at point a can be clamped between −0.7 and 5.7 V. Because there exists the inductance from the point a to the diodes, and the pins of the diode to VCC and GND, the junction capacitance of the diode and the response time of the diode, when the surge disturbance is actually applied, the potential at point a can only be clamped between±10 and ±20 V, and the actual measured voltage at this point with oscilloscope is shown in Figure 4.92.

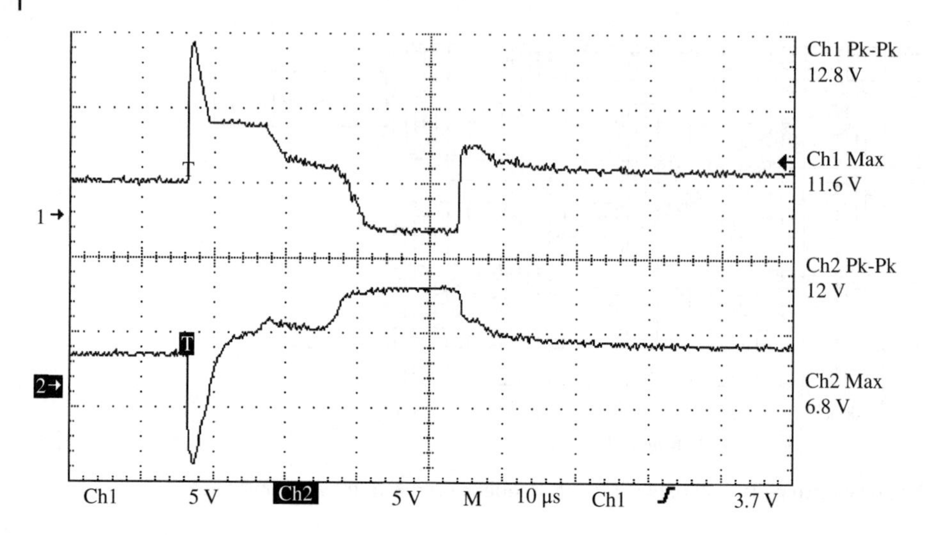

Figure 4.92 The measured clamp voltage waveform at point a.

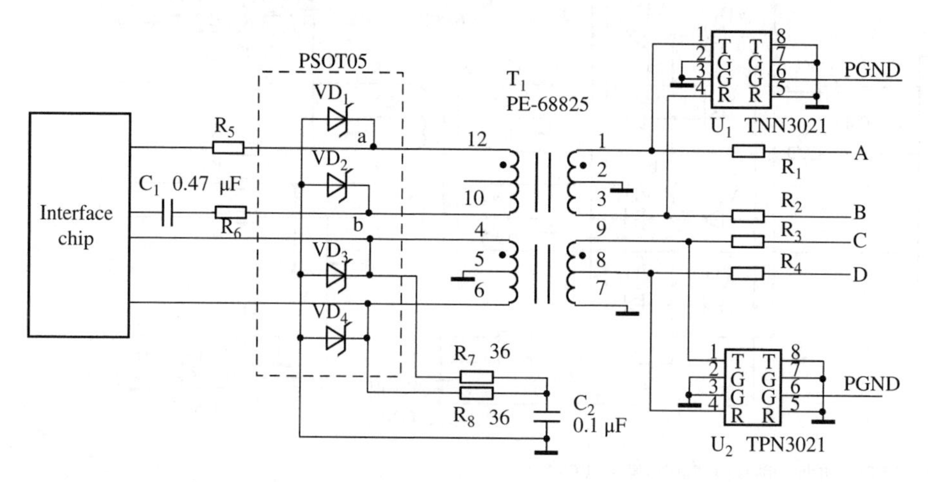

Figure 4.93 The principle diagram of interface circuit after modified.

In Figure 4.91, the clamping voltage of the clamping circuit consisting of DA108S1RL is not as low-as expected while a disturbance voltage is assigned. When looking at the datasheet of the interface chip, it is found that the interface chip can withstand maximum 18 V surge voltage. Visibly, when performing the surge test, the high-potential at point a is the main reason causing DS2154 being damaged. So if a lower clamping voltage (less than 18 V) protection circuit is found, the problem will be solved.

[Solutions]

Through the above analysis, the high potential at point a is the main reason causing DS2154 being damaged. Using the TVS PSOT05 from ProTek instead of DA108S 1RL can achieve a better clamping performance. The schematic is shown in Figure 4.93.

TVS is a kind of special diode for overvoltage protection by working in reverse breakdown mode. The surge protection capability of TVS is very strong, its clamping voltage and the breakdown voltage are low, while the transient response time is relatively fast and its junction

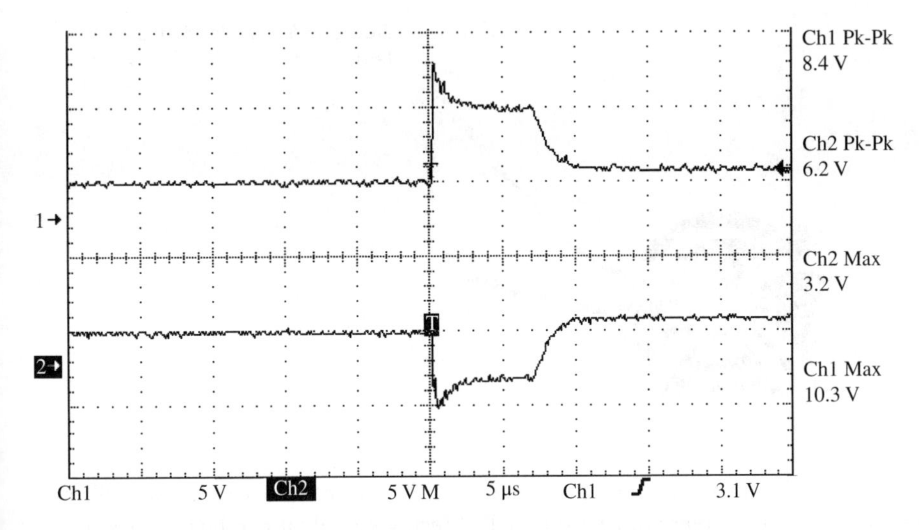

Figure 4.94 The measured clamp voltage waveform at point a after modified.

capacitance is low. Therefore, TVS becomes the ideal protection device for electronic equipment. A typical breakdown voltage of PSOT05 is about 6 V, its peak power is 300 W (for 8/20 μs pulse waveform), its junction capacitance is 5 pF, and its response time is 10×10^{-12} seconds. When the potential at point a is higher than the breakdown voltage of TVS, VD1 is in reverse breakdown state, and the surge current is discharged from VD1 to GND. By using this circuit, it can effectively protect the interface chip, and the waveform at point a is shown in Figure 4.94.

[Inspirations]

1) Diodes (including Schottky diode, high-speed diode, and PIN diode) and TVSs could be used as the protection against the transient disturbance, but their protection mechanisms are not the same. Diode uses the forward conduction characteristics; TVS uses reverse breakdown principle. Diode usually has lower junction capacitance compared with TVS. For the protection design, we must weigh their pros and cons.
2) If using diode as the protection against transient disturbance, it is recommended to use the diode with a fast response time.

4.2.17 Case 45: Ferrite Ring Core and EFT/B Immunity

[Symptoms]

For a certain kind of smart single-phase meter, when ±1 kV EFT/B immunity test voltage is injected on the power supply input port, it is found that the microprocessor inside the meter works abnormally, namely the meter is frequently reset, the display is wrong, the communication fails, and sometimes the MCU is crashed.

[Analyses]

Checking the circuit structure of the smart meter, it is found that the electric meter is supplied with the AC 220 V voltage, and the working voltage of the control circuit is converted from a linear power supply. As the linear power supply is not like a switching-mode power supply, which will produce high-frequency electromagnetic interference, so the design does not take into account the filter on the power supply input port, of course, the high cost of the filter is also another reason.

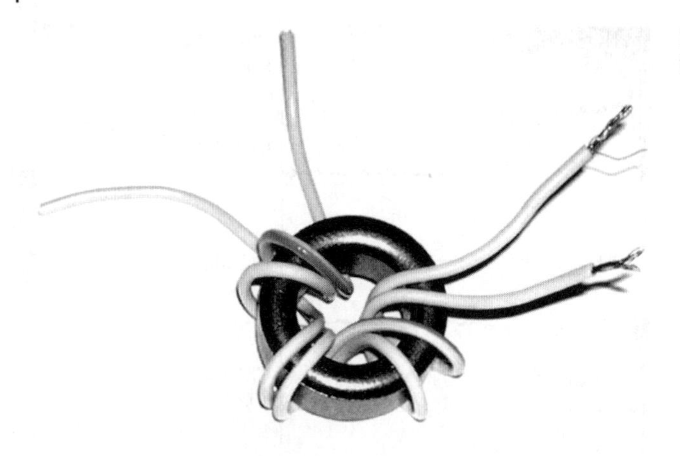

Figure 4.95 Double winded ferrite core.

The essence of the EFT/B disturbance has been described in other cases, so we will not repeat it here. To solve the problem during the EFT/B test, generally it can be achieved from the following three perspectives:

1) Change the flow of the EFT/B disturbance current and do not let it pass the sensitive circuits of the product.
2) Suppress the EFT/B disturbance current before it arrives at the sensitive circuits, such as adding the filter that has the suppression effect on EFT/B disturbance at the power supply input port.
3) Improve the immunity performance of the circuit itself, and it will not appear abnormal phenomena even the EFT/B disturbance current flows through it (this method has a high risk).

The way (1) and (3) should be considered during the product structure design and schematic design; The way (2) is the simplest and the most effective method to solve the problems due to the EFT/B disturbance during the late design period of the product. During the test, trying to use two high-temperature wires and bifilar wind them with three turns on the ferrite ring core (because the EFT/B disturbance appears in common mode). The outer diameter of the ferrite ring core is 25 mm, its inner diameter is 15 mm, its height is 12 mm, and its relative permeability is 800, which is shown in Figure 4.95.

Place this ferrite ring core in series with the power supply input (see Figure 4.96) of the meter, the EFT/B immunity performance of this product is greatly improved, namely it could pass ±4 kV EFT/B immunity test.

The magnetic material of this ferrite ring core can be represented with the chemical formula MFe2O4. In this formula, *M* represents the ions of manganese, nickel, zinc, copper, and other divalent metal. Ferrite is manufactured by sintering a mixture of these metal compounds. The main features of the ferrite is that the resistivity is much larger than metal magnetic material, which suppresses the generation of eddy current, and allow it widely used in high-frequency applications. Ferrites can be considered as a series of the inductance and the resistance in schematic. When used in products, it has significantly improvement for EFT/B immunity performance, mainly due to the equivalent resistance of the ferrite. The resistance is an energy-consuming component and it converts the energy of EFT/B disturbance into heat energy and dissipates it. So, usually this effect of ferrite on high-frequency disturbance is known as the absorption. The absorption mechanism of the ferrite is different from the inductor and capacitor, which are used to constitute a filter based on the principle of impedance mismatching (the filter with reflection effect will amplify the disturbance over a certain frequency band).

Figure 4.96 The ferrite core double winding is installed at the power supply input.

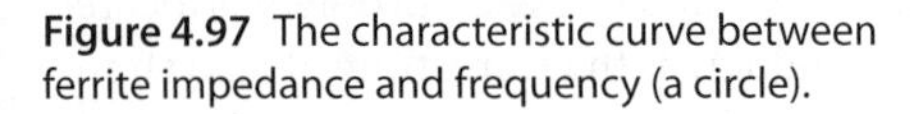

Figure 4.97 The characteristic curve between ferrite impedance and frequency (a circle).

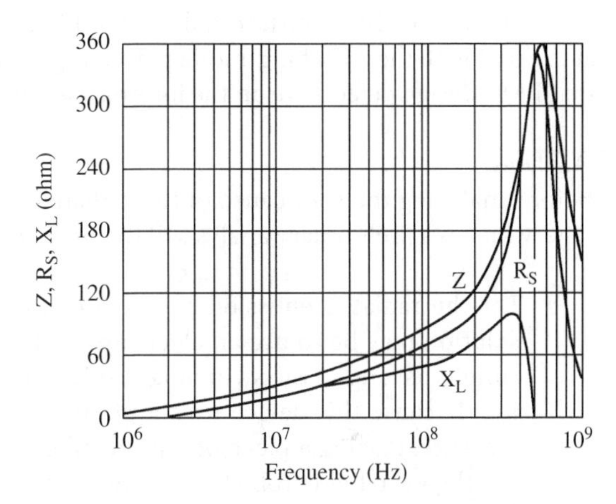

[Solutions]

According to the above tested results, bifilar winding three turns on the ferrite ring core and putting it in series with the power supply input port. Using nickel zinc ferrite ring core with 25 mm outer diameter, 15 mm inner diameter, and 12 mm height, the characteristic curve of the impedance vs. frequency is shown in Figure 4.97. The cost of this ferrite ring core is not more than 2 yuan in RMB.

[Inspirations]

1) In practice, it is also found that the ferrite is particularly effective on the suppression of EFT/B disturbance, especially for the thicker and larger ferrite ring core. If a better installation position can be found in the product, the ferrite ring core will be the best choice to improve EFT/B immunity performance of the product.
2) When winding the wire on the ferrite ring core, pay special attention to the turn number of the winding, winding on a single-layer as possible, and increasing the distance between the turns, because too many winding turns will increase the parasitic capacitance and lead the ferrite to be noneffective. The influence on the parasitic capacitance of the inductance from ferrite core is bigger than that from iron powder core and iron core.
3) When winding wire on the ferrite ring core, make the starting terminal of the coil away from the ending terminal (the angle between them shall be greater than 40°), in order to prevent big coupling between the input and the output.

4.2.18 Case 46: How Ferrite Bead Reduces the Radiated Emission of Switching-Mode Power Supply

[Symptoms]

During the radiated emission test on a controller product, it is found the test results are exceeding the standard limit. The original spectral plot of the radiated emission test results is shown in Figure 4.98.

Before determining the noise source, many kinds of filtering measures are tried on the external port, and the results show that there is no obvious effectiveness. Even disconnecting all the related external cables (only the power cord is left), the emission is still over the standard limit. In addition, while trying the measures such as filtering at the power supply port, it is found that the frequency with the problem is just shifted, but the result can't pass the radiated emission test. At the same time, these measures will also directly affect the conducted emission test results of the device (the conducted emission over low-frequency range), if the radiated emission test passes by this method, the conducted emission test should be re-performed with the risk of failure. After a series of tests and the spectrum analysis, for this product, its radiated emission problem is located at the part with the switching-mode power supply of the product, at the same time, it also shows that EMC design defects are on the internal circuits of the switching-mode power supply circuit.

[Analyses]

Before analyzing the root cause of the radiated emission, first, let's look at the following passage, which is a preliminary analysis of the product's radiated emission problem by an engineer:

> For the radiated emission problem caused by the switching-mode power supply, the solution may be to use an electromagnetic shielding. Electromagnetic shielding is very simple, as long as the switching-mode power supply circuit is covered by a metallic casing. Because this device uses a plastic casing, the first thought is to use the shielding measure. Placing a piece of flat metallic plate in front of the power supply section, and use the near-field probe to scan the noise, the tested noise amplitude is greatly reduced. It is recommended to implement a conductive painting or add a metallic foil on the casing. But in the next test with this modification, the test result is not as good as expected, because there are too many ventilation holes and seams around the connectors which are used to connect with the external cables, so, in the test, they cannot be completely shielded. And the feasibility of using metallic casing for this device is not possible.

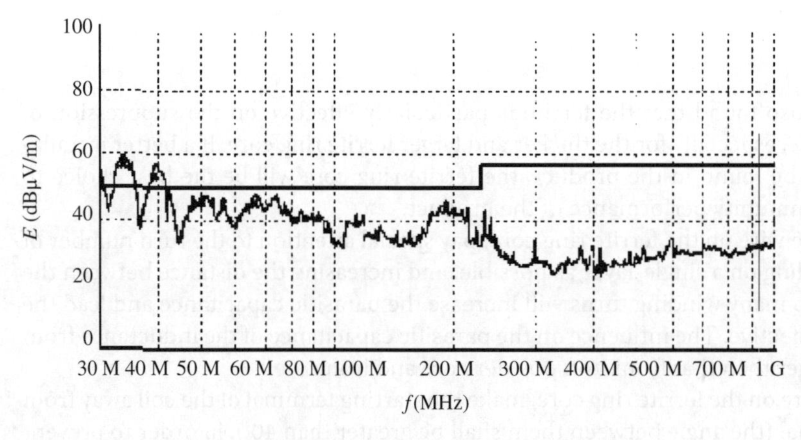

Figure 4.98 The spectrum of original radiation emission test results.

The analysis from this engineer actually exposed some misunderstanding on EMC problems, one of which is the misunderstanding on the electromagnetic shielding. Although the metallic casing is very helpful to improve the EMC performance of the product, but it is not the case as long as it is simply covering the product circuit with the metal (or painting the plastic casing with conductive material). Referring to case 13, the premise of using metallic casing (or painting the plastic case with conductive material) to reduce EMI is that the metallic casing (or the plastic casing painted with the conductive material) is correctly connected to the internal circuit of the product. This paragraph analysis exposes another misunderstanding about the locating of "the equivalent antenna," which actually radiates the emissions. This can be seen from the sentence that "because there are too many ventilation holes and seams around the connectors which are used to connect with the external cables, so, in the test, they cannot be completely shielded ..." The size of those holes and seams that are regarded as the radiation antenna by the engineer is not large (less than 1 cm), and they cannot become "the equivalent antenna" of the radiated emission. The real equivalent antenna is the power cord, which is longer than 1 m (including the grounding wire).

The most direct measure to solve the problem is to add the filtering circuit at the power port, but when the added filter cannot solve the problem or affects the conducted emission test result of the product, we need to analyze the switching-mode power supply circuit, through changing the operation of the switching-mode power supply or the PCB layout and routing to solve the problem.

There are mainly the following factors leading to EMI problems in the circuit:

1) The load of the switching transistors is the primary winding of the high-frequency transformer, which is the inductive load: when the switching transistor is switched on or switched off, large high-frequency current can be produced, which may be directly radiated through the loop or coupled to the power supply cord to be radiated.
2) The dU/dt caused by the switching on or off of the power transistor is also the main interfering source of the switching-mode power supply, and it acts as a voltage source and drives the power supply input/output cables to form the common-mode radiation through all kinds of coupling path.
3) Electromagnetic disturbance caused by the diode rectifying circuit, generally the reverse recovery current's di/dt of the rectifying diode in the main circuit is far smaller than that of the freewheeling diode. It also can cause the disturbance over a large frequency range.

According to the power supply circuit of the product, a tiny ferrite bead is in series with the collector of the switching transistor (in EMC, it is called as the hot spot), as shown in Figure 4.99; the radiated emission test can be passed.

The spectral plot of the tested radiated emission results after adding the ferrite bead is shown in Figure 4.100, and the contrastive schematics of the switching circuits before and after inserting the ferrite bead are shown in Figure 4.101.

It seems that the added ferrite bead reduces the high-frequency current of the switching circuit (shown in Figure 4.101), even if the resonance occurs in a certain frequency band, the Q value of the resonance circuit will be relatively low (due to the effect of the series resistance of the ferrite bead). Therefore, it can reduce the differential-mode radiation generated by the loop. In fact, combining with the electromagnetic field theory, in the switching circuit, the originally designed loop area is small (the area of the switching circuit is less than $0.5\,\text{cm}^2$ in the actual product), when the power is small, the emissions directly caused by the differential-mode radiation at 3 m far from the product (the site where the radiation emission is) is very limited and will not exceed the radiated emission limit. Readers may refer to case 25, and try to do some estimation.

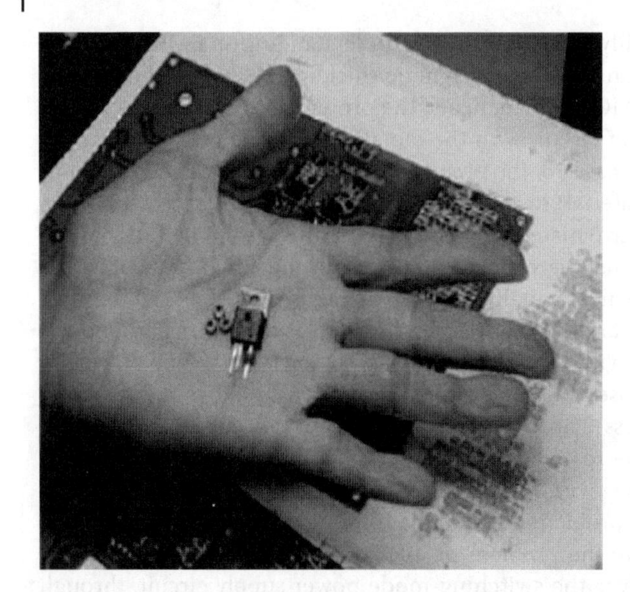

Figure 4.99 Magnetic beads.

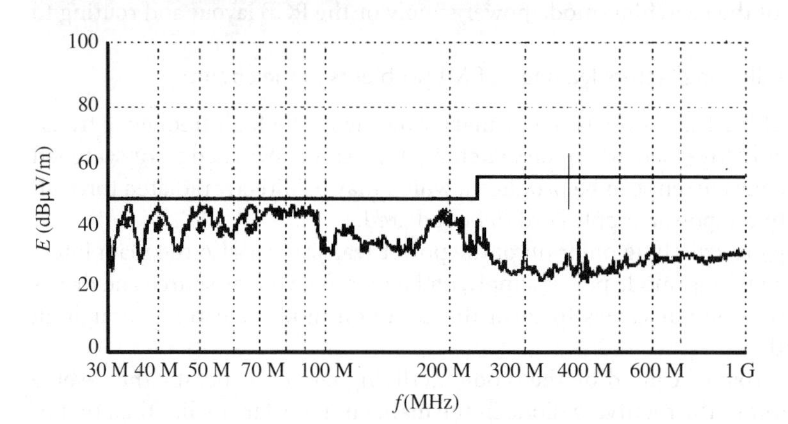

Figure 4.100 The spectrum of radiation emission test results after adding the magnetic beads.

In fact, the ultimate root cause of the power product lies in the power input cable. The switching circuit loop shown in Figure 4.101 is inside the large loop, which consists of the common-mode impedance of the power supply, the power supply input cable, the power supply module itself, the power supply output cable, the common-mode impedance on the load side, and the ground reference plate (as shown in Figure 4.102) in physical structure, which means there exists large inductive coupling or mutual inductance between these two. Although radiation formed by the switching circuit itself shown in Figure 4.101 is very limited, in the near-field range, its near magnetic field can be coupled to the common-mode large loop shown in Figure 4.102 by means of magnetic coupling. The relative longer power supply input cable can be equivalent to the common-mode radiation antenna, and as a part of the big common-mode loop, over the frequency range of the radiated emission test. As long as the common-mode current on it is more than a few micro-amps, it will cause the overall radiated emission of the product to exceed the standard limit. According to the electromagnetic theory, in the absence of any additional measures, the previously mentioned magnetic coupling phenomenon is easy to cause the excessive radiated emission of the power supply products.

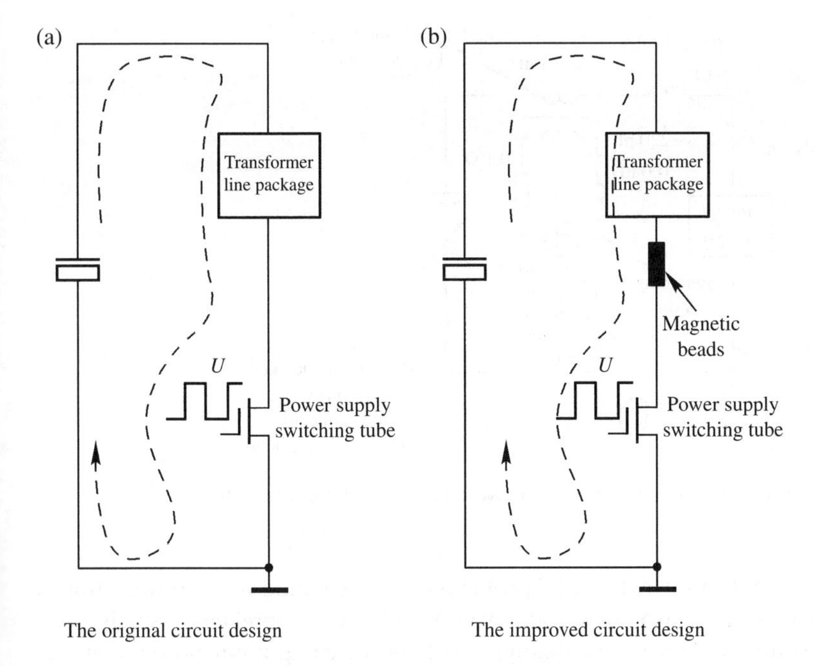

Figure 4.101 The principle diagrams of switch circuit before and after linking the magnetic beads.

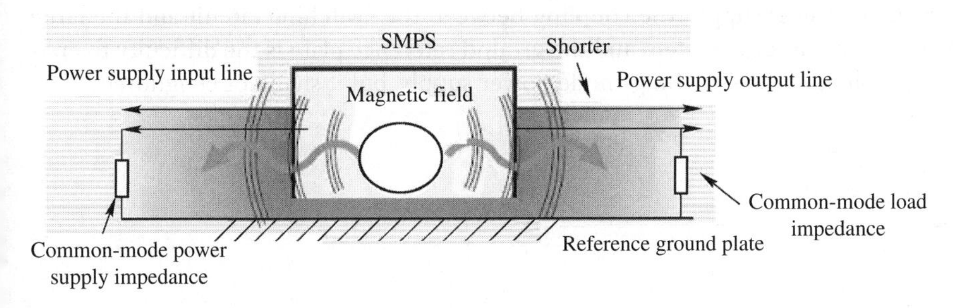

Figure 4.102 Common-mode circuit in product.

Figure 4.103 represents the coupling schematic between the switching circuit and the large loop. If readers have a simulation tool (such as PSPICE), you can also do the simulation. What is of concern in the simulation is the current in the power cord (the current will directly impact the radiated emission result). The measures taken in this case is to insert a ferrite bead in series with the switching circuit shown in Figure 4.103; its equivalent circuit is the series connection with a resistance and inductance.

[Solutions]

According to the above analysis and the test results, insert a tiny ferrite bead in series with the collector of the switching transistor in the switching loop, as shown in Figure 4.99 (such as nickel zinc ferrite beads, whose initial relative permeability is 700).

[Inspirations]

- In most cases, the radiated emission of low-power switching-mode power supply is not directly radiated from the internal circuit of the power supply, but from the input/output power cables.

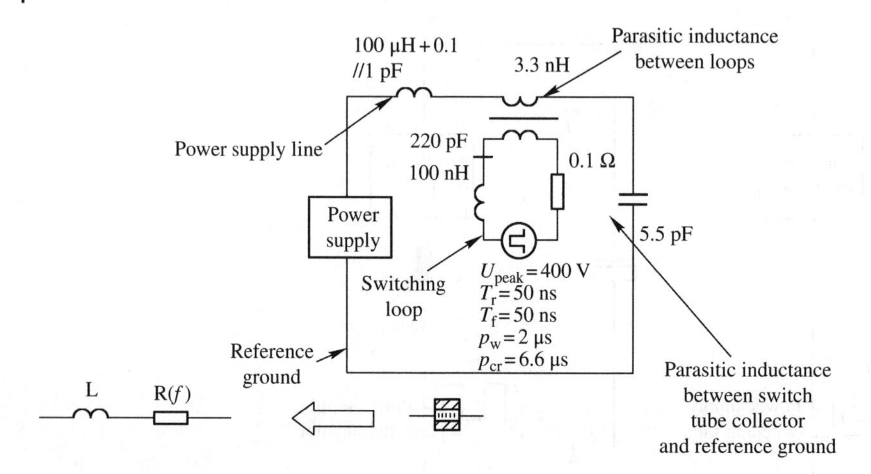

Figure 4.103 The coupling principle diagram between the switch circuit and the large loop.

- The filter is very important to solve the EMC problem of the switching-mode power supply, but it is not the only way and it also not absolutely safe. Only the comprehensive analysis on the EMI interfering source and the transmission path of the switching-mode power supply is the fastest, the most efficient and the lowest-cost solution to solve the EMI problems of the switching-mode power supply.
- In switching-mode power supply, the coupling between the switching circuit and the large loop consisting of the product system and the ground reference plate is the difficulty to analyze the EMI problem of the switching-mode power supply, but it shall not be ignored.

5

Bypassing and Decoupling

5.1 Introduction

5.1.1 The Concept of Decoupling, Bypassing, and Energy Storage

The bypassing and decoupling is to prevent useful energy from being transmitted from one circuit to another circuit, and to change the transmission path of the noise energy, so that the quality of the power distribution network is improved. There are three basic concepts: the power supply, the ground plane, the connection between the components and the power plane in inner layer.

Decoupling is that when the device is switched at high speed, the radio frequency (RF) energy is discharged from the power supply of the high-frequency components into the power distribution network. The decoupling capacitor also provides a local DC source for the component and the element, which is a very good effect on reducing the surge spike of the current on the printed circuit board (PCB).

The reason why the power supply needs decoupling is described in Case 47.

When the component is switched on and off and consumes the DC energy, there is a transient spike in the power distribution network without decoupling capacitor. This is because there is a certain inductance in the power supply network, while the decoupling capacitor can provide a local power supply with no inductance or very small inductance. By decoupling capacitors, the voltage is maintained at a constant reference point, which inhibits the wrong logic conversion. At the same time, the generated noise is reduced because it can provide a minimum loop area for the high-speed switching current to replace the large current return loop area between the element and the far-end power supply, as shown in Figure 5.1.

Another role of the decoupling capacitor is to provide the local energy storage source, which can reduce the power supply radiation path. The radiation of the RF energy in the circuit is proportional to the IAf, where I is the current in the circuit loop, A is the area of the circuit loop, and F is the frequency of the current. Because the current and frequency are determined when the logic device is selected, it is important to reduce the circuit area in order to reduce the radiation. In a circuit with a decoupling capacitor, the current flows in a small RF current loop, thereby the *RF* radiation is reduced. A small loop area can be obtained by careful placement of the decoupling capacitor.

In Figure 5.1, ΔU is the noise in the line induced on the ground by Ldi/dt; it is in the loop containing the decoupling capacitor and the IC. This ΔU drives the ground structure on the board and distributes the common-mode current flowing in the whole board. Therefore, the reduction of ΔU is related to the ground impedance as well as the decoupling capacitor.

Electromagnetic Compatibility (EMC) Design and Test Case Analysis, First Edition. Junqi Zheng.

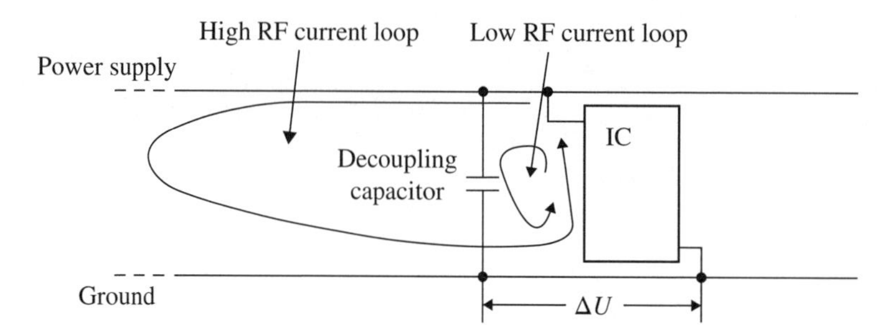

Figure 5.1 In the presence of the decoupling capacitor in the two layer, the area of the current loop can be greatly reduced.

Decoupling is also a method to overcome the physical and time sequence constraints, which is achieved by providing a low-impedance power source between the signal line and the power plane. Before the frequency rises up to the self-resonant point, with the increasing of the frequency, the impedance of the decoupling capacitor will be lower and lower, so that the high-frequency noise will be discharged effectively from the signal line, and then the remained low-frequency RF energy will not affect the circuit operation.

According to the operation principle of the decoupling capacitor, if the difficulty of absorbing the energy from the power line is increased, then most of the energy can be obtained from the decoupling capacitor, to completely play the role of the decoupling capacitor. Meanwhile, a smaller dI/dt noise will be generated on the power line. According to this thinking, the impedance of the power line can be artificially increased. A ferrite bead in series with the line is a common method, since the ferrite bead shows large impedance for high-frequency current, thus the decoupling effectiveness of the decoupling capacitor for the power supply can be improved.

Bypassing is to discharge the common-mode RF energy from the components or the cable that is not useful. The essence of bypassing is to set up an AC path to discharge the unwanted energy from the vulnerable areas. It also provides the filtering function (its bandwidth is limited), sometimes referred to the filtering in general.

Bypassing usually occurs between power source and ground, between signal and ground, or between different grounds. The essence of bypassing is different from that of decoupling, but the method used for the capacitor is the same, so the following descriptions of the capacitor's characteristics are applicable to the bypassing.

Energy storage is used to keep the constant DC voltage and current supplied to the device, when all the signal pins of the chip are switched with the maximum load capacity simultaneously. It can also prevent the voltage drop from being produced by the current surge di/dt of the component. If the decoupling belongs to the high-frequency domain, then the energy storage can be understood in the low-frequency domain.

5.1.2 Resonance

In fact, all the capacitors include a LCR circuit, where L is the inductance which is related to the lead length, R is the lead resistance, and C is a capacitance. Figure 5.2 shows the schematic of the actual capacitor.

At a certain frequency, the L in series with C will generate the oscillation to provide very low impedance. Above the self-resonant frequency, the impedance of the capacitor will be increased with the increasing of the inductive reactance, and then the capacitor will not play the role of

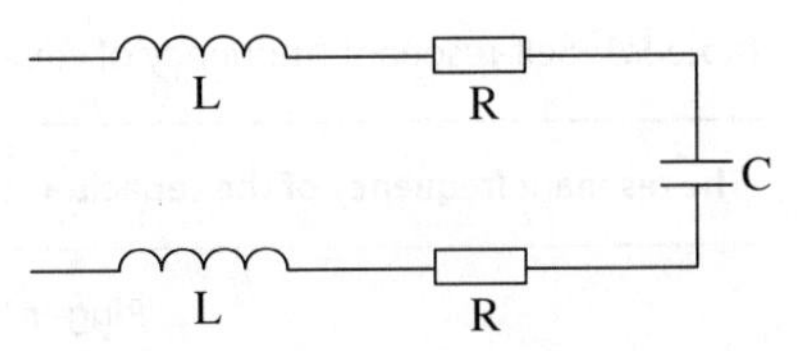

Figure 5.2 Actual physical characteristics of the capacitor with lead inductance and resistance.

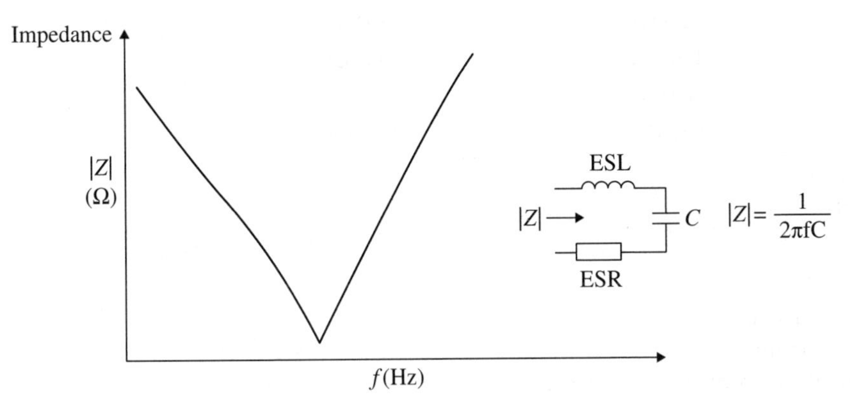

Figure 5.3 Capacitance frequency impedance.

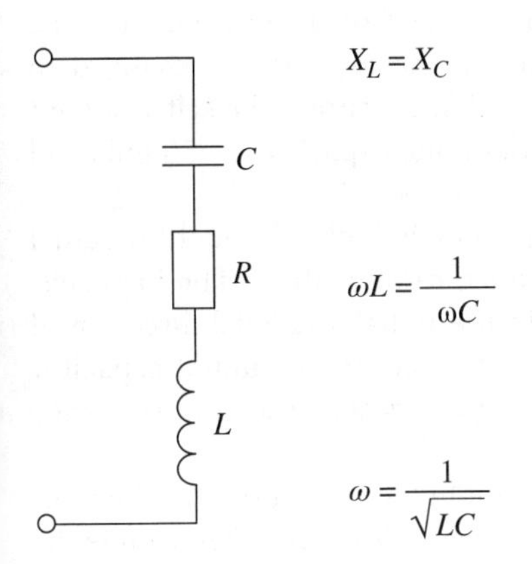

Figure 5.4 Series resonant circuit.

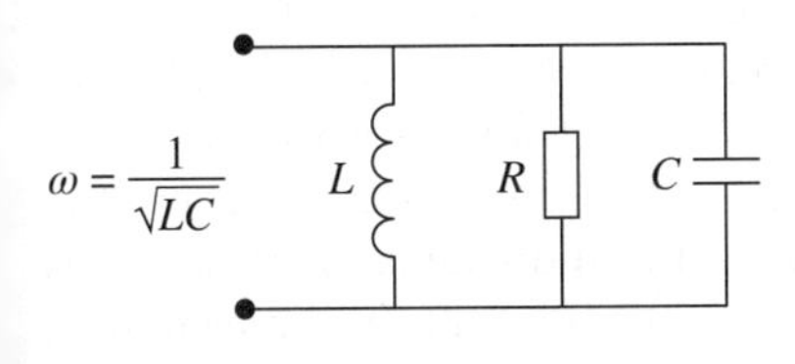

Figure 5.5 Parallel RLC resonant circuit.

bypassing and decoupling, as shown in Figure 5.3. Therefore, bypassing and decoupling are influenced by the lead inductance of the capacitor (including the SMD capacitor), the trace length between the components, the pads of the through-hole, and more.

Figure 5.4 shows the RLC serial resonant circuit.

The serial RLC circuit has the following characteristics at the self-resonant frequency:

- The minimum impedance
- The impedance equals to resistance
- The phase difference, 0
- The maximum current
- The maximum energy conversion (power)

When a parallel resonant circuit replaces the serial resonant circuit to be a load, the selected frequencies will not be effective anymore. Figure 5.5 shows the RLC parallel resonant circuit, in which the self-resonant frequency is the same as that of the serial RLC circuit.

The parallel RLC resonant circuit has the following characteristics as the self-resonant frequency:

- The maximum impedance
- The impedance equals to resistance

Table 5.1 Self-resonant frequency of capacitor.

The resonant frequency of the capacitor		
Capacitance	**Plug-in* 0.25-nch pin**	**SOT** 0805**
1.0 μF	2.6 MHz	5 MHz
0.1 μF	8.2 MHz	16 MHz
0.01 μF	26 MHz	50 MHz
1000 pF	82 MHz	159 MHz
500 pF	116 MHz	225 MHz
100 pF	260 MHz	503 MHz
10 pF	821 MHz	1.6 GHz

*Suppose ESL = 3.75 nH
**Suppose ESL = 1 nH

- The phase difference, 0
- The minimum current
- The minimum energy conversion (power)

The selection of the bypassing and decoupling capacitor does not depend on the size of the capacitance, but on the self-resonant frequency of the capacitor, which shall be matched with the required frequency for bypassing and decoupling. Below the self-resonant frequency, the capacitor is capacitive, and above the self-resonant frequency, the capacitor is changed to be inductive, which will reduce its RF decoupling function. Table 5.1 shows the self-resonant frequency of two types of ceramic capacitors: the lead length of one capacitor is 250 mil., and the other one is surface-mounted capacitor.

The self-resonant frequency of the surface-mounted capacitor is high, although in actual practice, the inductance of the connected wire can also reduce its advantage. The high self-resonant frequency is due to the smaller lead inductance in the radial and axial direction of the small package capacitor. According to the statistics, for the surface-mounted capacitor with different packages, with the changing of the lead inductance, its self-resonant frequency variation is ±2–5 MHz.

The through-hole capacitor can be regarded as the surface-mounted capacitor plus the pins. For a typical through-hole capacitor, the general lead inductance is 2.5 nH per 10 mil while the general lead inductance of the surface-mounted capacitor is 1 nH.

In conclusion, the most important point for using the decoupling capacitor is the lead inductance. Due to its low lead inductance, the surface-mounted capacitor has a good performance than the through-hole capacitor at high frequency.

Figures 5.6 and 5.7 respectively show the relationship between the frequency and the impedance of the commonly used through-hole and surface-mounted capacitors with different capacitance, from which the self-resonant frequency can be observed, for reference only.

The inductance is the main factor that the decoupling effect of the capacitor is lost above the self-resonant frequency. So in the actual circuit application, the inductance of PCB traces connected to the capacitor (including vias, etc.) must be taken into account. In some circuits, the operating frequency is high, and when the operating frequency is much higher than the

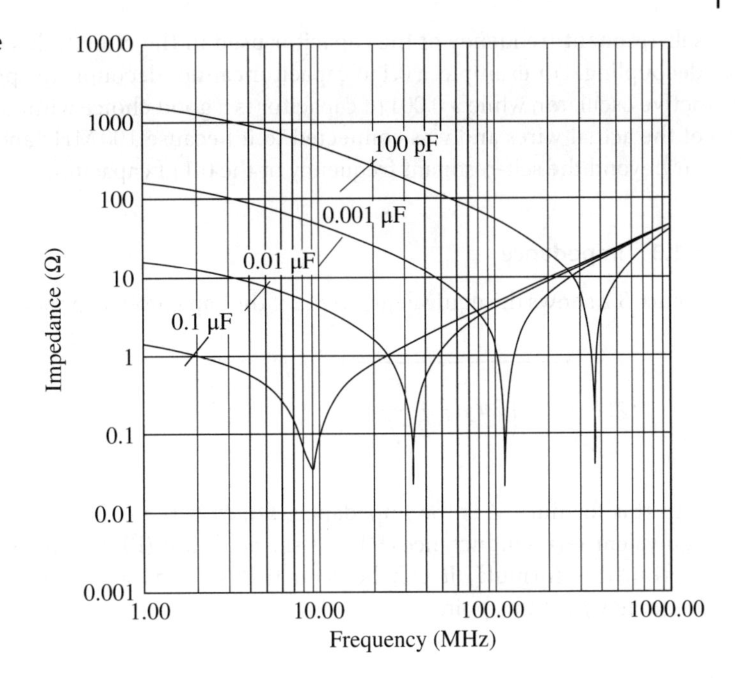

Figure 5.6 Frequency impedance diagram of capacitor used for different capacitance values of different values.

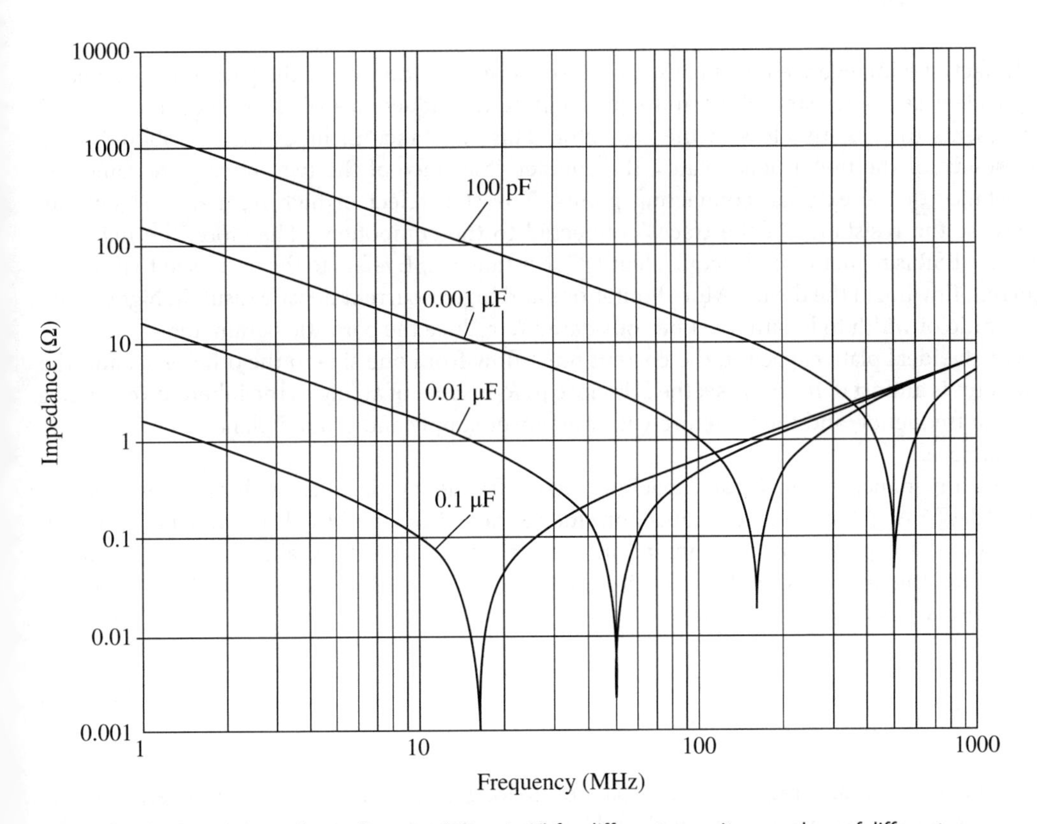

Figure 5.7 Frequency impedance diagrams (ESL = 1 nH) for different capacitance values of different capacitance values.

self-resonant frequency of the capacitor used in the circuit, this capacitor cannot be used for decoupling. For example, a 0.1 μf capacitor cannot decouple the power supply for the 100 MHz active oscillator, while a 0.001 μf capacitor is a good choice without considering the inductance of the actual wires and vias connected to it because 100 MHz and its harmonic frequency are far beyond the self-resonant frequency of the 0.1 μf capacitor.

5.1.3 Impedance

Figure 5.2 shows the equivalent circuit of the capacitor, its impedance can be expressed as (5.1):

$$|Z| = \sqrt{R_s^2 + \left(2\pi fl - \frac{1}{2\pi fC}\right)^2} \tag{5.1}$$

In this formula, Z is the impedance (Ω); R_s is the equivalent series resistance (Ω); L is the equivalent series inductance (H); C is the capacitor (F); F is the frequency (Hz).

From this formula, it can be seen that $|Z|$ has a minimum value at the self-resonant frequency f_0, at this point,

$$f_0 = \frac{1}{2\pi\sqrt{LC}} \tag{5.2}$$

In fact, the impedance formula (5.1) represents the situation while the parasitic parameters equivalent series resistor (ESR) and equivalent series inductance (ESL) are considered. ESR represents the resistive loss in the capacitor. This resistance includes the distributed plate resistance of the metal polar plates, the contact resistance of the two inner plates, and the resistance of the external connecting points. The skin effect of high-frequency signal can increase the resistance of the traces connected to the component. Therefore, the high-frequency ESR is higher than the equivalent DC resistance. ESL refers to the part used to limit the current flowing in the device. More limitations and higher current density result in higher ESL. The ratio of width to length must be considered to reduce the parasitic parameters.

For the ideal plate capacitor, the current must flow from one side to the other side, and the inductor is almost 0. In this case, the Z is close to R_s instead of acting as the inherent resonance at high frequency, and the structure with the power supply and ground plane in PCB belongs to this case.

The impedance of an ideal capacitor versus frequency declines with the slew rate of 20 dB dec^{-1}. And for the actual capacitor, due to the existence of the lead inductance, which prevents the capacitor from changing into the expected direction, above the self-resonant frequency, the impedance of the capacitor becomes inductive and increases with the slew rate of 20 dB dec^{-1}, as shown in Figure 5.8.

$$f_0 = \frac{1}{2\pi\sqrt{LC}} \tag{5.3}$$

At the self-resonant frequency, the capacitor no longer presents the capacitive characteristic. The ESR of the capacitor is very small, and has no impact on the self-resonant frequency of the capacitor.

Figure 5.8 Capacitance impedance frequency curve.

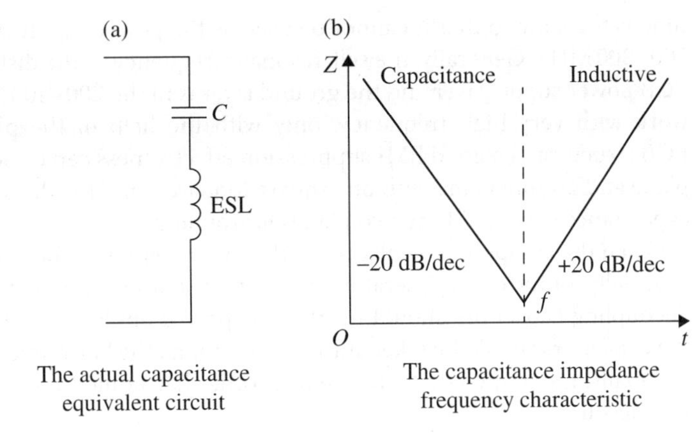

From the perspective of voltage distribution, at a specific frequency, the function that the capacitor can reduce the power supply distributed noise is shown in (5.4).

$$\Delta V(f) = |Z(f)| \cdot \Delta I(f) \tag{5.4}$$

Here, ΔV is the acceptable variation of the power supply, ΔI is the current supplied to the device, and f is the expected frequency. To optimize the power supply distribution system, it is necessary to ensure that the noise does not exceed the expected noise margin, $|Z|$ should be less than V/I for the required supply current. The maximum value of $|Z|$ is estimated from the maximum required ΔI. If $\Delta I = 1\,\text{A}$, $\Delta V = 3.3\,\text{V}$, the impedance of the capacitor must be less than $3.3\,\Omega$.

In order to make the capacitor work as expected, a high C is needed for the device to provide low impedance at the desired frequency. With a low L, the impedance of the capacitor will not be increased with the increasing of the frequency. Besides, a low R_s is required for the capacitor to get the lowest impedance.

The response of the decoupling capacitor is based on the transient current. The explanation to the impedance response in the frequency domain is very useful in the aspect that the capacitor provides a supply current. The charge conversion ability is an important index for selecting capacitor in time domain. The low-frequency impedance between the power supply and ground plane shows how many voltage variations there will be during the full speed transient change. This response represents the average voltage during the fast transient change. With the transient voltage change of the low-impedance voltage source, more current will flow into the component. The high-frequency impedance shows how much current can be supplied in the board during the rapid change. Above 100 MHz, with a few nanoseconds transient change, for a given voltage change, the lower impedance can provide more current.

5.1.4 The Selection of Decoupling Capacitor and Bypass Capacitor

In the actual circuit design, decoupling must be deployed for the clock and other periodic operating components. This is because the switching energy generated from these components is relatively concentrated, its amplitude is high, and it is injected into the distribution system of the power supply and ground. This energy will be transmitted to other circuits or subsystems in common mode and differential mode. The self-resonant frequency of the decoupling capacitor must be higher than the harmonic frequency of the clock to be suppressed. Typically, when the rising edge of the signal is 2 ns or more, the capacitor whose self-resonant frequency is 10–30 MHz can be chosen. The commonly used decoupling capacitors is the 0.1 μF capacitor and 0.001 μF capacitor in parallel, but since its inductance is too big, the charge and discharge time is too slow so that it cannot be used as the power supply source with frequency is above 200–300 MHz. Generally, the self-resonant frequency of the distributed capacitor between the PCB power supply layer and the ground layer is in the 200–400 MHz range. If the components work with very high frequency, only with the help of the plate capacitor constructed by PCB structure, the good EMI suppression effectiveness can be achieved. Usually for the power plane and ground plane with one square foot area, and the distance between them is 1 mil, the capacitance between these two planes is around 225 pF.

When the component is placed on the PCB, there must be sufficient decoupling capacitors, especially for the clock generator circuit, it must be ensured that the selection of the bypass and decoupling capacitors should meet the expected application. The self-resonant frequency of the capacitors should be taken into account for all the harmonics of the clock to be suppressed. Normally, the frequency up to the fifth-order harmonic frequency of the clock shall be taken into account.

A practical example is used to illustrate how to select the decoupling capacitor (although this method is not practical in the actual circuit design). Assume that the circuit has 50 driving buffers, which are switched on and off simultaneous, the edge of each output is 1 ns, the load is 30 pf, the power supply is 2.5 V, and the allowable power supply fluctuation is ±2% of the rated voltage (if the impedance of the PCB power plane is considered, the allowable fluctuation can be increased). The simplest method is to see the impact from the instantaneous current supplied to the load, and the calculation method is shown as follows:

1) Calculate the current I required by the load at first:

$$I = \frac{Cdu}{dt} = \frac{30\,\text{pF} \times 2.5\,\text{V}}{1\,\text{ns}} = 75\,\text{mA},$$

Then the total required current is:

$$50 \times 75\,\text{mA} = 3.75(\text{A})$$

2) The required capacitance can be worked out then:

$$C = \frac{Idt}{du} = \frac{3.75\,\text{A} \times 1\,\text{ns}}{2.5 \times 2\%} = 75(\text{nF})$$

3) Considering the temperature, aging and other impacts in the actual situation, an 80 nF capacitor can be selected to ensure a certain amount of margin. And it can be used with two 40 nF capacitors in parallel to reduce the ESR.

The above calculation method is simple, but the actual effectiveness is not very good. Especially in the high-frequency circuit, there will be a lot of problems. In the case of the previous example, even though the inductance of the capacitor is small, only 1 nH, but according to $dU = Ldi/dt$, a voltage drop of 3.75 V can be calculated, which is clearly unacceptable.

Therefore, in the design of high-frequency circuit, another more effective calculation method should be applied, which is mainly to see the impact from the circuit inductance. Just analyze the above example:

First, calculate the maximum allowable impedance of the power circuit X_{max}:

$$X_{max} = \Delta U/\Delta I = 0.05\,\text{V}/3.75\,\text{A} = 13.3(\text{m}\Omega)$$

Considering the working frequency range of the low-frequency bypass capacitor F_{BYPASS}:

$$F_{\text{BYPASS}} = X_{\max} / 2\pi L_0 = 13.3/(2\times 3.14\times 5) = 424(\text{kHz})$$

At this time, considering the decoupling capacitor of the power bus on the board, the larger electrolytic capacitor is selected generally. Here, assume that the parasitic inductance is 5 nH. It can be considered that, for the AC signal with the frequency less than F_{BYPASS} ac signal, the bypass is provided by the bulk capacitor on the PCB board.

The maximum effective frequency F_{knee} is also known as the cutoff frequency:

$$F_{\text{knee}} = 0.5/T_r = 0.5/1\,\text{ns} = 500(\text{MHz})$$

The cutoff frequency represents that most energy of the digital signal is concentrated in the frequency range below the cutoff frequency, and the part over the frequency F_{knee} will have less effect on the energy transmission of the digital signal.

At the maximum effective frequency F_{knee}, the maximum allowable inductance L_{TOT} is calculated:

$$L_{\text{TOT}} = \frac{X_{\max}}{2\pi F_{\text{knee}}} = \frac{X_{\max}\cdot T_r}{\pi} = \frac{13.3\,\text{m}\Omega\times 1\,\text{ns}}{3.14} = 4.24(\text{pH})$$

Assuming that the ESL of each capacitor is 15 nH (including the inductance of the soldering wire), the number of the required capacitors N can be calculated:

$$N = ESL/L_{\text{TOT}} = 1.5\,\text{nH}/4.24\,\text{pH} = 354$$

The impedance of the capacitors cannot exceed the allowable impedance range at low frequency, and the total capacitance C can be calculated:

$$C = \frac{1}{2\pi F_{\text{BYPASS}}\cdot X_{\max}} = \frac{1}{2\times 3.14\times 424\,\text{kHz}\times 13.3\,\text{m}\Omega} = 28.3(\mu F)$$

Finally, calculate the capacitance of each capacitor C_n:

$$C_n = C/N = 28.3\mu F/354 = 80(\text{nF})$$

The results show that, in order to achieve the best design effectiveness, 354 pieces of capacitors with value of 80 nF capacitors are needed to be distributed on the whole PCB uniformly. But from the actual situation, so many capacitors are often impossible to be implemented, if the number of simultaneous switching buffers can be reduced, the edge is not very fast, a greater range of voltage fluctuation is accepted, the calculated results will vary greatly. If the requirement of the actual high speed circuit is very strict, only selecting capacitors with small ESL can avoid using a large number of capacitors.

5.1.5 Capacitor Paralleling

Effective capacitive decoupling is achieved by properly placing the capacitors on the PCB. Arbitrary placement or excessive uses of capacitors is a waste of material. Sometimes strategically placing several capacitors will take a very good decoupling effect. In the practical applications, using two parallel capacitors can provide a wider suppression bandwidth. In order to achieve the best results, these two parallel capacitors must have different order capacitance (e.g. 0.1 and 0.001 μF) or 100 times difference between these two capacitors.

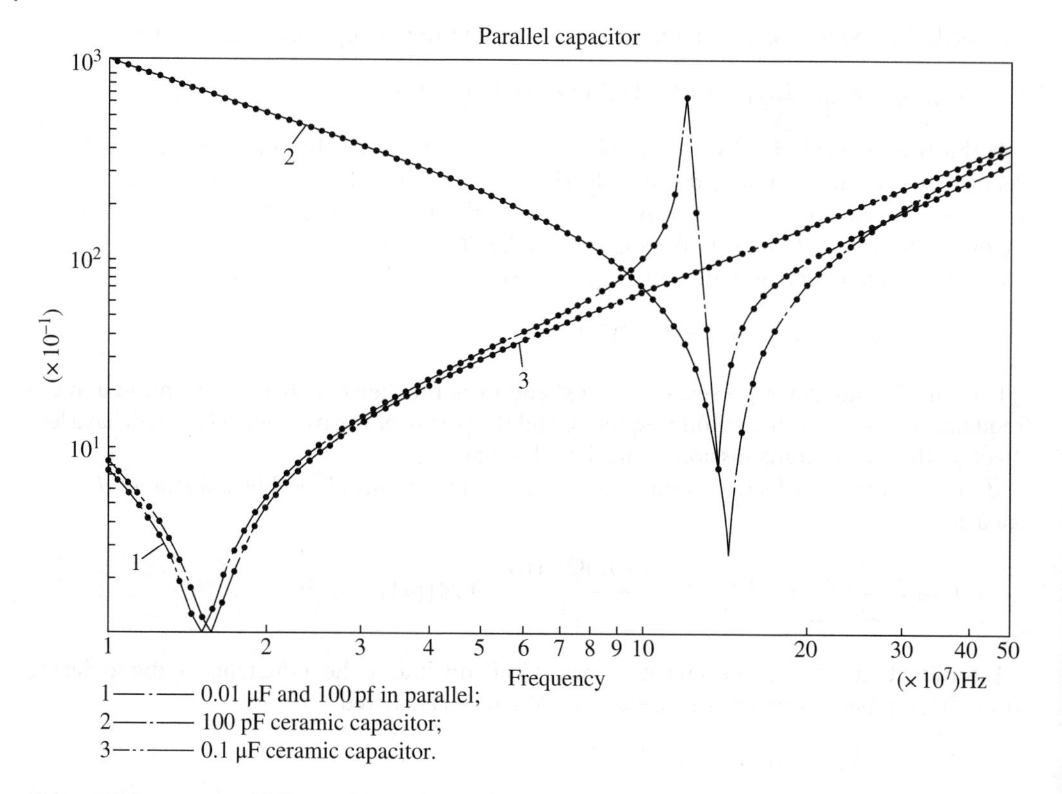

Figure 5.9 Resonance of different values of shunt capacitor.

Figure 5.9 shows the impedance-frequency curve of two decoupling capacitors 0.1 μF and 100 pF used separately and in parallel. The self-resonant frequency of 0.1 μF and 100 pF capacitor is, respectively, 14.85 and 148.5 MHz, at 110 MHz, the combination impedance of the parallel capacitors has a great increase, as the 0.1 μF capacitor becomes inductive, while the 100 pF capacitor is still capacitive. There is a parallel resonant LC circuit in this frequency range. In resonance, both inductance and capacitance exist. Therefore, an anti-resonant frequency appears around these resonant points, so the impedance of the parallel capacitors must be larger than that of the single capacitor. If the EMI requirements must be met around this frequency range, it will be a risk.

In order to eliminate high bandwidth noise, the common method is to place two parallel capacitors (such as 0.1 and 0.001 μF) near the power supply pins. If the parallel capacitors are used in the PCB layout, it must be ensured that the capacitors have different order of capacitance or 100 times difference of the capacitance. The total capacitance of the parallel capacitors is not the main factor, but the impedance of the parallel capacitors is.

In order to optimize the decoupling effectiveness of the parallel capacitors and allow the use of a single capacitor, the lead inductance of the capacitor needs to be reduced. The lead inductance may be different when the capacitor is placed on the PCB. This lead length includes the length of the vias interconnecting the capacitor to the plane. If the lead of single or parallel decoupling capacitors is shorter, the decoupling effect is better.

In addition, two capacitors that have the same capacitance in parallel can also improve the decoupling effectiveness and its application frequency range. This is because, with the capacitors in parallel, the parasitic resistance (ESR) and the parasitic inductance (ESL) are decreased.

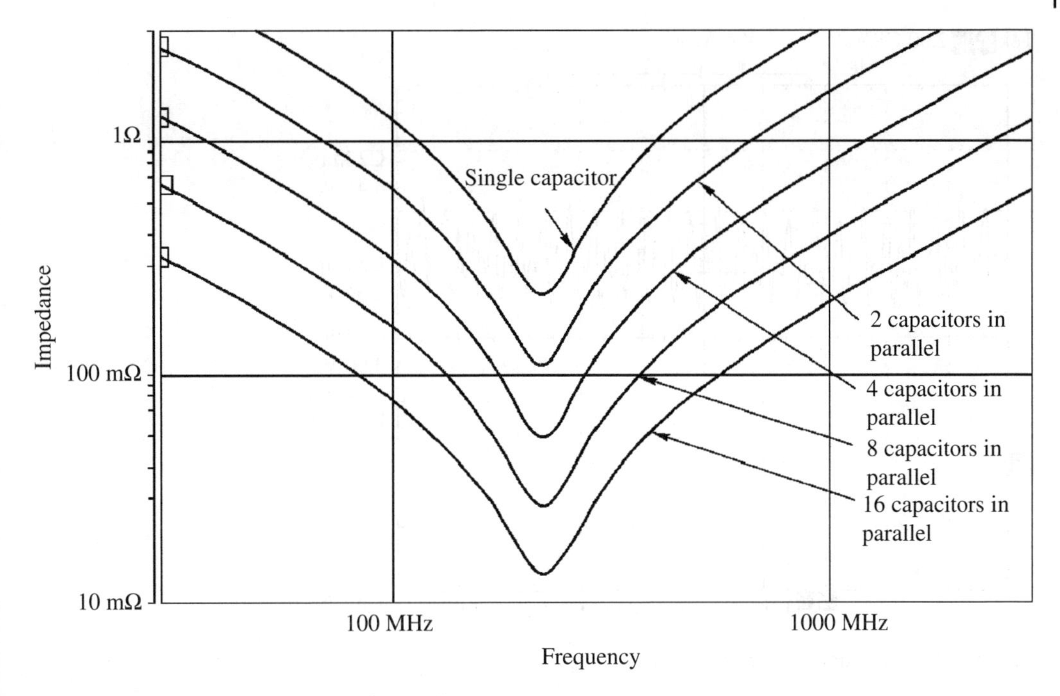

Figure 5.10 Parallel characteristics of equivalent capacitance.

For multiple (n) capacitors with same capacitance, after they are in parallel, the equivalent capacitance turns to **n***C from C, the equivalent inductance becomes L/n from L, the equivalent resistance turns to R/n from R, but the resonance frequency remains unchanged. From the perspective of energy at the same time, multiple capacitors in parallel will be able to provide more energy to the decoupled device (as shown in Figure 5.10).

5.2 Analyses of Related Cases

5.2.1 Case 47: The Decoupling Effectiveness for the Power Supply and the Capacitance of Capacitor

[Symptoms]

The clock driver chip **3807, which many digital circuit hardware design engineers are familiar with, is used in the circuit of one device (in this case, its power supply voltage is 3.3 V). It is found that the measured voltage waveform on its power supply pins by oscilloscope is shown in Figure 5.11.

As you can see from the waveform shown in Figure 5.11, the peak–peak voltage of the noise is 1.8 V, and its frequency is close to 100 MHz. Obviously, it does not meet the requirements of the power supply quality (usually the ripple shall be less than 5% of the DC voltage). The noise on the power supply directly affects the integrity of the power plane and the ground plane, and it also has a great impact on the common-mode radiated emission of the system. The common-mode radiated emission model of the signals on the PCB is shown in Figure 5.12.

In high frequency, the reference plane (including the ground plane and the power plane) acts as the current return conductor, and there may be the voltage with any frequency on it. This voltage drop is caused by the differential-mode current I_{DM} flowing on the reference plane

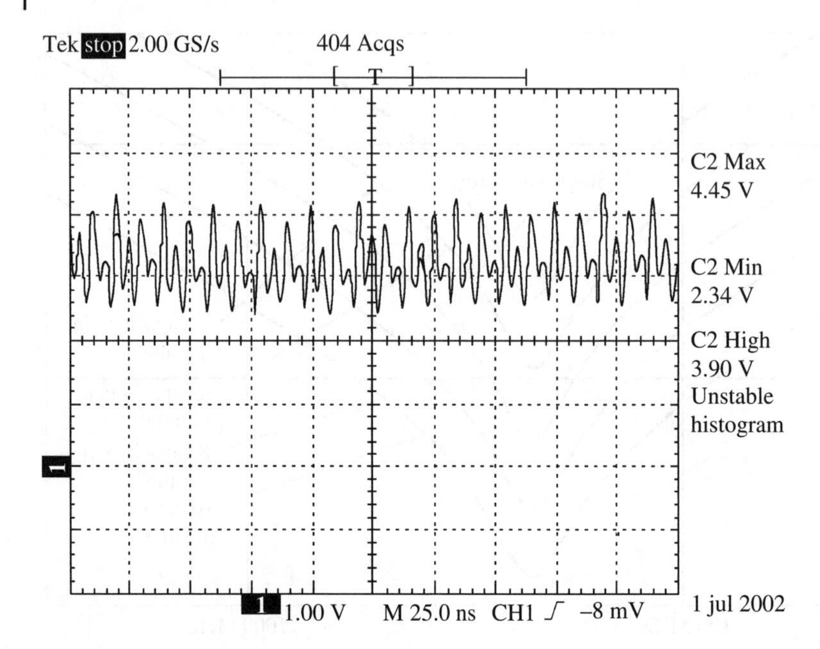

Figure 5.11 Power supply signal test result.

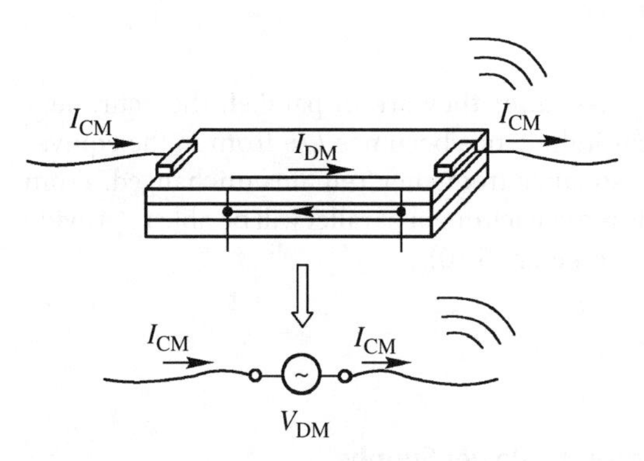

Figure 5.12 Equivalent model of common-mode radiation in multilayer PCB (electric cable shields connects to reference plane).

beneath the signal conductor, which is the common-mode voltage drop U_{CM}, as shown in Figure 5.12. This voltage drives the large peripheral structure and the common-mode current I_{CM} is generated; for example, the common-mode current on the cable shield connected to the reference plane with low impedance. This current and the long conductor it flows through (at this time, it is the transmitting antenna) together constitute the radiation source, and they cause a severe EMC problem. So this issue can raise a big risk on whether the EMC test will be successful, and it must be solved before the test.

[Analyses]

The schematic of the clock and its driving circuit are shown in Figure 5.13.

After analyzing the circuit schematic, it is found that there are three decoupling capacitors (all of them are surface-mounted capacitors) for V3.3_2 power supply. One is 10 μF, and it is placed at point A in Figure 5.14; Two are 0.1 μF, and they are placed at point B and point C in Figure 5.14. After a preliminary examination, the problem is identified: The PCB power supply trace is too long and the layout of the decoupling capacitors is not reasonable, so the

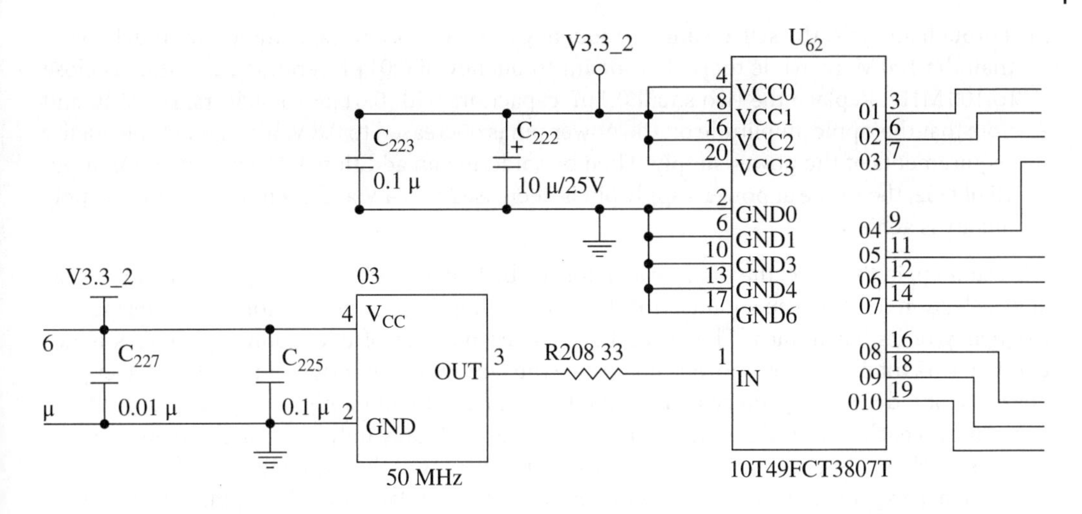

Figure 5.13 The principle diagram of clock and driver.

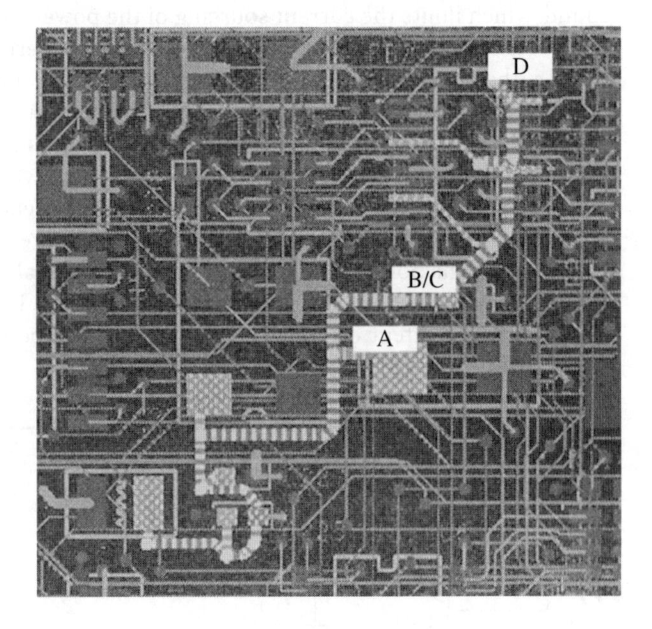

Figure 5.14 The diagram of power signal V3. 32 PCB network.

decoupling capacitors are not near the power supply pins, resulting in a larger lead inductance; the selected capacitance is unreasonable, the self-resonant frequency of 0.1 μF capacitor is far below 100 MHz.

In order to confirm this judgment, the following tests are performed:

1) Soldering additional three small 0.1 μF capacitors at each power supply pin of the component 3807 and V3.3_2 power net between point C and point D, as shown in Figure 5.14, in order that each power supply pin has a nearby decoupling capacitor. Then test it again. There is no improvement on the power supply waveform, which shows that the greatest part of the problem cannot be solved by only improving the lead inductance without changing the capacitance.

2) Through analysis, the self-resonant frequency of 10 and 0.1 μF capacitance are much lower than the 100 MHz, while the self-resonant frequency of 0.01 μF ceramic capacitors is close to 100 MHz. Replace the two small 0.1 μF capacitors with 0.01 μF capacitors, retest it, and note that the ripple amplitude on the power pin is decreased to 0.8 V. It can meet the quality requirements of the power supply. Then by soldering an additional 0.01 μF capacitor at pin 15 of U62, the ripple at power supply pin is decreased to 0.4 V and even better power supply quality is achieved.

To sum up, in this case, the main reason for the bad quality of power supply is the unreasonable selection of the capacitance for the decoupling capacitor, without considering the frequency of its harmonics. The second is that the number of decoupling capacitors is not enough, U62 totally has four power pins. Since the power consumption of the clock driver providing strong driving capability is high, the best thing is to add a high-frequency decoupling capacitor on each power pin. In this experiment, the 3807 only outputs four channels, and if it outputs 10 channels at the same time, the dynamic current of the load will be multiplied.

To further explain the above phenomenon, first, let's explain how the ripple on the power supply pins of the chip is caused, which is the reason why we need to implement the decoupling on it. Figure 5.15 shows a typical gate output circuit, when the output is high, Q3 is switched on and Q4 is switched off. On the contrary, when the output is low, Q3 is switched off and Q4 is switched on, for both states, there exists high impedance between the power supply and the ground, which limits the current sourcing of the power supply.

However, when the state is changed, there will be a period of time when Q3 and Q4 are conducted at the same time. At this period, a short-time low impedance between power supply and ground appears, and 30–100 mA peak current is produced. When the gate output is changed from low level to high level, the power supply provides not only the short-circuit current but also the charging current for the parasitic capacitance, which causes even greater peak current. As there is always the inductance on the power supply nets in different extent, when the current is changed transiently, there will be the induced voltage existing on the power supply nets. This is the noise on the power supply nets. When there is peak current flowing on the power supply nets, this peak current will also flow through the ground. Because there is also always the inductance in different extent on the ground, so there will be the induced voltage on ground, too.

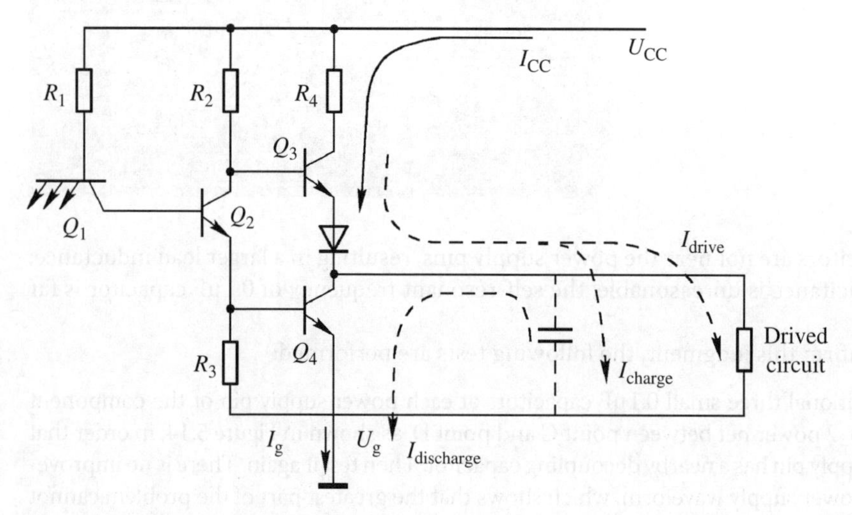

Figure 5.15 Typical gate circuit output level.

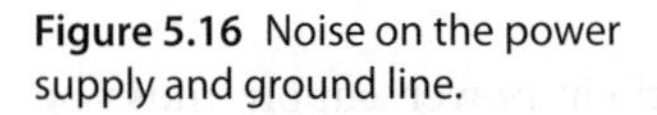

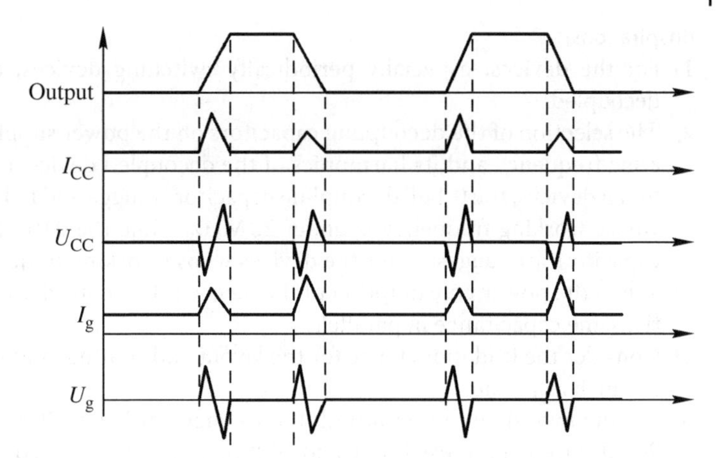

Figure 5.16 Noise on the power supply and ground line.

It appears as the ground noise, especially for the circuit with periodic signal, the noise peak is more concentrated, as shown in Figure 5.16.

Decoupling the capacitor is one way to overcome producing the noise peak. When all the signal pins work with the maximum load capacity and they are switched on and off at the same time, the decoupling capacitors also provide the dynamic voltage and current needed for the component to work normally while the clock and data is changing. Decoupling is achieved with a low-impedance power source provided between the signal and the power plane. Before the frequency is increased to the self-resonant frequency of the decoupling capacitor, with the increasing of the frequency, the impedance of the decoupling capacitor will be lower and lower, so that the high-frequency noise can be effectively reduced on the signals, then the remaining low-frequency energy has no impact on the signals.

0.1 μF capacitor and 0.01 μF capacitor are most commonly used as the decoupling capacitors in high-speed circuit design. Generally the self-resonant frequency of the surface-mounted capacitor is not more than 500 MHz, for 0.01 μF surface-mounted capacitor, its self-resonant frequency is from 50 to 150 MHz, and in the practical application on the PCB, the presence of the lead inductance and the via will further reduce the self-resonant frequency of the decoupling circuit. So, even the capacitance of the decoupling capacitor is small enough, the self-resonant frequency of the decoupling capacitor cannot be infinite. In the practical application, the lead inductance limits the decoupling frequency not more than 300 MHz no matter how small the capacitance is. This is also the reason why the minimum decoupling capacitor of a lot of circuits is only 0.01 μF capacitor even if the working frequency is very high. For two capacitors with the same capacitance in parallel, the lead inductance and its parasitic inductance will be decreased after parallel, which decreases the overall impedance, and it is beneficial for raising the working frequency of the decoupling capacitor. When the gate flips in the device, these two parallel capacitors with same capacitance can provide more energy at the same time. In addition, in a multilayer PCB design, the plate capacitor between the power plane and ground plane has the characteristics of low ESL, and it is an important mean to decouple the power supply in high-frequency circuit design.

[Solutions]

Replace the 0.1 μF capacitor with the 0.01 μF capacitor, and ensure that each power supply pin has more than one decoupling capacitor (the experienced value is 1.5 capacitors for each pin), and place them near the power supply pins on the PCB layout.

[Inspirations]

1) For the devices, especially periodically switching devices, their power supply must be decoupled.
2) The selection of the decoupling capacitors on the power supply needs to consider the operating frequency and its harmonics of the decoupled device, not to use the 0.1 μF capacitor for all devices, the 0.1 μF decoupling capacitor is suggested to be used for a variety of devices whose working frequency is under 20 MHz, while the 0.01 μF or even smaller decoupling capacitors are suggested for the devices whose working frequency is more than 20 MHz.
3) When the power consumption of the device is big, consider using multiple capacitors with the same capacitance in parallel.
4) Consider the lead inductance for the layout and routing, and minimize the lead inductance as much as possible.
5) For the hybrid circuit with both the frequencies below 20 MHz frequency and more than 20 MHz frequency, use 0.1 μF and 1000 pF in parallel to decouple the power supply.

5.2.2 Case 48: Locations of the Ferrite Bead and Decoupling Capacitor Connected to the Chip's Power Supply Pin

[Symptoms]

There is a clock driver chip in the PCB of a product, and there are 10 μF filtering capacitor C192 and 0.1 μF decoupling capacitor C202 (not drawn in Figure 5.17) in parallel with A5V1 near the power supply of the clock driver, and after a ferrite bead FB5 (17010145 ferrite – EMI magnetic bead – 60 Ω ± 25% – 4.0 A – 206). A5V1 is sent to the voltage current condenser (VCC) pin of the chip, as shown in Figure 5.17. It is found that the quality of the output clock signal is poor, and the duty ratio is changed. After further testing, it is found that the voltage on the VCC pin has severe oscillation and voltage drop phenomenon, the frequency of the oscillation is the same as the output frequency of the clock, and when there is the voltage drop on the VCC pin, the rising time of the output clock becomes very slow. The tested waveform is shown in Figure 5.18.

[Analyses]

In PCB layout, a power plane is poured under the chip, on the left and the right of the power plane, a 0.1 μF decoupling capacitor and a 10 μF filtering capacitor are, respectively, connected to it, and then through the ferrite bead FB5, the power supply plane is connected to VCC pins of the chip, which are pin 4, pin 8, pin 15, and pin 20 of the chip, respectively.

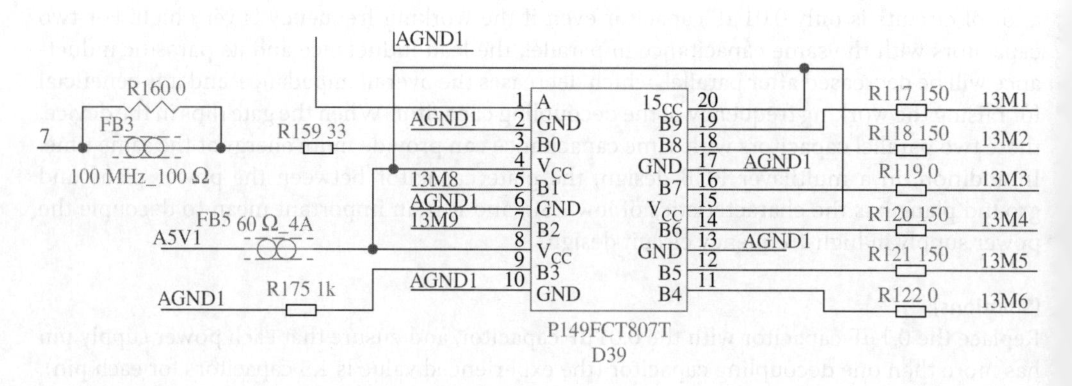

Figure 5.17 Chip connection diagram.

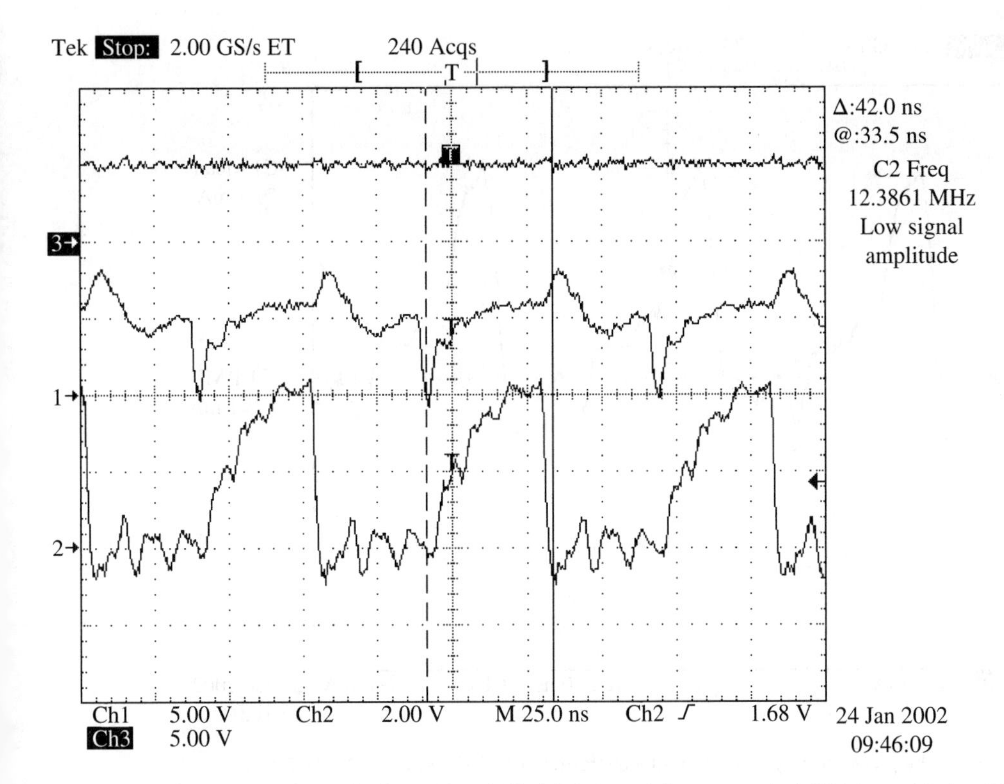

Figure 5.18 The test waveform of voltage U on A5V1, VCC pin and the output clock B5 (channel 3, 1 and 2).

The load of the high-speed clock driver chip is usually heavy. At the rising edge and the falling edge of the output clock, the power supply input current of the chip will be changed drastically and quickly, as shown in Figure 5.19.

The VCC pin of the clock chip is first connected to the ferrite bead, then to the capacitors. Due to the impedance characteristic of the ferrite bead, the high-speed variation on the power supply input current of the clock driver chip will cause great counter-electromotive force on the ferrite bead FB5, which is the only path the power supply input current flows through, and it causes the overshoot and undershoot on the voltage U of the VCC pin. For further detailed theoretical analysis, the equivalent circuit of the ferrite bead can be regarded as a resistor R and inductor L in series (sometimes regarded as L and R in parallel), in which the resistance of R and the inductance of L is relevant with the frequency, as shown in Figure 5.20. The curve R is the resistive impedance characteristic of the ferrite bead, the curve X is the inductive impedance characteristic of the ferrite bead, and the curve Z ($R + jX$) is the total impedance of the ferrite bead.

According to the equivalent circuit, the voltage drop Δu across the ferrite bead can be calculated, and the calculation formula is as follows:

$$\Delta u = A5V1 - L(f) \times dI/\mathrm{d}t - R \times I \cdots \tag{5.5}$$

In this formula, I is the current flowing through the ferrite bead FB5; and f is the frequency of the current I.

Use the current probe (TCP202 from TEK Co.) to measure some waveforms of the current I flowing through the FB5, and the voltage U and the current I of the power supply VCC pins, as shown in Figure 5.21.

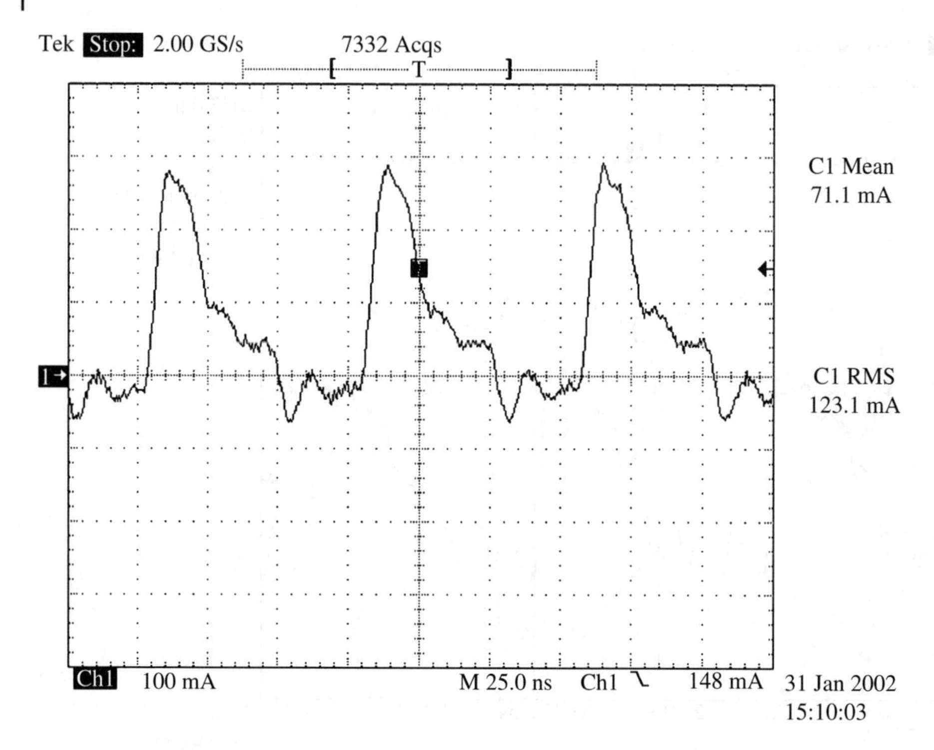

Figure 5.19 The power supply input current waveform of Clock driver chip P149FCT807T.

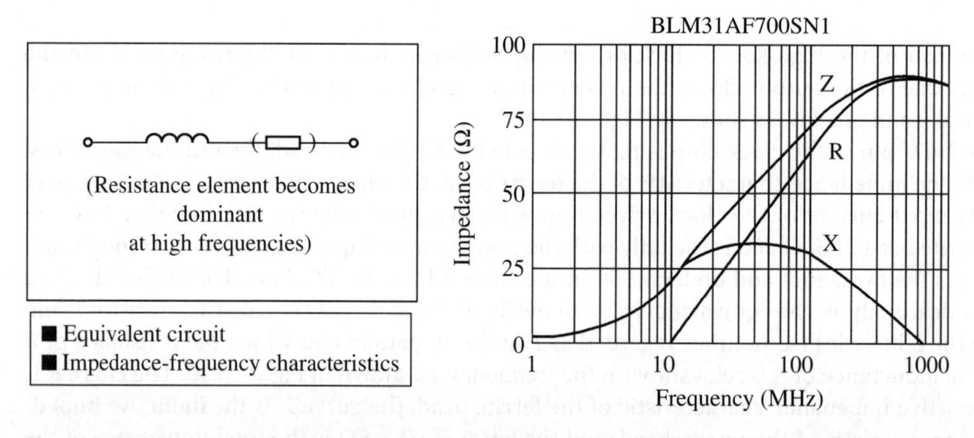

Figure 5.20 Magnetic beads equivalent circuit model and impedance frequency characteristic.

By the theory of digital signal, the energy of the current I is mainly distributed in the harmonic frequencies of 13 MHz within frequency F0, F0 = 0.5/t, t is the rising or falling time duration of the signal. From the waveform, it is about 5 ns, and its energy can be estimated mainly distributed from 13 to100 MHz. Within 13–100 MHz, the inductive impedance X of the ferrite bead varies from 25 to 35 Ω, according to the formula: L (f) = Z/(2Π f), the inductance L can be derived from 40 to 300 nH; The resistive impedance of the ferrite bead is between 25 and 60 Ω. From the waveform of the current, it can be seen that the maximum dI/dt is around 30 mA ns^{-1}, under such high changing rate of the current, due to the inductive characteristic of the ferrite bead, a counter electromotive force in volt level will be produced on the

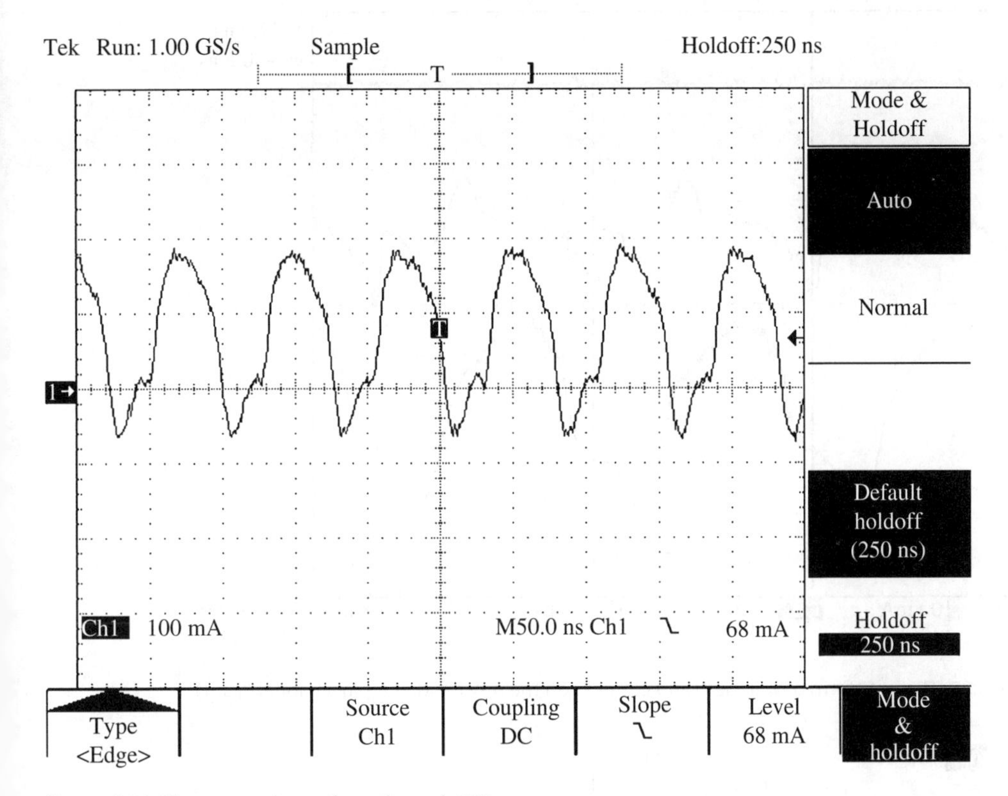

Figure 5.21 The current I waveform through FB5.

power input of the chip; And, due to the resistive impedance of the ferrite bead, the voltage $-RI$ (also in volt level) will be superimposed on the counter electromotive force, then there will be a high overshoot and undershoot on the voltage U of VCC pin, as shown in Figure 5.22.

[Solutions]

The right position and connection between the ferrite bead and the capacitor is to directly parallel the 10 μF capacitor and the 0.1 μF decoupling capacitor close to the VCC pin of the chip, in order to increase the ability to resist current changing. Then insert the ferrite bead FB5 between VCC pin and A5V1 plane to separate them with high impedance, in order to reduce the disturbance to the A5V1 plane. After moving these two capacitors to the VCC pin of the chip and inserting the ferrite bead FB5 between these two capacitors and A5V1 plane, the chip works well. The disturbance on the A5V1 plane is also reduced by the ferrite bead FB5, the voltage drop across the ferrite bead Δu and the voltage on A5V1 plane are shown in Figure 5.23.

[Inspirations]

1) We need to have a deep understanding on the operating principle and function of the ferrite bead, and recognize that adding a ferrite bead is not always a method to reduce the interference on the power supply of the chip.
2) In the filtering circuit design, the impedance of the source and the load must first be taken into account. The operating principle of the decoupling capacitor and the filtering capacitor is impedance mismatching. Table 5.2 provides the reference for selecting the filtering circuit.

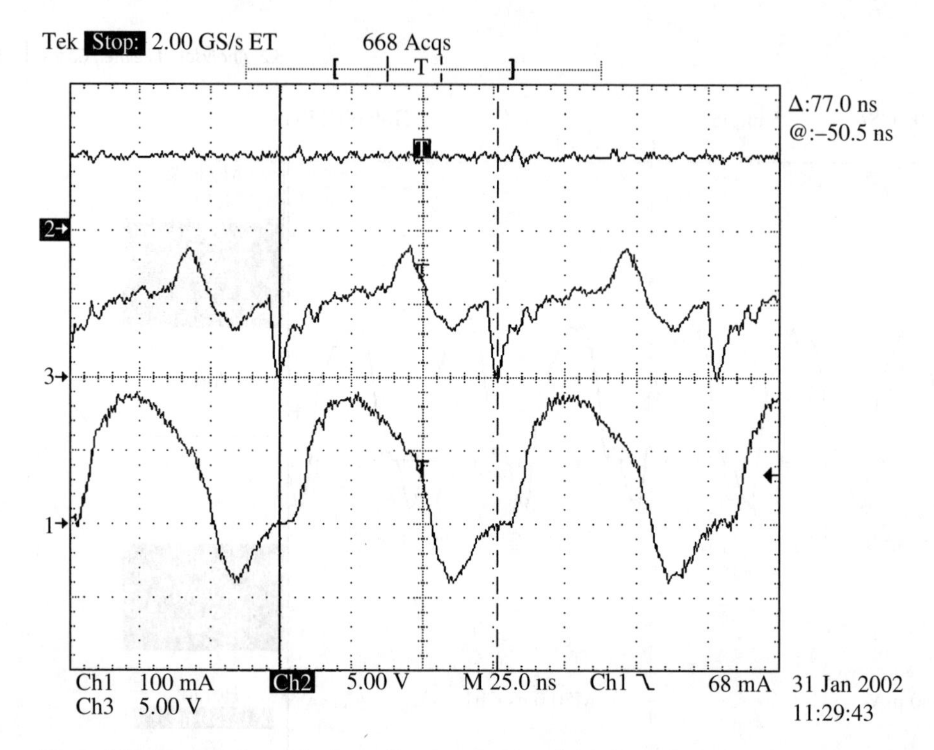

Figure 5.22 The waveform of voltage U and current I on the A5V1 and VCC pin (channel 2, 3, 1).

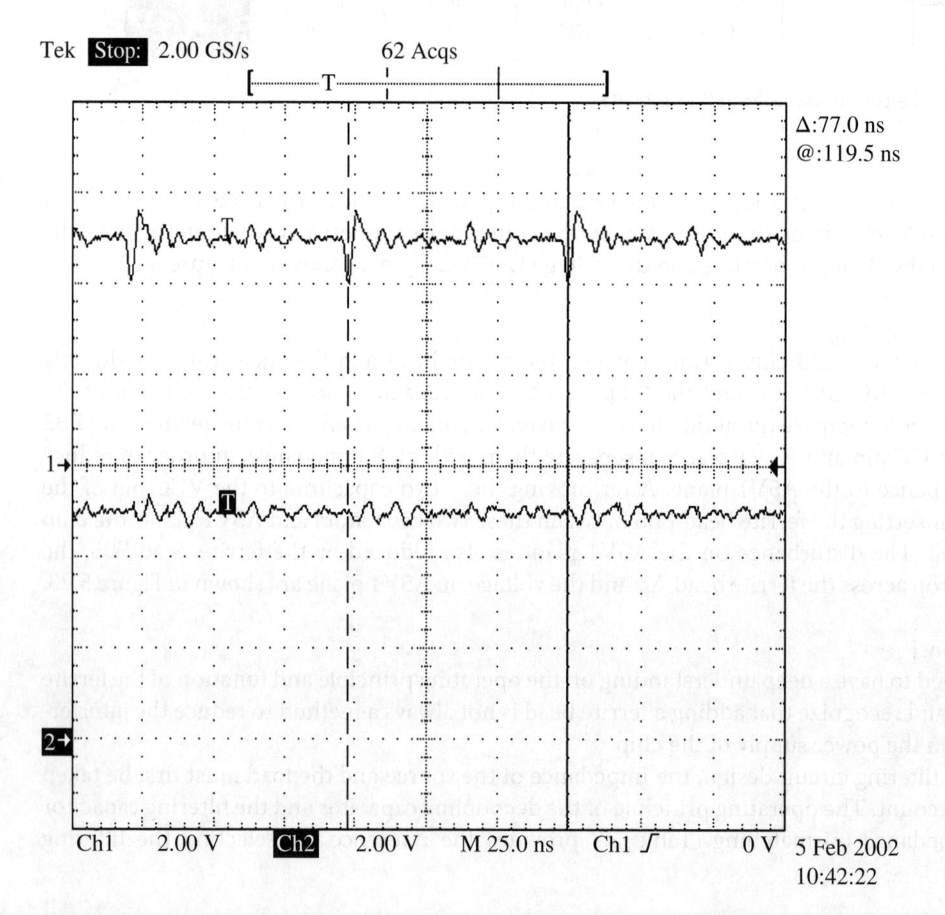

Figure 5.23 The voltage waveform of the end of the magnetic bead Δu and A5V1 (channel 1 and 2).

Table 5.2 Filter circuit selection.

Source impedance	Filter circuit	Load impedance
Low	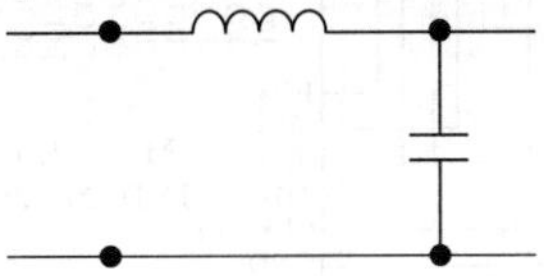	High
High	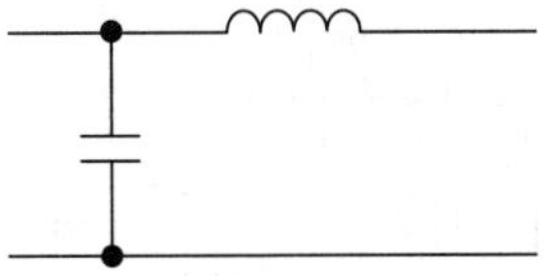	Low
High	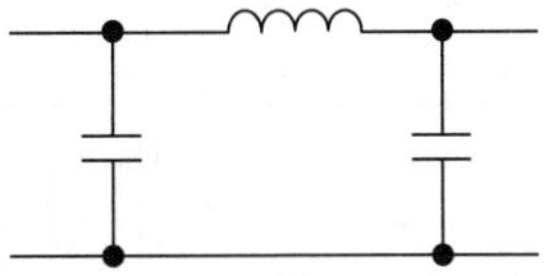	High
Low	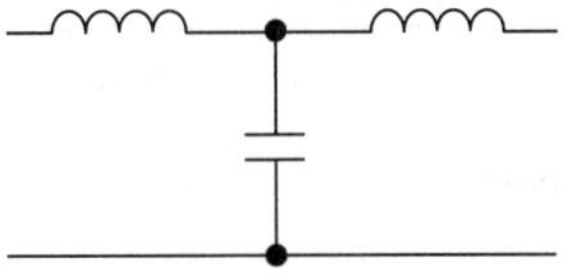	Low

5.2.3 Case 49: Producing Interference of the ESD Discharge

[Symptoms]

A product is a communication converter, of which the RJ-45 connector with metallic shell is used for one communication port, for the ESD test per IEC61000-4-2 standard, the ESD contact discharge is required for the RJ-45 metallic shell, and the test level according to the product standard is ±6 kV. When performing the contact discharge test on the RJ-45 connector, it is found that communication errors occur in the converter, the specific phenomenon is the transmission data errors.

Part of the schematic of the product is shown in Figure 5.24.

After checking the PCB layout, it is found, in the schematic, that the transmission distance of the interconnecting signals between pin 28, pin 27, pin 25 of U2, and pin 4, pin 1 of U5 and pin 30 of U4 is long, and because the PCB is four layer board, so the signal traces are routed on the external layers, in the experiment, respectively parallel 1 nF bypassing capacitors on the four signal traces (after the test, the capacitor's impact on the signal quality is accepted), the ESD test is passed, and there is no communication error.

[Analyses]

When ESD is discharged, it usually affects the electronic equipment through the following ways:

1) The initial electric field can be capacitively coupled to the circuits with large surface area, and the field strength where is 100 mm away from the ESD discharge position is up to a few kV m^{-1}.

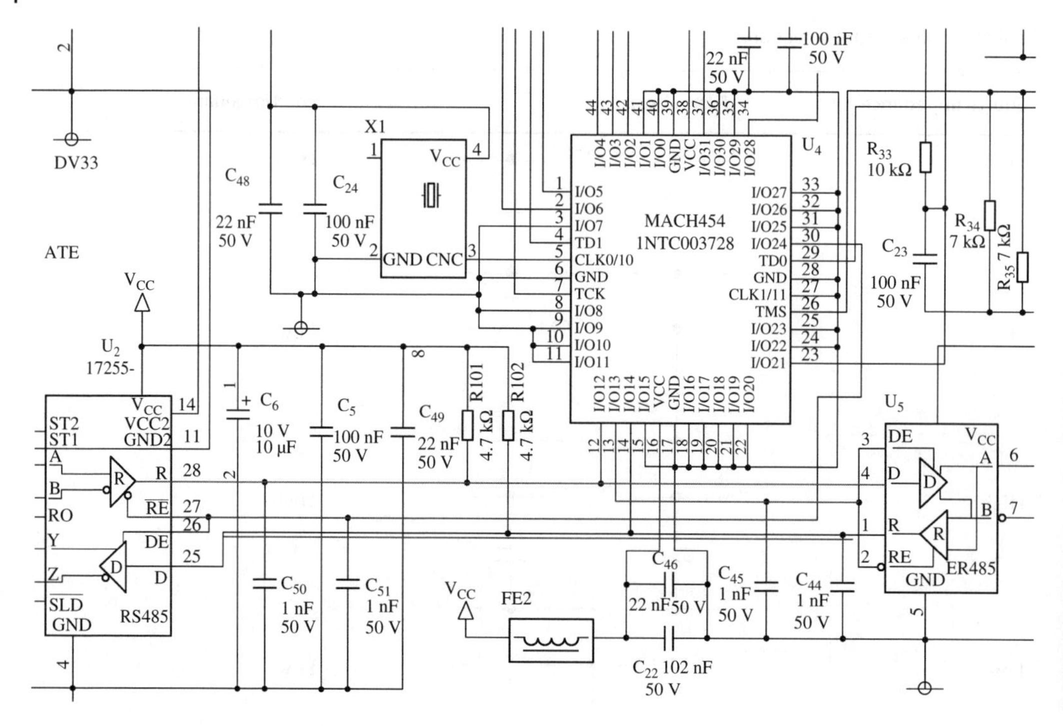

Figure 5.24 The principle diagram of product part.

2) The following damages and failures may be caused by the discharge current:
   a) Breakdown is caused in the internal thin insulating layer of the component, and damage the gate of MOSFET and CMOS components.
   b) The trigger is locked in CMOS components.
   c) The reverse biased p–n junction is short-circuited.
   d) The forward biased p–n junction is short-circuited.
   e) The soldering wires or aluminum wires inside the active device is welded.
3) ESD current may cause the voltage pulses ($U = L^* di/dt$) on the conductors, which may be a power supply or ground, signal traces, and the voltage pulses will enter each component connected to those conductors.
4) A strong magnetic field with the frequency range from 1 to 500 MHz will be caused by the ESD discharging current, which will be inductively coupled to each nearby circuit loop. The magnetic field, which is 100 mm away from ESD discharge position, is up to tens of $\mathrm{A\,m^{-1}}$.
5) The electromagnetic field radiated from the discharging current can be coupled with the long signal traces, which act as the receiving antenna.

In this product, when injecting contact discharge on the metallic shell of the RJ-45 connector, as the shell of the RJ-45 connector is connected to the reference ground through the grounding wire, it would produce a transient and large discharging current, and the current will induce a large electromagnetic field nearby the injection position. If the components or signals exposed in this electromagnetic field is sensitive, the system will operate abnormally. And the disturbed signal of this product is 3 cm away from the ESD discharge position on the RJ-45 connector, it can be seen that, in this case, the problem is the last kind of those problems described as above, namely the electromagnetic field radiated from the ESD discharge current is coupled to the long signal traces, these signal traces behave as the receiving antenna. After paralleling bypass

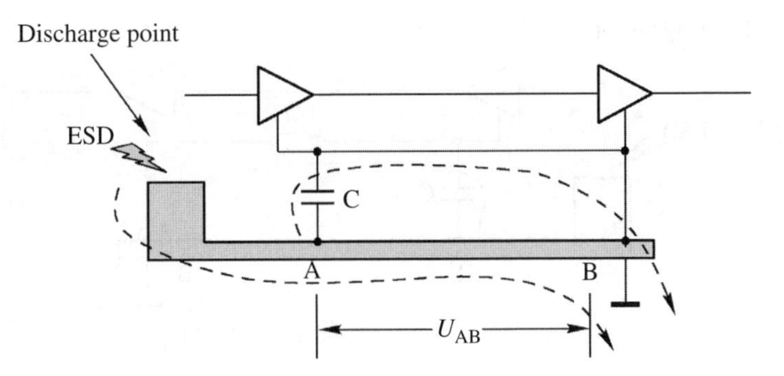

Figure 5.25 The principle diagram 1 of ESD analysis.

capacitors on the signals, a part of the coupled energy is filtered by these capacitors; thus, the signal at the receiver side of the device is protected.

If there is another possibility in theory, which is like (3) in several kinds of impacts on equipment by ESD discharge. It is hard to distinguish which kind of ways influences the internal circuit by tests, this possible situation will be analyzed, too. In this case, the product is equipped with the grounding terminal, the product is grounded during ESD test, the grounding terminal, the RJ-45 connector's shell and the reference ground of the internal circuit are connected together, as shown in Figure 5.25. In this figure, C represents the parasitic capacitance between the ESD injection position and the reference ground of the internal circuits, which is assumed as 2 pF; UAB represents the voltage drop between position A and position B when the ESD current flows through the grounding wire. Because of the structure limitation of the product and the inherent impedance of the grounding wire (about 1 m long), it is difficult to achieve zero ohm grounding impedance between position A and position B.

To help with the analysis, assume the ESD peak current is 20 A (in fact, it should be greater than this value, and the current path is represented by the dotted line) flowing through the grounding parts, and suppose the parasitic inductance is 10 nH between position A and position B of the grounding wire; then

$$\mathrm{U}_{\mathrm{AB}} = L * dI/dt = 10*10^{-6*}20/1*10^{-9} = 200(\mathrm{V})$$

In this formula, *dt* means the rising time of the ESD current, which is 1 ns.

When the peak voltage of UAB is 200 V, the current (this current path is represented by the upper blue dotted line) flowing through the reference ground of the internal circuit is

$$I = C * dv/dt = 2*10^{-12}*200/1*10^{-9} = 0.4(\mathrm{A})$$

When there is more than 0.4 A current flowing through the reference ground, if the ground plane is not very complete, and it has certain impedance due to the slot caused by the juxtaposed vias, as shown in Figure 5.26. Assuming that there is 1 cm long slot, its parasitic inductance L1 will be about 10 nH (regarding this original estimation, the relevant documents can be referred).

The voltage drop across this parasitic inductance is

$$\Delta U = L_1^* dI/dt = 10*10^{-9}*0.4/1*10^{-9} = 4(\mathrm{V})$$

This 4 V voltage is enough to cause abnormal operation for 5 V TTL level device, and this voltage is the lower value we estimate. In the actual situation, a greater voltage drop may be produced. After paralleling the bypass capacitors between the signals and the reference ground,

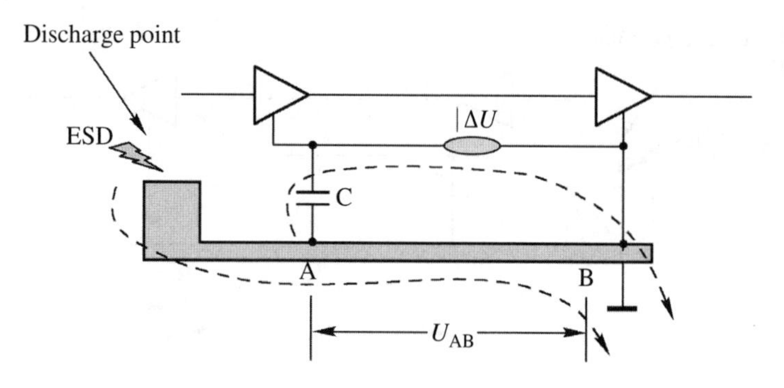

Figure 5.26 The principle diagram 2 of ESD analysis.

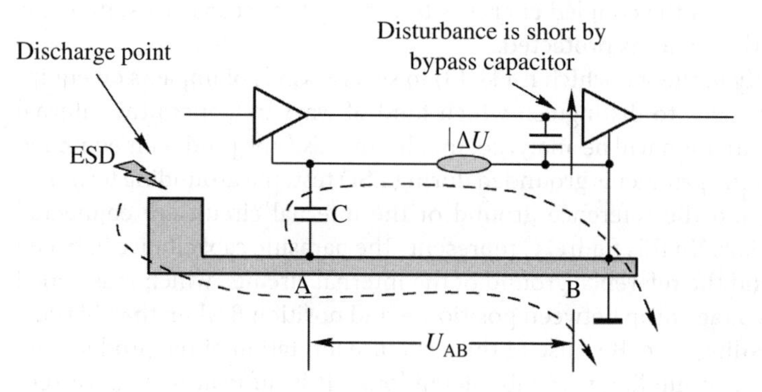

Figure 5.27 The function of bypass capacitor.

the capacitors approximately short-circuit the high-frequency noise from signals to ground; thus, the received signal of the device is protected, as shown in Figure 5.27.

[Solutions]

Paralleling the bypass capacitors with the signals, the bypass capacitors are placed on the PCB and close to the signal pins of the chip.

[Inspirations]

1) For sensitive signals with long transmission distance and near the ESD injection position, it is suggested to bypass and filter them, or route them in the inner layer of the PCB with more than six layers.
2) For the product structure design, it is necessary to avoid the common-mode disturbance current flowing through the reference ground plane of the internal circuit. If the common-mode current cannot be avoided flowing through the reference ground, we need to ensure that the reference ground plane is as complete as possible; generally, the impedance of the perfect ground plane without vias and slots is only 3 mΩ. For the 5 V TTL level circuit, it can withstand at least 200 A common-mode current. For 3.3 V TTL level circuit, it can at least withstand 130 A common-mode current.

5.2.4 Case 50: Using Small Capacitance Can Help Solve a Longstanding Problem

[Symptoms]

One medical product is used to receive the signals sent out by the auxiliary organ that is embedded inside the human body. We can supervise the operating status and the corresponding test data of the artificial organ. Because the signal is transmitted through the wireless electromagnetic field, the radiated immunity test is necessary for this device. The test level is $3\,\mathrm{V\,m^{-1}}$. The medical product will continuously receive a certain frequency signal that is generated by the analog signal source. We must assure that the product can work normally and receive the data without any data loss.

During the radiated immunity test, the received results of the product show that the product can't receive the right information at some certain frequencies. Repeating the trials, we can find that the frequency problem is not fixed.

[Analyses]

In terms of EMI, we must prevent the PCB wirings, coils, and cables, etc. from becoming antennas. Antenna effect will occur when the length of the PCB wirings or the cables can be comparative with the wavelength, which will cause the radiation. The radiation can radiate the energy out of the product through the free space or the cable. In terms of EMS, it is the same as EMI. Antenna effect (as shown in Figure 5.28) will occur when the length of the PCB wirings, the cables or the loops can be comparative with the wavelength, which may cause the circuit inside the equipment work abnormally.

According to Maxwell equations, the induced current will appear in the closed loop when the variable magnetic fields penetrate the closed loop. The magnitude of this current is related to the changing rate of the magnetic field.

The product structure is shown in Figure 5.29.

It can be seen from Figure 5.29 that the product consists of three parts, the transceiver, the processing unit and the shielded cable. The transceiver is used as a sensor to receive the signals that are sent out by the embedded auxiliary organ inside the body. The shielded cable transmits the received analog signals to the processing unit. Then the processing unit converts the analog signals into digital signals. Finally, the digital signals are sent to the computer by the USB interface.

The transceiver circuit is shown in the Figure 5.30. It can be seen in the figure, that the LC resonant circuit is used for the signal receiving part of the receiver. The inductor is a bigger coil. According to Maxwell equations, the coil will receive the high-frequency radiated field in radiated immunity test, which will be superimposed on the useful signal. When the induced voltage caused by radiated disturbance can be comparative with the useful signal voltage in amplitude and duration. The signal will be disturbed. The chip cannot judge the signals correctly. Thus, the above phenomenon that we have described appears.

Figure 5.28 Loop antenna.

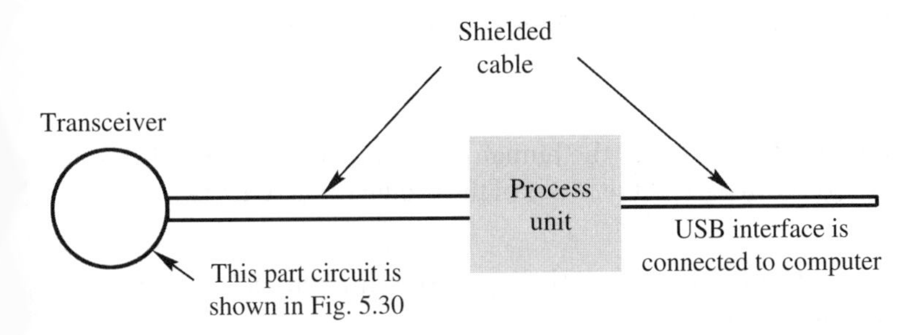

Figure 5.29 The product structure.

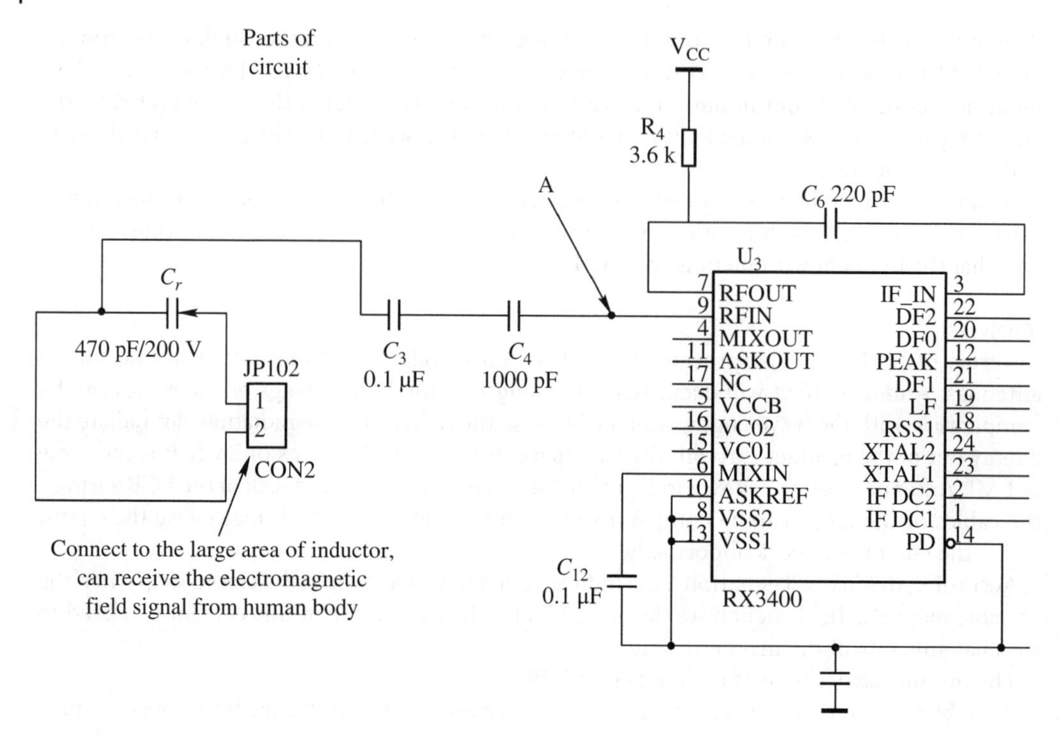

Figure 5.30 Transceiver circuit.

To solve this problem, we must get rid of the undesired signal. Because the frequency of the useful signal (29 Hz) is not same as the frequency of the undesired signal. The filtering method is feasible. The test results verify that, if a 100 pF capacitor is in parallel with position A (shown in Figure 5.30) after C4 in the schematic, the issue is solved.

[Solutions]

According to above analysis, we can connect a 100 pF bypass capacitor at position A in Figure 5.30.

[Inspirations]

The coil is the sensitive component. It can receive the electromagnetic disturbance easily. So we must make sure that the operating frequency of the coil is not same as the frequency of the disturbance.

5.2.5 Case 51: How to Deal with the ESD Air Discharge Point for the Product with Metallic Casing

[Symptoms]

An industrial product with metallic casing has the human interface on which there is a small hole for manipulating the toggle switch to set the status of the product. There are two problems during the ESD test:

1) Arc discharge phenomenon takes place around the toggle switch when ±8 KV ESD air discharge is injected on the toggle switch.
2) The component in the peripheral circuits of the toggle switch is damaged after multiple ±8KV air discharges.

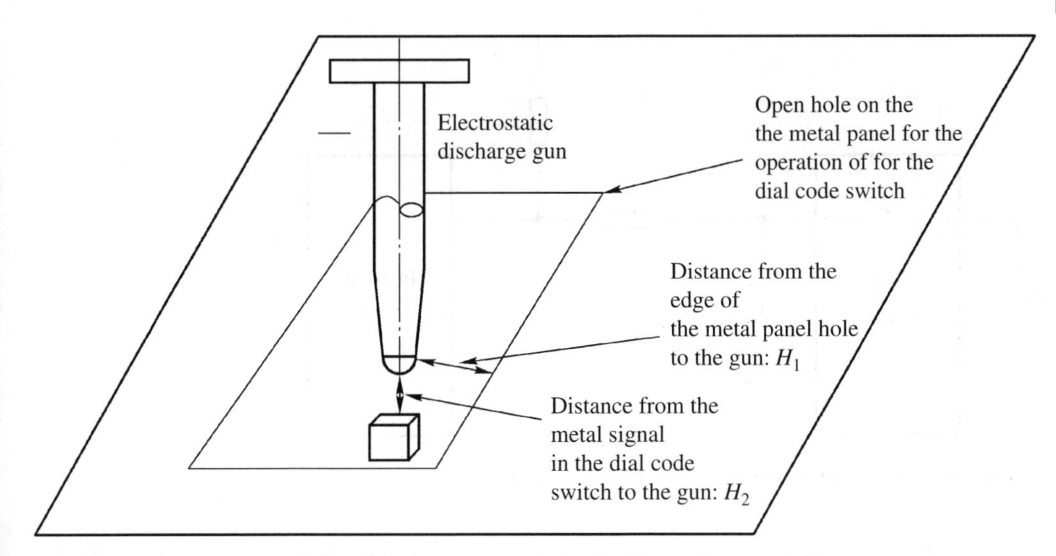

Figure 5.31 Schematic diagram of air discharge.

[Analyses]

ESD is a kind of high-energy discharge, which is easily charged to its nearby low potential conductor during the ESD test. For ESD air discharge, the discharge electrode shall be rapidly touched the injected position during the test operation. If the design of the product structure (as shown in Figure 5.31) causes that the distance H_2 from ESD injection position to the circuit of the toggle switch is smaller than the distance H_1 from ESD injection position to the metallic casing, then ESD discharge will take place between the ESD injection position and the toggle switch circuit.

After a further analysis on the structure design and the internal circuits of the human interface, it is found that:

1) The toggle switch is close to the metallic casing, so when air discharge test is performed, the distance H_2 from the ESD injection position to the circuit of the toggle switch is smaller than the distance H_1 from the ESD injection position to the metallic casing, which will cause the air discharge between the ESD injection position and the toggle switch.
2) The circuit design is not reasonable, there is no protection with the transient suppressing devices or proper ESD suppression circuits for the signals of the toggle switch, which will eventually cause the chip directly linked to the signals of the toggle switch damaged. The transient voltage suppressor (TVS) can suppress the transient overvoltage. The protection circuits, such as RC filtering circuit, can filter out the interference caused by ESD, so that it can protect the post-stage circuits.

[Solutions]

1) Taking the advantage of the characteristics of ESD that the ESD disturbance will be discharged to the nearby conductors, the opening area of the metallic panel is changed, and the distance between the toggle switch and the metallic panel is appropriately increased, so the distance H_2 from the ESD injection position to the circuit of the toggle switch is larger than the distance H_1 from the ESD injection position to the metallic panel, during ESD test, the discharge will not take place between the ESD injection position and the circuit of the toggle switch.
2) Add transient suppression devices or suppression circuits on the signals of the toggle switch, which are disturbed by the ESD discharge to suppress the high voltage and over current caused by ESD discharge.

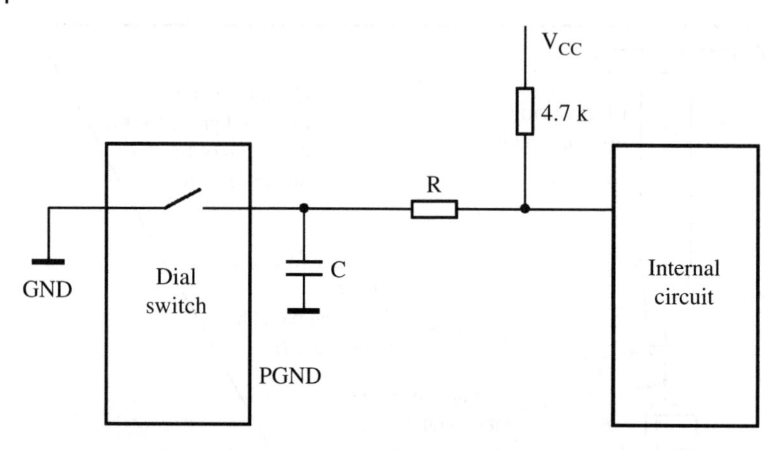

Figure 5.32 The principle of dial switching protection circuit.

Due to the time constraints of the product development, we add protection circuits in the circuit of the toggle switch in this product. The ESD protection schematics of the toggle switch and its internal circuits are shown in Figure 5.32.

In this protection circuit, the capacitance used to filter the high voltage caused during ESD discharge is from 1000 pF to 0.01 μF and a resistor about 50 Ω is in series with the circuit to suppress the overcurrent in the ESD discharge (while the ESD test level is high, the capacitor in Figure 5.32 can be replaced by TVS, and the recommended TVS is PSOT05). It is proven that this product can pass ±15 kV discharge test with this protection circuit.

[Inspirations]

1) For the design of the similar products, it is necessary to consider the relationship between H_1 and H_2 as mentioned in Figure 5.31 to make sure that H_2 is always bigger than H_1 or $H_2 > H_1$ while they are close to the breakdown distance of the ESD voltage.
2) It is necessary to add a transient suppression circuit or a filtering circuit in the circuit, which is easily disturbed by ESD discharge.

5.2.6 Case 52: ESD and Bypass Capacitor for Sensitive Signals

[Symptoms]

A product is constructed with the frame, PCB inserting cards, and the PCB backplane. Each of the PCB cards has a metal panel, while performing ESD immunity test for the panel of a PCB card (called VPU board), while ±4 kV ESD contact discharge is injected. The card is reset under the test, and the result is that it cannot meet the standard requirements for the ESD immunity test.

[Analyses]

The ESD discharge (regarding the description of ESD waveform, refer to Case 21 process is accompanied by the radiated noise and the conducted noise. The radiated noise includes the static magnetic field, the electric field, and the magnetic field caused by the discharge current. The conducted noise includes the direct charge injection and the current induced by the electric field and the magnetic field. Figure 5.33 illustrates the electric field and the magnetic field at the position that is 10, 20, and 30 cm away from the ESD discharge position when a certain ESD voltage is injected. It can be seen that the generated electromagnetic field strength

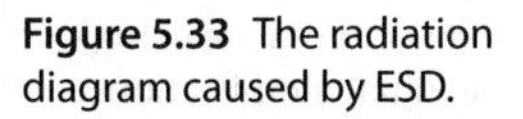

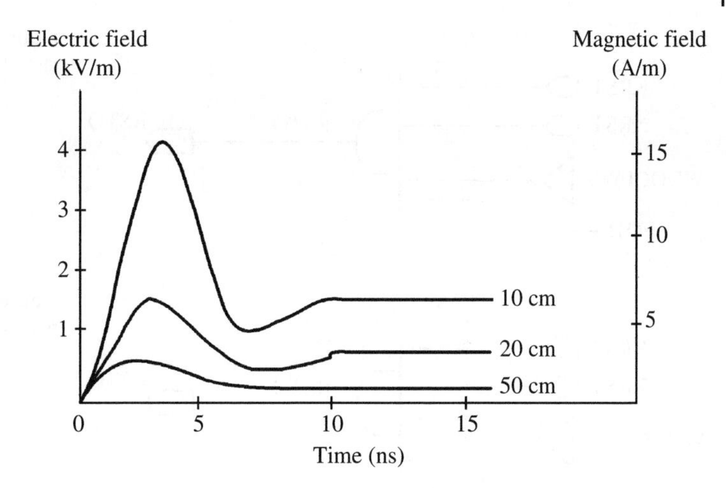

Figure 5.33 The radiation diagram caused by ESD.

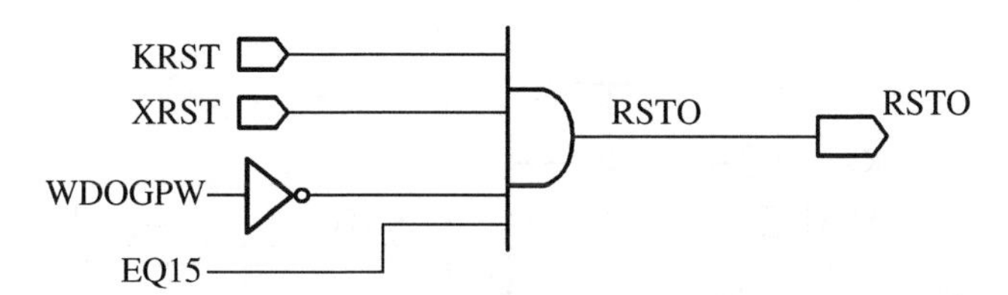

Figure 5.34 VPU board reset circuit.

is considerable while ESD discharge happens. Of course, in the actual situation, these effects are not independent from each other.

In this case, it is possible that the electromagnetic disturbance caused by ESD discharge is coupled with the control signals of the VPU board; therefore, the VPU board is reset during the ESD test. So, to analyze the root cause of why the VPU board is reset, we should start from its reset control circuit. The reset circuit of the VPU board is shown in Figure 5.34.

In Figure 5.34, KRST is the reset input signal from the manual reset button, and XRST is a reset signal from another PCB card named RPU board. When the RPU board is started or receives its reset command, the XRST signal level will be low; WDOGPW is the reset signal for Watchdog; EQ15 is the power on reset signal; RSTO is the reset output signal, which will be sent to the main processor of the RPU board. As shown in Figure 5.34, it can be seen from the reset circuit of the VPU board that the VPU board will be reset as long as any one reset input signal of KRST, XRST, WDOGPW, and EQ15 is enabled. When ESD contact discharge is applied on the VPU board, to locate the reset problem, we need to find out which one of the reset signals is disturbed in the test and then reset the VPU board.

There is no long wire inside EPLD in which the power on reset signal EQ15 is generated, so the EPLD will not cause the reset when it is subject to the disturbance, which should be excluded at first. As shown in Figure 5.34, because the implementation of the reset circuit for the VPU board is inside the EPLD, so that it is not necessary to cut off the signals or fly the signals to locate the other reset signals, as a result it is convenient for testing and finding the root cause through modifying the internal logic of the EPLD. The specific steps are as follows:

1) It is found that the trace of the reset signal KRST is near the edge of the VPU board and it is about 12 cm long, which is prone to be influenced by the electromagnetic interference radiated by the ESD discharge current. In order to confirm whether this reset signal is disturbed and causes the reset, the reset logic circuit inside the EPLD is modified, as shown in Figure 5.35.

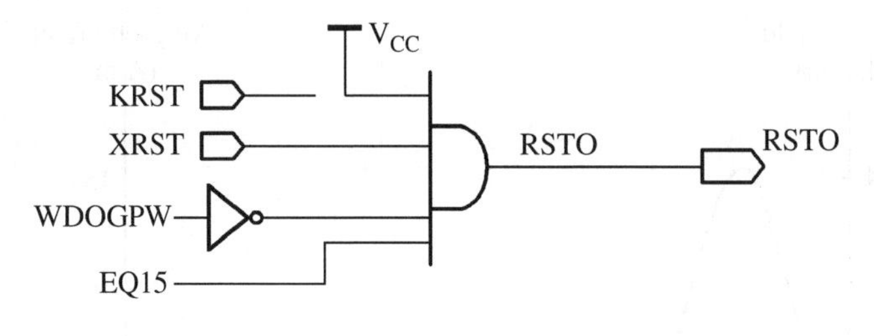

Figure 5.35 EPLD internal reset logic circuit after the initial modification.

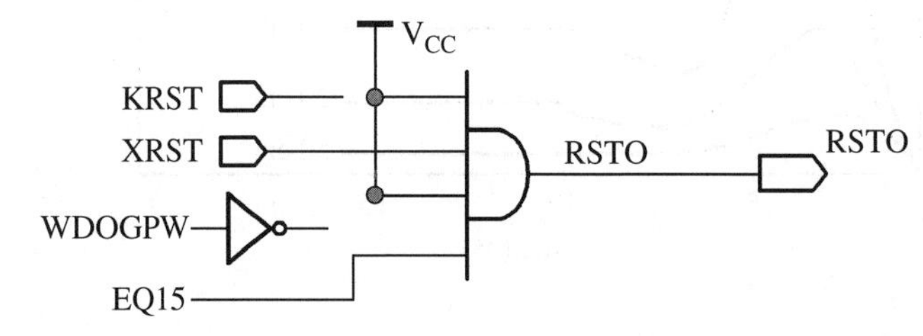

Figure 5.36 Re modified EPLD internal reset logic circuit.

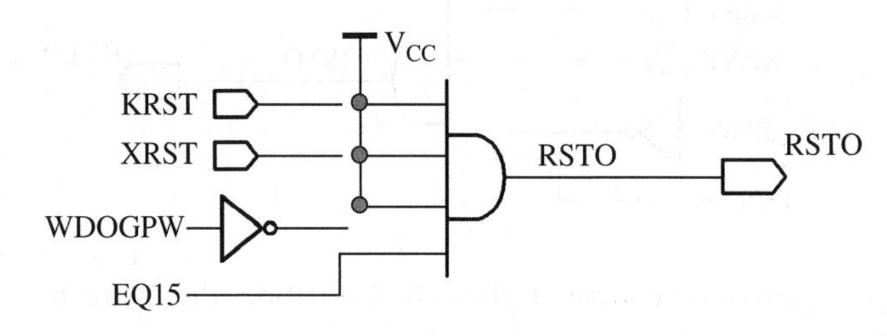

Figure 5.37 EPLD internal reset logic circuit after two times modification.

2) Disconnect the reset button to disconnect the reset signal KRST, and keep the reset signal XRST, the watchdog reset signal WDOGPW and the power on reset signal EQ15 in RPU board, then perform ±6 kV ESD contact discharge test, the VPU board is reset, it is preliminarily concluded that the reset is not caused by the reset signal KRST, which is not disturbed.
3) It is suspected that the program in the VPU board is disturbed to run away during the ESD test, which results in the watchdog reset signal is effective and resets the VPU board; therefore, the internal reset logic of the EPLD is modified, as shown in Figure 5.36.
4) Disconnect the watchdog reset signal WDOGPW, and maintain the reset signal XRST, the power on reset signal EQ15 of the VPU board, then perform ±6 kV contact discharge test, the VPU board is reset, this can be concluded that the reset is not caused by the program, which is not disturbed.
5) At this time it can be basically concluded that the reset of the VPU board is caused by the reset signal XRST from the RPU board, which is disturbed during the ESD test. In order to verify this conclusion, revise the internal reset control logic of the EPLD again, as shown in Figure 5.37
6) On the basis of (2), disconnect the reset signal XRST on the RPU board, and only maintain the power on reset signal EQ15; then perform ±6 kV contact discharge test. The VPU board is no longer reset, which indeed verifies that the XRST signal is disturbed and causes the board reset.

[Solutions]

The reset control signal XRST is passing from the chain of the RPU board→backboard→VPU boards, its wiring is very long, and it is not filtered before it enters the EPLD. Connect a 0.01 μF bypass capacitor between the XRST signal input pin on the EPLD and the reference ground on the VPU board, and then inject ±8 kV contact discharge ESD on the frame; the VPU board is no longer reset.

[Inspirations]

1) In the design of the reset circuit, the protection for it should be considered. A bypass capacitor should be connected between the reset signal input pin and the reference ground; its capacitance is recommended to be 0.01 μF.
2) The reset signal is the sensitive signal, so its wiring length should be reduced as small as possible in PCB design.
3) While locating the root cause for the ESD problems, the probe operating with the oscilloscope will pick up the strong disturbance radiated by the ESD discharge. It is difficult to identify whether it is the disturbance signal superimposed on the operation signals. The method of modifying the logic circuit may be considered to analyze and locate the problem.

5.2.7 Case 53: Problems Caused by the Inappropriate Positioning of the Magnetic Bead During Surge Test

[Symptoms]

When ±500 V surge test is performed for the interface circuit of one product, the interface circuit can't work well and the signal is interrupted; it can't be self-recovered after the test. After checking, it is found that the ferrite bead in series with the signal to suppress the high-frequency noise is broken and it is in open-circuit state.

[Analyses]

The highest data rate of the interface circuit in this product is 1.536 Mbps. The DC resistance of the ferrite bead is 1.3 Ω, and its impedance is between 10 and 20 Ω at 10 MHz, it reaches 600 Ω at 100 MHz. Therefore, it can absorb the high-frequency noise on the signals on the premise of less attenuation on the signal. The frequency versus impedance characteristics of the ferrite bead are shown in Figure 5.38

Ferrite bead is made of ferrite, which can convert the AC signal into heat energy. While the inductor can only store the energy and release it slowly. Ferrite bead has more impedance for high-frequency signal, and its resistance is much lower than its inductance at low frequency.

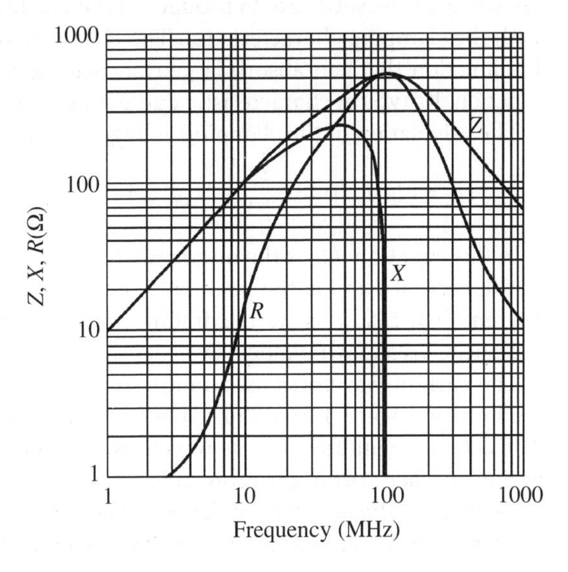

Figure 5.38 Frequency impedance characteristics of magnetic bead.

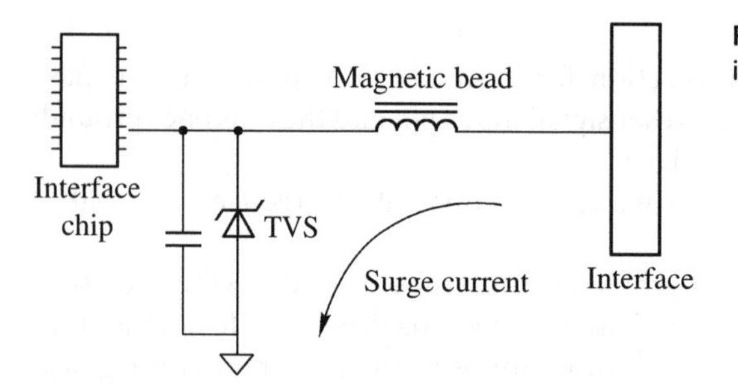

Figure 5.39 The principle diagram of product interface circuit.

When current goes through the wires, ferrite presents almost no impedance for low-frequency current while larger impedance for higher frequency current. The high-frequency current is dissipated on the ferrite in the form of heat. Its equivalent circuit is an inductor in series with a resistor, and the values of these two elements are in proportion to the length of ferrite bead. The ferrite bead can be used not only in the power supply circuit to filter the high-frequency noise (it can also be used in the DC and AC output port), but also in the other circuits of which the size needs to be small. Especially in digital circuits, the pulse signal includes higher harmonics, which are the main sources of high-frequency radiation in the circuit. And the ferrite bead can play its role in this condition. It has better high-frequency filtering properties than the ordinary inductors, and it behaves as resistance at high frequency. It is also able to maintain higher impedance over a quite wide frequency range, so the filtering effect is improved.

The schematic of the interface circuit in this product is shown in Figure 5.39.

It can be obviously seen from the schematic that the TVS is used to protect against surge. The ferrite bead together with the bypass capacitor constitutes a LC filter to suppress the high-frequency noise. When the surge test is performed on the interface, the surge current will flow though the ferrite bead first and then be released to the reference ground through TVS, which can protect the post-stage interface chip against the surge disturbance. TVS is a special element, which working principle is similar as the Zener diode. TVS is a PN junction made of polyhydrocarbon silicon. By controlling the doping concentration of the PN junction and the resistivity of the substrate to produce avalanche breakdown, it can clamp the transient voltage with the clamping characteristics. The characteristics of TVS are in proportion to the area of PN junction that can absorb large transient current by controlling the characteristics of the junction. Its typical characteristic curve is as shown in Figure 5.40.

The calculation of the clamping voltage is as follows:

$$U_{\mathrm{C}} = (I_{\mathrm{P}}/I_{\mathrm{PP}}) \times (U_{O\max} - U_{BR\max}) + U_{BR\max}$$

In this equation, U_{C} is the intermediate value of the clamped voltage (V), which is measured at the current I_{P}; I_{P} is the peak pulse current in the test (A); I_{PP} is its rated maximum peak current (A); U_{Omax} is its maximum clamping voltage (V) at I_{PP}; U_{BRmax} is the maximum avalanche breakdown voltage (V).

The current waveform used in the surge test is 8 μs/20 μs, and the voltage waveform is the combination wave 1.2 μs/50 μs, it means that most of the energy will be released in 20 μs. The actual magnitude and the waveform of the voltage and current that the circuit will suffer at the interface port of the product is related to the internal resistance of the generator and the impedance of the interface under test. The peak pulse power that the TVS at this interface can sustain is 500 W, and its clamping voltage is about 10 V. It can withstand about tens of amperes,

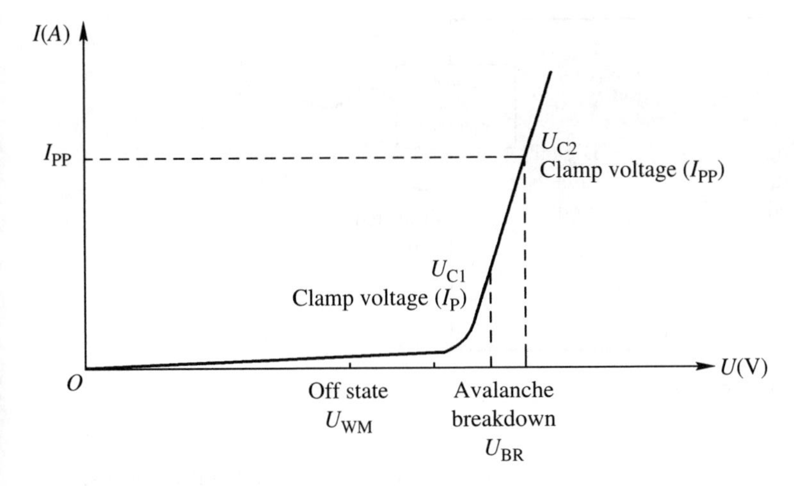

Figure 5.40 Voltage–current characteristic curve of TVS tube.

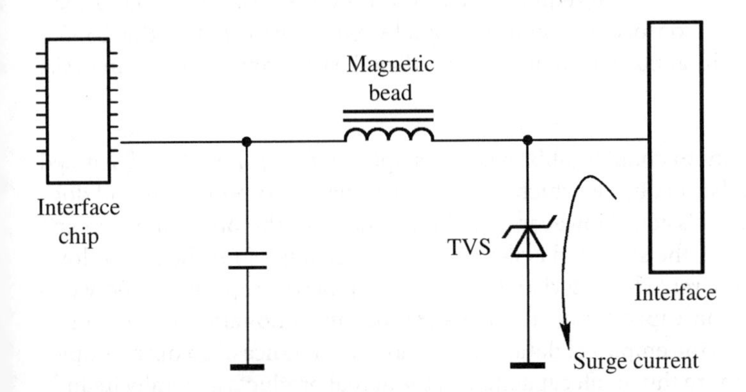

Figure 5.41 The principle diagram of TVS in front of the bead.

while the rated current of the ferrite bead is only 100 mA. In the surge test, the current flowing through the ferrite bead is too much higher than its rated current; consequently, it will be damaged.

[Solutions]

Move the TVS in front of the ferrite bead; thus, the large surge current won't go through the ferrite bead. The schematic is shown in Figure 5.41.

[Inspirations]

If the surge protection and the high-frequency noise suppression circuit or the bypass capacitor coexist, the design should first protect against surge and then suppress the high-frequency noise.

5.2.8 Case 54: The Role of the Bypass Capacitor

[Symptoms]

The structure diagram of a certain industrial product is shown in Figure 5.42.

There is only on PCB board in this product, its casing is made of plastics and there is a special grounding terminal in the power supply port. The circuit board is separated into the analog

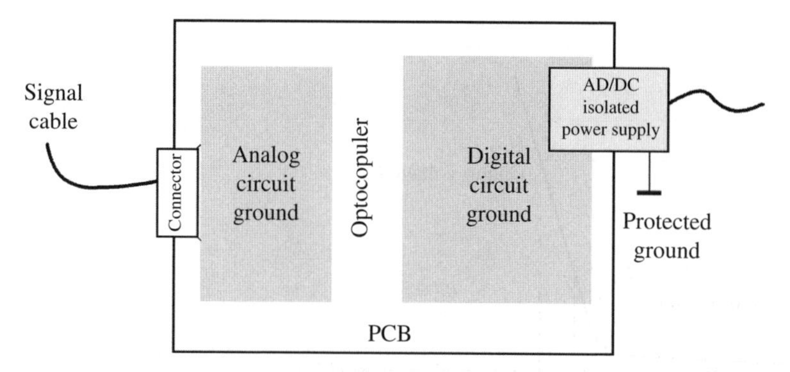

Figure 5.42 The diagram of product structure.

circuit and the digital circuit, and they are isolated by a photo coupler. The length of the signal cable is greater than 3 m, in addition to the power port, the signal ports must be subjected to the EFT/B test and other immunity tests. The required EFT/B test level is ±2 kV. It is found that, during the test, abnormal operation occurs when the signal cable is subjected to the ±500 V EFT/B disturbance in the test, the abnormal circuit is in the digital part through further analysis.

[Analyses]

To analyze this problem, the photo coupler (abbreviated as opto-coupler) should be discussed at first. Opto-coupler is an isolation device, which isolates the signal on its both sides and does not affect the DC signal transmission. However, it is important that the opto-coupler can't reach 100% isolation in any case, the so-called isolation is only meaningful for the DC or low-frequency signals, due to its inherent characteristics, there exists junction capacitance between both sides. Due to this junction capacitance, the isolation becomes noneffective for high-frequency signals. According to the empirical data, the junction capacitance of an opto-coupler is generally 2 pF. But do not ignore this small capacitance; the actual products generally include multiple signal channels for transmission and then multiple opto-couplers in parallel are required. In this case, the number of opto-couplers in this product is 5, so totally 2 pF × 5 = 10 pF capacitances exist between the digital ground and the analog ground. In the EFT/B test, the common-mode disturbance current is shown in Figure 5.43. (Regarding the essence of the EFT/B disturbance, you can refer to Case 11 and the explanations in Section 2.1.)

In Figure 5.43, the arrow curve indicates the flow of the common-mode current under EFT/B test, due to the presence of the junction capacitance of the opto-coupler, the common-mode current will pass through the opto-coupler, and flow through the digital circuit. It can be seen that the digital circuit isolated by opto-coupler is impacted by the EFT/B common-mode current, when the common-mode current flows through it, if the ground impedance of the digital circuit is large on its ground plane, due to the incomplete ground plane and too many via holes on it, a higher voltage drop may appear on it. If the voltage drop exceeds a certain level, the circuit will be affected.

According to the above analysis, it is approximately clear where the circuit is disturbed, i.e. the region the common-mode current passes by. If none or only a very small part of the common-mode current flows through the digital circuit, the probability of abnormal operation in the test could be reduced. According to this idea, a bypass capacitor is connected between the analog ground and the grounding terminal of this product, the capacitance is 10 nF, and then we perform the test again. The test is passed with ±1 kV EFT/B test voltage. Figure 5.44 shows the difference from the initial situation.

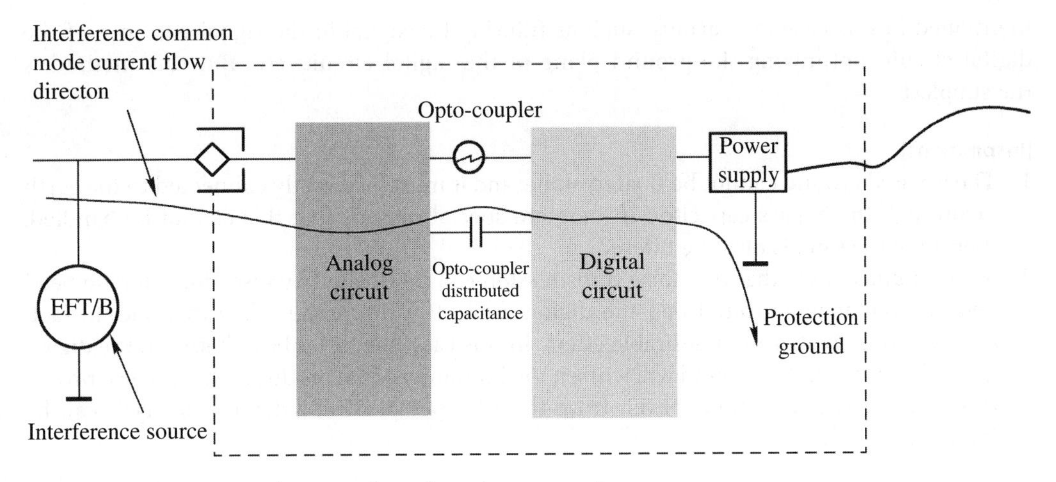

Figure 5.43 Common-mode current flows through opto-coupler.

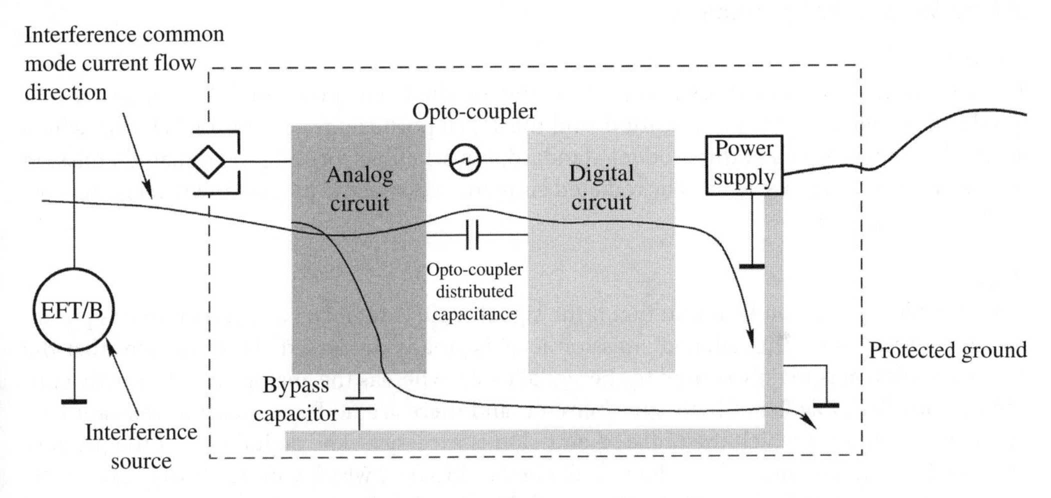

Figure 5.44 Common-mode current flow direction after connecting the bypass capacitor.

It can be seen clearly from Figure 5.44 that the common-mode current path under EFT/B disturbance is changed, i.e. another common-mode current path is added, and because the 10 nF bypass capacitor is much larger than the 10 pF junction capacitance and on the condition that the impedance of the grounding terminal where the bypass capacitor is connected is low enough, most of the common-mode disturbance current will flow from the bypass capacitor to the earth, so that the common-mode current flowing through the digital circuit is greatly reduced, the digital circuits is protected, and therefore the level of the EFT/B disturbance the product can pass is greatly increased. The grounding impedance of the bypass capacitor is very important, which must be ensured to be small enough. Repeating once again here, in the PCB layout, the PCB copper conductor with the ratio of length-to-width less than 5:1 presents very low impedance; it is about 3 mΩ at 100 MHz.

[Solutions]

According to the above analysis and the test results, a 10 nF bypass capacitor is connected between the analog ground and the protective ground. Certainly, the solutions to the problem mentioned in this case are various, such as filtering the signal in the signal entrance of the digital circuit, optimizing the ground plane of the digital circuit, etc. But this method is the simplest.

[Inspirations]

1) The isolated ground cannot be floated alone, and it must be directly connected to the earth or through the bypass capacitor. If there are special reasons that this cannot be handled, then all of the signals must be filtered.
2) Many engineers say that the noise transmission can be isolated by inserting a ferrite bead between the analog ground and the digital ground in the product like this one. In fact, however, this creates an undesirable effect; in this case, the series bead between the digital ground and the analog ground will worsen the immunity of the product. The test also proves this point. It is not desirable, even from the EMI perspective; a detailed analysis can be found in Case 55.

5.2.9 Case 55: How to Connect the Digital Ground and the Analog Ground at Both Sides of the Opto-Coupler

[Symptoms]

This case is the extension of Case 54 for the same product. The problem of EFT/B test is analyzed and solved, after the modification, and the signal port can pass ±1 kV EFT/B test, which meets the requirements of the product standard, but when performing the radiated emission test, the problem appears again. The radiated emission test spectral plot is shown in Figure 5.45; the test is not passed.

[Analyses]

From further tests, it can be found that if the signal cable is removed or a ferrite ring is put on the core on the cable, the radiated emission level is greatly decreased. This indicates that the radiated emission is mainly related to the signal cable, whereas the analog circuit directly connected with the cable is not high-speed circuit, and there are no frequencies or relevant harmonic frequencies at which the radiated emission is excessive. The digital part of the product has a high-speed circuit, and its clock frequency is 25 MHz, which can be clearly seen in the test spectral plot that the higher frequency with higher radiation is the multiplied frequency of 25 MHz. Thus, the noise is likely to be radiated from the digital circuit.

As mentioned in the other case, a necessary condition for forming the radiation is:

1) The driving source that can be a voltage source or a current source.
2) The antenna.

Obviously, the signal cable is the radiating antenna, then where is the driving source? Generally, it is believed that the noise in the digital circuit has been isolated by the opto-couple, there will be no noise transmitted to the signal cable, but it is not true in high-frequency condition. The principle of how the radiation is generated is shown in Figure 5.46.

Due to the existence of the junction capacitance of the opto-coupler (the number of optocouplers in the product is 5, so the capacitance in this case between the analog ground and the digital ground is 2 pF*5 = 10 pF), part of the noise in the digital circuit will be transmitted through the junction capacitance of the opto-coupler to the analog circuits. The noise voltage ΔU of the digital circuit is the driving source; thus, the two necessary conditions forming radiation are provided.

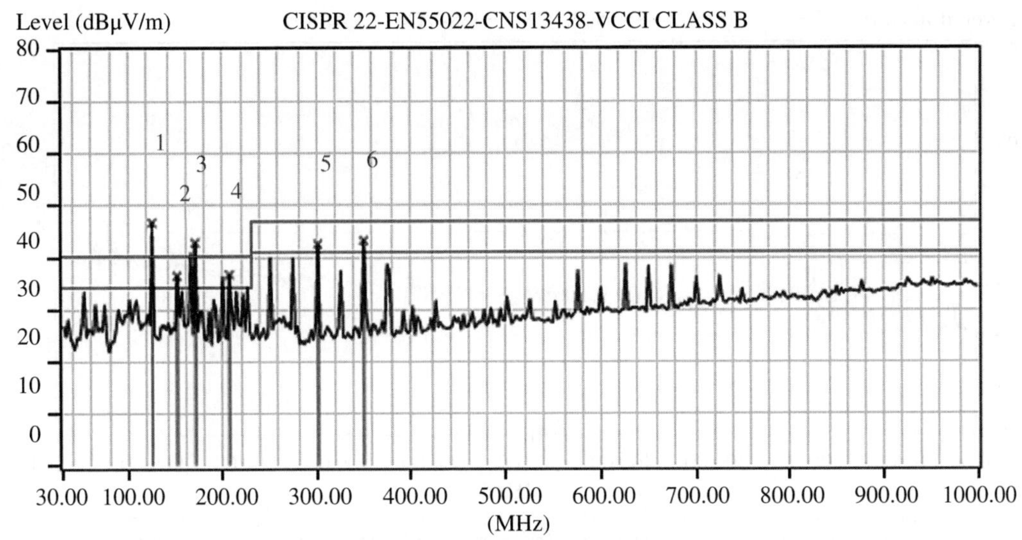

	No.	Frequency MHz	Factor dB	Reading dBμV/m	Emission dBμV/m	Limit dBμV/m	Margin dB	Tower cm	Table deg
*F	1	124.58	15.04	31.57	46.61	40.00	6.61	--	--
	2	151.25	16.98	19.39	36.38	40.00	–3.62	--	--
F	3	170.65	16.01	26.75	42.77	40.00	2.77	--	--
	4	207.03	13.07	23.54	36.60	40.00	–3.40	--	--
	5	301.60	16.58	25.88	42.46	47.00	–4.54	--	--
	6	350.10	17.47	25.71	43.18	47.00	–3.82	--	--

Figure 5.45 Spectrum of radiation disturbance test.

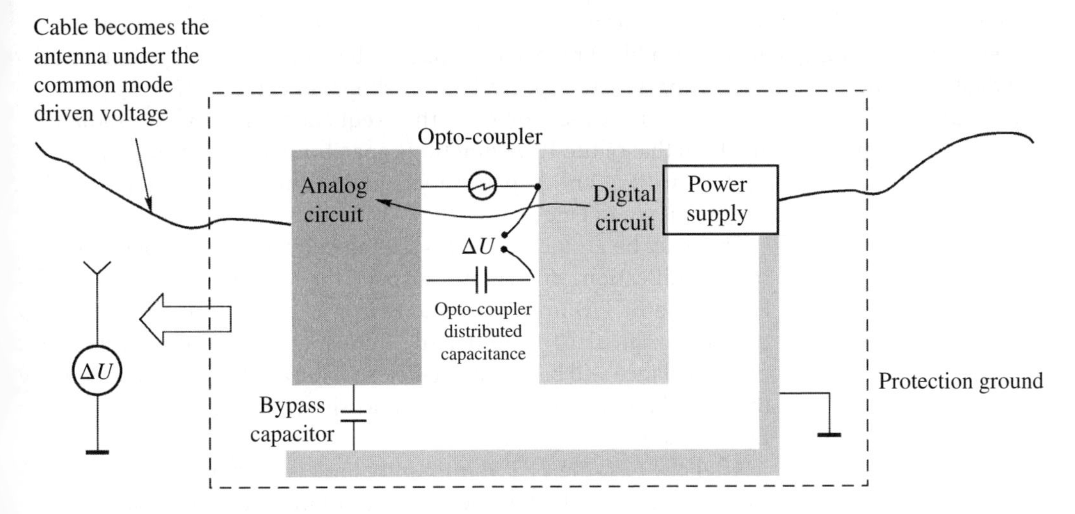

Figure 5.46 The principle of radiation.

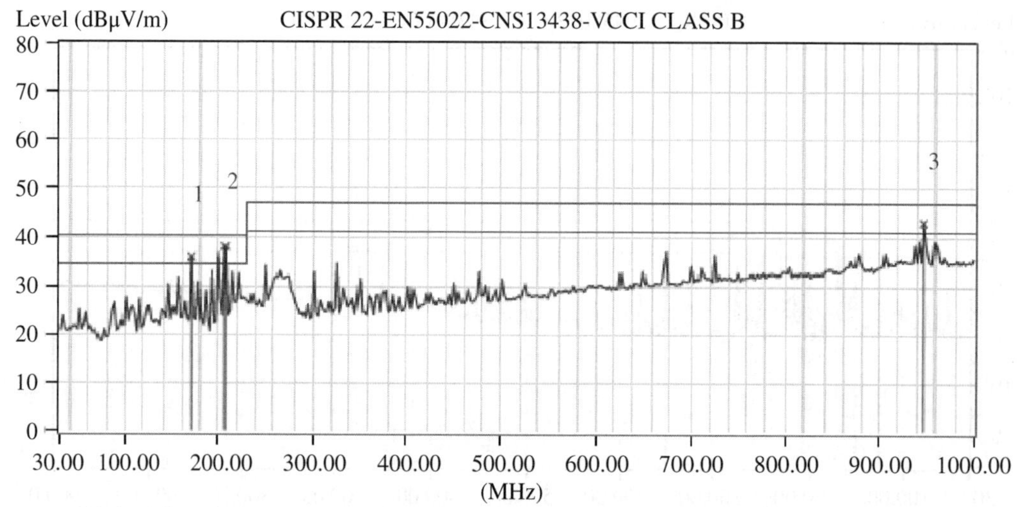

This data is for evaluation purposes only. It cannot be used for EMC approvals unless it contains the approved signature.
If you have any questions regarding the test data, you can write your comments to service@mail.adt.com.tw

	No.	Frequency MHz	Factor dB	Reading dBμV/m	Emission dBμV/m	Limit dBμV/m	Margin dB	Tower cm	Table deg
	1	170.65	16.01	19.51	35.53	40.00	−4.47	--	--
*	2	207.03	13.07	24.73	37.80	40.00	−2.20	--	--
	3	946.65	27.80	15.04	42.84	47.00	−4.16	--	--

Figure 5.47 The spectrum diagram after connecting the bypass capacitor.

Obviously, reducing ΔU is the best way to solve the radiation problem in this case. Maybe someone will try to suppress the noise transmission by the inserting ferrite bead between the digital ground and the analog ground, this method does not solve the problem, because the ferrite bead presents high impedance at high frequency, it cannot reduce the ΔU.

To reduce the radiated emission of this product, the driving voltage at the antenna port must be reduced. According to this idea, add a 1 nF bypass capacitor between the analog circuit and the digital circuit, then the test is passed, the spectral plot is shown in Figure 5.47.

Actually, the impedance of the 1 nF capacitor over the frequency range of the radiated emission test is much smaller than that of the 10 pF junction capacitor in this case, and the connection with a 1 nF bypass capacitor is equivalent to short-circuit ΔU, as shown in Figure 5.48.

This may be an incredible result, but the fact still happens. After this modification, one might suspect that the 1 nF capacitor will reduce the EFT/B immunity, the reason is that 1 nF is much larger than the original 10 pF, under the frequency of EFT/B disturbance, the impedance will be much smaller, so naturally the current through the digital circuit will be increased (as shown in Figure 5.49), so the EFT/B test may not be passed.

After the test, the EFT/B immunity level is decreased, on the contrary it is improved a lot, the product originally can only pass ±1 kV test for the signal port. Now it can pass the ±2 kV test (if 1 nF capacitor is removed, it only passes ±1 kV test).

Δ V
1 nF

Figure 5.48 The principle of bypass capacitor.

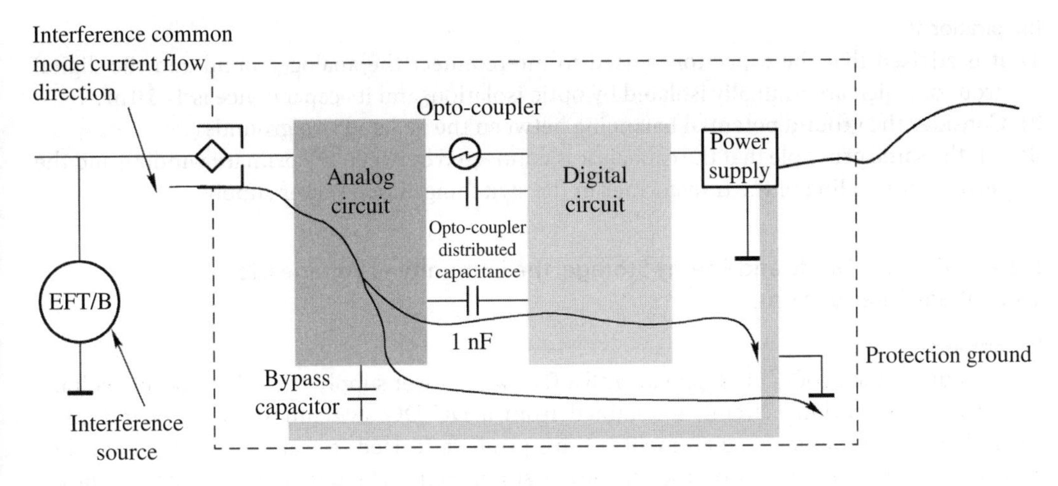

Figure 5.49 Common-mode current flow direction after connecting bypass capacitor.

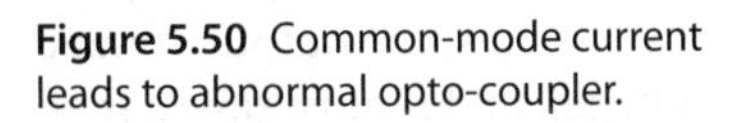
Figure 5.50 Common-mode current leads to abnormal opto-coupler.

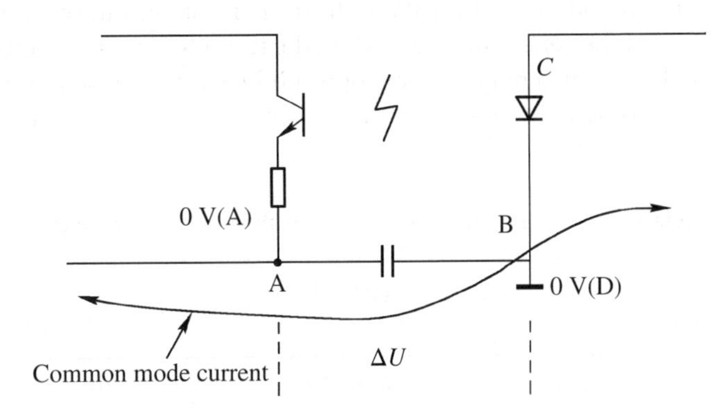

The following is the interpretation of how the EFT/B immunity is improved after adding 1 nF bypass capacitor between the digital circuit and the analog circuit.

In Figure 5.50, supposing that the common-mode disturbance current flows starting from the left to the right, since the high impedance between A and B (the distributed capacitance) leads to a part of the common-mode disturbance current flowing from A to C, and then through the diode B, the most sensitive part of the opto-coupler, that is the light emitting diode, is disturbed, and it cannot work normally.

The same is true for the radiation problem, after connecting the 1 nF bypass capacitor, the voltage between A and B is reduced, and then the situation is improved a lot. (*Note:* The usually used bypass capacitor is the high voltage capacitor, of which the rated voltage is above 1 kV.)

[Solutions]

According to this analysis and the test results, the capacitance of the bypass capacitor between the analog circuit and the digital current is 1 nF. The test results are shown in Figure 5.47.

Remember that the common-mode current flowing through the digital circuit is indeed increased after connecting the 1 nF bypass capacitor between the digital circuit ground and the analog circuit ground, which is also a verification for the digital circuit. In this case, the reason for the overall improved immunity performance is that the sensitive level of the opto-coupler is relatively low. It is necessary to take the overall consideration for product design, and the EMC design is not only to publicize design rules, but also to deeply understand the circuit characteristics.

[Inspirations]

1) It is advised that the capacitor is used to interconnect the analog ground and the digital ground, which are mutually isolated by optic isolation, and its capacitance is 1–10 nF.
2) Consider the ground potential balancing between the isolated two grounds.
3) It is the same principle that the capacitor is connected between the primary winding and the secondary winding of the transformer in the switching-mode power supply.

5.2.10 Case 56: Diode and Energy Storage, the Immunity of Voltage Dip, and Voltage Interruption

[Symptoms]

A communication product is supplied with a DC-48 V power supply, the 3.3 V working voltage of the internal working circuit is obtained from a DC/DC switching-mode power supply. According to the standard requirements of this product, it needs to carry out the voltage dips and voltage interruption test on the DC power supply port. The test requirements are shown in Tables 5.3 and 5.4.

It is found that, for the 0% of the normal voltage interruption test in which the output impedance of the generator is low, when the test interruption duration is 1–10 ms, the output voltage of the DC switching power supply module is off (i.e. no output voltage), and it must take a long time to be recovered after the power supply input is switched off rather than automatically

Table 5.3 The measurement level, duration time, and performance criterion for the voltage drop test.

Test	Test level %U1	Duration(s)	Performance criterion
Voltage drop	40 and 70	0.01	A
		0.03	B
		0.1	B
		0.3	B
		1	B

Table 5.4 The measurement level, duration time, and performance criterion for the voltage interruptions test.

Test	Test condition	Test level %U1	Duration(s)	Performance criterion
Short interruption	High impedance/Low impedance	0	0.001	A
			0.003	A
			0.01	A
			0.03	B
			0.1	B
			0.3	B
			1	B

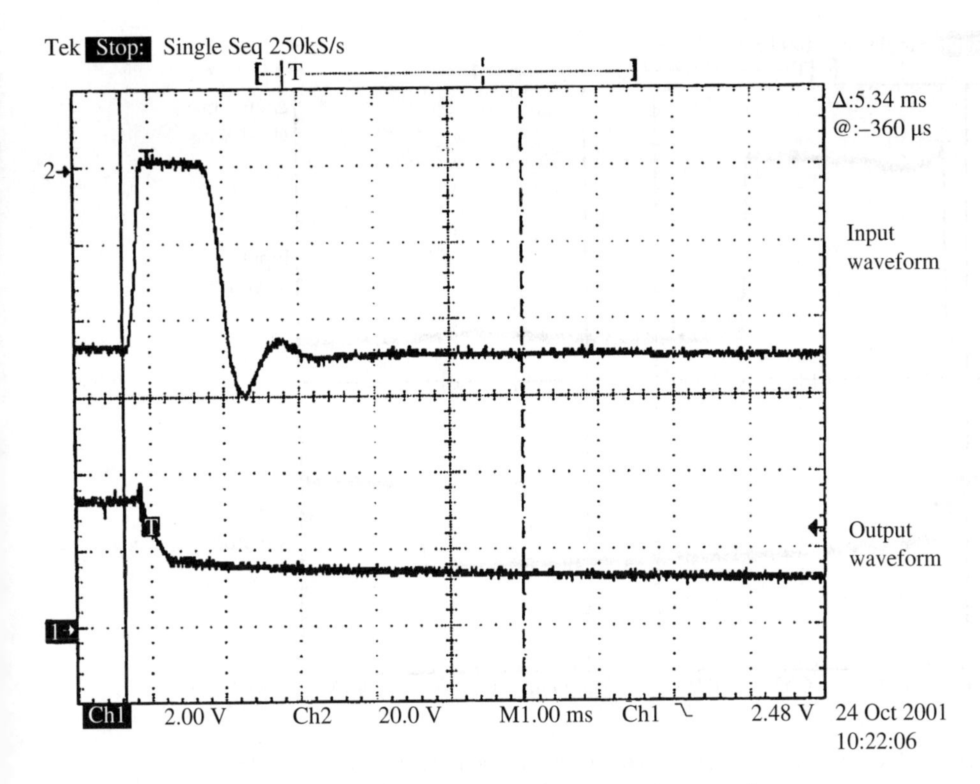

Figure 5.51 DC/DC power input/output voltage waveform at 1 ms interrupt time.

recovered (this phenomenon is similar as the over current protection phenomenon of the switching-mode power supply, which is abbreviated as "protection" phenomenon in the following description of this case); the test cannot be passed. In order to solve this problem, we try to increase the capacitance of the energy storage capacitor by paralleling more capacitors in the DC port. However, there is no obvious effect when an additional 200 μF capacitor is added. This kind of phenomenon does not appear when the interruption time is greater than 10 ms, while only system reset phenomena occurs. As the accepted performance is criteria B for the voltage dip and voltage interruption tests on the DC power supply port, it is passed. Figures 5.51 and 5.52, respectively, show the input and the output voltage waveforms of the DC/DC power supply in the test during which the output impedance of the generator is low and the interruption duration is, respectively, 1 and 14 ms.

For the 0% of the nominal voltage interruption test in which the output impedance of the generator is high, for all the combined tests, the "protection" phenomenon of the DC/DC module's output does not appear. Only in the interruption test with longer than or equal to 10 ms duration will the reset phenomenon appear in the product. The input and output voltage waveform of the DC/DC switching-mode power supply is measured, from which it can be found that this reset phenomenon is caused by the reset of the DC/DC power supply when its power supply input is off. Figure 5.53 is the working schematic of the input and output voltage at the port of DC/DC power supply when the reset phenomenon appears. It is also found in the test that the problem can be solved by increasing the capacitance of the energy storage capacitor at the power input port (originally, it is a 47 μF capacitor, and then another 47 μF capacitor is in parallel with it).

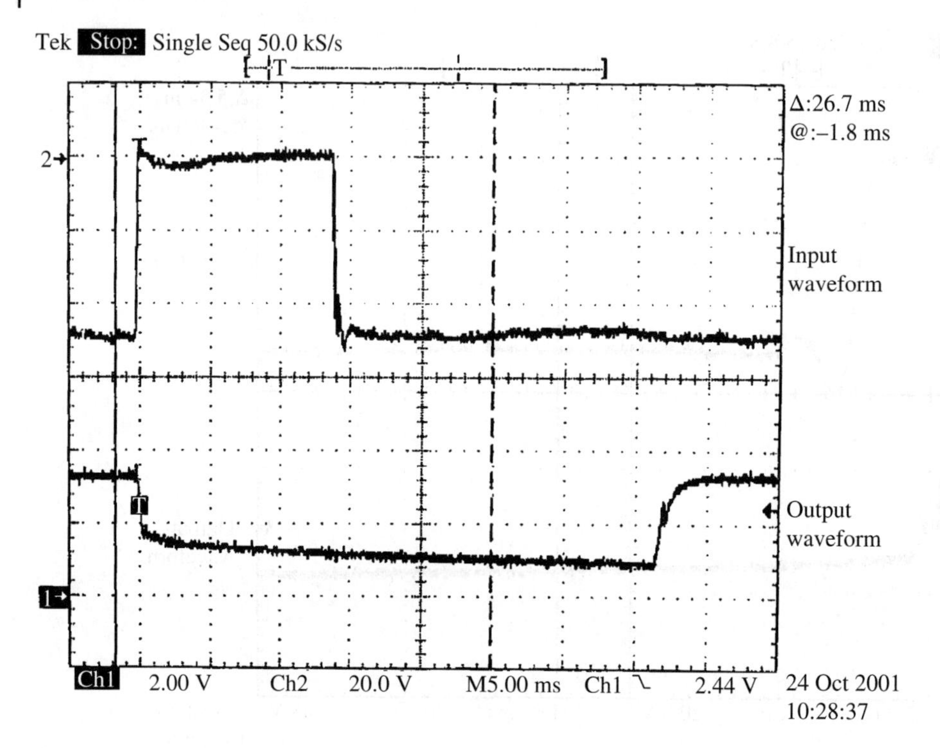

Figure 5.52 DC/DC power input/output voltage waveform when interruption time is 14 ms.

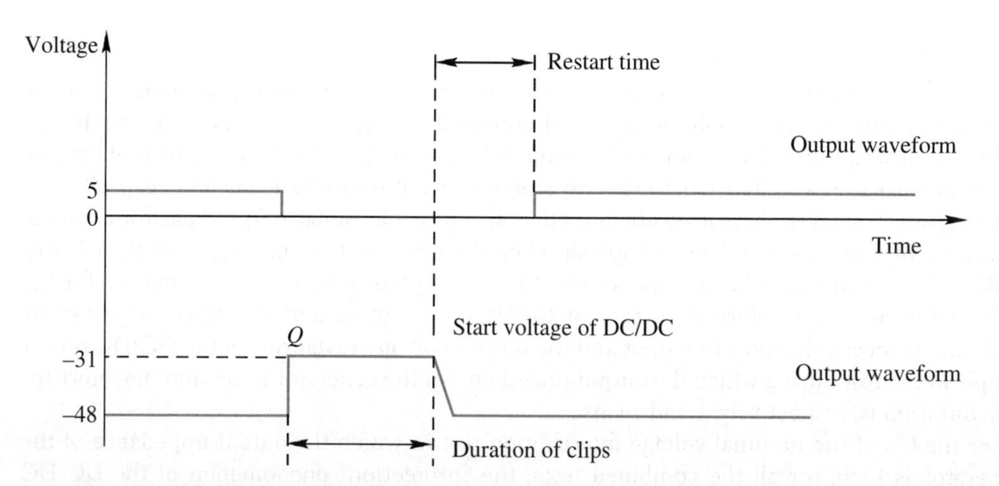

Figure 5.53 DC/DC power input/output voltage waveform when 0% interruption time is more than 14 ms under high-impedance state.

For the above test results, the following questions arise:

1) In the 0% of the nominal voltage interruption test, why does the "protection" phenomenon appear on the power port of the DC/DC switching-mode power supply in the product when the interruption duration is less than 14 ms, and no "protection" phenomenon appears when the interruption duration is greater than 14 ms?

2) In the test during which the internal impedance of the generator is high, why does the voltage waveform shown in Figure 5.53 appear at the input and output port of the DC/DC power supply, namely the voltage dip generator outputs zero voltage, however the voltage at the output port of the DC /DC power supply is not zero, but −31 V?
3) Why is the protection phenomenon not obviously improved when increasing the capacitance of the energy storage capacitor when the internal impedance of the generator is low, however it is obviously improved when the internal impedance of the generator is high?

[Analyses]

Someone who understands the operation principle of the DC /DC switching-mode power supply points out that there is a capacitor inside the separate-excitation type power supply circuit (called capacitor A in this paper), when the power supply is off, the voltage of the capacitor A drops to a certain voltage (called voltage B in this paper). If we want the power supply to work normally while it is restarted, we must first wait for the voltage across the capacitor A to fall to zero, and then power on it again, which can make sure the normal starting of the DC/DC module (the module can always work properly before the voltage across capacitor A is not decreased to voltage B).

The protection phenomenon appears in the 0% interruption test during which the internal impedance of the generator is low, while no "protection" phenomenon appears in the 40% voltage interruption test (the voltage across the capacitor A does not fall to the voltage B). The DC/DC power module in the product, in this case, belongs to this type of power supply; thus, the protection phenomenon of DC/DC power module is caused by the inherent characteristics of itself, but the inherent characteristics results in the low immunity performance for voltage dip and interruption of the DC power supply in this product.

So, why does the voltage waveform shown in Figure 5.53 occur at the input and output port of the DC/DC power supply in the test, during which the impedance of the generator is high; that is, the output voltage of the voltage dip generator is zero. However, the voltage at the output port of the DC/DC power supply is not zero, but −31 V?

Now, let's analyze how the input and output waveform of the DC/DC module is generated in the 0% and 40% drop test in which the internal impedance of the generator is high. The principle of the supply system for this product in this case is shown in Figure 5.54.

For the test in which the output impedance of the generator is high, at the moment of power off, because there is the residual voltage across the capacitor C_1, the energy will continue to power DC/DC power module to work for a period of time, while the energy in the capacitor C_1 is consumed rapidly for the working of the DC/DC power module, namely the voltage across the capacitor C_1 drops rapidly, until the voltage across the capacitor C_1 cannot power the DC/DC power module working properly. As shown in Figure 5.53, starting from −31 V, the DC/DC

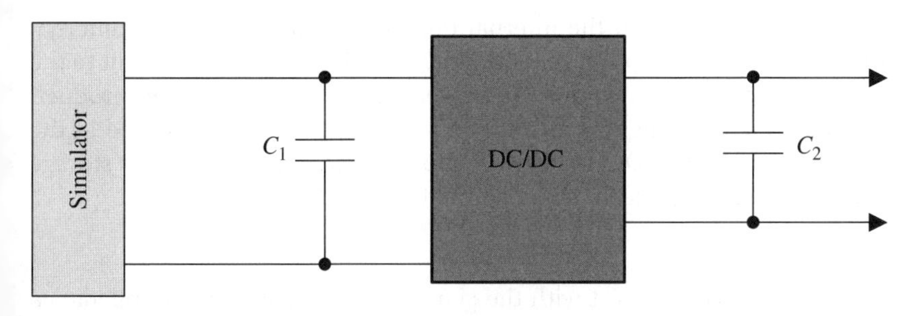

Figure 5.54 Schematic diagram of power supply system of the product.

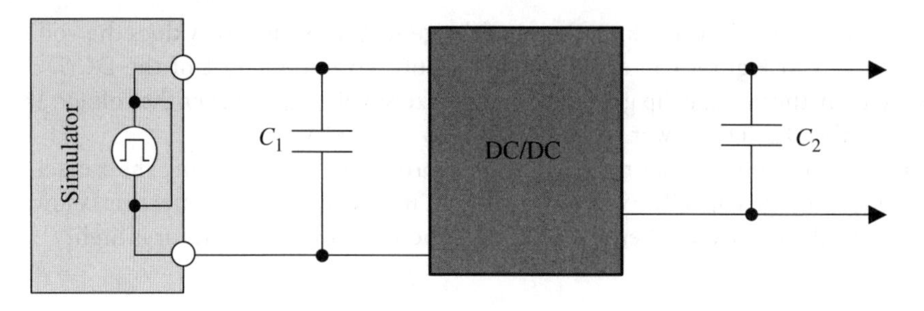

Figure 5.55 Schematic diagram of voltage break test under low-impedance state.

power supply is no longer in normal working condition, so the consumption is greatly reduced, the voltage across C_1 drops very slowly. This is why the product in the test during which the impedance of the generator is high, the voltage across C_1 doesn't quickly drop to zero, and it will be kept as −31 V for a period of time.

In addition, because the DC/DC power module is off due to the voltage drop across the capacitor C_1, the post-stage integrated circuit originally supplied by that DC/DC power module can only rely on C_2 to maintain its working for a period of time, but the integrated circuit isn't like the DC/DC power supply module, which has a wide working voltage range. According to the statistics, the normal working voltage of the integrated circuit is generally ±5% of it rated voltage, which is very difficult for C_2 to supply for a longer period of time. Usually, the output voltage of the DC/DC power module is zero. The integrated circuit quickly consumes the energy of C_2 when the voltage across capacitor C_2 is less than 95% of the normal voltage; the integrated circuit is equivalent to power down, and there will be a system reset phenomenon.

Why, for the test during which the impedance of the generator is low, increasing the capacitance of the energy storage capacitor, is the protection phenomenon not obviously improved, but the opposite is true when the impedance of the generator is high?

Actually, in the test with the generator of which the output impedance state is low, the internal impedance of the voltage dip and interruption generator is low, which can be equivalent to the short circuit of the power supply, as shown in Figure 5.55. At the moment of the voltage dip, on the one hand, the voltage across the energy storage capacitor C_1 continues to supply the DC /DC power module. On the other hand, it will be discharged through the short circuit on the side of the generator, and quickly be 0 V. According to the actual test, while C_1 = 47 μF, this discharging time is about 50 μs, far less than 1 ms. The discharging time constant of the short circuit from the capacitor C_1 to the side of the generator is far less than that of the DC/DC power module. Therefore, increasing the capacitance of C_1 (such as 100, 200 μF) cannot significantly improve the test results.

In the test with the generator of which the internal impedance state is high, the internal impedance of the voltage dip and interruption generator is high, which can be equivalent to the disconnection between the voltage source and the power supply entrance of the product. As shown in Figure 5.56, at the moment of the voltage interruption, the energy stored in the capacitor C_1 is consumed only by the DC/DC power supply. Increasing the energy storage capacitor's capacitance will have a significant effect on the test results.

[Solutions]

Since the equipment can easily pass the test with the generator in which internal impedance is set high, then a diode is in series with the entrance of the power supply input, the principle is

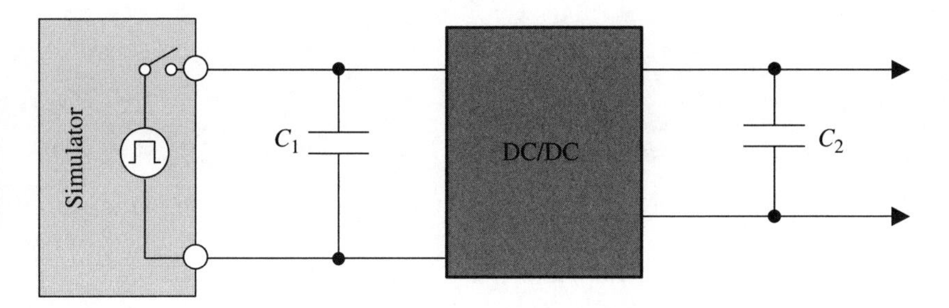

Figure 5.56 Schematic diagram of voltage break test under high-impedance state.

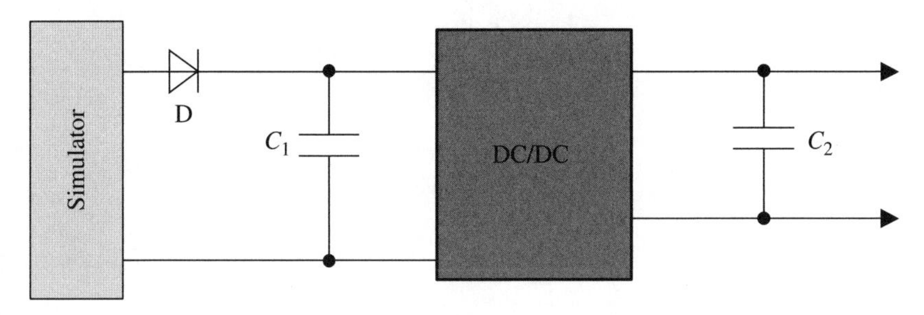

Figure 5.57 Diode in series at the entrance of power.

shown in Figure 5.57, which can achieve the change for the generator from low internal impedance state to a high internal impedance state. According to the test results, after the diode is in series, and adding another 47 μF capacitor, all the required tests on the DC power supply port are passed.

[Inspirations]

1) Understand the meaning of the voltage interruption test comprehensively. The voltage interruption test is to simulate two scenarios of the actual voltage interruption, that is, the high-impedance state and the low-impedance state. The high-impendence state is caused when the power supply is changed from one source to another source; The low-impendence state is caused when the overload or a defect on the power network is removed, and it can cause the reverse current (the negative impulse current) generated from the load side. In the high-impendence state, this reverse current is blocked; when in the low-impendence state, the low-impendence absorbs the negative impulse current from the load side.
2) From an energy storage perspective, the test is easier to be passed in a high-impedance state, so the series-connected diode at the power supply entrance is equivalent to change the low-impedance source to the high-impedance source, which is beneficial for the test to be passed.
3) In the premise of the series-connected diode at power supply entrance, the capacitance of the energy storage capacitor in the product or in a local circuit of the product can be calculated by

$$1/2C\left(U_1 - U_2\right)^2 = Pt$$

where, U_1 is the normal working voltage; U_2 is the lowest voltage on which the circuit can work normally; P is the total power of the energy storage capacitor can deliver for the circuit; t is the dropping time; and C is the required capacitance for the energy storage capacitor.

6

PCB Design and EMC

6.1 Introduction

6.1.1 PCB Is a Microcosm of a Complete Product

A printed circuit board (PCB) is a miniature of a complete product. It is the most important part worth being discussed regarding electromagnetic compatibility EMC techniques, the part of the equipment with highest operating frequency, and also the most sensitive part with the lowest voltage level. The EMC design of the PCB, in fact, already includes the design of the grounding and bypassing. For the PCB with a good ground plane, not only the voltage drop caused by the common-mode current but also the loop area can be reduced. A PCB board with good decoupling and bypassing design is equivalent to a robust body.

PCB is the most basic component in the electronic product, and also the carrier of most electronic components. When the PCB design of the product is completed, it can be said that the disturbance and immunity characteristics of its core circuit have been basically determined. After the PCB design is finished, in order to improve the EMC, a filter must be added on the interface circuit or the cable must be shielded. These options not only greatly increase follow-up product costs but also increase the product complexity and reduce the product reliability.

It can be said that a good PCB can solve most electromagnetic noise problems. As long as, in the design of the interface circuit layout, we add the transient suppression device and the filtering circuit, we can solve most of the noise problems.

In the PCB design, electromagnetic compatibility problems will cause additional costs to the final completion of the product. If, in the PCB design, product designers only focus on increasing the product density, reducing the occupied space, chasing an artistic vision, and unifying the layout, but ignore the impact caused by the layout on the electromagnetic compatibility problems, they cause radiation disturbance. This lack of forethought regarding potential compatibility issues can result in a large number of EMC problems. In many cases, even if we add filters and other noise-reducing components, we cannot solve these problems. At the end, we have to reroute the entire PCB. Therefore, it is important to start with a good PCB layout.

6.1.2 Loops Are Everywhere in PCB

From the schematic of the digital circuit, it can be seen that the digital signal is transferred in the gate circuits. These signals are transferred in the form of current. The current is always circulated, but there is no the return path for the signal current in the schematic.

Many digital circuit engineers think the return path is not relevant to the current. And in the actual signal transmission, the return current always exists, which is the reason for the current

Electromagnetic Compatibility (EMC) Design and Test Case Analysis, First Edition. Junqi Zheng.

electromagnetic interference (EMI) problem. Because the transmission of a signal implies the existence of a current loop, the main radiation source is the current flowing through the circuit (the clock, the video and data driver and the other oscillators) on the PCB. The loop formed by the signal transmission path and the return path is one of the reasons for the PCB radiation. It can be described by the model of small loop antenna. The small loop refers to its size being less than 1/4 wavelength ($\lambda/4$, when the frequency is 75 MHz, the length is 1 m). When the frequency is a few hundred MHz, most PCB loops are still considered as small loops. When the dimensions are close to $\lambda/4$, the phase of the current at different points of the loop is different. This effect can cause the field strength at a specified point to be low and may be high. In free space, the radiated field strength declines with the decreasing of the distance between the radiation source and the observation position. When the distance is fixed at 10 m, the radiated emission can be estimated. In the worst case, because of the reflection of the ground, the radiated field strength may be increased twice. This also meets the test configuration requirements specified in the test standard. Once a loop exists on the ground, considering the ground reflection, the maximum electric field strength at the position with 10 m away from the loop can be obtained by the formula (6.1):

$$E = 263\times10^{-12}\left(f^2 A I_S\right)\mathrm{V\,m^{-1}} \tag{6.1}$$

In this formula, A is the loop area (cm^2), f (MHz) is the frequency of the current source I_S (mA).

The loop area in this formula is known. The loop is composed of the signal transmission path and the return path. I_S is the current component at a single frequency. Because the square wave includes rich harmonics, I_S must be calculated by the Fourier series.

You can use (6.1) to roughly predict the differential-mode radiation. For example, if $A = 10\,\mathrm{cm}^2$, $I_S = 20\,\mathrm{mA}$, $f = 50\,\mathrm{MHz}$, then the electric field strength is 42 dBμV m^{-1}, the radiation exceeds the Class B limit regulated in the standard EN55022 by 12 dB. If the frequency and working current are fixed, and the loop area cannot be reduced, the shielding will be needed. Note, the conclusion in return is not true. That is, according to the formula (6.1), we predict that the differential-mode radiation of the PCB does not exceed the standard, but we cannot say the device does not require the shielding design, because the differential-mode current on the small loop of the PCB is not the only radiation source. The common-mode current flowing in the PCB, especially the common-mode current flowing in the cable, plays a big role in the radiation. The common-mode current on the PCB, compared with the differential-mode current (Kirchhoff current law), is very difficult to be predicted. The common-mode current return path is often from the stray capacitance (displacement current) to other nearby objects. Therefore, to get a complete prediction scheme, the mechanical structure of PCB and its casing, and the closeness between PCB and the ground and other equipment must be taken into account.

The existence of the loop is one of the reasons causing the immunity problems, because all these loops are the receiving antenna. The induced voltage of single turn loop can be calculated by the formula (1.17) or (1.18).

So, for the digital circuit design engineers, reducing the loop area as much as possible is very important.

6.1.3 Crosstalk Must Be Prevented

In the PCB design, crosstalk is very important. A PCB with good EMC design can avoid the common-mode noise current flow through the internal circuit of the product and make the current flow to the low-impedance casing, the earth, or the nonsensitive circuit in the PCB.

The crosstalk problem must be considered, which exists between the region through which the common-mode current flows, and the sensitive region through which the common-mode current does not flow. If the crosstalk is not considered, the electric-field coupling and the magnetic-field coupling exist between the two regions, which will lead to design failure. Similarly, for the internal PCB noise source circuit, such as the clock-generating circuit, the clock transmission lines, the switching circuit of the switching-mode power supply, the high-frequency signal lines, and the EMI noise and the common-mode voltage should be isolated and controlled within the circuit, to avoid the formation of radiation.

6.1.4 There Are Many Antennas in the PCB

When the length of the high-speed signal PCB trace is comparable with the signal wavelength, PCB can radiate the energy directly to the free space, and the common-mode voltage on the ground plane can drive the cable and cause radiation like an antenna. Even the printed trace length is smaller than the signal wavelength. See Section 2.1.3. When the PCB is equalized with the antenna, what is the specific meaning? The antenna is specifically used for radiating the energy outward, and most of the PCB is the unintentional antenna, i.e. the PCB is not designed to be an antenna, unless it is specifically used for energy transmission. If the PCB unintentionally becomes an ideal antenna and the effective measures cannot be found, it is necessary to take some shielding measures. Sometimes, PCB is not an antenna, but because the common-mode noise drives the cable, the cable naturally becomes the antenna.

The efficiency of the antenna is a function of frequency, whether it is intentional or accidental. When an antenna is driven by a voltage source, its impedance will be significantly changed. When the antenna is in a state of resonance, its impedance will be very high and it will mainly present in the inductive state. The impedance equation is $Z = R+J(R)$. In the equation, R is called as the radiation resistance. The radiation resistance is a measurement of the radiation tendency of the antenna at a certain frequency.

The radiation efficiency of most antennas is relatively in a specific spectrum. These frequencies are generally lower than 200 MHz, because the length of the I/O line is about 2–3 m, which is longer than the wavelength. If the frequency is even higher, the radiation can be clearly seen from the seam of the casing.

For the common-mode radiation driven by the common-mode voltage, reducing the driving voltage is the most simple and feasible way. There are three reasons for the existence of radiofrequency (RF) driving voltage:

1) The impedance of PCB routing
2) Ground bounce
3) The bypass or shielding used to reduce the driving voltage for the unintentional antenna

In order to reduce the radiation efficiency on the PCB, we need to use EMC design and control measures. Besides the shielding and the reasonable layout, appropriately selecting the filter can reduce the radiation from the unwanted RF signal and obtain the best expected results.

6.1.5 The Impedance of the Ground Plane in PCB Directly Influences the Transient Immunity

The transient disturbance always flows into the circuit through the parasitic capacitance and the distributed capacitance. Just as shown in Figure 6.1, for ungrounded device, when the common-mode disturbance electrical fast transient/burst (EFT/B) is applied on the power supply line, due to the presence of distributed capacitance between the signal cables and ground, the

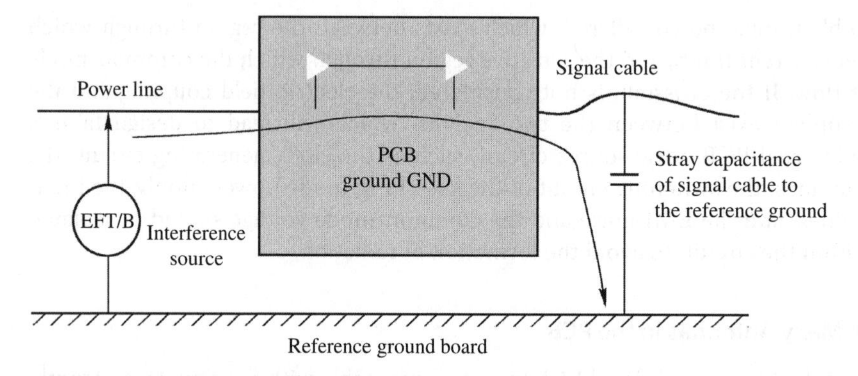

Figure 6.1 EFT interfere with the work in the PCB board.

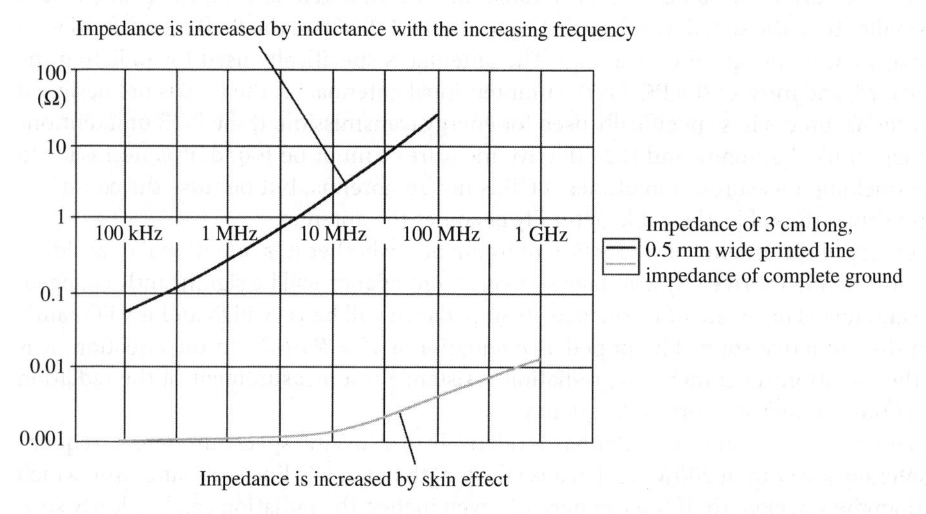

Figure 6.2 The relationship between the complete ground plane frequency and resistance in PCB.

EFT/B common-mode disturbance current flows into the ground through the distributed capacitance of the cable (shown in Figure 6.1).

On the PCB, because the ground impedance is always low, most of the current will flow through the ground on the PCB. When the common-mode disturbance current flows through the PCB, if the ground impedance between two logic circuits is too large, the logic circuits will be disturbed. Figure 6.2 shows the relationship between the frequency and the impedance of the PCB ground plane.

We know from Figure 6.2 that, for a complete (without holes and slots) ground plane, at 100 MHz, its impedance is only 3 mΩ. When the current flowing through it is 100 A, only 0.3 V voltage drop will be produced, which can be tolerable for the 3.3 V TTL circuit, because 3.3 V TTL circuit will switch the logic when the voltage is higher than 0.4 V, which has good anti-interference ability. The logical relations of 3.3 V TTL is shown in Figure 6.3

If there is a 1 cm slot in the ground plane, through which the disturbance current flows, the inductance of the slot is around 10 nH. When 100 A EFT/B common-mode current passes through this slot, the generated voltage drop is:

$$U = L \cdot \mathrm{d}I/dt = 10\,\mathrm{nH} \times 100\,\mathrm{A}/5ns = 200\,\mathrm{V}$$

Figure 6.3 TTL level logic relationship of 3.3 V.

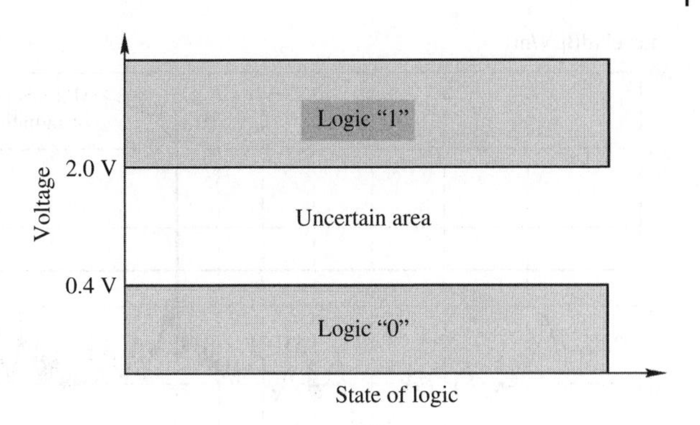

The 200 V voltage drop is very dangerous for 3.3 V TTL circuits. In the PCB, the importance of ground impedance is visible. Practice has proven that, for the 3.3 V TTL logic level circuit, the common-mode disturbance current will be safe in the ground plane if the IR drop is less than 0.4 V. If the voltage drop is more than 2 V, it will be dangerous. For the 5 V TTL logic circuit, the threshold voltage would be a bit higher (1 and 2.2 V). 5 V TTL circuit has higher anti-interference ability than 3 V TTL circuit (this method can be used in the design of product and the EMC risk assessment for the product).

6.2 Analyses of Related Cases

6.2.1 Case 57: The Role of "Quiet" Ground

[Symptoms]

A product has the shielded cabinet structure, there is a RS-232 serial port on the panel for monitoring purpose, a RJ-45 connector is used, and the serial port circuit is on the control board. This serial port is connected with the AC/DC power module of this product through a meshed shielded cable for warning purpose. The length of the serial port cable is 1.5 m; however, the radiated emission at 100 MHz is significantly high, and beyond the Class B limit of the standard EN55022 by 15 dB. Figure 6.4 shows the spectral plot of the tested radiated emission.

[Analyses]

The majority of the EMI problems in the I/O circuit are caused by five reasons:

1) The common-mode noise is generated from the components of the I/O circuit or the I/O signal itself.
2) The noise on the power plane is coupled to the I/O circuit and wirings.
3) The clock signal is coupled to the I/O cable or signal wires through capacitive coupling or inductive coupling.
4) The grounding of the casing, the digital ground, and the analog ground are interconnected improperly.
5) The hybrid using of multiple kinds of connectors (the metallic connector is connected to plastic connector, etc.).

In this case, the overall PCB structure of the serial port board can be shown in Figure 6.5.

Disconnect the interconnection between the serial port and the control board. The radiated emission at 100 MHz is greatly reduced, which proves the interference source is from the control board rather than the power supply cable of the AC/DC power supply module. We suspect

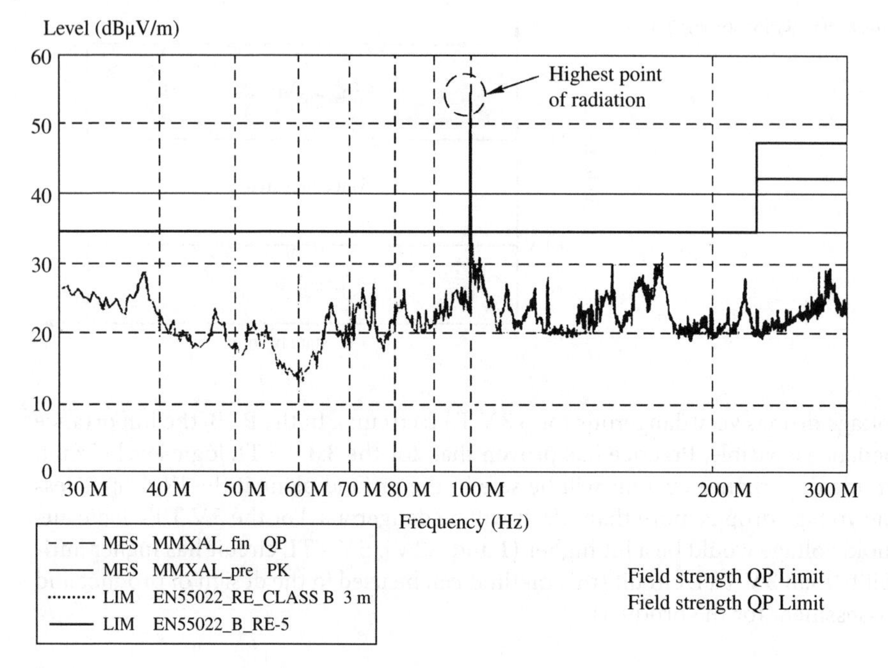

Figure 6.4 Spectrum of radiation harassment test.

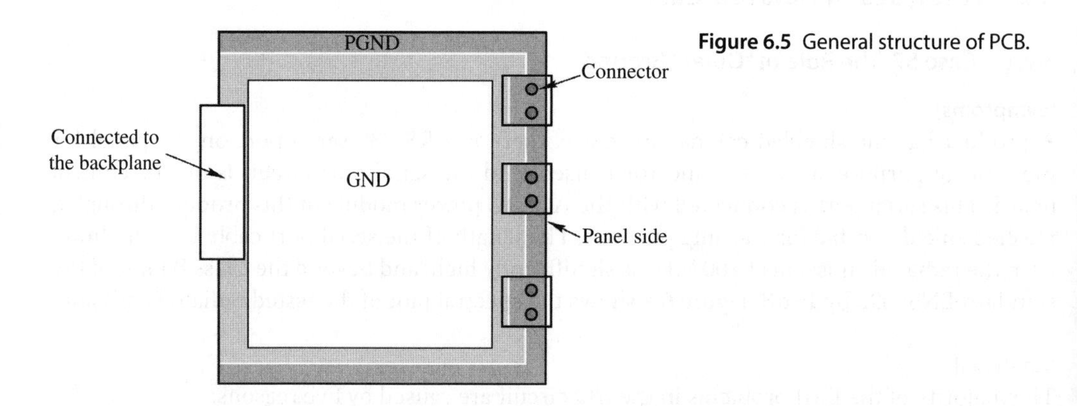

Figure 6.5 General structure of PCB.

that the poor contact between the metallic structure and the shield layer of the shielded cable causes the radiated emission, because this poor contact (disconnection) is a common EMC defect in the structure design. If a poor contact exists, high-impedance at this contact position appears; it eventually causes the radiated emission. Using a conductive copper foil to connect the control panel and the shield layer, the radiated emission is not reduced.

In the serial port cable, its shield layer near the RJ-45 connector is cut with a small opening, and receiving, transmitting, and ground (RX, TX, and GND) signal wires inside the cable are cut off, while the shield layer of the shielded cable is still connected to the metallic panel of the product through the RJ45 connector. The tested radiated emission at 100 MHz is reduced to 30 dBμV m^{-1}, indicating that the radiation in the shielded serial port cable is caused by the coupled signal from the inner signal wires of the cable to its shield layer. Connecting the GND wire that was cut off before, the radiation at 100 MHz increases to 62 dBμV m^{-1}, which proves the GND on the PCB of the control board is not "clean" but full of common-mode noises. Cut the connection between the GND inside the cable and the GND of the control board PCB, while using the copper foil to connect the GND with the protective ground (PGND) in the PCB board (PGND is routed as a round along the edge of the control board, and the GND is connected to the PGND with four zero ohm resistors and six capacitors in this product, so the circuit schematic is not changed), the tested radiated emission at 100 MHz is 61 dBμV m^{-1}, showing that the noise on the PGND of the control board PCB is considerable, too.

The GND in the control board is connected to the PGND by four 0 Ω resistors and six capacitors. While those components are removed during the test, the tested radiated emission is not improved. As the area of the PGND is big in the control board, it may pick up the noises from near field inside the control board. Remove the RJ-45 connector, cut 2 GND pins and extend pin 4 (GND) by 3 cm wire, and this wire is kept floating inside the cabinet, the radiated emission at 100 MHz is 46 dBμV m^{-1}. Connect this wire to the metallic panel, the radiated emission at 100 MHz is reduced to 37 dBμV m^{-1}, showing that most of the emissions at 100 MHz are inside the cabinet. Connect the wire to the GND of the PCB board, the radiated emission goes back to 57 dBμV m^{-1}. Change the wire to a 100 Ω/100 MHz ferrite bead, then connect it to the PCB's GND, the radiated emission is reduced to 51 dBμV m^{-1}. Adding another 200 pF capacitor cascading the ferrite bead to the metallic panel (i.e. the casing of the product), as shown in Figure 6.6, the radiated emission at 100 MHz is decreased to 32 dBμV m^{-1}.

Remove the ferrite bead shown in Figure 6.6, and then the radiated emission at 100 MHz goes back to 49 dBμV m^{-1}. Remove the capacitor and connect the ferrite bead again, which is directly connected to the metallic panel. The radiated emission at 100 MHz is reduced to 34 dBμV m^{-1}. As we can know, the connection methods shown in Figure 6.7 can meet the Class B requirements for the radiated emission test.

After dealing with the GND as above, and the RX and TX signals are connected again, the radiated emission is back to 54 dBμV m^{-1}, showing that the RX and TX wires need to be filtered and decoupled. Connect the same ferrite beads in series with RX and TX signal, the radiated emission is reduced by 4 dB to 50 dBμV m^{-1}. Add the ferrite beads, and then connect 220 pF capacitor between RX/TX signals and the metallic panel, which results in the filtering on the TX/RX wire. As shown in Figure 6.8, the radiated emission is reduced further by 17 dB to 33 dBμV m^{-1}.

For the circuit connection shown in Figure 6.8, if the ferrite bead is replaced by a short wire, the radiated emission is increased to 46 dBuV m^{-1}, meaning that the ferrite bead cannot be removed.

Separating PCB into different areas is an important technique to reduce RF coupling in between the sub-systems of the PCB. In this case, there is common-mode noise on the I/O signal and the GND, at the beginning of the design, the separation between circuits is already

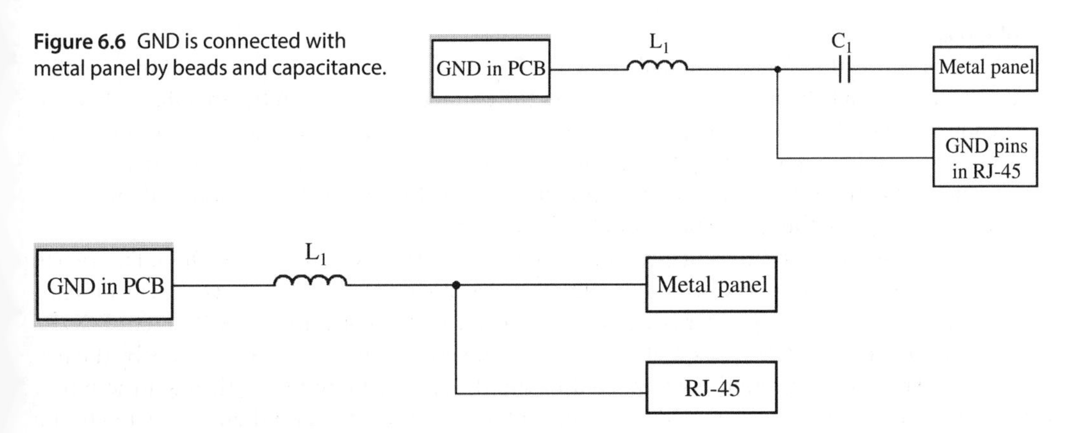

Figure 6.6 GND is connected with metal panel by beads and capacitance.

Figure 6.7 GND is connected with metal panel by beads.

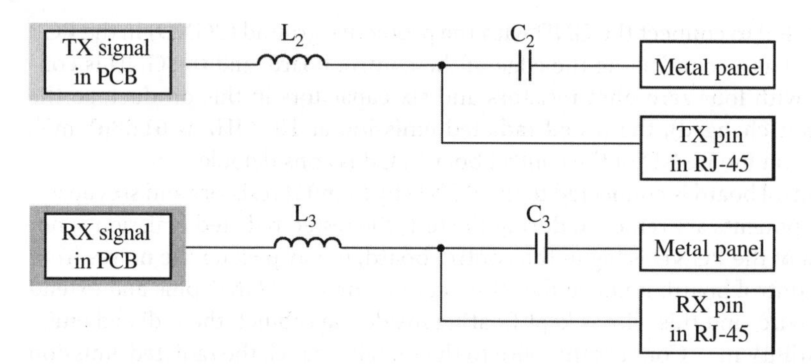

Figure 6.8 Filtering on TX, RX signal line.

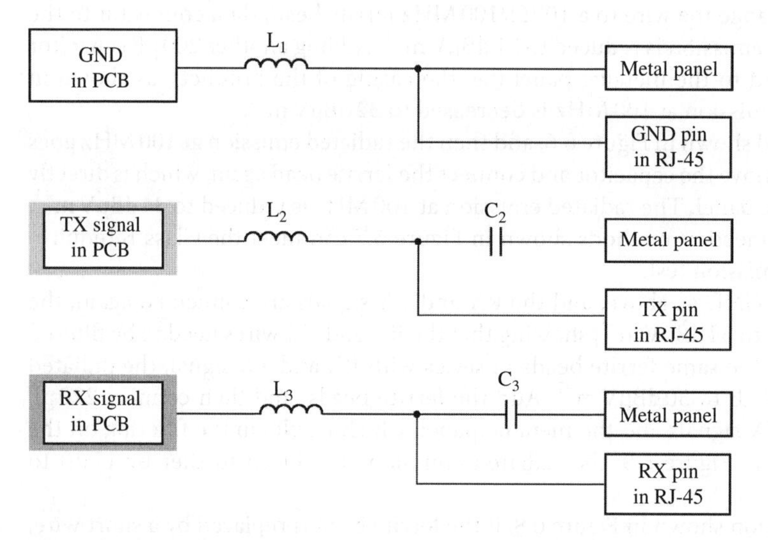

Figure 6.9 Easiest filter circuit of RS-232 serial port.

taken into account; there is an area for the "quiet" ground, in which there is no noise. However, as there is no filtering and decoupling circuit for the signals entering and outgoing the quiet ground area, including the RXD/TXD signals, the separated quiet ground is not effective.

[Solutions]

In this procedure, the "metallic panel" shown in Figures 6.6–6.8 is the extension of the metallic panel in the cabinet through a copper tape. As the connection between the metallic panel and the cabinet is good, the metallic panel can be regarded as an area without noise, so the extension of the metallic panel is a "quiet" ground for EMC. Moreover, the serial port needs surge protections, as the protection is the first priority and the filtering is the second, TVS needs to be connected before the ferrite bead and the capacitor.

In a word, the simplest filtering circuit for the RS-232 serial port can be shown in Figure 6.9.

In Figure 6.9, the ferrite bead with the characteristics of 100 Ω/100 MHz–0.1 Ω–500 mA is chosen for L_1, L_2 and L_3, the capacitor is a ceramic capacitor with the feature of 100 V–2200 pF-X7R.

In the design of the PCB, the "quiet" ground is connected with the PCB GND only by the ferrite bead. The "quiet" ground is connected to the RJ-45 metal connector through the pin of RJ45 connector, which is connected to the shell of RJ-45 connector, and then connected to the

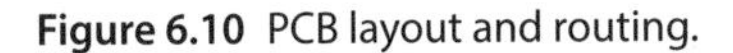

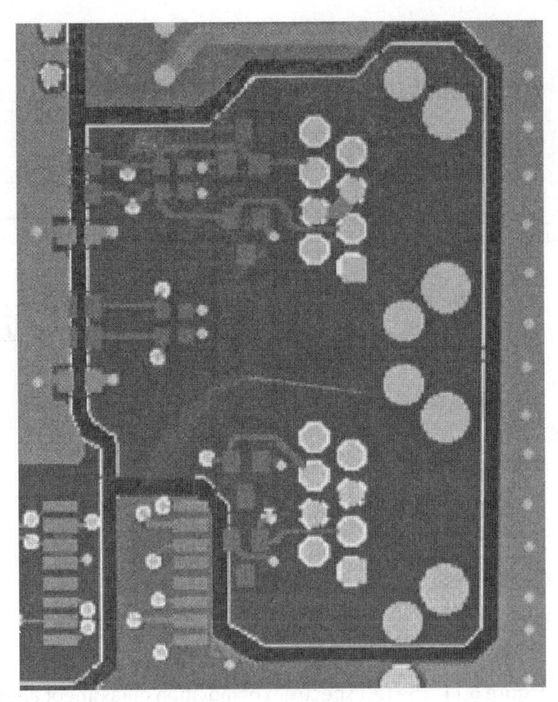

Figure 6.10 PCB layout and routing.

metallic panel through the reed of the RJ45 connector. The ferrite bead crosses the segment line between the PCB GND and the "quiet" ground. The actual PCB design can be shown in Figure 6.10.

After the modified PCB board is installed, the spectral plot of the tested radiated emission can be shown in Figure 6.11.

As can be seen from Figure 6.11, the radiated emission at 100 MHz is now reduced by 25 dB.

[Inspirations]

The schematic of the filter, the surge protection and the PCB design, which are finally used, can be a good reference for the serial port design of some products with large size casing, meanwhile note these eight items:

1) The RJ-45 connector must be a connector with metallic shell, and the places where the RJ-45 connector is installed shall be metallic.
2) The way with the quiet ground can only be used when there is strong high-frequency noise on the inner PCB GND; i.e. there is large noise on the inner GND of the serial port.
3) The area of the quiet ground should not be too large; otherwise, it may couple the noise from the internal PCB board and then work inefficiently.
4) In the quiet ground area, there shall be no power plane, GND plane, or any other signals that are not in the serial port.
5) The quiet ground should be connected to the metallic panel and the shell of the RJ-45 connector.
6) The ferrite bead interconnecting the quiet ground and the PCB GND must cross the segment line between the quiet ground and the PCB GND, and the ferrite beads on TX and RX signals should cross this segment line, too.
7) We must pay attention to the design of the quiet ground. If there are communication signals between the circuits in which these two grounds are separated, the way with the quiet

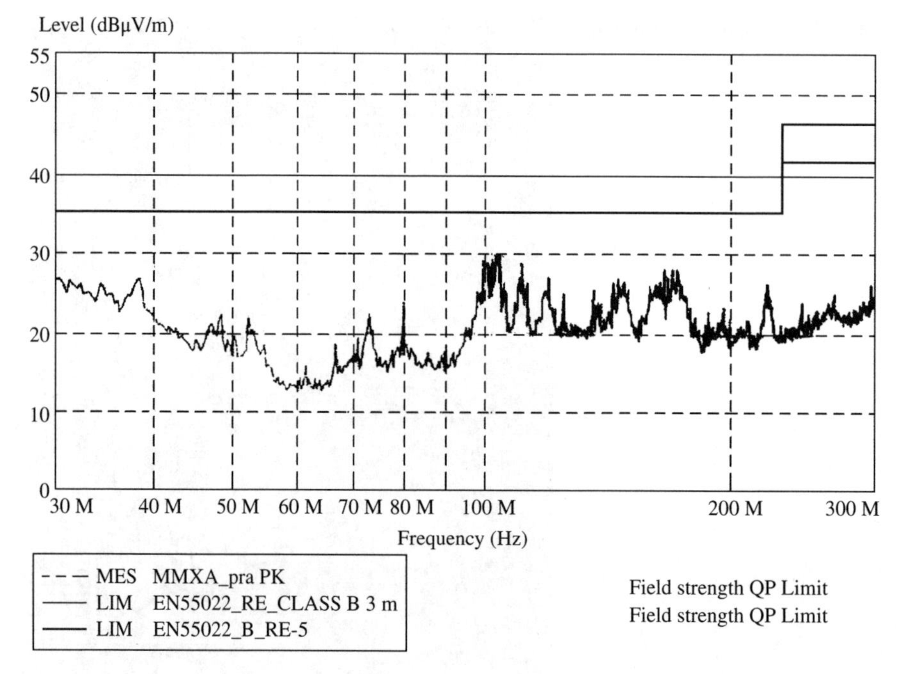

Figure 6.11 The test spectrum of radiation harassment after modifying.

ground and the ferrite bead crossing these two grounds is not suitable. In this case, the best way is to design these two GNDs as an equipotential, which may decrease the common-mode voltage, and only a capacitor interconnecting these two GNDs can compensate for the potential difference caused by the separated grounds at high frequency.

8) There must be no signal return current on the quiet ground, so the quiet ground can be regarded as the ground without noise. The quiet ground should be properly connected with the metallic casing, and the impedance (mainly the inductance) should be as small as possible. A multipoint connection can be used to ensure the quiet ground and the metallic casing is in equipotential. The quiet ground should be connected to the digital ground. A high-frequency bypass capacitor shall be paralleled between each I/O wire, including the signal wire and its return wire, and the quiet ground, the inductance in the decoupling circuit, should be as small as possible. The surface mounted capacitor can be used; therefore, the external disturbance like ESD, surge pulse, etc., which can be conducted to the circuit through the I/O wires, will bypass the metallic casing of the product by the bypass capacitor before arriving at the components area. Eventually, the internal circuit is protected. Meanwhile, the noise on the I/O wires, which is coupled with the PCB, is bypassed before being transmitted to the outside.

6.2.2 Case 58: The Loop Formed by PCB Routing Causes Product Reset During ESD Test

[Symptoms]

Figure 6.12 is the frame structure of a certain device. Various kinds of PCB boards are inserted in the frame box, and two of them are the same, which is called board A. These two boards are, respectively, used as the main board and the backup board for the master controlling. In this

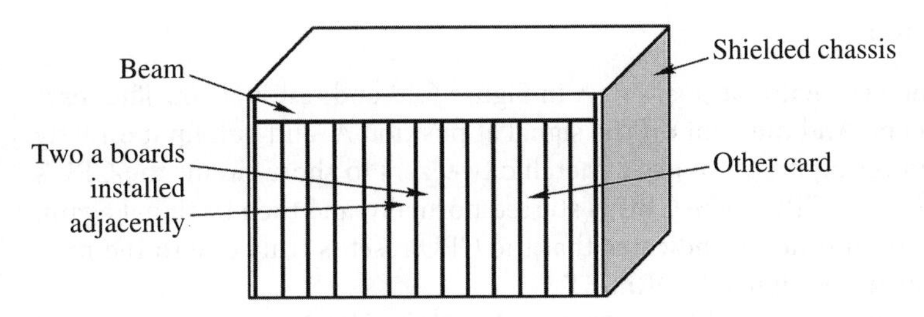

Figure 6.12 Frame structure diagram of equipment.

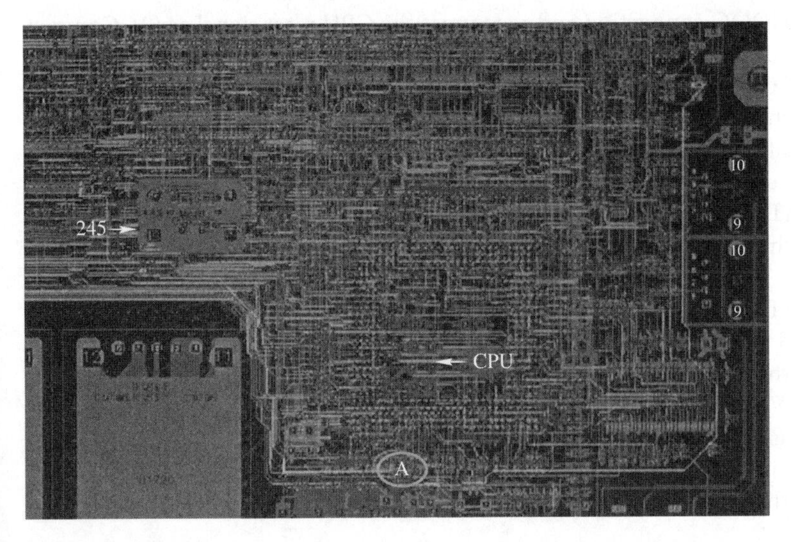

Figure 6.13 PCB routing.

device, an ESD contact discharge test is performed near the position where board A is installed. When the test voltage is ±6 kV, it is found that the reset phenomena appear on one board A, but the other one is in normal operation. Exchanging these two boards and testing them again, the result is that the prior problematic one is still reset and the other is still normal. The result shows that it is not due to the installation position of the PCB boards to cause the different immunity performance. Compare these two boards and find out that the PCB versions of these two boards with the same function are different, the version of the board with the reset phenomenon is version 0.2 and the version of the other one is version 0.1. Replace the PCB with reset phenomenon in the device with another version 0.2 PCB, the result is the same. So now we can get the conclusion that the problem is not accidental, but it is because there is the weakness in the version 0.2 PCB under the ESD disturbance test.

[Analyses]

As can be seen from the results, it is possible that the reset signal is disturbed. Then check the layout of the reset signal in this PCB. It can be seen from the PCB layout that the reset signal trace PRORESET is indeed long. The white highlight trace in Figure 6.13 is the PRORESET, which is routed under the CPU and then connected to the chip 245, and it is connected through a resistor to CPU from the intermediate position of the trace PRORESET to be the reset signal for CPU (inside the circle shown in Figure 6.13).

Carry out the following tests:

1) Add a 0.01 μF decoupling capacitor at position A in Figure 6.13 and test it again. The reset phenomenon still appears. And then cut off the signal at position A, and pull up it directly near the CPU. After powering it up, we use a metallic tweezers to short circuit the CPU's reset signal (namely reset the CPU). The CPU is started normally, and then testing it again, the reset phenomenon disappears. It indicates that the CPU reset is truly due to the mis-trigger caused by the disturbed signal PRORESET.
2) Connect PRORESET again and cut off the output of the WATCHDOG. Test it again, and the CPU is still reset. It indicates again that the reset is independent of the WATCHDOG output, but is caused by the interference coupled on the reset signal wire.
3) Recover the WATCHDOG connection and reconnect the CPU reset signal as (2). Connect PRORESET to an indicator light. If PRORESET is in low-voltage level then the light should lighten, so the signal can be detected. Then perform the test again. The CPU is not reset but the light is on; it shows that there is a low-level signal on PRORESET. Interrupt the WATCHDOG output again; the light is not on anymore. So there is indeed an output from the WATCHDOG, but why? To find the answer, we continue the test.
4) Disconnect the WATCHDOG MR signal and test it again. There is no output from the WATCHDOG, which means the reason for the output from WATCHDOG is that it receives a reset signal.
5) There is a SOFTREST signal (the I/O pin controlled by software) from the CPU in the input of the WATCHDOG MR signal, which is controlled by a reset button. Switch them off one by one to perform the test. It proves that the reset signal controlled by the reset button causes the reset of WATCHDOG, which means the output of the reset button (the bright light in Figure 6.14) is also disturbed.
6) Remove the resistor in series with the reset button and test it again. WATCHDOG is still reset. That is to say, the signal trace itself is disturbed, independent on the reset button. The

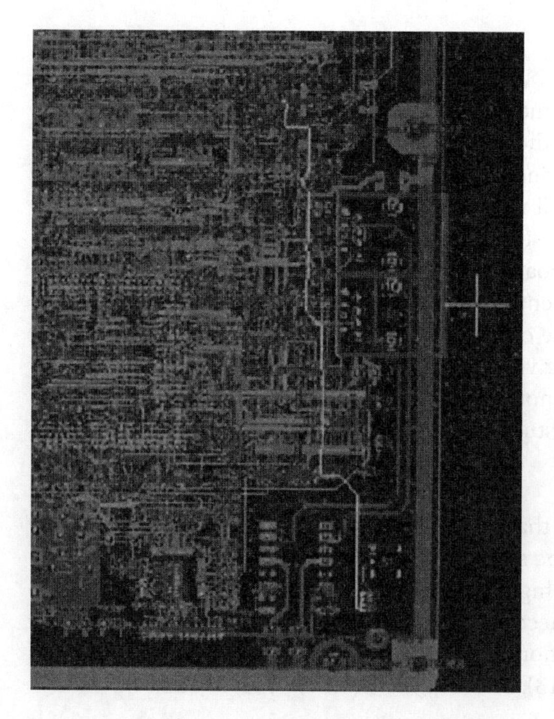

Figure 6.14 Routing of reset switch output lines.

output signal of the CPU reset button is close to the edge of the board, and passes through the hollow area of the ground plane in the ethernet interface circuit, which can easily couple the radiated disturbance during ESD test.

According to the above tests, the reset signal output of the reset button and the PRORESET output of WATCHDOG (the highlight traces shown in the above two figures respectively) are both disturbed by ESD disturbance. During the test, if any of these two signal traces is disturbed by ESD, the reset of the PCB board will appear.

The rising time of ESD current is shorter than 1 ns. When the discharge current flows through conductors, voltage pulse will appear on them ($U = L \cdot dI/dt$). These conductors can be power, ground, or signal traces. Those voltage pulses will be conducted into every components connected with these conductors. At the same time, the discharge arc and the discharge current flowing through the conductors will produce a strong magnetic field over the frequency range of 1 MHz–1 GHz, which can be inductively coupled to the nearby circuit loop nearby it. According to the test, at the position that is 100 mm away from the ESD discharge arc, the magnetic field strength can be up to tens of $A\,m^{-1}$. The electromagnetic field radiated from the discharge arc can also be coupled to the long signal traces. These signal traces act as the receiving antenna. In the ESD test for this device, the reset signal exactly couples the above-mentioned interference. That is why the device is reset.

As regards to the PCB layout of the reset signal, the following tests are added to explain the impact on ESD immunity performance of this product versus the position of the reset signal trace on the PCB.

First, use a fly-wire to replace PRORESET signal in the PCB, and put it far away from the edge of the PCB, as shown in Figure 6.15. Test it in this condition and there is no abnormal phenomenon. The test is passed.

As shown in Figure 6.16, use a fly-wire to replace the former PRORESET trace in the PCB and put it near the edge of the PCB. Test it again, and the product is reset again.

In the condition shown in Figure 6.16, maintain the others, only adjust the routing near the ethernet interface to keep it away from the hollow area of the ground plane in the ethernet

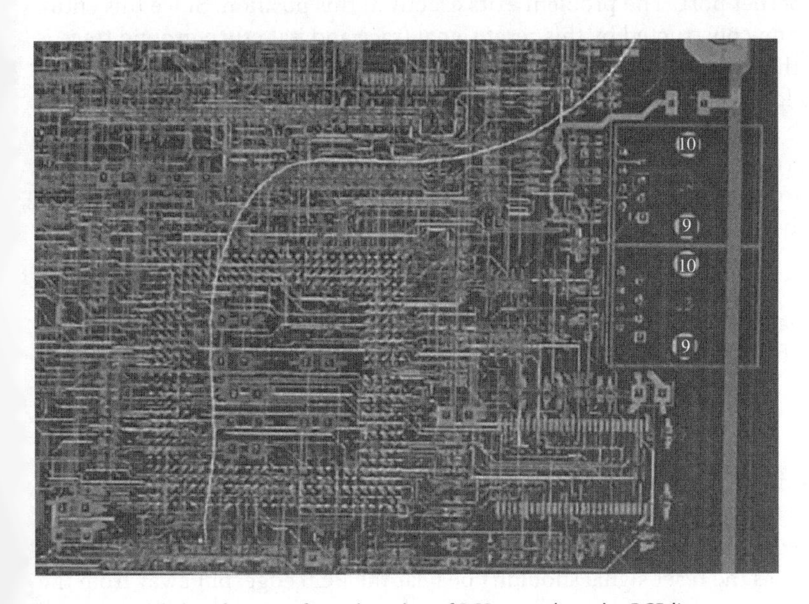

Figure 6.15 Fly line far away from the edge of PCB to replace the PCB line.

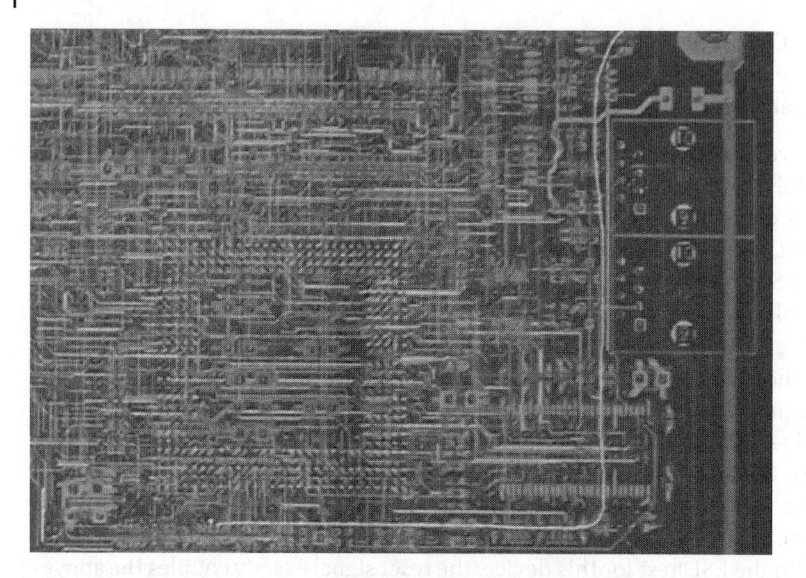

Figure 6.16 Fly line close to the edge of PCB to replace the PCB line.

interface circuit, which is shown in Figure 6.17. And perform the ESD test again. During the test, there is not any abnormal phenomenon. The test is passed. It indicates that the routing of the PRORESET signal traces that are close to the ethernet interface circuit is the main reason for this ESD problem.

In fact, in the condition shown in Figure 6.17, the problem can be solved if the distance from the signal trace to the ethernet interface is longer than 1 cm.

Now let's see the routing of the PRORESET signal at the ethernet interface, shown in Figure 6.18.

The white trace in Figure 6.18 is the output of WATCHDOG, which goes through the segment line under the ethernet port. The problem exits exactly at this position. Since this entire place is hollowed, the loop constructed by this reset signal trace and its return ground trace is too big. According to the principle of the electromagnetic induction in the closed loop, the induced voltage on this loop will be increased, i.e. the coupled disturbance is big. As long as this wire is routed out of the segment area and above the DGND plane, the problem will be solved. There exists the same problem for the reset signal from the reset button. The reset signal from the reset button goes through the hollowed area, as shown in Figure 6.19. This is why these two signals are easily disturbed.

[Solutions]

Just move the reset signal out of the segment area and route it on the area above the complete DGND plane and away from the edge of the PCB farther than 1 cm, and then the problem can be perfectly solved.

[Inspirations]

1) In a multilayer board, the current return plane of the signal traces should be the ground plane. Routing across the segmented ground area will increase the area of the signal loop, which will cause the circuit more sensitive to the external disturbance.
2) Sensitive signals such as the reset signal shouldn't be near the PCB edge, but away from the edge over 1 cm.

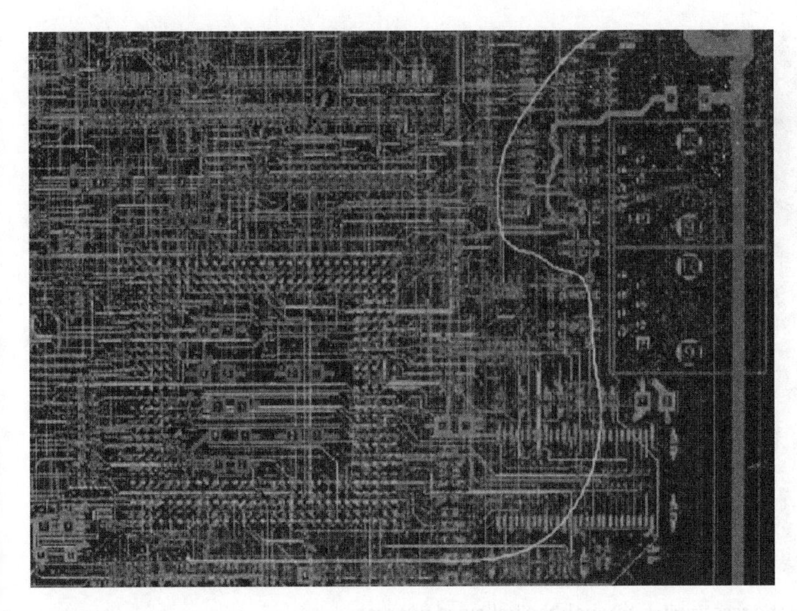

Figure 6.17 Fly line close to the edge of PCB but away from the network port to replace the PCB line.

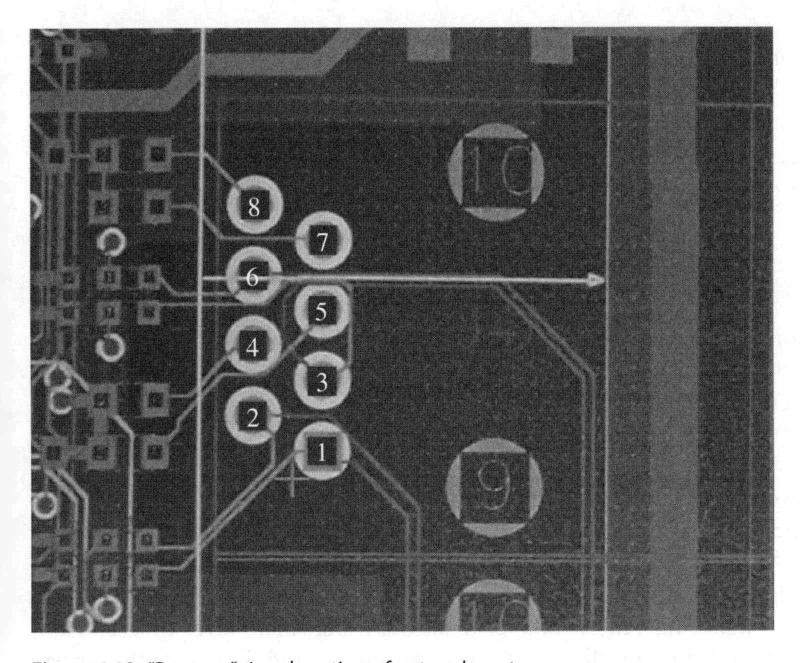

Figure 6.18 "Proreset" signal routing of network port.

6.2.3 Case 59: Unreasonable PCB Wiring Causes the Interface Damaged by Lightning Surge

[Symptoms]

After suffering one lightning surge event on site, the ethernet communication of a router becomes abnormal. After further checking, we find that the physical layer of the ethernet interface does not work normally. The test result shows that the receiving circuit of the ethernet

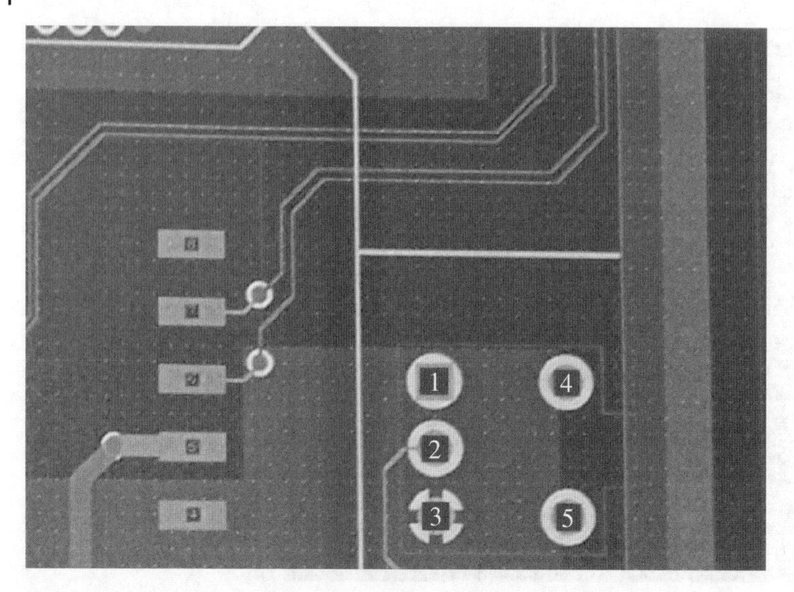

Figure 6.19 Reset signal of the reset switch through the hollowed out area.

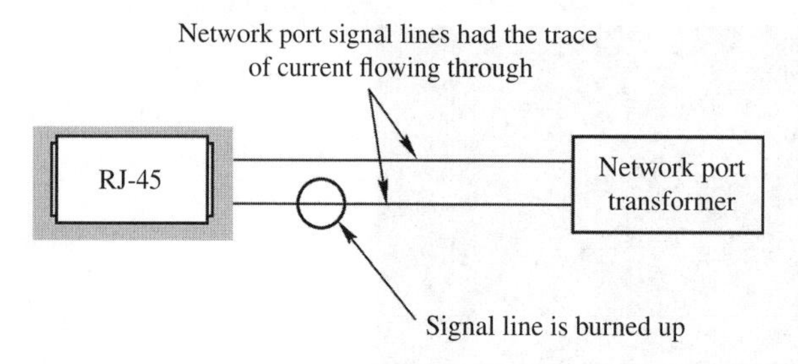

Figure 6.20 Connection diagram of ethernet receive line after it burned up.

PHY chip is already short-circuited to the ground. Besides, the ethernet receiving signal traces on the PCB top layer is burned out, and there is an obvious overcurrent track on this part of the PCB traces. The connection schematic after the burnout is shown in Figure 6.20.

[Analyses]

The PCB routing of the ethernet interface in this product is that the receiving signals between the RJ-45 connector and ethernet transformer are routed on the TOP layer, the transmitting signals are on BOTTOM layer, and the RJ-45 connector is equipped with a metallic shell. According to the investigation result of the lightning strike phenomenon, the overvoltage and overcurrent generated by lightning surge will appear on the cable shield layer first, and will then affect the internal circuit through the shield layer and the ground. But since the air gap between the metallic shell of the RJ-45 connector and the PCB is very small, the big current caused by the lightning surge causes a very high voltage drop on the cable shield layer, and the cable shield and the shell of RJ-45 connector are interconnected. The insulation between the ethernet receiving signal traces on the PCB and the metallic shell of the RJ-45 connector breaks down. The impedance of the loop that is broken down is small, so large lightning current passes through the signal traces, and then the signal traces are burned out. Meanwhile, the insulation between the signal wires (and the ground wires) and the shield layer of the cable will be broken down and the high common-mode voltage on the cable shield layer can be transformed into the different-mode voltage between ethernet signals; therefore, the PHY chip at the ethernet interface is damaged.

In the ethernet interface, since the different-mode voltage propagation characteristics of the ethernet transformer can be used as the isolation from the common-mode lightning overvoltage and overcurrent. When it is suffered from lighting strike, very high common-mode voltage will appear on the shielded cable connected to the ethernet port. If the dielectric strength between the high-voltage signal part and the other low-voltage signal wires is not enough, once the voltage between these two exceeds the maximum voltage, which the insulation can sustain, the insulation will be broken down. The insulation breakdown means that the voltage and the current will propagate from the high-voltage point to the low-voltage point, large current will be generated on the signal wires. The breakdown phenomena, including the voltage, the current and the impedance, are different on different signal wires. Exactly because of these differences, the common-mode voltage is transformed into the different-mode voltage, and for the different-mode signals at certain frequency, the isolation function of the transformer is lost. Then, high voltage will smoothly pass through the transformer to the internal circuit and damage the post-stage circuit.

The metallic shell of the RJ-45 connector covers around the ethernet interface. If the signal wiring is on the TOP layer, the insulation gap between the signal wires and the metallic shell of the connector is not enough to sustain the required withstanding voltage. Thus, if we use metallic RJ-45 connector, the signal wires linking the connector and the ethernet transformer must not be on the TOP layer. Otherwise, the common-mode protection based on the voltage-clamping technique or special insulation treatment is needed.

[Solutions]

Route the primary signal linking the ethernet connector and transformer on the inner layer of the PCB, in order to strengthen the dielectric strength between signal traces and the shell of the RJ-45 connector.

[Inspirations]

If the product is possible to suffer lightning strike and the ethernet connector with metallic shell needs to be used, we suggest that the signal wires linking the ethernet connector and the transformer should not be routed on the top layer and the bottom layer.

6.2.4 Case 60: How to Dispose the Grounds at Both Sides of Common-Mode Inductor

[Symptoms]

The spectral plot of the tested conducted emission on the 24 V DC power supply port in one communication module is shown in Figure 6.21. It can be seen from the plot that the average value over the low-frequency range exceeds the Class A limit.

[Analyses]

The schematic of the power supply input section of the product is shown in Figure 6.22.

In Figure 6.22, U_5 is the controller of the unisolated switching-mode power supply. It is considered as one noise source from the EMI point of view, in which the switching signal and the harmonics will directly influence the magnitude of the tested conducted disturbance. L_1 is the common-mode choke, which can filter the common-mode noise on the power supply port, and prevent the common-mode noises from being transmitted between its both sides mutually. For EMI, it's mainly used to suppress the disturbance from the switching-mode power supply, including the power supply and the 0 V, so as to achieve good EMC effect.

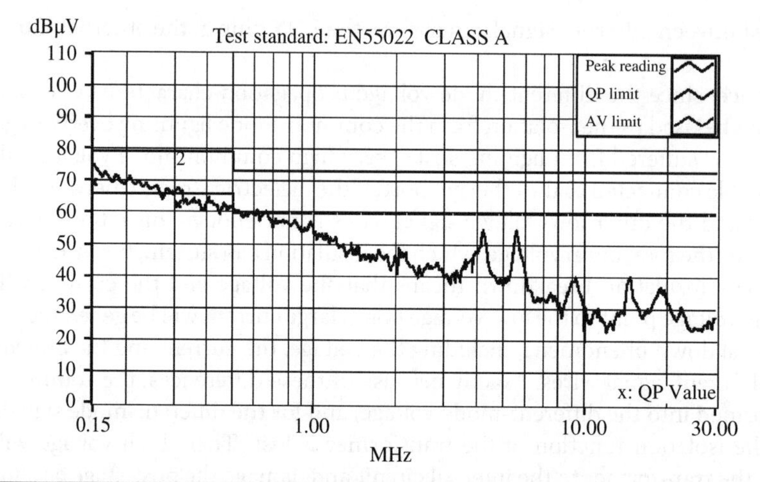

No.	Frequency MHz	Corr. factor dB	Reading dBμV QP	Reading dBμV AV	Emission dBμV QP	Emission dBμV AV	Limit dBμV QP	Limit dBμV AV	Margins dB QP	Margins dB AV	Notes
+ 1X	0.15000	1.96	68.43	65.51	70.39	67.47	79.00	66.00	−8.61	1.47	
2	0.31016	0.75	61.52	58.15	62.27	58.90	79.00	66.00	−16.73	−7.10	

Figure 6.21 Spectrum of conduction harassment test.

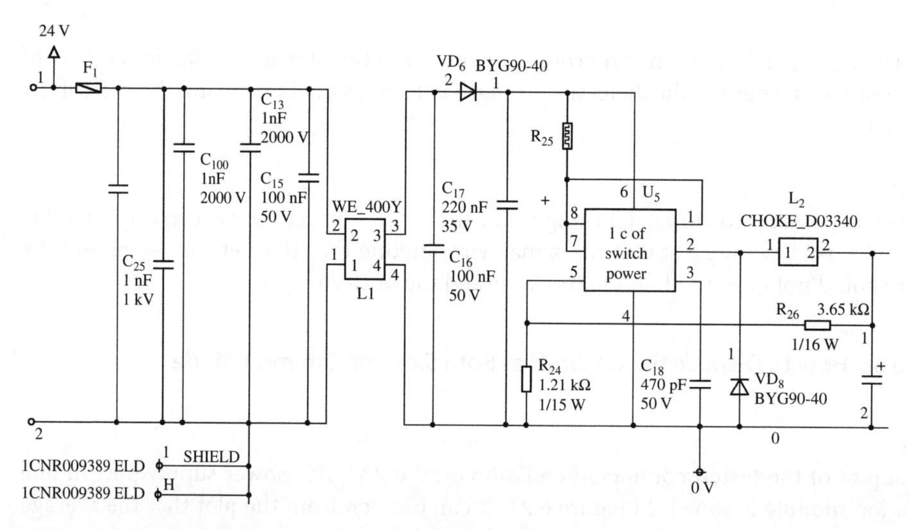

Figure 6.22 The principle diagram of the power entry.

The PCB layout of the power supply input port in this product is shown in Figure 6.23. In this figure, the white box represents the position of the common-mode choke L_1.

After careful analysis, it is found that it is not necessary to pour 0 V copper plane under the common-mode choke. Due to the small size of the common-mode choke, the area of the poured copper plane is also small, which is about 1 cm^2. Although this area is small, it will also play the role of capacitive coupling on the two separated sides of the common-mode choke, which causes the inductance of the common-mode choke to be lost to a certain extent.

The equivalent schematic illustrating how this coupling is formed is shown in Figure 6.24, where C_{s1} and C_{s2} present the distribution capacitors caused by the poured copper plane, which

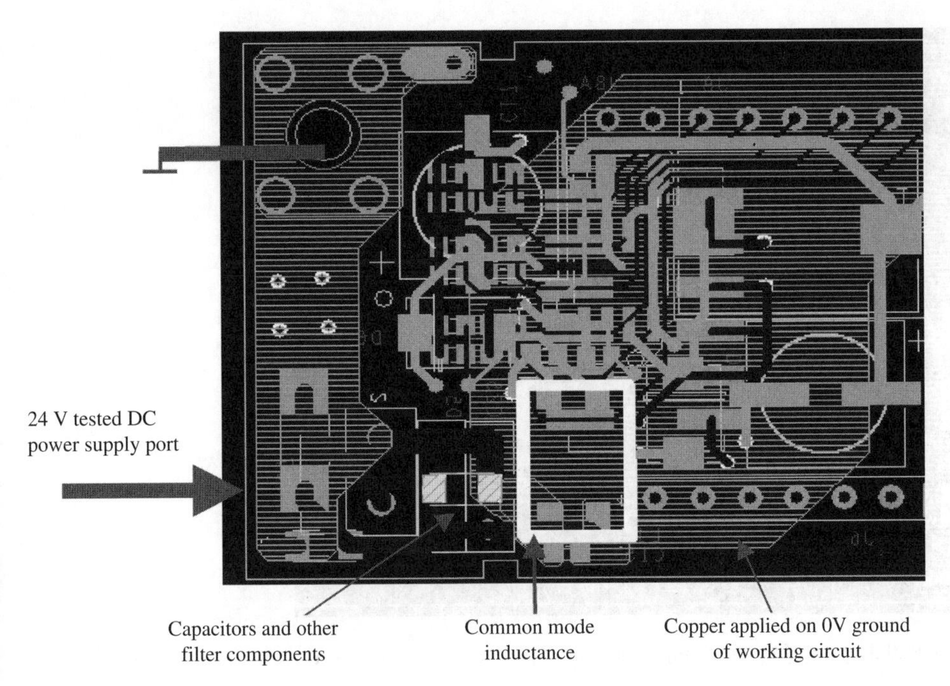

Figure 6.23 PCB diagram of the power entry.

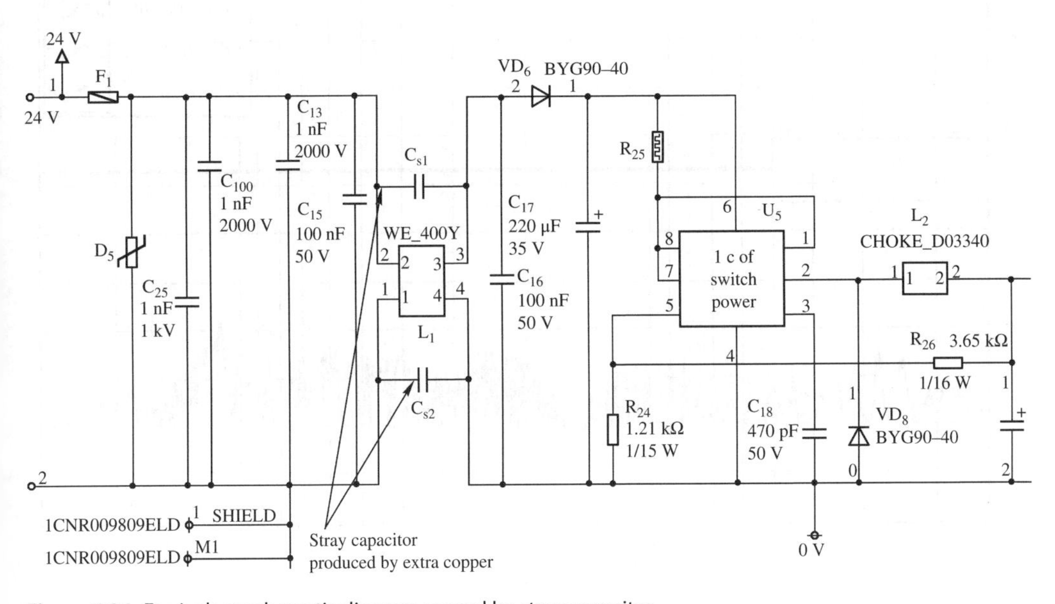

Figure 6.24 Equivalent schematic diagram caused by stray capacitor.

short-circuit the two sides of the common-mode choke at a certain frequency, so the noise from the post-stage power supply will pass through the distributed capacitor directly to the conducted disturbance test instrument.

In order to confirm the correctness of the analysis, modify the PCB, and remove the excessive poured ground plane, and the PCB layout after removing the excessive copper plane is shown in Figure 6.25.

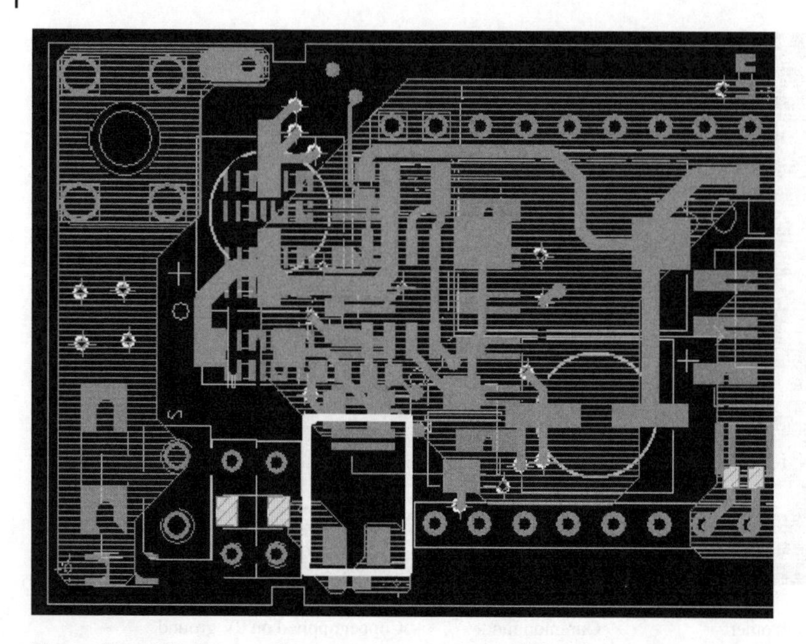

Figure 6.25 PCB diagram after removing extra ground plane.

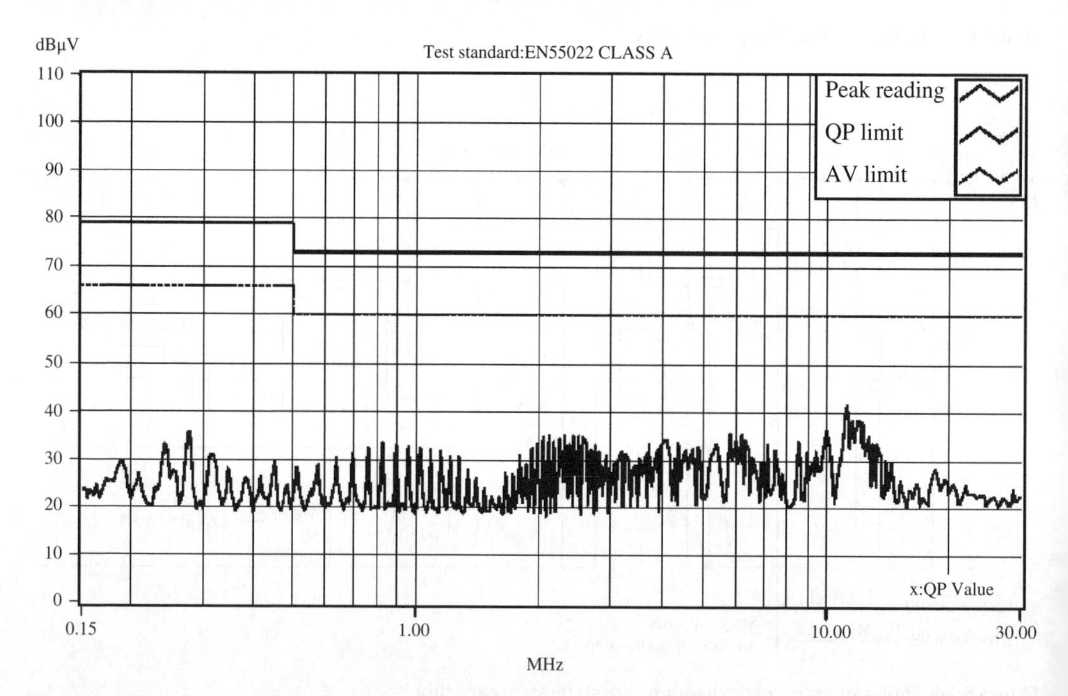

Figure 6.26 Test result after removing extra ground plane.

The test results after removing the excessive poured ground plane are shown in Figure 6.26. The tested conduced disturbance is greatly improved, which is incredible.

[Solutions]

Remove the excessive 0 V plane under the common-mode choke, and then the two sides of the common-mode choke are well separated.

[Inspirations]

1) The ground plane and power plane cannot be poured unintentionally, even if it is only a very small area.
2) Good separation between the input and output of the filtering circuit must be ensured in order to maximize the expected effectiveness of the filtering circuit.

6.2.5 Case 61: Avoid Coupling When the Ground Plane and the Power Plane Are Poured on PCB

[Symptoms]

A product uses the backplane structure, and other PCBs are inserted on the backplane and interconnected through the backplane. The backplane PCB is fixed on the front panel, and other PCB is vertically connected with the backplane. The product structure diagram is shown in Figure 6.27. The whole frame is supplied by the −48 V DC power supply. The −48 V power supply is transferred to each PCB inserted on the frame and connected to the backplane via the backplane. Among them, the main control board is the overall control system of the frame.

When performing the radiated emission test, it is found the radiated emission is high at 32.76 MHz, and its quasi-peak value is 53.8 dBuV m^{-1}, which is nearly 4 dB more than the Class A limit, as shown is Figure 6.28.

During the problem-locating test, it is found that the excessive emission will disappear if the main control board is not inserted into the slot, as long as the main control board is inserted, regardless of the configuration of the other PCB boards, the excessive radiated emission at that frequency appears. It is also found that the radiated emission at that frequency will also disappear when the ferrite ring core is suited with the power supply cable, which indicates that the emission at that frequency is radiated by the power supply cable, the noise source at that frequency is from the control board, and the coupling path may exist on the main control board or the backplane.

[Analyses]

To determine the coupling path of the radiation source, first, the backplane of the frame and the PCB on the main control board are examined in detail. After checking the PCB routing of

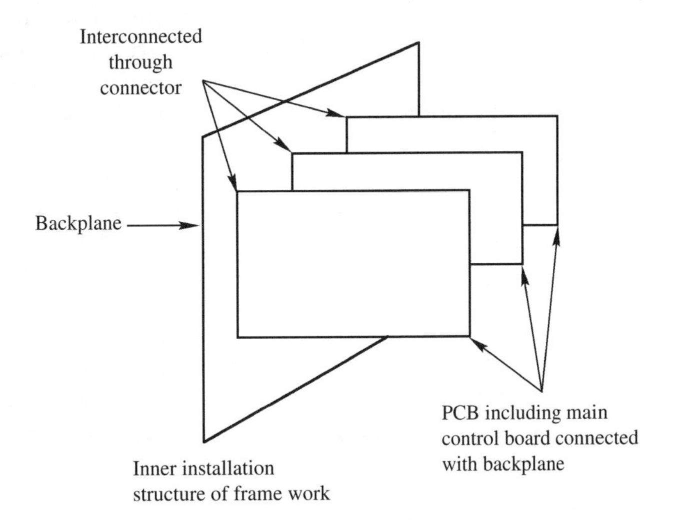

Figure 6.27 Structure installation diagram of the product.

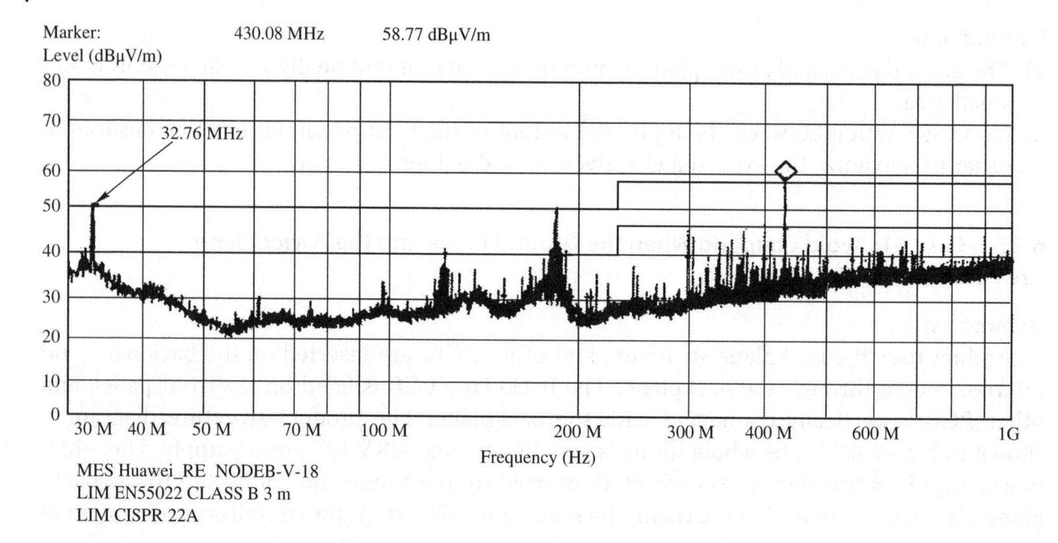

Figure 6.28 Spectrum of radiation emission test.

the backplane board and the main control board, there are two possibilities why the interference is coupled to the power supply cable:

1) The clock trace of the slot through which the main control board is inserted on the backplane is close to the −48 V power supply wires, at the same time, the separation distance from the backplane DGND is 50 mil, so the clock noise may be coupled to a power supply cable.
2) The impedance matching is implemented at both ends of the clock trace, which is pulled up to the VTT power plane through the resistor. The schematic of the clock output is shown in Figure 6.29

If the selection of the filtering capacitor for the VTT plane is not reasonable, the clock noise may be transmitted to the VTT plane, which is overlapped with the −48 V power supply plane with a large area on the main control board, and then the −48 V power supply plane is likely to be coupled with the interference.

After the above preliminary analysis, the following steps are followed to locate the problem:

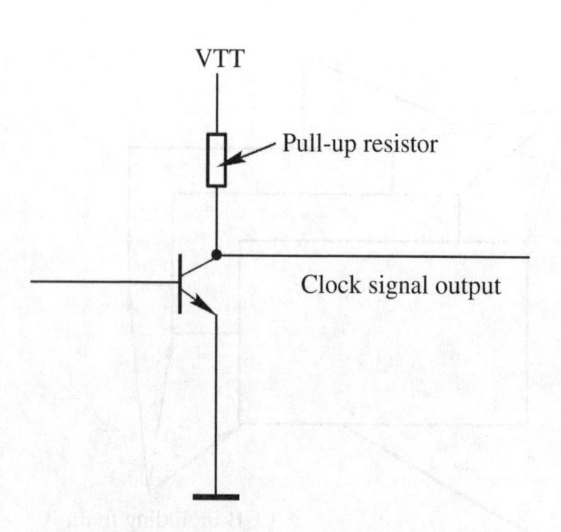

Figure 6.29 Schematic diagram of clock signal output.

Step one:

Optimize the filtering capacitors of the power plane to which the clock impedance matching resistor is pulled up on the frame backplane, and change them to a 0.1 μF capacitor in parallel with a 0.022 μF capacitor.

The impedance characteristic curve of the capacitors in Figure 6.30 shows that, the filtered frequency band after two capacitors are in parallel is around tens of megahertz. After modification, the test results with two capacitors in parallel are shown in Figure 6.31.

The result shown in Figure 6.31 is an improvement over the prior test result, which shows that the interference is related with the VTT power plane, but it needs further locating test to determine whether the coupling occurs in the backplane or the main control board.

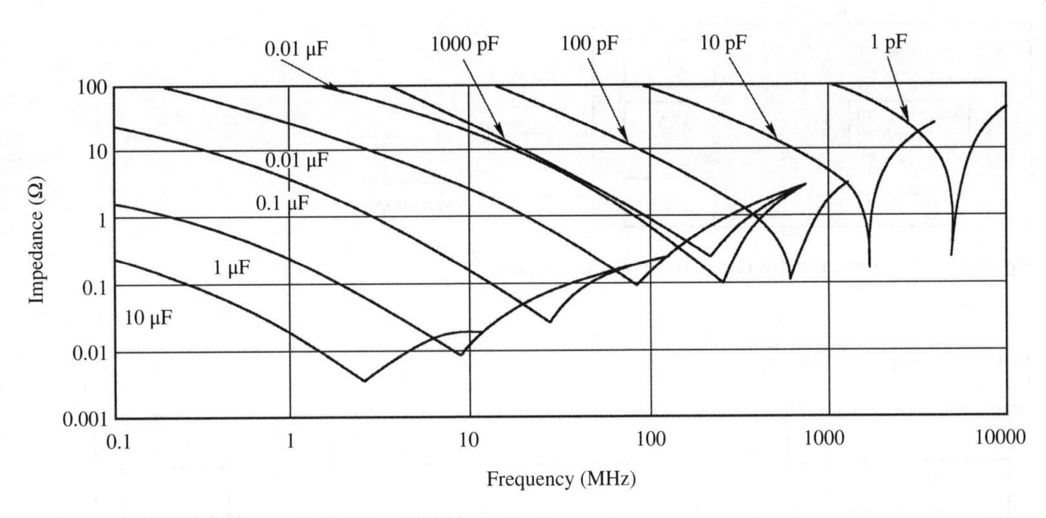

Figure 6.30 Capacitive impedance characteristic curve.

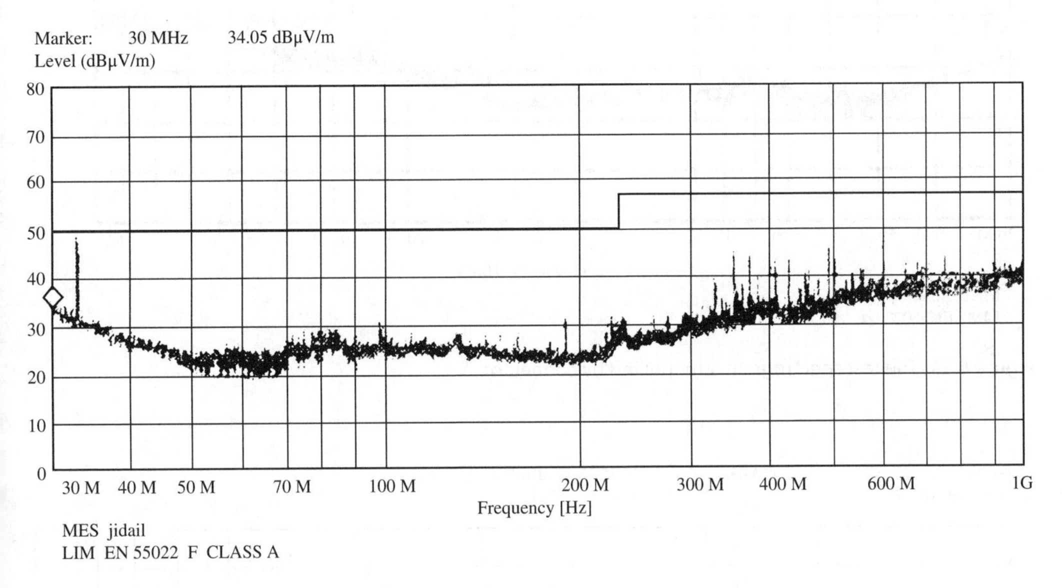

Figure 6.31 Test result after two capacitances are in parallel.

Step two:

Pull up the 32.768 MHz clock output of the main control board to the VTT plane through a special disposed connector, and then power on the main control board. The schematic is shown in Figure 6.32.

The test is done again after pulling up the clock to the VTT plane through the disposed connector. The test result is shown in Figure 6.33. The test is repeated after putting the ferrite ring core on the power cable, and the test result is shown in Figure 6.34.

So far, it appears that the problem is on the main control board, rather than on the backplane board, i.e. there exists the coupling inside the control board. It needs further locating test to determine whether the coupling is directly caused by the clock trace or the VTT power plane.

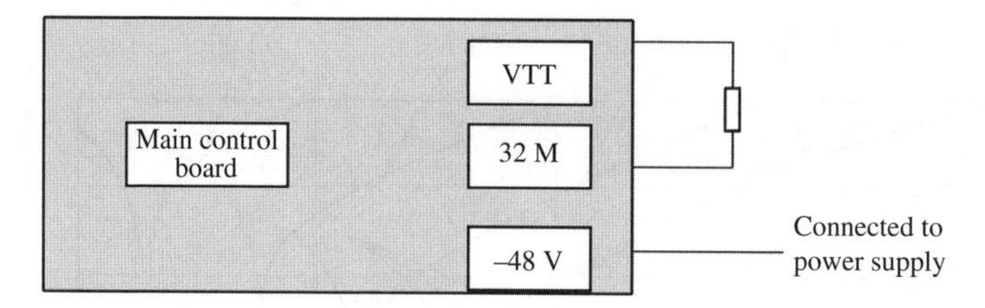

Figure 6.32 The principle diagram of pulling by connector.

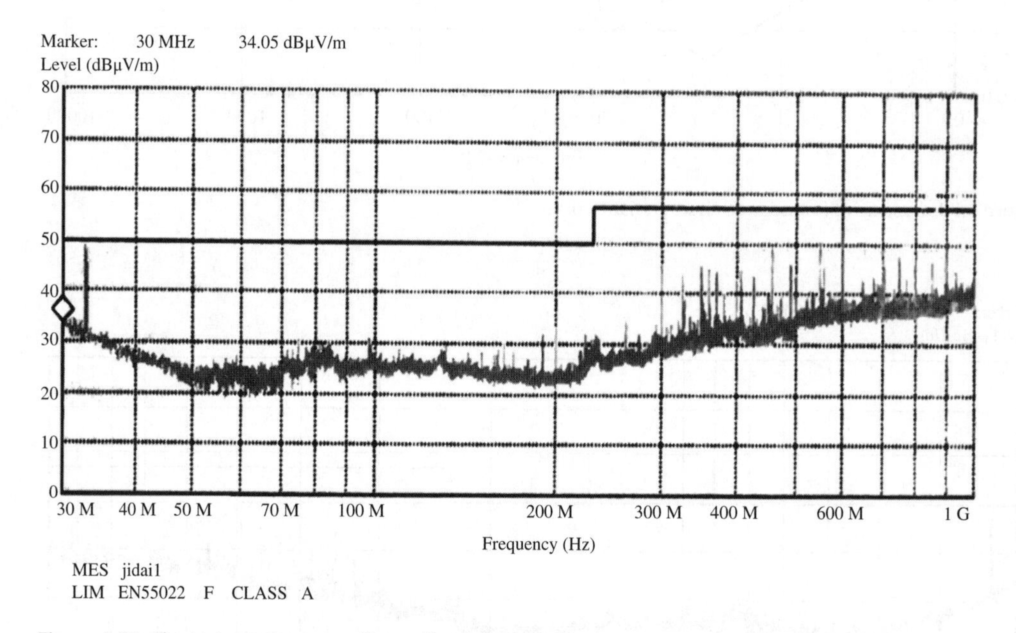

Figure 6.33 The test spectrogram after pulling by connector.

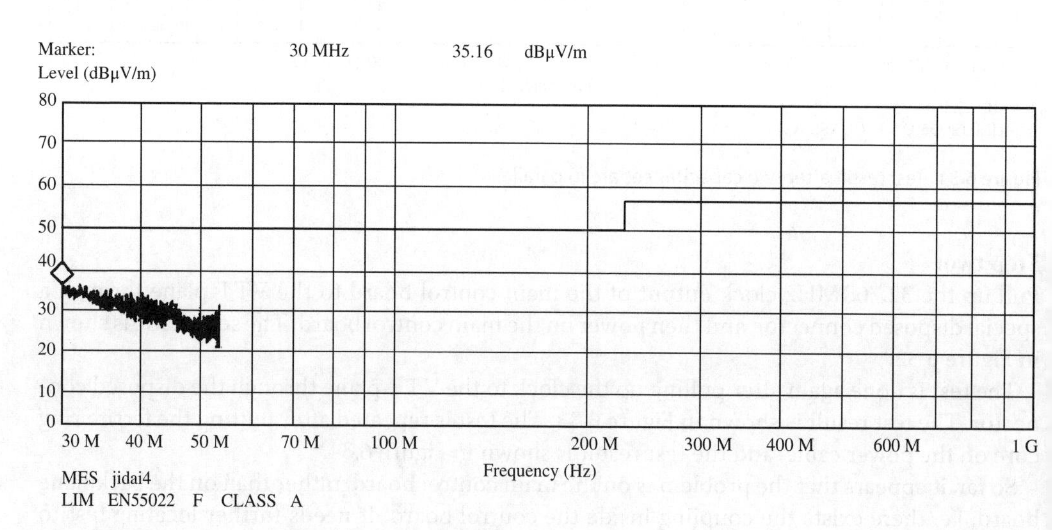

Figure 6.34 The test result after putting the ring on the power cord.

Step three:
Modify the main control board, and turn off its VTT power supply, and then the VTT plane is powered by and external linear power supply, as shown in Figure 6.35.

Do the radiated emission test again after power on the main control board. The result is shown in Figure 6.36.

The radiated emission at the 32.768 MHz almost disappears, which shows that the excessive radiated emission is not caused by the fact that the clock signal is directly coupled to the −48 V power supply plane, but the fact that the VTT power plane for the clock signal is interfered with by the clock signal and then the −48 V power plane is disturbed.

The results show that the radiation from the 32.768 MHz clock is caused by the fact that the clock noise is coupled to the main control board, and then coupled to the −48 V power plane through the VTT plane. After checking the main control board, it is found that there is a large overlapping area between the VTT power plane and the −48 V power plane, so the clock noise on the VTT plane is coupled, through the capacitive coupling, to the −48 V and the −48 −GND wires, and then the power supply cable becomes a good radiating antenna, which is directly connected to the −48 V and the −48 −GND power supply of the frame. The schematic how the clock noise is coupled to the power supply is shown in Figure 6.37.

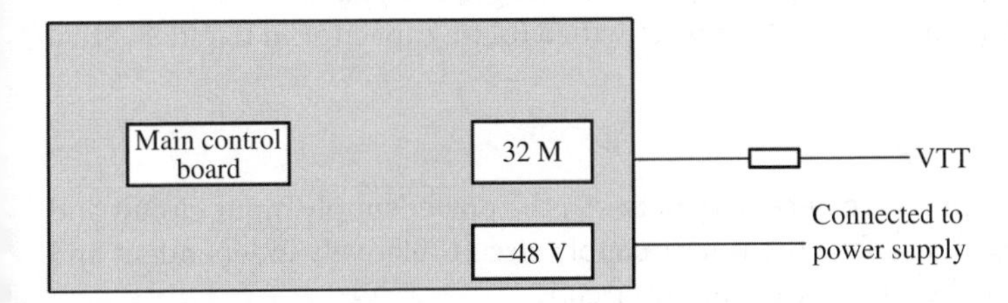

Figure 6.35 VTT supplied by external power.

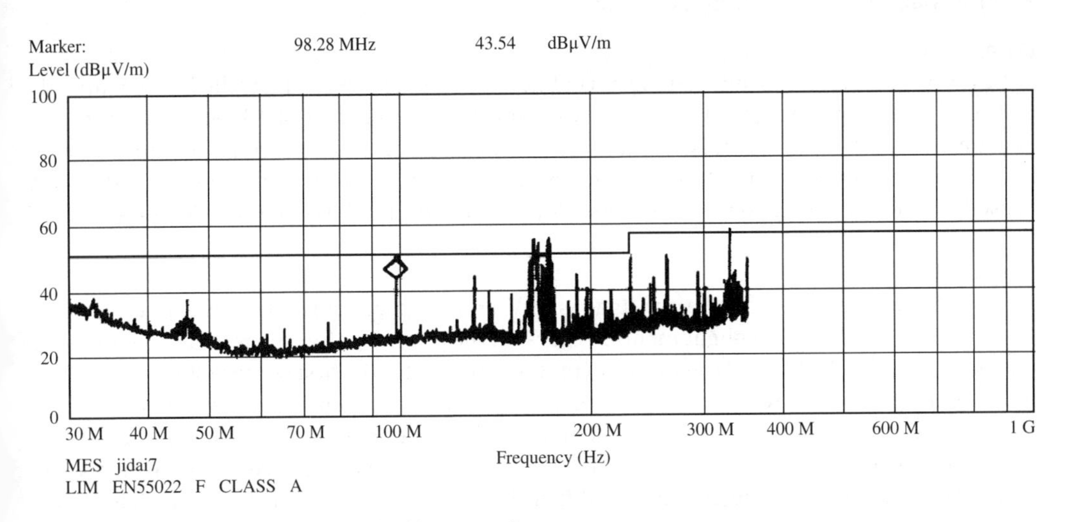

Figure 6.36 Test result after VTT supplied by external power.

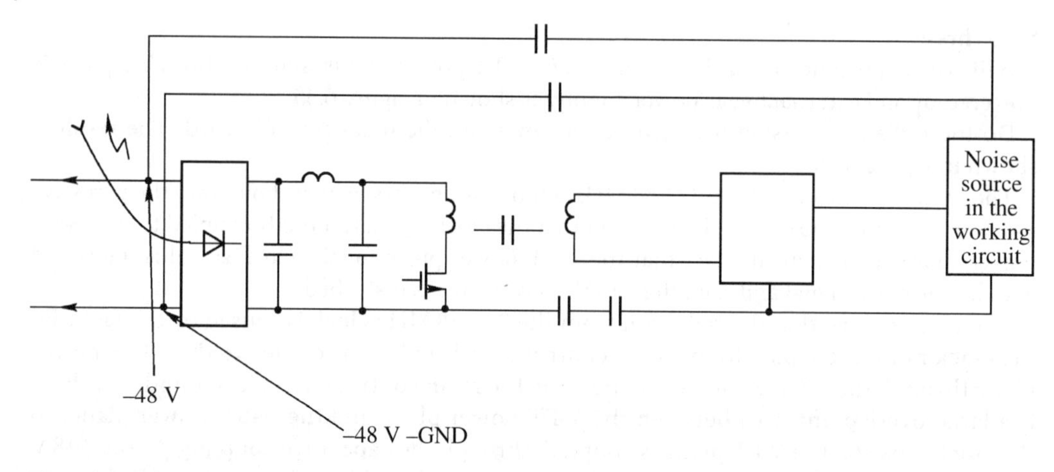

Figure 6.37 The principle diagram of clock noise coupling to power supply.

[Solutions]

1) Change the distribution of the VTT power plane in the main control board, and route it away from the –48 V power plane, so that there is no other plane on the area where the –48 V power plane is, except the –48 V power plane and its ground plane.
2) Optimize the filtering capacitor of the VTT power with a 0.1 μF capacitor in parallel with a 0.022 μF.

[Inspirations]

1) Good separation and decoupling are necessary between the power supply input circuit and the other circuits in the PCB, to make the power supply circuit relatively independent and avoid the PCB signal coupled to the power supply circuit.
2) For the isolated power supply, it is necessary to well isolate the signals and the 0 V trace.

6.2.6 Case 62: The Relationship Between the Width of PCB Trace and the Magnitude of the Surge Current

[Symptoms]

A 1.5 kV surge test is performed on the port of the clock signal in a product (the internal impedance of the surge generator is 12 Ω in the test, and the surge current of 100 A is shown on the instrument). After the tests, it is found that the interface under test cannot work normally no matter in the differential-mode test or in the common-mode test, and then the test is not passed. The RJ-45 connector is used in the clock interface, and the signal is differential.

[Analyses]

During the surge test, the first test voltage is 1 kV, and the surge current shown on the instrument is 70 A. No abnormal phenomenon occurs on the interface after the test. The second test voltage is raised to 1.5 kV, and the surge current is 100 A. After the first surge voltage is injected, it is found that no current data are displayed on the instrument in the subsequent tests. However, when it is turned to the test on another interface (not clock interface), the current data appear again, which shows that the problem may exist in the interface circuit. Doing some measurements with multimeter, it is found that the link between the pin of the clock interface connector and the protection device is broken up, so it can be concluded that it is the problem

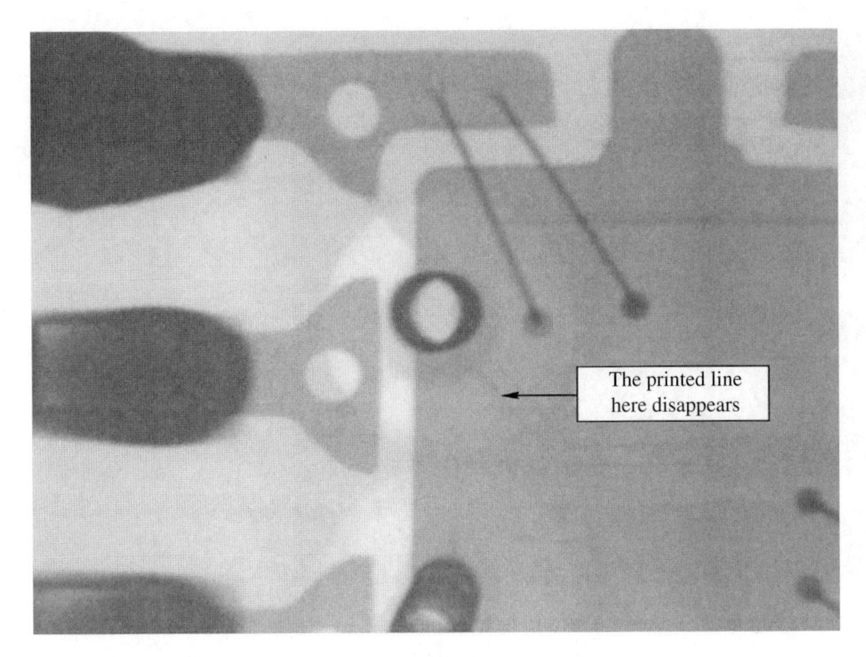

Figure 6.38 The X-ray scanning result after the first test finishes.

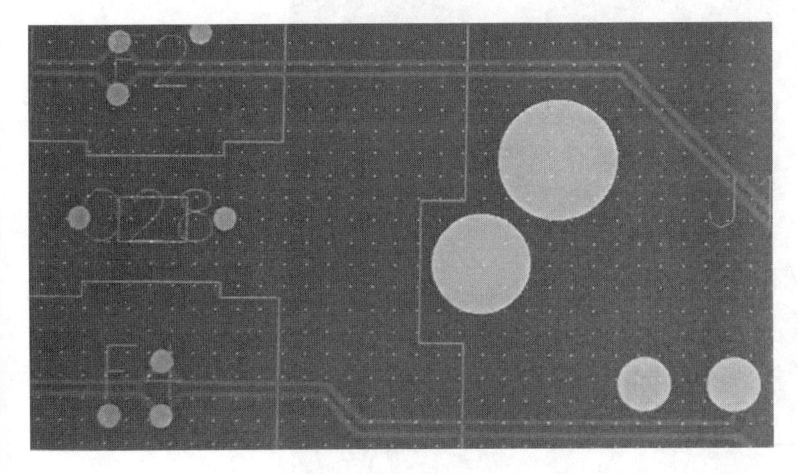

Figure 6.39 PCB diagram of interface circuit.

of PCB routing. In order to confirm the specific situation, X-ray scanning is performed on the interface board of the failed device. The X-ray scanning result is shown in Figure 6.38 after the first test.

It can be seen from Figure 6.38 that the routing in PCB, namely the printed wiring close to the protection device, which is near the corner in this figure, has been burned out, as shown by the black line. The PCB layout of the interface circuit is shown in Figure 6.39, the two pairs wirings shown above and below are the clock signals under test.

After checking, the width of this wiring is 5 mil, which meets the CAD design requirements. Furthermore, in order to verify whether the protection circuit can achieve the design goal, run the test again after replacing this wiring with a fly wire. It is found that the interface is in good

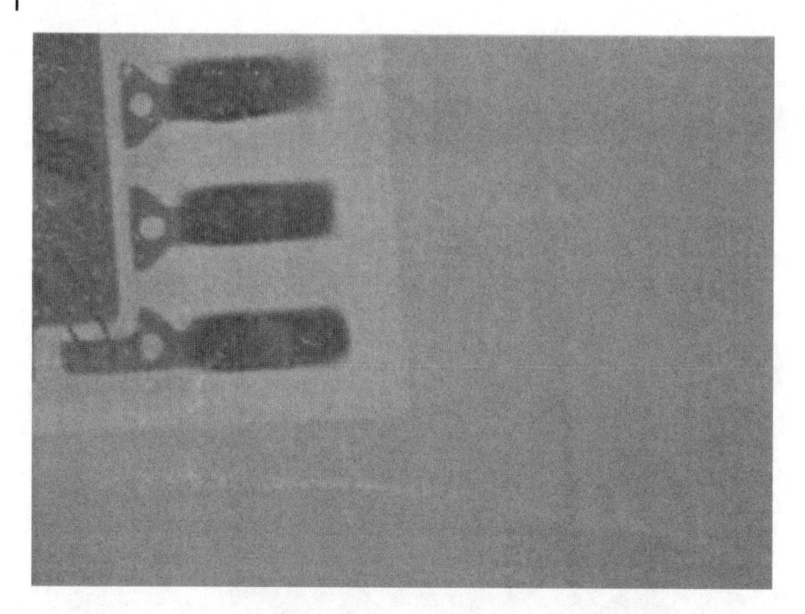

Figure 6.40 The X-ray scanning result 1 after the second test finishes.

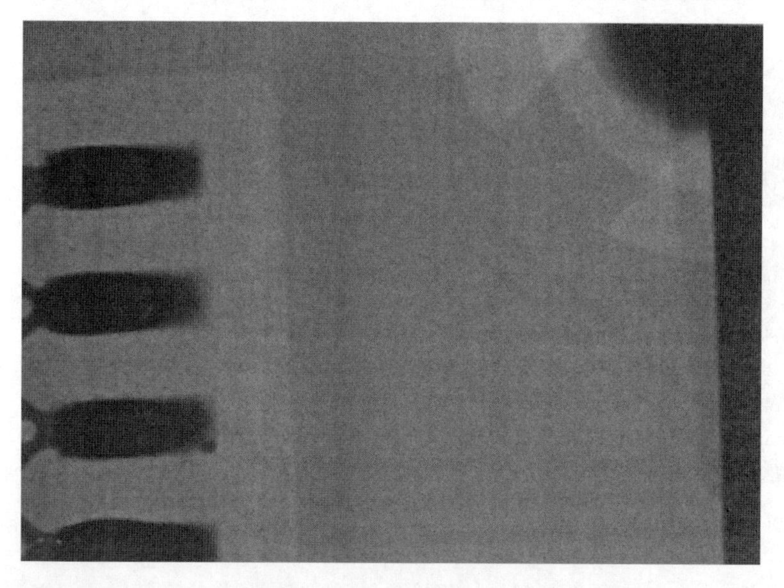

Figure 6.41 The X-ray scanning result 2 after the second test finishes.

condition and all operations are normal after applying the 1.5 kV surge voltage on the differential clock signal lines. In addition, it is found that the test results of the receiving side in the clock signal port is same as that of the transmitting side, and the PCB wiring is also burned out when 1.5 kV surge voltage is applied. After the second test, the X-ray scanning results are shown in Figures 6.40 and 6.41.

It can be seen that the width of the PCB wiring is too small and its current handling capability is not enough, which causes the PCB wiring is burned out when the transient large current passes through it in the surge test.

[Solutions]

Based on these experimental results, the small width of the PCB wiring and the insufficient current handling capability of the clock interface circuit cause the clock wirings are burned out when the transient large current passes through them. So it is easy to solve this problem, just increasing the width of the PCB wiring, such as to 10 mil.

[Inspirations]

The width of the PCB wiring from the interface connector to the protection devices should be as wide as possible, more than 10 mil is recommended.

6.2.7 Case 63: How to Avoid the Noise of the Oscillator Being Transmitted to the Cable Port

[Symptoms]

The radiated emission test is performed for a certain medical equipment. The spectral plot of the tested radiated emission is shown in Figure 6.42.

The measured interval of the frequencies at which the emission exceeds the standard limit is the crystal frequency, the same as the crystal frequency of the control board in this system. After testing, it is also found that the radiation is not directly radiated by the shell of the crystal but from the serial port line connected to the control board.

[Analyses]

The PCB layout of the control board, the routing path of the CPU, the serial port signal driver chip, the crystal, the serial port connector, and the serial port signals are shown in Figure 6.43. Obviously, this problem is because the serial port signal is routed under the body of the crystal.

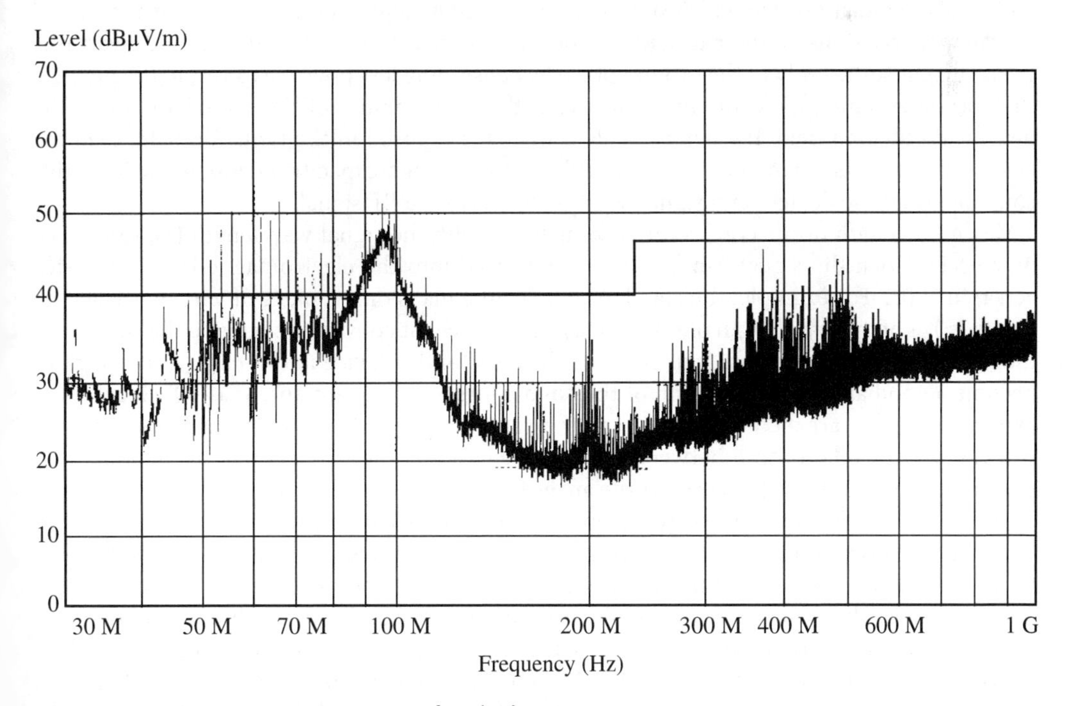

Figure 6.42 Radiation emission spectrum of medical equipment.

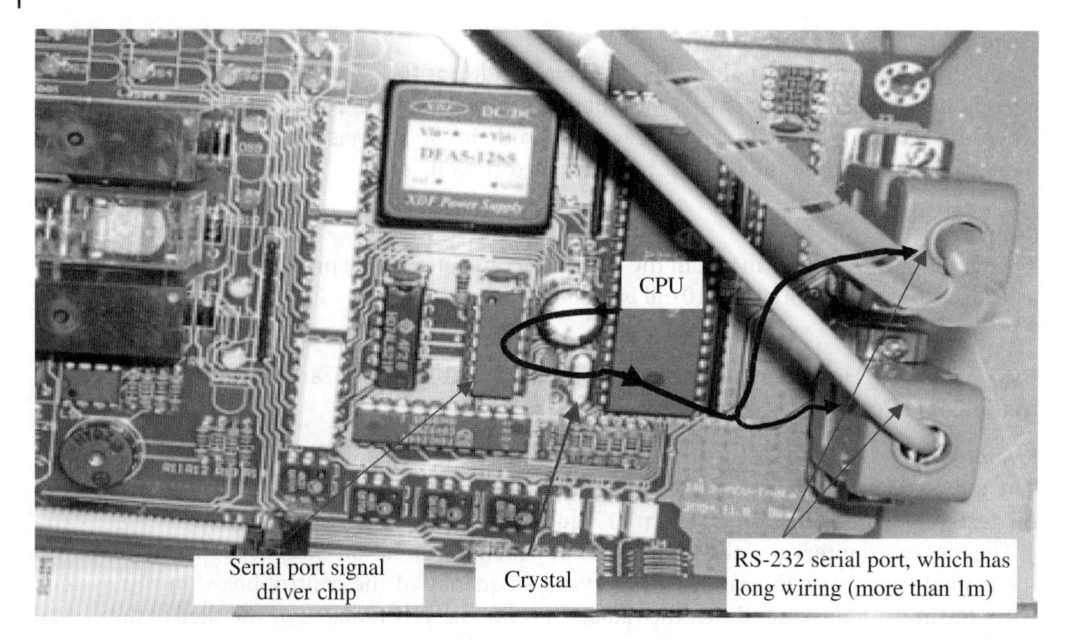

Figure 6.43 Part of PCB diagram of the control board.

The harmonic signal generated by the crystal is directly coupled to the serial port signal wirings, the serial port signal wirings become the carrier of the harmonics of the crystal, and the serial port signal wiring is very long, including the serial port cable. Thus, they have become a good radiating antenna, which brings the crystal's harmonics out of the PCB.

Crystal is a radiation source. The internal crystal circuit generates the RF current and the RF current generated inside the package may be large, so that the crystal pins cannot divert this great current sufficiently to the ground plane by consuming less power, as a result, the metallic shell becomes a monopole antenna. Therefore, the surroundings of the crystal are full of the near-field radiation field. If a device or PCB wiring is exposed in the radiation field, then the clock signal and its harmonics will be coupled on it through capacitive coupling or inductive coupling, i.e. these devices and signal wirings also carry the RF signal.

When the length of the conductor is comparable with the signal wavelength (λ), resonance may occur. Then the signal can be almost converted into the electromagnetic field (or vice versa) in 100%. For example, the standard dipole antenna is only a piece of wire, but when its length is 1/4 of the wavelength of the signal, it is an excellent converter changing the signal into the field. This is a very simple fact. For the cable used in the device in this case, it is enough to become a good antenna. In fact, all of the conductors are resonant antennas. Obviously, it is expected that they are inefficient antennas. If we assume that the conductor is a dipole antenna, you can use the relationship between the cable length and the efficiency of the antenna (the plot shown in Figure 6.44) to support the analysis.

In Figure 6.44, the vertical axis represents the conductor length (in units of m). In order to facilitate the observation, the spectrum is copied. The rightmost skewed line shows the relationship between the conductor length and the frequency when the conductor becomes an ideal antenna. Obviously, in the generally used frequency band, even a short conductor can cause the emission and immunity problems. It can be seen that, at 100 MHz, 1 m long conductor is a very effective antenna, and at 1 GHz, 100 mm conductor becomes a good antenna. In Figure 6.44, the middle skewed line shows that, although the conductor does not become an

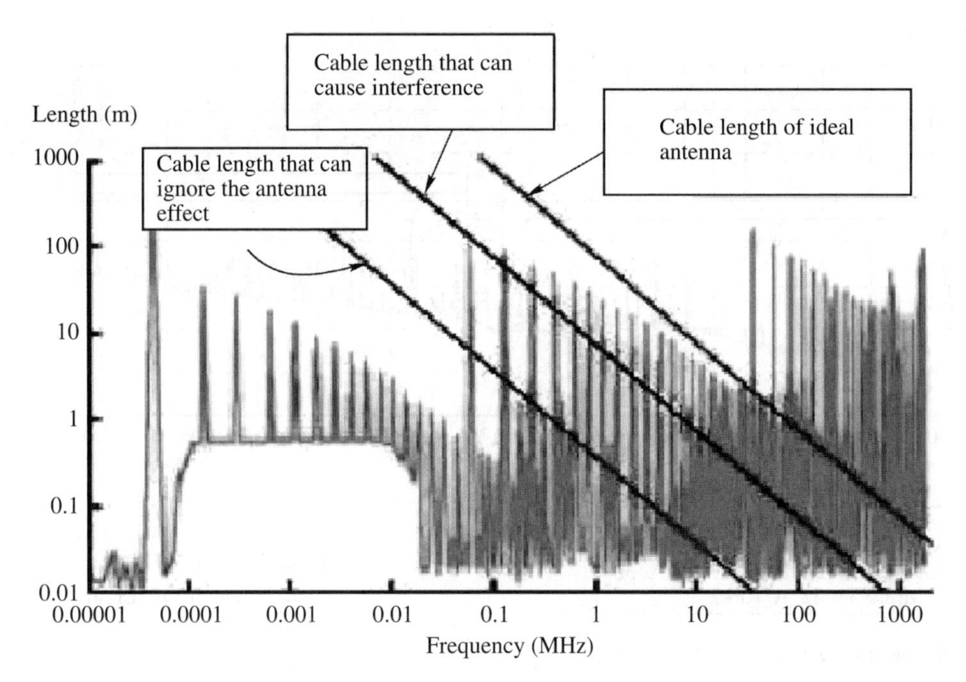

Figure 6.44 Relation between cable length and antenna efficiency.

efficient antenna, this conductor may still cause problems. The left skewed line shows that, when the conductor is very short, its antenna effects can be ignored (except for the particularly strict product).

[Solutions]

Based on the above analysis, the crystal is a noise source, which cannot be changed. The RS-232 serial port cable is an antenna, which also cannot be changed. But the coupling and driving relationship between the noise source and the antenna can be changed or eliminated, which should be taken into account in the design. Therefore, in this case, as long as the wiring of the serial port signal is away from the crystal (practice has proven that a distance of more than 300 mil can meet the basic requirements), the results will be satisfied, which is confirmed in the test.

[Inspirations]

1) Crystal has strong radiation, so the PCB wiring should be inhibited under its body and should be placed more than 300 mil from the crystal to avoid the crosstalk.
2) When there is no way to change the elements (the noise source and the antenna) forming the radiation, changing the driving relationship between the noise source and the antenna is also a feasible way to solve the radiation problem.

6.2.8 Case 64: The Radiated Emission Caused by the Noise from the Address Lines

[Symptoms]

It is found that there is large radiated emission at 37.5 MHz in the radiated emission test of a product. The spectral plot of the tested radiated emission is shown in Figure 6.45.

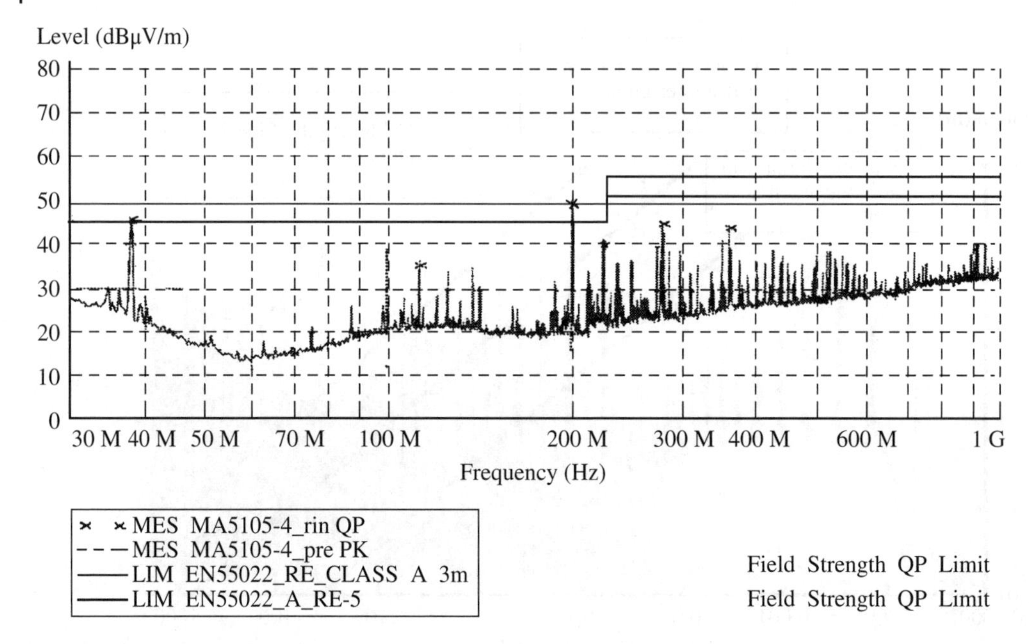

Figure 6.45 Spectrum of radiation emission test.

In order to find the root cause of the problem, some preliminary tests are carried out as follows:

1) Determine whether the signal cable is relevant. It is found that there is no change of the radiated emission at 37.5 MHz without the cable, which excludes the possibility of the radiation from the signal cable.
2) Determine whether the radiation from the power cable is relevant. On the condition that the signal cable is not connected, the decoupling clamp is put on the power cord, which does not influence the test results too, which eliminates the possibility of the radiation from the power cable.
3) Since 37.5 MHz is three times 12.5 MHz, it is suspected that it is related to the 25 MHz crystal of a PCB in the product (although it is not 12.5 MHz). Then disconnect the 330 Ω resistor in series with the crystal's clock output, and carry on the test again. It is found the radiated emission at 37.5 MHz disappears in the spectral plot, and the amplitude of the radiated emission at the frequencies near 37.5 MHz becomes smaller, too.
4) For further verification, the 330 Ω resistor is recovered, and the high emission at 37.5 MHz appears again, which shows that the radiated emission at 37.5 MHz is related to the 25 MHz clock.

[Analyses]

Since the device has the 37.5 MHz clock signal and the 25 MHz crystal has no direct relationship with 37.5 MHz, the general view is that it is caused by the resonance of the cabinet cavity.

According to the phenomenon in the test, it is proved that the radiation at 37.5 MHz is related to the 25 MHz clock signal indeed. And putting some microwave absorbing materials into the cabinet does not influence the test result (using absorbing material is to verify whether it is caused by the resonance of the cavity, because the absorbing material can change the original characteristic of the resonance, which can be used in the locating of the radiation). So it is concluded that the resonance is not related to the radiation. After further analysis, the propagation diagram of the 25 MHz clock signal is shown in Figure 6.46.

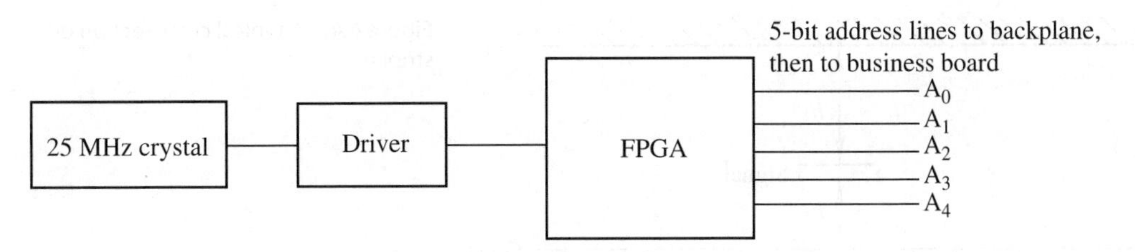

Figure 6.46 Flow diagram of 25 MHz clock signal.

The address lines of A_0, A_1, A_2, A_3, and A_4 come out from the FPGA. While there is no service, there will be a regular signal as 01010101 on A_3/A_4 according to the requirement of the protocol. It is triggered by the rising edge of the 25 MHz clock, and its frequency is 12.5 MHz, and 37.5 MHz is just its third harmonic. In the protocol, once the state of A_0, A_1, and A_2 is changed, a data with 1 F is added. Then this signal is not a periodic square wave.

It is generally considered that the signal on the address signal lines is not periodic, so the frequency spectrum of the address signal is continuous and the energy is relatively scattered, then its radiation is generally lower. While the frequency spectrum of clock and other periodic signal is discrete and its energy is relatively centralized, the radiation is higher at its harmonic frequencies. Considering that the frequency spectrum of the aperiodic address signal is continuous and low-emissive, the address signals of the product are routed on the external layer of the PCB, and the wiring length is long. However, due to the particularity of this product, the address signal becomes a periodic signal same as the clock signal. Thus, the harmonic of the address signal causes a large radiation due to the periodic address signals and the long distance of their wiring. In the test, it is also found that the radiation at 37.5 MHz disappears while the matching resistor at the starting terminal of A_3/A_4 address lines is removed. So the inference is confirmed correct.

[Solutions]

In the test, the root cause of the 37.5 MHz radiation is that the frequency of the FPGA address signal is half of the clock frequency 25 MHz, which can be equivalent as a 12.5 MHz periodic clock signal, and its long wiring on the external layer leads to the larger radiation. Since it is difficult to change the signal in the product, only the routing of the address signal can be changed, namely, the address signal wiring of the PCB is changed to the inner layer (this is a six-layer PCB).

[Inspirations]

1) Address signals are not always nonperiodic, and sometimes it can become a periodic signal and act as a high-radiation source.
2) The data signal is the same as the address signal, and a similar problem may also occur. So in the design, the clock signal, the address signal, and the data signal should be taken into account as a whole.
3) High-frequency signal can be routed in the inner layer of the six-or-more-layer PCB. Strip lines routed in between reference planes are used for the routing of transmission lines, which is the most common routing way in PCB design. The following figure shows the comparison relationship of the return current density on these two reference planes (the magnitude of the return current density impacts the common-mode radiation) when the distances from the signal line to the upper plane and the lower plane are different, and this result is calculated with the electromagnetic field numerical simulation tool, which can be a reference for the readers.

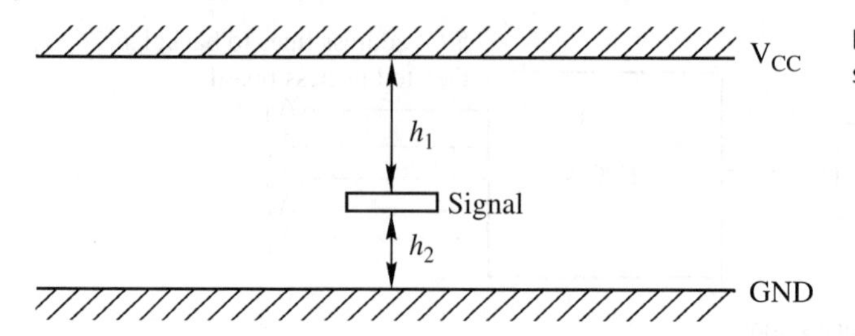

Figure 6.47 A typical cross section of stripline.

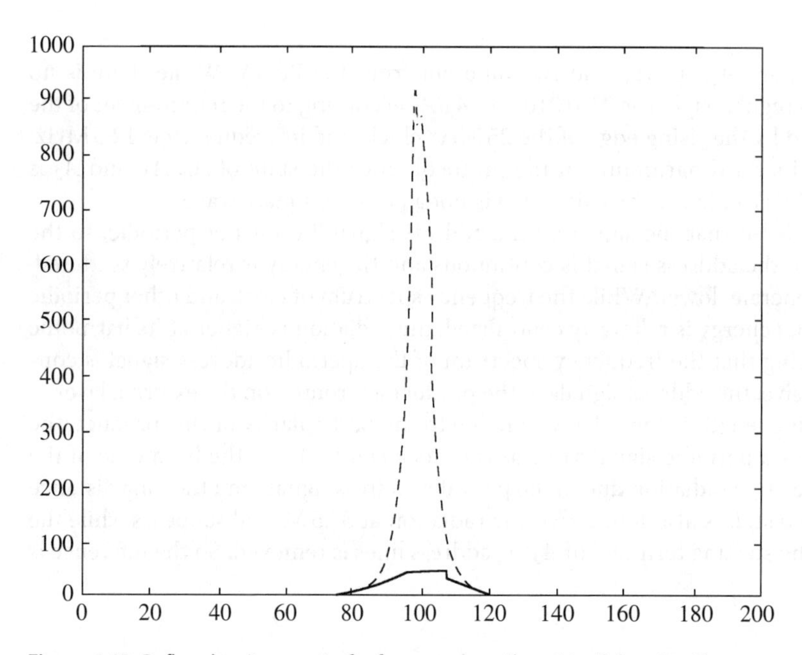

Figure 6.48 Reflux density map 1 of reference plane ($h_1 = 14\,\text{mil}$, $h_2 = 2\,\text{mil}$).

Figure 6.47 is the cross-sectional view of a typical strip line.

For ease of description, the lower reference plane is named as GND, the upper reference plane is named as VCC, and the signal line is named as signal. The distances from signal to VCC and GND are named h_1 and h_2, respectively. The width of the signal is 5 mil, and the dielectric of the PCB is FR4, which is commonly used.

Simulate the following four cases:

1) $h_1 = 14\,\text{mil}$, $h_2 = 2\,\text{mil}$
2) $h_1 = 12\,\text{mil}$, $h_2 = 4\,\text{mil}$
3) $h_1 = 10\,\text{mil}$, $h_2 = 6\,\text{mil}$
4) $h_1 = 8\,\text{mil}$, $h_2 = 8\,\text{mil}$

In these four cases, the comparison relationship of the return current density in these two reference planes, GND and VCC, are shown in Figures 6.48–6.51, respectively. Among them, the dotted line is the return current density in GND. The solid line is the return current density in VCC. Since the comparison relationship is considered only, it does not need to care about the specific units.

By comparing the return current density of the reference plane shown in Figure 6.48–6.51, it can be seen the signal return current is concentrated on its closer reference plane. Among

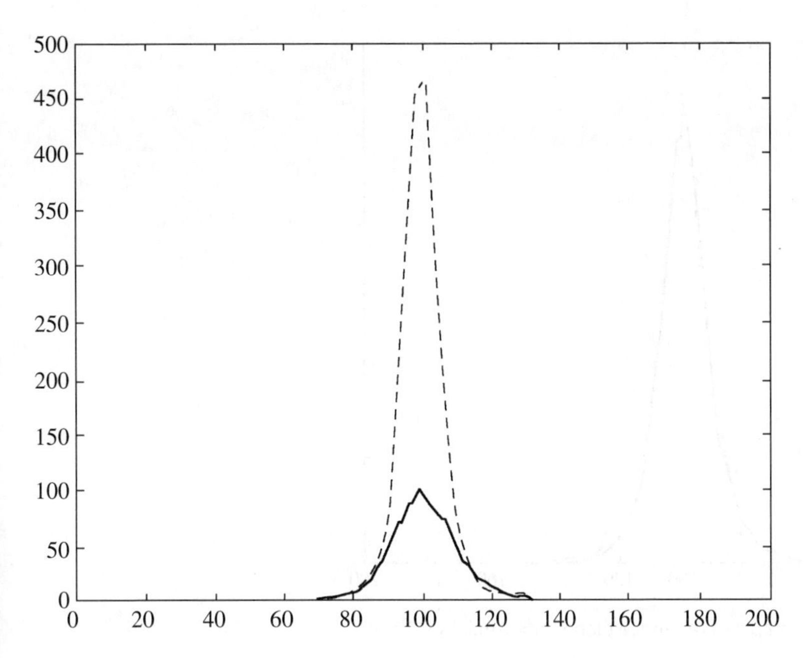

Figure 6.49 Reflux density map 2 of reference plane ($h_1 = 12$ mil, $h_2 = 4$ mil).

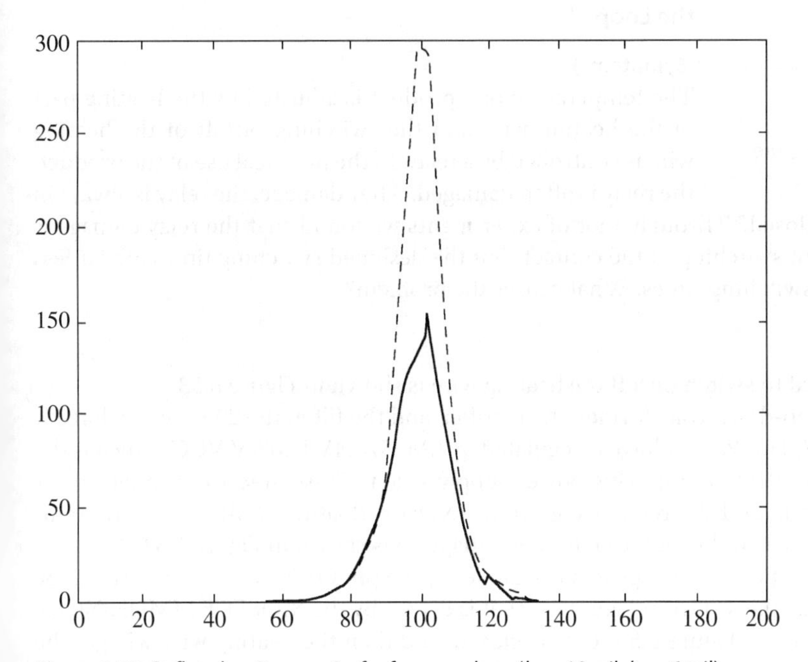

Figure 6.50 Reflux density map 3 of reference plane ($h_1 = 10$ mil, $h_2 = 6$ mil).

them, the situation illustrated in Figure 6.48 shows that in the actual routing, with the PCB stack-up shown in Figure 6.52, the return current of Signal1 is concentrated on VCC reference plane, and the return current of Signal2 is concentrated on the GND reference plane according to the simulation result. The VCC power plane is often incomplete in practical applications, so we must pay special attention on the high-speed Signal1.

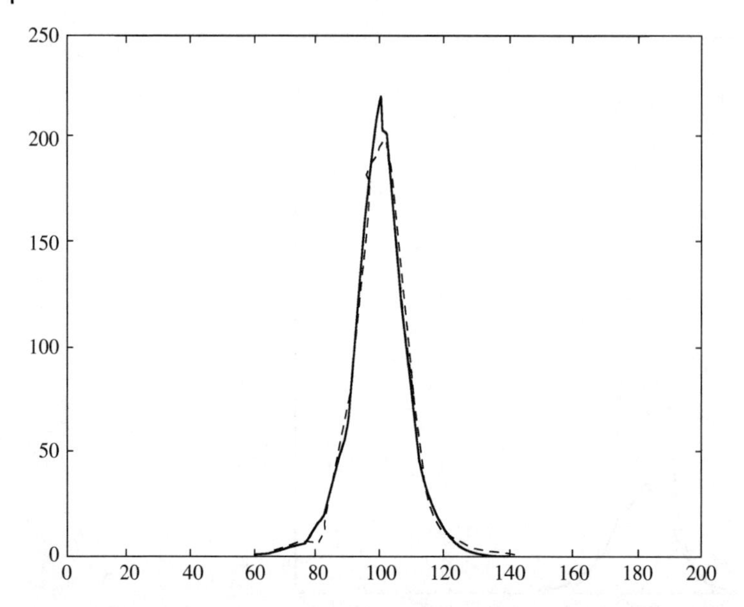

Figure 6.51 Reflux density map 4 of reference plane (h_1 = 8 mil, h_2 = 8 mil).

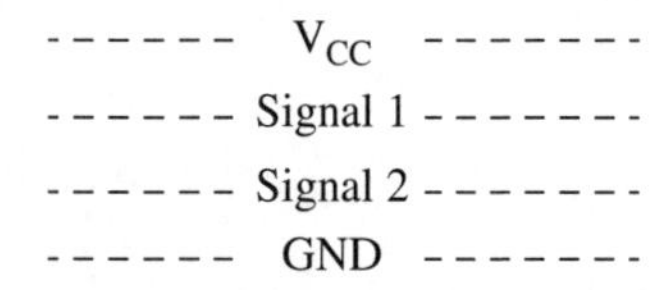

Figure 6.52 PCB layer structure.

6.2.9 Case 65: The Disturbance Produced by the Loop

[Symptoms]

The temperature of a product is adjusted by the heating part of the heating wire and the switching on/off of the heating wire is controlled by a relay. In the practical use of the product, the relay is often damaged. After damage, the relay is always in the state of "contact closed." Through a lot of experiments we found that the relay damage is due to the too-frequent switching of the contact. But the designed switching times are far less than the relay's rated switching times. What causes the problem?

[Analyses]

The relay control board to switch on/off the heating wire is shown in Figure 6.53.

After the isolated step-down transformer, the rectifier, and the filter, the 220 V AC voltage is converted to U_{IN}=28 V. This 28 V voltage is regulated by LM2576HVT to 5 V VCC, which is the power supply of the control circuit. This power supply system is isolated from other power supply of the equipment, and the reference ground GNDH is floating. TMP01 is a programmable temperature sensor, and its internal function diagram is shown in Figure 6.54.

When the outside temperature drops to a certain level, the pin OVER of the TMP01 will be set as a logic level signal (i.e. CTL1 in Figure 6.53). It is driven by the MOSFET MMDF3N03D, the contacts of the relay (in Figure 6.53) can be closed, and then the heating wire will get the power supply. When its temperature rises to a certain temperature, the pin OVER of the TMP01 will be set as an opposite logic level signal. The contacts of the relay will be opened. The heating of the heating wire will be stopped. In order to prevent the unstable output of the pin OVER at the critical temperature, a temperature hysteresis loop is set in TMP01. The hysteresis curve of TMP01 is shown in Figure 6.55.

In the problem locating test, a hair dryer is used to blow the temperature sensor, and the temperature of the sensor is measured using the infrared meter. It is found that, when blowing

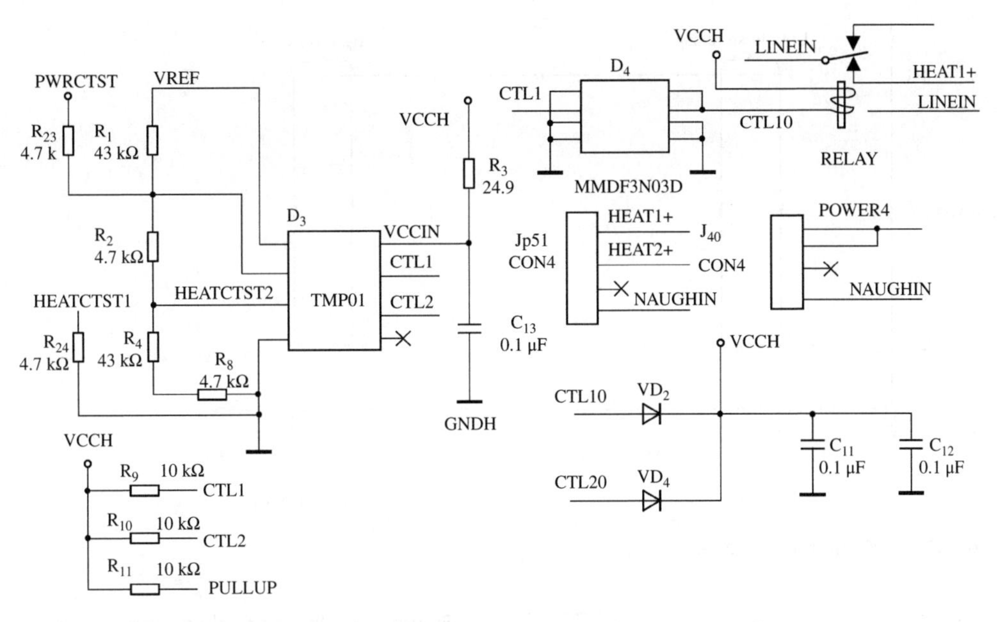

Figure 6.53 Schematic diagram of control board when switch electric wire on/off.

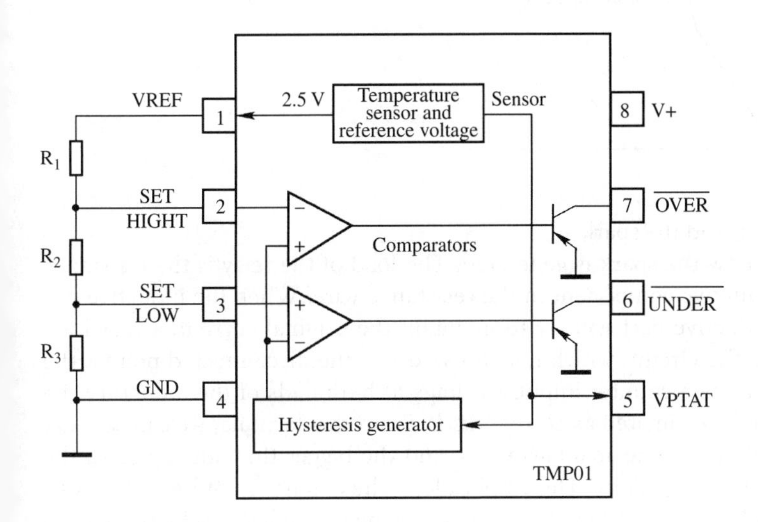

Figure 6.54 Internal function diagram.

the hot air, the temperature rises, and when blowing cold air, the temperature continues decline. There is no big temperature fluctuation, which means the temperature sensor has no problem. And measure the power supply VCC; nothing is wrong with it. The output of the MOSFET is opposite to the output of the temperature sensor, which is consistent with the design, and it indicates that the MOSFET has no problem. In the locating test, we also found that almost every time the relay is switched, there are considerable occurrences of fast switching phenomenon, i.e. the contacts are close – open – close, oscillating back and forth. At both contact pins of the relays, we can see sparks. When the temperature drops, the phenomenon will be severe. The more slowly the temperature drops, the more severe the spark is. The long-term oscillation and the spark damage the relay.

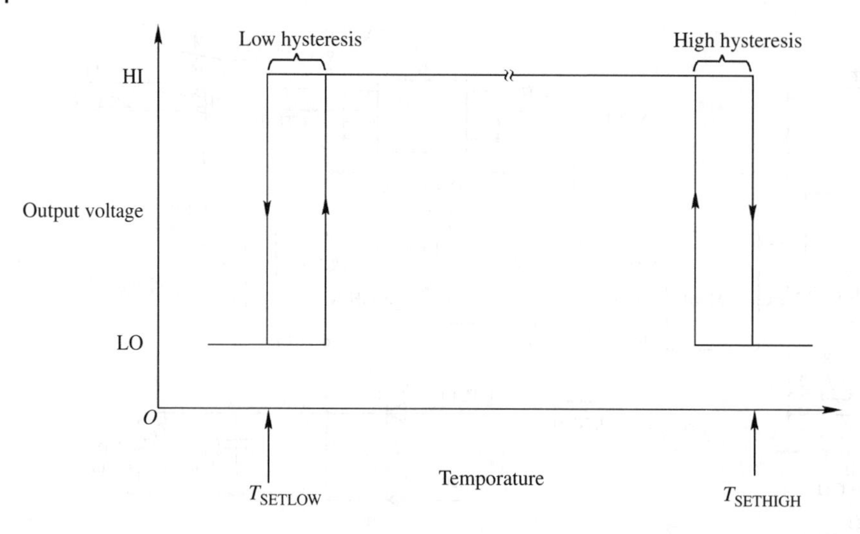

Figure 6.55 TMP01 hysteresis curve.

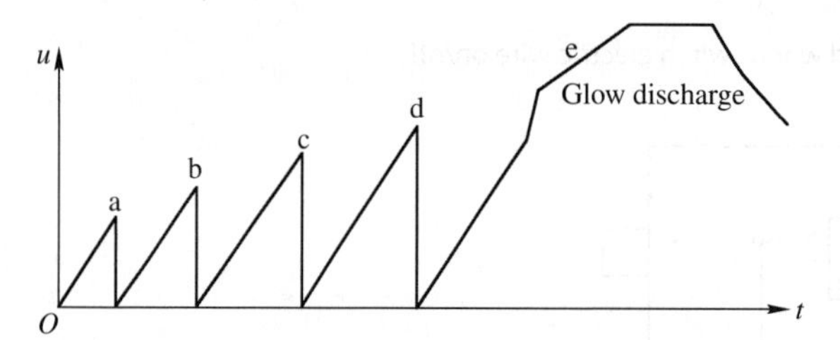

Figure 6.56 Discharge voltage waveform of contact spark.

What causes the oscillation and the spark?

Let's first have a look on how the spark is generated. The load of the relay is the resistance wire. There must be the inductive component in the resistance wire. When the inductive circuit is disconnected, the inductive part will try to maintain the original current, a very high voltage will be generated in the circuit, which may break down the disconnected points (the contacts of the relay) and create a counter impulse voltage at both ends of the inductor. The counter impulse voltage can be estimated as $U = -L\mathrm{d}i/dt$. From this formula, we can see that the greater the changing rate of the normal current is, and the bigger the inductance is, the higher the counter impulse voltage will be. The amplitude of the counter impulse voltage can be 10–200 times of the supply voltage. When the current flowing through the inductive load is suddenly interrupted by the relay contacts, the stored energy of the inductor will be dissipated in the discharge of the circuit and the contacts (which is also a strong noise). When the relay, which is generally used in the circuit, is switched off, as long as the voltage across the opened contacts is above 15 V and the current in the circuit is above 0.5 A, there will be spark discharge across the opened contacts. This discharge also produces high-frequency radiation. Glow discharge occurs when the voltage across the opened contacts is higher than 300 V. Figures 6.56 and 6.57, respectively, show the spark discharge voltage waveform and the spark discharge current waveform of the contacts.

In the figure, the spark discharge starts at point a, the voltage of the contacts drops to zero instantly, but due to the energy cannot be fully released, so the voltage will be immediately increased and the discharge happens again when the voltage is increased to point b. The discharge is repeated when the voltage increases to point c and point d. As the distance between

Figure 6.57 Discharge circuit waveform of contact spark.

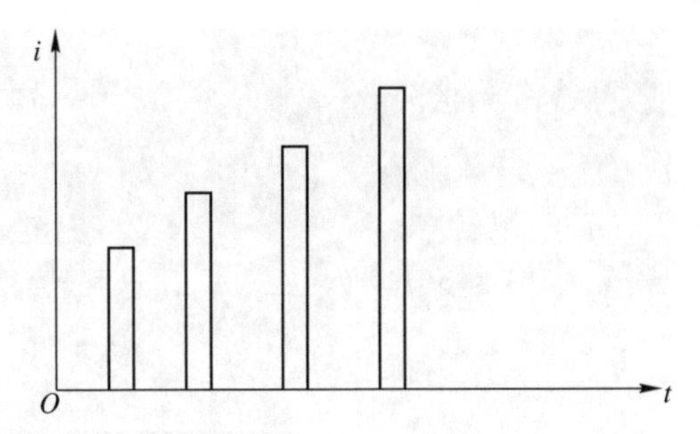

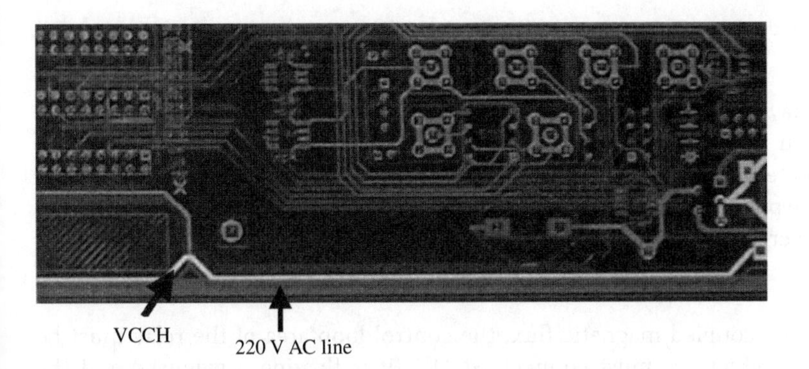

Figure 6.58 PCB routing of power supply.

the contacts is gradually increased, the discharge voltage will be gradually increased. When the voltage between the contacts is greater than 300 V, glow discharge occurs at point e. The discharge will be stopped until the energy in the inductor is exhausted. The contacts of the relay are opened completely at this time. It can be seen that when the relay is switched off, its contacts will suffer the high voltage and the high current, which will affect the service life of the relay.

What causes the oscillation? After checking the layout and the routing of this PCB, we found two problems. The BOTTOM layer and TOP layer are both poured with copper plane, which are the digital ground, and there is no direct connection between these planes and the working circuit of the relay. When the PCB circuit works, the GNDH can be equivalent as floating, but this is not shown in Figures 6.58 and 6.59.

1) There are long and close parallel routings between the high-voltage AC 220 V and the power supply VCC 5 V of the relay, which are respectively routed in the adjacent two layers, there will exist severe coupling. Moreover, the AC line is the power supply of the heating wire. The spark of the relay contacts is inside this loop. The highlighted line in Figure 6.58 is VCC, the one near the VCC is the line of AC 220 V.
2) Large loop exists in the working circuit of the relay. When the spark discharge occurs between the relay contacts, strong high-frequency radiation produced by this discharge will be coupled into this closed loop through magnetic coupling (the highlight line in Figure 6.59 is the loop). The electromagnetic energy is similar as the voltage source in the loop. The amplitude of the voltage source is proportional to the area of the loop. How much interference is received also depends on the loop area.

In the experimental test, we found that if we connect the ground of the power supply circuit of the relay to the poured copper plane on BOTTOM layer and TOP layer, the oscillation

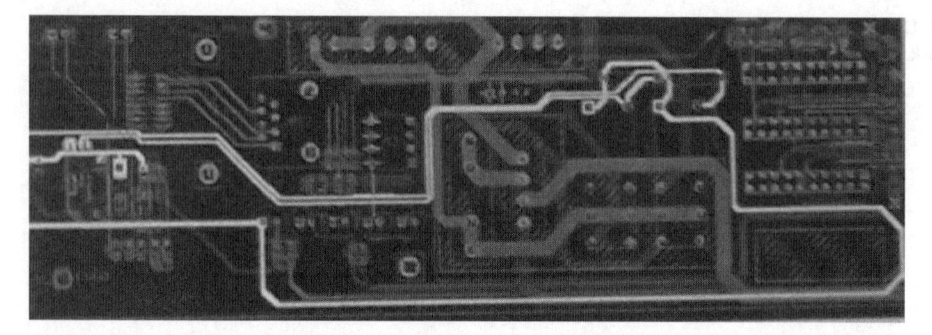

Figure 6.59 The actual loop diagram of relay coil operating circuit in PCB routing.

phenomenon of the relay will disappear. It is easy to understand that the disappearing of the oscillation phenomenon is due to the working loop area of the relay coil is greatly reduced when GNDH is connected with the poured plane on BOTTOM layer and TOP layer. In the process the interference phenomenon takes place, the loop and TMP01 act as a sensitive circuit when it works with the critical temperature (inside the hysteresis loop).

[Solutions]

1) In order to reduce the coupled magnetic flux, the control loop area of the relay must be minimized. In this product, we must connect GNDH (it is floating previously) and the poured plane on TOP layer and BOTTOM layer.
2) In order to further improve the reliability of the circuit. We must route 220 V AC and VCC separately on the PCB and add energy absorbing circuit to consume the arc across the contacts of the relay, such as RC, C, and varistor.

[Inspirations]

1) The signal loop will cause more problems than the interference of the power supply system. From the immunity point of view, because the disturbance can be injected directly into the loop and then to the inputs of components, in order to alleviate the consequences from the disturbance, reducing the loop area is the simplest way. If the loop area of the PCB is large enough, the immunity performance of the circuits on the PCB will be decreased, due to the antenna effect of the loop, at the same time the loop can also act as a transmitting antenna and transmit the radiation to the free space. The EMI problem will occur. After all, the loop is an antenna, if it is able to receive radiation, it will be able to emit radiation.
2) In this case, the signal in the loop is not a high-speed signal. If the signal is a high-speed signal, the EMI problem will appear, just as the situation that the return current loop area is increased by the slot on the ground plane (the current image plane). Taking this case as example, let's illustrate more. According to the analysis, in the high-speed digital circuit board, when there exists slot on the ground plane, why will the field strength around the printed trace be significantly increased? Because the return current image in the ground plane must flow along the slots and thus the high-frequency current loop area is increased HF. Because this impacts the characteristic impedance of the micro strip line a little, the output current of the high-frequency source is not changed too much, but because the slot on the ground plane can be regarded as a capacitor in this return current loop, the total output current will be slightly reduced. The radiated field will be increased because the high-frequency current loop is enlarged. When the signal passes through the slot, the RF return current at the slot is shown in Figure 6.60.

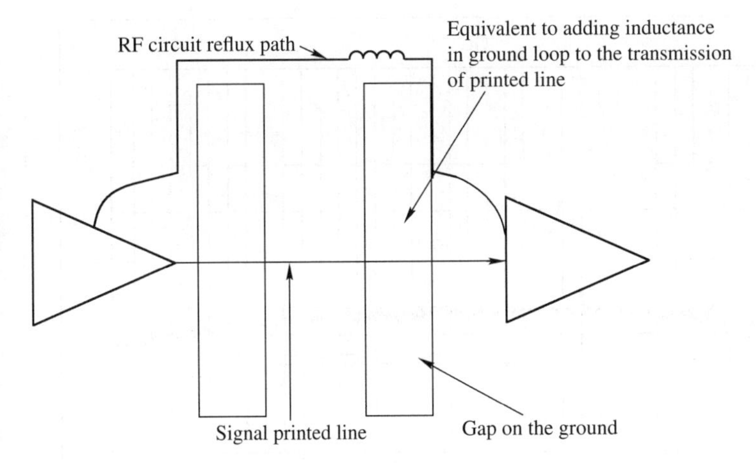

Figure 6.60 RF loop current at trench.

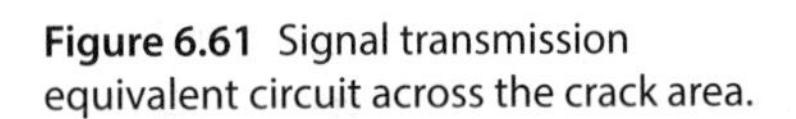

Figure 6.61 Signal transmission equivalent circuit across the crack area.

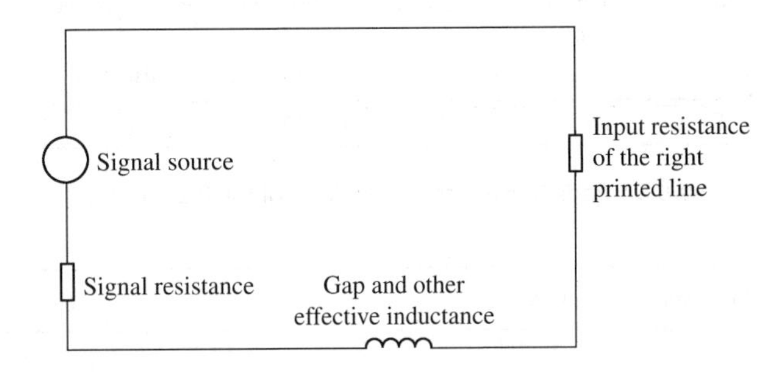

The existence of the slots can be regarded that an inductance is in series with the signal transmission circuit. This inductance and the input impedance of the right transmission line (the printed wire is designed with impedance matching, so the input impedance of the right transmission line is equal to the characteristic impedance of the printed wire.) constitute a voltage dividing circuit, so only part of the voltage from the high-frequency voltage source is divided on the printed wire. The equivalent signal transmission circuit across the slot can be shown in Figure 6.61.

To analyze the design of the printed wire across the slot, we can use the equivalent circuit shown in Figure 6.61. When the signal trace crosses the slot of the ground plane, the electromagnetic field near the printed wire will be slightly reduced because of the existence of the lot on the ground. However, because the high-frequency radiation loop area is increased, the noise on the ground will be greatly increased, so the radiation from the product to the outside is increased, and the amplitude of the radiation depends on the structure of the specific PCB.

6.2.10 Case 66: The Spacing Between PCB Layers and EMI

[Symptoms]

A certain product is the floating device (which is not grounded). The radiated emission test results of two batches (the first batch and the second batch are called in the following description.) of the products are, respectively, shown in Figures 6.62 and 6.63.

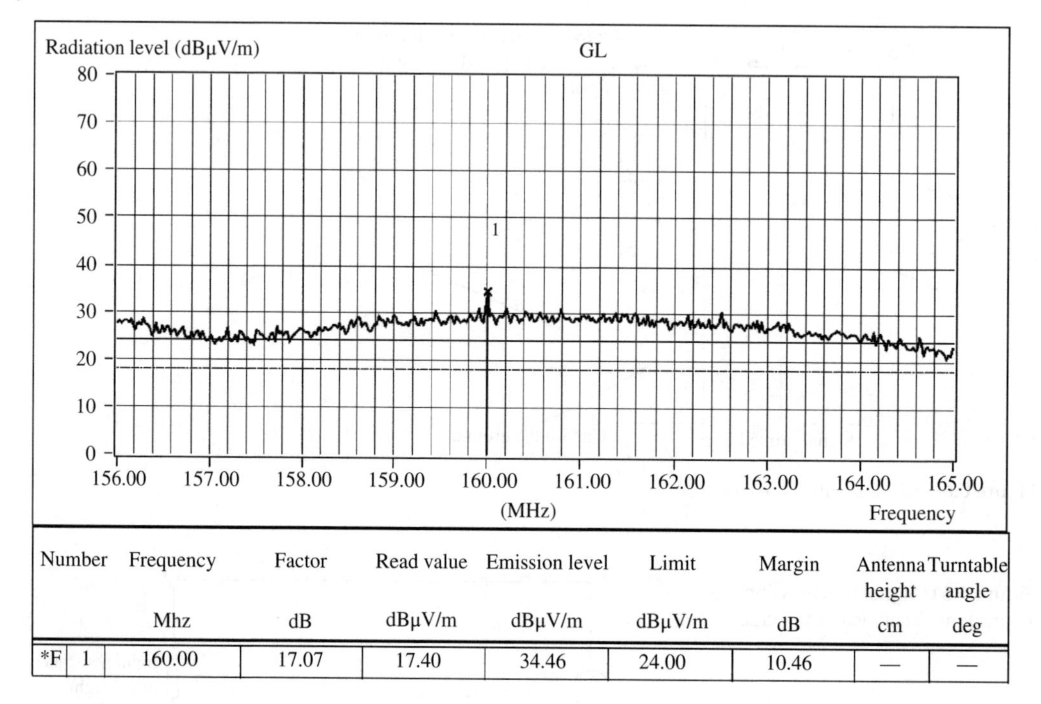

Number		Frequency	Factor	Read value	Emission level	Limit	Margin	Antenna height	Turntable angle
		Mhz	dB	dBμV/m	dBμV/m	dBμV/m	dB	cm	deg
*F	1	160.00	17.07	17.40	34.46	24.00	10.46	—	—

Figure 6.62 Radiation emission spectrum of the first product.

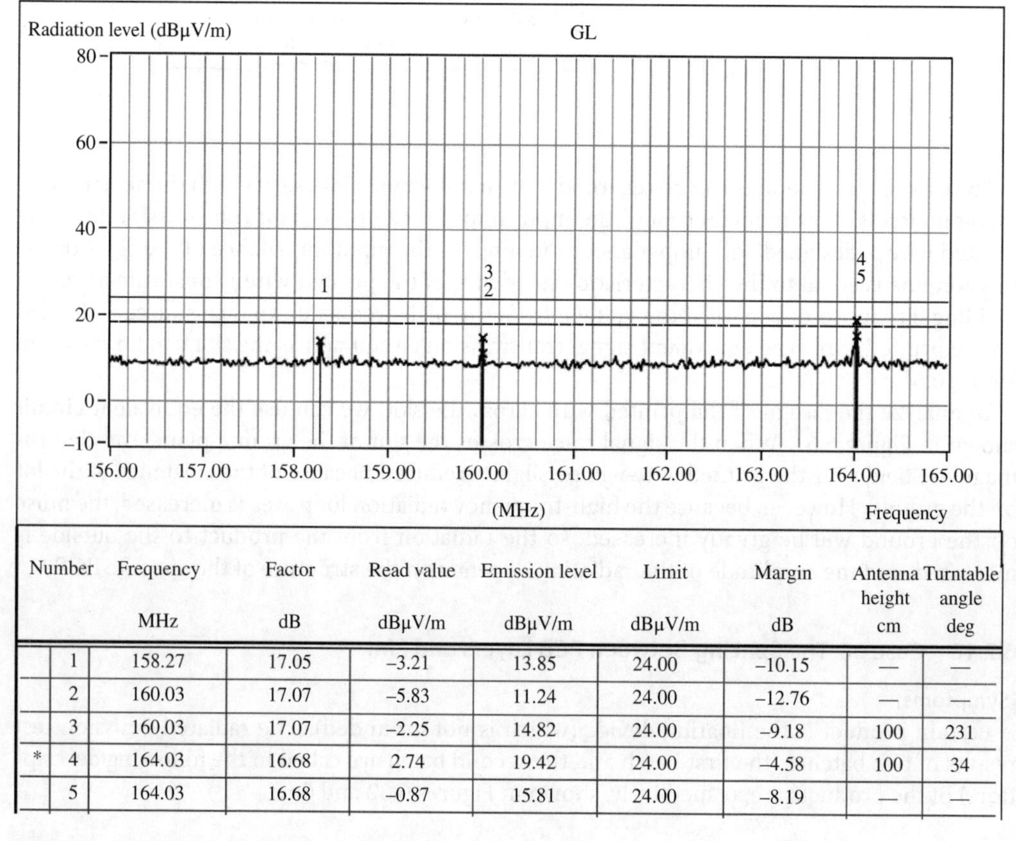

Number		Frequency	Factor	Read value	Emission level	Limit	Margin	Antenna height	Turntable angle
		MHz	dB	dBμV/m	dBμV/m	dBμV/m	dB	cm	deg
	1	158.27	17.05	−3.21	13.85	24.00	−10.15	—	—
	2	160.03	17.07	−5.83	11.24	24.00	−12.76	—	—
	3	160.03	17.07	−2.25	14.82	24.00	−9.18	100	231
*	4	164.03	16.68	2.74	19.42	24.00	−4.58	100	34
	5	164.03	16.68	−0.87	15.81	24.00	−8.19	—	—

Figure 6.63 Radiation emission spectrum of the second product.

[Analyses]

Compare the PCB designs of the two batches of products and find that their stackups are different, i.e. the spacing between PCB layers is different. The PCB stackup of the first batch product is shown in Table 6.1 and that of the second batch product are shown in Table 6.2. Among them, the signal (the clock) related with the frequency at which the emission exceeds the standard limit is routed in Lay2, the signal layer.

When the high-speed signal propagates on the PCB, the signal return current will cause a voltage drop on its return path. If the cable is driven by this voltage, the common-mode current (in microampere level) will be produced on the cable. This is a basic driving-mode for

Table 6.1 Cascading settings of the first PCB.

PCB layer setting	Signal/ground layer setting	Thickness (mm)
Signal layer	Top-signal layer	0.035
Dielectric layer 1		0.28
Cupreous layer2	Lay1-ground layer	0.035
Dielectric layer 2		0.28
Cupreous layer3	Lay2-signal layer	0.035
Dielectric layer 3		0.27
Cupreous layer4	Lay-3-power layer	0.035
Dielectric layer 4		0.28
Cupreous layer5	Lay4-ground layer	0.035
Dielectric layer 5		0.28
Cupreous layer6	Signal/ground layer	0.035
	Aggregate thickness	1.6

Table 6.2 Cascading settings of the second PCB.

PCB layer setting	Signal/ground layer setting	Thickness (mm)
Signal layer	Top-signal layer	0.035
Dielectric layer 1		0.2
Cupreous layer2	Lay1-ground layer	0.035
Dielectric layer 2		0.15
Cupreous layer3	Lay2-signal layer	0.035
Dielectric layer 3		0.69
Cupreous layer4	Lay-3-power layer	0.035
Dielectric layer 4		0.15
Cupreous layer5	Lay4-ground layer	0.035
Dielectric layer 5		0.2
Cupreous layer6	Signal/ground layer	0.035
	Aggregate thickness	1.6

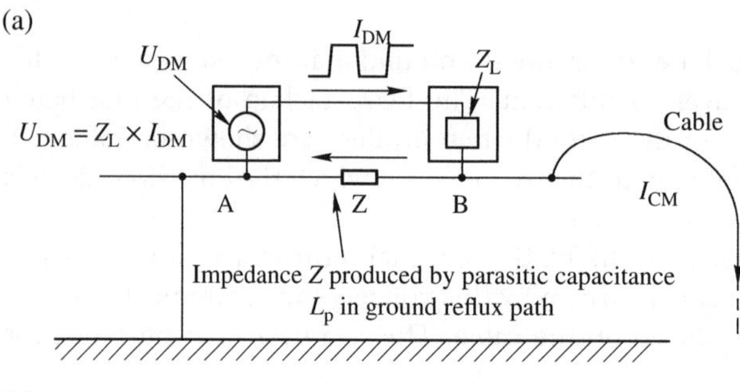

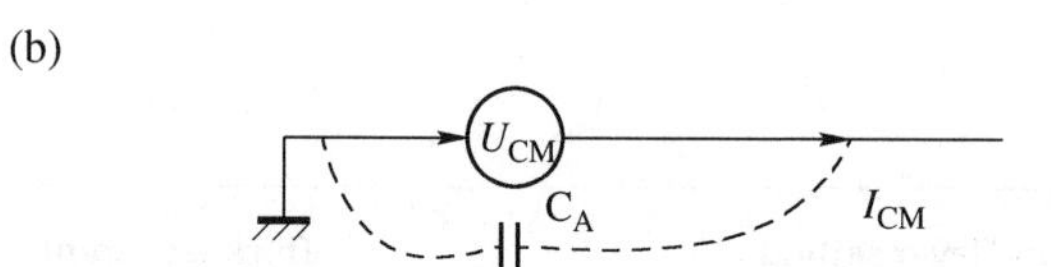

Figure 6.64 Common-mode radiation schematic diagram of current drive mode.

common-mode radiation, which is called current driving mode. Figure 6.64 is the common-mode radiation schematic with the current driving mode.

In Figure 6.64a, U_{DM} is the differential-mode voltage source. There are many such sources in the PCB, such as all kinds of digital circuits, high-frequency oscillation source, etc. Z_L is the load of the loop and I_{DM} is the differential-mode current flowing through the load of the loop. This current flows through the current return ground between the point A and B (such as the PCB ground wire). If there is a certain impedance Z existing between A and B (such as the parasitic inductance L_P caused by the connector pins or the incomplete ground plane between point A and B). The voltage drop on the impedance Z is: $U_{CM} = ZI_{DM} = I_{DM} \times (J\omega L_P)$

U_{CM} is the driving voltage of the common-mode radiation. To form the radiation, besides the driving source, there must be an antenna. This antenna is composed of the ground on the right of the point B and the external cable, shown in Figure 6.64a. The equivalent circuit of the radiation system composed by them is shown in Figure 6.64b. This is actually a pair of asymmetric dipole antenna.

According to the specific conditions in this case, the limit of the radiated emission is 24 dBμV (at 3 m measuring distance). We need control the radiated emission $E_{dB\mu V} < 16\,\mu V\,M^{-1}$ (the linear value of 24 dBμV). For the narrowband disturbance signal, when the length of the cable, which is the equivalent radiated emission antenna meets $L_M \geq \lambda/2$, the radiation can be estimated by the formula (6.2). The estimated results show that we must ensure the common-mode current flowing on this cable meet $I_{CM} < 0.8\,\mu A$ (if we consider the reflection from the ground reference plate, the common-mode current I_{CM} will be even smaller). If the characteristic impedance between the cable and the ground is 150 Ω, the noise voltage of the product at 160 MHz on the PCB must be less than 120 μV.

$$E_{\mu v\,m^{-1}} \approx 60 \times I_{\mu A} / D_m \tag{6.2}$$

$E_{\mu V\,m^{-1}}$ is the field strength of the radiation source at the measuring distance and its unit is ($\mu V\,m^{-1}$). I_{CM} is the common-mode current flowing through the cable and its unit is (μA). F_{MHz} is the signal frequency of the radiation source and its unit is (MHz). L_m is the cable length and its unit is (M). D_m is the distance from the radiation source to the measuring antenna and its unit is m.

For the high-speed transmission line, the current image plane originally provides a current return path for the signal (physically, the return path is exactly the image of the signal trace, so the plane is an image plane). When the complete image plane provides the image path to the

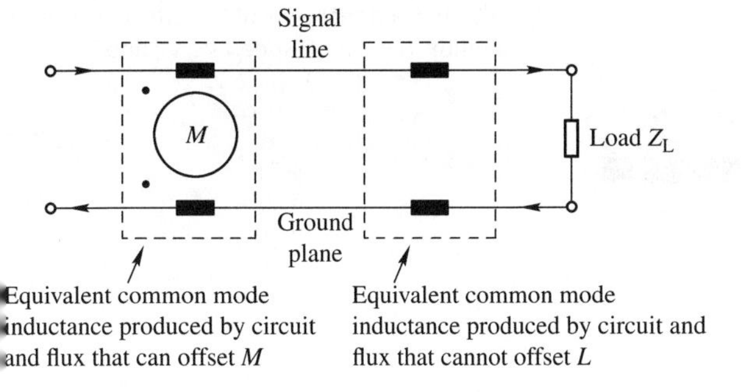

Figure 6.65 Equivalent circuit diagram of the high-speed signal reflux principle in PCB.

high-speed signal trace, because each pair of the paths (the signal current path and its image path) are very close to each other, the amplitude of the current in the signal trace is the same as that of the RF return current path and they are in opposite direction, the differential-mode RF current will cancel each other, which is just like the common-mode choke. If the cancelation ratio of the current and the magnetic flux is not 100%, most of the remained current will be converted to the common-mode voltage, which will be superimposed on the cable of the signal path. It is this common-mode current that produces the excitation source and then causes the EMI radiation. In order to minimize this common-mode current, we must maximize the mutual inductance between the signal trace and the image plane to confine the magnetic flux inside the loop and suppress the unwanted RF energy.

In the actual PCB design, because there is a certain distance between the signal trace and the signal return path (on the ground), the above mentioned "current cancelation" or "magnetic flux cancelation" cannot be 100%. Therefore, when the return current flows through the image plane, the "leakage inductance" L cannot be avoided. The equivalent circuit is shown in Figure 6.65. When the signal return current flows through this inductance L, a voltage drop $E{:}E = |L(dI/dt)|$ will appear.

It is worth noting that the effectiveness of the "current cancelation" and the "magnetic flux cancelation" is related with the physical structure of the signal trace and the PCB image ground plane. When the signal trace is closer to its image ground plane, this kind of cancelation effect will be more obvious. That is, the leakage inductance L in Figure 6.65 will be small, and it will result in a voltage drop when the signal return current flowing through the ground plane becomes even smaller. This is the main reason for the different radiation results of these two PCBs in this case. Because the clock trace of the second batch of the product is closer to the ground plane, the common-mode disturbance voltage on the ground plane is lower than that of the first batch of the product, and then the radiated emission level of the second batch of the product is lower than that of the first batch of product.

[Solutions]

According to this analysis, in the PCB stackup design, we must ensure that the high-speed signal trace is as close to the complete ground as possible.

[Inspirations]

In addition, the return current density of the high-speed signal is not completely concentrated in a small thin printed wire, which is similar as the signal trace, but is distributed right below the signal trace and drops quickly from the center of the signal wire on the two sides of the signal wire. Usually, 90% of the signal return current density is distributed within a distance

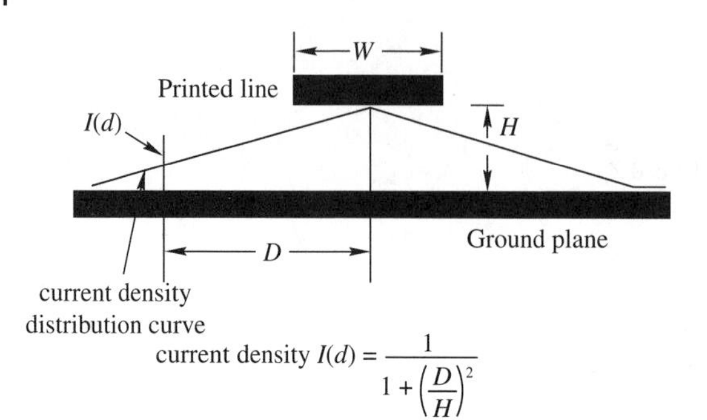

Figure 6.66 Sectional view of the effective area of mirror reflux in the high-speed signal line.

of ±5 times h (h is the distance between the signal trace and the image plane), which is shown in Figure 6.66.

This area is called the effective area for the high-speed signal image current.

So, if there are via holes in the image ground plane, the return path will be interrupted, and the return current on the image plane has to flow along the vias. This will also greatly reduce the magnetic flux cancelation between the signal trace and the RF current return path, and greatly increase the excessive inductance in the return path. It is visible that, for the multilayer PCB, any vias or slot on a large ground area (the width of the area in Figure 6.66 is $10h$) under the high-speed signal trace will impact the integrity of the image plane and causes EMC problems. In some cases, even if the signal trace can be routed across the vias or the slot, EMC problem could take place if the image ground area under the signal trace is not enough. In order to avoid this problem, for the signal routed between through-hole components, the distance between the signal trace and the blank area around the through-hole components shall be not small than $10h$. For example, for a four-layer PCB, its thickness is 1.6 mm, and the spacing between two layers is 0.4 mm, there shall be no vias on the ground area with a distance ±2 mm from the signal trace.

6.2.11 Case 67: Why the Sensitive Trace Routed at the Edge of the PCB Is Susceptible to the ESD Disturbance

[Symptoms]

A tabletop product system has a ground terminal, when ESD contact discharge voltage (the test voltage is ±6 kV) is injected on its ground terminal, system reset appears. In the test, we try to disconnect the Y capacitor, which is connected between the ground terminal and the internal digital reference ground, but the results are not significantly improved.

[Analyses]

There are many paths through which the ESD disturbance can enter the internal circuit. For the product under test in this case, the test point is on the ground terminal, so most of the ESD disturbance energy shall flow away through the grounding cable. That is to say the ESD current does not flow directly into the internal circuit. This tabletop product under test is configured according to the ESD test standard IEC61000-4-2, the grounding cable is about 1 m, so there is a large inductance (which can be estimated by $1\,\mu\text{H}\,\text{m}^{-1}$). When the ESD discharge occurs (i.e. the switch K is closed in Figure 6.67), due to the high-frequency ESD discharge current (the rising time is less than 1 ns) the voltage on the ground terminal of the product is zero (i.e. the voltage on point G is not zero when the switch K is enclosed; Figure 6.67). This voltage, which is not zero on the ground terminal, will propagate into the internal circuit of the product. Figure 6.67 shows the schematic how the ESD disturbance goes into the PCB of the product.

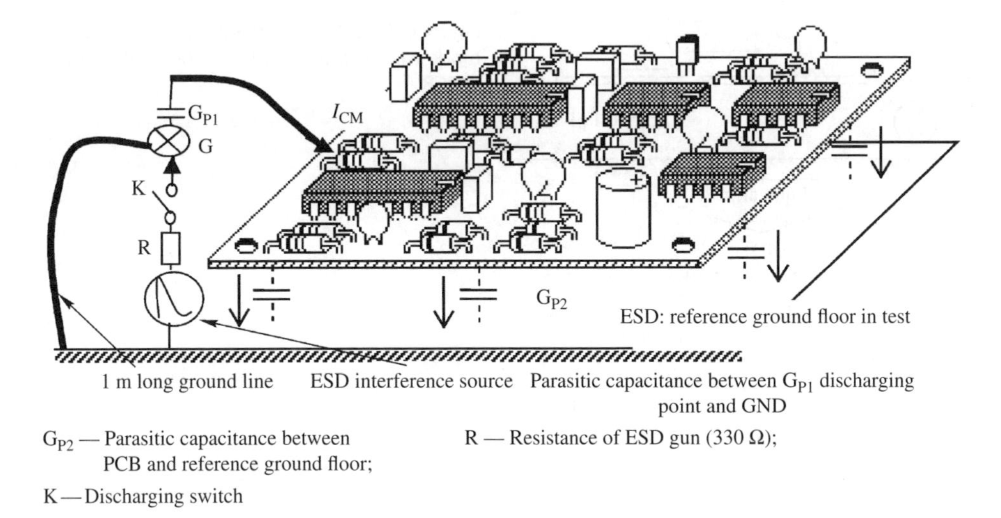

Figure 6.67 Schematic diagram of ESD interference into the internal PCB of product.

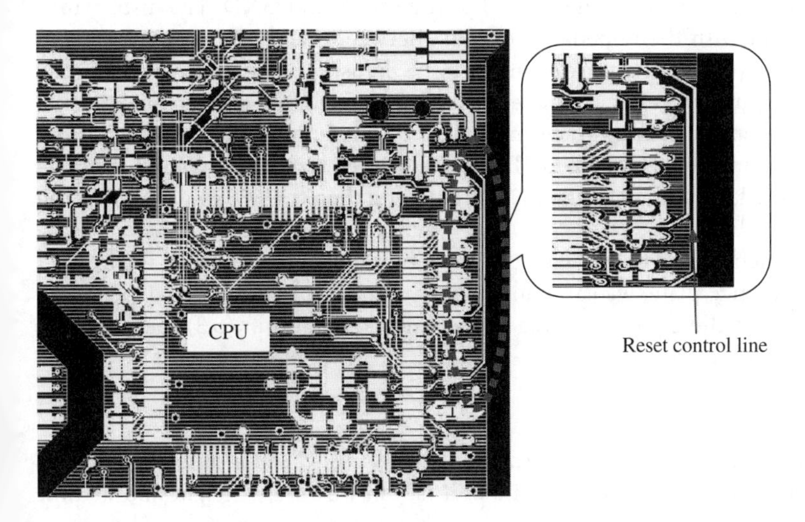

Figure 6.68 Partial PCB layout of tested product.

From Figure 6.67, we can see that, C_{P1} (the parasitic capacitance between the ESD discharge point and GND) C_{P2} (the parasitic capacitance between PCB and the ground reference plate), the PCB reference ground GND and the ESD gun (including the grounding cable of the ESD gun) constitute the disturbance path and the disturbance current is I_{cm}. The PCB is also inside this disturbance path and it is clear that the PCB suffers the ESD disturbance at this time. If there are some other cables in this product, this kind of disturbance will be even more severe.

After the careful examination of the product, it is found that the CPU reset control signal is routed on the edge of the PCB and not covered by the GND plane, which is shown in Figure 6.68.

Let's illustrate why the printed trace routed on the edge of PCB is sensitive to be disturbed, we can start this illustration from the parasitic capacitance between the PCB printed trace and the ground reference plate. Due to this parasitic capacitance, the PCB printed trace is disturbed. The schematic that the printed signal trace is disturbed by the common-mode voltage is shown in Figure 6.69. From this figure, we can see that when the common-mode disturbance

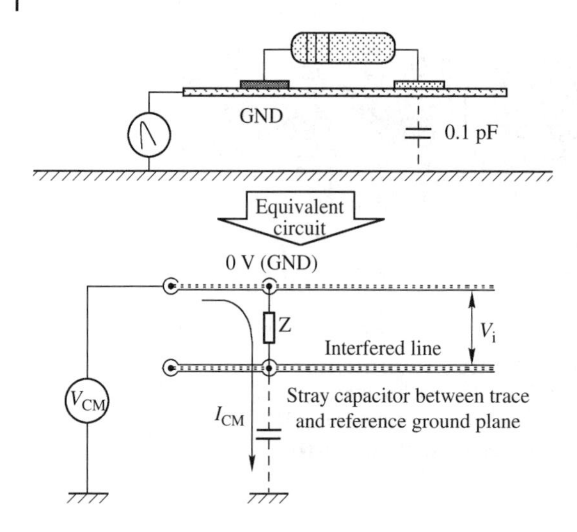

Figure 6.69 Schematic diagram of common-mode voltage interfering with the printed line in PCB.

(the common-mode disturbance voltage relative to the ground reference plate) flows into GND, there will be a disturbance voltage between the PCB printed trace and GND. The disturbance voltage is not only relevant with the impedance existing between the PCB printed trace and GND (Z is shown in Figure 6.69), but also relevant with the parasitic capacitance between the PCB printed trace and the ground reference plate. Assume Z is fixed, the greater the parasitic capacitance existing between the printed trace and the ground reference plate, the higher the disturbance voltage U_i existing between PCB GND and the printed trace. This voltage is superimposed on the normal working voltage of the PCB circuit, and then the working circuit in the PCB will be directly affected.

From formula ((6.3), which is used to calculate the parasitic capacitance between the printed trace and the ground reference plate, we can know that the parasitic capacitance depends on the distance (H in (6.3)) and the equivalent area of the electric field (S in (6.3)) between the printed trace and the ground reference plate.

$$C_P \approx 0.1 \times S/H \tag{6.3}$$

In this formula, C_P is the parasitic capacitance and its unit is (pF); S is the equivalent area of the electric field between the printed trace and its unit is (cm^2); H is the height and its unit is (cm).

Because the electric field between the printed trace inside the PCB and the ground reference plate will be squeezed by other printed traces, but the electric field between the printed trace on the edge of the PCB and the ground reference plate is scattered, when the printed trace is routed on the edge of the PCB, a relatively large parasitic capacitance will exist between the printed trace and the ground reference plate. The schematic of the electric field distribution between the printed trace and the ground reference plate is shown in Figure 6.70.

Obviously, for the circuit design in this case, because the PCB reset signal is routed on the edge of the PCB and outside the GND plane, the reset signal trace will be largely disturbed, leading to the system reset in the ESD test.

[Solutions]

According to the above analysis, it is easy to get the following two disposals:

1) Move the reset signal trace on the PCB to the left side, and ensure the trace is covered by the GND plane and far away the edge of the PCB. In order to reduce the parasitic capacitance

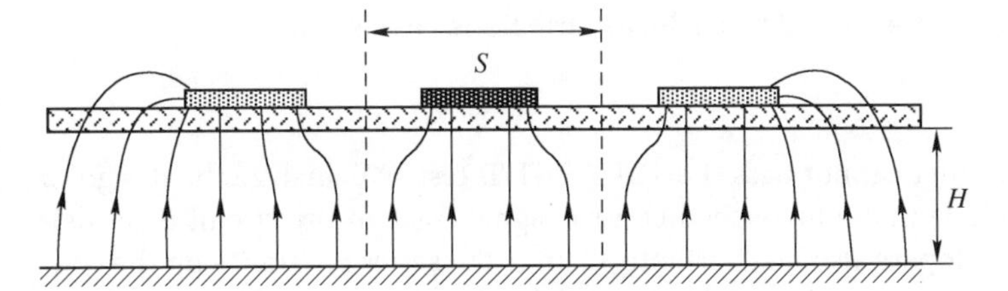

Figure 6.70 Electric field distribution diagram between the printed line and the reference ground floor.

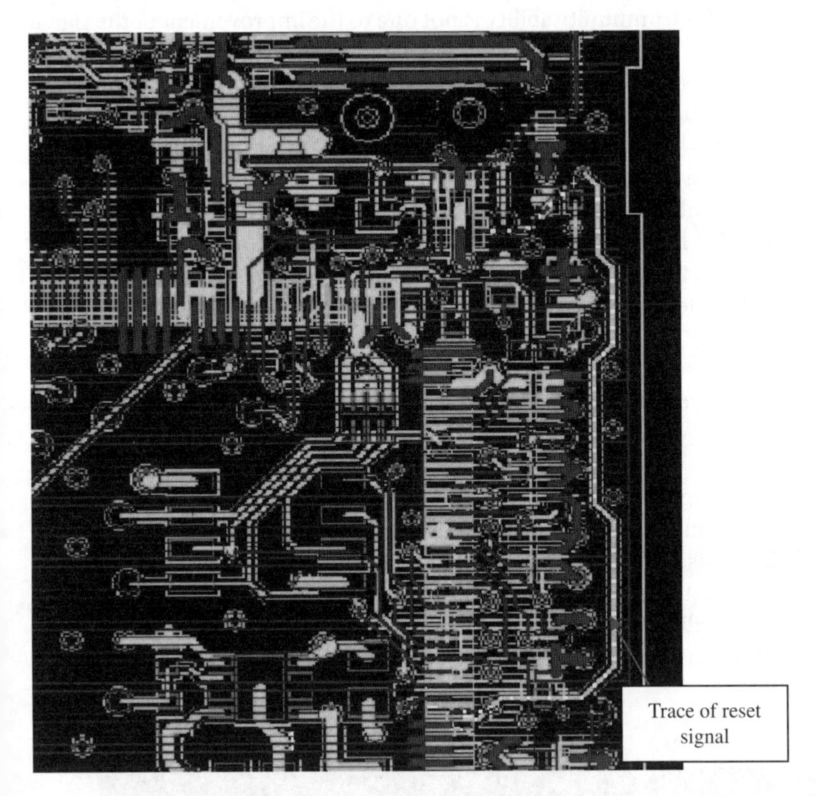

Figure 6.71 PCB layout of reset signal line after modifying.

between the printed trace of the reset signal and the ground reference plate, we can pour copper plane on the blank place of PCB layer on which the reset signal trace is (the reset signal trace is routed on the external layer in this case). The design is shown in Figure 6.71.

2) On the printed trace of the reset signal, we can connect a capacitor in parallel with the reset trace and near the CPU reset pin. The capacitance can be selected from 100 to 1000 pF.

[Inspirations]

1) We cannot route the sensitive signal trace on the edge of PCB.
2) Carefully analyzing the ESD current path is helpful for the analysis of the ESD test problem.
3) Surrounding the printed trace routed on the edge of the PCB with GND guard ring can reduce the parasitic capacitance between the printed trace and the ground reference plate or the metallic casing.

6.2.12 Case 68: EMC Test Can Be Passed by Reducing the Series Resistance on the Signal Line

[Symptoms]

The I/O signal port of a product cannot pass the ±2 kV EFT/B test. We analyze the design of the circuit and finally conclude that a low-speed control signal trace in the circuit is in malfunction. In the test, the developers also inadvertently change the series resistance on the control signal, i.e. the resistance is changed from 1 kΩ to 100 Ω (the resistance in series with the signal is allowed to be 1 kΩ, and the signal input port of the chip is a high-impedance input), and the test can be passed. The signal quality with these two kinds of resistors is the same. This shows that the improvement of the immunity ability is not due to the improvement of the signal quality itself.

From the schematic of the circuit, the resistor is in series with the signal, and it can limit the current. If the current is the disturbance current coming from the external side to the internal side, the disturbance voltage on the pin connected with the signal of the chip can be reduced to a certain extent. In addition, the resistor and the filtering capacitor in parallel with the signal constitute the RC filtering circuit. For EMI, the resistor can also limit the EMI current on the signal trace and reduce the EMI level of the product system. In this way, on the condition that the signal quality is not affected, the larger the resistance in series with the signal, the better the circuit performance. But why does the opposite result occur in this case?

[Analyses]

We carefully check the PCB layout of the product and find there are two signal traces with similar function (the source, the load, and the logic of these traces are the same, and the resistance in series with these two signals are also the same in resistance). The signal routing in the PCB layout is shown in Figure 6.72. From this figure, we can clearly see that a long section of one signal trace is routed on the edge of the PCB (the shape of the PCB looks like L). According to the analysis of case 67, this is a severe design defect. A large parasitic capacitance exists between the signal trace on the edge of the PCB and the ground reference plate or the metallic casing. According to the principle how the common-mode disturbance impacts the product circuit, this kind of signal trace will suffer a big disturbance. Case 67 explains this principle in detail.

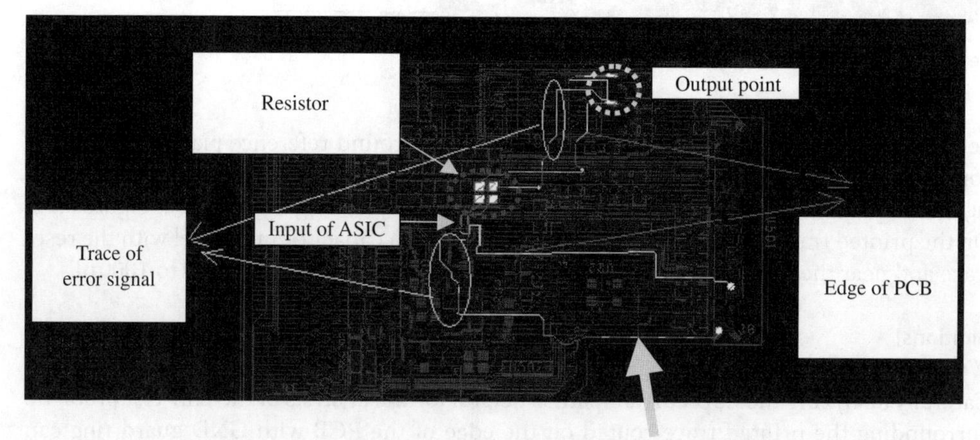

Figure 6.72 Signal line layout in PCB.

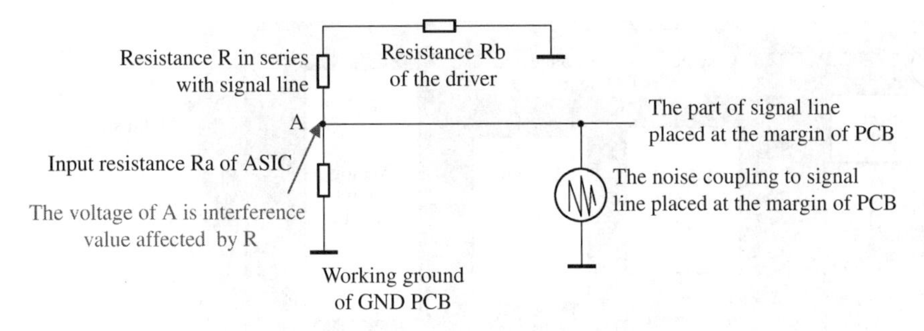

Figure 6.73 Working schematic diagram of interfered signal line.

Although the reason the signal trace is disturbed seems clear, why can the chip and signal work normally after we reduce the resistance in series with the signal? If we draw the circuit schematic, we will find the reason. Figure 6.73 shows the schematic of the disturbed circuit. R_b represents the output impedance of the signal driver. R_a is the input impedance of the signal input port, and R is the resistance in series with the signal trace (originally, it is 1 kΩ).

Because a long section of the signal trace is routed on the edge of the PCB, this trace section will pick up the common-mode noise injected into the internal side of the product, and then the coupled common-mode noise will be transformed into the differential-mode disturbance (i.e. the disturbance between the input of the ASIC and the reference ground). According to the disturbing principle described in Case 67, the differential-mode disturbance voltage is not only related with the parasitic capacitance between the printed trace and the ground reference plate, but also related with the impedance between the signal input port of the ASIC and its reference ground which is related with R_a, R, and R_b shown in Figure 6.73. R is the paralleled resistance between R_a and the series resistance of R and R_b. R_a is large (in kΩ level), but R_b is very small. R directly determines the voltage at point A. When R is changed from 1 kΩ to 100 Ω, the noise voltage at point A is greatly reduced by around 90% of the original one, so the test is passed.

[Solutions]

According to the above analysis, we can move the signal trace on the edge of the PCB to the internal side of the PCB, or change the series resistance on the signal trace from 1 kΩ to 100 Ω.

[Inspirations]

1) Do not route the sensitive signal of the circuit on the edge of the PCB.
2) The noise on the signal port of the chip is related to its input impedance. The signal port with high-input impedance is more susceptible to the noise. Therefore, in the circuit design, avoid the unused signal port of the chip floating (especially for the high-input impedance port of the chip, such as the input port of the CMOS components), connect them to the reference ground with low impedance.

6.2.13 Case 69: Detailed Analysis Case for the PCB Design of Analog-Digital Mixed Circuit

[Symptoms]

In the design of the audio equipment used in the car, which has the metallic shell, we found that the high-order harmonic noise of the digital circuit affects the quality of the radio signals. One local PCB layout of the product is shown in Figure 6.74. When we check the specific design of the product, we find that the analog radio circuit is connected with the metallic casing by screws.

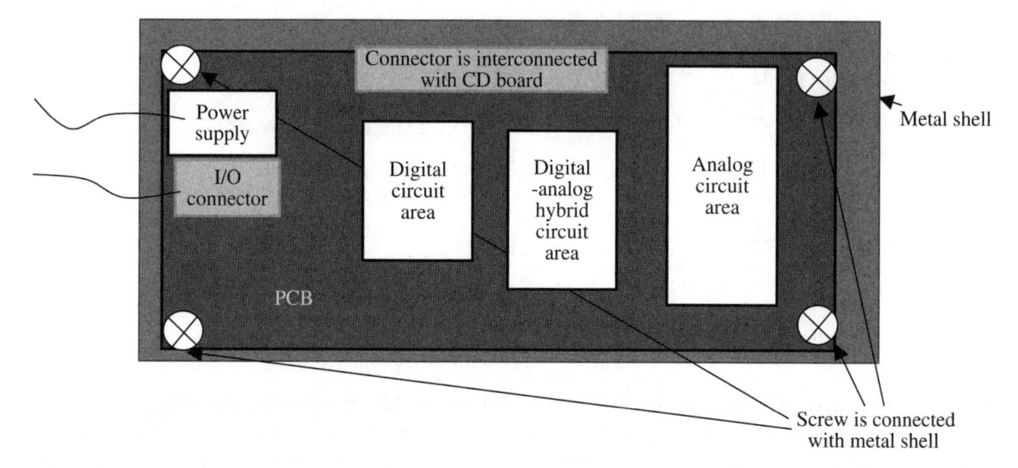

Figure 6.74 Partial PCB layout of the product.

[Analyses]

This is a typical case about the mutual interference of the analog-digital mixed circuit, and in the PCB board with the coexisting of the digital circuit and the analog circuit, when the operating voltage level of the digital circuits is high-(such as the signal voltage is 3.3–5 V), the operating frequency is high (e.g. frequency is up to tens of megahertz, and its harmonic frequency is up to hundreds of MHz), and the operating voltage level of the analog circuits is low (e.g. lower than 1 mV), the operating frequency of the analog circuit is right in the operating frequency range of the digital circuit or its harmonic frequency range. If the circuit isn't designed properly, the mutual interference occurs between the digital circuit and the analog circuit. Then how is this interference caused?

There are three ways through which the noise of the digital circuit affects the noise of the analog circuit:

The first one is the power supply, mainly the noise from the switching-mode power supply.
The second one is the crosstalk from the digital signals to the analog signals.
The third one is the noise on the reference ground of the digital circuit.

Two of them are from the power supply, and the first one is the output noise of the switching-mode power supply itself, which is usually a differential-mode noise existing between the power supply and 0 V ground, and the characteristics of this noise is low frequency, high amplitude, but relatively easy to be filtered by the composition of capacitors and inductors. The second one is the noise on the power supply pins of the chip used in the digital circuit, and this noise is inherent when the digital chip works. It can be reduced by the reasonable selection of the decoupling circuit, and its propagation can be minimized by separating the power supply plane from the others, as shown in Figure 6.75.

For the noise caused by the crosstalk between signals, the propagation of this noise will be more direct and its amplitude will be much higher. For example, when the crosstalk occurs between the digital signal and the analog signal, the analog signal will be directly affected by the noise, as shown in Figure 6.76.

For example, the voltage amplitude of signal1 $U_c = 5\,\text{V}$, the rising time of this signal is 1 ns, and its load $Z_0 = 100\,\Omega$, the source impedance and the load impedance of signal2 Z_1Z_2, both are $100\,\Omega$, the parasitic capacitance between Signal1 and Signal2 C_p is 10 pF, then,

$$I_c = U_c/Z_0 = 5/100 = 0.005\,\text{A}; Z = Z_1 \bullet Z_2/(Z_1 + Z_2) = 50\,\Omega.$$

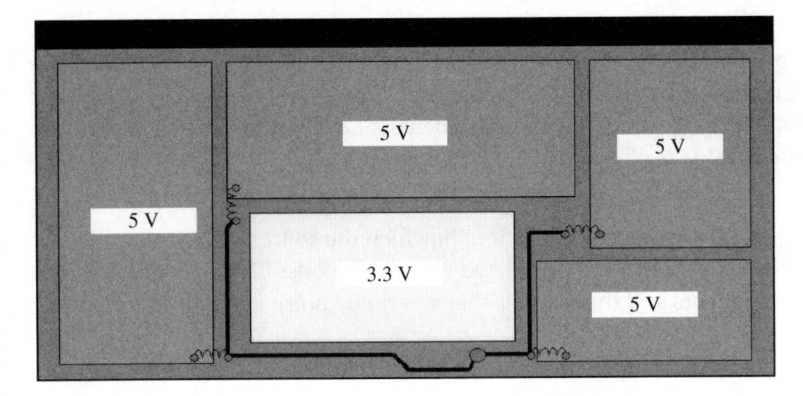

Figure 6.75 Power is divided to reduce power noise transmission.

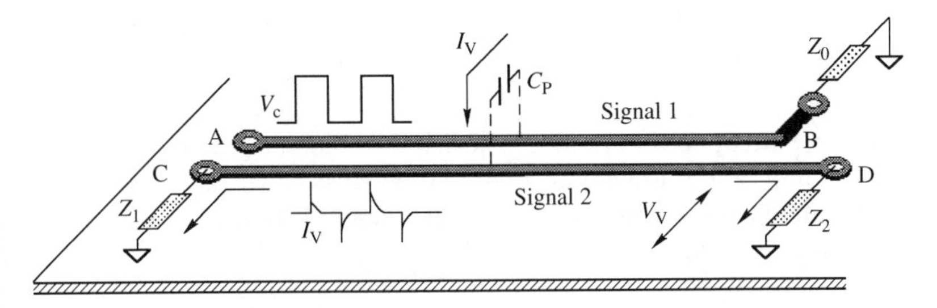

Figure 6.76 Schematic diagram of capacitive crosstalk.

The voltage on Signal2 induced by the crosstalk (the near-end crosstalk and the far-end crosstalk are not taken into account) is:

$$U_U = Z \cdot I_U = Z_1 Z_2 / (Z_1 + Z_2) \cdot C_p \cdot \Delta U_c / \Delta t = 50\Omega * 10\text{pF} * 5\text{V}/1\text{ns} = 2.5\text{V}$$

We can see that this kind of crosstalk between the analog signal and the digital signal is relevant with the parasitic parameters between the digital signal trace and the analog signal trace. A kind of capacitive crosstalk is shown in Figure 6.76, and the crosstalk between two traces is directly related to the parasitic capacitance between these two traces, so long as the parasitic capacitance between the digital signal trace and the analog signal trace is reduced, this problem can be solved. How can the parasitic capacitance between these signal traces be reduced? Let's see the following experimental results.

The following results tell us how to prevent such noise propagation caused by crosstalk.

The essence of the experiment is that, when the signal with a certain amplitude and frequency propagates in one printed trace, the other printed trace will couple (crosstalk) this signal, the coupling (crosstalk) effectiveness is related to the PCB layout of these two PCB traces.

The experiment is carried out with the following four scenarios. The experimental conditions are set with the same source and the same test way, but the signal is injected into four different PCBs on which the signal layouts are different.

Regarding the experimental results, measure the coupled noise on the other signal trace, and compare this measured noise on the other signal trace in these four test scenarios.

The same source in the four experiments is a 40 MHz square wave clock, and its peak-to-peak voltage is 5 V.

Experiment 1:
Figure 6.77 illustrates the experimental configuration, in which the connector BNC1 is injected with the source signal mentioned as above; BNC2 and BNC4 are connected with 50 Ω loads; BNC3 is connected with the oscilloscope for measurements, and the result is that a periodic noise with 2 V peak-to-peak value is measured.

Experiment 2:
Figure 6.78 illustrates the experimental configuration, in which the source signal is injected to the connector BNC5; BNC6 and BNC8 are connected with 50 Ω loads; BNC7 is connected to the oscilloscope for measurements and the result is that a periodic noise with 100 mV peak-to-peak value is measured.

Experiment 3:
Figure 6.79 illustrates the experimental configuration, in which the source signal is injected to the connector BNC1'; BNC2' and BNC4' are connected with 50 Ω loads; BNC3' is connected to the oscilloscope for measurements, and the result of the measurement is that a periodic noise with 1.2 V peak-to-peak is measured.

Experiment 4:
Figure 6.80 illustrates the experiment configuration, in which the source signal is injected to the connector BNC5'; BNC6' and BNC8' are connect with 50 Ω loads; BNC7' is connected to the oscilloscope for measurements and the result of the measurement is that a periodic noise with 1 V peak-to-peak value is measured.

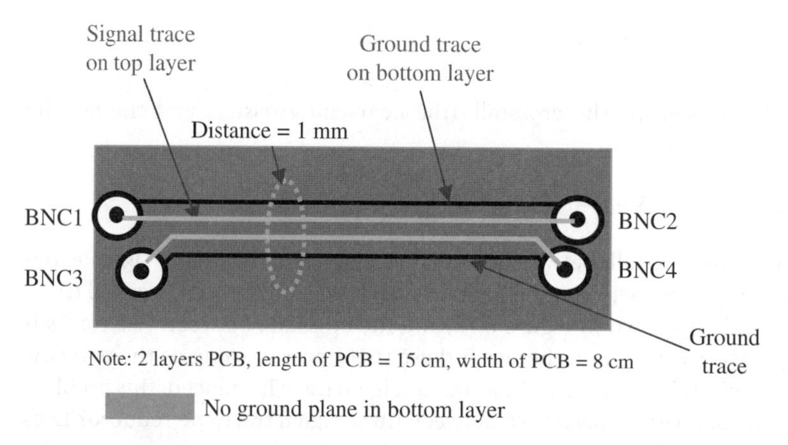

Figure 6.77 Construction diagram of experiment 1.

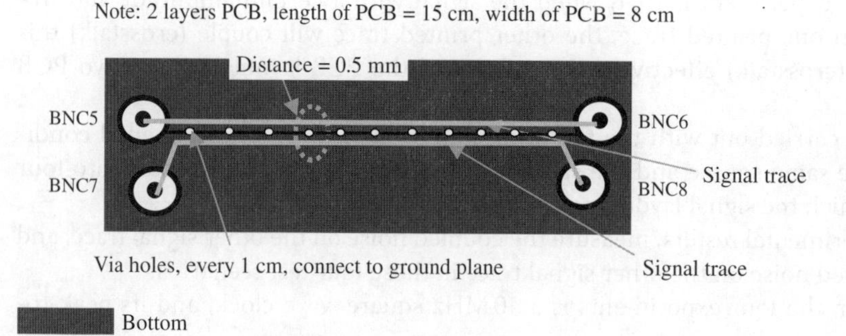

Figure 6.78 Construction diagram of experiment 2.

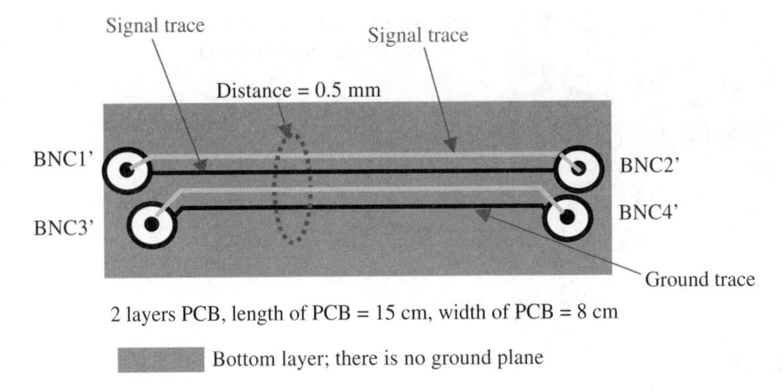

Figure 6.79 Construction diagram of experiment 3.

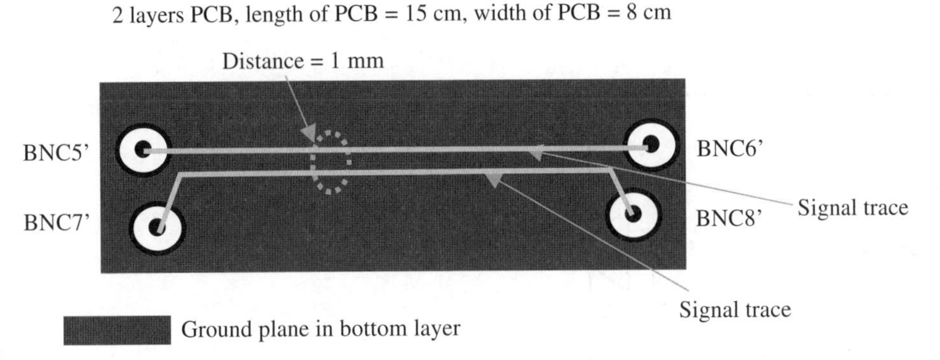

Figure 6.80 Construction diagram of experiment 4.

Per the test results, the PCB layout used in the experiment 2 is the best way. Visibly, in order to prevent crosstalk between the digital signal trace and the analog signal trace, in addition to avoiding the parallel routing in the same layer, the more important thing is to pour the reference ground (0 V) plane under the signal trace, and use the ground guard trace, at the same time the ground guard trace is connected to the reference ground plane at multipoints. Or place the digital signal traces and the analog signal traces on different PCB layers, between which there is the ground (0 V) plane for isolation. Figure 6.81 is a good PCB layout for an analog-digital mixed circuit. In Figure 6.81, the part inside the dotted lines is the low-voltage analog circuit, the dotted line on the top layer is the ground guard trace, and the copper shapes around the dotted line are connected to the ground (0 V) plane through vias. Similarly, the other signal layers are treated with the similar way.

As to the ground noise, first, the noise transferred from the ground to the chip is due to the common impedance coupling, as shown in Figure 6.82.

That is, when the noise current I_{ext} produced in other circuits flows through the ground between IC_1 and IC_2, due to the ground impedance Z between IC_1 and IC_2, there will be a voltage drop on the impedance Z, and this voltage drop will be superimposed on the transferred normal signal between IC_1 and IC_2, then the chips will be disturbed. If the normal operating voltage of the signal transferred between IC_1 and IC_2 is U_s, the actual signal voltage is the U_s plus the coupled interference voltage (Z multiplied by I_{ext}) when the common-mode interference current I_{ext} flows through the common impedance. In the study of the mutual interference

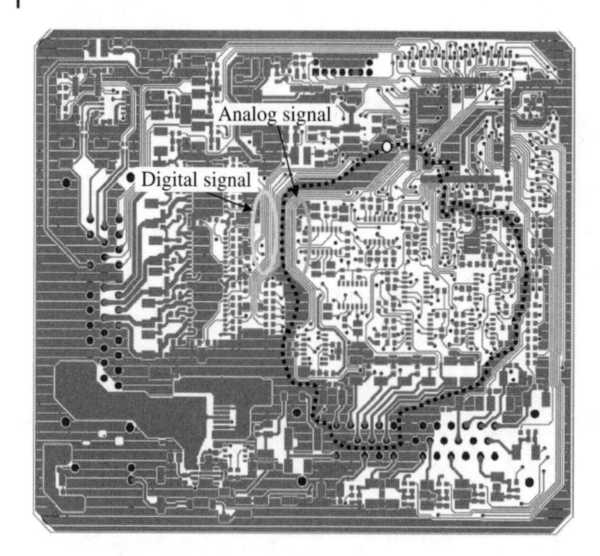

Figure 6.81 PCB crosstalk processing example of digital-analog mixed circuit.

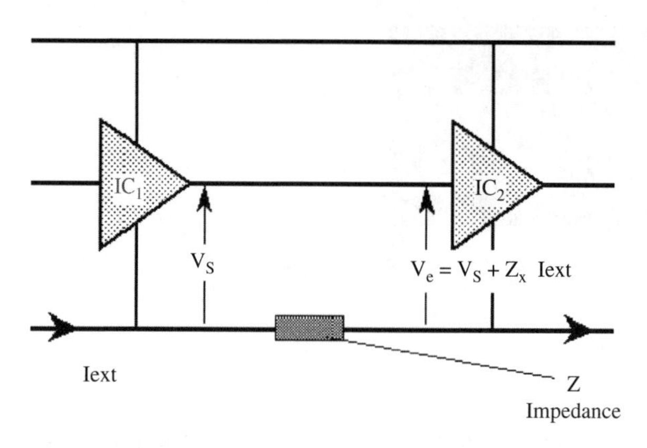

Figure 6.82 Coupling schematic diagram of public ground impedance.

in analog-digital mixed signal circuit, this common-mode interference current I_{ext} is not from the outside but from the inside of the product or the inside of the PCB, and the return current of high-speed digital signal on the ground is the main source of this common-mode interference current I_{ext}. If the noise on the digital signals and the power supply noise are called "active" noise, the noise on the digital ground can be the "passive" noise because there is no noise source on the digital ground, and the ground noise is caused when the return current of the digital signal flowing on the ground passes by the common impedance of the ground. The voltage of this kind of noise is generally lower than that of the noise on the signal line, but ground noise is a kind of common-mode noise, and it will be easier to be transferred to other circuits of the product. Generally, when there are some connections between the analog ground and the digital ground, such digital ground noise may enter the analog circuit, and affects the low-voltage analog circuits.

The PCB layout of an analog-digital mixed circuit is shown in Figure 6.83, where I_{DM} is the high-speed digital signal current, ΔU_{AB} is the ground noise voltage when the return current of this digital signal flows on the digital ground (which is also called as the ground bounce); Z_1 is the ground impedance (i.e. the impedance of the signal return path between the ground pin of the digital device and the ground pin of digital-analog mixed device), through which the

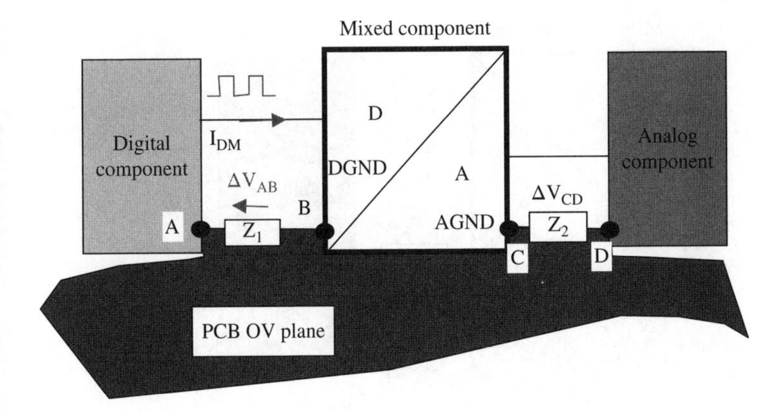

Figure 6.83 PCB layout of digital-analog mixed circuit.

high-speed digital signal current I_{DM} flows. For example, the frequency of a square wave clock signal is 50 MHz, its current $I_{DM}$is 15 mA, and then its fundamental current I_0 is:

$$I_0 = 0.64 \bullet I_{DM} = 0.64 * 15\,\text{mA} = 9.6\,\text{mA}$$

The third-order harmonic (150 MHz) current I_3 is:

$$I_3 = I_0/3 = 9.6\,\text{mA}/3 = 3.2\,\text{mA}$$

The length of the 50 MHz square wave signal trace on the PCB is 10 cm, then the voltage drop due to the third harmonic current flowing on the following different ground return paths are assessed as below.

When the return path of the clock signal is a thin printed trace, the voltage drop ΔU_{AB} generated by the third harmonic return current on this thin printed trace is as follows:

$$\Delta U_{AB} = Z_3 \bullet I_3 = 105\,\Omega * 0.0032\,\text{A} \approx 0.336\,\text{V}$$

where Z_1 = 105 Ω, which is the approximated impedance of a thin PCB trace at 150 MHz.

When the return path of the clock signal is a square ground plane with vias, the voltage dropΔU_{AB} generated by the third harmonic return current on this ground plane is as follows:

$$\Delta \text{U}_{\text{AB}} = \text{Z1} * 0.0032\,\text{A} \approx 2.65\,\text{mV}$$

where, Z_1 = 0.83 Ω, which is the approximated impedance of the square ground plane at 150 MHz. When the return path of the clock signal is a meshed ground plane on which the side length of the grid is 5 mm, the voltage drop ΔU_{AB} generated by the third harmonic return current on this ground plane is as follows:

$$\Delta U_{AB} = Z_1 \bullet I_3 = 1.1\,\Omega * 0.0032\,\text{A} \approx 3.5\,\text{mV}$$

where Z_1 = 1.1 Ω, which is the approximated impedance of the ground grid plane at 150 MHz.

When the return path of the clock signal is a ground plane without via holes, and the ground plane area is much larger than the area of the traces (i.e. 100% of the signal return current is confined in the ground plane), which is perfect square ground plane, the voltage drop ΔU_{AB} generated by the third harmonic return current on this ground plane is as follows:

$$\Delta U_{AB} = Z_1 \bullet I_3 = 0.005\,\Omega * 0.0032\,\text{A} \approx 16\,\mu\text{V}$$

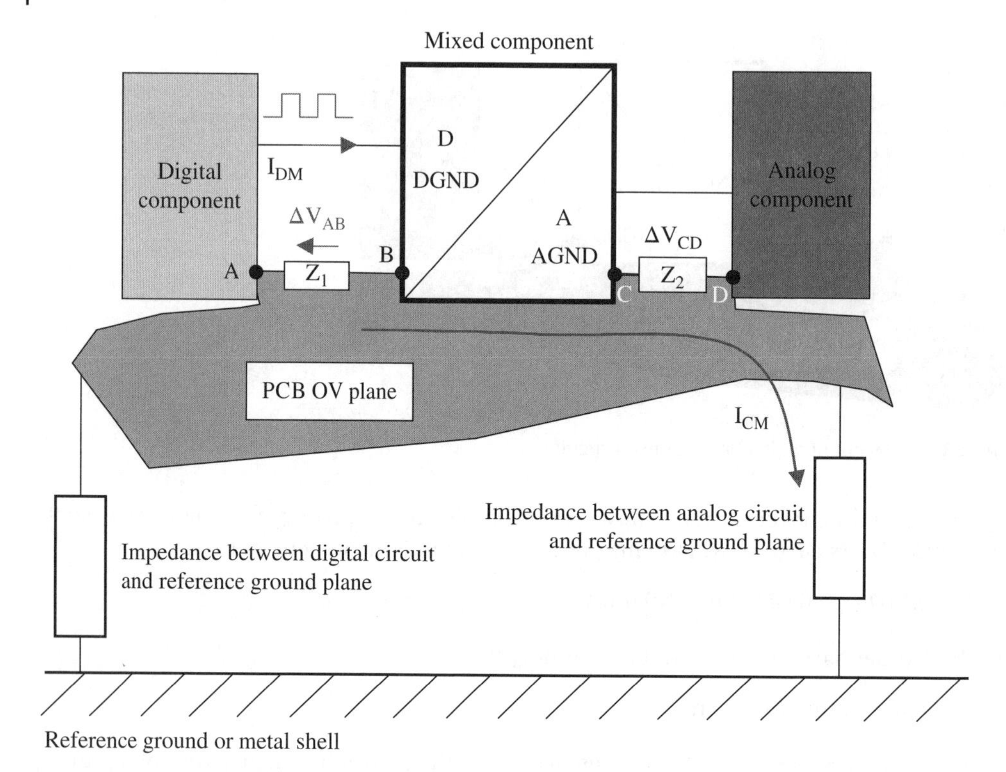

Figure 6.84 Working schematic diagram of digital circuit ground noise affecting analog circuit.

where $Z_1 = 0.005\,\Omega$, which is approximated impedance of the ground plane without vias at 150 MHz.

The above examples show that reducing the ground impedance of the digital circuit can help reduce the ground noise.

We will analyze how the noise (voltage) from the digital circuit affects the analog circuit. For analytical purposes, it can be assumed that there are two extremities. Assume that, for the product shown in Figure 6.84, the noise voltage on the digital ground when the return current of the digital signal flows on the digital ground is ΔU_{AB}, and the impedance Z_A between the analog circuit and the ground reference plate or the metal casing is infinite. Under this condition, obviously there will be no common-mode current flowing through the analog circuit, i.e. $I_{CM} = 0$, and the voltage difference between the analog ground position C and position D is $\Delta U_{CD} = 0$, i.e. to say the point B, point C, point D shown in Figure.6.84 are equipotential, so the voltage difference from these points to point A is ΔU_{AB}. Since points C and D are equipotential, $\Delta U_{CD} = 0$; therefore, this ΔU_{AB} will not disturb the transferred analog signals between points C and D (the normal operation signal is transferred in differential mode).

Further assume that, for the product shown in Figure 6.84, the noise voltage on the digital ground when the return current of the digital signal flows on the digital ground is ΔU_{AB}, and the impedance from the analog circuit to the ground reference plate or the metallic casing is $Z_A = 0$. Under this condition, clearly there is a common-mode current flowing through the analog circuit, i.e. $I_{CM} \neq 0$, and when the nonzero common-mode current flows from the position C to the position D of the analog circuit, the voltage difference $\Delta U_{CD} \neq 0$ arises. Because the potential difference between the point C and point D is directly superimposed on the analog signals transferred between the point C and point D, ultimately the analog circuit is disturbed.

Suppose the impedance of analog circuit ground between point C and point D shown in Figure 6.84, $Z_2 = 0$, and no matter the impedance Z_A is infinite or zero, ΔU_{CD} is equal to 0. There will be no disturbance on the analog circuit.

As we can see from the analysis, the noise voltage from the digital circuits does not directly affect the normal operation of the analog circuits. When the noise voltage of the digital circuit is not transformed to the current flowing through the analog circuit (the common-mode current), the mutual interference phenomenon between the analog circuit and the digital circuit through ground will not occur. Only when the noise voltage from the digital circuit is transformed to the current that flows through the analog circuit (the analog ground) will the voltage directly impact the operation of the analog circuit can be produced. It can be seen that it is the voltage rather than the current that directly impacts the normal operation of the analog circuit, but this voltage is not the noise voltage on the digital ground or the common-mode noise. However, this ground noise voltage drives the analog circuit and produces the common-mode current flowing through the analog circuit, and when the common-mode current flows through the analog ground, there will be another ground noise (the noise voltage on the analog ground) in the analog circuit, and this voltage can directly affect the normal operation of the analog circuits.

These assumptions are several extreme situations, which hardly happen in the actual circuit, but it can be seen that, from the analysis of these assumptions, the digital noise affecting the analog circuit is not only related with the impedance between the analog ground and the ground reference plate or the metallic casing, but also related to the ground impedance of the circuit ground. To reduce the impact on the normal operation of the low-voltage (mV or less) analog circuit, from a product design perspective, we should try to increase the impedance between the analog circuit section (or the analog ground) and the ground reference plate or the metallic casing, and reduce the analog ground impedance (the impedance of a well-designed ground plane could be 3.7 mΩ at 100 MHz). But in the actual product, there will be the following situations:

- For some products, the impedance between the analog circuit section (or the analog ground) and the ground reference plate or the metallic casing must be very low, i.e. the impedance Z_A shown in Figure 6.84 is very small, which generally occurs in the following four design scenarios:
  1) At the far end of the analog circuit, the analog circuit is connected with the ground reference plate or the metallic casing (directly or through a capacitor), as shown in Figure 6.85.

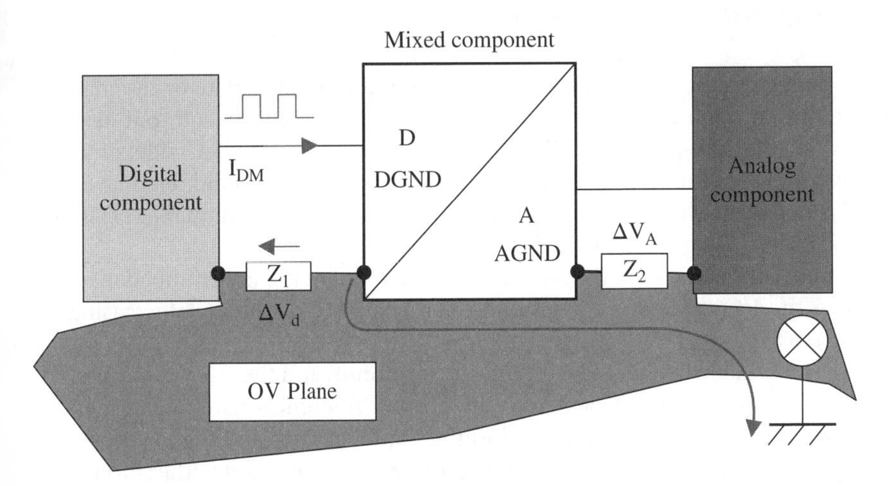

Figure 6.85 Schematic diagram of analog circuit working ground interconnecting with reference ground or metal shell.

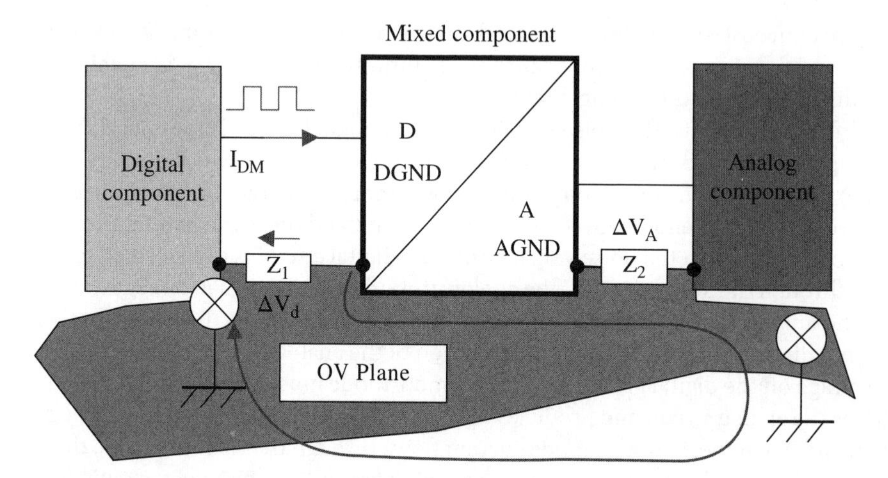

Figure 6.86 Schematic diagram of analog circuit working ground interconnecting with reference ground or metal shell.

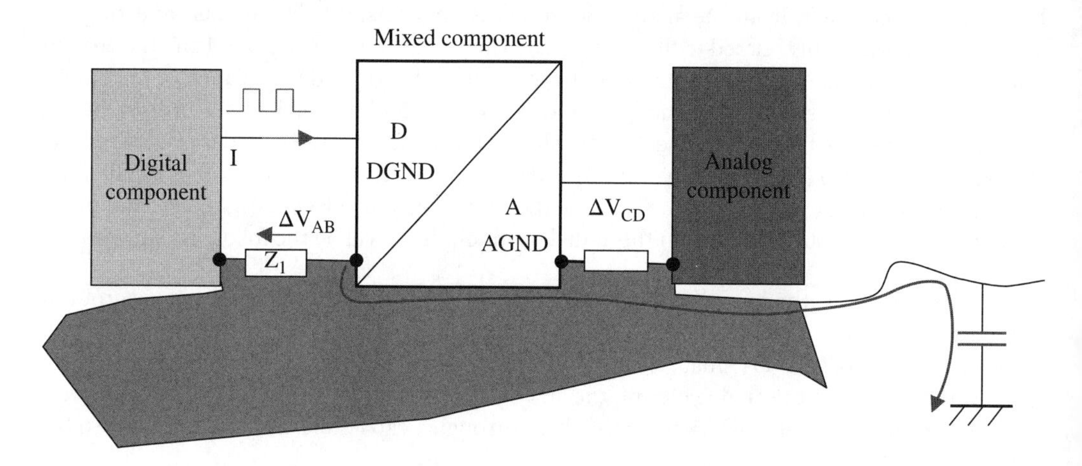

Figure 6.87 Schematic diagram of I/O cable of analog circuit.

2) Not only the analog circuit is connected with the ground reference plate or the metallic casing (directly or through capacitive) at the far end of the analog circuit, but also the digital ground at the low-potential side of the digital circuit (the left side of Z_1 in Figure 6.84) is connected (directly or through a capacitor) with the ground reference plate or the metallic casing, as shown in Figure 6.86. In this scenario, the mutual interference will be greater in the analog-digital mixed circuit.
3) There is I/O cable connected with the analog circuit, and there is a lower impedance (i.e. 150 Ω) between the I/O cable and the ground reference plate or the metallic casing, as shown in Figure 6.87.
4) Not only there is I/O cable connected with the analog circuit, and there is a lower impedance (i.e. 150 Ω) between the I/O cable and the ground reference plate or the metallic casing, but also there is I/O cable connected to the digital circuit, and the low-potential side of the digital circuit (the left side of Z_1 in Figure 6.84) is connected (either directly or through a capacitor) with the ground reference plate or the metallic casing, as shown in Figure 6.88.

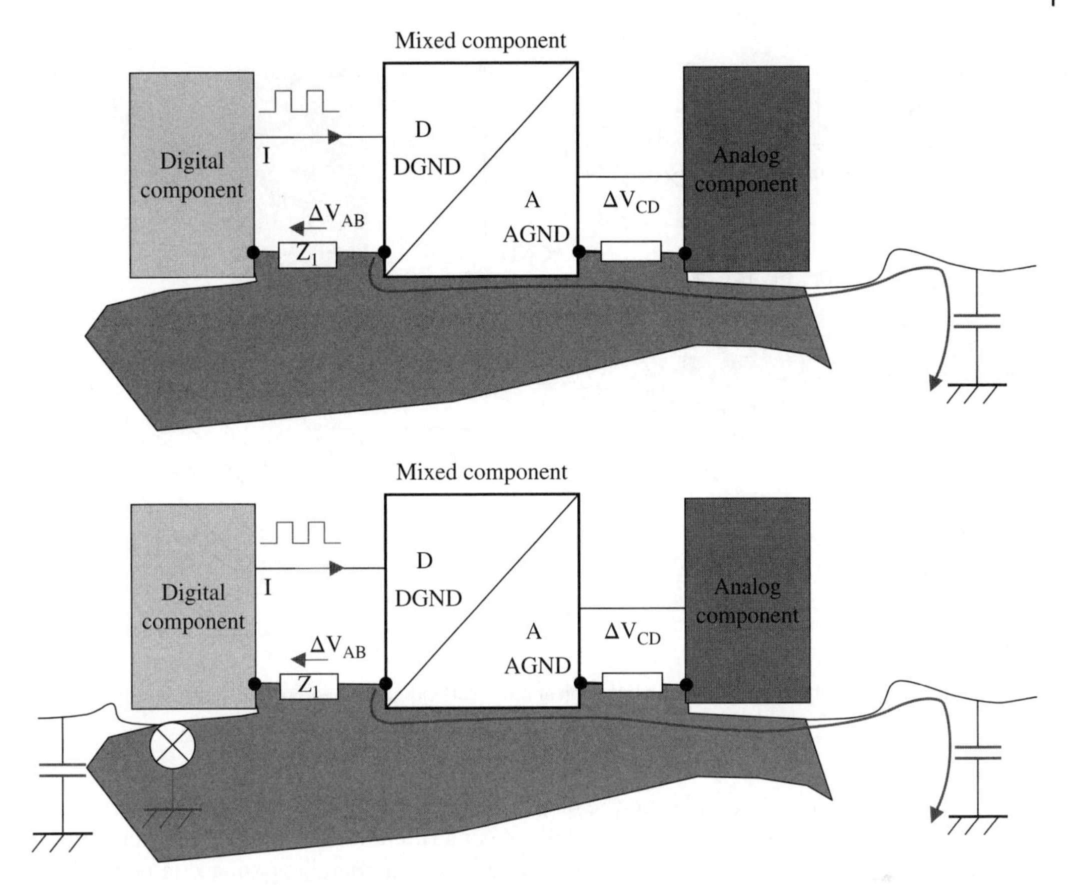

Figure 6.88 Schematic diagram of I/O cable of analog circuit.

In the four scenarios respectively shown in Figures 6.85–6.88, the common-mode noise voltage on the ground can be converted to the common-mode current, which will flow through the analog circuit, because, in these four scenarios, the ground noise voltage ΔU_{AB} can cause a current, and this current will flow through the analog circuit area. At this time, the common-mode current is multiplied with the impedance of the analog ground, i.e. a noise voltage appears on the analog ground. If this noise voltage exceeds the precision of the analog circuit or the lowest voltage of the analog circuit, then the analog circuit will be disturbed.

- For the product designed with single layer or double layer PCB, because there is no complete plane, the ground impedance of the digital circuit must be high, and then the common-mode noise voltage on the digital ground will be high. Meanwhile, the ground impedance of the analog circuit is also high, and then the noise voltage on the analog ground will be high, which is caused when the current driven by the noise voltage on the digital ground flows through the analog ground, so the mutual interference in the digital-analog mixed circuit will be more severe. If, at the far end (the right side of Z_2 in Figure 6.84) of the analog circuit, the analog ground is connected with the ground reference plate or the metallic casing, the mutual interference will be even more severe.

In this case, you can use the low-impedance characteristics of the metallic casing, as long as the connection point between the PCB ground and the metallic casing is chosen reasonably, and the connection way is correct, good design results can be achieved. For the product shown

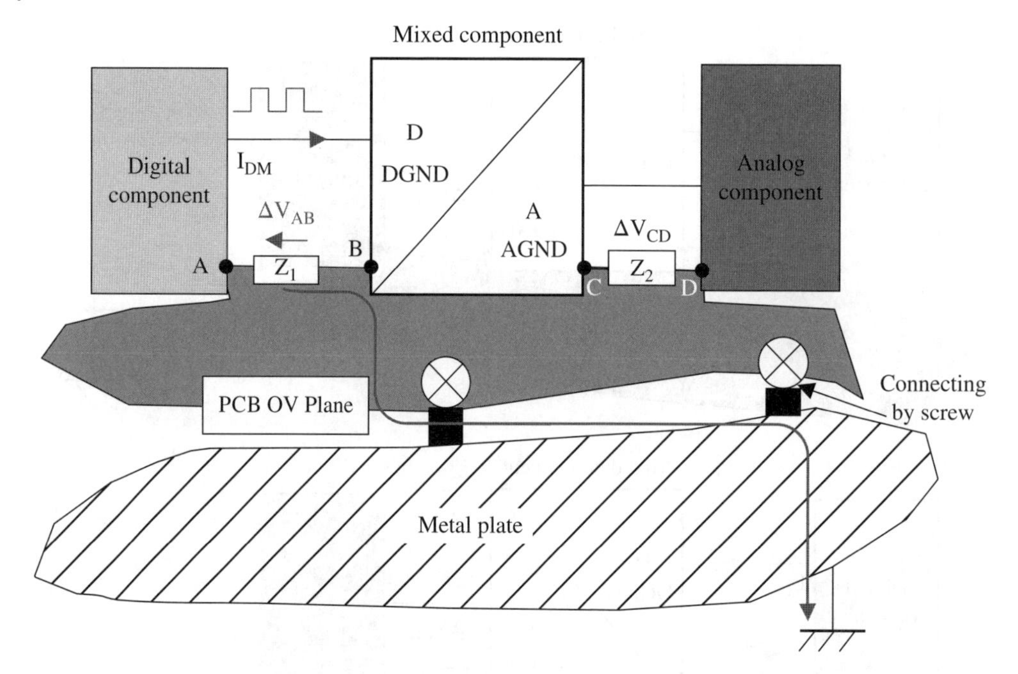

Figure 6.89 Correct multipoint connection between analog working ground and metal shell.

in Figure 6.89, near the position D and the position between B and C, when the PCB ground is connected with the good and complete metallic casing (no holes and no slots) by screws, the common-mode interference current can be bypassed to the metallic casing due to the low-impedance of the metallic casing, and then the interference voltage ΔU_{CD} will be greatly reduced, which directly impacts the operation of the analog circuit.

It is worth noting that the position selection of the connection between the PCB reference ground and the metallic casing is not random. As shown in Figure 6.90, when the PCB reference ground is connected with the metallic casing near point A and point D, the noise voltage ΔU_{AB} generated in the digital circuit is not reduced, the converted common-mode current flowing through the analog circuit will not be reduced (in contrast, it will be increased), and then the interference voltage ΔU_{CD} on the analog circuit will not be reduced (in contrast, it will be increased).

On the basis of the product design shown in Figures 6.90 and 6.89, if there is an additional connection between the PCB ground and the metallic casing (as shown in Figure 6.91), the situation will be greatly changed, due to the metallic casing between point D and point B, point C, the analog circuit is less affected by the common-mode interference current (the common-mode interference current is bypassed by the metallic casing between point D and point B, point C). In addition, the noise voltage U_{AB} on the digital circuit is also bypassed by the metallic casing between point A and point B. This is a perfect connection way.

[Solutions]

From this analysis, at the border area between the analog circuit and the digital circuit, another connection between the PCB reference ground and the metallic casing is added. In addition, for the product discussed in this case, disconnecting the link between the analog ground on the PCB and the metallic casing is certainly helpful for solving the interference problem.

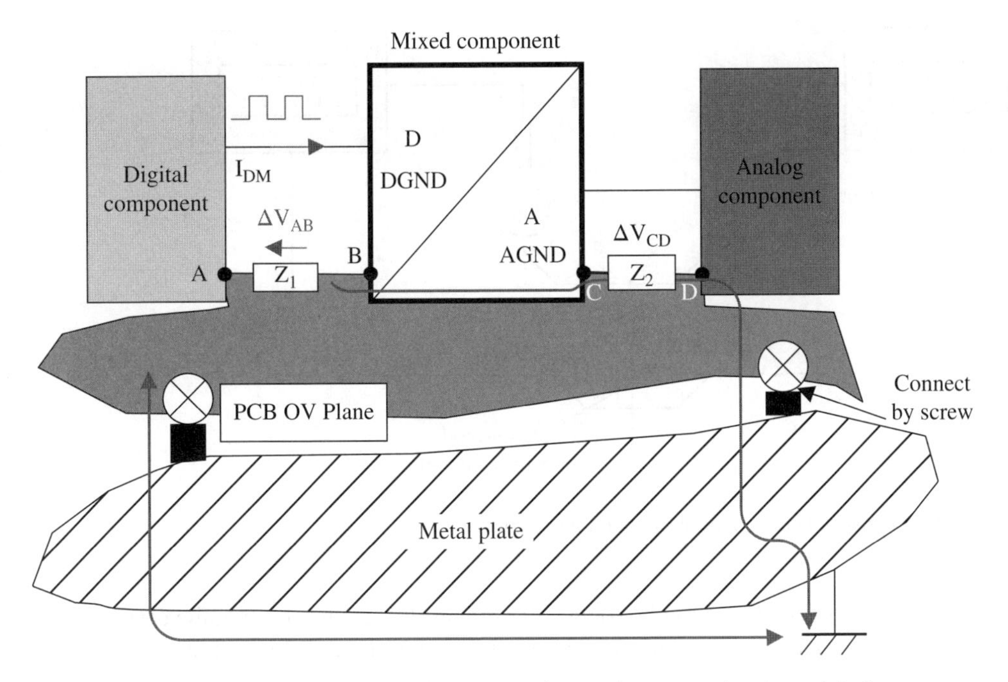

Figure 6.90 Wrong multipoint connection between analog working ground and metal shell.

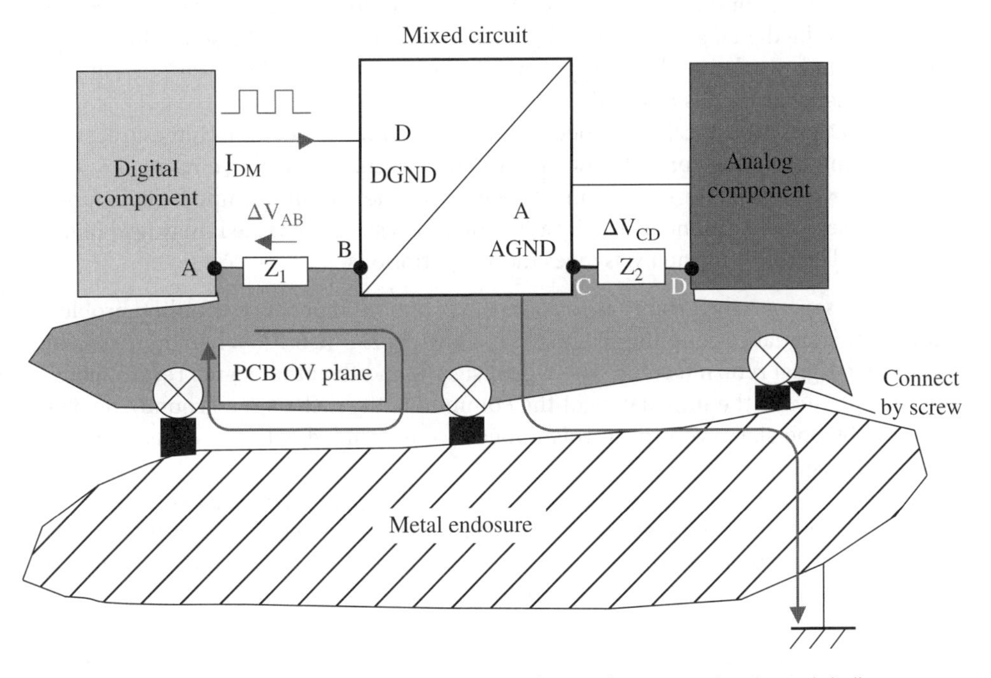

Figure 6.91 Prefect multipoint connection between analog working ground and metal shell.

[Inspirations]

For products with metallic casing, any kind of mutual interference described as above occurs, the metallic casing can be used to solve the problem. By reasonably choosing the connections between the PCB reference ground and the metallic casing, the mutual interference problem in

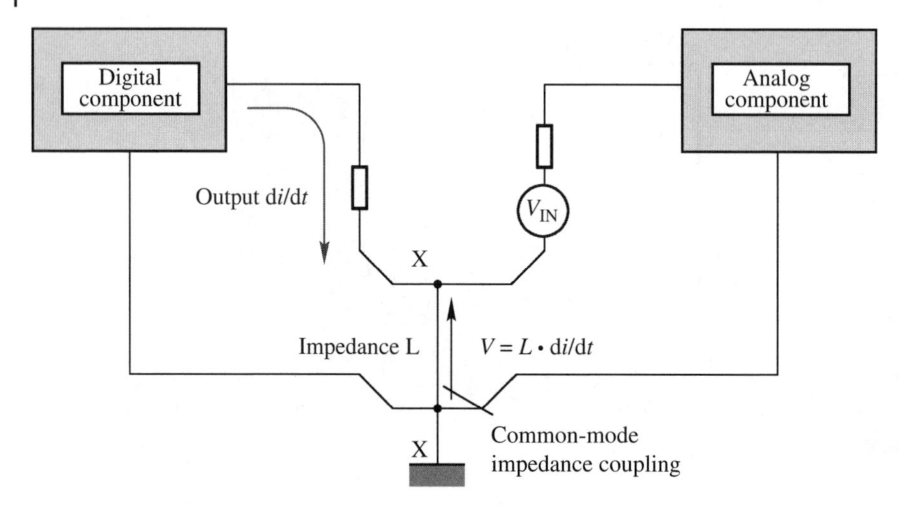

Figure 6.92 Schematic diagram of public impedance coupling.

the analog-digital mixed circuit can be solved. For the products with nonmetallic casing, naturally there are fewer alternative solutions. The general disposals are listed as follows,

1) Increasing the number of PCB layer, the main purpose is to enlarge the ground plane and reduce the impedance of the ground plane. With this solution, not only the common-mode noise caused by the digital signal on the digital ground is reduced, but also the interference voltage generated on the analog ground when the common-mode noise current flows through the analog ground is reduced.
2) Change the location of the cable connector and the original product architecture, so that there is no common-mode current flowing through the sensitive analog circuit.
3) In the PCB design, separate the ground to eliminate the common impedance coupling. However, the separated ground is usually a "banded" measures, because it may lead to more EMC defects. The following analysis gives more illustrations.

The purpose to separate the ground is to solve the common impedance coupling problems, and the common impedance coupling schematic is shown in Figure 6.92. When there is a common path for one signal return (such as the digital signal) and the other signal return (such as the analog signal), due to the impedance of the common path (such as the ground), a voltage drop generated by one signal return on this impedance is coupled with the other signal and then the interference occurs.

In the design of PCB, if these two signal return paths are separated, known as "separate the ground" as shown in Figure 6.93, you can solve the common impedance coupling problem shown in Figure 6.92. In this way, these two signal loops are independent, and even if they are interconnected at only one point, we can also achieve a common reference potential.

We can see that "separate the ground" seems to be a relatively simple and a good solution for solving the interference problem of the analog-digital mixed circuit. However, some severe EMC defects are possible:

1) After the ground is split or separated, the length-to-width ratio of the ground conductor or the ground plane will be inevitably increased. For certain circuit (such as high-speed digital circuit), it means that the impedance of its signal return path is increased, and eventually the common-mode noise voltage on the ground is increased. Meanwhile, the immunity of these circuits to the external noise (mainly the common-mode noise) is decreased.

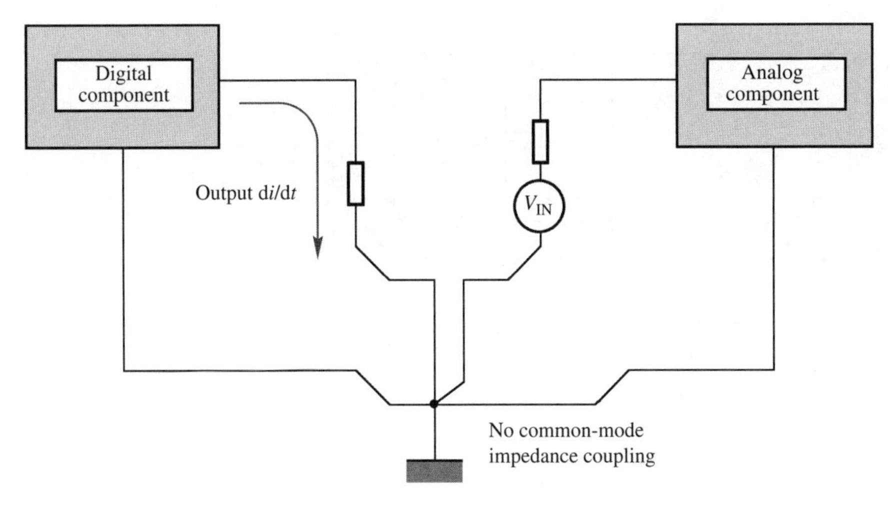

Figure 6.93 Schematic diagram of the solution of public impedance coupling by separating signal reflux ground.

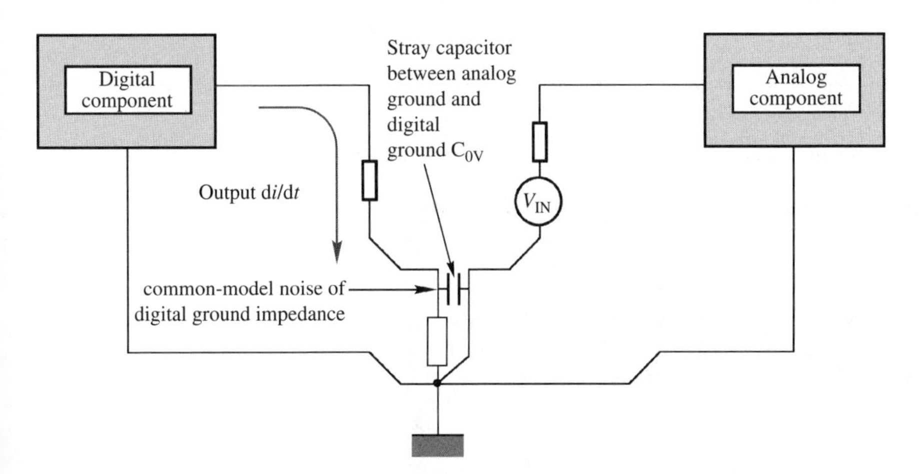

Figure 6.94 Schematic diagram of ground noise coupling caused by parasitic capacitance between ground lines.

2) Splitting the ground or separating the ground easily leads to the signal trace crosses the boundary between the separated grounds, and then the area of the signal loop is greatly increased.
3) At high frequencies, splitting ground or separating ground does not completely solve the problem of noise coupling between different circuits. Figure 6.94 shows the ground noise coupling because of the parasitic capacitance between ground traces.

It can be seen that the design of separating the ground can't be easily implemented, for the actual product design. Separating the ground can be considered, which can be divided into two general types of conditions:

1) It usually occurs in single-layer board and double-layer board. When the single-layer board and the double-layer board cannot be designed with a more complete ground plane, separating the ground can be used to reduce the mutual interference of the analog-digital mixed circuit by partially scarifying the overall EMC performance.
2) It occurs in the ground of four-layer PCB board, which could be designed with a ground plane. Figures 6.95 and 6.96, respectively, show how the digital noise flows without separating ground, and how the digital noise flows with separating ground on the PCB.

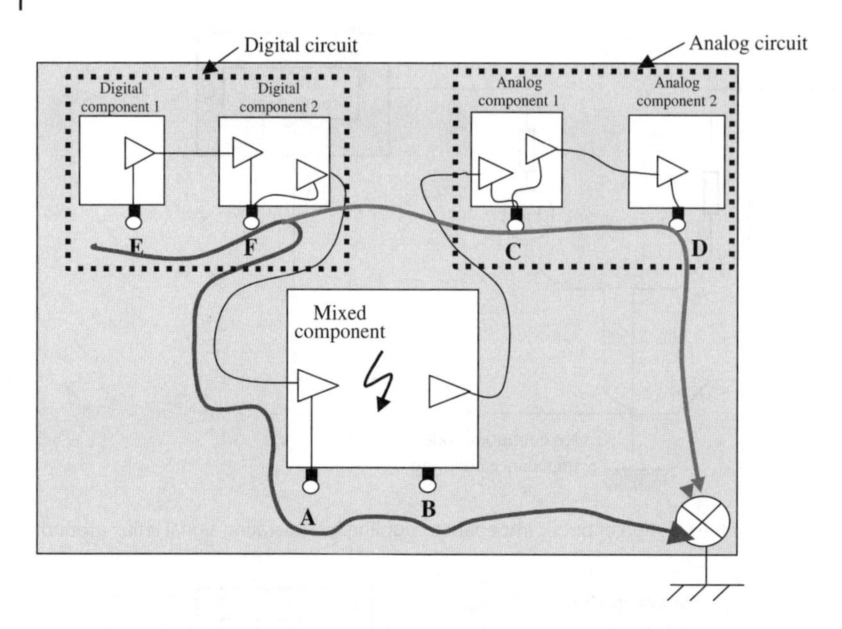

Figure 6.95 Digital circuit noise interference path when PCB do not divide ground.

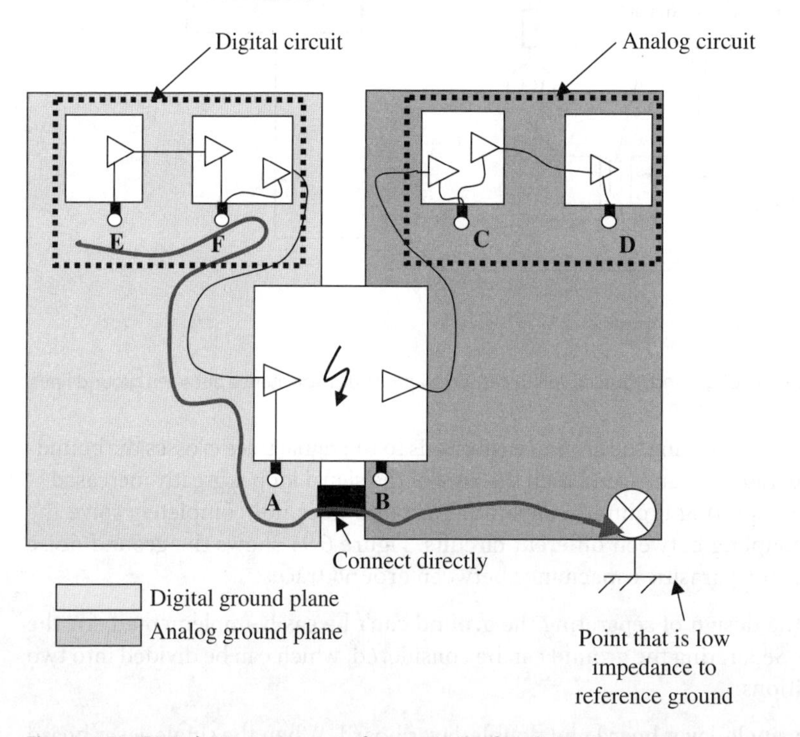

Figure 6.96 Digital circuit noise interference path after dividing ground.

From the arrows (the flowing direction of the common-mode interference current caused by the digital ground noise) in Figures 6.95 and 6.96, it can be clearly seen that the current path of the common-mode interference current originally flowing into the analog circuit in Figure 6.95 is altered by the separated ground, so that the common-mode interference current does not

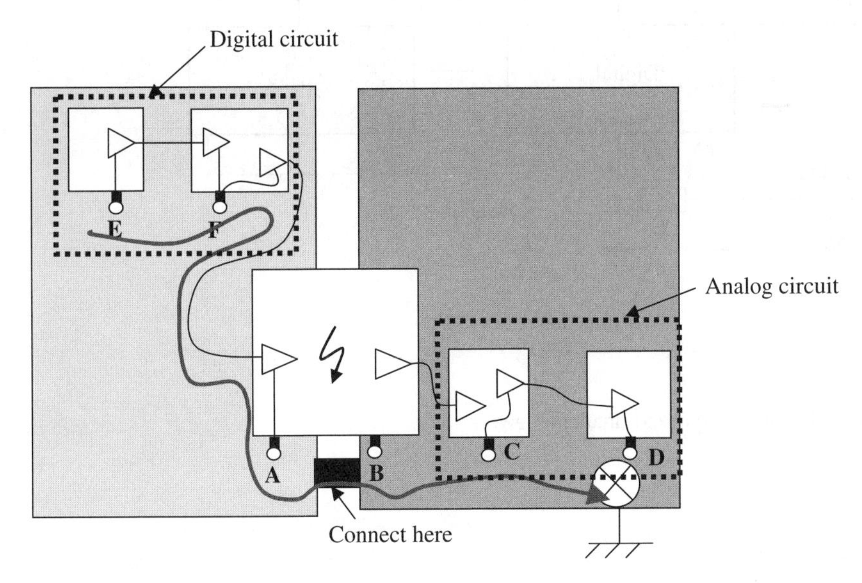

Figure 6.97 Digital circuit noise interference path still flows through analog sensing circuit area after dividing ground.

pass through the sensitive analog circuit (the area between point C and point D in Figure 6.96). However, it is worth noting that, in the PCB shown in Figure 6.96, if there is an I/O cable on the analog sensitive circuit (there will be a low-impedance between the analog ground and the ground reference plate, the common-mode current will flow through the sensitive analog circuit), then with the way of separating ground, the effect of improving the mutual interference in the analog-digital mixed circuit will be deteriorated, and it may cause the immunity test problems for the product.

Consider the situation shown in Figure 6.97, after separating the ground, the noise current in the digital circuit still flows through the sensitive analog circuit. Therefore, the purpose of separating ground is not to prevent the transmission of the common-mode voltage, but to change the path of the common-mode current generated by the common-mode voltage noise, to achieve the effect of separating ground, we must start from the analysis of the common-mode current path. The separating ground is effective only when it causes the common-mode interference current path is away from the sensitive analog circuit.

The main concern for analog-digital mixed circuit designers is how the digital signal will affect the low-level analog signal. How should an analog-digital mixed circuit be designed so that the digital signal doesn't impact the analog signal? Here are some considerations:

1) Try to construct the overall structure as shown in Figure 6.98 as much as possible. Separate the components and the signal traces, ensure that the digital components are in one area and the analog components or signal traces in another area, and insert ground guard trace between the digital circuit and the analog circuit on PCB external layers.
2) Use the metallic casing to bypass the common-mode noise of the digital ground and the common-mode current, which is caused by the common-mode noise of the digital ground and flows through the analog circuit.
3) From the flowchart shown in Figure 6.99, you can determine whether the digital and analog ground need to be separated. This flowchart shows the design thought of the analog-digital mixed circuit.

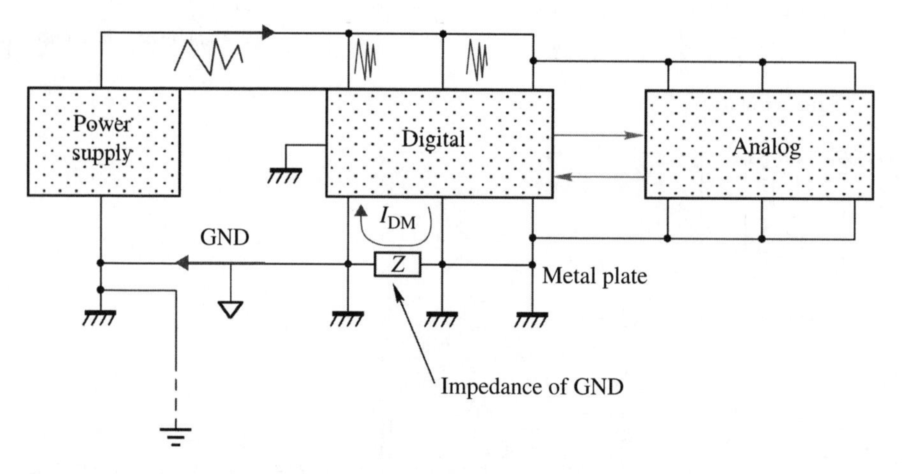

Figure 6.98 Very good digital-analog mixed circuit framework.

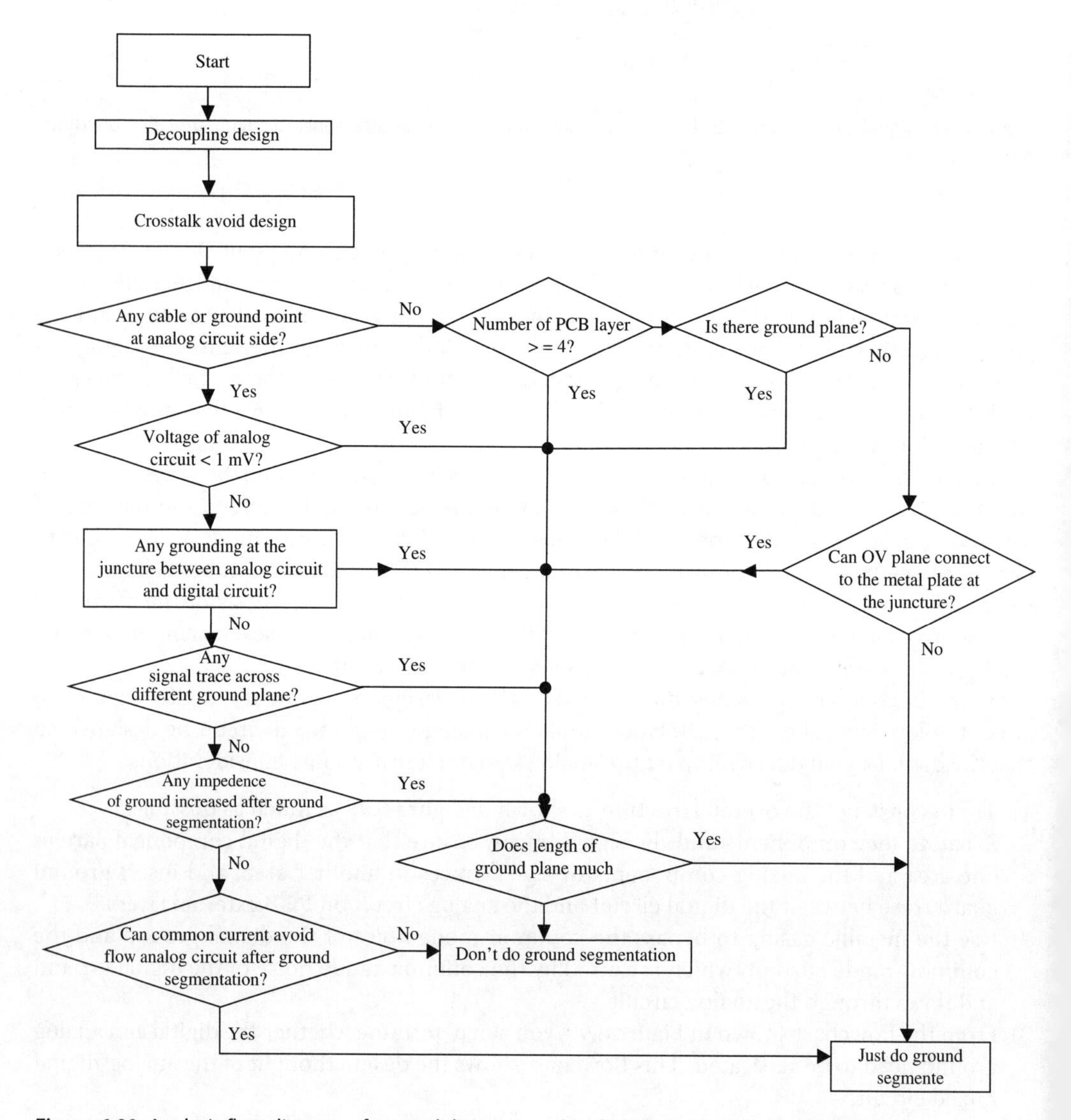

Figure 6.99 Analysis flow diagram of ground division need in digital-analog mixed circuit design.

6.2.14 Case 70: Why the Oscillator Cannot Be Placed on the Edge of the PCB

[Symptoms]

For a plastic product with one I/O cable, when the radiated emission test is performed per the EMC standard for marine product, it is found that the radiated emission is over the standard limit, and the specific frequency is 160 MHz at which the emission exceeds the standard limit. We need to analyze the root cause for its excessive radiation and find the corresponding countermeasures. The spectral plot of the tested radiated emission is shown in Figure 6.100.

[Analyses]

There is only one PCB board in this product, in which there is one 16 MHz oscillator. Thus, the radiated emission at 160 MHz should be related to the oscillator. (Note: It does not mean the excessive emission is directly radiated by the oscillator.) Figure 6.101 shows the partial PCB layout of the product, and from Figure 6.101, we can clearly see that the 16 MHz oscillator is placed just at the edge of the PCB.

When the product is placed in the test environment for the radiated emission test, there is the capacitive coupling between the high-speed signal traces or the high-speed components of the product under test and the ground reference plate in the laboratory, which means there is the electric field distribution or the parasitic capacitance between the high-speed signal trace or the high-speed components of the product and the ground reference plate in the laboratory, and the parasitic capacitance is very small (such as less than 0.1 pF), but it will still lead to the common-mode radiation from the product. The schematic illustrating how the common-mode radiation is produced is shown in Figure 6.102. In Figure 6.102, the voltage on the oscillator shell (its shell is not connected to 0 V ground) or the voltage on the clock signal pin of the oscillator

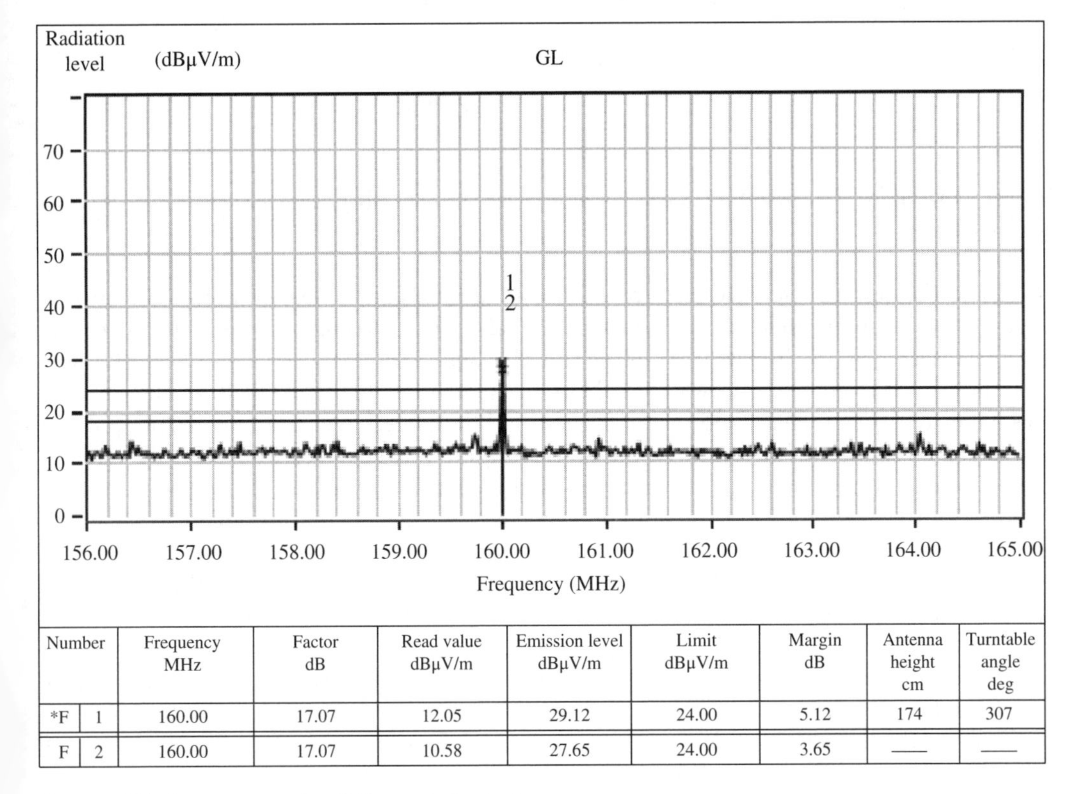

Number		Frequency MHz	Factor dB	Read value dBμV/m	Emission level dBμV/m	Limit dBμV/m	Margin dB	Antenna height cm	Turntable angle deg
*F	1	160.00	17.07	12.05	29.12	24.00	5.12	174	307
F	2	160.00	17.07	10.58	27.65	24.00	3.65	——	——

Figure 6.100 Spectrogram of radiation emission test.

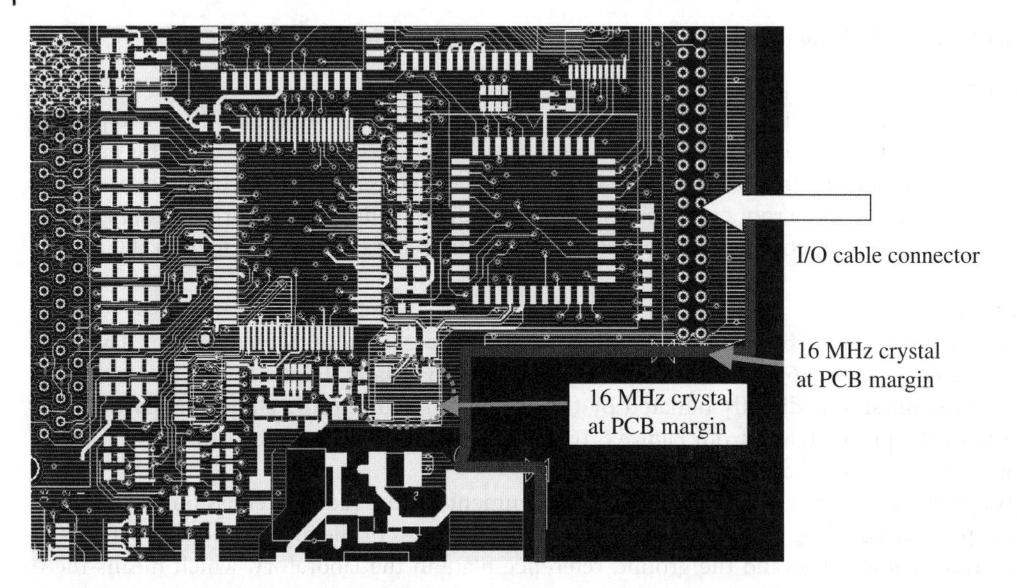

Figure 6.101 Partial PCB layout of the product.

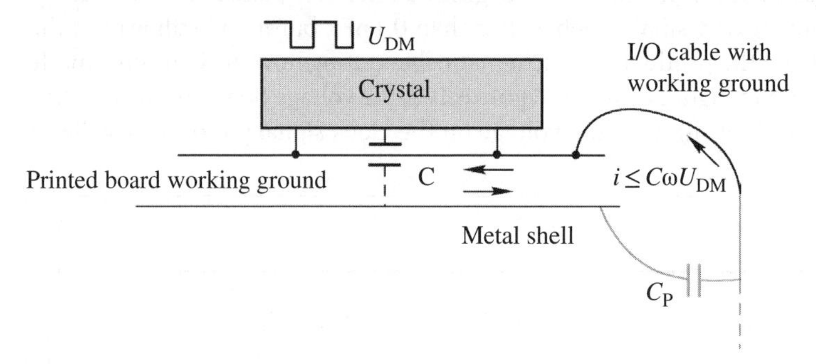

Figure 6.102 Principle of capacitive coupling between crystal and reference ground board leading to radiation emission.

U_{DM} and the ground reference plate constitute a parasitic path. The common-mode current leads to the common-mode radiation through the cable, and this common-mode current can be estimated as $I_{CM} \approx C \cdot w \cdot U_{DM}$, where C is the parasitic capacitance between the PCB signal traces and the ground reference plate, which is about one-tenth of pF to several pF; C_P is the parasitic capacitance between the ground reference plate and the cable, which is about 100 pF; and ω is the angular frequency of the signal. So the common-mode current I_{CM} is from several microamperes to tens of micro amps. The common-mode radiation formula (6.2) (refer to Case 66) shows that the magnitude of this common-mode current flowing on the cable is sufficient to cause the radiated emissions beyond the standard limit.

Then to analyze why, the radiated emission is over the standard limit when the oscillator is placed at the edge of the PCB, and if the oscillator is moved to the internal side of the PCB board, the radiation emission is passed.

It can be seen from this analysis that the essence of the coupling between the oscillator and the ground reference plate is the parasitic capacitance between the oscillator and the ground reference plate, which leads to the common-mode radiation from the cable. It means the larger the parasitic capacitance is, the stronger the coupling between the oscillator and the ground reference plate is, the greater the common-mode current flowing through the cable is and then the

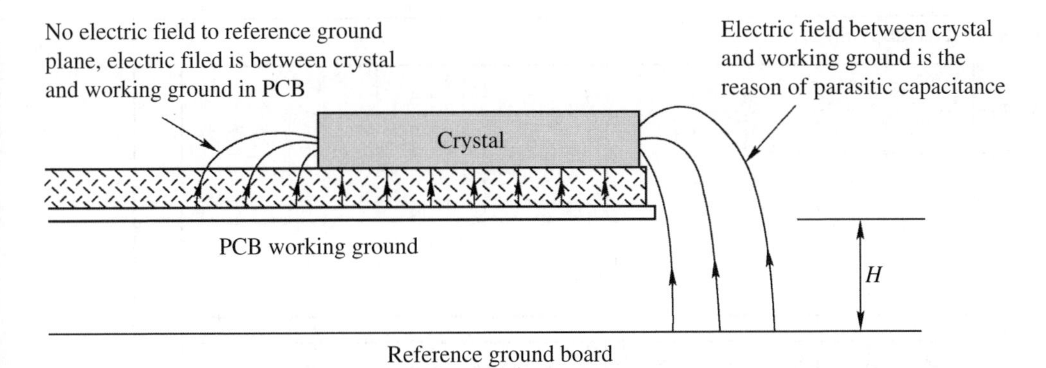
No electric field to reference ground plane, electric filed is between crystal and working ground in PCB
Electric field between crystal and working ground is the reason of parasitic capacitance
Crystal
PCB working ground
H
Reference ground board

Figure 6.103 Electric field distribution diagram between crystal at the margin of PCB and reference ground board

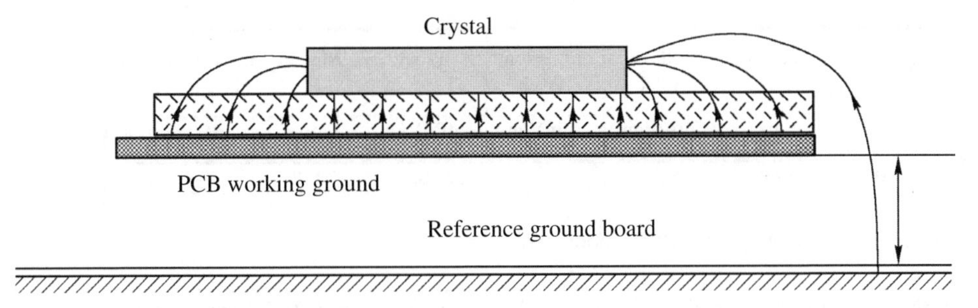
Crystal
PCB working ground
Reference ground board

Figure 6.104 Electric field distribution diagram between crystal in the middle of PCB and reference ground board.

greater the common-mode emission radiated by the cable is, and vice versa. What is the essence of the parasitic capacitance? In fact, the parasitic capacitance between the oscillator and the ground reference plate means the electric field distribution between the oscillator and the ground reference plate, and when the voltage difference between these two is constant, the more the distributed electric field is, the greater the electric field strength between these two will be and the greater the parasitic capacitance between these two will be. When the oscillator is placed at the edge of the PCB, the distribution of the electric field between the oscillator and ground reference plate is shown in Figure 6.103. When the oscillator is placed in the middle of the PCB, or far away from the edge of the PCB, the electric field distribution is shown in Figure 6.104.

From the comparison between Figures 6.103 and 104, we can see that, when the oscillator is placed in the middle of the PCB, or far away from the edge of the PCB, the reference ground (GND) plane in the PCB confines the most of the electric field between the oscillator and the reference ground (GND), i.e. in the middle the PCB, the electric field distributed to the ground reference plate is greatly reduced and then the parasitic capacitance between the oscillator and the ground reference plate is greatly reduced. Then it is not difficult to understand why an oscillator placed at the PCB edge will lead to the excessive radiation, and after it is moved to the middle of the PCB, the radiated emission is reduced.

[Solutions]

Move the oscillator to the middle of the PCB, place it at least 1 cm away from the edge of the PCB ground plane, and pour copper plane in the surrounding area which is within 1 cm distance from the oscillator on the external layer of the PCB, meanwhile this poured copper plane on the external layer needs to be connected to the PCB ground plane. After this modification,

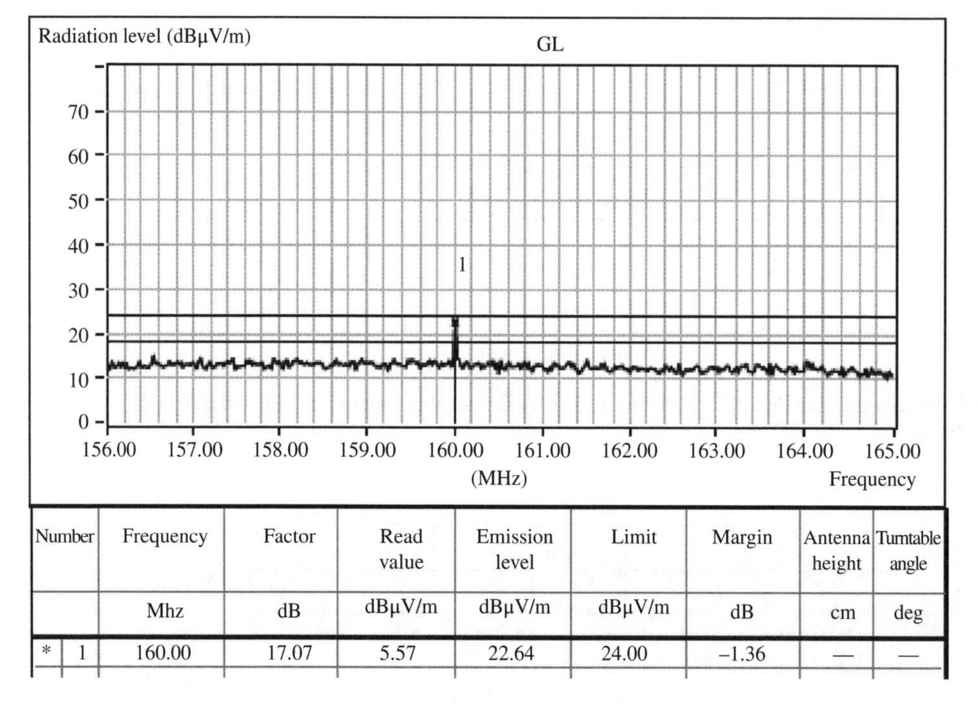

Number		Frequency	Factor	Read value	Emission level	Limit	Margin	Antenna height	Turntable angle
		Mhz	dB	dBμV/m	dBμV/m	dBμV/m	dB	cm	deg
*	1	160.00	17.07	5.57	22.64	24.00	−1.36	—	—

Figure 6.105 Spectrogram of test result after modifying.

the spectral plot of the tested radiated emission is shown in Figure 6.105. From Figure 6.105, we can see that the radiated emission is significantly reduced.

[Inspirations]

1) The capacitive coupling between high *dU*/*dt* printed traces or components and the ground reference plate will cause EMI problems, and the immunity problem may occur if the sensitive trace or components is placed at the edge of the PCB.
2) Avoid placing high *dU*/*dt* traces or components at the edge of the PCB, if, in the design of PCB, they must be placed at the edge of the PCB for some reasons, a reference ground (GND) trace can be placed between these traces or components and the edge of the PCB, and it shall be connected to the reference ground (GND) plane with vias.
3) Eliminate a kind of misunderstanding. Do not think the emission is directly radiated by the oscillator, the oscillator is actually small, which directly impacts the near-field radiation (it is represented as the parasitic capacitance between the oscillator and the other conductors, such as the ground reference plate), the direct factor causing the far field radiation is the biggest part in size, i.e. the cable or the conductors inside the product, of which the size is comparable with the wavelength of the radiation frequency.

6.2.15 Case 71: Why the Local Ground Plane Needs to Be Placed Under the Strong Radiator

[Symptoms]

For a household electrical appliance with plastic casing, its radiated emission exceeds the standard limit. In the process of locating the problem, a simple near-field probe is made with the oscilloscope and its probe to locate the problem, and the specific type of the oscilloscope and locating test schematic are as follows.

The test equipment: the oscilloscope is TDS784D (1 GHz), and the probe is P6245 (1.5 GHz).

Test method: The signal pin and the ground of the probe are connected to construct a ring, like a receiving antenna loop, which is shown in Figure 6.106, and this ring is put close to the circuit in order to find the position with the maximum radiation.

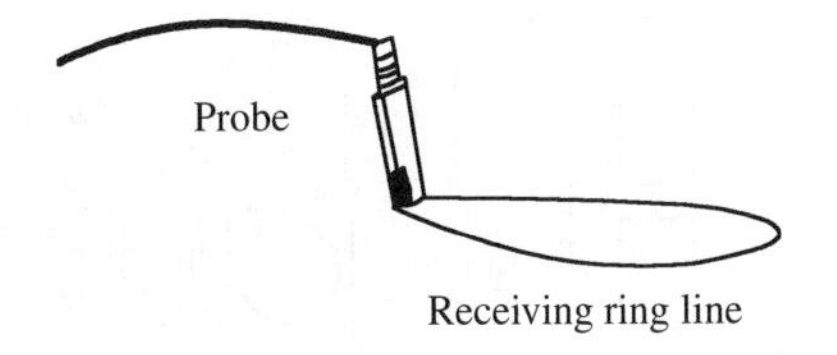

Figure 6.106 Receiving loop composed of probe and ground line.

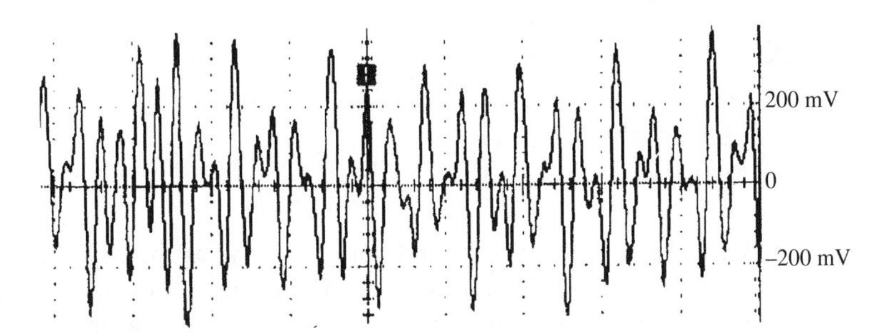

Figure 6.107 Noise peak-to-peak value when probe is close to crystal.

In the test, we found that when the probe is close to the oscillator clock or its clock output trace on the PCB, the radiation is the maximum. The peak-to-peak voltage of the maximum radiation when the probe is near the oscillator is shown in Figure 6.107, and the peak-to-peak voltage exceeds 400 mV. According to experience, this amplitude is too large.

Thus, we preliminarily determine that the design of the clock trace or the oscillator in this PCB is unreasonable. After observing further, we can see that the clock trace is too long and it is routed on TOP layer and BOTTOM layer, and the layout of the oscillator is not specially treated.

[Analyses]

The oscillator and its corresponded clock signal is one of the main interference sources because of the characteristics of the cyclic nature and fast rising edge, on the PCB; meanwhile, it includes rich harmonics.

The internal circuit of the oscillator, due to its characteristics, will produce RF current, and the RF currents produced inside its package may be large, so that the oscillator pins can't sufficiently drain this great Ldi/dt to the ground plane by consuming less power; therefore, the metallic shell becomes a monopole antenna. Sometimes there are two or three PCB layers between the shell of the oscillator and its nearest ground plane, thus, the coupling path from the RF current to the reference ground is insufficient. If the oscillator is a surface-mounted device, since the surface-mounted devices are often plastic packaged, at this time the situation will become even worse. The RF current generated within the package will be radiated into the space and coupled to the other devices, and the impedance of PCB material relative to the ground pin of the oscillator is very high, which will prevent the RF current flowing into the ground plane. These reasons cause high-frequency noise on the oscillator pin or its adjacent space.

Placing the oscillator, the crystal, and all the applicable clock circuits (such as buffers, drivers, etc. which usually have high-speed and high slew-rate characteristics) above a local ground plane is a simple and effective method of reducing the common-mode radiation from the oscillator, the crystal, and the clock circuits. The local ground plane is a local copper plane of the PCB, and it is usually directly connected from the PCB external layers whose components are

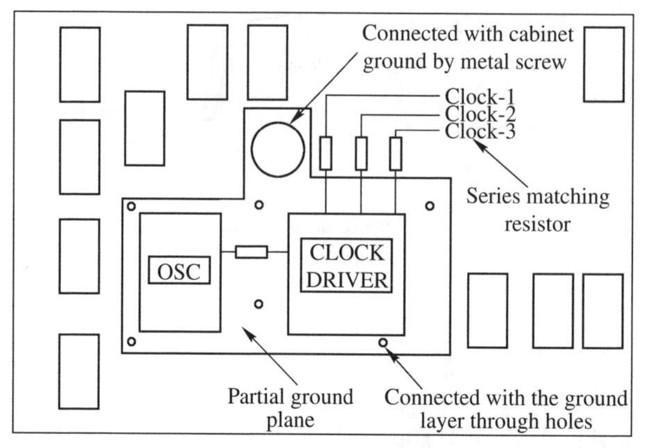

Figure 6.108 Example of partial ground plane below crystal.

on the main ground plane in the PCB inner layer by the ground pin of the crystal and at least two vias. In addition, the clock drivers, the buffers, etc. must be placed near the oscillator. The local ground plane should be extended under the circuits supporting the logics. One example of the local ground plane under oscillator is shown in Figure 6.108.

The main reason to pour the local ground plane below the clock generating circuits is that the local ground plane under the oscillator and the clock circuits provides a return path for the common-mode RF current generated inside the oscillator and its associated circuits. The RF field is confined in a much smaller range, so that the RF emission is minimized. In order to sustain the differential-mode RF current flowing on the local ground plane, we need connect the local ground plane to the other internal ground planes at multipoints. The local ground plane on the external layers is connected to the main ground plane in the PCB inner layer through the low-impedance vias. Sometimes, in order to improve the performance of the local ground plane, the clock generation circuit is placed near the chassis ground connection, which can take even better result.

At the same time, it should be avoided the signal traces go through the local ground plane. Otherwise it will deteriorate the effect of the local ground plane. If there are traces going through the local ground plane, the local ground plane will not be continuous, resulting in a small ground loop voltage. In high frequency, this ground loop will cause some problems.

[Solutions]

1) Pour the local ground plane under the oscillator, which is on the external layers of the PCB, and connect it to the ground layer through multiple vias.
2) Route the clock signal originally routed on external layers in the third layer (six-layer PCB board).

Install the modified PCB in the product, and then test it with the above mentioned near-field probe. The measured results are shown in Figure 6.109. The peak-to-peak voltage is about 40 mV.

[Inspirations]

1) In the design of multilayer PCB, it is recommended to pour a local ground plane under the oscillator. For the double-layer board, this method is even more important.
2) It is not recommended to route long clock traces on the external layers of a six-layer or more than six-layer board. The maximum recommended length of the clock signal on the external layers is 1/20 of the signal wavelength.
3) It is not recommended to route signals under the oscillator and its driving circuit, and the area less than 300 mil away from these circuits.

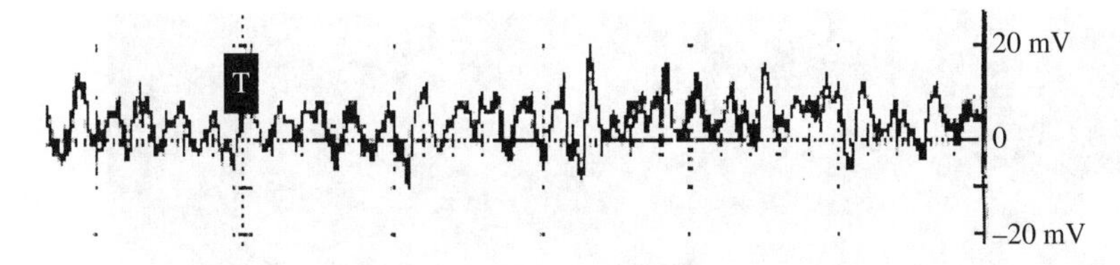

Figure 6.109 Noise peak-to-peak value after modifying.

6.2.16 Case 72: The Routing of the Interface Circuit and the ESD Immunity

[Symptoms]

There is a maintenance window on one module of a product, which contains ethernet ports, 2S clock signal port, serial port, and other external communications interface. These communication interfaces are used for monitoring and maintenance in the actual applications. During the ESD test, when ±2 kV ESD voltage is injected on the metallic socket of the 2S clock signal port (connected with a subminiature assembly (SMA) connector), the device works properly. When the ESD voltage is increased to ±4 kV, an alarm appears in this module and there are many errors on the transmitted data. When performing ±6 kV ESD contact discharge test, the data errors in this module are increased dramatically, and eventually the module is reset.

[Analyses]

The 2S clock signal of this module communicates with the outside through a SMA coaxial connector. The metallic shell of this connector is connected to the PCB reference ground, and the reference ground is connected to the metallic casing of this module. There is no independent protective ground on the PCB. The 2S signal is connected with a TVS and a NAND gate after coming into the PCB, and then it goes to the clock chip after the conversion of the NAND gate. The 2S clock is the reference clock of the PCB, and all clock signals are synchronized with this reference clock. Therefore, if the signal error occurs for this clock, the clocks of the entire module will be wrong, which results in the loss of function for this module, and since this module is the basis of the device, the loss of its control function will lead to the entire module crashed.

From the test results, it can be seen that when ESD voltage is injected on the shell of the coaxial connector, three situations can cause the disturbance to the 2S clock signal. The first one is that the coaxial connector is not connected to the ground, or the connection is bad, and then a strong electromagnetic disturbance is produced when ESD is discharged on it and the clock circuit will be disturbed. The second one is that the ESD current may be directly injected into the device pins, although the coaxial connector has been connected to the ground, but this is not a protective ground, and then the disturbance will directly come into the PCB and affect the 2S signal, and the 2S signal has no ESD protection after coming into the PCB. The third one is that, although the 2S signal is in parallel with TVS, the TVS does not play its protective role in the circuit.

In the first situation, do some measurements for the metallic shell of the SMA connector and confirm that the shell has been connected to the ground, so this possibility can be excluded.

In the second situation, the metallic shell of the SMA connector is directly connected to the PCB reference ground; thus the ESD disturbance can directly disturb the operation signal on the PCB. In this case, if ESD protective treatment is not implemented for the interface components, it is easy to impact the normal operation of the PCB, so in the design of the interface circuit, the signal paralleled with TVS is protection against ESD. The PCB is treated as this; therefore, this situation is also excluded.

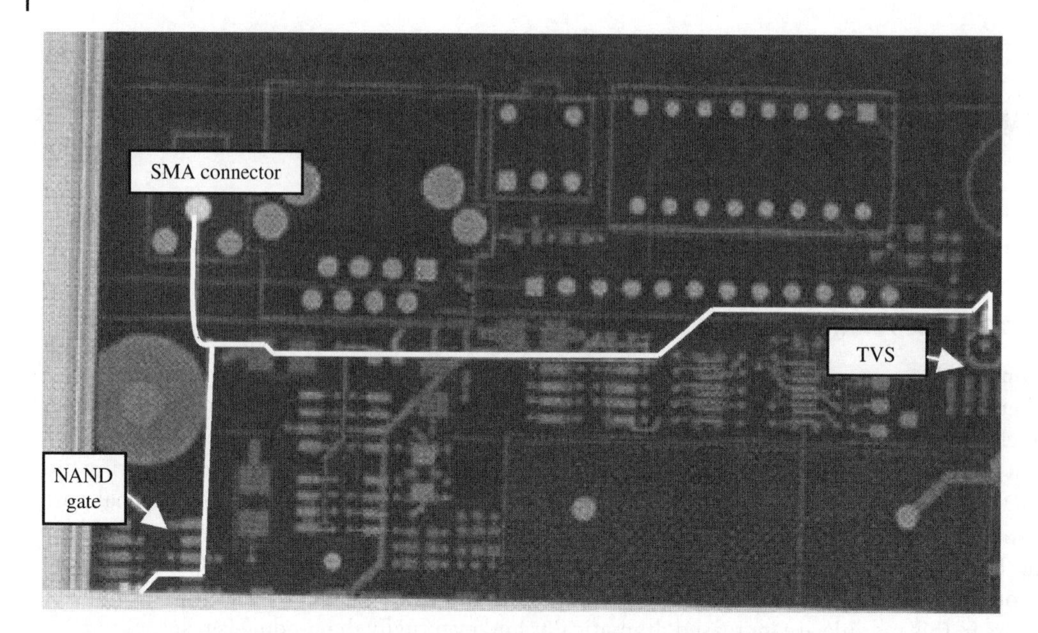

Figure 6.110 The wiring of 2S signal and the relative position of TVS and NAND in PCB.

For the third situation, the metallic shell of the SMA connector is connected to the PCB reference ground, and the signal is paralleled with TVS for protection, but in the ESD test, a large number of data errors and resets still occur. It means that the TVS does not take effect, which is the root cause of the issue in this case, but why does the TVS not play a protective role?

Although there is a proper connection of the TVS achieved in the PCB design, the circuit schematic shows that the signal has been connected by a TVS in parallel. According to the basic requirement for signal protection, the protection device must be placed at the entrance of the interface, so during the PCB routing, it should be considered that the signal is connected to the TVS first when it comes into the PCB, and then connected to the other internal circuits, starting from the TVS. While analyzing the PCB routing, it is found that the requirement is not satisfied in the PCB design. The routing of the 2S signals on the PCB and the relative position of the TVS and the NAND gate are shown in Figure 6.110.

In Figure 6.110, the white wire is the routing of the 2S signal. From the previous illustration, the 2S signal is divided into two branches after coming out of the SMA connector. One is directly connected to the NAND gate and the other one is routed from the right side and connected to the TVS through a relatively long trace. After measurement, the trace length from the TVS to the SMA connector is 60 mm + 10 mm = 70 mm, while the trace length from the SMA connector to the NAND gate is 30 mm, which is much less than the distance to the TVS. Therefore, when applying disturbance to the metallic shell of the 2S signal, the disturbance coupled to the signal will first come into the NAND gate, and then arrive at the TVS, so, at this time the TVS does not well play the role of restraining the ESD disturbance.

It can be explained that the printed trace from the SMA connector to the TVS pin is long. Since there is a certain impedance of this printed trace (this impedance is the end-to-end impedance of the printed trace, not the characteristic impedance of transmission line), so the impedance depends mainly on the length of the printed trace and then on the width of the printed trace. Figure 6.111 shows the impedance versus frequency relationship of a group of printed traces with 10 cm length and different width. It can be seen that the printed trace will behave as high impedance at higher frequencies.

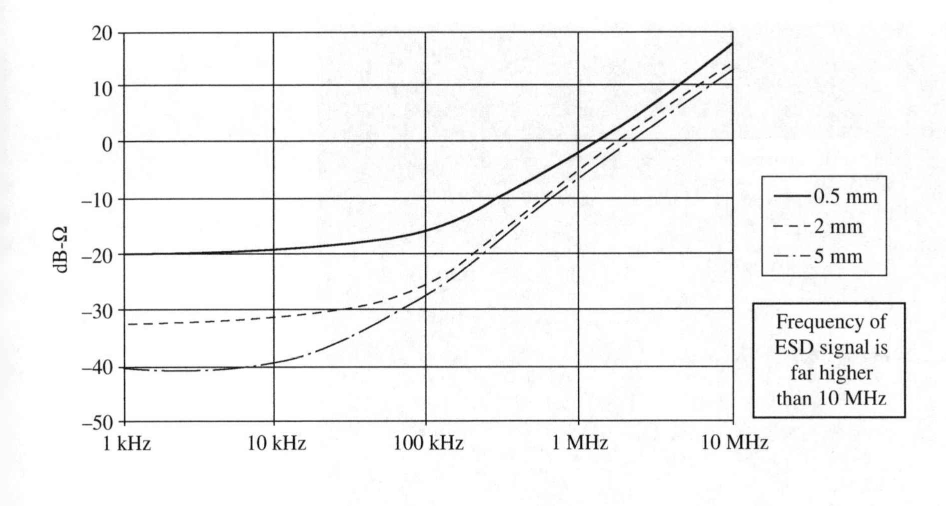

Equivalent circuit of 10 cm long, 0.5 mm wide printed line $R = 100\ \text{m}\Omega$ $L = 60\ \text{nH}$

Figure 6.111 The relation between the printed line width and impedance.

In this case, the trace from the SMA connector to the NAND gate is long (about 7 cm). The ESD problem analytical schematic of the interface circuit is shown in Figure 6.112. The trace from the SMA connector to the NAND gate is long, leading to large R_1, L_1, and $L_1 > L_2$, so, it is possible to cause 1 kV voltage drop on the L_1 when ESD current (transient, with high di/dt) flows through the TVS, so that the voltage on point A shown in Figure 6.112 cannot be reduced and the poststage logic components cannot be protected.

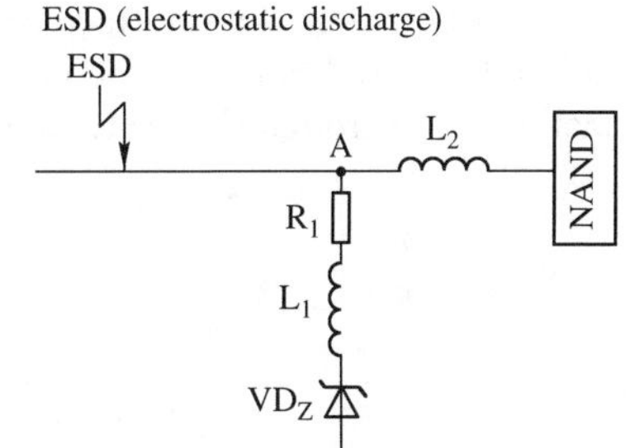

Figure 6.112 Analysis principle diagram of ESD problem of interface circuit.

When modifying the PCB layout in this part of the interface circuit, the TVS is placed between the connector and the NAND, and the PCB layout is shown in Figure 6.113.

In Figure 6.113, the white wire is the 2S clock. The clock signal is first connected to the TVS and then connected to the NAND gate after coming out of the SMA connector, and the distances are respectively 18 and 22 mm. This kind of routing conforms to the requirement of the ESD protection. The test result also proves that the module works properly without any data errors, under the ESD voltage from ±2 to ±6 kV.

[Solutions]

From the previous analysis, the declined ESD protection performance is mainly caused by the unreasonable layout of the components in the PCB, which leads to the TVS not playing a protective role. After modifying the layout, which is to place the TVS between the connector and the NAND gate, ensuring that the signal passes through a protection device first and then enters the logic components, the ESD test is passed.

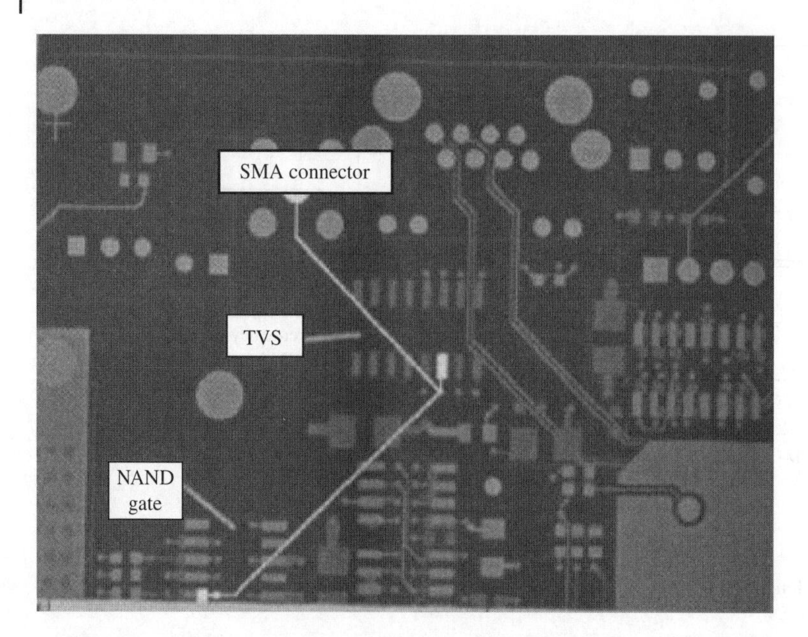

Figure 6.113 PCB routing and layout when TVS is placed between connector and NAND.

[Inspirations]

In the layout of the PCB, TVS and the similar protection devices should be placed at the signal entrance and after the interface connector and close to the connector. The protective device should be placed between the protected components and the interface connector; the signal shall first go through the protection device, and then be connected to the protected devices.

7

Components, Software, and Frequency Jitter Technique

7.1 Components, Software, and EMC

As we know, the circuit is made up of components, but the electromagnetic compatibility (EMC) performance of the components is often neglected. The EMC performance of the component is very important in the design of the circuit. For example, for a CPU chip with strong anti-interference ability, because the chip itself has the ability to create anti-interference, when we use this component in the circuit, we can save some additional peripheral protection and filtering components. Similarly, for a CPU chip with low EMI, when using this device in the circuit, we can also save some additional peripheral suppression and filtering components. At the same time, it can save a lot of time for the PCB design.

When the circuit design engineers select the digital components, they only pay attention to the function, the operating speed, and the propagation delay data of the component, but do not pay much attention to the real edge of the input/output signal. There is an inverse relationship between the working speed and the EMI. The operation speed of the component is faster and faster, so the problem of EMI will be more and more prominent. That is to say, the low-speed component is better than the high-speed component for the EMI problem. On the PCB, the circuit design engineers always focus on the key points, such as the layout of the components, the wiring, the bus structure and the decoupling capacitor. However, they often dismiss the package (such as silicon materials, plastic, or ceramic materials) of the component. Designers often only take the function and price of the component into account, but not control the parameter requirements for the package. Why shall we care about the package of the component? Although the operation speed is considered to be the most important parameter in the high-speed circuit design, in fact, the package of the component plays a major role of increasing or decreasing the RF current. The independent lead inside the package may cause some EMC problems. The biggest problem is the inductance of the lead, which will cause some abnormal operating states, the ground bounce, and the IC pins are driven by the signal noise, which may lead to a large radiation problem.

From an EMI perspective, while selecting the components, it is helpful to reduce the RF energy produced by using the logic components (especially the digital logic components), if the following recommendations are considered:

1) While selecting the components, choose the component that will sink a small current when its logic stage is switched. Here, the current refers to the maximum inrush current when all the pins of the component are switched at the same time and their capacitive loads are the maximum, not the average or static current.
2) In the condition that the circuit function is ensured, choose a component with as low a speed as possible. Although the low-speed components are now more and more difficult to be found, but for some common logic function, do not use the component with the speed in ns level.

Electromagnetic Compatibility (EMC) Design and Test Case Analysis, First Edition. Junqi Zheng.

3) Select the components on which the power pins and the ground pins are located in the center of the package and they are adjacent to each other.
4) Use the components with the metallic shielding shell (crystal) and use the low impedance vias as many as possible to connect the metallic shell to the ground.
5) For the component with ceramic package and metal embedded on the top, we should provide a grounded heat sink for the component. It seems that this structure should be incorporated in one product, but it may be difficult to achieve this. If there is no other way, we can only do as this.

From EMS (the electromagnetic immunity) perspective, while selecting the components, it is helpful to improve the immunity level of the product if the following recommendations are considered:

1) The component with high ESD immunity should be preferentially chosen.
2) Select the components on which the power pins and the ground pins are located in the center of the package and are adjacent to each other.
3) For the digital-analog hybrid components, the component with good isolation between the digital part and the analog part should be preferentially chosen.

Software itself does not belong to the category of EMC, but it can be used as a kind of fault-tolerant technique used in EMC domain. Its function is mainly reflected on the immunity of the product, such as preventing the CPU program against "running away," which can be realized by the disturbance through the software trap, eliminating the signal noise to improve the precision of the system by digital filtering, and avoiding the presence of the interference effect by reasonable software time-sequence mechanism.

7.2 Frequency Jitter Technique and EMC

Switching frequency jitter technique has become a popular technique in recent years, and it can reduce the conducted disturbance and the radiated disturbance. The difference between the dithering frequency and the fixed frequency is that the frequency of the ordinary periodic signal is very stable, but the frequency of the jittering signal's cycle changes according to a certain rule. That is, frequency jitter is produced artificially. This technique is also used in the digital periodic signal circuit, in which the technique is called the spread spectrum modulation technique.

We should pay attention to the problem that, the effect of the frequency jitter is only to distribute the disturbance of the equipment in a wide spectrum so it can easily pass the EMC test. The whole disturbance energy is not changed. Because this technique has been recognized by all the administrative agencies, it is indeed a simple way to make the device successfully pass the EMC test. This technique is particularly suitable for civilian equipment but does not play a significant role in military applications due to strict compliance. Military equipment mainly relies on the shielding and the filtering techniques.

7.3 Analyses of Related Cases

7.3.1 Case 73: Effect on the System EMC Performance from the EMC Characteristics of the Component and Software Versus Cannot Be Ignored

[Symptoms]

A product has E_1 communication interface, when 6 kV ESD contact discharge is injected on the aluminum framework and the metallic shell of the SMB connector, which is on the front panel of E_1 communication interface board and, the E_1 communication link is interrupted immediately and cannot be automatically recovered, using software to reset E_1 chip and manually resetting the product do not work, then the link can be recovered after the communication interface board is plugged-out and plugged-in and the software is reloaded. At the same time, the EFT/B test is carried out for the power port of the product, when the test voltage is ±1 kV, the phenomenon that appears in the test is same as that in ESD test. The connection diagram of the product in the ESD test and the electrical fast transient/burst (EFT/B) test is shown in Figure 7.1. The input and output of the E_1 communication interface in the product are, respectively, connected to the input and output of the data error measurement meter for the E_1 interface in order to simulate the normal communication of the product.

[Analyses]
The E_1 interface chip of the communication interface board in this product is DS2154. In order to study the root cause of the problem, we continue to do the following three tests and measure the E_1 transmission signal:

1) Modify the program and some simple hardware connections, so that the E_1 signal doesn't pass through the other components after going to the interface chip DS2154, and loop back directly; the signal flow is shown in Figure 7.2. Perform the test again, and simultaneously measure the signal on the control circuit of the E_1 interface chip DS2154, at the point A shown in Figure 7.2.
2) The E_1 signal waveform at the point A is measured during the test. Under normal circumstances, the received E_1 data on the data-error measuring meter is normal (9B, 000000..., DF, 000000..., 9B, 000000...). The waveform at the point A is shown in Figure 7.3 (because it needs a very long wire for measurement, so the measured signal quality is not very good). When 6 kV ESD voltage is injected to the framework, the received E_1 data on the data error measuring meter (9B, 0000000... 0001, DF, 000... 0001) becomes abnormal; that is, the last bit in the frame is changed to 1, and the waveform at the test point A is shown in Figure 7.4.
3) Do some simple hardware connections so that the E_1 signal passes from the SMB connector into the product, and the input and output signals are short-circuited before the E_1 signal goes into the interface chip DS2154; namely, the loopback is achieved on the interface chip DS2154. The schematic is shown in Figure 7.5. Then, performing the ESD test, the received E_1 data on the data error measuring meter is normal.

These experiments show that the problem lies in the DS2154 chip.

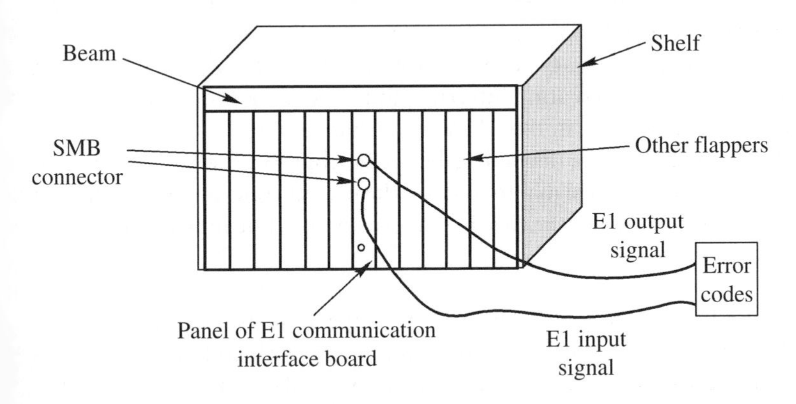

Figure 7.1 Test equipment connection diagram.

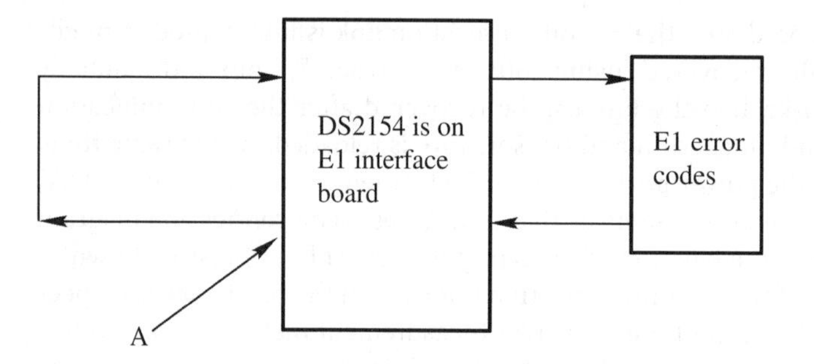

Figure 7.2 Test schematic diagram.

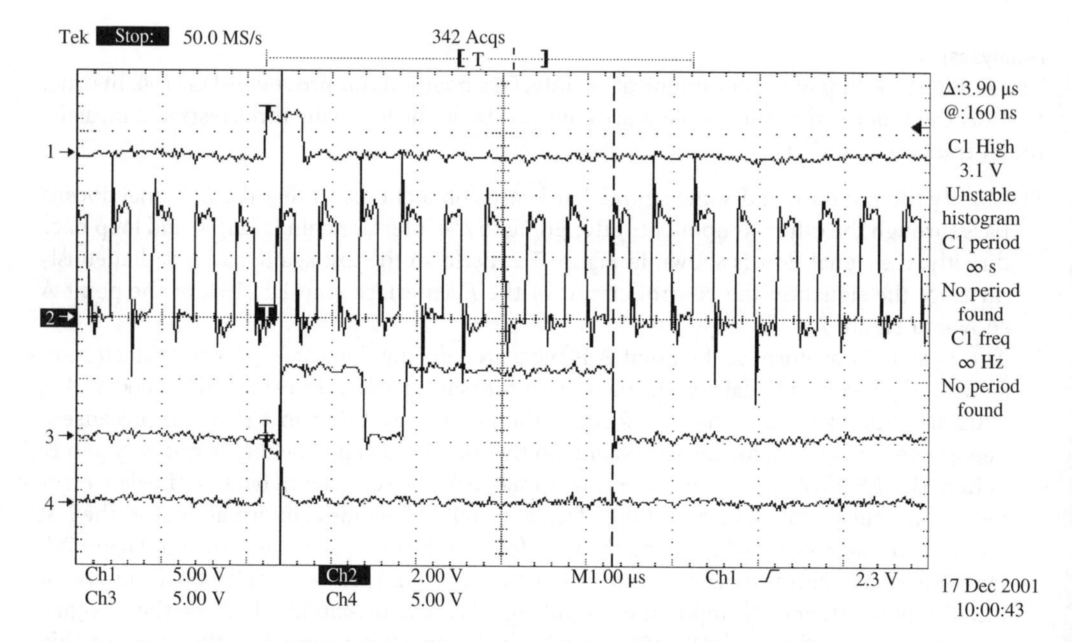

Figure 7.3 Normal waveforms (each channel are frame head, 2 M, data, clock distribution frame head).

[Solutions]

Look up the datasheet of the E_1 interface chip, the ESD immunity of the interface chip DS2154 is ±1 kV, while the ESD immunity of DS21554 is ±2 kV. Replace the interface chip DS2154 with DS21554 in this product, and then perform ±6 kV ESD test. The phenomenon of communication interrupt and system crash no longer appear.

The EFT/B test is performed after replacing the interface chip with DS21554, the test voltage is +1 kV, and the phenomenon of communication interrupt and system crash no longer occurs, but the communication link is interrupted in the −1 kV test. Change the CPU software of the E_1 communication interface board, increase the fault-tolerant capability of the E_1 port, namely, increase data error rate threshold of the E_1 signal to twice of the original one, and then perform the ±1.5 kV EFT/B test. The phenomenon of the E_1 signal interrupt does not appear at this time.

[Inspirations]

1) The EMC performance of the component itself has great impact on the system EMC performance, it should be paid attention to choose the components with strong immunity in the early EMC design; The EMC characteristics of the component itself must be considered when choosing the components, especially for the ESD immunity (ESD immunity is also

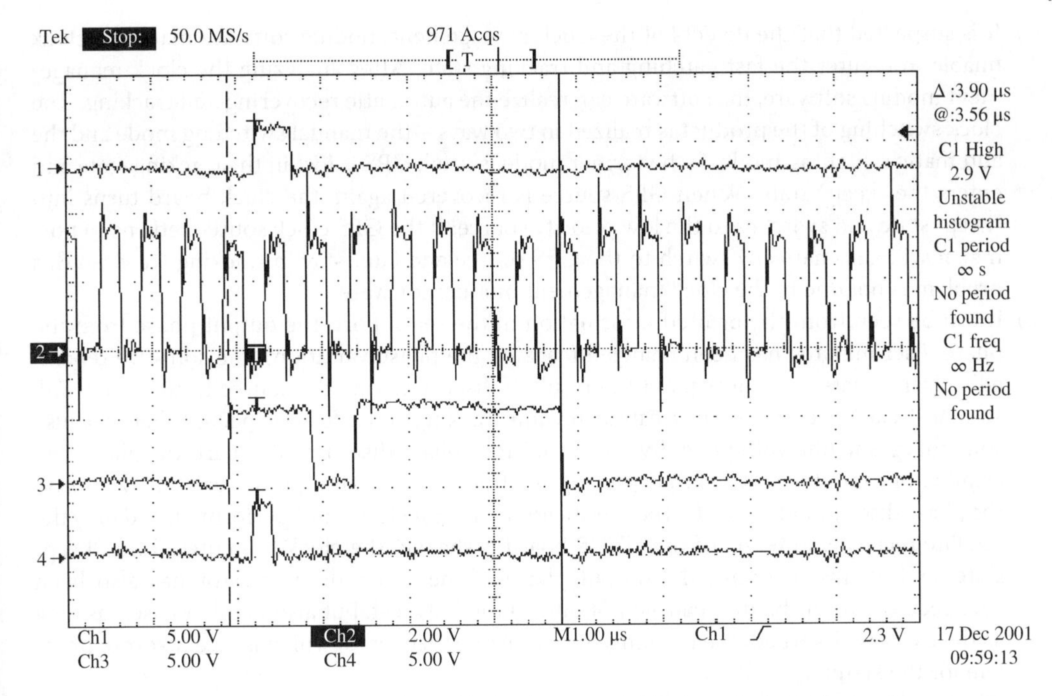

Figure 7.4 Exception waveforms (each channel is the frame head, 2 M, data, clock distribution frame head).

easier to be found in the datasheet of the components).

2) Software is also closely related to the EMC performance of products, a certain fault tolerant ability should be set up in the product software.

7.3.2 Case 74: Software and ESD Immunity

[Symptoms]

1) When the −6 kV ESD test is performed for the clock input port of a product, the traced clock status is changed to "keep" state and cannot be recovered except resetting or replugging in the product. This does not meet the requirements of the product standard.

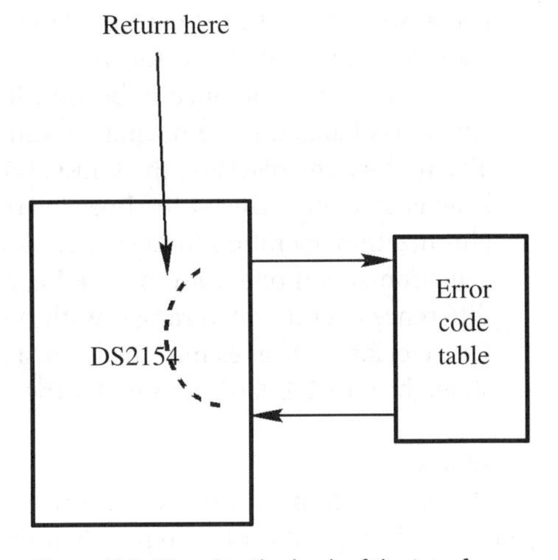

Figure 7.5 Signal in the back of the interface before entering the interface chip.

[Analyses]

1) In the ±6 kV ESD test, according to the product standard, the clock is allowed to be interrupted but must be able to be recovered automatically. The GPS2 clock is traced in the test, when the ESD test voltage is −6 kV, the test point is on the shell of the clock port, the GPS2 clock source goes to "keep" state, through querying the clock information, GPS2 clock source is immediately recovered to "USABLE" state, which indicates that the clock source itself has returned to the normal state and the hardware circuit is able to be self-recovered at this time.

2) It is suspected that the defects of the clock management module software make the clock unable to reenter the fast catching and tracking state. After analyzing the clock management module software, the software can realize the automatic recovering and tracking. The clock switching of the product is realized in two ways – the manual switching mode and the automatic switching mode. If the former mode is used, GPS is lost in the tracking state and enters the "keep" state. When GPS source is recovered again, the clock board turns into "keep" state and switches to the low priority source; If the GPS clock source returns to normal, it will automatically switch to the GPS source and later stay in tracking state. So this problem is not due to the clock management module software.
3) It can be seen from the printed information in the serial port, the output phase from the phase discriminator has been changed (causing the phase variation), the software repeatedly discriminates the source, but the result of discriminating the source is not successful, so that it cannot enter the fast capturing and tracking state. It is suspected that, because something is wrong with the software of the clock phase discriminator part, the phase discriminator cannot enter the fast capturing and tracking state after phase variation occurs on the phase discriminator. For further verification, the ESD test is not performed, and only the satellite receiving antenna of the UTCP board is shaken, the clock has entered the "keep" state, and at this moment, the output phase of the phase discriminator has also been changed. So this problem is caused not only in the ESD test, but also in other cases, as long as the clock is disturbed, the tracking state won't be recovered, which is a very severe problem for the system.
4) After analyzing the software, it is found that, when the input of the phase discriminator is beyond the range the software can manage, the CLKC unit cannot re- enter the fast capturing and tracking state, and will stay in discriminating the source state repeatedly. The reason is that when the CLKC board tracking the GPS clock source is in the "keep" state, the software hasn't executed the forced synchronization (aligned) operation to the logic as the detected phase difference is beyond its rated range, which means that even if the clock source is changed to be normal, it still cannot enter the fast capturing and tracking state. The reason why resetting the CLKC board permits reenter the fast capturing and tracking state is that when the CLKC board is restarted, its reset signal implements the forced synchronization operation to the logic. Modify the software so that in the keep state, the forced synchronization operation on the logic is implemented after the software detects a phase difference over its rated range. With this modification, when the CLKC board tracking the GPS clock source goes into the keep state, when the clock source is recovered to the stable state, the CLKC board can reenter the fast capturing and tracking state.

[Solutions]

This problem can be improved from the software perspective, i.e. the forced synchronization operation is applied on the logic when the software detects a phase difference over than its rated range and the CLKC board is in keep state. After modifying the software, the ±8 kV ESD contact discharge and ±15 kV ESD air discharge test are carried out, and the clock can be recovered automatically.

[Inspirations]

EMC design is a kind of system design, involving all aspects, software, hardware, structure, and cable. When EMC problems are found, do not rush to solve the problem from the perspective of interference isolation, but assess the severity of the problem and find out the root cause of the problems which may cause severe defects to the system and solve them from the root.

7.3.3 Case 75: The Conducted Emission Problem Caused by Frequency Jitter Technique

[Symptoms]

An industrial product is supplied by 24VDC power supply, the 5V working voltage of the internal circuit is obtained by the DC/DC switching mode power supply, and the schematic of the switching mode power supply is shown in Figure 7.6 (in this figure, the filtering circuit and its post-stage circuit of the power supply output are omitted).

For the first design of the product, the load of the power supply is large (500 mA), the conducted disturbance on the product power port is tested, and the test results are shown in Figure 7.7.

Each frequency harmonic of power supply switching can be clearly seen from frequency spectrum, and the results are less than the limit value requirement of conduction disturbance in power port shown in Figure 7.7, in which the disturbance allowance in 0.17457MHz frequency is only 2.37dB.

Later, some designs are changed so that the power supply load becomes 300 mA when the product works normally. At this time, the conducted disturbance on the power port is tested; the test results are shown in Figure 7.8.

As seen from Figure 7.8, the frequency curve becomes smoother, unlike Figure 7.7, in which there are many frequency spikes, and the conduction disturbance levels of most frequency points are low, but accidentally it is found that the radiation becomes higher at 150 kHz, exceeding the limit by 3.64 dB.

This raises two questions:

1) Why does the conduction disturbance level of most frequency points become low and the spectral curve become smoother when the power supply load is small (300 mA) (i.e. the switching frequency jitter) are transmitted?
2) Why does the conducted disturbance become higher and exceed the limit at 150 kHz?

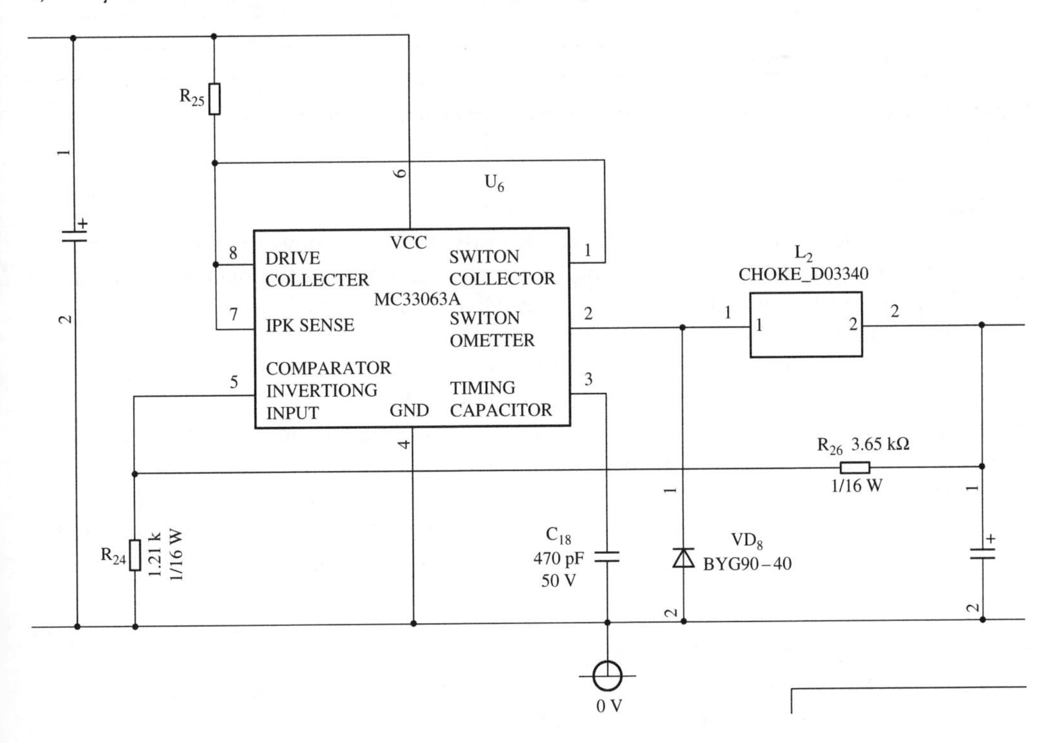

Figure 7.6 Switching power supply schematic diagram.

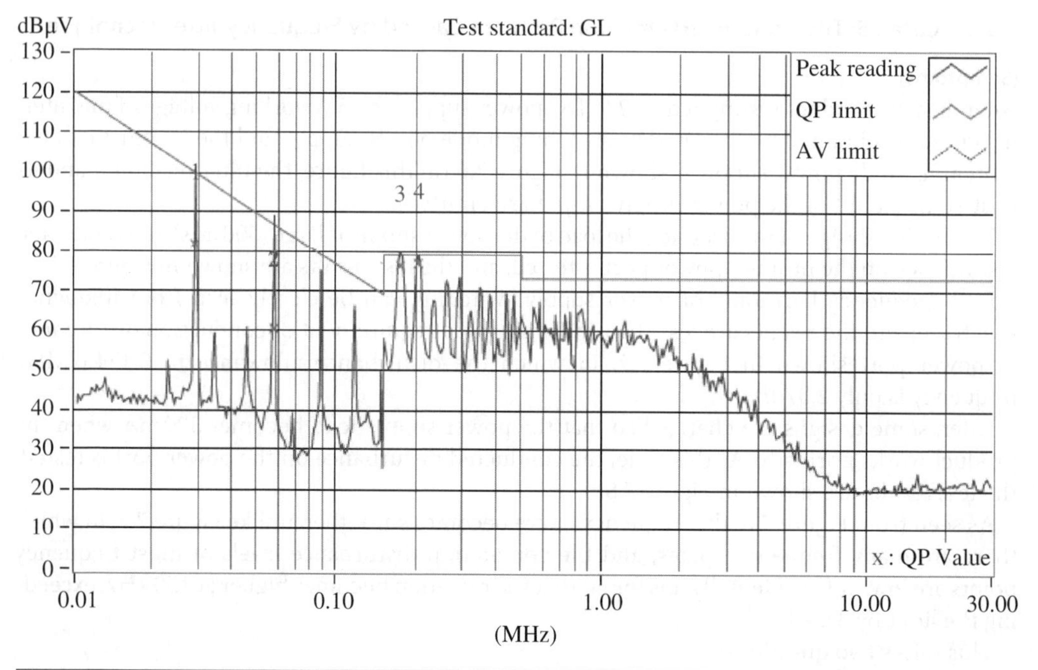

	Frequency	Corr. factor	Reading dBμV	Emission dBμV	Limit dBμV	Margins dB	Notes
No.	MHz	dB	QP	QP	QP	QP	
1	0.02915	9.87	80.75	90.62	99.85	−9.23	
2	0.05842	4.95	59.59	64.54	86.76	−22.22	
+3	0.17457	1.45	75.18	76.63	79.00	−2.37	
4	0.20480	0.89	75.64	76.53	79.00	−2.47	

Figure 7.7 500 mA load when the power port of the transmission disturbance spectrum.

[Analyses]

It is found through looking up the information of the switching power supply chip, the frequency jitter technique (i.e. the switching frequency is variable in the operation) is applied when the power supply load is small (e.g. 300 mA), and the switching frequency is fixed at a certain frequency when the load is large (e.g. 500 mA).

To explain the benefits of the frequency jitter technique for the conducted disturbance test to the switching power supply, the nature of the signal shall be discussed at first. Signal can be divided into two groups, nonperiodic signals (such as data signals, address signals and some randomly generated signals) and the periodic signal (such as switching signal of power supply, digital periodic signal (CLK). As described in Section 1.3.1, the frequency spectrum of each sampling period of the periodic signal is the same, so its frequency spectrum is discrete, but the amplitude is large, which is usually referred to as narrow band noise. The spectrum of each sampling period of the nonperiodic signal is different, the spectrum is very wide, and the amplitude is weak, which is often referred to wide band noise. In the switching mode power supply, the PWM signal is usually a rectangular pulse with fixed frequency, the frequency spectrum contains higher harmonics, so the disturbance level of the fundamental and harmonic frequency in the PWM signal is relatively high.

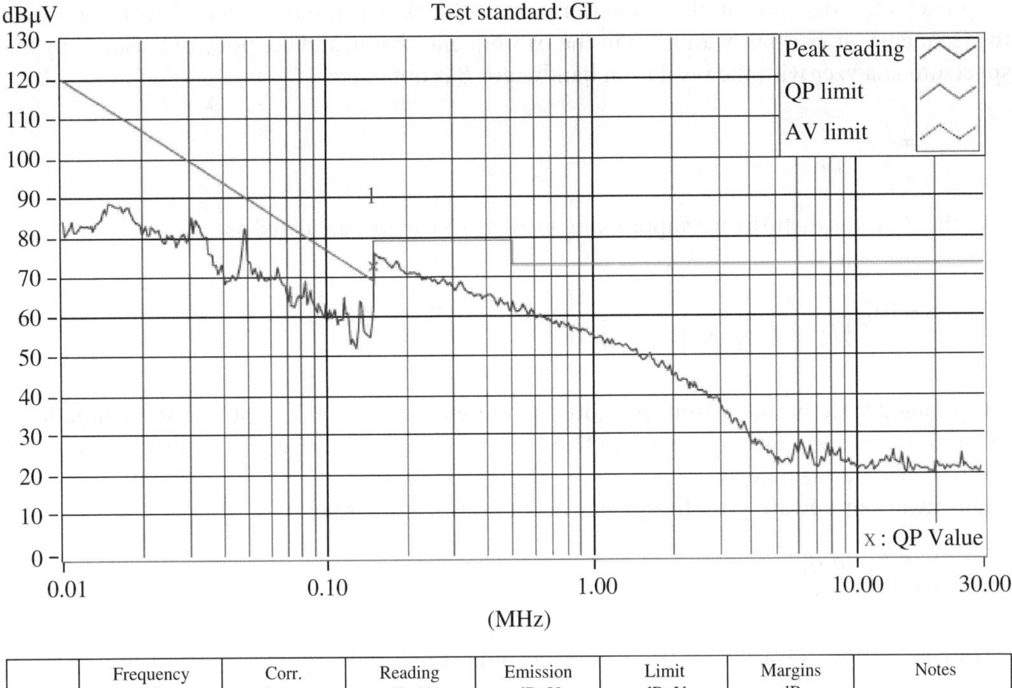

	Frequency MHz	Corr. factor	Reading dBμV	Emission dBμV	Limit dBμV	Margins dB	Notes
No.	MHz	dB	QP	QP	QP	QP	
+1X	0.15000	1.98	70.66	72.64	69.00	3.64	

Figure 7.8 300 mA load when the power port of the transmission disturbance spectrum.

Why can the frequency jitter reduce the level of the conducted disturbance? The frequency jitter technique can reduce the disturbance by about 7–20 dB. The larger the range of frequency jitter, the more obvious the effect of reducing the disturbance becomes.

The effect of frequency jitter on the conduction disturbance and radiated disturbance can be estimated by the following ways. In order to simplify the analysis, only the fundamental harmonic wave of the periodic signal is considered. The signal with fixed frequency can be expressed as follows:

$$A\sin\left(2\pi f_c t\right)$$

Frequency jitter can be expressed as

$$A\sin\left[2\pi\left(f_c + w(t)\right)t\right]$$

in which, $w(t)$ is the modulated waveform, when the frequency jitters, the frequency is changed by Δf and is back to the initial frequency in a certain period, as if it were a signal modulation, and the modulation waveform represents the relationship of the frequency curve versus time; usually, it is a sawtooth wave.

The frequency spectrum without jitter is a spectral line at f_c, and its amplitude is $A^2/2$.

Since the frequency spectrum is just a spectral line, the amplitude has nothing to do with the resolution bandwidth B of the spectrum analyzer. However, the spectral amplitude with the frequency jitter depends on the resolution bandwidth B. Because the power distribution with the frequency jitter is fairly uniform in the Δf frequency band, and the power measured by the spectrum analyzer with the resolution bandwidth B is obtained:

$$P = \frac{1}{2}A^2 \frac{B}{\Delta f}$$

In this way, the disturbance suppressing rate (dB) S can be calculated as:

$$S = 10\lg\left[\frac{1/2A^2}{\left(1/2A^2\,B/\Delta f\right)}\right] = 10\lg\frac{\Delta f}{B}$$

Combined with the above frequency jitter parameters, the frequency jitter rate δ (sometimes called the expansion rate), the original fixed frequency f_c and the frequency jitter mode, the S can be calculated as follows:

Downward or upward jitter (frequency becomes small or large):

$$S = 10\lg\frac{|\delta| \bullet f_c}{B}$$

Downward and upward jitter (frequency becomes small and large at the same time):

$$S = 10\lg\frac{2|\delta| \bullet f_c}{B}$$

Among them, the frequency jitter rate is the ratio of jitter (or extension) range (Δf) and the fixed frequency (f_c). The type of jitter includes downward jitter, upward jitter, or downward and upward jitter at the same time. If the jitter range is Δf, then δ is defined as, for downward jitter, $\delta = -\Delta f/f_c \times 100\%$, for center jitter, $\delta = \pm 1/2\Delta f/f_c \times 100\%$, and for upward jitter: $\delta = \Delta f/f_c \times 100\%$.

It should be noted that when $f_{SW} << f_m << f_c$, the disturbance suppressing rate S is not related to the modulation rate f_m. f_{SW} is the scan rate of the spectrum analyzer. f_m is the modulation rate, which is used to determine the cycle of frequency jitter, in which the frequency is changed by Δf and is back to the initial frequency. The modulated waveform represents the curve that the frequency varies with time, and it is usually a sawtooth wave. Figure 7.9 shows the relationship between the frequency jitter modulation rate f_m and the frequency jitter rate δ.

Through the above explanation, it is now possible to clearly answer the first question; namely when the power load is small (300 mA) (switching frequency jitter operating), the conduction disturbance levels of most of the frequency will be lower. Except for the reason that small power consumption of the power supply itself causes decreased disturbance levels, the more important reason is that after the frequency jitter, the occurrence number of pulses in some particular frequency (harmonic frequency) in the original fixed frequency is reduced (here, the pulse is the pulse in spectrum, an impact energy at one frequency point). In order to reduce the disturbance level of the harmonic frequencies at the fixed frequency, the radiation spectrum in the same frequency band is scattered into more frequency bands. Because the spectrum is a bundle distribution, there is a lot of space between the beams. The result of the oscillation frequency jitter is that the spectrum bandwidth becomes wider and the peak value decreases. While the receiving bandwidth of the receiving device used for the conduction disturbance test is constant, when the spectral line becomes wider, a portion of the energy is outside the receiving bandwidth of the receiver, which can also make the measurement value small. The reason

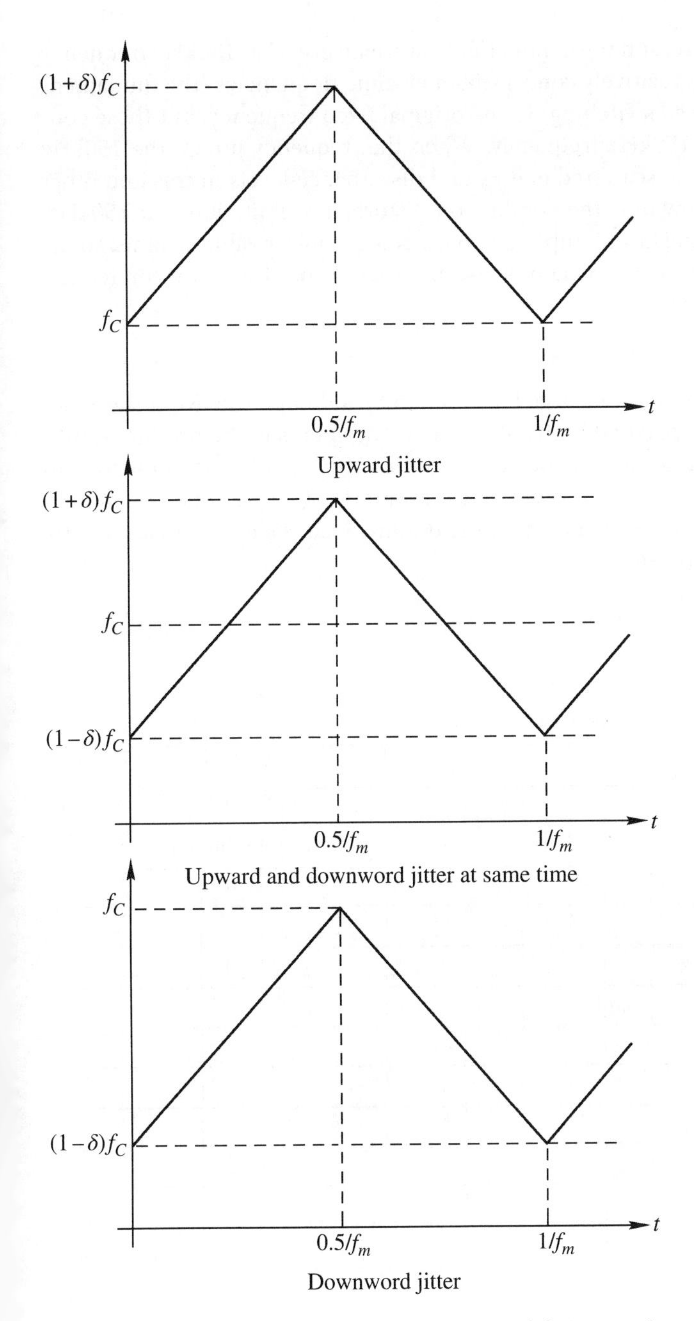

Figure 7.9 Relation between frequency shaking modulation rate f_m and frequency shaking rate.

the spectrum curve becomes smoother is that the frequency jitter make the spectrum bandwidth wider and scatters the energy. But when the frequency bandwidth is larger than the scanning step (stepwide) in the conduction disturbance test, a relatively smooth spectrum curve will appear.

Figure 7.9 shows the relationship between frequency shaking modulation rate f_m and frequency shaking rate.

As for the second question, the reason the conduction disturbance level at 150kHz frequency becomes high is that the energy is relatively concentrated in some frequencies (the fundamental and harmonic frequency of fixed switching) in the original fixed frequency, but these concentrated frequencies are not at 150kHz frequency. When the frequency jitters, the 150kHz frequencies are also assigned to the scattered energy and disturbance levels in this and other similar points become high. Meanwhile, the conduction disturbance limits jump at 150kHz, according to the standard, when the limit jumps at a point, take the lower value as limit, so the limit value at that point is relatively low, which is also the reason that leads the conduction disturbance exceeding the standard.

[Solutions]

When the switching power supply is in the large load (500 mA), although the overall conduction disturbance level is high, the specified limit value line requirements in the product standard can be met, so in the product design, the power supply is connected to a dummy load to make the product work properly, and the power consumption is 500 mA. The conduction disturbance test spectrum diagram after connecting the dummy load is shown in Figure 7.10, and the data show that the test is passed.

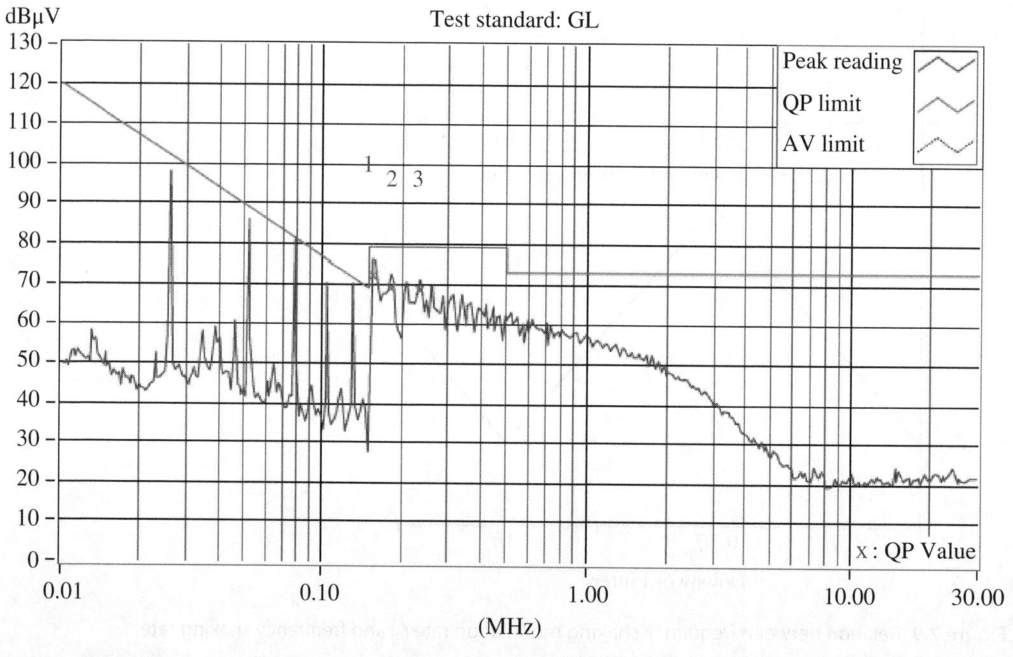

No.	Frequency MHz	Corr. factor dB	Reading dBμV QP	Emission dBμV QP	Limit dBμV QP	Margins dB QP	Notes
1X	0.15487	1.87	71.14	73.01	79.00	–5.99	
2X	0.18168	1.30	68.29	69.59	79.00	–9.41	
+3X	0.23457	0.84	67.72	68.56	79.00	–10.44	

Figure 7.10 Conduction disturbance test spectrogram after dummy load is connected.

[Inspirations]

1) The frequency jitter technology is usually beneficial to the pass of EMI test, but this is a special case, which also shows that the product design needs trade-off.
2) The effect of frequency jitter is only to make the device easy to pass the EMI test. The disturbance energy over the entire frequency range has not changed. It just scatters the more concentrated energy over a wide frequency band.
3) Both frequency jitter and low pass-filtering technologies can reduce the disturbance of periodic signal, but it is not absolute to evaluate which one is better. A comparison of them is done below, and the selection from developers depends on the situation in the actual project:
   - *Different principles.* Frequency jitter technology broadens the periodic signal spectral line. Using the condition that the receiving bandwidth is constant in measuring method, a portion of the spectral line energy is received so that smaller measured values are obtained. The filtering technology is to filter out the energy and reduce the amplitude of the disturbance. Therefore, it can be considered that the frequency jitter is proposed to pass the test easily, and the filter is the real solution to suppress the energy of electromagnetic disturbance. Of course, the effect of the frequency jitter technology is effective for solving the interference of the periodic signal to the narrow band receiver.
   - *Different impacts on the waveform.* The effect on periodic signal waveform from frequency jitter technology is the frequency jitter, and the rising/falling edge of the pulse remains same, as steep as the original ordinary periodic signal. The effect on the periodic signal waveform from filter is to make the corner of the pulse passivation and extend the rising edge of the pulse. The wider rising edge of the pulse will lead to a decrease in the circuit operation speed.
   - *Different effective frequency ranges.* The filter can only reduce the harmonic amplitude of periodic signal in higher harmonics (in order to ensure the fundamental waveform of the signal cycle, generally retain 15 harmonics), while there isn't any inhibitory effect on the lower harmonics (especially the fundamental frequency). The low frequency, even the fundamental frequency, also has the function of reducing the amplitude of the frequency, which depends on whether the range of the frequency jitter is larger than the receiving bandwidth of the receiver. For example, if the frequency modulation of frequency jitter is 0.5%, for 120kHz periodic switching signals, for 10 times harmonic, the range of frequency change is 12kHz, which is larger than the 9kHz bandwidth of the receiver in the conduction disturbance test, so it can obtain the smaller measurement value.

7.3.4 Case 76: The Problems of Circuit and Software Detected by Voltage Dip and Voltage Interruption Tests

[Symptoms]

For a communication product, its rated working voltage is 48 V dc. In the voltage dip and voltage interruption tests on the power port, it is found that the voltage dips from 48 V to 70% of the 48 V, the duration for voltage dip is respectively 0.3 and 1 seconds, the reset phenomenon appears. When the test is finished, the product is also continuously reset repeatedly and cannot be self-recovered to work normally, in the test, the bit-error tester which is used to monitor whether the product works properly displays "NO SIGNAL." This indicates that the product is not in normal working condition, only replugging the PCB can achieve the normal reset and recover the operations; When the voltage dips from 48 V to 40% of the 48 V, the duration for the voltage dip is, respectively, 0.1, 0.3 and 1 seconds, the bit-error tester shows "NO SIGNAL," namely the product is not in normal mode. Software reset cannot recover the operations, and it can return to normal operating after forced reset or power down and power on again.

[Analyses]

There are two kinds of power supply systems in the product, the 5 V and the 3.3 V, which are obtained by two DC/DC switching mode power supply modules. The output voltage of these two modules in the product is not necessary to response the dip change of the input voltage at the same time. The watchdog circuits in the product only monitor the output voltage of the 5 V power supply module while the output voltage of the 3.3 V power module is not monitored. When the voltage dip occurs, since the 5 V power supply circuit doesn't drop, the watchdog circuit monitoring the supply voltage doesn't detect the voltage change, but in fact the circuit supplied by 3.3 V loses its power supply in an instant and is powered on again, the main program cannot judge that the restored information in the memory supplied by the 3.3 V power supply is changed and lost at the moment its 3.3 V power supply is lost and the memory is reinitialized, and it still thinks the stored information is correct, which must impact the normal operations.

In addition, among the two types of memories in the product, one kind completely recovers the information restored before the drop occurs, while there is more random information in the part of storage units of the other memory. Through checking the software of the product, it is found that the program check is only implemented for one memory when the program is initialized, and it is assumed that, if the program check of one register is passed, all the registers will have no problem. This overgeneralization mistake leads to the fact that a portion of memory information cannot be recovered in the product reset and the program initialization, so the operation is affected and many "NO SIGNAL" or bit errors appear.

[Solutions]

The watchdog circuit is used not only to reset the chip powered with 5 V but also to provide a reset signal to the chip supplied by other power supply voltage. The power supply voltage monitoring circuit is used to monitor the voltage fluctuation of all the power supply networks without any exceptions. The software of the product needs to implement the initialization and the check for all the registers and memories.

[Inspirations]

There are several reasons the DC voltage dip and voltage interruption tests results in the product operation is interrupted and cannot be recovered automatically, greatly related to the hardware design and software design. Therefore, it is more complex and needs to be considered case-by-case. But the basic rule is that it is generally related to the reset circuit design and the check program for the memory and the register.

Appendix A

EMC Terms

Auxiliary equipment, AE: when doing EMC test, the device that ensures the EUT to be in normal operation.
Cabinet radiation: the radiation generated by the casing of device, excluding the radiation caused by the antenna or cable connected with the casing.
Conduct emission: the energy conducted from source to another medium via conductive medium in the form of voltage or current, also known as conducted disturbance.
Conducted interference: the energy causing the performance degradation of device, transmission path, or system in the form of voltage or current interference.
Coupling path: the path, via which portion or all electromagnetic energy is transferred to another circuit or device.
Coupling: in a given circuit, electromagnetic quantity (generally voltage or current) are transferred from an assigned position to another one in the form of magnetic field, electric field, voltage or current.
Degradation (of performance): undesired deviation between the working performance of equipment, device or system and its normal performance.
Disturbance suppression: measure to alleviate or eliminate disturbance.
Disturbance: any electromagnetic phenomenon that may cause performance degradation of equipment, device or system or may harm the organic or abiotic matter. Note: electromagnetic disturbance can be electromagnetic noise, useless signal, or the change of media itself.
Electromagnetic compatibility, EMC: the equipment should not be affected by external electromagnetic sources and should not itself be a source of electromagnetic noise that can pollute the environment.
Electromagnetic compatibility level: the expected highest electromagnetic disturbance level applied to working device, equipment, or system in specified conditions.
Electromagnetic compatibility margin: the difference value between immunity limit of device, equipment or system, and limit of disturbance.
Electromagnetic environment: the summation of all electromagnetic phenomena that exists in the given place.
Electromagnetic interference, EMI: the performance degradation of device, transmission path, or system caused by disturbance.
Electromagnetic screen: the screen to reduce the penetration of alternative electromagnetic field to designated area using conductive material.
Electromagnetic susceptibility: capacity for not guaranteeing performance of equipment, device or system against electromagnetic disturbance.
Electrostatic discharge, ESD: charge-transfer caused by the mutual nearness or direct contact of objects with different electrostatic potentials.

Electromagnetic Compatibility (EMC) Design and Test Case Analysis, First Edition. Junqi Zheng.

Emission level (of a disturbance source): the certain electromagnetic disturbance level generated by specified device, equipment, or system that be measured by specified measuring method.

Emission: the phenomenon that the electromagnetic energy is transmitted from source to outside.

Equipment under test, EUT: the equipment under test.

Far field: region where the power flux density from an antenna approximately obeys an inverse square law of the distance. For a dipole this corresponds to distances greater than $\lambda/2\pi$, where λ is the wavelength of the radiation.

Field strength: the term "field strength" is applied only to measurements made in the far field. The measurement may be of either the electric or the magnetic component of the field and may be expressed as V/m^{-1}, A/m^{-1} or W/m^{-2}; any one of these may be converted into the others. For measurements made in the near field, the term "electric field strength" or "magnetic field strength" is used according to whether the resultant electric or magnetic field, respectively, is measured.

Immunity (to a Disturbance disturbance): capacity for guaranteeing performance of equipment, device or system against electromagnetic disturbance.

Immunity level: the highest electromagnetic disturbance level applied to a certain device, equipment or system and it is still in normal operation and keep the required performance.

Immunity limit: the specified lowest immunity level.

Immunity margin: the difference value between immunity limit of device, equipment or system, and the compatibility level.

Impulsive disturbance: a series of distinct pulses or transient electromagnetic disturbance presenting on a certain equipment or device.

Impulsive noise: a series of distinct pulses or transient noise presenting on a certain equipment or device.

Interference suppression: measure to alleviate or eliminate interference.

Limit of disturbance: the highest sufferable electromagnetic disturbance level for specified measuring method.

Line impedance stabilization network, LISN: the main function of an LISN is to provide precise impedance to the power input of the EUT, in order to get repeatable measurements of the EUT noise present at the LISN measurement port. This is important because the impedance of the power source and the impedance of the EUT effectively operate as a voltage divider. The impedance of the power source varies, depending on the geometry of the supply wiring behind it.

Mains decoupling: the ratio between the voltage applied to a certain position in the power supply and the voltage applied to the assigned input of the device that can cause the same disturbed effect on the device.

Mains immunity: immunity against mains-borne disturbance.

Mains-borne disturbance: the electromagnetic disturbance transferred from power supply line to device.

Noise: unintentional or useless signal in the environment or circuit. High frequency part is designated in this book.

Pulse: a physical quantity that suddenly changes in a short time and then return to its initial value rapidly.

Radiate emission: the energy radiated from source to space in the form of electromagnetic waves, also known as Radiated radiated Disturbancedisturbance.

Radiated interference: the energy causing the performance degradation of device, transmission path, or system in the form of electromagnetic interference.

Rise edge: the time duration for a quantity from 10% to 90% of peak value.

Rise time (of a pulse): the time for the pulse instantaneous value from the given lower limiting value up to upper limit value.
Screen: the measure to reduce the penetration of field to designated area.
Semi-anechoic chamber: shielded enclosure where all internal surfaces are covered with anechoic material with the exception of the floor, which shall be reflective (ground plane).
Transient: the variational physical quantity and phenomenon during mutual transformation between two adjacent steady state. The variation time is less than the concerned time scale.

Appendix B

EMC Tests in Relevant Standard for Residential Product, Industrial, Scientific, and Medical Product, Railway Product, and Others

The International Special Committee on Radio Interference (CISPR) and International Electrotechnical Commission (IEC) standards such as CISPR11, CISPR13, CISPR14, CISPR15, CISPR22, IEC61000-4-2, IEC61000-4-3, IEC61000-4-4, IEC61000-4-5, IEC61000-4-6, IEC61000-4-8, IEC61000-4-11, IEC61000-3-2, IEC61000-3-3, and so on, regulate the required EMI and EMS tests for industrial, scientific, and medical equipment, broadcasting receiver, household appliances, hand tools, lamps and lanterns, and IT products.

B.1 Radiated Emission Test

B.1.1 Radiated Emission Test Purpose

Since EMC design and the analyses of EMC problem are all based on EMC tests, it is necessary to simply elaborate the EMC tests. It is to test the radiated emission from electronic, electrical, and electromechanical equipment and its assembly units, including all modules, cables and interconnection. It is used to authenticate whether the radiation meets the standards, in order to avoid affecting other devices that are in the same environment while in normal operation.

B.1.2 Radiated Emission Test Equipment

According to the standards of Emission Test Standard CISPR16 and EN55022, the devices needed in the radiated emission test are as follows:

1) EMI automatic measurement control system (computer and interface elements)
2) EMI test receiver
3) Various antennas (active and passive rod antenna, the loop antenna with all sorts of sizes and shapes, power gain bi-conical antenna, log-spiral antenna, horn antenna) and antenna control unit
4) Semi-anechoic chamber or open area

EMI test receiver is the most common basic test device in EMC test. Based on the requirements for the frequency response characteristic of the test receiver, as CISPR16 requires, the test receivers should have four basic detection modes: quasi-peak detection, root-mean-square detection, peak detection, and average detection. However, most electromagnetic disturbance is impulse, whose objective effects on RF frequency are bigger as the repetition frequency increases and the output characteristics of the quasi-peak detector with specified time constant can approximately reflect this effect. So in the wireless broadcast frequency domain, the

Electromagnetic Compatibility (EMC) Design and Test Case Analysis, First Edition. Junqi Zheng.

EMC standard recommended by CISPR adopts quasi-peak detection. Since quasi-peak detection will not only use the amplitude of the disturbance signal but also reflect its temporal distribution, the charging time constant is bigger than that of the peak detector but the discharge time constant is smaller than that of the peak detector. For different frequency bands, there should be different charge and discharge time constants. These two detection ways are mainly used in impulse disturbance test.

Antennas are the sensors of the radiated emission test. The frequency range of the radiated emission is from tens of kHz to tens of GHz. To test in such a wide frequency range, various antenna types are used, and all kinds of detection antenna must be used to convert field strength to the voltage. In the frequency range of 30–300 MHz, dipole antenna and bi-conical antenna are commonly used. In 300 MHz–1 GHz, dipole, log-periodic and log-spiral antenna are generally used, and in 1–40 GHz we use horn antenna. The relevant parameter and theory of these antennas can be referred to the documentation provided by the manufacturer. The antenna used in the radiated emission tests has the following characteristics: broadband antenna is widely used in order to improve the testing speed except the tests at only several known interfering frequency points. Manufacturer will provide calibration curve before broadband antenna is released, and during it is used, the antenna curve should be inputted. Many testing antennas work in near field region, where the test results are greatly sensitive to test distance. Therefore, the test must be performed strictly by the test rules. Second, in near-field region, the ratio of electric field and magnetic field (wave impendence) is not a constant any more. So, though the correction coefficients of electric field and magnetic field are provided for some antennas, they are only valid when these antennas are used for far-field test. When testing near field EMI, the test results of the electric field and magnetic field cannot be converted in this way. These problems can be easily ignored during the test.

Open test area is a professional radiated emission test site, which meets the requirements for the test distance required in the standard. In the standard, there should be no overhead wiring, construction or reflection object in the testing area (obstacle-free zone), and underground cable should be avoided. If necessary, there should be climate protection mask. The site should meet the related requirements about site attenuation defined in CISPR16, ANSI63.4, and EN50147-2. Semi anechoic chamber is a space that simulates open test area. Except the reflective plane is installed on the ground, other five internal surfaces are all installed with absorbing material. This site also meets the related requirements about site attenuation defined in CISPR16, ANSI63.4, and EN50147-2.

Control units are only used to test the coordinated action in between every device during the test, in order to auto-complete the radiated emission test.

B.1.3 Radiated Emission Test Methods

Figure B.1 shows a radiated emission test configuration diagram according to the requirements in CISPR16 and EN55022. When doing the test, EUT is placed inside the semi-anechoic chamber and rotated on the turntable to find the maximum radiation. The radiated emission is received by the antenna and then transmitted to the receiver outside the semi anechoic chamber via the cable.

The layout of tabletop EUT is shown in Figure B.2. The detailed requirements are as follows:

1) The distance from interconnection I/O cable to the ground should be not less than 40 cm.
2) Besides connecting the actual load, the EUT can also be connected with artificial load. But the artificial load should conform to the impedance relationship, and can represent the actual situation of the disturbance at the same time.

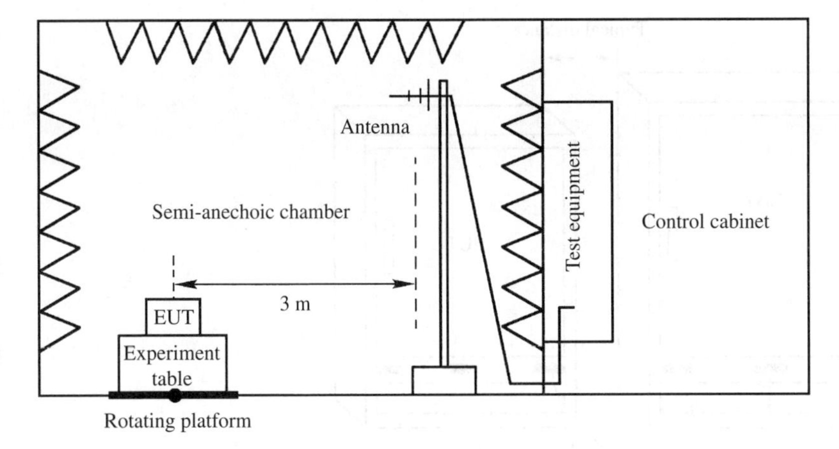

Figure B.1 Arrangement diagram for radiated emission test.

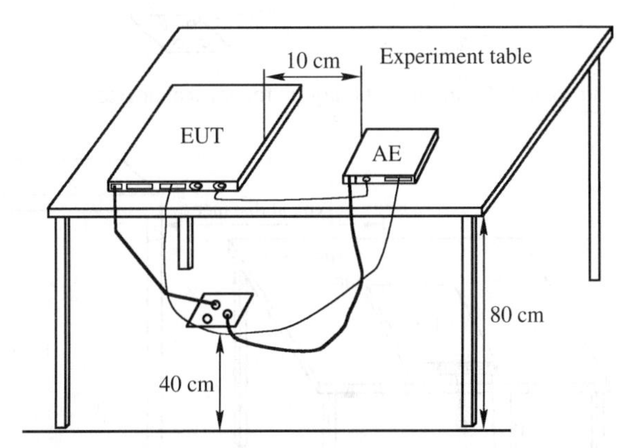

Figure B.2 Arrangement diagram for bench device.

3) The power lines of EUT and AE shall be directly plugged to the socket on the ground, and the lines connected to the socket should not be lengthened.
4) The distance between EUT and AE is 10 cm.
5) The control component of the EUT itself (such as keyboards) should be set as the normal use.
6) If there are many cables connected with the EUT itself, the cables should be carefully settled, disposed respectively, and recorded in test reports in order to gain the repeatability for another test.

The layout of floor-standing EUT is shown in Figure B.3. The detailed requirements are as follows:

1) The interconnecting I/O cables between equipment cabinets should be placed naturally. If they are so long and can be bundled as a 30–40 cm harness, they must be bundled.
2) Place the EUT above the metallic plane and the insulation distance away from the metallic plane insulation is 10 cm; the insulation from the metallic plane should be noticed when the artificial load or the cable going outside the chamber are connected.
3) If the power line of the EUT is too long, it should be bundled as a 30–40 cm harness, or shortened to the exact length for working.
4) If there are many cables connected with the EUT itself, the cables should be carefully settled, disposed respectively, and recorded in test reports in order to gain the repeatability for another test.

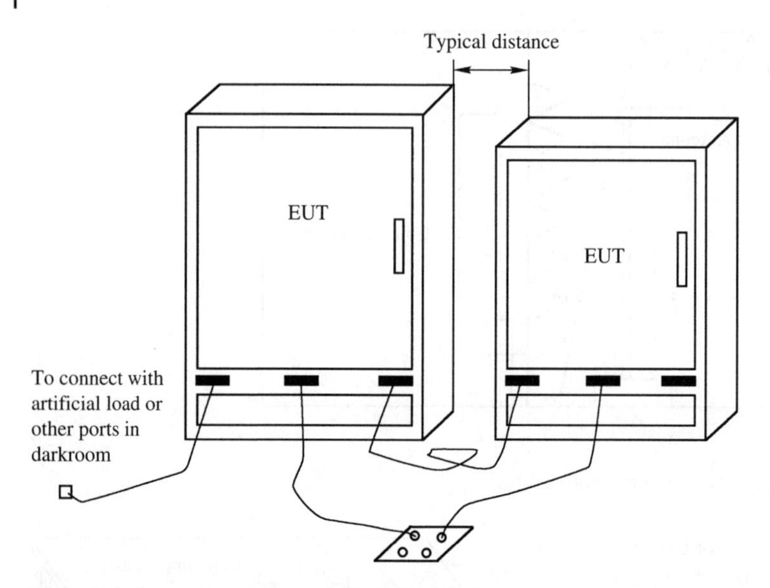

Figure B.3 Arrangement diagram for vertical device.

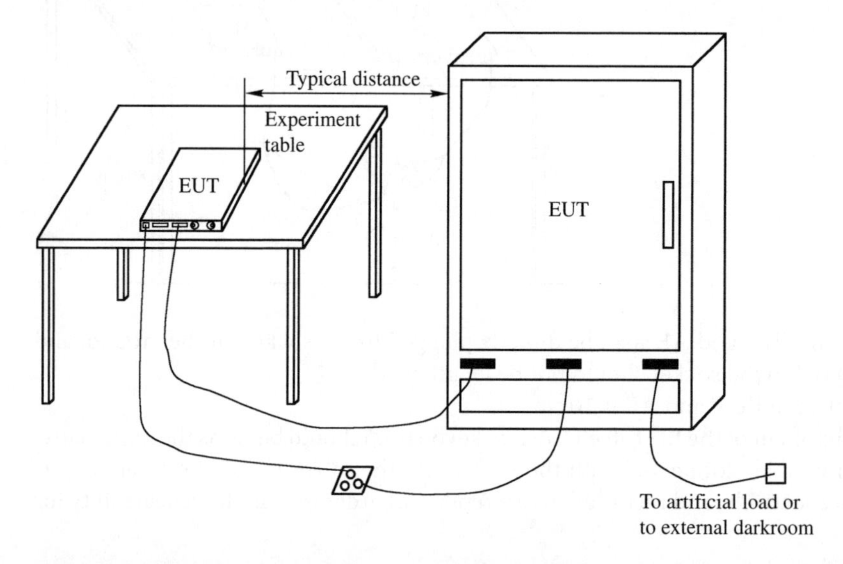

Figure B.4 Arrangement diagram for vertical and bench device.

The layout of the tabletop plus floor-standing EUT is shown in Figure B.4. The detailed requirements are as follows:

1) The interconnecting cables should be bundled as a 30–40 cm harness and placed on the metallic plane; if the length of the interconnecting cable is suitable and not touched with metallic plane, it should be hung up.
2) Power line should be placed naturally.
3) The insulation from the metallic plane should be noticed when connecting the artificial load or the cable going outside the chamber.
4) If there are many cables connected with the EUT itself, the cables should be carefully settled, disposed respectively, and recorded in test reports in order to gain the repeatability for another test.

B.2 Conducted Emission Testing

B.2.1 The Testing Purposes of Conducted Emission

Conducted emission testing measures the disturbance transmitted from power port and signal port of the equipment to the power network and signal network.

B.2.2 Commonly Used Conducted Emission Test Equipment

According to the commonly used conducted emission test standard CISPR16 and EN55022, conducted emission test mainly needs the following equipment:

1) EMI automatic test control system (computer and its interface unit)
2) EMI test receiver (or spectrum analyzer)
3) Line impedance stabilization network (LISN), current probe

LISN is a kind of coupling and decoupling circuit, mainly used to provide a clean DC/AC power quality and block the disturbance of the equipment under test back to the power supply and the RF coupling. It also provides specific impedance characteristics at the same time. The internal circuit architecture and impedance characteristic curve are shown in Figure B.5.

Current probe uses the principle that the current flowing through a conductor produces the magnetic field, and this magnetic field is picked up by another induction coil. It is usually used to perform the conducted emission test for the signal lines.

B.2.3 Conducted Emission Test Method

Compared with the radiated emission test, the conducted emission test requires less instruments, but a very important condition is that a reference ground plane with more than 2 × 2 m area is needed, and it must be extended beyond the EUT boundary at least 0.5 m because the environmental noise of the shielding room is very low. At the same time, inside the metallic shielding room, the metallic walls and can be used as the ground reference plate, so the conducted emission test is usually performed in the shielding room. Figure B.6 shows the test configuration for the conducted emission test on the power port of the table top equipment.

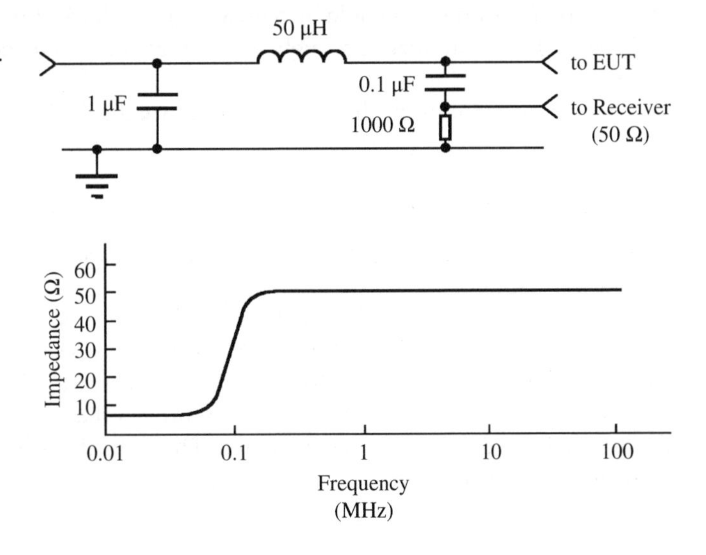

Figure B.5 LISN equivalent circuit and impedance characteristics curve.

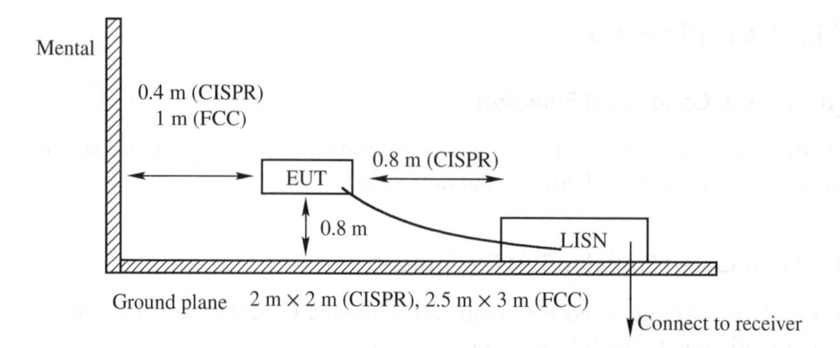

Figure B.6 Radiated emission test configuration for power port of bench device.

LISN picks up the conducted emission signals and implements impedance matching, then transmits the signals to the receiver (the specific test schematic diagram is described in Case 1). For floor standing equipment, in the test, the device is placed on the insulation support with 0.1 m height from the ground. Also, the power supply port needs to be performed with the conducted emission test, signal and communication port are also needs to performed with the conducted emission test. The test method for the signal port is relatively complicated. There are two ways to test them, i.e. to measure the voltage and current. The test results are compared with the current limit and voltage limit specified in the standard to determine whether the test is passed.

B.3 Electrostatic Discharge Immunity Testing

B.3.1 Purpose of an Electrostatic Discharge Test

This text checks the ability of a single device or system to resist electrostatic disturbance. It simulates (i) the discharge of an operator or an object in contact with the device; (ii) the discharge of a person or object to an adjacent object. Electrostatic discharge may have the following consequences:

1) Semiconductor devices could be damaged through the direct energy exchanging.
2) Electric field and magnetic field caused by discharge could cause the equipment to malfunction.
3) The noise current of the discharge could cause the device to a false action.

B.3.2 Electrostatic Discharge Test Device

The basic schematic and the discharge current wave form of the ESD generator are presented in Figures B.7 and B.8.

In B.8 I_m is peak current, the rise time $t_r = (0.7 - 1)$ ns. Energy storage capacitor C_s in discharge circuit stands for the body capacitance, it is acknowledged that 150 pF is more appropriate. Discharge resistance R_d is 330 Ω, used to represent the human resistance when a human holds a key or other metal tools. It has been proven that using this discharge model to reflect the human body discharge is harsh enough. The test voltage is gradually increased from low to high, until the specified value.

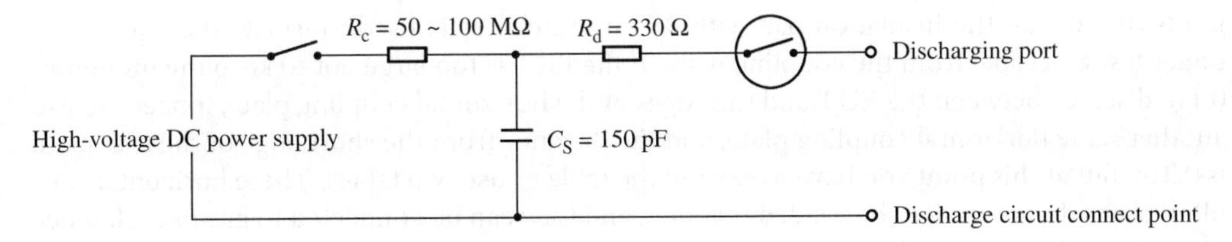

Figure B.7 Electrostatic discharge generator.

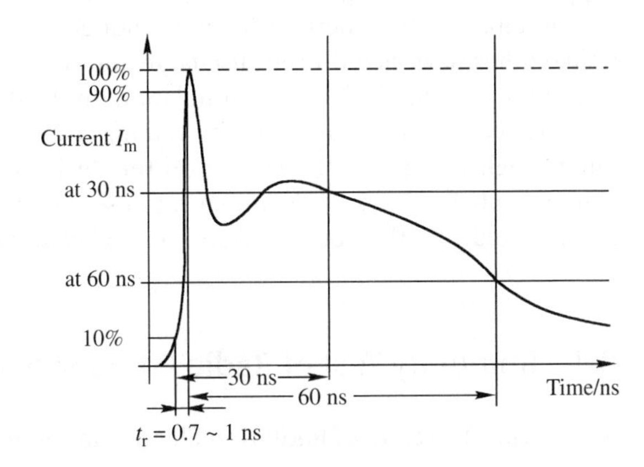

Figure B.8 Current waveform for electrostatic discharge.

Table B.1 Level of ESD test.

Level	Contact discharge(kV)	Air discharge(kV)
1	2	2
2	4	4
3	6	8
4	8	15

B.3.3 Electrostatic Discharge Test Method

ESD testing includes contact discharge and control discharge; contact discharge includes direct discharge and indirect discharge. The discharge point includes all accessible surfaces. Air discharge is performed for insulating surfaces, the maximum voltage can be up to 15 kV; contact discharge is performed for metal surface, the maximum voltage can be up to 8 kV (including vertical and horizontal coupling test). The electrode of the ESD gun is usually vertical to the surface of the device under test, at least 10 times of discharges for each polarity, the positive and negative polarity. The test interval is generally about one second. Before and after the ESD test, the function of the EUT is monitored to see whether it is normal to determine whether the test is passed.

The severity of the test is shown in Table B.1.

The test level selection depends on the environment and other factors, but for specific products, it is already specified in the corresponding product or product family standard.

For tabletop devices, the test equipment includes a wooden table that is above the reference plane with 0.8 m height. The area of the horizontal coupling plate (HCP) on the top of the table is 1.6 × 0.8 m, and the insulation pad with 0.5 mm thickness is used to isolate the equipment under test and cable from the coupling plate. If the EUT is too large not to keep the minimum 0.1 m distance between the EUT and the edges of the horizontal coupling plate, it needs to use another same horizontal coupling plate, and the distance from the short edge of the first plate is 0.3 m. But at this point you have to expand the table or use two tables. These horizontal coupling plates do not have to be welded together, and they can be connected to ground reference plane through a cable with resistors.

For floor standing equipment, the EUT and the cables shall be isolated with an insulation support with 0.1 m height from the ground reference plane.

Regarding the test method for nongrounded equipment, as the nonground equipment is unlike other devices with self-discharge ability. If the charge is not removed before the next ESD pulse is applied, charge accumulation on the device will make the voltage two times that of the expected test voltage in the test. It may cause accidental insulation breakdown discharge due to high energy. So the charge on the device of the nongrounded equipment should be eliminated before each ESD pulse, i.e. to use the cable with 470 KΩ discharge resistors, which is connected with the horizontal and vertical coupling plates.

B.4 Immunity Test of Radio Frequency Electromagnetic Field

B.4.1 The Objective of Radiation Electromagnetic Field Immunity Test

The disturbance of radio frequency electromagnetic field on equipment is always produced when the equipment is in operation and maintenance and the safety inspection person uses mobile phones. Others such as radio station, television transmitting station, mobile radio transmitters and various industrial electromagnetic radiation sources (the above ones belong to intentional radiation), and the parasitic radiation produced by the electric welding machine, the thyristor rectifier and the fluorescent lamp (these are nonintentional radiation), they will also produce radio frequency radiated disturbance. The purpose of the test is to establish a common criterion for evaluating the electrical and electronic equipment about the ability of anti-RF radiated electromagnetic field disturbance.

B.4.2 Test Instrument

1) *Signal generator*. The main index includes bandwidth, modulation function, automatically or manually scanning, the dwell time on scanning point can be set, the signal amplitude can be automatically controlled, and so on.
2) *Power amplifier*. It is required to achieve the field strength specified in the standard with either 3 m test distance or 10 m test distance. Also, a 1 m test distance can be used for small products, when there is a dispute for the test results between 1 m test distance and 3 m test distance, 3 m test distance is the criterion.
3) *Antenna*. The biconical antenna and log-periodic antenna are used in different frequency bands. The composite antenna has been used in the whole frequency band abroad.
4) *Field strength test probe*. The probe connects to field strength recording equipment.
5) *Field strength testing and recording equipment*. When additional equipment such as a power meter, computer (include special control software), or automatic walking mechanism of the field strength probe are added based on the basic instrument, a complete automatic test system can be constructed.

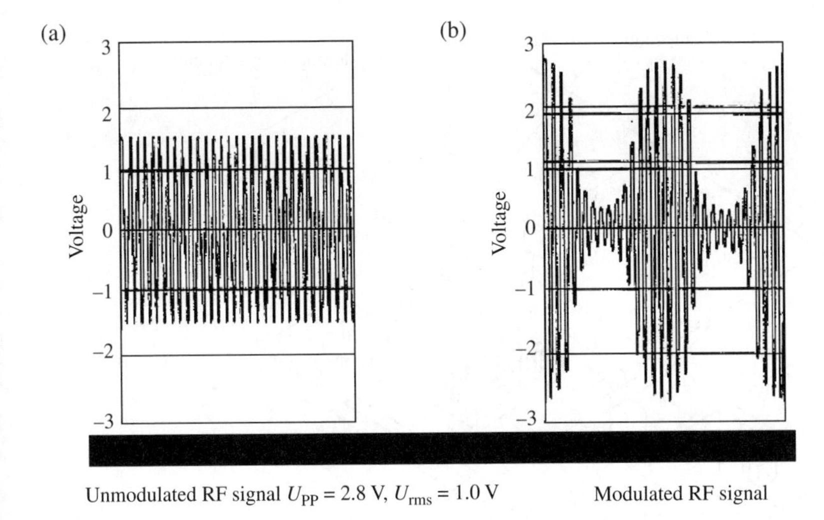

Figure B.9 Output voltage waveform of signal generator.

6) *Anechoic chamber*. The uniformity of the site is main considered. If, in the anechoic chamber, the electromagnetic waves generated from the product itself are also considered, it needs to compare the anechoic chamber compared with the open field. Verify the uniformity of the test site in order to ensure the comparability and repeatability of the test results.

B.4.3 Radiation Electromagnetic Field Immunity Test Method

The test with 1 kHz sine wave amplitude modulation, modulation depth is 80%, see Figure B.9 (modulation is not needed in early testing standards). It may be possible to add a new key to control FM in the future (EC standards have been adopted). Modulation frequency is 200 Hz, duty cycle is 1 : 1.

Test in anechoic chamber (shown in Figure B.10), monitor the EUT working with the monitor (or connect the signal from the EUT to test room and check its working status, it is judged by special instrument.). There are antennas (include the antenna's lift tower), turntable, test sample, and monitor in anechoic chamber. Test engineer, instrument, signal generator, power meter, computer, and other equipment are in the measurement room. The high-frequency power amplifiers are placed in the amplifier room. The wiring is very important and should be recorded during the test in order to reproduce the test results if necessary.

The relationship between field strength, test distance, and power amplifier is shown in Table B.2 (for reference only).

B.5 Electrical Fast Transient/Burst Immunity Test

B.5.1 Purpose Electrical Fast Transient/Burst Test

In the circuit, the switching of the inductive load by mechanical switch usually produce disturbance to the other electrical and electronic equipment in same circuit. The characteristics of this kind of disturbance are: transients occur with burst, high repetition rate of pulse, short rise time of pulse waveform, and low energy of single pulse. In fact, the probability of equipment malfunction caused by the electric fast transient/burst is less, but the false action of the equipment is

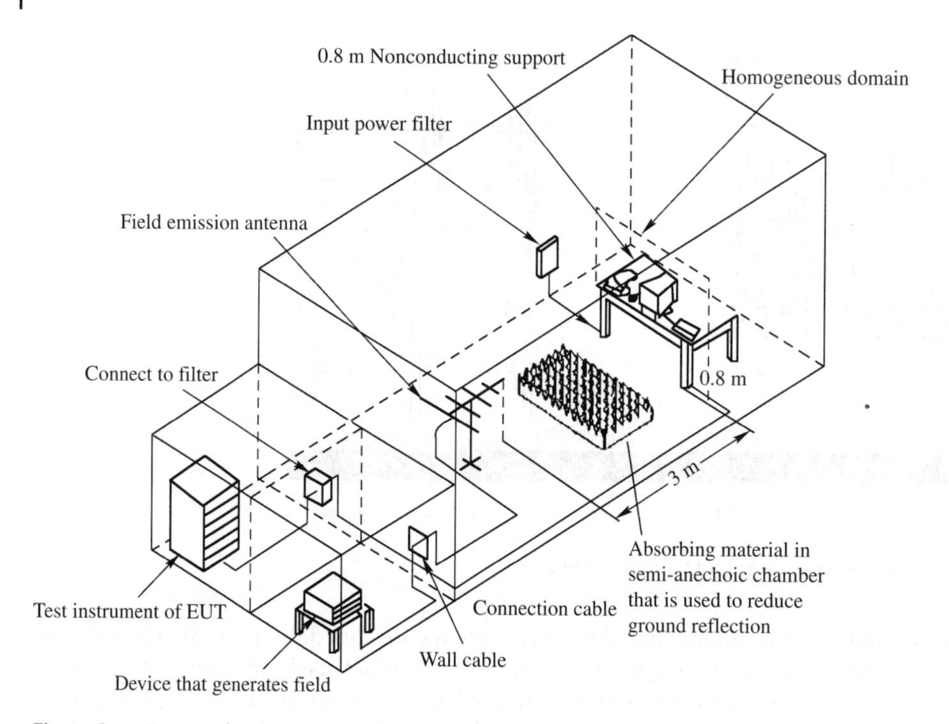

Figure B.10 Test configuration for RF radiated electromagnetic field immunity.

Table B.2 The relationship of field intensity, test distance, and power amplifier.

Amplification power	Field intensity and test distance
25 W	3 V m^{-1} field intensity can be generated using 1 m method. If frequency is high than 200 MHz, 10 V m^{-1} field intensity can be generated using 1 m method.
100 W	3 V m^{-1} field intensity with 80% modulation depth can be generated using 3 m method. 10 V m^{-1} field intensity can be generated using 1 m method.
200 and 500 W	10 V m^{-1} field intensity can be generated using 3 m method on the virtual plane 1.5 × 1.5 m. If reduce the distance, 30 V m^{-1} field intensity can be generated.

often seen. Unless the appropriate measures are taken, the immunity test is difficult to pass. The purpose of the electric fast transient/burst test is to establish a common basis for the evaluation of the electric and electronic devices to immunize the fast transient/burst. The mechanism of the test is to use the accumulation effect of the energy of the line distributed capacitance from the pulse group, and when the energy accumulates to a certain level, it may cause the line (even equipment) malfunction. Once the line in the test goes wrong, it will continue to make mistakes. Even if the pulse voltage is slightly reduced, the error still occurs.

B.5.2 Equipment for Electrical Fast Transient/Burst Test

Figure B.11 gives the basic circuit diagram of the generator of electrical fast transient/burst. The pulse waveform is seen in Figure B.12. The basic requirements for the electric fast transient/burst are:

Rise time (refer to 10–90%): 5 ns ± 30%

Pulse width (50% of rising edge to 50% of falling edge): 50 ns ± 30%

Repetition frequency: 5 or 2.5 kHz
Burst duration: 15 ms
Burst period: 300 ms
Open circuit output voltage (peak) of generator: (0.25–4) kV
Dynamic output impedance of generator: 50 Ω ± 20%
Polarity: positive/negative
Relationship to power supply: asynchronous

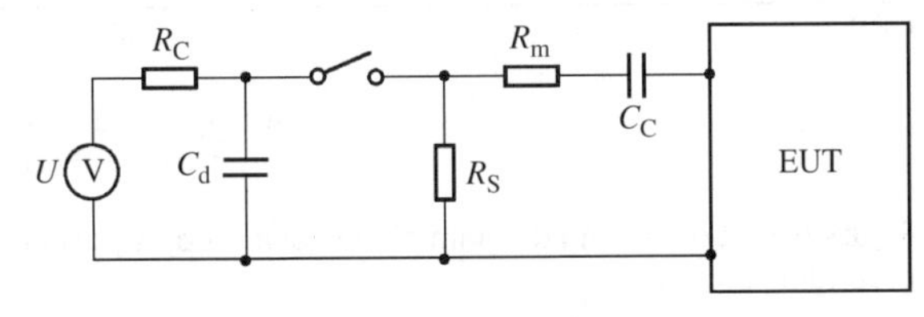

Figure B.11 The basic circuit diagram of the generator of electrical fast transient pulse burst.

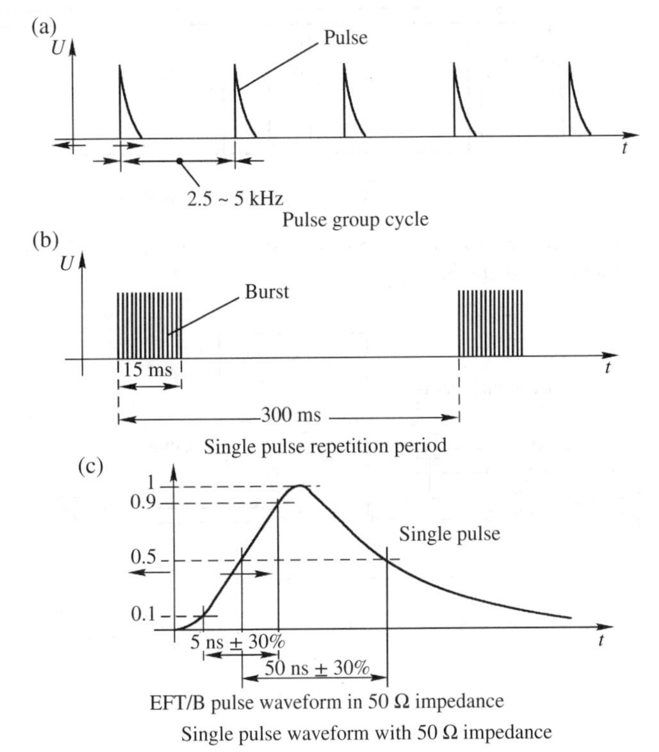

Figure B.12 The electrical fast transient pulse group waveform.

B.5.3 Method for Fast Transient/Burst Test

There are two types of tests: The laboratory test and the field test after the installation of the equipment. The first test is to be a priority for the test; the second test is adopted only when the manufacturer and the user come to an agreement. The laboratory configuration of electrical fast transient/burst is similar to ESD test, a ground reference plane, ground floor on the ground, and the material of plane requirements are same; but for the desktop equipment, it shouldn't lay metal plate on the table. As shown in Figures B.13 and B.14.

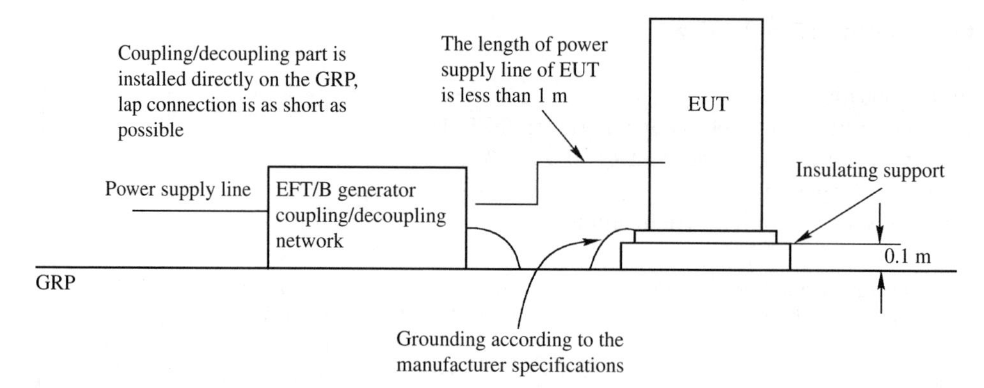

Figure B.13 Connecting diagram of the power supply port of the landing equipment and the protection of grounding port EFT/B.

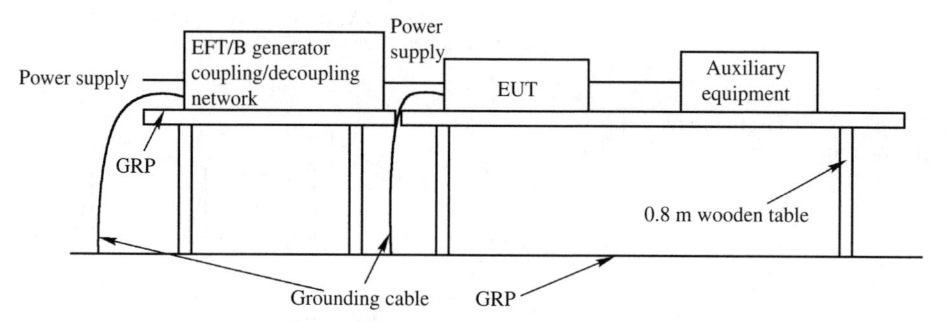

Figure B.14 Connecting diagram of the power supply port of the desktop equipment and the protection of grounding port EFT/B.

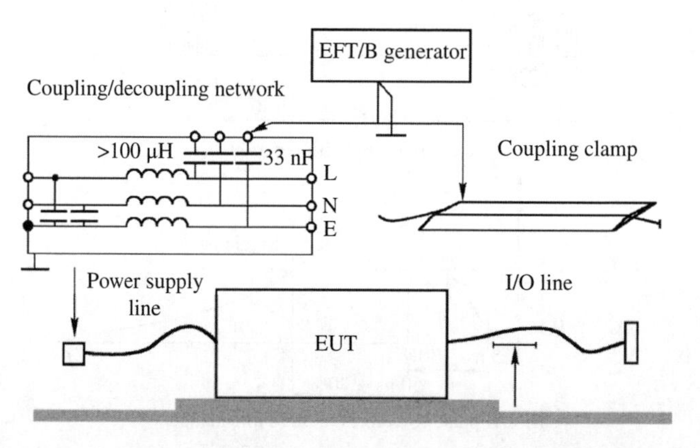

Figure B.15 The coupling principle in the test.

The coupling and decoupling principle in the test is shown in Figure B.15:

1) Power supply line test (include AC and DC) is used with common mode by coupling and decoupling network. The test voltage is injected between each power supply terminal and nearest protection grounded point, or between each power supply terminal and the ground reference plane.
2) For control line, signal line and communication equipment, the test voltage is applied by capacitance coupling clamp in common mode.

Table B.3 Severe degree level.

Open circuit output test voltage (±10%),repetition frequency of pulse (±20%)				
	On the power supply port and PE		On the I/O (output/input), signal, data, and control ports	
Level	Voltage peak (kV)	The repetition frequency of pulse (kHZ)	Voltage peak (kV)	The repetition frequency of pulse (kHZ)
1	0.5	5 and 100[a]	0.25	5 and 100[a]
2	1	5 and 100[a]	0.5	5 and 100[a]
3	2	5 and 100[a]	1	5 and 100[a]
4	4	5 and 100[a]	2	5 and 100[a]
X[b]	undetermined	undetermined	undetermined	undetermined

[a] The custom is to use 5 kHz, but 100 kHz is closer to the actual. Product technical committee may determine the frequency for a particular product or product type.
[b] *X* is an open level, with special requirements for special equipment.

3) For the protection grounded terminal of the equipment, the test voltage is added between the terminal and the ground reference plane. Test time is at least one minute each time, and the positive/negative polarity are the necessary items. Table B.3 is a severity level table for the test.

Note: In the table the voltage refers to the voltage on the signal energy storage capacitor of the/burst generator. The frequency refers to the repetition frequency of the pulse group

B.6 Surge Immunity Test

B.6.1 Purpose of Surge Immunity Test

Surge immunity tests are designed to check to two primary things: damage from lightning strikes and switching disturbances.

1) *Lightning strikes* (mainly indirect lightning). This could include lightning strikes to outdoor circuit injecting high currents producing voltages by either flowing through ground resistance or flowing through the impedance of the external circuit, produces the interference voltage; another example, an indirect lightning stoke (i.e. a stoke between or within clouds) that induces the voltage and current on the line; another example, a stroke to nearby objects that produce electromagnetic fields, when outdoor line passes through the electromagnetic field, it can induce the voltage and current; also, when the lightning stroke to the near ground, the interference can be introduced when the ground current passes through the common grounding system.
2) *Switching transients.* One example is major power system switching disturbances (such as the switching of the compensation capacitor group); another example, in the same power grid, the interference is formed by larger switch switching near the device; another example, switching the thyristor device with resonant circuit; also, various system faults include short circuits or arcing faults in grounding network or between grounding system. A common standard for evaluating the anti-surge interference ability of electrical and electronic equipment is established by the method of simulation testing.

B.6.2 Equipment for Surge Simulation

In accordance with the requirements of the IEC61000-4-5 (T/GB 17626.5) standard, it is required to simulate the surge test on the power supply line and the communication line. As the line impedance is not the same, the surge waveforms are different. Hence, it needs two simulations. Figure B.16 is a diagram of integrated wave generator.

The waveform of integrated wave generator is shown in Figure B.17.

Figure B.17 (a) shows a 1.2/50 μs open circuit voltage waveform (according to IEC60-1), the wave front time: $T_1 = 1.67 \times T = 1.2\,\mu s \pm 30\%$;

Time to half value: $T_2 = 50\,\mu s \pm 20\%$; Figure B.17 (b) is a 1.2/50 μs short-circuit voltage waveform (according to IEC60-1), the wave front time: $T_1 = 1.25 \times T = 8\,\mu s \pm 30\%$; time to half value: $T_2 = 20\,\mu s \pm 20\%$. In addition to the characteristic of generating the waveform shown in

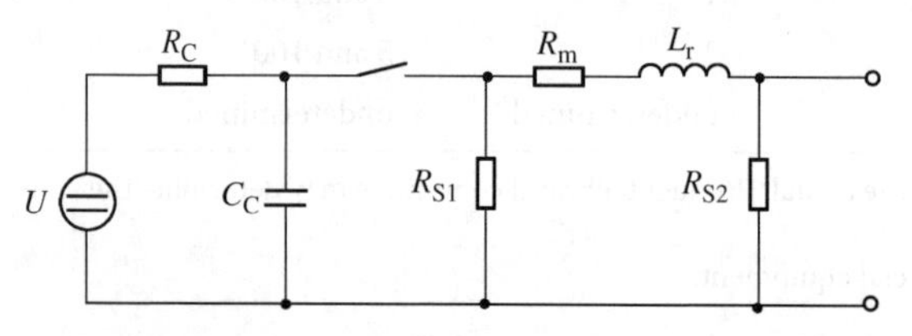

Figure B.16 The simple diagram of integrated wave generator.

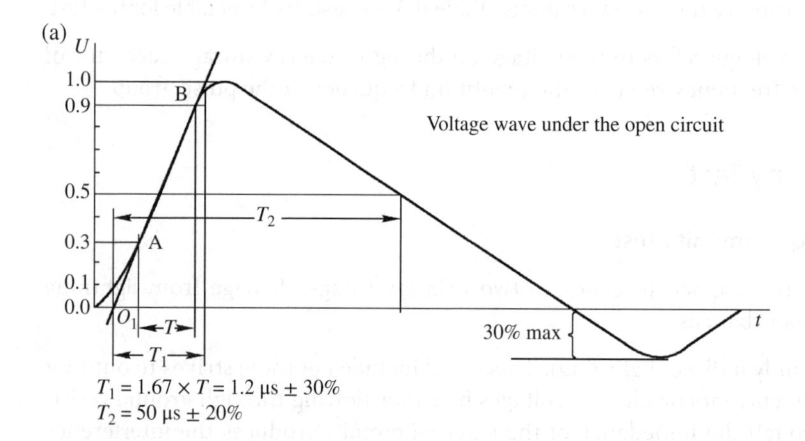

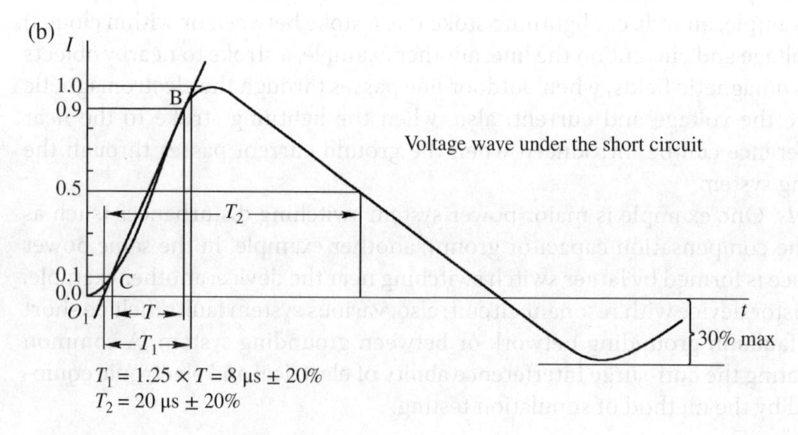

Figure B.17 Integrated waveform.

Figure B.18 The basic line of the 10/700 μs surge voltage generator.

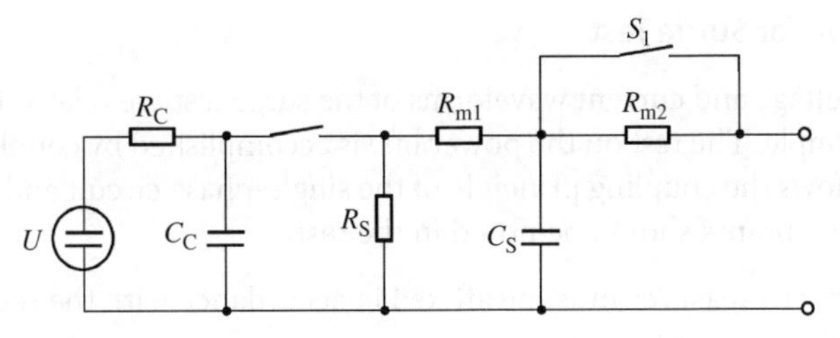

U–High-voltage power supply;
R_m–Resistor for impedance match (R_{m1} = 150 Ω; R_{m2} = 25 Ω);
R_C–Charge resistor; C_C–Stored energy capacitor (20 μF);
C_S–Capacitor for rise time (0.2 μF); R_s–Resistor for pulse duration (50 Ω); S_1–Switch

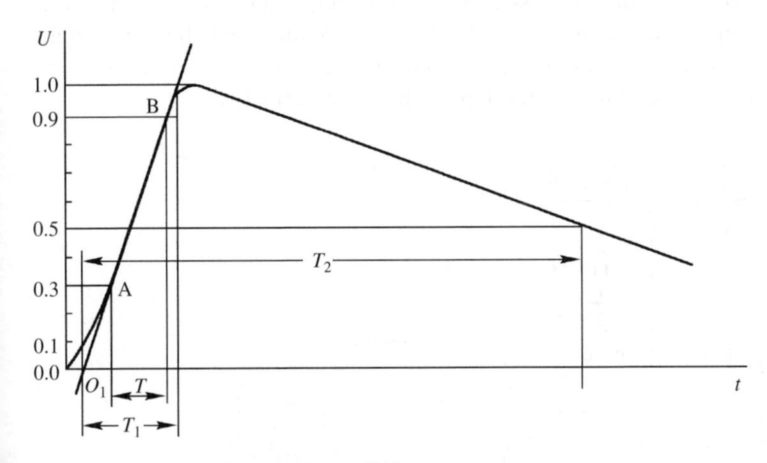

Rise time: $T_1 = 1.67 \times T = 10$ μs ± 30%;
Pulse duration: $T_2 = 700$ μs ± 20%.

Figure B.19 CCITT voltage surge waveform.

Figure B.17, the integrated wave generator shall have the following basic performance requirements:

Open-circuit output voltage (peak): 0.5–4 kV
Short-circuit output current (peak): 0.25–2 kA
Impedance of generator: 2 Ω, (additional resistance of 10 Ω or 40 Ω, so as to form 12 Ω or 42 Ω impedance of generator)
Polarity: positive/negative
Phase shift range: 0°–360°
Maximum repetition frequency: at least one time per minute
The circuit diagram of the 10/700 μs surge voltage generator used for communication line test is shown in Figure B.18. The corresponding International telephone and consultative committee (CCITT) voltage surge waveforms are shown in Figure B.19.
In addition to the characteristic of generating the waveform shown in Figure B.19, the 10/700 μs surge voltage generator shall have the following basic performance requirements:
Open-circuit output voltage (peak): 0.5–4 kV
Dynamic resistance: 40 Ω
Output polarity: positive/negative

B.6.3 Method for Surge Test

Because the voltage and current waveforms of the surge test are relatively slow, the configuration of the test is simple. The test on the power line is accomplished by coupling/decoupling network. Figure B.20 shows the coupling principle of the single-phase circuit and the signal line surge test.

The following points should be noted in the test:

1) The protective measures must be affixed in accordance with the requirements of the manufacturer before the test.
2) Test rate is one time per minute, not too fast, so that the protection device has a performance recovery process. In fact, in the natural world, lightning strikes and large switching station switching do not have a very high repetition rate.
3) Test times, five times for positive/negative, respectively.
4) The test voltage should be gradually increased from low to high, and avoid artifacts due to the volt-ampere nonlinear characteristics of EUT. Also, be sure that the test voltage does not exceed the requirements of the product specification, so as to avoid unnecessary damage. The specified severity level in the standard is shown in Table B.4.

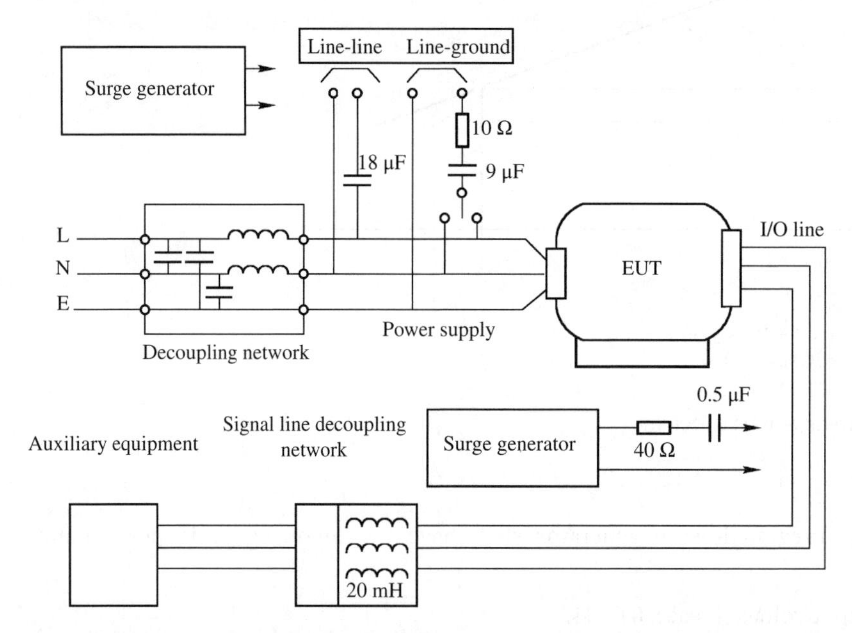

Figure B.20 The coupling principle of surge test.

Table B.4 Severe degree level.

Level	Line-line	Line-ground
1	—	0.5
2	0.5	1
3	1	2
4	2	4
X	Undetermined	Undetermined

B.7 Conducted Immunity Testing

B.7.1 Purpose of Conducted Immunity Test

Under normal circumstances, the dimensions of the disturbed equipment are assumed to be small compared with the wavelengths of the disturbance frequency, but the length of the lead of equipment (including mains, communication lines and the interface cable) is comparable with several wavelengths of interference frequency. In this way, the conducted disturbance to the equipment can be generated from the lead wire. The test is to evaluate the electrical and electronic equipment for the conducted immunity caused by the radio frequency field. Note: For devices without cables (such as power supply line, signal line or ground wire), this test is not required.

B.7.2 Equipment for Conducted Immunity Test

The block diagram of the conducted immunity test equipment is shown in Figure B.21.

1) RF signal generator (with bandwidth 150 kHz–230 MHz, amplitude modulation function, automatic or manual scanning, the dwell time in scanning points can be set; the amplitude of the signal can be automatic controlled).
2) Power amplifier (depending on the test method and the severity level).
   a) Used in coupling between lines and lines, the output of the generator is floating.
   b) Used in coupling between lines and lines, the output of the generator is grounded.
3) Low-pass and high-pass filter (to avoid signal harmonics interference to product).
4) Fixed attenuator (it is fixed to 6 dB. In order to reduce the mismatch between the power amplifier and the coupling network, install it closed to coupling network).

Along with a computer, electronic millivolt meter, and the instruments above, the combination can form an automatic test system.

B.7.3 Method for Conducted Immunity Testing

The simulation test frequency range is 150 kHz–80 MHz. When the product size is small, the upper limit frequency can be extended to 230 MHz. In addition, in order to increase the difficulty of the test, a 1 kHz sine wave can be used for amplitude modulation in the test. The amplitude modulation depth is 80%.

The classification of severity level (shown in Table B.5) is the same with IEC61000-4-3 (GB/T 17626.3). The test can be carried out in a shielded room, shown in Figure B.22. There are two interference injection methods:

1) Coupling/decoupling network (which is often used in power line test, as well as the number of signal lines is less).
2) Current clamp and electromagnetic coupling clamp (especially suitable for test of multi core cable. The electromagnetic coupling clamp has good reproducibility for the test results at the frequency above 1.5 MHz. When test frequency is higher than 10 MHz, electromagnetic

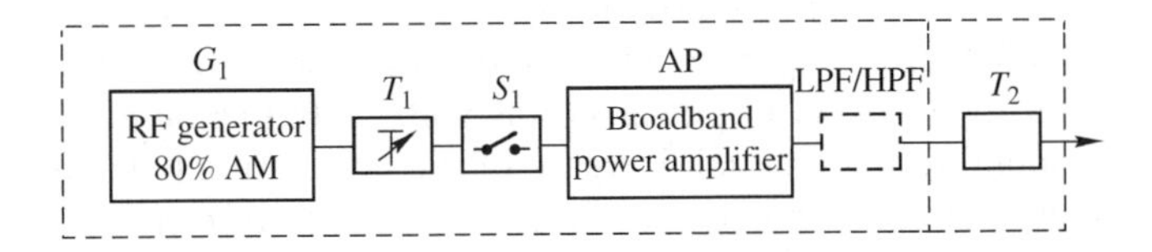

Figure B.21 Test equipment composition block diagram of conductive immunity test.

Table B.5 Severe degree level.

Level	Test voltage
1	1
2	3
3	10
X	Undetermined

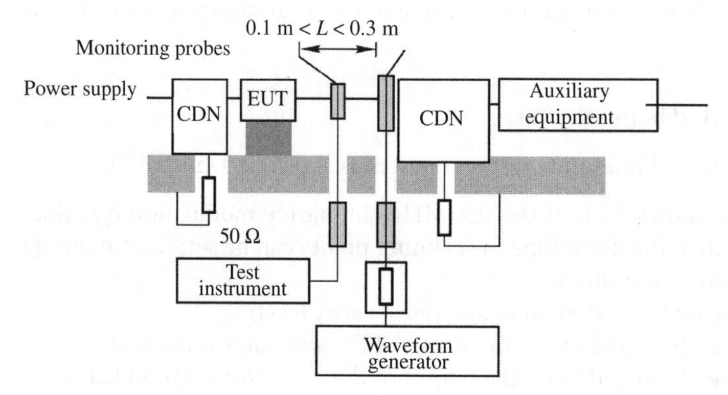

Figure B.22 The test instrument configuration diagram of conduction immunity test.

coupling clamp has better direction than conventional current clamp, and there is no need for special impedance between auxiliary equipment signal reference and reference grounded plane. Therefore, it is more convenient.

B.8 Voltage Dips, Short Interruptions, and Voltage Variations Immunity Testing

B.8.1 Purpose of Immunity Test of Voltage Dips, Short Interruptions, and Voltage Variations

With voltage instantaneous dips, short interruptions are caused by the fault of the power grid, the electric facilities, or a sudden change of the load. In some cases, two or more consecutive dips or interruptions occur. Voltage change is caused by continuous change of the load. These phenomena are essentially random, and their characteristics are shown to deviate from the rated voltage and continue for a period of time. Voltage dips and short interruptions are not always abrupt, because the rotating motor and protective device connected with the power supply network has a certain reaction time. If the large power network is disconnected (a large area in a factory or in a region), the voltage will be reduced by a lot of rotary motor connected to the power grid. These rotating motors will be run as a motor in the short term, and the electricity will be delivered to the power grid. As with most data processing equipment, there are generally built-in power failure detection devices, so that the device can start in the correct way after the recovery of power supply voltage. But some power detection devices fail to quickly respond to the gradual decrease of the power supply voltage, so the DC voltage in the integrated circuit has been reduced below the lowest operating voltage level before the power detection device triggering, which results in data loss or change. In this way, when the power supply voltage is recovered, the data-processing equipment cannot be properly started again. IEC61000-11-4/29 standard specifies the different tests to simulate the voltage mutation effect, in order to establish a general immunity evaluation of electrical and electronic equipment in this effect. Voltage variations could perform as a routine test, according to the product or the relevant standards, and is used in particular and reasonable condition.

B.8.2 Equipment for Immunity Test of Voltage Dips, Short Interruptions and Voltage Variations

Main performance requirements:

Output voltage: precision ±5%.
Output current capability: < 16 A while test level is 100%U_T, and it can maintain constant power for other test level.
Peak starting current capability: no more than 500 A (220 V voltage); 250 A (100–120 V voltage).
Rise or fall time for mutation of voltage: 1–5 μs (with 100 Ω load).
Phase: 0°–360° (accuracy ±10°).
The output impedance is resistive, and should be as small as possible.

Testing equipment to achieve the above functions: There are two basic formats, shown in Figures B.23 and B.24. Figure B.23 is a relatively cheap test generator, when the two switches are off at the same time, the output voltage is interrupted (interrupt time can be set in advance);

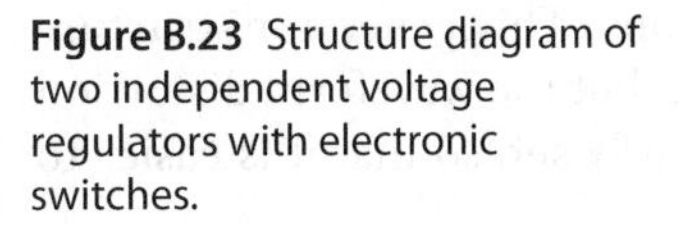
Figure B.23 Structure diagram of two independent voltage regulators with electronic switches.

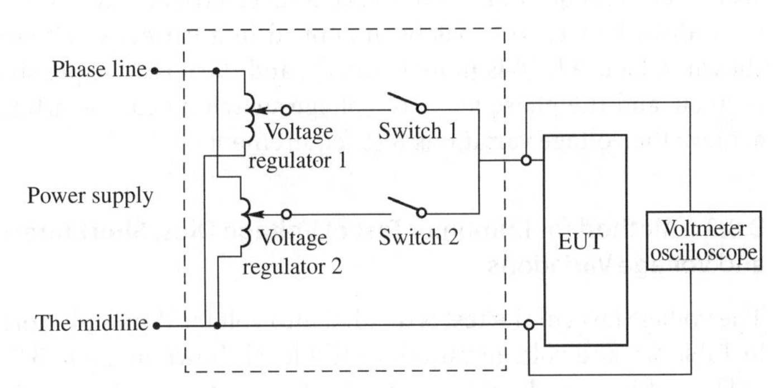

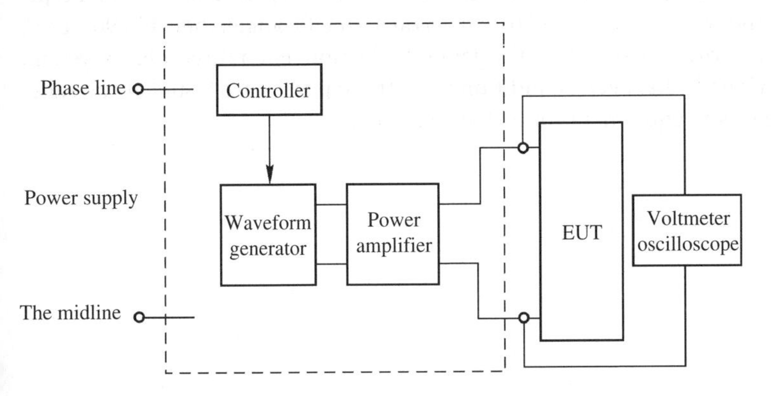

Figure B.24 The form of test generator made up of waveform generator and power amplifier.

Table B.6 Voltage sag and short time interrupt test level.

Test level	Voltage sag and short time interrupt	Duration time (period)
0 40 70	100 60 30	0.5 1 5 10 25 50 X

Table B.7 Voltage gradient test level.

Test level	Falling time	Keeping time	Rising time
40%UT	2 s ± 20%	1 s ± 20%	2 s ± 20%
0%UT	2 s ± 20%	1 s ± 20%	2 s ± 20%

when the two switches are on alternately, it can simulate the voltage dips or rise. The switch of the generator can be composed of a thyristor or bidirectional thyristor. The control circuit is usually on at zero voltage and off at zero current, so the circuit can only simulate the initial angle 0° and 180° of the voltage switch. Even so, because of the low cost of the instrument, it can also meet the requirements of the general electrical and electronic products to the power network disturbance, so it has been applied in a variety of situations. This generator structure shown in Figure B.24 is more complex, and the cost is expensive, but the waveform distortion is small, and the phase angle of voltage switching can be arbitrarily set, so that it is easier to achieve the voltage variations test requirements.

B.8.3 Method for Immunity Test of Voltage Dips, Short Interruptions, and Voltage Variations

The voltage level of the test is divided into voltage dips and short interruption test levels shown in Table B.6 and voltage variations test level shown in Table B.7.

The test is carried out according to the selected test level and duration time. The test is done for three times, and the interval for each time is 10 seconds. The test is performed under a typical working condition. If the voltage is required to switch on a specific angle, then 45°, 90°, 135°, 180°, 270°, 315°, 225° are priorities. 0° or 180° is the general selection. For a three-phase system, each phase is tested individually. In special conditions, the three phases are tested at the same time, and there are three sets of equipment for test at this time.

Appendix C

EMC Test for Automotive Electronic and Electrical Components

For standards of automotive electronics and electrical components, EMC tests include ISO11452, CISPR25, ISO7637, J1113 SAE, etc., which provide the specification also for automotive electronics, electrical components EMC, and EMI test. At the same time, in order to emphasize automotive, in the automotive electronics, electrical components EMC test, immunity test is more important. ISO11452 and ISO7637 are standard for the immunity performance of automotive electronics.

C.1 Automotive Electronics and Electrical Components Radiation Emission Test

C.1.1 Purpose of Automotive Electronics and Electrical Components Radiation Emission Test

The purpose is to test the radiation emission produced by automotive electronics and electrical components, including from the shell, all parts, cables, and connection lines. It is used to identify whether the radiation of the automotive electronics, electrical parts, and components comply with the related standards, so that the normal use of the car does not affect other electronic, electrical equipment.

C.1.2 Equipment for Automotive Electronics and Electrical Components Radiation Emission Test

The purpose is to test the radiation emission produced by car interior electronics electrical components, including from the shell, all parts, cables, and connection lines. It is used to identify whether the radiation of the automotive electronics, electrical parts, and components comply with the related standards, so that the normal use of the car does not affect other devices in the same environment (such as the car interior).

According to the provisions of automotive electronics, electrical equipment radiation emission test standard of CISPR25 (domestic equivalent use, corresponding to the national standard is GB18655-2002 the radio disturbance characteristics limits and measurement methods for car receiver protection for), radiation emission test mainly needs the following equipment:

1) EMI automatic test control system (computer and software)
2) EMI test receiver
3) Antenna control unit

Electromagnetic Compatibility (EMC) Design and Test Case Analysis, First Edition. Junqi Zheng.

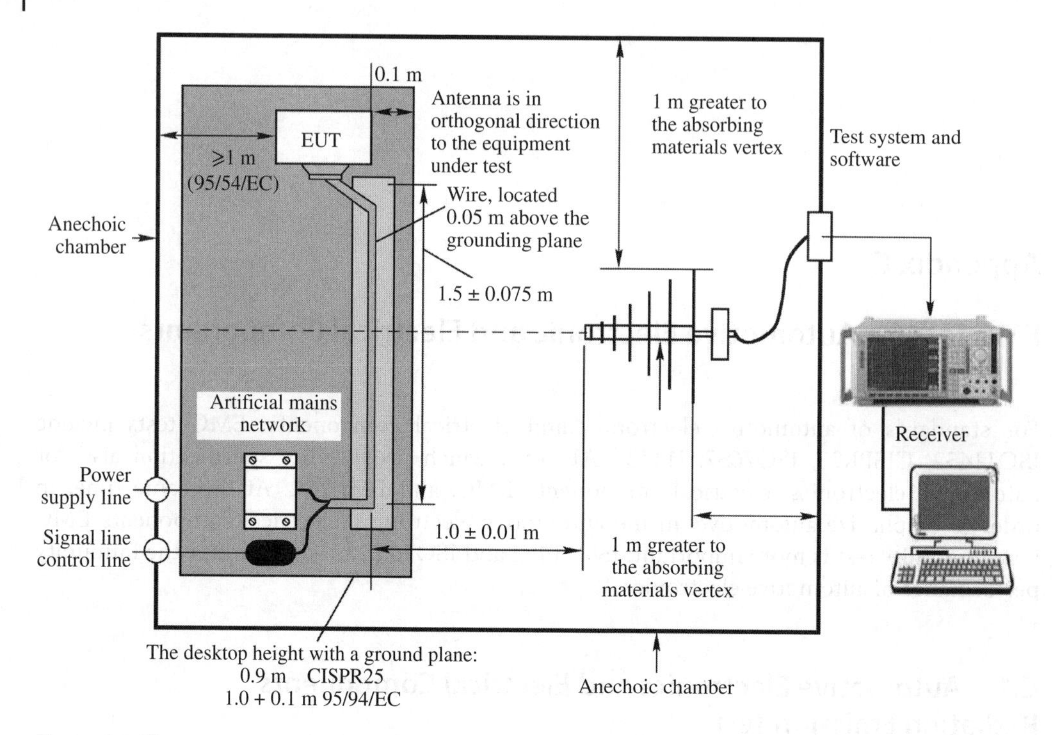

Figure C.1 The requirements for the radiation emission test arrangement in CISPR25 standard.

4) Semi-anechoic chamber
5) Artificial mains network (AMN, sometimes called linear impedance stabilization network, LISN), in the laboratory.

Artificial power network is used to replace the wire impedance, in order to determine the work status of the measured device. The parameters of artificial power supply network have strict requirements, which provide the basis for the comparability of test results in different laboratories.

C.1.3 Automotive Electronics and Electrical Components Radiation Emission Test Method

For radiation emission test of automotive electronics and electrical components, the layout should be in accordance with CISPR25 standard requirements for the radiation emission test, as shown in Figure C.1. During radiation emission test, the equipment under test (EUT) of automotive electronics is placed in semi-anechoic chamber, the distance from receiving antenna to the EUT wire is 1 m, respectively, in case of receiving antenna in vertical and horizontal polarization plane, to find the maximum radiation point. The radiation signal is received by the receiving antenna, and passes to the receiver outside the anechoic chamber through the cable.

C.2 Automotive Electronics and Electrical Components Conducted Disturbance

C.2.1 Purpose of Automotive Electronics and Electrical Components Conducted Disturbance

The purpose is to evaluate the conducted disturbance transmit to inner power supply or signal network from power port, signal port of automotive electronics, electrical components.

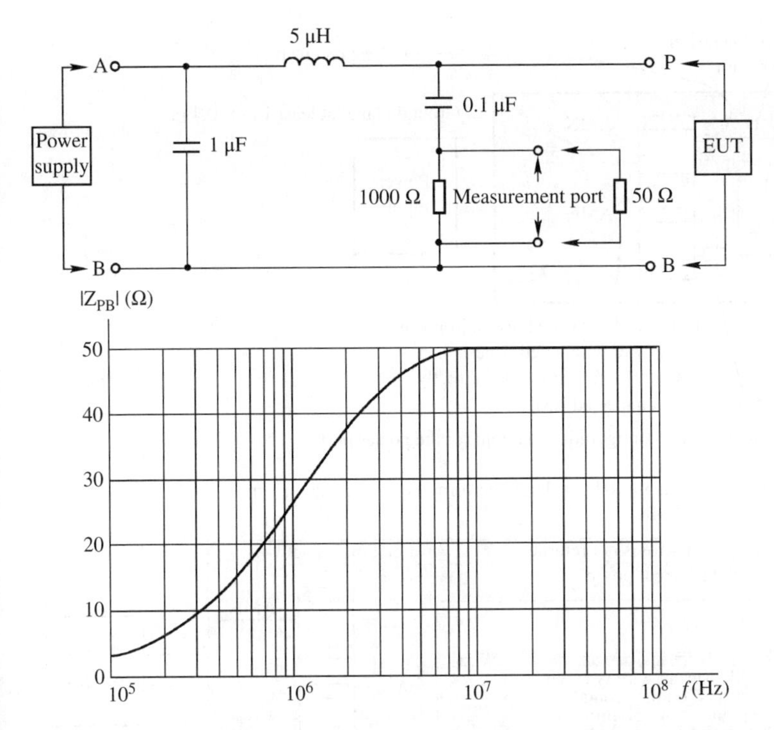

Figure C.2 The internal circuit architecture of the automotive electronics, electrical components products EMC test LISN standard, and the impedance characteristic curve.

C.2.2 Equipment for Conducted Disturbance Test of Automotive Electronics and Electrical Components

According to the provisions of automotive electronics, electrical components conducted disturbance test standard of CISPR25, conducted disturbance test mainly needs the following equipment:

1) EMI automatic test control system (computer and software)
2) EMI test receiver
3) Power supply LISN or AMN, the internal circuit architecture of the CISPR25 LISN standard and the impedance characteristic curve is shown in Figure C.2.
4) The current probe is use of magnetic field generated by current flowing the conductor, and other coil induce this magnetic field. This is usually used to measure conducted disturbance on the signal line.

C.2.3 Automotive Electronics and Electrical Components Conducted Disturbance Test Method

A very important condition for the conduction disturbance test is required a more than $1 \times 0.4\ \text{m}^2$ reference ground plane, and beyond the EUT boundary at least 0.1 m. Conducted disturbance tests are usually performed in a shielded room. Figure C.3 shows the configuration of conducted disturbance test, in which (i) is the conducted disturbance test configuration diagram for the power port, and (ii) is the conducted disturbance test configuration diagram for the signal port, artificial mains network achieves impedance matching and pickup conductive signal, then transmit the signal to the receiver.

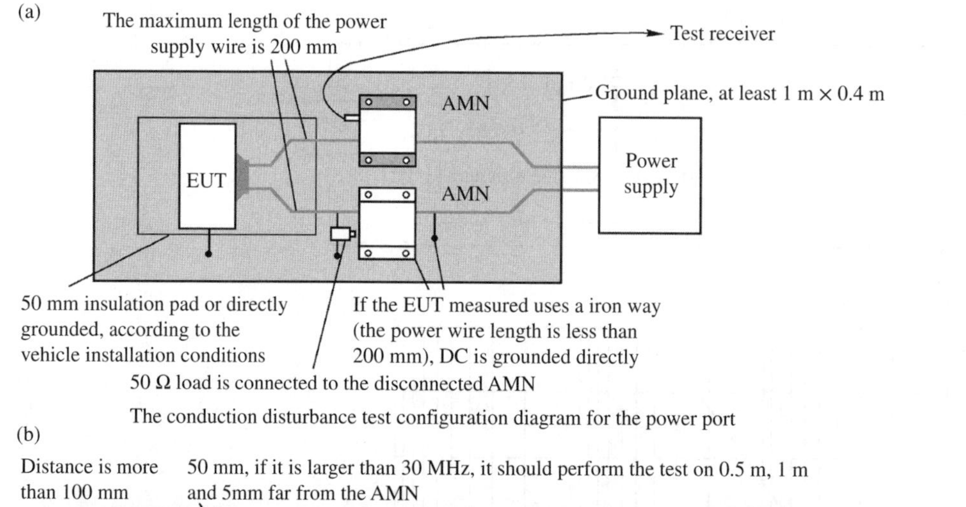
(a)
The maximum length of the power supply wire is 200 mm
Test receiver
Ground plane, at least 1 m × 0.4 m
AMN
AMN
EUT
Power supply
50 mm insulation pad or directly grounded, according to the vehicle installation conditions
If the EUT measured uses a iron way (the power wire length is less than 200 mm), DC is grounded directly
50 Ω load is connected to the disconnected AMN
The conduction disturbance test configuration diagram for the power port
(b)
Distance is more than 100 mm
50 mm, if it is larger than 30 MHz, it should perform the test on 0.5 m, 1 m and 5mm far from the AMN
The power wire is not returned from the current probe
Artificial mains network
AMN
EUT
Measuring wire is 1.5 m, placed on the ground plane above 50 mm
Power supply
AMN
Current probe
Auxiliary equipment
Distance is more than 100 mm
Test receiver
Grounded plane
The conduction disturbance test configuration diagram for the signal port

Figure C.3 The configuration diagram of conductive disturbance test.

C.3 Automotive Electronics and Electrical Components ESD Immunity Test

C.3.1 Purpose of Automotive Electronics and Electrical Components ESD Immunity Test

The purpose is to evaluate the ESD immunity capability of the automotive electronic, electrical components. It simulates: (i) the discharge between the operator or object and the electronic and electrical components invehicle; and (ii) the discharge between a person or object and the close object.

C.3.2 Equipment for Automotive Electronics and Electrical Components ESD Immunity Test

To meets the ISO10605 standard, the electrostatic discharge needs human body electrostatic discharge model in two cases:

1) Electrostatic discharge phenomena occurs when passenger is in compartment.
2) The electrostatic discharge phenomena occurred when the passenger from the outside into the passenger compartment.

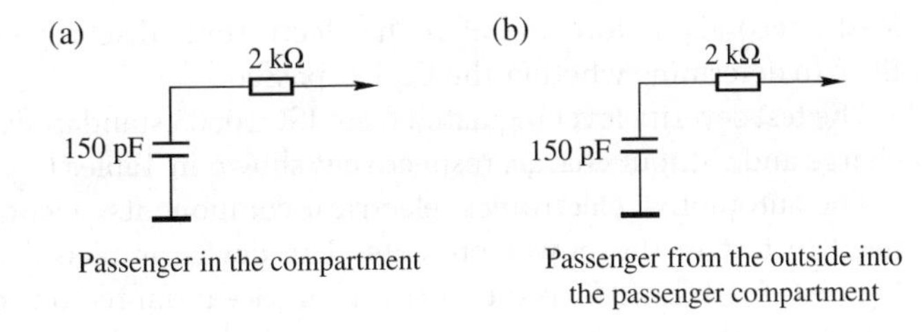

Figure C.4 Automotive electrostatic discharge gun resistance capacity network.

These two discharge models correspond to different electrostatic discharge gun resistor capacitor network, respectively, as shown in Figure C.4a and b. On the other hand, the electrostatic discharge gun of the automobile electronic and electrical components requires that the output voltage range is at least −25 to +25 kV. The discharge current waveform should be in accordance with the waveform shown in Figure C.5, the direct contact discharge current waveform verification parameters are shown in Table C.1.

For the air discharge waveforms, the standard required that the waveform parameters of the discharge voltage at ±15 kV need to be verified and the rise time should be less than 5 ns.

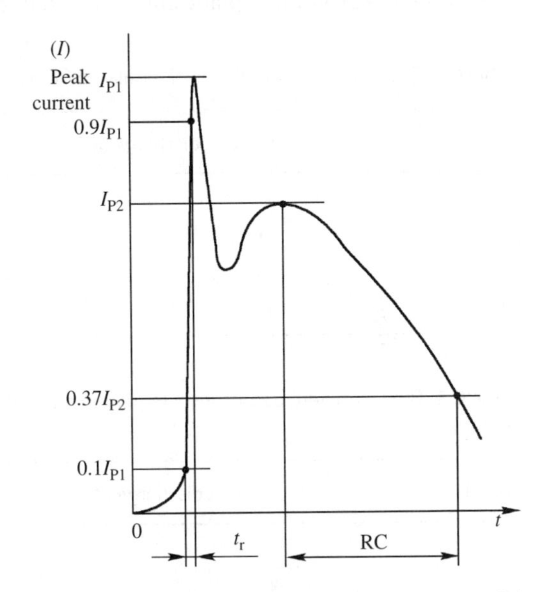

Figure C.5 The electrostatic discharge current model that meets the ISO10605 standard.

C.3.3 Automotive Electronics and Electrical Components ESD Immunity Test Method

The electrostatic discharge test of the automotive electronics and electrical components includes the contact discharge and the air discharge. The contact discharge includes the direct discharge and indirect discharge. The discharge point includes all the contact surfaces, while the air discharge is used for the insulation surface, and the contact discharge is used for the metal surface. During the electrostatic discharge test, the electrode head of the electrostatic discharge generator is usually perpendicular to the surface of the device, and the test times are positive and negative polarity. The discharge is required to be done three times at positive and negative polarity respectively according to ISO10605 standard and the discharge interval is at

Table C.1 The direct contact discharge current waveform verification parameter that meets the ISO10605 standard.

Level	Display voltage (KV)	Peak current for the first time (A)	The rise time of the discharge switch operation (nS)
1	2 ± 0.5	7.5	
2	4 ± 0.5	15	0.7–1
3	6 ± 0.5	22.5	
4	8 ± 0.5	30	

least 5 seconds. Before and after the electrostatic discharge test, monitor the function of the EUT to determine whether the EUT is passed.

The test severity level regulated from ISO10605 standard is divided into the level of test with charge and without charge, respectively, shown in Tables C.2 and C.3

The automotive electronics, electrical components, according to the ISO10605 standards, need to test in the power on state. The device is placed on the grounded plane (seen in Figure C.6). If the EUT is the electronic device mounted on vehicle chassis, it should be placed directly on the grounding plane and connected to each other. If the equipment is insulated with

Table C.2 The test severe degree level regulated from ISO 10605 standard (with charge).

Level	Contact discharge (KV)	Air discharge (KV)
1	±4	±4
2	±6	±8
3	±7	±14
4	±8	±15

Table C.3 The test severe degree level regulated from ISO 10605 standard (without charge).

Level	Contact discharge (KV)	Air discharge (KV)
1	±4	±4
2	±6	±15
3	±8	±25

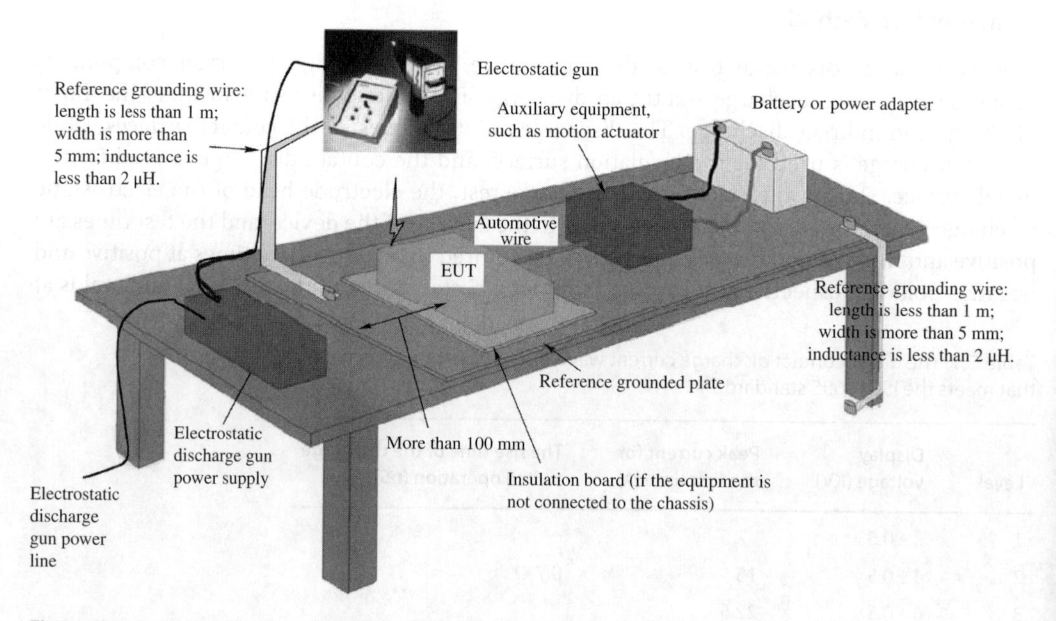

Figure C.6 Automotive electronic electrostatic discharge configuration diagram.

ground when properly installed, during the test, an insulation board can be placed between the EUT and grounding plane. When a test is performed under the condition of no power on, the EUT needs to be installed on the electrostatic dissipative material between the ground plane and the EUT to release the charge accumulated during the test.

For the equipment without grounding, it cannot discharge by itself like other equipment. In the test, if the charge is not removed before the next electrostatic discharge pulse is applied, the charge accumulation on the device may cause two times of the expected voltage, which may cause a high-energy insulation breakdown discharge accidently, so for the equipment without grounding need remove the charge before the next electrostatic discharge pulse is applied. According to the IEC61000-4-2, a cable with 470 kΩ discharge resistance can be applied, which is used in the horizontal coupling plate and the vertical coupling plate. And ISO10605 regulates the use of 1 MΩ dissipation resistor.

C.4 Immunity Test of Radiated, Radio Frequency Electromagnetic Field of Automotive Electronics and Electrical Components

C.4.1 Purpose of Immunity Test of Radio Frequency Electromagnetic Field of Automotive Electronics and Electrical Components

In order to measure the state of the electronic and electrical components in a variety of electromagnetic fields, the immunity test must be carried out in the electromagnetic field.

C.4.2 Equipment for Immunity Test of Radio Frequency Electromagnetic Field of Automotive Electronics and Electrical Components

1) Signal generator
2) Power amplifier
3) Transmitting antennas, such as broadband transmitting antennas, strip line antennas, parallel plate antennas, etc.
4) Field strength testing probe
5) Field strength testing and recording equipment. On the basis of the basic instrument, adding some additional such as power meter, computer (including special control software), self-moving mechanism of the electric field probe, can constitute a complete automatic test system.
6) Anechoic chamber
7) Transverse electromagnetic cell (TEM cell)

C.4.3 Method for Immunity Test of Radio Frequency Electromagnetic Field of Automotive Electronics and Electrical Components

For automotive electronics and electrical components, the method for radiation electromagnetic field immunity test includes:

1) Free field test method (from ISO11452-2 standard)
2) Transverse electromagnetic mode cell, TEM cell test method (from ISO11452-3 standard)
3) Tri-plate test method (from SAE J1113-25 standard)
4) Strip line test method (from ISO11452-5 standard)

5) Parallel plate antenna test method (from ISO11452-6 standard)
6) Helmholtz coil test method (from SAE J1113-22 standard)

In the free field test method, the transverse electromagnetic wave cell test method and the guided wave test method are adopted in the ECE and EEC regulations.

C.4.3.1 Free field test method

Because the space of the anechoic chamber (absorber lined chamber) is large, free field test method generally does not limit the size of the measured equipment, and it can accommodate equipment with larger size to take the test. It is also easy to use CCTV or other monitoring device to observe the action characteristics of the measured equipment in the test. General automotive electronics and electrical components, such as electric rearview mirrors, are available to free field test method. The frequency range of the free field test method is from 200 MHz (or 20 MHz) to 18 GHz, and the configuration diagram of free field test method is shown in Figure C.7a and b.

C.4.3.2 Transverse electromagnetic mode cell, TEM cell test method

According to the ISO11452-3 and SAE J1113-24 standards, the TEM cell is a simple closed transmission line, one end is inputted RF power, and the other end is connected to load impedance. With the propagation of electromagnetic waves in the transmission line, an electromagnetic field is established between the conductors. TEM (transverse electromagnetic wave) is the dominant electromagnetic field generated in this type of cell effective area. When the length of transmission line is given, the electric field intensity is uniform and easy to be measured or calculated in a certain area. EUT is placed in the effective area of the TEM cell. The TEM unit is usually in the form of a box, with a built-in isolation surface, the wall of the box is used as one end of the transmission line, and the isolation surface (or called septum) is the other end. The geometrical configuration of the TEM unit has a decisive influence on the characteristic impedance of the transmission line. Its main drawback is that it has an upper frequency limit, which is inversely proportional to its physical size. When the frequency is higher than the upper limit, the field uniformity begins to become worse. The maximum EUT size that can be measured by the TEM unit is limited by its internal available field strength uniform region size, so the maximum EUT size is directly related to and the maximum frequency of the unit. The lowest measuring frequency of the TEM cell is DC. The transverse electromagnetic wave cell test can apply to the radiation immunity test of small equipment. The box is closed, in addition to a small leak, and there is no electromagnetic field outside cell, so the cell can be applied in any environment without external shielding. The general applicable frequency range is 0.01–200 MHz, or higher. The configuration diagram of transverse electromagnetic wave cell test method is shown in Figure C.8.

C.4.3.3 Strip line test method

The strip line contains 150 and 800 mm two kinds of height specification. The test object for 150 mm is limited to the line, while test equipment can be measured in the strip line with 800 mm. The limit of the strip line test method is that the maximum diameter of the EUT itself or line under test is 1/3 or less of maximum diameter of the strip line's height, and it must be tested in the shielded room. The general applicable frequency range is 0.01–200 MHz. The configuration diagram of strip line test method is shown in Figure C.9.

C.4.3.4 Parallel plate antenna test method

For the test object, the parallel plate antenna test method is similar to the free field test method. But it is suitable for the low frequency band, especially for the low frequency electric field test. The general applicable frequency range is 0.01–200 MHz. The configuration diagram of

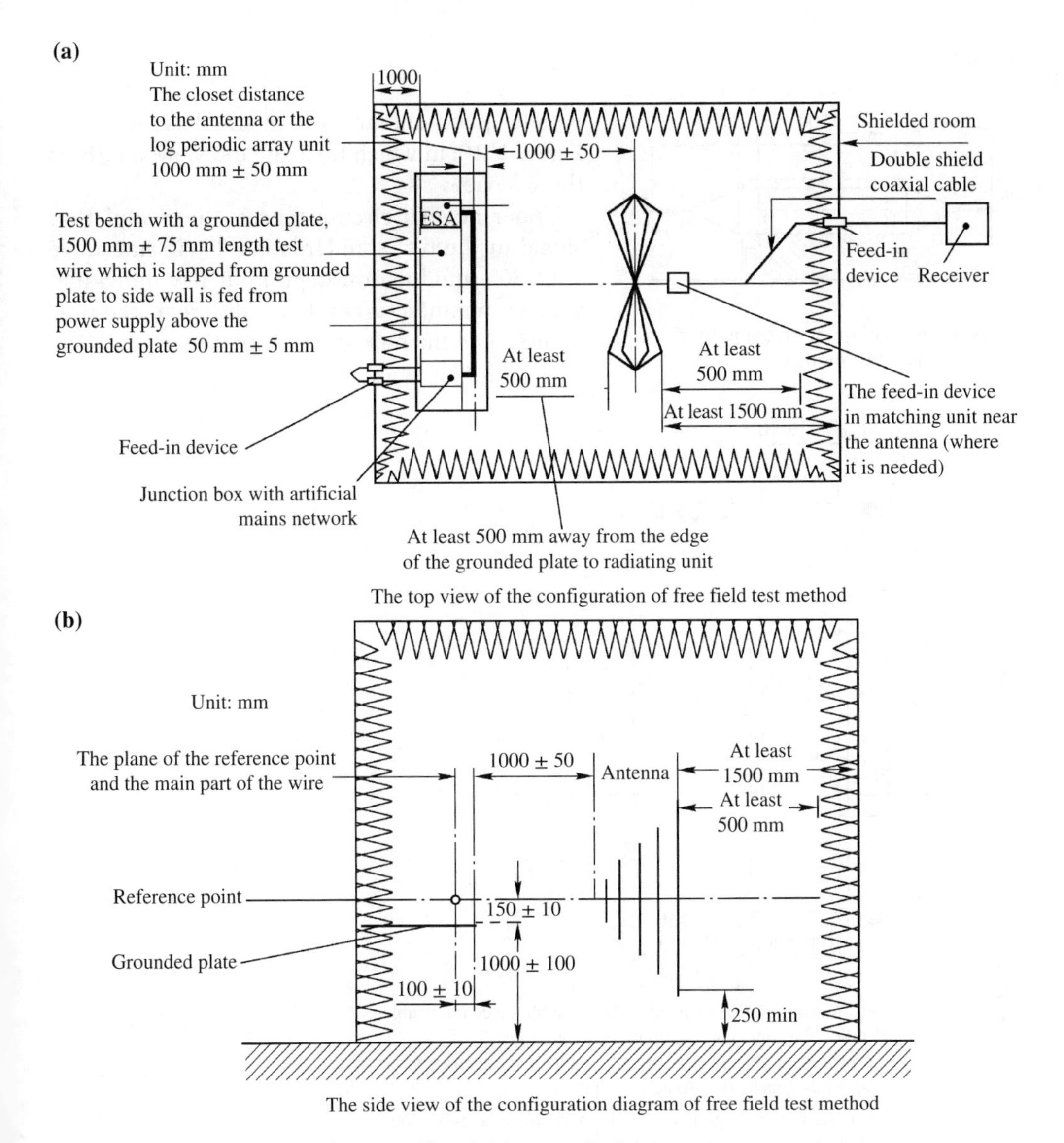

Figure C.7 The configuration diagram of free field test method.

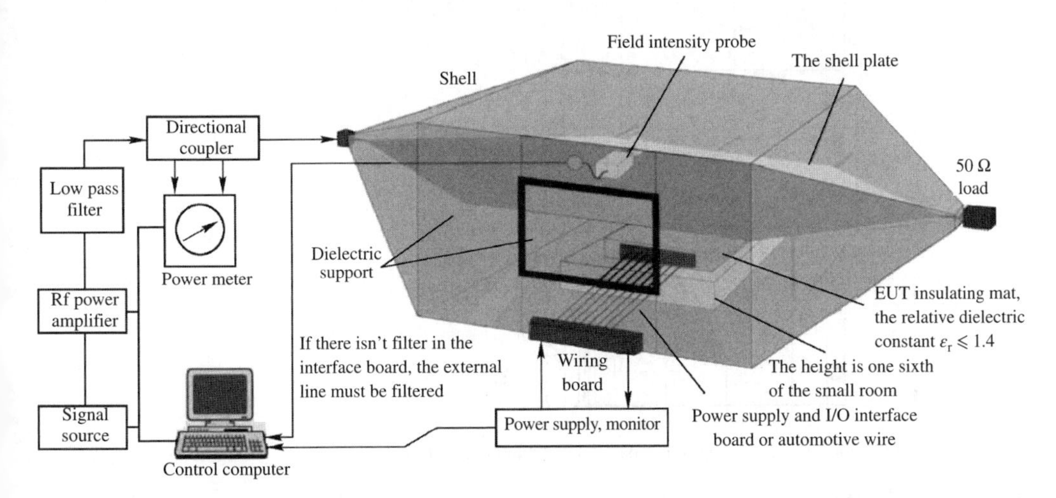

Figure C.8 The configuration diagram of transverse electromagnetic wave cell test method.

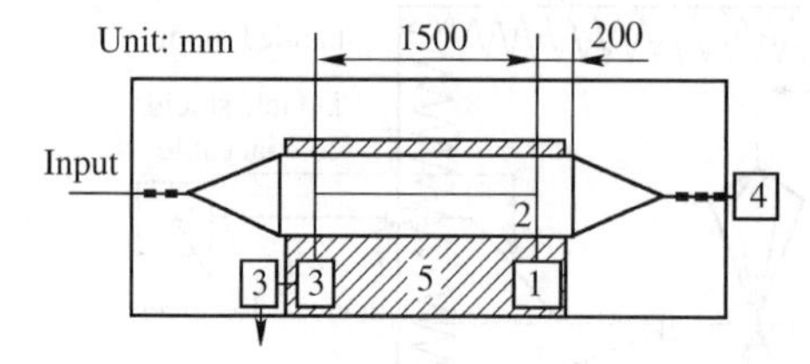

Figure C.9 The configuration diagram of 150 mm strip line test method.

parallel plate antenna test method is shown in Figure C.10, in which (a) is the top view and (b) is the side view.

Under normal circumstances, the modulation signal in the test standards is a 1 kHz sine wave with 80% modulation depth, but few individual vehicle manufacturers have different requirements. The purpose of defining the modulation

(a)

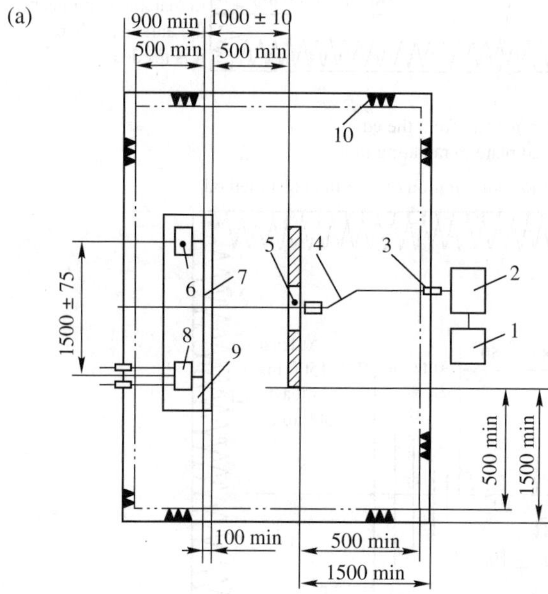

1—Signal generator; 2—Amplifier; 3—Connector; 4—double lie coaxial cable; 5—Parallel plate antenna; 6—Equipment under test; 7—Circuit under test (power supply and signal line); 8—Artificial mains network; 9—Test bench; 10—Anechoic chamber

The top view of the configuration diagram of parallel plate antenna test method

(b)

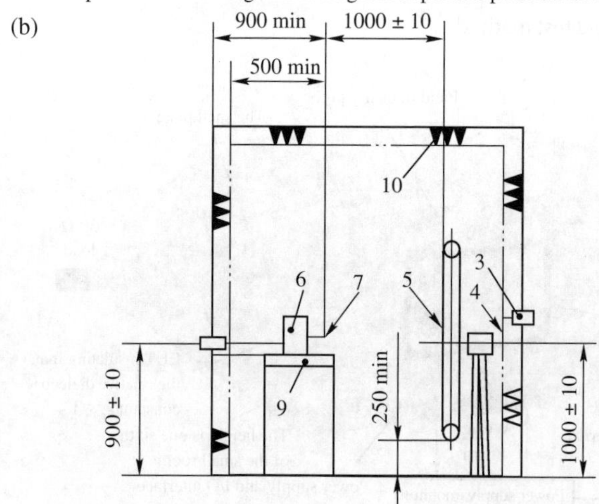

The side view of the configureation diagram of parallel plate antenna test method

Figure C.10 The configuration diagram of parallel plate antenna test method. (a) is the top view and (b) is the side view.

parameter is to set a constant peak value. This is not the same as the immunity test regulated in the IEC61000-3-4. In IEC61000-3-4 standard, peak power of the modulated signal is higher 5.3 dB than unmodulated signal in the immunity test. In the test with the constant peak voltage, the power of modulated signal with 80% modulation depth is 0.407 times of the unmodulated signal. The application process of this signal is clearly defined in ISO11452:

- At each frequency point, use the linear or logarithmic method to increase signal intensity until it meets the demand (the net power meet the requirements for the open loop method, the voltage level of the test signal meet the requirements for the closed loop method), according to +2 dB criteria to monitor the forward power.
- Apply modulated signal by the requirement and keep the test signal time equal to the EUT minimum response time.
- Slowly reduce test signal strength, and then go to the next frequency of testing.

C.5 P3a, P3b Transient Pulse Immunity Test in ISO2-7637 Standard

C.5.1 Purpose of P3a, P3b Transient Pulse Immunity Test in ISO2-7637 Standard

For automotive electronics and electrical components, the test is used with P3a and P3b transient pulse waveforms in ISO2-7637 standard. In all kinds of switch, relay and fuse during the process of opening or closing from automotive electronics system, the P3a is used to simulate the fast transient pulse group generated by the electric arc. P3b is used to simulate the fast transient pulse group generated during the switching process of the driving unit of the electric door and window, the speaker, or the central gating system. Same with the test in the IEC61000-4-4 standard; the purpose is to provide a common basis for evaluating the immunity for the electronic and electrical components of the vehicle in the P3a, P3b transient pulse test.

C.5.2 Equipment for P3a, P3b Transient Pulse Immunity Test in ISO2-7637 Standard

The principle of test pulse P3a, P3b in ISO7637-2 standard (as shown in Figure C.11): the test pulse P3 occurs in the switching moment. The characteristics of this pulse are affected by the wire distributed capacitance and inductance. Since the value of the wire distributed capacitance and inductance is usually small, the P3 pulse is a series of high speed and low energy pulse in the whole ISO2-7637 standard, which often leads the device controlled by a microprocessor or digital logic control to generating an error action.

The test generator for P3a and P3b transient pulse test should have the parameter characteristics shown in the Table C.4 and the Figure C.12.

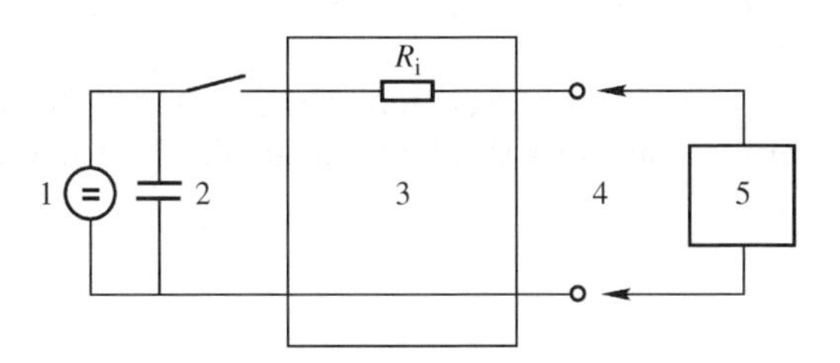

Figure C.11 P3 waveform generator has the advantages of simple circuit diagram.

Table C.4 P3 pulse parameter calibration table.

	P3a pulse		P3b pulse	
Parameters	**No load**	**50 Ω load**	**No load**	**50 Ω load**
U_S	−200 V ± 20 V	−100 V ± 20 V	+200 V ± 20 V	+100 V ± 20 V
t_r	5 ns ± 1.5 ns	5 ns ± 1.5 ns	5 ns ± 1.5 ns	5 ns ± 1.5 ns
t_d	150 ns ± 45 ns	150 ns ± 45 ns	150 ns ± 45 ns	150 ns ± 45 ns

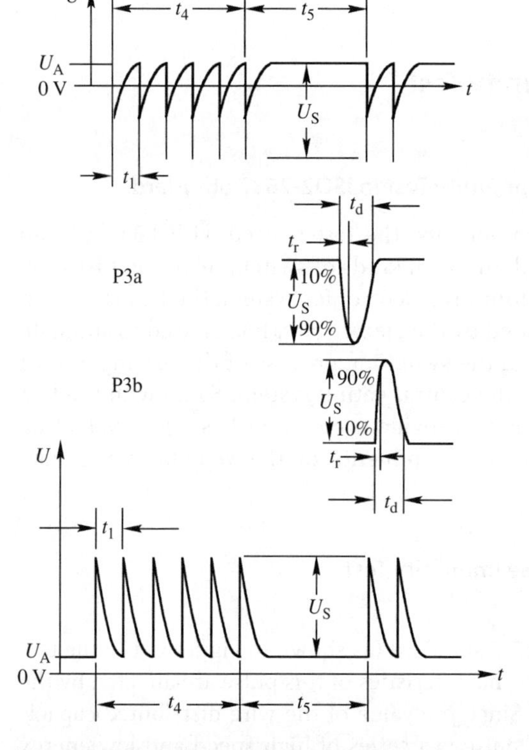

Parameters	12 V system	24 V system
U_S	−112 to −150 V	−150 to −200 V
R_i	50 Ω	
t_d	0.1 ~ 0.2 μs	
t_r	5 ns ± 1.5 ns	
t_1	100 μs	
t_4	10 ms	
t_5	90 ms	

Parameters	12V system	24V system
U_S	+75 to +100 V	+150 to +200 V
R_i	50 Ω	
t_d	0.1 μs to 0.2 μs	
t_r	5 ns ± 1.5 ns	
t_1	100 μs	
t_4	10 ms	
t_5	90 ms	

Figure C.12 P3 pulse waveform parameters.

C.5.3 Test Method for Impulse Immunity of P3a and P3b in the Standard ISO 7637–2 and ISO 7637-3

The schematic diagram of the test configuration for transient pulse of P3a and P3b in the standard ISO 7637–2 and ISO 7637-3 is shown in Figure C.13. The 12 V system test level is shown in Table C.5.

The standards ISO 7637-2 and ISO 7637-3 using pulse P3a and P3b for equipment immunity test are very similar with standard IEC61000-4-4, which also uses pulse P3a and P3b for

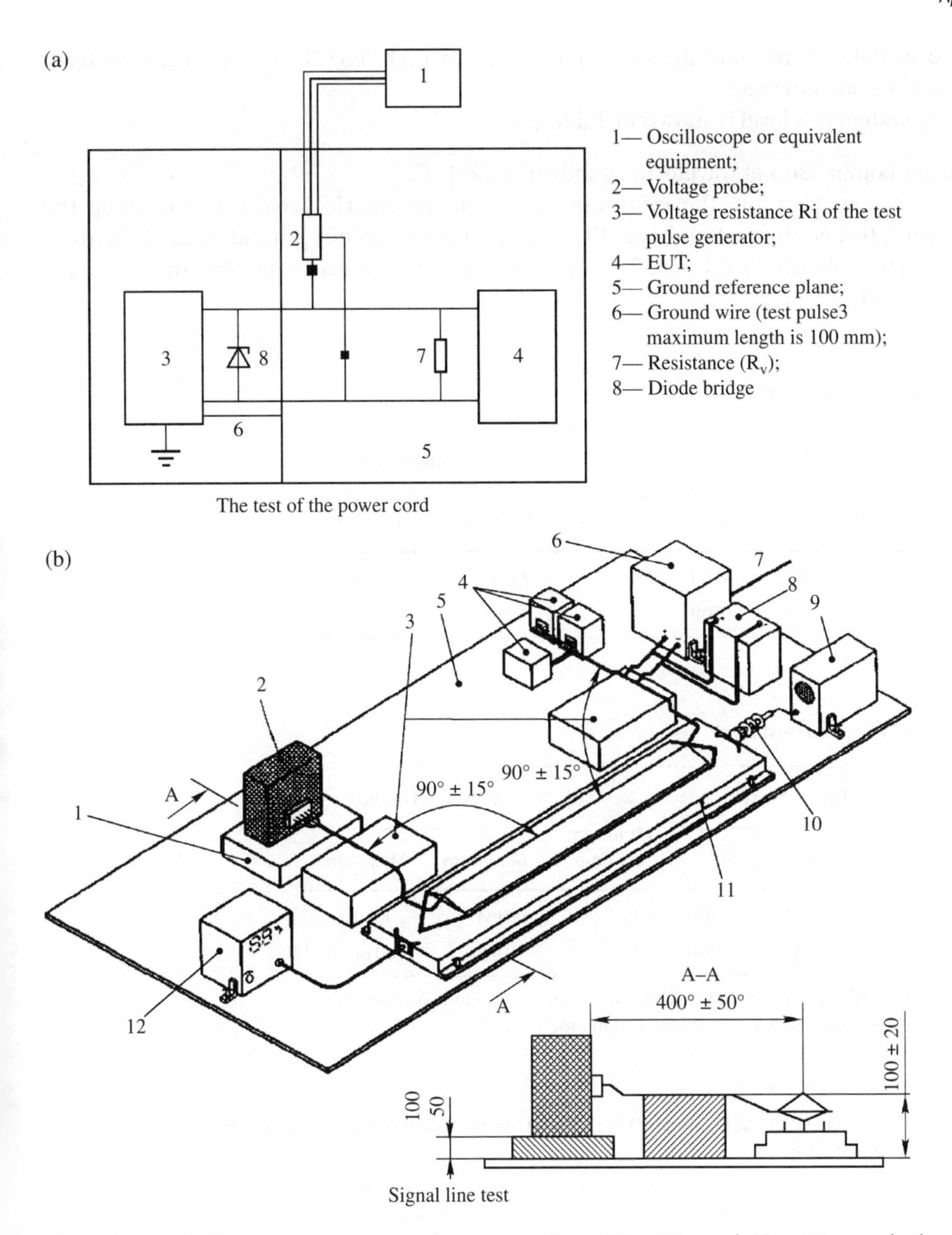

Figure C.13 P3a, P3b test circuit diagram of transient pulse in ISO 7637-2 and ISO 7637-3 standard.

equipment immunity test to represent high-frequency EMC test of automotive electronic and electrical components. So the method of EMC design risk analysis shown in the book can also be used in automotive electronics and electrical components. Also, the best point of standards ISO 7637-2 and ISO 7637-3 for configuration testing is that the test layout is clear, especially test configuration should adopt mechanical fixing to obtain repeatable test results as mentioned in standards ISO 7637-2 and ISO 7637-3. Due to space reasons, these are not described in detail here; only the following points in ISO 7637/3 that are different from IES61000-4-4-2 are covered:

The 24 V system test level is shown in Table C.6.

- Parameter comparison of waveform as shown in Table C.7.
- In the test configuration, the minimum size from geometric projection including the instrument, test equipment and coupling clamp to the reference ground edge of the floor is not given explicitly in ISO 7637-2, ISO 7637-3, which must be greater than 0.1 m in IEC61000-4-4.

Table C.5 The 12 V system test level.

Test pulse	Level of test				Minimum test time	Group pulse cycle	
	I	II	III	IV		Minimum	Maximum
3a			−112	−150	1 h	90 ms	100 ms
3b			+75	+100	1 h	90 ms	100 ms

Table C.6 The 24 V system test level.

Test pulse	Level of test				Minimum test time	Group pulse cycle	
	I	II	III	IV		Minimum	Maximum
3a			−150	−200	1 h	90 ms	100 ms
3b			+150	+200	1 h	90 ms	100 ms

Note: The test level of I and II is not given, because the test level is too low to ensure that there is enough immunity for the vehicle equipment.

Table C.7 P3a, P3b, ISO in ISO 7637-2 and ISO7637-3 and electric fast transient pulse group parameters in IEC61000-4-4.

	ISO 7637	IEC61000–4 -4
Pulse front t_r (frontier 10% to front 90%)	5 ns ± 30%	5 ns ± 30%
Pulse duration t_d	150 ns ± 30% (10% of front to 10% the back porch)	50 ns ± 30% (50% of front to 50% the back porch)
Impedance of generator R_i	50 Ω	50 Ω
pulse interval time t_1	100 μs	200 μs–400 μs
Burst duration t_2	10 ms	15 ms
Pulse train time interval t_3	90 ms	285 ms

- The insulation support of the EUT is set at 50–100 mm in standard ISO 7637-2 and ISO 7637-3, but it is set at 0.1 m in IEC61000-4-4.
- The distance that wire harness is left between the test pulse generator and coupling clamp or among the load, the sensor and the coupling clamp is 0.45 m in standard ISO 7637-3, which is 1 m in the IEC61000-4-4.
- The biggest difference in the test method is that in the standard ISO 7637-3, one of the coupling clamps is connected to a test pulse generator and the other end is connected with a 50Ω coaxial attenuator. But in the IEC61000-4-4 50Ω coaxial attenuator is not mentioned. And the structures of coupling clamps are not the same in two standards: in ISO 7637-3, the distance between the bottom plate of coupling clamp and the coupling plate is 27 mm, which is 100 mm in IEC61000-4-4.

C.6 P1, P2a, P2b, P5a, P5b Impulse Immunity Test in Standard ISO7637-2

C.6.1 The Purpose of P1, P2a, P2b, P5a, P5b Pulse Immunity Test in the Standard ISO 7637-2

ISO 7637-2 standard consist waveform P1, P2a, P2b, P5a, P5b immunity test, which has the relatively large energy (pulse width is above 50 μs and the amplitude is high), so the interference signals contained spectrum is relatively narrow (pulse rise time is in microsecond and millisecond), and it can be classified as the surge test. These waveforms are tested in order to simulate the following several phenomena occurring in the car-generated pulse.

P1 pulse is generated from the moment when the power of the inductive load releases. It can directly affect the work of the device in parallel with this inductive load. As the standard does not present the inductance range of inductive load, it generally refers to the interference by the switching of inductive load. After statistics and optimization, P1 pulse has large internal resistance and high voltage, leading faster and larger width of the negative pulse. It belongs to medium speed and medium energy pulse interference in the standard ISO 7637-2, has two effects for interference (causes the faulty operation) and damage (damage to the equipment components) to the EUT.

The P2a pulse is induced transient on wiring harness when the device connected in parallel with the tested device is suddenly powered off. Consider using a wiring harness with a small inductance value, so the pulse amplitude is not so high, frontier faster, smaller width, smaller internal resistance, and positive pulse. In the whole, ISO 2-7637 belongs to the pulse interference with fast speed and small energy and has the similar function with P1 pulse, but it is positive pulse.

The P2b pulse is caused by the transient phenomena due to DC motor working as generator when the ignition is cut off. This is a pulse whose voltage is not high, frontier is slow, width is large and internal resistance is very small. In the whole ISO 7637-2 belongs to pulse interference with low speed and high energy, focus on the destructive assessment of equipment (components). This effect of P2b pulse is similar with P5, but the voltage is lower and the pulse is wider.

P5 pulse occurs when the battery in the discharge is released, while the AC generator is charging the battery, and at the same time, other loads are still connected to the circuit of AC generator. The magnitude of the discharge pulse depends on the speed of the generator and the excitation of the generator at the moment of release of the battery. The duration of the discharge pulse is mainly determined by the time constant of the excitation line and the magnitude of the pulse. P5 pulse is composed by P5a and P5b. Described previously, it is the formation of P5a pulse. However, in most of the new AC generator, the magnitude of the discharge pulse is suppressed by the addition limiting diode (clamp), which forms P5b pulse. Thus, the difference between P5a and P5b pulse is that one is the pulse before the clamp, and the other is the pulse after clamp. P5 pulse has higher amplitude (100–200 V, relative to the power supply voltage of the system, it is high voltage), wide width (up to a few hundred milliseconds), extremely low resistance (few ohms, even a few tenths of ohm). So in the ISO 7637-2, the P5 pulse belongs to relatively large energy pulse, in addition to evaluate the tested equipment anti-interference ability in the P5 test, to a considerable extent, it is to evaluate destructive ability to the tested equipment components in this test.

C.6.2 Simulation Equipment of P1, P2a, P2b, P5a, P5b Impulse Immunity Test in Standard ISO 7637-2

Parameter	12 V system	24 V system
U_S	−75 to −100 V	−450 to −600 V
R_i	10 Ω	50 Ω
t_d	2 ms	1 ms
t_r	0.5–1 μs	1.5–3 μs
t_1	0.5–5 s	
t_2	200 ms	
t_3	<100 μs	

P2a pulse waveform and parameter is shown in Figure C.15.

Parameter	12 V system	24 V system
U_S	+37 to +50 V	
R_i	2 Ω	
t_d	0.05 ms	
t_r	0.5–1 μs	
t_1	0.2–5 s	

P2a pulse correction parameter is shown in Table C.9.
P2b pulse waveform and parameter is shown in Figure C.16.

Parameter	12 V system	24 V system
U_S	10 V	20 V
R_i	0–0.05 Ω	
t_d	0.2–2 s	
t_{12}	1 ms ± 0.5 ms	
t_r	1 ms ± 0.5 ms	
t_1	1 ms ± 0.5 ms	

P2b pulse correction parameter is shown in Table C.10.
P5a pulse waveform and parameter is shown in Figure C.17.

Parameter	**12 V system**	**24 V system**
U_S	+65 to +87 V	+123 to +174 V
R_i	0.5–4 Ω	1–8 Ω
t_d	40–400 ms	100–350 ms
t_r	5–10 ms	

P5b pulse waveform and parameter is shown in Figure C.18.

Parameter	**12 V system**	**24 V system**
U_S	+65 to +87 V	+123 to +174 V
U_S*	Specified by user	
t_d	Values for the same not being suppressed	

P5a pulse correction parameter is shown in Table C.11.
Table C.11 P5a pulse correction parameter.
In accordance with the requirements of the ISO 7637-2, the P2a, P2b, P1, P5a, P5b pulse immunity test equipment for the automotive electronics and electrical components products output waveform should have the following characteristics.

The waveform and parameters of pulse P1 are shown in Figure C.14.
The correction parameters of pulse P1 is shown in Table C.8.
The waveform and parameters of pulse P2a are shown in Figure C.15.
The correction parameters of pulse P2a is shown in Table C.9.
The waveform and parameters of pulse P2b are shown in Figure C.16.
The correction parameters of pulse P2b is shown in Table C.10.
The waveform and parameters of pulse P5a are shown in Figure C.17.
The waveform and parameters of pulse P5b are shown in Figure C.18.
The correction parameters of pulse P5a are shown in Table C.11.

C.6.3 Immunity Test Method of P1, P2a, P2b, P5a, P5b in Standard ISO7637-2

Because P1, P2a, P2b, P5a, P5b, waveforms' rising edge is relatively slow corresponding to the transient pulse test voltage and current waveform (microsecond or millisecond), the interference waveform contains a lower spectrum, which leads to a smaller parasitic parameters. The waveform of P1, P2a, P2b, P5a, P5b in ISO 7637-2 standard, corresponding to the transient pulse test configuration requirements, is similar with P3 pulse test on power supply port. Here it is no longer repeatable, but during the pulse 5b test, a suppression diode must be used.
As to the test level, the test level of 12 V system is shown in Table C.12.
The test level of 24 V system is shown in Table C.13.

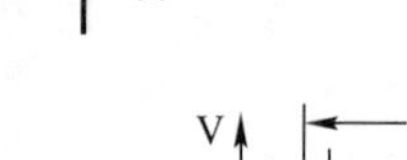

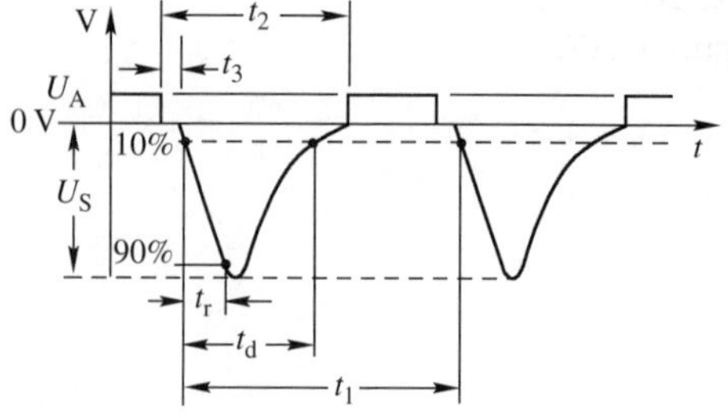

Parameters	The 12 V system	The 24 V system
U_S	–75 to –100 V	–450 to –600 V
R_i	10 Ω	50 Ω
t_d	2 ms	1 ms
t_r	0.5 to 1 μs	1.5 to 3 μs
t_1	0.5 to 5 s	
t_2	200 ms	
t_3	<100 μs	

Figure C.14 Waveforms and parameters.

Table C.8 Impulse correction parameters of P1.

	12 V system		24 V system	
Parameter	**No-load**	**10 Ω load**	**No-load**	**50 Ω load**
U_S	−100 V ± 10 V	−50 V ± 10 V	−600 V ± 60 V	−300 V ± 30 V
t_r	0.5–1 μs	—	1.5–3 μs	—
t_d	2000 μs ± 400 μs	1500 μs ± 300 μs	1000 μs ± 200 μs	1000 μs ± 200 μs

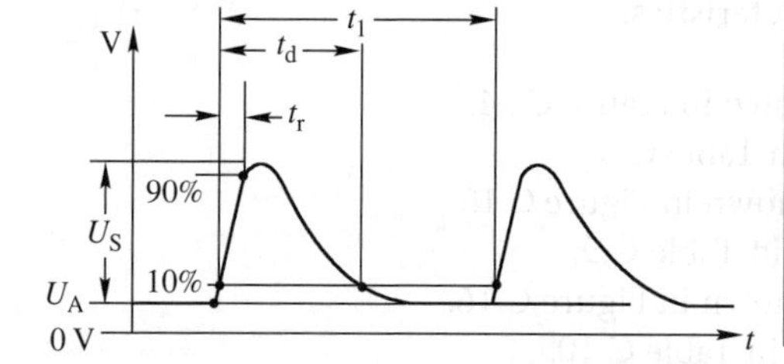

Parameters	The 12 V system	The 24 V system
U_S	+112 to +150 V	
R_i	2 Ω	
t_d	0.05 ms	
t_r	0.5 to 1 μs	
t_1	0.2 s to 5 s	

Figure C.15 P2a pulse waveform and parameter.

Table C.9 Impulse correction parameters of P2a.

	12 V and 24 V system	
Parameter	**No-load**	**2 Ω load**
U_S	+ 50 V ± 5 V	+ 25 V ± 5 V
t_r	0.5–1 μs	—
t_d	50 μs ± 10 μs	12 μs ± 2.4 μs

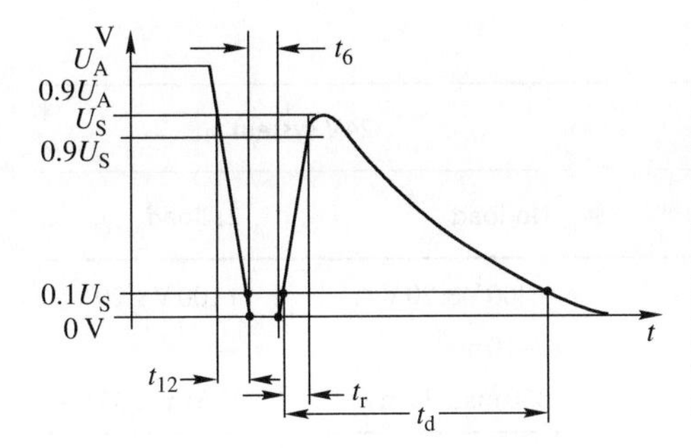

Parameters	The 12 V system	The 24 V system
U_S	10 V	20 V
R_i	0 to 0.05 Ω	
t_d	0.2 to 2 s	
t_{12}	1 ms ± 0.5 ms	
t_r	1 ms ± 0.5 ms	
t_6	1 ms ± 0.5 ms	

Figure C.16 P2b pulse waveform and parameter.

Table C.10 Impulse correction parameters of P2b.

	No-load and 0.5 Ωload	
Parameter	**12 V system**	**24 V system**
U_S	+10 V ± 1 V	+20 V ± 2 V
t_r	1 ms ± 0.5 ms	
t_d	2 s ± 0.4 s	

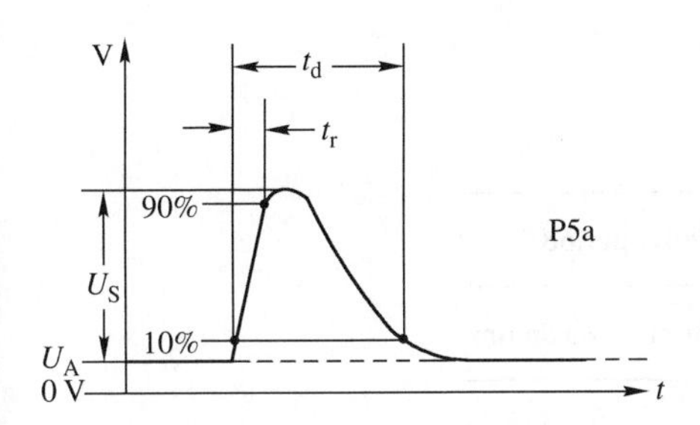

Parameters	The 12 V system	The 24 V system
U_S	+65 to +87 V	+123 to +174 V
R_i	0.5 to 4 Ω	1 to 8 Ω
t_d	40 to 400 ms	100 to 350 ms
t_r	5 to 10 ms	

Figure C.17 P5a pulse waveform and parameter.

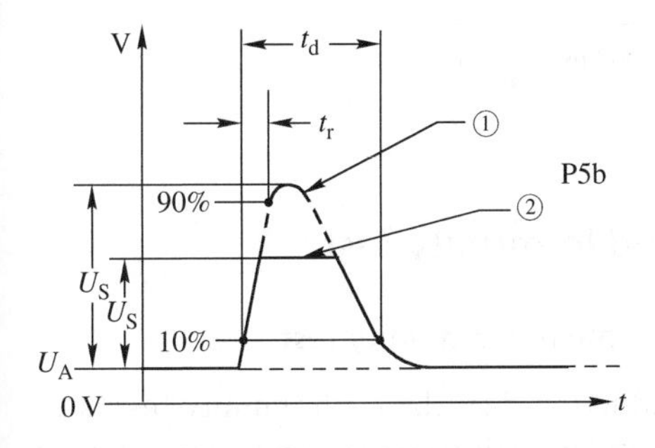

Parameters	The 12 V system	The 24 V system
U_S	+65 to +87 V	+123 to +174 V
U_s*	Specified by user	
t_d	Values for the same not being suppressed	

Figure C.18 P5b pulse waveform and parameter.

Table C.11 Impulse correction parameters of P5a.

	12 V system		24 V system	
Parameter	**No-load**	**2 Ω load**	**No-load**	**2 Ω load**
U_S	+100 V ± 10 V	+50 V ± 10 V	+200 V ± 20 V	+100 V ± 20 V
t_r	5–10 ms	—	5–10 ms	—
t_d	400 ms ± 80 ms	200 ms ± 40 ms	350 ms ± 70 ms	175 ms ± 35 ms

Note that the standard has calibration parameters only for P5a, no data for P5b.

Table C.12 The 12 V system test level.

	Level of test					Pulse period	
Test pulse	**I**	**II**	**III**	**IV**	**Minimum pulse number**	**Minimum**	**Maximum**
P1			−75	−100	5000	0.5 s	5 s
P2a			+37	+50	5000	0.2 s	5 s
P2b			+10	+10	10	0.5 s	5 s
P5			+65	+87	1		

Table C.13 The 24 V system test level.

	Level of test					Pulse period	
Test pulse	**I**	**II**	**III**	**IV**	**Minimum pulse number**	**Minimum**	**Maximum**
P1			−450	−600	5000	0.5 s	5 s
P2a			+37	+50	5000	0.2 s	5 s
P2b			+20	+20	10	0.5 s	5 s
P5			+123	+173	1		

Note: The test level of I and II is not given because the test level is too low to ensure that there is enough immunity for the vehicle equipment.

C.7 BCI and Direct Injection of Conducted Immunity Test

C.7.1 The Purpose of BCI and Direct Injection of Conducted Immunity Test

The purpose of BCI and direct injection of the conduction disturbance immunity test is to evaluate the conduction disturbance of the electronic and electrical components caused by the radio frequency field.

C.7.2 The Way of BCI and Direct Injection of Conducted Immunity Test

ISO11452-4 and ISO11452-7 provide two kinds of conducted immunity testing method for automotive electronics and electrical components, bulk current injection (BCI) and direct injection method. The former method needs to inject interference current into EUT and control the amount of the injected current. The latter needs to inject power into EUT and controls the amount.

1) Bulk current injection (BCI)
   General vehicle routing arrangements are made of a variety of wiring harness, and each wire harness has a respective current signal. Because the wiring harness is bundled with each other and the opportunity of disturbance is larger, vulnerable wire harness is easy to be affected, resulting in the signal change and influencing electrical equipment at end of the wire harness. BCI method in ISO 11452-4 and SAE J1113/4 were described BCI testing methods. When we use this method, a current injection probe is put on cable harness devices, which is connected to the tested equipment (such as wire harness of audio and video systems, CD-ROM, electric rearview mirror and automotive electronics, electrical components), then RF interference is injected into the probe. At this point, the probe is the first current converter and the cable device is the second current converter. Therefore, the RF current first flows through the cable device by common mode (the current is in the same way flowing through all conductors of the device), then flows into the EUT port.

   The real current is determined by the common mode impedance of the device where the current flows in, but at low frequency it is almost determined by the EUT and the ground impedance of the device is connected with the other end of the cable. Once the cable length is up to 1/4 wavelengths, the change of impedance becomes very important, which may reduce the repeatability of the test. In addition, because the current injection probe can bring loss, larger driving capability is needed to establish a reasonable interference level on the EUT. In spite of this, the BCI method has a great advantage, which is noninvasive, because the probe can simply clamp on any diameter not more than the maximum acceptable cable diameter, and does not need any direct cable conductor connection, and does not affect the circuit working function that is connected with cable. BCI test method should be carried out in the shielded room in order to obtain the correct test results. General BCI test method is applicable to the frequency range from 1 to 400 MHz (or up to 1000 MHz), and the BCI method test configuration is shown in Figure C.19.
2) Direct injection method
   BCI test method has high requirements for driving capability, and in the test isolated with related equipment is not good. Direct injection method in ISO 11452-7 and J1113 SA/3 standard is to overcome the two shortcoming of BCI method. The specific approach is directly connected test equipment to the EUT cable, and RF power is injected into the EUT cable through a broadband artificial network (BAN). The RF energy is directly coupled with the tested device, without interfering with the EUT and its sensor and load interface, so the RF impedance of BAN in the test frequency range can be controlled. BAN can provide at least 500 W block impedance in the current flow direction to the auxiliary equipment. The interference signal is directly coupled with the measured line through a DC blocking capacitor. It can be targeted at the individual power line or signal line to carry out immunity test. The direct injection method is also applied in shielded chamber, and the frequency range is 0.25–400 MHz (or extend to 500 MHz), and the direct injection method test configuration is shown in Figure C.20.

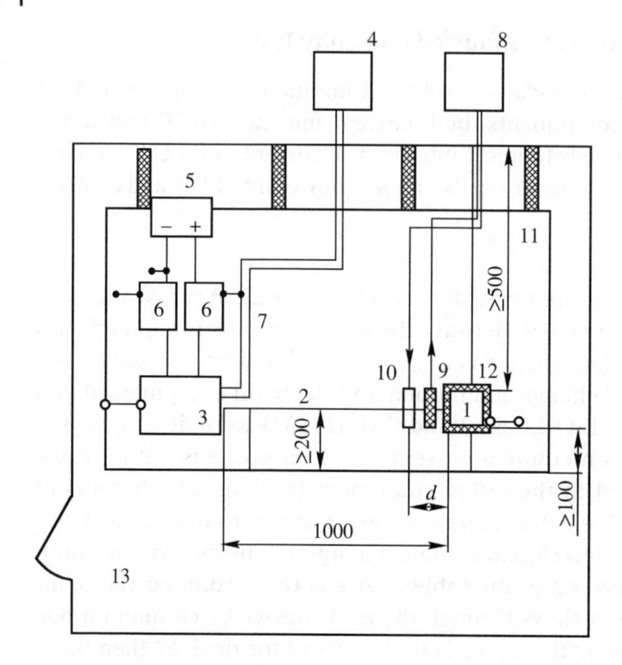

Figure C.19 BCI test configuration diagram.

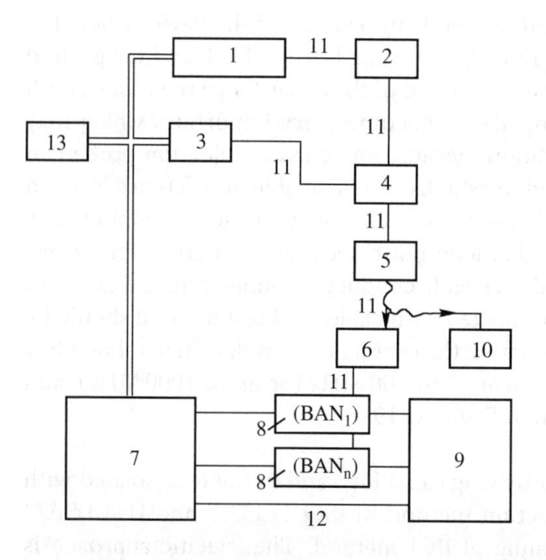

Figure C.20 Direct injection method test configuration diagram.

C.8 The Immunity Test of P4 Transient Pulse in the ISO 76372-2 Standard

C.8.1 The Purpose of the Immunity Test of P4 Transient Pulse in the ISO 76372-2 Standard

The immunity test of automotive electronics, electrical components in standard ISO76372-2 specify a similar voltage dip test, which is the transient pulse P4 immunity test. It simulates the voltage dip of the vehicle power supply system due to turning on the starting circuit of the

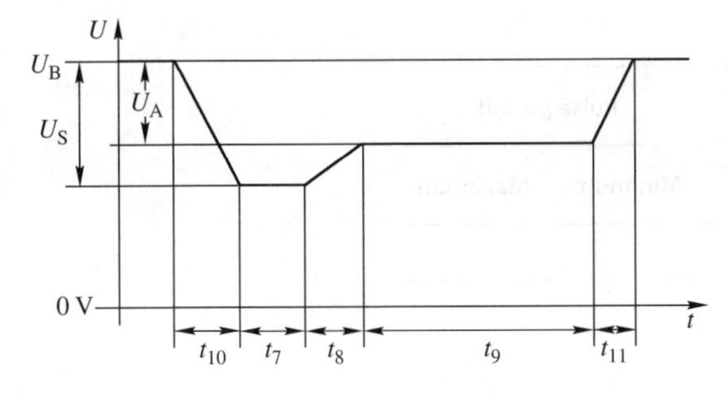

Parameter	The 12 V system	The 24 V system
U_s	–6 to –7 V	–12 to –16 V
U_A	–2.5 to –6 V I Ua I ≤ I US I	–5 to –12 V I Ua I ≤ I US I
R_i	0 to 0.02 Ω	
t_7	15 to 40 ms	50 to 100 ms
t_8	≤ 50 ms	
t_9	0.5 to 20s	
t_{10}	5 ms	10 ms
t_{11}	5 to 100 ms	10 to 100ms

Figure C.21 P4 waveform and parameter.

internal combustion engine. This is a voltage dip more than half and the duration last a few seconds to tens of seconds. In the ISO7637-2, the main test aspect is EUT's malfunction in the dip process, especially the situation of data loss and program disorder in the case of EUT with the microprocessor.

C.8.2 The Immunity of P4 Transient Pulse Test Equipment in the Standard ISO 76372-2

In accordance with the requirements of the ISO 7637-2 standard, the immunity test equipment (i.e., P4 waveform generator) for the electronic and electrical components of the vehicle (i.e., P4) should have the waveform and parameters as shown in Figure C.21.

C.8.3 The Immunity of P4 Transient Pulse Test Method in the ISO 76372-2 Standard

P4 immunity transient pulse test configuration requirements in ISO 7637-2 standard is similar to the other immunity test configuration requirements in ISO 7637-2 standard.

The test level of 12 V system is shown in Table C.14.

The test level of 24 V system is shown in Table C.15.

Parameter	12 V system	24 V system
U_s	-6 to -7 V	-12 to -16 V
U_A	-2.5 to -6 V | Ua | ≤ | US |	-5 to -12 V | Ua | ≤ | US |
R_i	0–0.02 Ω	
t_7	15–40 ms	50–100 ms
t_8	≤50 ms	
t_9	0.5–20s	
t_{10}	5 ms	10 ms
t_{11}	5–100 ms	10–100 ms

Table C.14 The 12 V system test level.

Test pulse	Level of test				Minimum pulse number	Pulse period	
	I	II	III	IV		Minimum	Maximum
P4			−6	−7	1		

Table C.15 The 24 V system test level.

Test pulse	Level of test				Minimum pulse number	Pulse period	
	I	II	III	IV		Minimum	Maximum
P4			−12	−16	1		

Note: The test level of I and II is not given because the test level is too low to ensure that there is enough immunity for the vehicle equipment.

Appendix D

Military Standards Commonly Used for EMC Test

GJB 151A-97: Military equipment and subsystems electromagnetic emission and susceptibility requirements (equivalent to US military standard MIL – STD – 461D) and GJB 152A-97: Military equipment and subsystems electromagnetic emission and susceptibility measurements (equivalent to US military standard MIL – STD – 462D) Standard specifies EMI and EMS testing for military products.

D.1 Special Requirements of Military Products

For naval equipment, from the view of the control EMC, the filter should be used as little as possible between line and ground, because this kind of filter can provide a low-impedance path through the ground plane for the structure current (common mode). The current may be coupled to other equipment connected on same ground plane. Thus, it may be a major cause of electromagnetic interference in system, platform or device. If this type of filter must be used, the capacitance between each line and the ground should be less than $0.1\,\mu F$ for 50 Hz equipment; for 400 Hz equipment, it should be less than $0.02\mu F$. For the DC power supply on submarine and aircraft equipment, in the user interface, each line to ground filter capacitor should be not exceed $0.075\mu F/kV$ of the connection load. If the load is less than 0.5 kW, the filter capacitor should not exceed $0.03\mu F$.

D.2 Military Specifications for EMC Test Project

This section describes GJB151A – specifically, electromagnetic emission and susceptibility test method. The test project and the name are shown in Table D.1. The test method is suitable for the whole frequency range. However, a specific device or device type should be tested as GJB151A on a test project and frequency range according to the electromagnetic environment of the platform. The applicability of each test item to each platform is described in Table D.2.

D.3 Military Standard EMC Test Basic Configuration

Equipment under test (EUT) should be installed on the ground reference plane to simulate the actual conditions. If the actual situation is unknown, or must be installed in a variety of forms, it should use the metal ground plate of reference. Unless otherwise provided, the area of the

Electromagnetic Compatibility (EMC) Design and Test Case Analysis, First Edition. Junqi Zheng.

Table D.1 Military product EMC test project list.

Project	Designation
CE101	power line transmission of 25 Hz–10 kHz
CE102	power line transmission of 10 Hz–10 kHz
CE106	the antenna terminal conducted emission of 10 kHz–40 GHz
CE107	the power cord (time domain) conduction emission peak signal
CS101	The power cord conduction sensitivity of 25 Hz–50 kHz
CS103	The antenna terminal intermodulation conduction sensitivity of 15 kHz–10 GHz
CS104	The antenna terminal useless signal sensitivity for restraining of 25 Hz–20 GHz
CS105	The antenna terminal/sensitivity of adjustable conduction of 25 Hz–20 GHz
CS106	The power cord sensitivity peak signal transduction
CS109	Shell conduction current sensitivity of 50 Hz–100 kHz
CS114	Cable beam injection conduction sensitivity of 10 kHz–400 MHz
CS115	Cable transmission beam injection pulse sensitivity
CS116	Cable and power cord damped sinusoidal transient conduction sensitivity of 10 kHz–100 MHz
RE101	Magnetic field radiation emission of 25 Hz–100 kHz
RE102	Electric field emission of 10 kHz–18 GHz
RE103	Harmonic and spurious output antenna radiation emission of 10 kHz–40 GHz
RS101	Radiation sensitivity of magnetic field of 25 Hz–100 kHz
RS103	Radiation sensitivity of electric field of 10 kHz–40 GHz
RS105	Radiation sensitivity of transient electromagnetic field

reference ground plate shall not be less than $2.25m^2$, and its section edge shall not be less than 760 mm. When there is no the reference ground plane in the EUT installation, the EUT should be placed on the insulation plane.

When the EUT is mounted on a metal reference plate, the surface resistance of reference is not greater than $0.1m\Omega$ each block (minimum thickness: copper 0.25 mm; brass 0.63 mm; aluminum plate 1 mm). The DC overlap resistance between the ground plate and the shield room is not greater than $2.5m\Omega$. The metal reference plate shown in Figures D.3 and D.4 should be in the 1 m spacer overlap to the shield room wall or floor. The metal overlap joint grounding should be solid, and the ratio of the length and width should not be greater than 5 : 1. In the outside of shielded room test the area of metal reference ground floor shall be at least 2 m × 2 m, and shall exceed the test configuration boundary at least 0.5 m. Unless otherwise stated in a single test method, all test methods in the standard GJB151A use line impedance stabilization network (LISN) to isolate the power supply interference and provide the required power source impedance for the EUT. The LISN circuit should be in accordance with Figure D.1, and its impedance should be in accordance with Figure D.2. LISN impedance characteristics are measured at least every year under the following conditions:

- Impedance should be tested between power output line on the load side of LISN and the LISN metal housing.
- The LISN signal output port should be connected to 50 Ω resistor.
- LISN power input port should be vacant.

Table D.2 The applicability of the test items for each platform.

		Surface ship	Submarine	Army aircraft (airline security equipment)	Naval aircraft	Air Force planes	Space system (including carrier rocket))	Army ground	Navy ground	Air force ground
Applicability of test project	CE101	A	A	A	L					
	CE102	A	A	A	A	A	A	A	A	A
	CE106	L	L	L	L	L	L	L	L	L
	CE107			S	S	S				
	CS101	A	A	A	A	A	A	A	A	A
	CS103	S	S	S	S	S	S	S	S	S
	CS104	S	S	S	S	S	S	S	S	S
	CS105	S	S	S	S	S	S	S	S	S
	CS106	S	S	S	S	S	S	S	S	S
	CS109		L							
	CS114	A	A	A	A	A	A	A	A	A
	CS115			A	A	A	A	L		A
	CS116	A	A	L	A	A	A	L	A	A
	RE101	A	A	A	L					
	RE102	A	A	A	A	A	A	A	A	A
	RE103	L	L	L	L	L	L	L	L	L
	RS101	A	A	A	L			L	L	
	RS103	A	A	A	A	A	A	A	A	A
	RS105	L	L	L	L				L	

Note: Table D.2 lists the test items required for the pr installation of the equipment and the subsystem of the equipment and the platform or the military platform or device. If a device or a system is expected to be installed in a multitype platform or device, a class of the most stringent requirements shall be required. A indicates that the item is required to be used in the form of L, which means that the item requirements shall be limited according to the relevant provisions of this standard. The blank column indicates that the item is not applicable. For army equipment, there are five main EMC test requirements; see Table D.3.

Table D.3 The five general requirements of the army.

Test project	Designation
RE102	Electric field emission of 10 kHz–18 GHz
CE102	power line transmission of 10 Hz–10 kHz
CS101	The power cord conduction sensitivity of 25 Hz–50 kHz
CS114	Cable beam injection conduction sensitivity of 10 kHz–400 MHz
RS103	Radiation sensitivity of electric field of 10 kHz–40 GHz

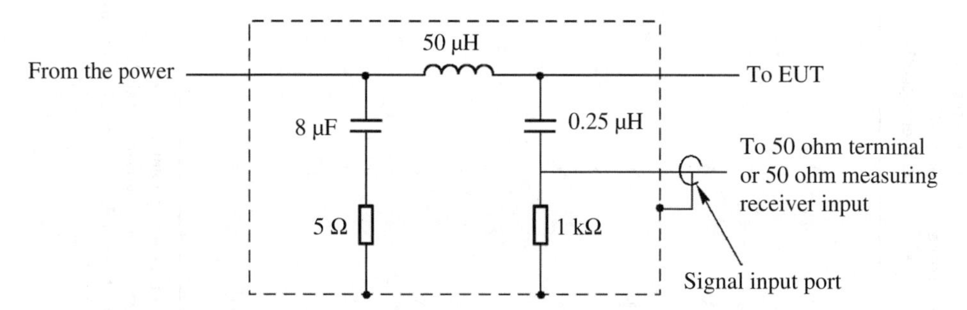

Figure D.1 The principle diagram of the LISN.

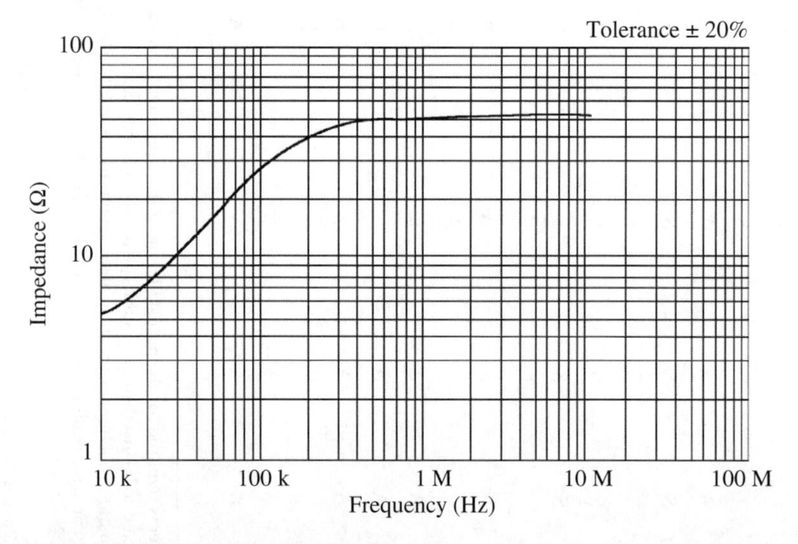

Figure D.2 LISN impedance characteristic curve.

The test configuration of EUT should be in accordance with the requirements of the general test configuration of Figures D.3–D.6. In the whole test device, the above configuration should be maintained unless there are explicitly directed to a specific test method. Any changes to the general test configuration should be specifically described in the single test method. Only when the EUT design and installation instructions are provided can the housing of equipment be directly overlapped to the reference ground plane. When the actual installation needs overlap,

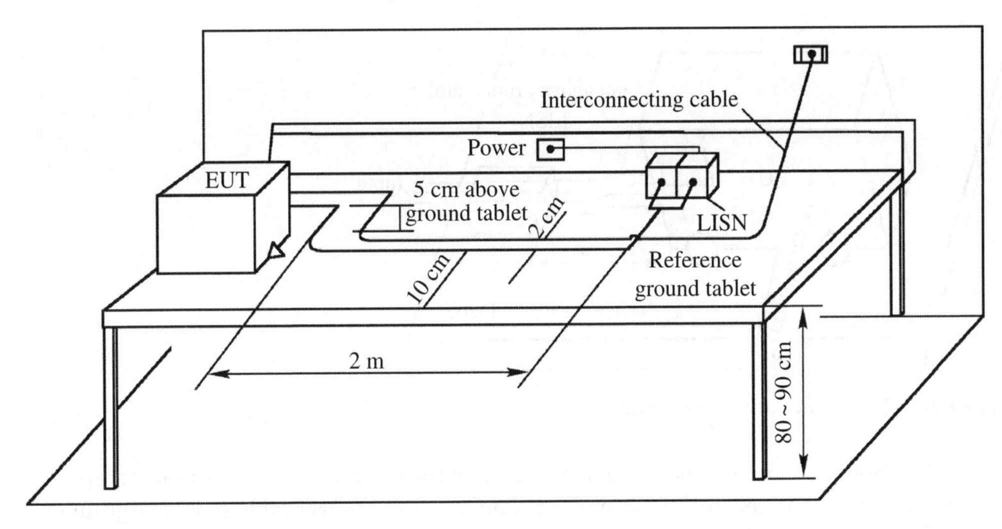

Figure D.3 General test configuration.

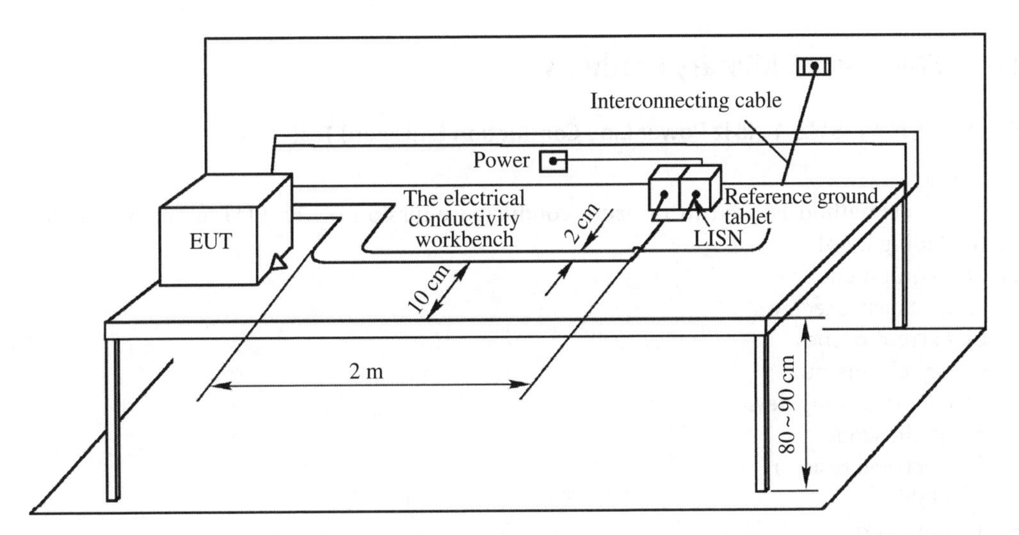

Figure D.4 The conductive surface test configuration of the EUT while the test table surface is unconductive.

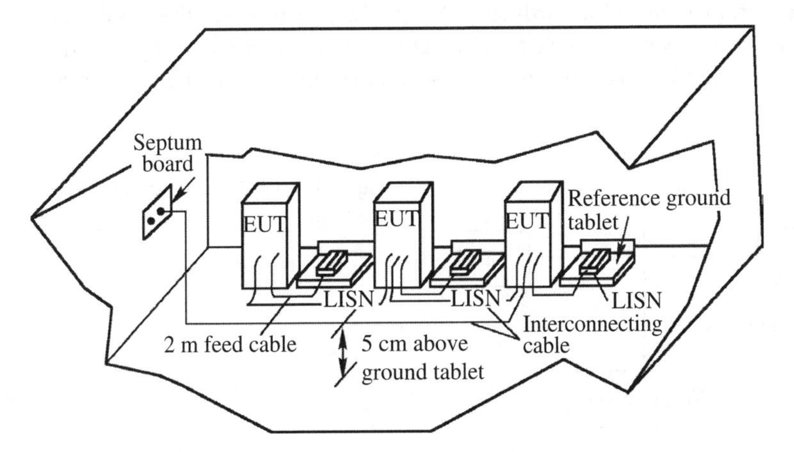

Figure D.5 Independent of the EUT and multiple EUT shielding chamber test configuration.

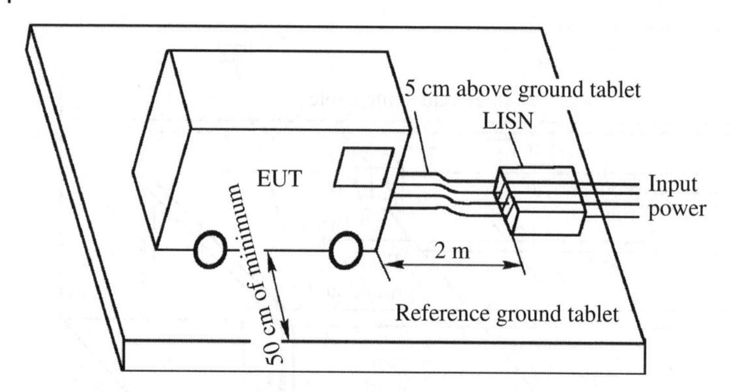

Figure D.6 Independent EUT test configuration.

the overlap should be same as the actual installation of the provisions. The portable equipment that is connected to the ground by the safety grounding wire of power cable should be grounded in accordance with the relevant test method.

D.4 EMC Test of Military Products

D.4.1 CE101 (25 Hz–10 kHz Power Line Conduction Emission) Test

1) Purpose
   This test method is used to measure conducted emission of the EUT input power line (including loop).
2) Test equipment
   a) measuring receiver
   b) current probe
   c) signal generator
   d) data recording device
   e) oscilloscope
   f) electrical resistor
   g) LISN
3) Test method
   According to the general requirements of Figures D.3–D.6, on the basic test configuration of EUT, the configuration is tested according to Figure D.7. Determine the conduction emission of the EUT input power line (including the return), EUT power on and preheating, so that it can reach a stable working state, and choose a power line clamped by the current

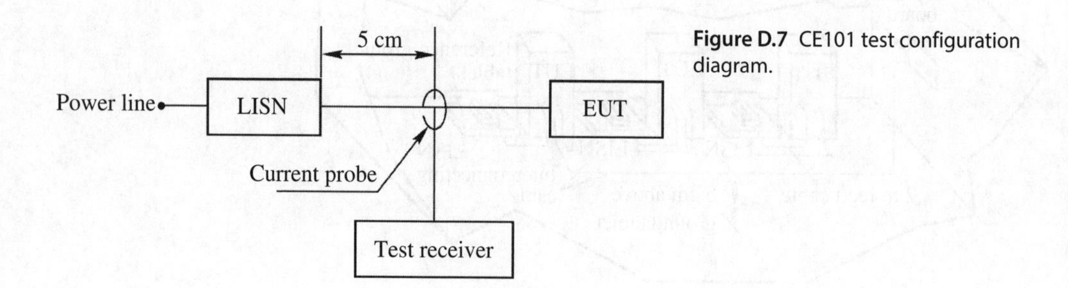

Figure D.7 CE101 test configuration diagram.

probe. The current probe is placed at the distance of 50 mm from the LISN in testing. The measurement of the receiver scans is within the applicable frequency range. After testing a power line, the test will be performed on the other line.

D.4.2 CE102 (10 kHz–10 MHz Power Line Conduction Emission) Test

1) Purpose
   This test method is used to measure conducted emission of the EUT input power line (including loop).
2) Test equipment
   a) measuring receiver
   b) data recording device
   c) signal generator
   d) attenuator, 20 dB
   e) oscilloscope
   f) T coaxial connector
   g) LISN
3) Test method
   According to the general requirements of Figures D.3–D.6, maintain the basic test configuration of EUT, test the configuration as Figure D.8, and connect the receiver to attenuator of 20 dB at the LISN signal output port. Determine the conduction emission of the EUT input power line (including the return), EUT power on and preheating, so that it can reach a stable working state, and choose a power line clamped by the current probe. The measurement of the receiver scans within the applicable frequency range. After testing a power line, the test will be performed on the other line.

D.4.3 CE107 (Power Line Spike (Time Domain) Conducted Emission

1) Purpose
   It is adequate to equipment and subsystem that may produce spike, measuring the amplitude of the spike in the time domain.
2) Test equipment
   a) $10\mu F$ of feed-through capacitor
   b) Current probe: 10 kHz–50 MHz frequency range amplitude uniformity within $\pm 3dB$
   c) LISN
   d) Voltage probe: 10 kHz–50 MHz frequency range amplitude uniformity within $\pm 3dB$
   e) Memory oscilloscope (bandwidth $\geq 50MHz$) or peak memory voltmeter (bandwidth $\geq 50MHz$)

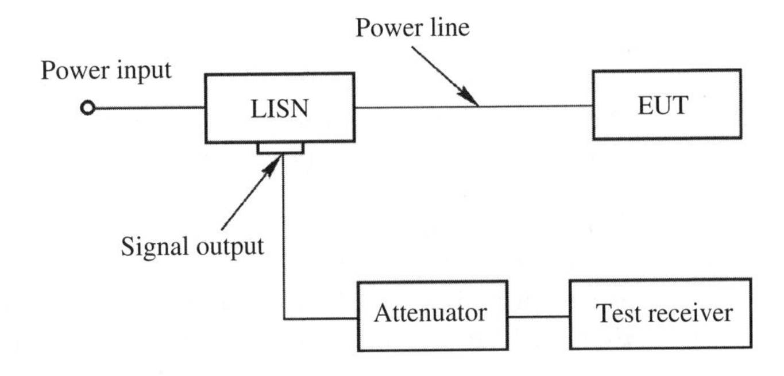

Figure D.8 CE102 test configuration diagram.

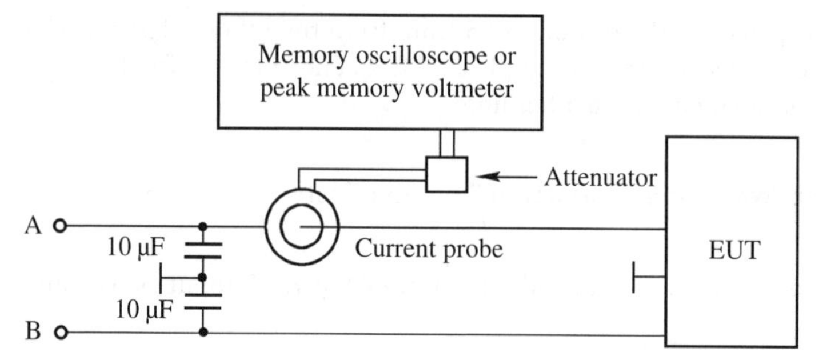

Figure D.9 CE107 cord closed-circuit current rush test.

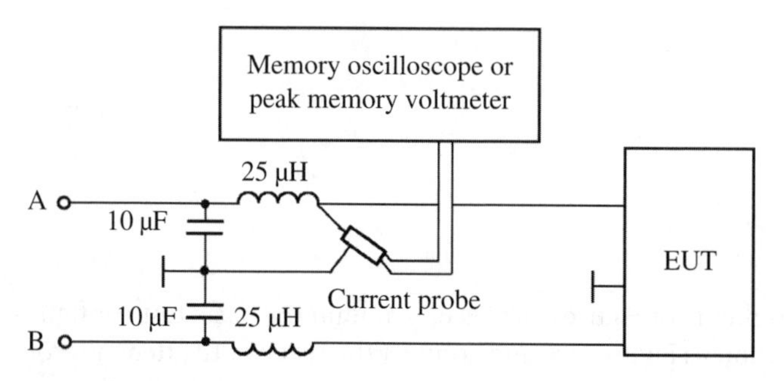

Figure D.10 CE107 cord open-circuit current rush test.

3) Test method

According to the general requirements of Figures D.3–D.6, maintain the basic test configuration of EUT, and test configuration as Figures D.9 or D.10. It can also use LISN instead of $10\mu F$ feed-through capacitor or $10\mu F$ feed-through capacitor group with $25\mu F$ inductor. The current or voltage probe is placed near the $10\mu F$ feed-through capacitor or $10\mu F$ feed-through capacitor group with $25\mu F$ inductor.

1) Closed-circuit peak current testing
   a) As shown in Figure D.9, the current probe is placed near the EUT power line $10\mu F$ feed-through capacitor. Output terminal of current probe is connected to the memory oscilloscope or peak memory voltmeter.
   b) All kinds of working state of the EUT for each state should be repeated at least five times, including on-off switches, read the maximum value in all kinds of working state. When it is possible to synchronize, EUT switch conversion should be adjusted in the power line appearing peak or zero value.
   c) The bandwidth of memory oscilloscope or the peak memory voltmeter is greater than the reciprocal of 50% of the peak current amplitude width. The voltmeter reading is narrow band, according to formula (D.1) to calculate the spike wave:

$$I_{dB\mu V} = V_{dB\mu V} - Z_{dB\Omega} \tag{D.1}$$

   In the equation, $Z_{dB\Omega}$ is transformation impedance of current probe loaded by memory oscilloscope or peak memory voltmeter. If the input impedance of the memory oscilloscope or peak memory voltmeter is in parallel with a 50Ω resistor, the impedance of the current probe can be used.

2) Open circuit voltage spike test
   a) As shown in Figure D.10, the current probe is placed on inductor near the power line $10\mu F$ feed-through capacitor. The inductor is at least $25\mu H$ and connected with a

10 μF feed-through capacitor providing corner or notch below 10 kHz. And the resonant frequency is higher than 50 MHz, the current of EUP can pass. The output of the voltage probe is connected to a memory oscilloscope or a peak memory voltmeter.

b) The same as b of closed-circuit peak current tests.

3) LISN, peak voltage or current test

a) In Figures D.9 and D.10 10 μF feed-through capacitor and 25 μH inductor is replaced by the line impedance stabilization network. When the line impedance stabilization network is connected to a 50Ω resistor, 30~50Ω impedance is provided in the frequency range of 10 kHz–50 MHz. The current probe is close to LISN, located between the EUT and the line impedance stabilization network. Voltage probe is connected to 50Ω output resistor of the line impedance stabilization network (50Ω resistor is removed).

b) The same as b of closed-circuit peak current tests.

D.4.4 CS101 (the Power Line Conduction Susceptibility) Test

1) Purpose
This test method is used to test EUT capability to withstand signal coupled to the input power line.
2) Test equipment
   a) signal generator
   b) power amplifier
   c) oscilloscope
   d) coupling transformer
   e) capacitor, 10 μF
   f) isolation transformer
   g) resistor, 0.5Ω
   h) LISN
3) Test method
According to the general requirements of Figures D.3–D.6, the basic test configuration of EUT is maintained, the configuration according to Figures D.11, D.12, and D.13 is tested, respectively. The configuration of Figure D.11 is corresponding DC or single-phase AC power supply; the configuration of Figure D.12 is corresponding three-phase Δ type connection power supply; the configuration of Figure D.13 is corresponding Y-type connection power supply (four power lines). At the same time, in order to protect the power

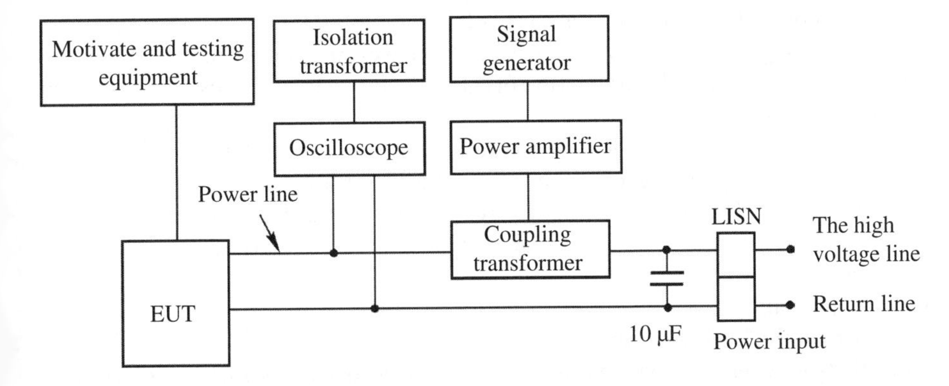

Figure D.11 DC or the single-phase AC power supply cord of CS101 test configuration diagram.

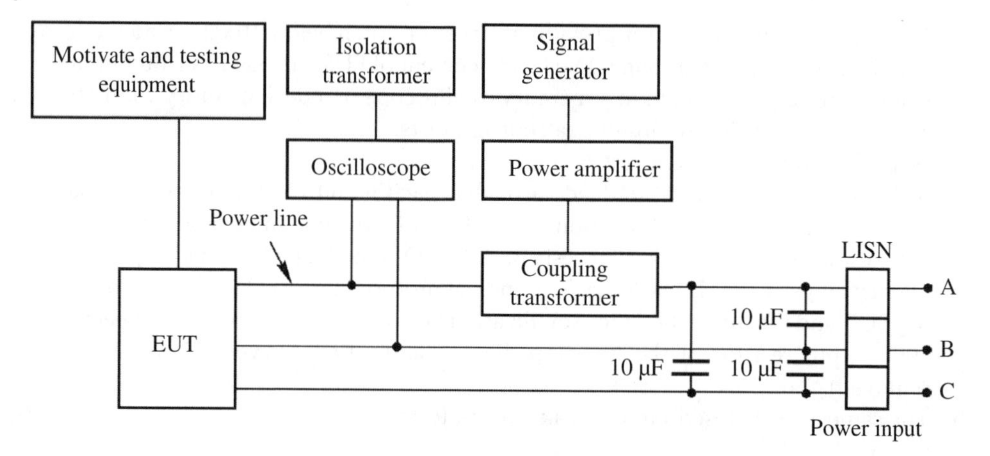

Figure D.12 Three-phase triangular connection the CS101 test configuration diagram of the power cord.

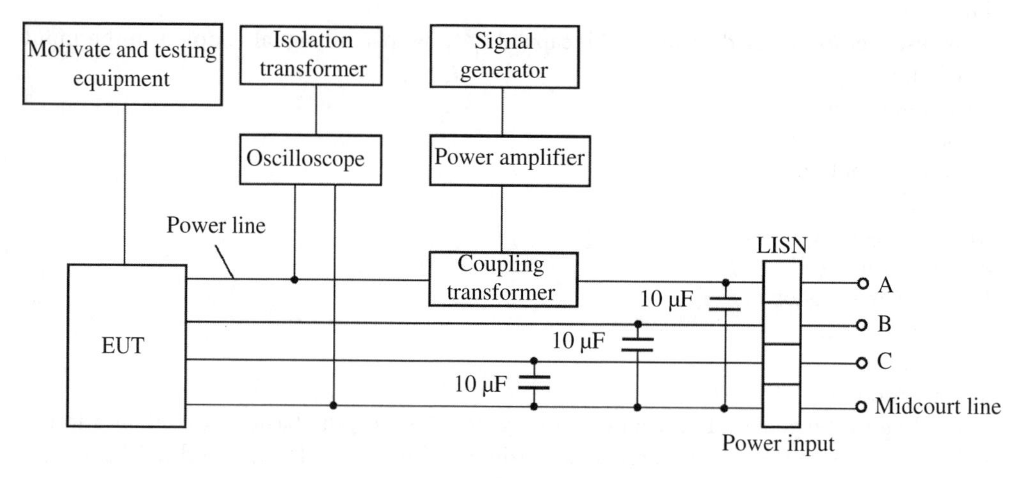

Figure D.13 Three-phase Y connection the configuration diagram of the power cord of CS101 test.

amplifier, an equivalent EUT dummy load and an additional coupling transformer can be used to make its induction voltage equal to the induction voltage of injected to transformer, but its phase is opposite. Detailed test steps are as follows:

a) Preheat EUT after power on, so as to achieve a stable working state. The test should be particularly careful, because disconnection of safety ground wire of the oscilloscope may create electric shock hazard.
b) The signal generator is set to the lowest test frequency, increasing the signal level until the required voltage or power level is reached on the power line.
c) Maintain the required signal level, and the scan rate is not greater than the requirements of standard GJB152A. Get the scanning measurement at required entire frequency range.
d) Sensitivity appraisal:
   i) Monitoring if the EUT is sensitive
   ii) If sensitivity is generated, the sensitivity threshold level need to be further determined
e) Repeat the test for each power line b–d if it is necessary.

The following requirements shall be measured for three-phase Δ type connection power supply:

Coupling transformer's line	Voltage measurement location
A	A to B
B	B to C
C	C to A

The following requirements shall be measured for three phase Y-type connection power supply (four power lines):

Coupling transformer's line	Voltage measurement location
A	A to neutral line
B	B to neutral line
C	C to neutral line

D.4.5 CS106 (Power Line Spike Conducted Susceptibility) Test

1) Purpose
   In the equipment, the subsystem of all nongrounded AC and DC input power line test equipment, the subsystem susceptibility by injection spike to the power supply.
2) Test equipment
   a) The spike generator, with the following features:
      A. pulse width: $0.15\mu s$, $5\mu s$, $10\mu s$
      B. pulse repetition rate: 3–10 PPS
      C. voltage output: not less than 400 V (peak)
      D. output control: from 0 to 400 V (peak) is adjustable
      E. output spectrum:$160 dB\mu V/MHz$ at 25 kHz; reduce to $115 dB\mu V/MHz$ at 30 MHz
      F. phase adjustment: 0° – 360°
      G. signal source impedance (with injection transformer): 0.06Ω
      H. transformer (current capacity): 30A
      I. external synchronization 50–1000 Hz
      J. external trigger 0–20 PPS
   b) $10\mu F$ feed-through capacitor
   c) 100 MHz bandwidth, any scope meets sweep frequency requirements
   d) suppression filter (suppress power frequency should be at least 40 dB)
3) Test method
   According to the general requirements of Figures D.3–D.6, the basic test configuration of EUT is maintained, performing the test according to the configuration of Figure D.14, respectively, and the test procedure is as follows:
   a) For EUT with AC and DC power supply is tested according to the Figure D.14a configuration, for EUT with DC power supply is tested according to the Figure D.14b configuration.
   b) The output level of the spike signal generator is slowly increased to provide the required peak voltage, but cannot exceed the output level of the pre calibrated spike signal generator.
   c) Adjust the synchronization and trigger, so that the peak signal will produce the maximum susceptibility on EUT specific position.

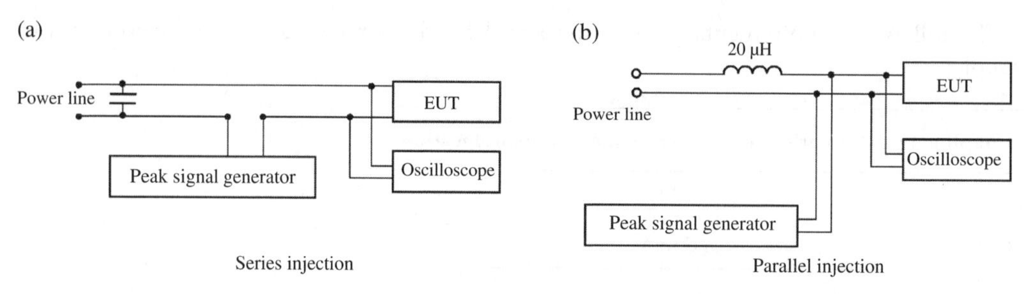

Figure D.14 Test configuration of signal conduction sensitivity of CS106 power supply line.

d) The positive, negative, single, and repetitive (6–10 PPS) spike signal is applied to the nongrounded input of EUT, and the injection time is not more than 30 minutes. The peak signal should be synchronized with the power supply, and the injection time is not less than 5 minutes at every 90-phase angle. In addition, it is also required to adjust the trigger phase of the spike signal, so that it is in 0–360 V range of the power frequency. Change the peak signal synchronization frequency (from 50 to 1000 Hz), and pay attention to the impact of the device susceptibility. The trigger spike signal should be generated in any time within the logic circuit of any open time and generated any pulse time in digital circuit.
e) If EUT is found to be sensitive to the peak signal, it should be measured and recorded at the threshold level, considering repetition frequency, phase position of the AC waveform and the time in the digital gate circuit.
f) The suppression filter is inserted between the power line and oscilloscope if necessary.

D.4.6 CS106 (Shell Conduction Current Susceptibility) Test

1) Purpose
   This test method is used to test the capability of EUT to withstand shell conduction current.
2) Test equipment
   a) signal generator
   b) oscilloscope or voltmeter
   c) resistor, 0.5Ω
   d) isolated isolation transformer
3) Test method
   This test does not need to be configured as the requirement of Figures D.3–D.6 for a basic test configuration. Its test configuration is shown in Figure D.15 and meets the following requirements:
   a) Set the EUT and test equipment as the Figure D.15 (including signal generator, test current monitoring equipment and equipment for making EUT work and monitoring the performance of EUT), ensure that the test configuration is single and point grounded.
   b) EUT is isolated with all test equipment AC power supply by isolated transformer, the isolated transformer is not applicable to the power supply of the DC power supply of EUT,
   c) Place EUT and test equipment on the nonconductive plane and disconnect all the input EUT power line safety ground wire.
   d) The choose of EUT test point depends on the type of EUT and final assembly or installation method, and test point should be connected to the crossover through all face diagonal endpoints of EUT.
   e) Connect signal generator and resistor to a set of test points.

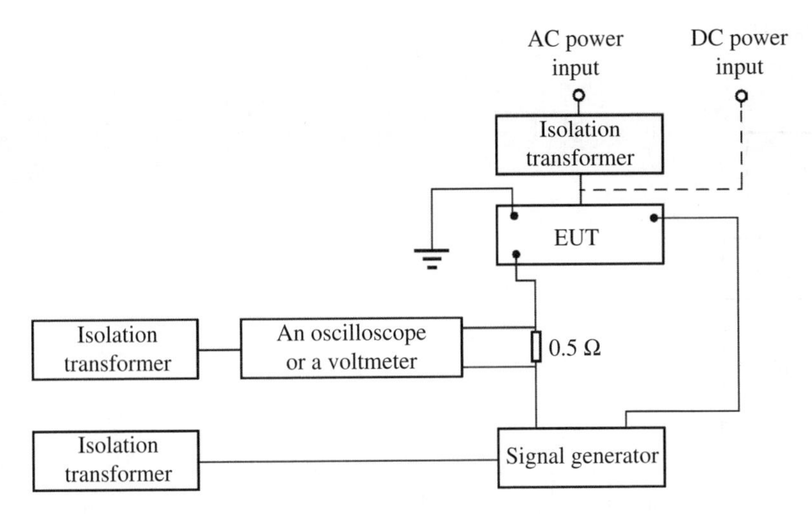

Figure D.15 CS109 (shell conduction current sensitivity) test configuration diagram.

CS109 (shell conduction current susceptibility) test method is as follows:

a) Preheat EUT, so as to achieve a stable working state.
b) The signal generator is set to the lowest frequency, and then adjusted to the required level. Monitor the current by measuring the voltage across the resistor.
c) Maintain the current level according to the applicable limit, at the same time, according to this standard, the general requirements scan in the 50 Hz–100 kHz frequency range, and monitoring if EUT is sensitive.
d) If EUT appears sensitive, it is to determine the susceptibility threshold level (at the level, EUT do not appear to response that does not want) and whether the level meets the requirements of GJB 151A.
e) Test on the other surface diagonal endpoint of the EUT, repeat b–d.

D.4.7 CS114 (10 kHz–400 MHz Cable Injection Beam Injection Susceptibility) Test

1) Purpose
   This test method is used to test the ability of the EUT to withstand the radio frequency signal coupled to the cable of EUT.
2) Test equipment
   a) receiver
   b) injection probe
   c) current probe
   d) calibration device: 50Ω a characteristic impedance at both ends of the head with coaxial connectors and provides enough space for the coaxial transmission line around the center conductor to calibrate the injection probe. The typical CS114 calibration device is shown in Figure CS114-3.
   e) directional coupler
   f) signal generator
   g) plotting instrument
   h) attenuator, 50Ω
   i) coaxial load, 50Ω
   j) power amplifier
   k) LISN

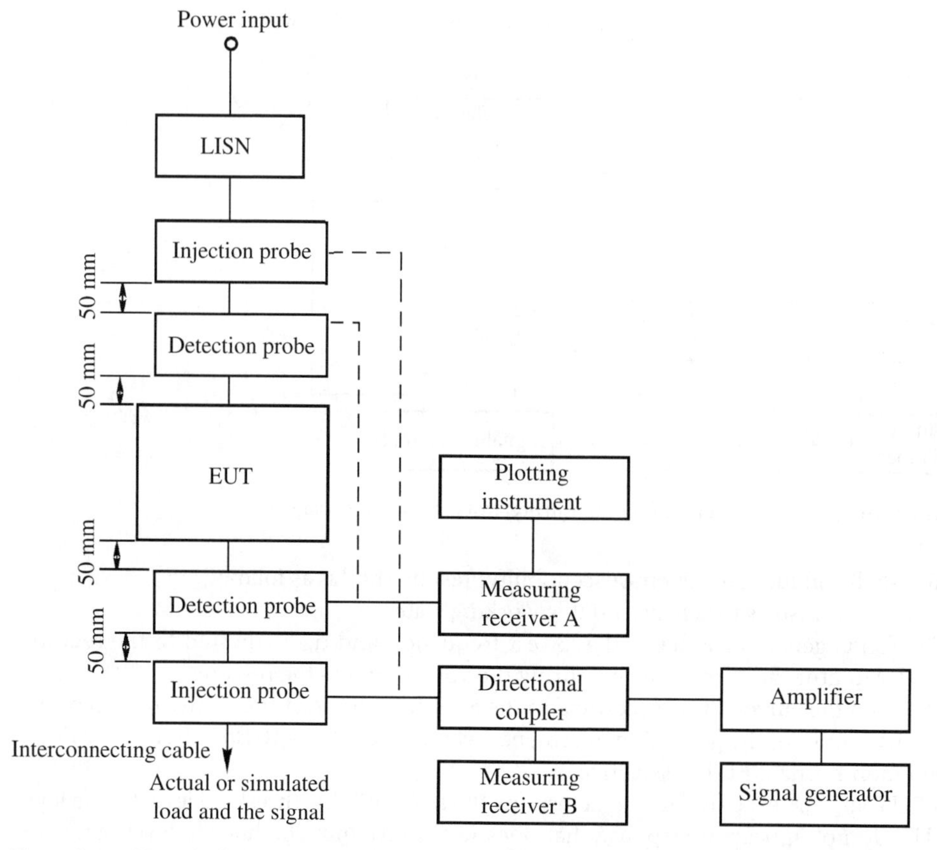

Figure D.16 CS114 (10 kHz–400 MHz cable transmission beam injection sensitivity) test configuration diagram.

3) Test method
   According to the general requirements of Figures D.3–D.6, maintain the basic test configuration of EUT and configure the test setup according to the requirements of Figure D.16. Meet the following requirements:
   a) Clamp inject and monitor probe on the cables that connected to the EUT connector.
   b) Place the monitor probe at 50 mm from the EUT connector, and if the connector and the base housing is more than 50 mm, the probe should be close to the connector's base housing.
   c) Place current injection probe 50 mm from monitor probe.

Each cable bundle that includes a full power cable (high line and return line) on the EUT shall be tested in accordance with the following steps. Containing the power supply cable tested the following requirements:

1) Preheat EUT, to get the stable working state.
2) Determination of loop impedance:
   a) Set signal generator to 10 kHz, without modulation.
   b) Apply the approximately 1 mW power level signal to the injection probe, and record the power level of the B indicator of the receiver (considering injection probe into the injection probe insertion loss converted into the injection probe output interface) and the induced current level of the measuring receiver A.

c) Scan at 10 kHz–400 MHz frequency range, and the applied power level and the induced current level are recorded.
d) Normalize measure the results to the watt per ampere (A/W).

3) Susceptibility evaluation
a) Set signal generator to 10 kHz, with 1 kHz duty cycle of 50% pulse modulation.
b) Feed the incident power level that determined in 4.2 d to injection probe, while monitoring the induced current.
c) In the frequency range of 10 kHz–400 MHz, scan testing as the general requirements of this standard, while the incident power is maintained at the corresponding level of calibration or the maximum current level in GJB 151A (select low level of both).
d) Monitor whether EUT performance degradation during the test.
e) If EUT appear sensitive, determine the susceptibility threshold level (under this level, EUT just does not appear undesired response), and determine the level does not meet the requirements of GJB 151A.
f) Because of security reasons for EUT, we can use multiple cables simultaneously in the injection method.

D.4.8 CS115 (Cable Transmission Beam Injection Pulse Susceptibility) Test

1) Purpose
This test method is used to test the ability of EUT to withstand the pulse signal coupled to the cable of EUT.

2) Test equipment
a) pulse signal generator, 50Ω
b) current injection probe
c) Excitation cable, 2 m long, 50Ω characteristic impedance, no more than 0.5 dB insertion loss at 500 MHz
d) current probe
e) Calibration device: characteristic impedance at both ends of the coaxial connector to provides enough space for the coaxial transmission line around the center conductor to calibrate injection probe
f) oscilloscope, 50Ω input impedance
g) attenuator, 50Ω
h) coaxial load, 50Ω
i) LISN

3) Test method
According to the general requirements of Figures D.3–D.6, maintains the basic test configuration of EUT, and configure the test setup according to the requirements of Figure D.17. It meets the following requirements:
a) Clamp injection and monitor probe on the cable, which connected to the EUT connector.
b) Place the monitor probe at 50 mm from the EUT connector, and if the connector and the base housing is more than 50 mm, the probe should be close to the connector's base housing.
c) Place current injection probe 50 mm from monitor probe.

Before the test, the EUT is in stable condition. Each cable bundle that includes a full power cable (high line and return line) on the EUT shall be tested in accordance with the following steps. Containing the power supply cable loop should be tested according to the following requirements:

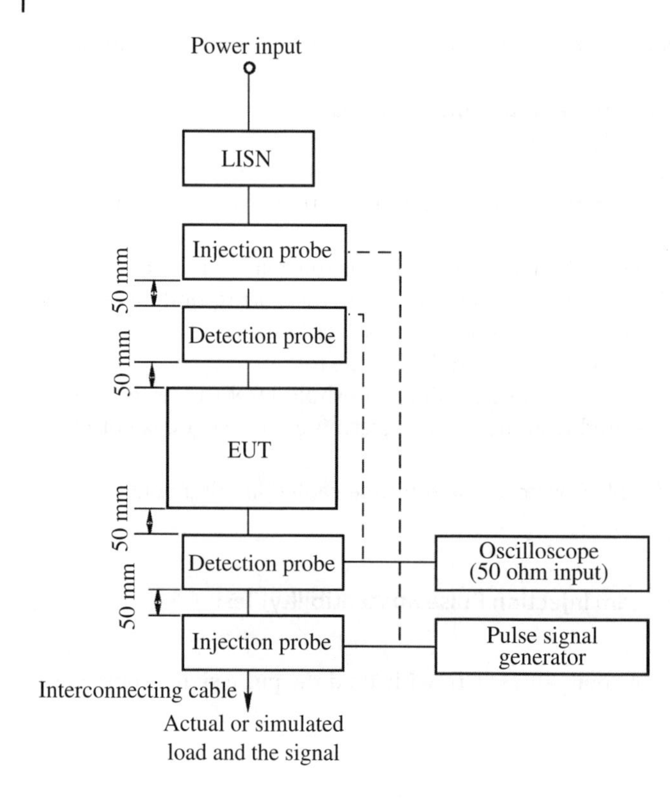

Figure D.17 CS115 (cable transmission beam injection pulse sensitivity) test configuration diagram.

i) Preheat EUT to get the stable working state.
ii) Conduct susceptibility evaluation.
   a) Adjust the pulse signal generator position from the minimum value to the determined amplitude.
   b) Apply the pulse with the repetition frequency and duration specified in GJB 151A.
   c) Monitor whether EUT performance degradation during the whole testing.
   d) If the EUT is sensitive, the susceptibility threshold level (under which the EUT does not appear undesired response) is determined, and the level of the GJB 151A is confirmed.
   e) Record the peak induced current on the cable from the indicator value of the oscilloscope.
   f) For each cable bundle of each connector on the EUT, repeat the above test.

D.4.9 CS116 (10 kHz–100 MHz Damped Sinusoidal Transients Conducted Susceptibility Test on Cable and Power Supply Line)

1) Purpose
   This test method is used to test the ability of EUT to withstand the sinusoidal transient signal coupled to the related cable and power supply line of EUT.
2) Test equipment
   a) Sinusoidal transient signal generator, output resistance <100 Ω
   b) Current injection probe
   c) Memory oscilloscope, input impedance is 50 Ω

d) Calibration device: with 50 Ω characteristic impedance, the coaxial connector is in the two ends. Provides enough space for the coaxial transmission line around the center conductor to calibrate the injection probe
e) Current probe
f) Wave form recorder
g) Attenuator
h) Measuring receiver
i) Power amplifier
j) Coaxial load
k) Signal generator
l) Directional coupler
m) LISN

3) Test method

According to the general requirements of Figures D.3–D.6, maintain the basic test configuration of EUT. The configuration is shown in Figure D.18 and meets the following requirements:

a) Test configuration according to Figure D.18.
b) The injection and monitoring probe are connected to the EUT connector cable beam.
c) The probe is placed near the EUT connector, and if the length of connector and the base housing is over 50 mm, the probe should be close to the base housing of the connector.
d) The probe is placed 50 mm away from monitoring probe.

Before the test, EUT works in a stable state. The EUT with complete power cable (high line and return line) of the EUT and each connector should be tested according to the following steps:

1) Preheat EUT, to get the stable working state.
2) Loop impedance characteristic is fixed.
   a) According to Figure D.18, signal generator is adjust to 10 kHz without modulation.
   b) Apply the approximately 1 mW power level signal to the injection probe, and record the power level of the B indicator of the receiver (considering injection probe insertion loss converted into the injection probe output interface) and the induced current level of the measuring receiver A.

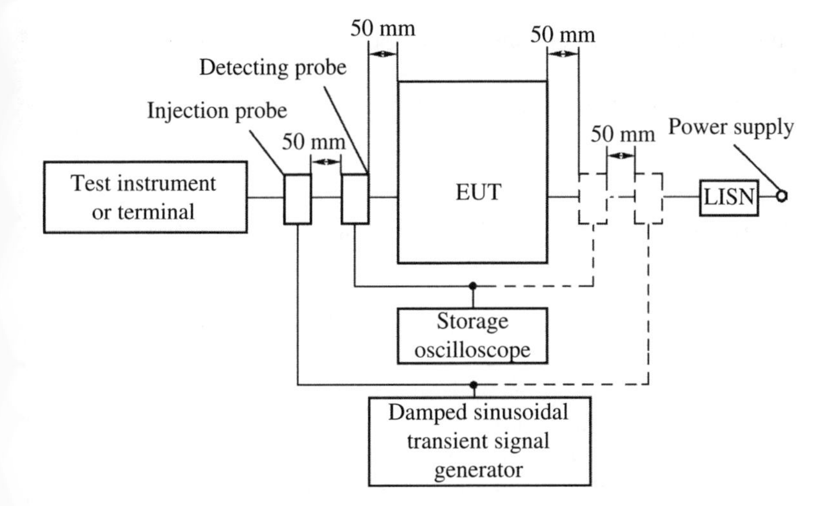

Figure D.18 10 kHz–100 MHz cable and power line damped sinusoidal transient conduction sensitivity test apparatus.

c) Scanning in the frequency range of 10 kHz–100 MHz. Record the power level and the induced current level.
d) The measurement results are normalized to amps per watt (A/W).
e) Marked resonance frequency at maximum and minimum impedance is indicated.

3) Susceptibility evaluation.
   a) Preheat EUT and to stable state.
   b) When the damped sinusoidal transient signal source is connected well but is not triggered, EUT should not be affected.
   c) The damped sinusoidal transient signal generator is transferred to the test frequency.
   d) Sequentially applying test signal to EUT each cable or power line. Slowly increasing the output level of the damped sinusoidal transient signal generator to provide the required current. But not more than damped sinusoidal transient signal generator. Precalibrated output level recording obtained peak current.
   e) Monitor whether there is EUT performance degradation.
   f) If EUT appears sensitive, determine the susceptibility threshold level important (under this level, EUT does not appear undesired response;), and determine if the level does not meet GJB151 requirement.
   g) Repeat b–f to the test each frequency and resonance frequency specified in the GJB151A.

   In addition, it is necessary to repeat the test in the case of power-off.

D.4.10 RE101 (25–100 kHz Magnetic Field Radiation Emission Test)

1) Objective

   This test method is used to test whether the magnetic field emission from the EUT and its related wires or cables exceeds the requirements.
2) Test equipment
   a) Measuring receiver
   b) Data recording device
   c) Ring sensor
      1) Diameter: 133 mm
      2) Turns: 36
      3) Lead specification: 7 × φ0.07 mm strand insulation line
      4) Shielding: electrostatic
      5) Correction factor: the measurement receiver's reading dBμV plus correction coefficients covert to dBpT
   d) LISN
3) Test method

   According to the general requirements of Figures D.3–D.6, maintain the basic test configuration. The configuration is shown in Figure D.19 and meets following the requirements:
   a) Preheat EUT to the stable state.
   b) The distance between ring sensor located and the EUT or cable is 70 mm and ring sensor plane is paralleled to the EUT plane and the axis of the cable. Using the bandwidth and minimum measurement time specified in standard. Enable the receiver to be in the entire application of the frequency range scan and find out the maximum radiation frequency point.
   c) Change the measuring receiver to the frequency or frequency range to the step c.
   d) When moving the ring sensor (keeping 70 mm distance) along the house plane or along the cable, monitor and measure the output of the receiver. Recorded the maximum points for each frequency at the step d.

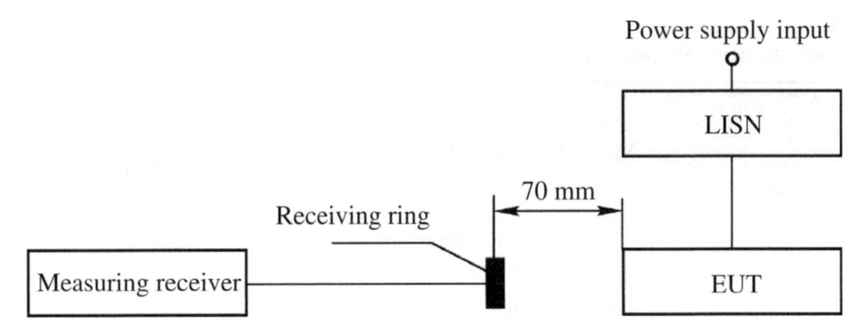

Figure D.19 25–100 kHz field emission test device.

e) At 70 mm distance from the maximum radiation point, adjust the direction of the ring sensor plane, find a maximum reading of the test receiver, and write down the reading.
f) Remove the ring sensor and place it 500 mm from the EUT or cable; record the test receiver reading.
g) When the frequency is below 200 Hz, at least two maximum radiation frequency points are per octave frequency selected; repeat steps d–g. When the frequency is higher than 200 Hz, at least three maximum radiation frequency points are selected; repeat steps d–g.
h) For each plane and cable of the EUT, repeat test b–h.

D.4.11 RE102 10 kHz–18 GHz Electric Field Radiation Emission Test

1) Objective
   This test method is used to test whether the electric field radiation emission from EUT and its related wires and cables exceeds the specified requirements.
2) Test equipment
   a) Measuring receiver
   b) Data recording device
   c) Antenna
      1) 10 kHz–30 MHz, 1040 mm antenna with impedance matching network
      2) When the impedance matching network includes a preamplifier (active rod antenna), pay attention to overload protection.
      3) Using a square grid, at least 600 mm per side
      4) 30–200 MHz, double cone antenna, top to top distance is about 1370 mm
      5) 200 MHz–18 GHz, double ridge horn antenna
   d) Signal generator
   e) Short rod radiator
   f) Capacitor, 10 pF
   g) LISN
3) Test method
   According to the general requirements of Figures D.3–D.6, based on EUT basic test configuration, the configuration is shown in Figure D.19 and meets following requirements:
   a) For all configurations, the antenna should be 1 m away from the edge of the test configuration.
   b) In addition to the 1040 mm rod antenna, the antenna should be higher than the ground plate 1200 mm.
   c) To ensure that the distance between any part of the antenna and wall of the shielded room is not less than 1 m and the top plate not less than 0.5 m.

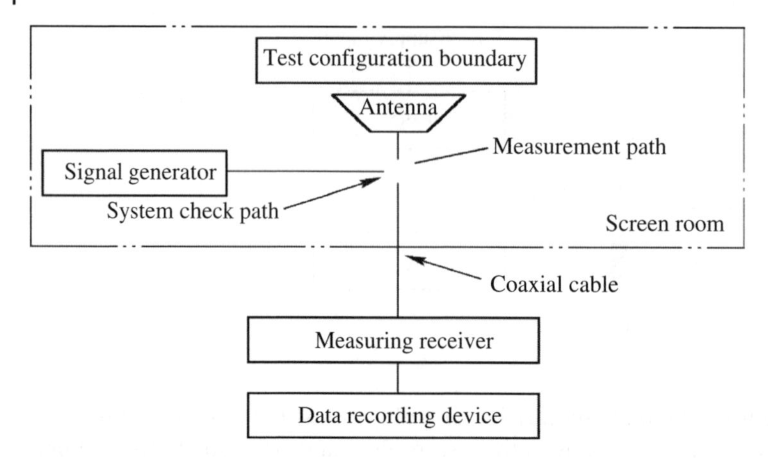

Figure D.20 10 kHz–18 GHz electric field emission test device.

d) On the use of the test bench configuration, the positioning requirement to the pull rod antenna and the test bench ground distance is shown in Figure D.20.
e) To the large EUT unfixed installed on the shield room floor, overlap and install the 1040 mm rod antenna matching the network on the ground plane. Do not use ground grid.

According to the measurement path shown in Figure D.20, determine the radiation of EUT and its relevant cable.

a) Using the specified bandwidth and minimum measurement time in this standard, so that the measurement receiver will scan the entire frequency range.
b) For frequency above 30 MHz, the antenna should take two directions of horizontal polarization and vertical polarization.
c) Test each antenna position described below.

The required number of antenna position quantity depends on the size of the EUT test boundary and the EUT extension number and the direction diagram of the antenna.

The following criteria are used to determine the specific location of the antenna for testing below 200 MHz:

a) When the test boundary edge is less than 3 m, the antenna should be located at the position where bisector perpendicular of the corresponding edge.
b) When the test boundary edge is large than 3 m, using multiple antennas intervals just as shown in Figure D.21. Divide the distance from one edge to the other edge (unit is m) by three and take it as the integer. Which is the number of antenna position.

To the test from 200 MHz to 1 GHz, the antenna should be placed in a sufficient number of locations. In order to make the entire width of each EUT shell and 350 mm of the wire and cable connected to the shell end of the EUT is exposed to the 3 dB of the bandwidth.

To the test above 1 GHz, the antenna should be placed in a sufficient number of locations. In order to make the entire width of each EUT housing and 70 mm wire and cable connected to the terminal of the EUT housing are exposed to the 3 dB beam width range.

Specific test procedures are shown as follows:

1) Preheat the EUT to reach a steady state.
2) Check the path of the system in Figure D.20 for each antenna used its highest frequency to evaluate the entire measurement system from each antenna to the data output device.

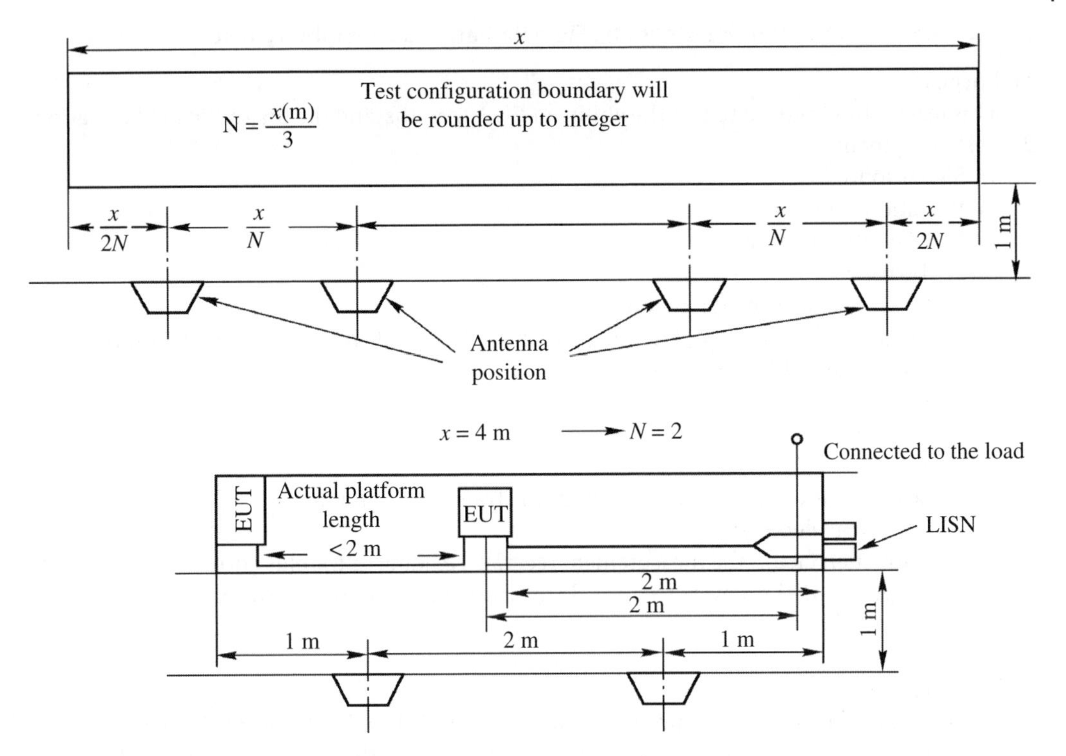

Figure D.21 Multiple antenna positions in RE102 test.

To the tie-rod antenna with a passive matching network evaluated it by the center frequency of each frequency band.

a) Applying a calibration signal to the coaxial cable of the antenna connection point. Its level is lower6 dB than the GJB151A limited value subtract antenna; coefficient.
b) The measuring receiver is scanned by normal method, and check the indicator level of the data-recording device is in the range of ±3 dB of the injection signal level.
c) To the 1040 mm rod antenna, remove the 10 pF capacitor of rod and add signal to antenna matching network.
d) If the deviation of the reading is more than ±3 dB, find out the cause of the error and correct.

3) Use the measurement path shown in Figure D.20. Assess each antenna to according to following requirement to confirm that the antenna can be used in good condition.
   a) Use antenna or short rod radiator to radio signal in the highest frequency measurement point of each antenna.
   b) Adjust the measurement receiver to the added signal frequency, and check whether the received signal is appropriate.
4) Preheat EUT power to achieve a stable working state.
   According to the measurement path shown in Figure D.20, find out the EUT and its related cable radiation emission.
   a) Use the standard requirement bandwidth and minimum measurement time, so that the measurement receiver is in the entire application frequency range of scan. For frequencies above 30 MHz, the antenna should take two directions of horizontal polarization and vertical polarization.
   b) Test each antenna's determined location.

D.4.12 RS101 25 Hz–100 kHz Magnetic Field Radiation Susceptibility Test

1) Purpose
   This test method is used to test the ability of EUT to withstand the magnetic field radiation.
2) Test equipment
   a) Signal source
   b) Radiation ring
      1) Diameter: 120 mm
      2) Turns: 20
      3) Lead specification: φ1.25 mm wire
      4) Magnetic flux density: The magnetic flux density at the distance of 50 mm from the ring plane is 9.5×10^7 pT/A
   c) Ring sensor
      1) Diameter: 40 mm
      2) Turns 51
      3) Lead specification: 7 × φ0.07 mm multi strand insulation line
      4) Shielding: electrostatic
      5) Correction factor: the measurement receiver's reading dBμV plus correction coefficients in figure RS101-1 covert to dBpT.d.; measuring receiver or narrow band voltmeter
   d) Current probe
   e) LISN
3) Test method
   According to the general requirements of Figures D.3–D.6, based on EUT basic test configuration, the configuration is shown in Figure D.22 and the detailed test procedure is as follows:
   1) Preheat EUT power preheating, so as to achieve a stable working state.
   2) Selected test frequency is shown as follows:
      a) The radiation ring is placed 50 mm away from the surface of EUT, and the plane of the ring should be parallel to the EUT surface.
      b) A sufficient current is applied to the radiation ring to produce a magnetic field intensity of at least 10 dB more than GJB151A, but not more than 15A (183dBpT).
      c) Scan within the frequency range of the GJB151A; the scan rate is three times faster than the speed in Table 3 of GJB151A.

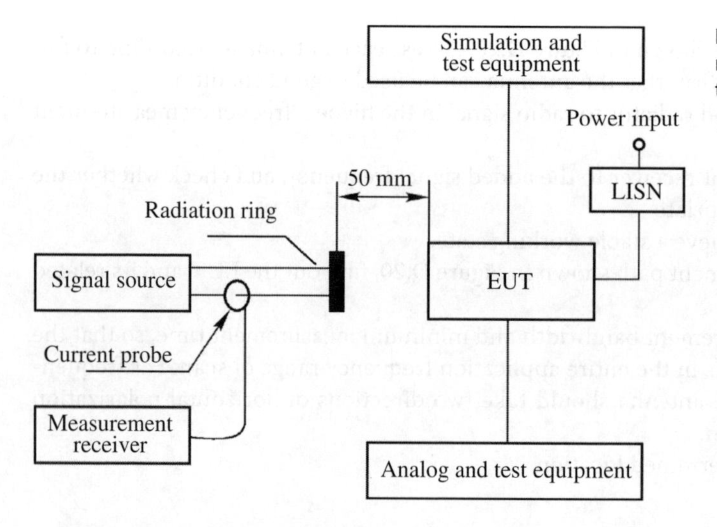

Figure D.22 25 Hz–100 kHz magnetic field radiation sensitivity test device.

d) If EUT appears as sensitive at the frequency, at least three test frequencies per octave frequency should be conducted.
e) Change the position of the ring, and make the ring sequentially align to each of the 300 regions on each side of the EUT and each connector interface. Repeat c–d at each position to determine the sensitive position and frequency;
f) From all the sensitive frequency in c–d, select three per octave frequency according to entire applicable frequency range of GJB151A.

3) For each frequency point determined in f, a current should be applied to the radiation ring that reaches the limited level of GJB151A is. Keep the spacing between the ring plane and EUT surface. The cable or the electrical connector is 50 mm; meanwhile, move the radiation ring. Pay attention to the determined location line and detect possible sensitive location. Determine whether the sensitive situation occurs.

D.4.13 RS103 10 kHz–40 GHz Electric Field Radiation Susceptibility Test

1) Purpose
   This test method is used to test the ability of EUT and related cables to withstand electric field radiation field radiation.
2) Test equipment
   a) Signal generator
   b) Power amplifier
   c) Receiving antenna:
      1) 1–10 GHz, double ridge horn antenna
      2) 10–40 GHz, other antennas approved by purchasing department
   d) Transmit antenna, or GTEM (0–18 GHz) alternative
   e) Electric field sensor
   f) Measuring receiver
   g) Power meter
   h) Directional coupler
   i) Attenuator
   j) Data recording device
   k) LISN
3) Test method
   According to the general requirements of Figures D.3–D.6, the configuration of a EUT basic test is shown in Figure D.23 and meets the requirements shown as follows, with the transmit antenna located according to the following requirement placed 1 m from the boundary of the configuration:
   1) The frequency is 10 kHz–200 MHz:
      a) Test configuration bounds are less than or equal to 3 m, the antenna is on the center line of the test configuration boundary, the boundary includes all the EUT extension housing, and the 2 m exposed power lines and interconnects are specified in this standard. If the actual installation of the platform interconnector is less than 2 m, it is also acceptable.
      b) Test configuration boundary is greater than 3 m. The antenna's position number (N) should be determined by dividing the distance (unit:m) from one edge to the other edge by 3 and selecting a rounded value.
   2) When the frequency is above 200 MHz, the antenna position number (N) should be determined as follows:
      a) To the test from 200 MHz–1 GHz, the antenna should be placed with sufficient number of locations. In order to make the entire width of each EUT shell and 350 mm of the wire and cable connected to the shell end of the EUT is exposed to the 3 dB of the bandwidth.

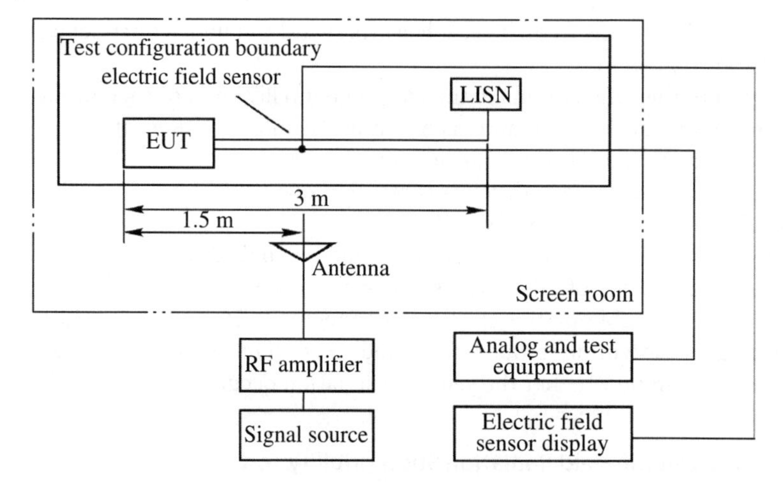

Figure D.23 10 kHz–40 GHz electric field radiation sensitivity test device.

b) To the test above 1 GHz, the antenna should be placed in a sufficient number of locations. In order to make the entire width of each EUT housing and 70 mm of the wire and cable connected to the housing end of the EUT is exposed to the antenna 3 dB beam width range.

D.4.14 RS105 Radiation Transient Electromagnetic Radiation Susceptibility Test

1) Purpose
   This test method is used to test the ability of EUT housing to withstand transient electromagnetic field.
2) Test equipment
   a) GTEM chamber, parallel plate, transverse electromagnetic cell or equivalent device
   b) High-voltage pulse adapter
   c) Transient single pulse generator
   d) Storage oscilloscope: Single sampling bandwidth is at least 200 MHz. Variable sampling rate is 1 GSa/s
   e) Protector device
   f) High-voltage probe
   g) Electric field broadband time-domain probe or B sensor and integrator or D sensor and integrator
   h) LISN
3) Test method
   Maintain the basic test configuration of EUT. Note: if the test uses an open radiation system, the test should be extra cautious.
   a) As shown in Figure D.24, if no more than GTEM available space, place the EUT housing on the GTEM base plate or on the ground plate, and the distance between GTEM core plate and the bottom plate is at least three times EUT.
   b) The GTEM base plate is overlapped on the ground reference point.
   c) When using open radiation system, as shown in Figure D.25, place test instrument in a shielded room. Keep the distance between roof (such as flat plate device) and near metal ground (including the ceiling, the building frame, metal air pipe and shielded room wall, etc.) at least two times h, which is the maximum vertical distance between the roof and the floor of the open radiation system.

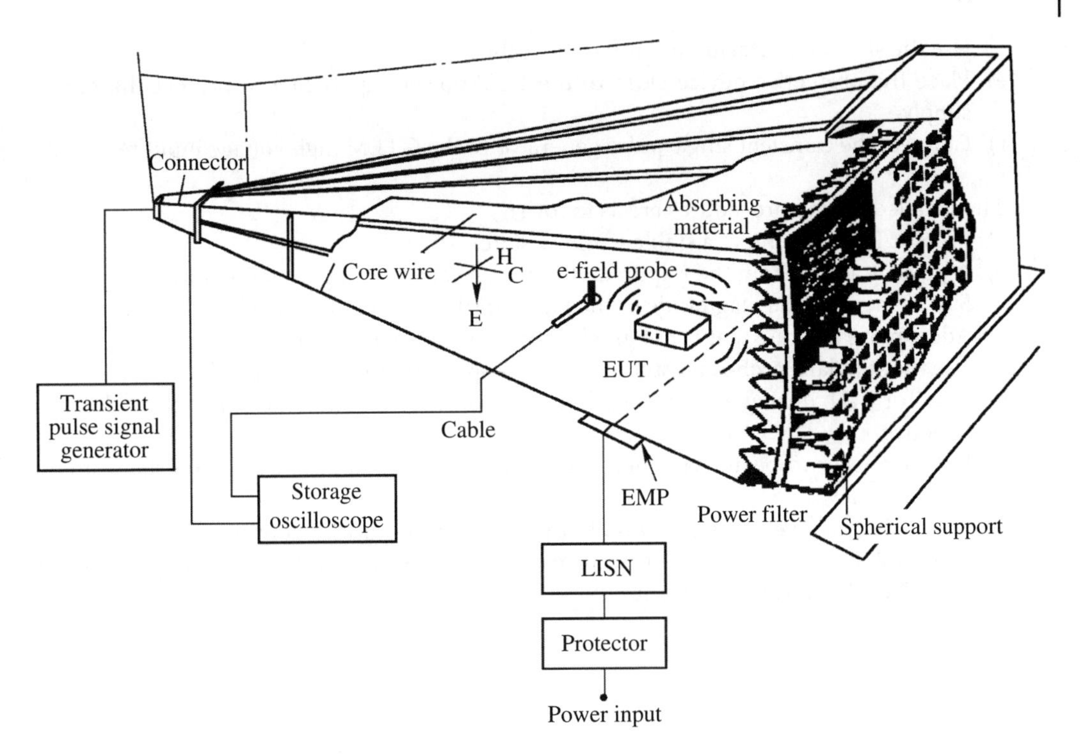

Figure D.24 RS105 test using GTEM device.

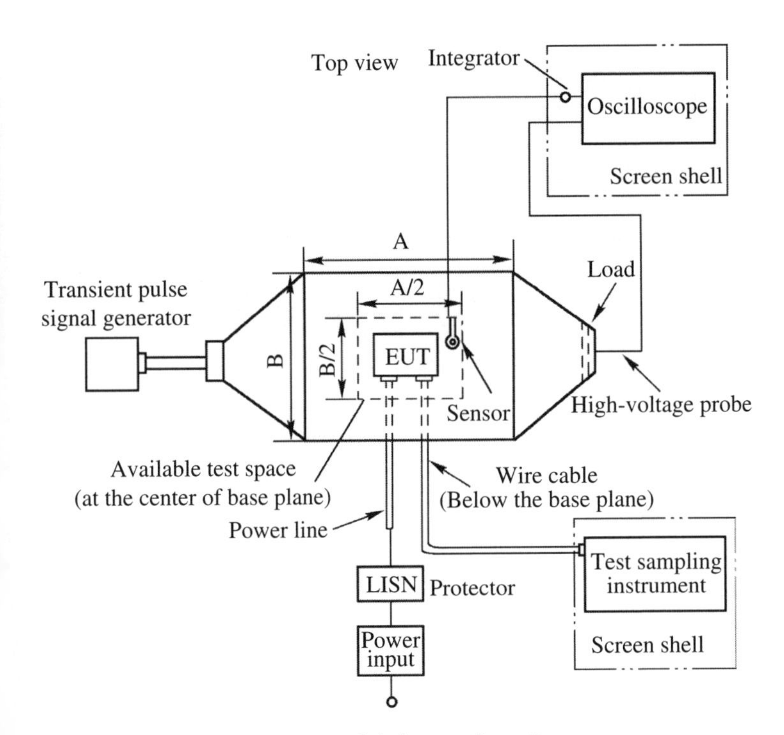

Figure D.25 RS105 test using parallel plate configuration.

d) Use the shielding measure to protect the cable.
e) Place the protective device close to the EUT power supply line to protect the power supply.
f) Connect the transient single pulse generator to the GTEM high voltage input terminal.

The detailed testing procedures are as follows:
a) Preheat EUT to achieve a stable working state.
b) A pulse is added from the required 50% peak level according to the required waveform. Slowly increase the magnitude of the pulse until reach required level.
c) Add a pulse at a rate that is not greater than one pulse per minute.
d) Use the standard probe and storage oscilloscope to monitor the applied pulse.
e) Monitor EUT during and after each applied pulse and determine whether the EUT is sensitive.
f) If EUT appear to be an error or fault below the specified peak level, stop the test and record the level.
g) If EUT is sensitive, the susceptibility threshold level must be determined (under level, undesirable response does not occur on the EUT). Confirm that the voltage does not meet the GJB151A requirements.

Appendix E

EMC Standards and Certification

E.1 The Origin for EMC Technical Standards

In 1934, The International Electrician Commission (IEC) has set up a special committee on radio interference (CISPR). The mechanism specializes research on radio interference problem and formulates relevant standards and aims to protect the broadcast reception effect. At the beginning only a small number of countries participated in the Commission, such as Belgium, France, Holland, and the United Kingdom. After many years of development, people's understanding of electromagnetic compatibility (EMC) has changed profoundly. In 1989, the European Commission issued an 89/336/EEC. The directive expressly provides that, since January 1, 1996, all electronic or electrical products must pass EMC performance certification; otherwise, they will be banned in the European Community market sales. This has caused great repercussions in the world, so that EMC has become an important indicator in the international trade. With the development of technology, the scope of the work of CISPR has also expanded from the original protection of broadcasting and receiving services to all operations involved in the protection of radio reception. The IEC has two technical committees specializing in the work of EMC: one is CISPR, which was founded in 1934; the other is EMC TC77, which was founded in 1981. CISPR was initially concerned with the issue of radio noise in the broadcast receiving frequency band, and it was followed by a tireless effort work on EMC standardization.

Today, CISPR has a total of seven subtechnical committees. Branch A is involved in the radio noise and immunity noise measuring equipment and measuring method; Branch B is involved in the industrial, scientific, medical RF equipment EMC; Branch C is related to the EMC of overhead power lines and high-voltage equipment; Branch D is related to the EMC problem of vehicle, motor vehicle, and spark ignition engine; Branch E is related to the radio and television receivers and the related equipment EMC; Branch F is related to the EMC issue of household appliances, electric tools, and fluorescent lamps and lighting fixtures; Branch G is related to the information technology equipment EMC problem.

CISPR has basically taken the industry and consumer products into account in the standards of EMC. CISPR has also drafted international standards for radio frequency harassment. To those newly developed products that are temporarily unable to correspond to the existing CISPR product standard, it can be limited by the radio frequency noise limitation. A few years ago, CISPR extended its operating frequency to 0–400 GHz; the actual operating range is 9 kHz–18 GHz now. The previous CISPR standards were mainly related to radio interference limits and measurement methods, but in recent years, the anti-disturbance ability has been studied and new criteria have been developed. TC77 was primarily concerned with the EMC problem of the low-voltage power system (below 9 kHz). Then the scope of its work was expanding to the whole frequency range and the entire EMC and products.

Electromagnetic Compatibility (EMC) Design and Test Case Analysis, First Edition. Junqi Zheng.

Table E.1 EMC standard number of countries or related organizations.

State or organization	Setting unit	Standard code
IE	CISPR	CISPR Pub.**
IEC	TC77	IEC*****
European Community	CENLEC	EN*****
USA	FCC, DOD	FCC Part**, MIL-SID.***
Japan	VCCI	VCCI
China	Quality and technical supervision, Defense departments	GB****-**** GJB***-**

Countries (in particular, developed countries) have their own national EMC standards. Table E.1 is the national EMC standard number of some country or related organizations.

China's EMC test and standardization work started in 1960s. At that time, some of the domestic institutes established a relatively simple test room and take research on radio interference (disturbance). While the Soviet Union and European and US national standards were being developed, China developed its own EMC standards and technical conditions. Since the establishment of the national radio interference Standardization Committee in 1986, China has begun organizing a systematic and corresponding CISPR/IEC work for the domestic EMC standardization. At present, the national radio interference Standardization Committee has set up eight subtechnical committees, the first seven branches corresponding with the international branches. S branch is set up according to the national conditions of China, it mainly involves the EMC problem between the radio system and the nonradio system. At present, China has developed 60 EMC national standards.

E.2 EMC Standard Structure and Classification

According to the work procedure of International Standardization Organization, publications on EMC have many forms, generally including standards, recommendations, technical specifications, and technical reports, etc. Standards and recommendations are used to repeat and continuous use and is a set of technical specifications approved by the approved recognized standards organizations. These technical specifications are only recommended, not compulsory. Technical specification specifies the characteristics of product requirements, such as performance, security, or size and requirements for the product, such as the term, symbol, and test program. Technical reports, in addition to the consensus view of the technology, the rest of the content is usually in the technical development stage, are not suitable for publication as an international standard.

EMC standard is the EMC design guidance documents for the product, and is the most important guarantee for the realization of the system. Especially when the product enters the domestic or international market, only comply with the relevant EMC standards, the product can be accepted is possible to be outside. "Several Provisions on Mandatory Standards for the Implementation of Mandatory Provisions" published by National Bureau of Quality and Technical Supervision provides for EMC standards for mandatory standards as, mandatory requirements for compliance.

The standard system framework of most organizations is adopts the standard method of IEC (International Electrical Commission). All the standards are divided into basic standards/publications, general standards/publications, and product standards/publications. Among them, the product standards can be divided into a series of product standards and special products. Each type of standard includes two aspects of the emission and immunity.

E.2.1 Basic EMC Standards

Basic EMC standards provide the general and basic conditions or rules of EMC. They are related to all of the series products, systems, or facilities involved in the EMC problem. They apply to these products, but they do not specify the product's emission limits or the degree of immunity. They are the basis of the development of other EMC standards (such as common standards or product standards) or the reference documents. The contents of the basic standards include the terms, the description of the electromagnetic phenomena, the level of compatibility, the general requirements, measurement, testing techniques and methods, the description and classification of the testing techniques and methods, the test level, and the environment.

E.2.2 General EMC Standards

General EMC standards are the EMC standards in a specific environment. They provide a minimum set of basic requirements and measurement/testing procedures. They can be applied to all products or systems that work in this particular environment. If a specific product does not have a series of product standards, the product can use general EMC standards. General EMC standards specific environment are divided into two major categories of specific environments:

1) Residential, commercial, and light industrial environment. Living environment, such as residential, apartments, etc. Business environment, such as shops, supermarkets, and other retail outlets, office buildings, banks and other commercial buildings, cinemas, internet cafes, and other public places of entertainment. Light industrial environment, such as small factories, laboratories, etc.
2) Industrial environment, such as the large inductive load or capacitive load frequent switch place, large current, and a place with a strong magnetic field, etc.

Developing general EMC standards must refer to the basis EMC standard, because they do not contain detailed measurement and test methods, and equipment required for measurement and testing. General EMC standards include emission (limit) and immunity (performance determination) requirements and corresponding measurement and test requirements. Only a limited number of requirements and measurement/testing methods are provided to achieve the best technical/economic results. But this does not prevent the product be designed to work normally in all kinds of electromagnetic noise.

E.2.3 Product EMC Standards

According to the size of the product and the characteristics of the product, product EMC standards can be further divided into series product EMC standards and special product EMC standards.

Series products are a group of similar products, systems, or facilities that use the same EMC standards. For specific product categories, series EMC product standards set up special

EMC (including emission and immunity) requirements, limits, and measurement/testing procedures. Product standard has more specific and specific detailed performance requirements and product operating conditions than the general standard. Range of product categories can be very wide or very narrow.

The EMC standard of the series products shall be based on the measurement and test method provided by the EMC standard. Its test and limit or performance criteria must be compatible with the general EMC standards. System products EMC standards is preferred than general EMC standards. Series product standards have more professional and detailed performance criteria than general standards.

E.2.4 EMC Standards for Special Products

EMC standards for special products are the standards for a particular product, system, or facility. According to the characteristics of these products, some special conditions must be considered. They have the same rules as the series EMC standard. Specialized product EMC standards should be preferred over the series EMC standards. Only in exceptional circumstances, different limits can be used. The special function of the product must be considered when deciding the immunity of the product. Special product EMC standard has accurate performance criteria. Therefore, the product standard and the series product standards or common standards are different.

E.3 International EMC Standards Organization

In the 1930s, many international organizations began studying EMC technology and released a number of standards and normative documents, such as International Electrical Commission (IEC), the International Telecommunication Union (ITU), the International Union of Railways (UIC), the International Conference on international electric power (CIGRE), and the European Telecommunications Standards Association (ETSI), the European Commission for electrical and technical standardization (CENELEC). IEC, ITU, and the European EMC standards have important influence and have their own characteristics.

E.3.1 International Electrical Commission (IEC)

The International Electrical Commission was established in 1906 is the world's first international electrical standardization agency, and it is headquartered in Geneva. According to the agreement reached between ISO and IEC in 1976, the two organizations are independent of each other. IEC is responsible for the international standardization work of electrical and electronic fields and other areas are responsible by ISO. IEC's aim is to promote international cooperation and standardization in related issues in the field of electrical and electronic.

IEC is equipped with three certification committees, namely the electronic component quality assessment committee (IECQ), the electronic safety certification board (IECEE), the explosion-proof electrical Certification Board (IECEX) and in 1996 they also set up a qualified assessment committee was set up, specifically responsible for the development of a series of certification and accreditation standards including system certification.

IEC has a very important role in the international standardization activities of EMC. The main research work in this area is the EMC Advisory Committee (ACEC), the Special Committee on radio interference (CISPR), and the EMC Technical Committee (TC77). Among them, CISPR has published a number of publications and amendments reached 38. TC77 organizations including the TC77 plenary and SC77A, SC77B, SC77C three branches of technical committee. SC77A is mainly responsible for the low frequency phenomenon; SC77B is responsible for the high frequency phenomenon; SC77C is responsible for the high altitude nuclear electromagnetic pulse. TC77 EMC standard is the IEC61000 series standards.

E.3.2 International Telecommunication Union (ITU)

The ITU is the standard organization in the field of international telecommunication, and is an international organization in the world. It has been developed for more than 130 years. In 17 May 1865, International Telegraph Union was founded in Paris, France by 20 European countries, signed an international telecommunications convention. In 1906, 27 national representatives signed an "international radio telegraph convention" in Berlin, Germany. The purpose is to develop standards for the telegraph network to exchange. In 1932, representatives of more than 70 countries in Spain decided to merge the two conventions (International Telecommunication Convention, and the International Telegraph Union) as the "International Telecommunication Union."

ITU includes three major sectors, namely, the Telecommunications Standardization Sector (T – ITU – T), the radio communications sector (R – ITU – R), and the telecommunications development department (D – ITU – D). The telecommunication standardization department is composed of the original CCITT (International Telegraph and Telephone Consultative Committee) and the CCIR (International Radio Consultative Committee). Its main responsibility is to study technology, operation, and tariff issues, and develop a global telecommunications standards. The results of the study are published in the form of proposals. ITU – R research radio communication technology and operation of radio communication, publishing proposals, and the exercise of the functions of the World Radio Administrative Convention (WARC), CCIR and frequency registration board. The telecommunication development department is composed of the original Telecommunication Development Bureau (BDT) and the telecommunication development center (CDT). Its responsibility is to encourage the participation of developing countries in the study of the electricity, and encourage the international cooperation.

ITU's Fifth Research Group is a research group that studies the EMC problem of telecommunications equipment and networks, the research area is EMC of the communication system and human body safety. The research group is the most experienced organization in the research of the telecommunications system of the EMC. In particular, the work in the area of overvoltage (overcurrent) protection is the most authoritative.

E.3.3 European Commission for Electrical and Technical Standards (CENELEC)

The European electrical technology standardization committee was established in 1973 and is headquartered in Brussels, Belgium. CENELEC is an organization holding standardization activities in the field of electrical engineering and in accordance with the EC 83/189/EEC. It is responsible for the coordination of all the standards for all the member states in the electrical field (including EMC), and is responsible for the development of European standards. In 1996, CENELEC and IEC signed a cooperation agreement in Dresden in Germany (Dresden Agreement). The contents of the agreement include: speeding up the publication and the common use of international standards, ensuring the rational use of resources, ensuring that the standard content of technology is an international level; meeting market needs to accelerate the development of standards for the program setting process, planning new projects together, and so on.

CENELEC is engaged in the technical committee TC2110 (before is TC110). It is responsible for the development transformation of EMC standards or conversion work. TC210 changes the available standards of IEC relevant technical committees and other EMC standards into the European EMC standards. The organizational structure of TC210 consists of five working groups, and the responsibilities of the working groups are as follows:

WG1: general standard
WG2: basic standards
WG3: power facilities and telephone line
WG4: anechoic chamber
WG5: responsible for military equipment for civilian use

Similarly, EMC will be divided into four types by TC210, i.e. basic EMC standards, general EMC standards (for residential, commercial and light industrial environment and industrial environment), product EMC standards, and professional products EMC standards.

E.3.4 European Telecommunication Standards Institute (ETSI)

ETSI is a nonprofit telecommunication standardization organization established by the European Commission in 1988. The headquarters are in Nice in southern France. The recommended standards developed by ETSI are often used as a technical basis for the European regulations and are required to be implemented. ETSI standardization is mainly in the field of the telecommunications industry, but also cooperates with other organizations in the field of information and broadcasting technology.

ETSI technical institutions can be divided into four kinds: technical committee and subtechnical committee, ETSI project team, and ETSI cooperation project team. The technical committee and the subtechnical committee are set according to the research field and the research content. The ETSI project team includes a set of issues that are very clear in a certain period of time.

ERM TC in the ETSI technical institution is mainly responsible for the problem of the EMC and the radio spectrum technology, including the study of EMC parameters and test methods, the use and allocation of the radio spectrum, and the relevant wireless and electromagnetic equipment for the standard to provide expert opinion on the EMC and wireless frequency.

E.4 EMC Standard System in China

China's EMC standard system is gradually improving. Referring to international classification methods, combined with China's actual situation, EMC standards can be divided into the following four categories: basic standards, generic standards, product family standards, and standards of intersystem compatibility. Basic standards are mainly related to the EMC term, the electromagnetic environment EMC measurement equipment specification and EMC measurement methods, such as GB/T 4365-95 "electromagnetic compatibility term." General standards are mainly involved in the protection of the human body in a strong magnetic field environment, as well as the radio business requirements of signal/interference protection ratio. Product standards are more, up to 38. Intersystem EMC standard specify EMC requirements between different systems after coordination. Most of these criteria are based on years of research structure and provide for the protection of the distance between the different systems. Most of the Chinese EMC standards come from the international standard. Its sources includes the international radio interference (CISPR) publication, the IEC standards, and the ITU recommendation. Because most of China's national standards come from the international standard, its products easily export to the international market. Commonly used EMC national standards are described in Tables E.2–E.5.

Table E.2 Basic standards.

Standard code	Standard name
GB/T 4365-1995	Electromagnetic compatibility terms
GB/T 6113-1995	Specification for radio disturbance and immunity measuring equipment
GB/T 3907-1983	Basic measurement method for industrial radio interference
GB/T 4859-1984	Basic measurement method of anti-interference degree for electrical equipment
GB/T 15658-1995	Measurement method of urban radio noise
GB/T 17626.1-1998	EMC immunity test general test and measurement technology
GB/T 17626.2-1998	Magnetic compatibility testing and measurement techniques for electrostatic discharge immunity test
GB/T 17626.3-1998	Electromagnetic compatibility testing and measurement techniques for RF electromagnetic field radiated immunity test
GB/T 17626.4-1998	Electromagnetic compatibility testing and measurement techniques for fast transient pulse group immunity test
GB/T 17626.5-1998	Electromagnetic compatibility testing and measurement techniques for surge (impact) immunity test
GB/T 17626.6-1998	Electromagnetic compatibility testing and measurement techniques of RF field induced conduction immunity
GB/T 17626.7-1998	Electromagnetic compatibility testing and measurement techniques for power supply systems and connected devices – a guide for measurement and measurement of harmonics and harmonics
GB/T 17626.8-1998	Electromagnetic compatibility testing and measurement techniques of power frequency magnetic field immunity test
GB/T 17626.9-1998	Electromagnetic compatibility testing and measurement techniques of pulse magnetic field immunity test
GB/T 17626.10-1998	Electromagnetic compatibility testing and measurement techniques for damping of oscillating magnetic field immunity test
GB/T 17626.11-1998	Electromagnetic compatibility testing and measurement techniques for voltage dips, short interruptions, and voltage variations immunity test
GB/T 17626.12-1998	Electromagnetic compatibility testing and measurement techniques for oscillating wave immunity test

Table E.3 General standard.

Standard code	Standard name
GB 8702-1988	Electromagnetic radiation protection
GB/T 14431-1993	Signal/interference protection required by radio service
GB/T 17799.1-1999	Immunity test in the environment of commercial and light industry
GB/T 15658-1995	Measurement method of urban radio noise

Table E.4 Product standard.

Standard code	Standard name
GB 4343-1995	Household and similar electrical appliances, electric heating appliances, electric tools, and similar radio interference characteristics measurement methods and allowable values
GB 9254-1998	Limits and measurement method of radio disturbance for information technology equipment
GB 4824-1996	Measurement methods and limits of electromagnetic disturbance characteristics of industrial, scientific, and medical RF equipment
GB/T 6833.1-1986	Electromagnetic compatibility test specification for electronic measuring instrument
GB/T 6833.2-1987	Electromagnetic compatibility test for electronic measuring instruments
GB/T 6833.3-1987	Electromagnetic compatibility test for electronic measuring instruments – Specification for electrostatic discharge sensitivity test
GB/T 6833.4-1987	Test of electromagnetic compatibility of electronic measuring instruments
GB/T 6833.5-1987	Test of electromagnetic compatibility of electronic measuring instruments – test of radiation sensitivity
GB/T 6833.6-1987	Test of electromagnetic compatibility of electronic measuring instruments
GB/T 6833.7-1987	Electromagnetic compatibility test for electronic measuring instruments – nonworking state magnetic field interference test
GB/T 6833.8-1987	Electromagnetic compatibility testing of electronic measuring instruments and the test of the working state of magnetic field interference
GB/T 6833.9-1987	Test of electromagnetic compatibility of electronic measuring instruments
GB/T 6833.10-1987	Test of electromagnetic compatibility of electric measuring instrument
GB/T 7343-1987	10 kHz–30 MHz passive radio interference filter and the method of measuring the suppression characteristics of components
GB/T 7349-1987	Measurement method for radio interference of high voltage transmission line and substation
GB 9254-1988	Limits and measurement methods for radio interference of information technology equipment
GB 9383-1995	Noise and television broadcast receivers and related equipment for the disturbance rejection limits and measurement methods
GB 13421-1992	Limits and measurement methods for stray transmit power levels of radio transmitters
GB/T 13836-1992	30 MHz–1 GHz sound and television signal cable distribution system equipment and components of the radiation interference characteristics and measurement method
GB 13837-1997	Audio and television broadcast receivers and associated equipment, radio interference characteristics, limits, and measurement methods
GB/T 13838-1992	Allowable value and measurement method for the radiated immunity characteristics of sound and television broadcast receivers and associated equipment
GB 13839-1992	Voice and television broadcast receivers and associated equipment – immunity and measurement methods
GB 14023-1992	Measurement method and allowable value of the radio interference characteristics of a vehicle, motor vehicle, and device driven by a spark ignition engine

Table E.4 (Continued)

Standard code	Standard name
GB 15540-1995	Requirements and measurement methods of electromagnetic compatibility for land mobile communication equipment
GB 15707-1005	Radio interference limits for high voltage AC transmission lines
GB/T 15708-1995	Measurement method of radio interference from the operation of AC electric railway electric locomotive
GB/T 15709-1995	Measurement method for the radio interference of AC electrified railway contact net
GB 15734-1995	Limits and measurement method of radio disturbance characteristics of electronic dimming equipment
GB 15949-1995	Cable distribution systems for sound and television signals – the limits and measurement methods for equipment and components
GB/T 16607-1996	Microwave oven at 1 GHz above the radiation interference measurement method
GB 16787-1997	30 MHz–1 GHz sound and television signal cable distribution system radiation measurement methods and limits
GB 16788-1997	30 MHz–1 GHz sound and television signal cable distribution system to measure method and limit value of immunity

Table E.5 System standard.

Standard code	System standard
GB 6364-1986	Electromagnetic environment requirements for aeronautical radio navigation station
GB 6830-1986	Allowable danger on telecommunication lines from power lines
GB/T 7432-1987	Index of anti-radio broadcasting and communication interference in the coaxial cable carrier communication system
GB/T 7433-1987	Index of anti-radio broadcast and communication interference in the symmetrical cable carrier communication system
GB/T 7434-1987	Open wire carrier system anti-interference from the radio broadcasting and telecommunication index
GB 7495-1987	The protective distance between overhead power line and receiving station of AM broadcasting
GB 13613-1992	Electromagnetic environment requirements for remote radio navigation stations in the sea
GB13614-1002	Electromagnetic environment requirements for shortwave radio direction finder
GB 13615-1992	Electromagnetic environment protection requirements for earth station
GB 13616-1992	Electromagnetic environment protection requirements for microwave relay station
GB 13617-1992	Shortwave radio receiving station (station) electromagnetic environment requirements
GB 13618-1992	Electromagnetic environment protection requirements for air intelligence radar station
GB/T 13620-1992	Determination and interference calculation method of coordinated area between earth station and ground station in satellite communication

E.5 EMC Certification

With the development of electrical and electronic technology, household electrical appliances are becoming increasingly popular, and electronic, broadcast television, post and telecommunications, and computer networks are becoming more and more advanced. The electromagnetic environment is increasingly complex and deteriorating, which makes it important for governments and business to pay more attention to the EMC of electrical and electronic products (EMI and EMS). Electromagnetic compatibility is a very important quality index for electronic and electrical products. It is not only related to the work reliability and safety of the product itself, but also may affect the normal operation of other equipment and systems and the protection of the electromagnetic environment. EC government regulates that from 1 in January 1996, all electrical and electronic products in the European Community market must pass EMC certification, demonstrated with an affixed CE logo. The movement has caused widespread impact. All the governments have had to take measures. Compulsory management has been implemented for electrical and electronic products EMC performance. International Comparison such as the EU89/336/EEC directive and the Federal Communications Commission (FCC) regulations have a clear requirement. China's 3C certification also gives a clear request to the EMC performance of the related products.

E.5.1 CE Authentication

CE is a certification mark and is deemed to be a passport for the manufacturer in entering the European market. Products affixed with a CE mark can be sold in EU domestic and are not required to meet the requirements of each member, allowing the free flow of goods in the EU member states within the scope. In the EU market, "CE" is a mandatory certification mark. Whether it is the product of the EU's internal enterprise or the production of other countries, in order to freely flow in the EU market, it is necessary for the product to add the CE rule. This shows that the product is in line with the European Union's "Technical Coordination and Standardization of New Methods," the basic requirements of the directive.

CE is the abbreviation of Communate Europpene, which is the European community (European Union). CE signs are used more and more popular in the market for goods sold in the European Economic Area (the European Union, the European Free Trade Association). CE logo shows the product is in line with EMC, safety, health, environmental protection, consumer protection, and a series of European requirements.

In the past, EU countries' requirements varied on the import and sale of product. According to a national standard, the manufacture of goods to other countries may not be listed. As part of the effort to eliminate trade barriers, CE came into being. Therefore, CE is on behalf of European unity. In fact, CE is an abbreviation for the "European Union's economic community" in many countries. Italian is COMUNITA EUROPEA; Portuguese is COMUNIDADE EUROPEIA. Of course, it might as well regard CE as CONFORMITY WITH EUROPEAN (DEMAND).

E.5.2 FCC Certification

FCC is an independent agency of the United States government established in 1934, directly responsible to Congress. The FCC coordinates domestic and international communications through the control of radio, television, telecommunications, satellite, and cable, which involves more than 50 states, the District of Columbia, and the United States. To protect life and property, the Office of Engineering and Technology is responsible for the technical support of the committee, and is responsible for the affairs of the equipment approval. To enter the US market, many radio applications products, communications products, and digital products are required to be recognized by the FCC. The FCC investigates and studies the various stages of product safety in order to find out the best way to solve the problem, while the FCC also includes a radio device and aircraft detection, for example.

In accordance with the provisions of FCC regulations (CFR47), the United States of America's electronic products must carry EMC certification (except for some special provisions of the relevant provisions of the product). There are three common ways of authentication: certification, doc, and verification. These three kinds of authentication methods and procedures have big differences. Different products can choose the relevant authentication method provided by FCC. For these three kinds of authentication, the FCC committee of the testing room also has relevant requirements.

E.5.3 China Compulsory Product Certification (3C Certification)

In order to protect the safety of consumers, protect animals and plants, protect the environment, and protect national security, compulsory product certification system is an assessment system by the competent authorities of all countries in accordance with the relevant laws and regulations. The development of a mandatory product certification catalog and mandatory product certification program provides for the inclusion of "directory" in the implementation of mandatory product testing and review. The products listed in the directory without the certification of the designated institution or the certification mark are not allowed to be manufactured, imported, sold, and used in the business premises.

China returned to the international standard organization member state in 1978. The work of establishing China's product certification system in accordance with international standards was actively carried out. Compulsory product certification, voluntary product certification, import and export food hygiene registration, management system certification, laboratory accreditation and certification, and personnel registration and other works have been developed, which improves the overall level of product quality and competitiveness in the international market, protects national economic interests and economic security, protects the health and safety of the people, animals, and plants, and protects the environment.

Since China's certification and accreditation, work has begun in the early stage of reform and opening up. Product quality certification, licensing, registration, and other systems are established in accordance with the relevant departments according to their respective administrative work. This promotes the development of China's certification and accreditation work. The certification and accreditation system have many disadvantages. For a long time, China's compulsory product certification has two sets of certification management system for internal and external. The former State Bureau of Quality and Technical Supervision is responsible for the implementation of safety certification for domestic products. The former State Inspection and Quarantine Bureau is responsible for the implementation of the safety and quality licensing system for import and export commodities. These two systems make some import products into catalog of mandatory certification. Thus, some imported products led by the two administrations and certified twice with different marking. Over the years many enterprises, through different ways and channels, reflect the situation to the relevant departments. This problem has become the focus of China's entry into the WTO negotiations.

With China's accession to the World Trade Organization, in order to better integrate with the international market, to honor our commitment to WTO, the State Administration of Quality Supervision, Inspection and Quarantine and the national certification and Accreditation

Administration Commission issued a notice: China implemented the new mandatory product certification system in 1 August 2003, i.e. China Compulsory Certification (3C certification).

China Compulsory Certification's implementation is based on four major regulations Chapter file ("Mandatory product certification management" "Compulsory product certification mark management approach" "The first batch of products to implement mandatory product certificationlist") and "Notice of compulsory product certification related issues" released by the Quality Supervision and Inspection and Quarantine Bureau and the national certification and Accreditation Administration Commission. China Commodity Inspection Bureau (CCIB) or China Commission for Conformity Certification of Electrical Equipment (CCEE) will be replaced by the new 3C logo. The former "Safety quality licensing system for imported goods" and "Mandatory supervision and management system for product safety certification" were abolished 1 August 2003.

The use of the 3C logo has the following requirements:

1) Applicants certified after the product within the certificate is valid; the logo can be used on the certified product.
2) In special circumstances, the applicant can apply for temporary use of the logo, to provide a certificate of the relevant certification authorities. The applicant can temporary use the logo before obtaining the certification.
3) When the applicant passes the authentication, it can only use the logo on certified products. If the applicant needs add nameplate printing/molding logo on certified product, it shall implement this according to "CCC compulsory certification mark printing/molding control program."
4) The applicant shall always be in accordance with the provisions of the use of signs. Effective control of the use of the logo is needed. Reserved for this use.

Further Reading

1 Williams, T. (2001). *EMC for Product Designers*, 3e. Newnes. www.newnespress.com.
2 Johnson, H. and Martin, G. (1993). *High-Speed Digital Design:A Hand Book of Black Magic*. Englewood Cliffs, NJ: Prentice Hall.
3 Clayton, P. (1996). *Introduction to Electromagnetic Compatibility*. New York: Wiley Series.
4 Mardigian, M. (1998). *EMI Control Methodology and Procedures*, vol. 8. Interference Control Methodologies Inc.
5 Ott, H. (1976). *Noise Reduction Techniques in Electronic Systems*. New York: Wiley.
6 CISPR, pub1.22, Limits and method of radio disturbance characteristics of information technology equipment.IEC1993.
7 Poon, F.N.K. and Pong, B.M.H. (November 2003). *High Frequency EMI for Switching Power Supplies*. PowerElab. Ltd.
8 IEC 61000-4-2. Immunity requirements and test methods for equipment subjected to static electric discharges. IEC, 1995+A1:1998.
9 IEC 61000-4-3.Immunity requirements and test methods for equipment subjected to radiated electromagnetic energy. IEC, 1995+A1:1998.
10 IEC 61000-4-4. Immunity requirements and test methods for equipment subjected repetitive electrical fast transients. IEC, 1995.
11 IEC 61000-4-5. Immunity requirements and test methods for equipment subjected to unidirectional surges. IEC, 1995.
12 IEC 61000-4-6. Immunity requirements and test methods for equipment subjected to RF transmitters in the frequency range 9 kHz to 80 (230) MHz. IEC, 1996.
13 CISPR 22 Third edition. Information technology equipment – Radio disturbance characteristics – Limits and methods of measurement.CISPR22, 1997.
14 Junqi, Z. (2006). *EMC Design And Test Case Analysis*. Beijing: Electronics Industry Press.
15 Paul, C.R. (1992). *Introduction to Electromagnetic Compatibility [Z]*. New York: Wiley.
16 Mardiguian, M. (2001). *Controlling Radiated Emissions by Design*, 2e. Boston: Kluwer Academic Publishers.
17 Dr. Eric Bogatin Bogatin Enterprises. Rules of Thumb I Have Known and Loved. Bogatin Enterprises, October 31, 1999.
18 John R. Barnes. Designing Electronic Systems for ESD Immunity. http://www.dbicorporation.com/esd-anno.htm, 2002.
19 Technical Note of Micron "Bypass Capacitor Selection for Highspeed Designs" 1996.
20 Golumbeanu, V., Suasta, P., Leonescu, D. The decoupling efficiency of power for low voltage circuit [Z]. IEEE, Elec. Tech-Med'98, 1998, 1:120.
21 Keith Armstrong. Design Techniques for EMCPart 3: Filtersand Surge Protection Devices[M/OL]. http://www.compliance-club.com/archive1/990609.Html, 1999.

Electromagnetic Compatibility (EMC) Design and Test Case Analysis, First Edition. Junqi Zheng.

22 Franco Fioro. Electromagnetic Emissions of Integrated Circuits [M/OL]. www.bolton.ac.uk/technology/mind/corep/emissions/emissions.html, 1999.
23 A EMC design training documents Paris: AEMC2006.
24 ISO 11452-4. Road vehicles – Electrical disturbances by narrowband radiated electromagnetic energy – Component test methods – Part 4: Bulk current injection (BCI). ISO, 2001.
25 ISO 11452-7. Road vehicles – Electrical disturbances by narrowband radiated electromagnetic energy – Component test methods – Part 7: Direct radio frequency (RF) power injection. ISO, 2001.
26 SAEJ1113-2, Electromagnetic compatibility measurement procedures and limits for vehicle components (except aircraft)-conducted immunity, 30 Hz to 250 KHz – all leads. ISO, 2001.
27 ISO 11452-2, Road vehicles – Electrical disturbances by narrowband radiated electromagnetic energy – Component test methods – Part 2: Absorber-lined chamber. ISO, 1995.
28 ISO 11452-3. Road vehicles – Electrical disturbances by narrowband radiated electromagnetic energy – Component test methods – Part 3: Transverse electromagnetic mode (TEM) cell. ISO, 2001.
29 SAEJ1113-25. Electromagnetic compatibility measurement procedures for vehicle components- immunity to radiated electromagnetic fields, 10KHz to 1000 MHz, trip-plat line method. ISO, 2001.
30 ISO 11452-5. Road vehicles – Electrical disturbances by narrowband radiated electromagnetic energy – Component test methods – Part 5: Stripline. ISO, 2001.
31 ISO 11452-6, Road vehicles – Electrical disturbances by narrowband radiated electromagnetic energy – Component test methods – Part 6: Parallel plate antenna. ISO, 2001.
32 SAEJ1113–22, Electromagnetic compatibility measurement procedures for vehicle components- immunity to radiated magnetic fields from power lines. ISO, 2001.
33 ISO 7637-0, Road vehicle-electrical disturbance by conduction and coupling -part 0:definitions and general. ISO, 2001.
34 ISO 10605. Road vehicle-electrical disturbance from electrostatic discharges. ISO, 2001.
35 ISO 7637-1. Road vehicles – Electrical disturbancesfrom conduction and coupling Part1: Definitions and general considerations. ISO, 2002.
36 ISO 7637-2. Road vehicles – Electrical disturbances from conduction and coupling – Part 2: Electrical transient conduction along supply lines only. ISO, 2004.
37 ISO7637-3. Road vehicles – Electrical disturbances from conduction and coupling Part 3: Vehicles with nominal 12 V or 24 V supply voltage – Electrical transient transmission by capacitive and inductive coupling via lines other than supply lines. ISO, 1995
38 IEC/CISPR 25. Limits and methods of measurement of radio disturbance characteristics for the protection 0f receivers used on board vehicles. ISO, 1995
39 W. Michael King. Common-Mode Architectural Current Flow Paths: Impact to Functional Reliability and Performance Stability of Systems-Products. Design News, April 2004

Index

Electromagnetic Compatibility (EMC) Design and Test Case Analysis, First Edition. Junqi Zheng.